Introductory
ALGEBRA

INTRODUCTORY ALGEBRA WITH P.O.W.E.R. LEARNING

1 2 3 4 5 6 7 8 9 0 DOW/DOW 1 0 9 8 7 6 5 4 3

ISBN 978–0–07–340626–8
MHID 0–07–340626–0

ISBN 978–0–07–748364–7 (Annotated Instructor's Edition)
MHID 0–07–748364–2

Senior Vice President, Products & Markets: *Kurt L. Strand*
Vice President, General Manager, Products & Markets: *Marty Lange*
Vice President, Content Production & Technology Services: *Kimberly Meriwether David*
Director, Developmental Mathematics: *Dawn R. Bercier*
Director of Development: *Rose Koos*
Director of Digital Content Development: *Nicole Lloyd*
Development Editor: *Liz Recker / Elizabeth O'Brien*
Market Development Manager: *Kim M. Leistner*
Lead Project Manager: *Peggy J. Selle*
Buyer: *Nicole Baumgartner*
Senior Media Project Manager: *Sandra M. Schnee*
Senior Designer: *David W. Hash*
Cover/Interior Designer: *Rokusek Design, Inc.*
Cover Image: *Power button icon © tkemot*
Lead Content Licensing Specialist: *Carrie K. Burger*
Compositor: *Aptara®, Inc.*
Typeface: *10/13 Times New Roman MT Std*
Printer: *R. R. Donnelley*

Library of Congress Cataloging-in-Publication Data

Messersmith, Sherri.
  Introductory algebra with P.O.W.E.R. learning / Sherri Messersmith, College of DuPage, Lawrence Perez, Saddleback College, Robert S. Feldman, University of Massachusetts.
    pages cm
  Includes index.
  ISBN 978–0–07–340626–8 — ISBN 0–07–340626–0 (hard copy : alk. paper) — ISBN 978–0–07–748364–7— ISBN 0–07–748364–2 (annotated Instructor's edition : alk. paper) 1. Algebra–Textbooks. 2. Study skills.
I. Perez, Lawrence. II. Feldman, Robert S. (Robert Stephen), 1947- III. Title.

  QA154.3.M474 2014
  512.97–dc23
                                    2012020394

Introductory
ALGEBRA

SHERRI MESSERSMITH
College of DuPage

LAWRENCE PEREZ
Saddleback College

ROBERT S. FELDMAN
University of Massachusetts Amherst

With contributions from William C. Mulford, *The McGraw-Hill Companies*

About the Authors

Sherri Messersmith
Professor of Mathematics, College of DuPage

Sherri Messersmith began teaching at the College of DuPage in Glen Ellyn, Illinois in 1994 and has over 25 years of experience teaching many different courses from developmental mathematics through calculus. She earned a Bachelor of Science degree in the Teaching of Mathematics at the University of Illinois at Urbana-Champaign and taught at the high school level for two years. Sherri returned to UIUC and earned a Master of Science in Applied Mathematics and stayed on at the university to teach and coordinate large sections of undergraduate math courses as well as teach in the Summer Bridge program for at-risk students. In addition to the P.O.W.E.R. Math Series, she is the author of a hardcover series of textbooks and has also appeared in videos accompanying several McGraw-Hill texts.

Sherri and her husband are recent empty-nesters and live in suburban Chicago. In her precious free time, she likes to read, cook, and travel; the manuscripts for her books have accompanied her from Spain to Greece and many points in between.

Lawrence Perez
Professor of Mathematics, Saddleback College

Larry Perez has fifteen years of classroom experience teaching math and was the recipient of the 2010 Community College Professor of the Year Award in Orange County, California. He realized early on that students bring to the classroom different levels of attitude, aptitude, and motivation sometimes accompanied by a tremendous fear of taking math. Confronted by this, he developed a passion for engaging students, demanding him to innovate traditional and online pedagogical techniques using architecture created with student feedback as the mechanism of design. He is the creator of the award-winning online learning environment Algebra2go® and has presented his work and methodology at conferences around the country.

Larry is a Veteran of the United States Navy Submarine Force and is a graduate of California State University Fullerton earning degrees in Electrical Engineering and Applied Mathematics. In his spare time he enjoys mountain biking and the great outdoors.

Robert S. Feldman
Dean and Professor of Psychology, University of Massachusetts Amherst

Bob Feldman still remembers those moments of being overwhelmed when he started college at Wesleyan University. "I wondered whether I was up to the challenges that faced me," he recalls, "and although I never would have admitted it then, I really had no idea what it took to be successful at college."

That experience, along with his encounters with many students during his own teaching career, led to a life-long interest in helping students navigate the critical transition that they face at the start of their own college careers. Bob, who went on to receive a doctorate in psychology from the University of Wisconsin-Madison, teaches at the University of Massachusetts Amherst, where he is the Dean of the College of Social and Behavioral Sciences and Professor of Psychology. He also directs a first-year experience course for incoming students.

Bob is a Fellow of both the American Psychological Association and the Association for Psychological Science. He has written more than 200 scientific articles, book chapters, and books, including P.O.W.E.R. Learning: *Strategies for Success in College and Life,* 6e and *Understanding Psychology,* 11e. He is president-elect of the FABBS Foundation, an umbrella group of societies promoting the behavioral and brain sciences.

Bob loves travel, music, and cooking. He and his wife live near the Holyoke mountain range in western Massachusetts.

Table of Contents

Section 1.3 gives students the opportunity to relearn and practice geometry before getting to applications later in the book that will require this knowledge. It helps them make connections to other math skills they will need and breaks down the information into smaller pieces for easier understanding.

Section 2.7 will begin by teaching students the arithmetic before relating it to algebra. If students can't understand the arithmetic, they will find it next to impossible to solve an algebra problem.

Rules of Exponents is one of the more difficult topics for students. We give the material its own chapter and then break the topic out into more manageable pieces. Students are given a chance to *practice* the easier parts first before progressing to harder topics.

Division tends to be a struggle for students but it is easier to digest if it is broken up, as we have done in **Sections 6.3 and 6.4**.

Sections 7.2 and 7.3 help students make the connection by starting off with exercises that are purely arithmetic.

Consistent Integration of Study Skills

In *Introductory Algebra,* strategies for learning are presented alongside the math content, making it easy for students to learn math *and* study skills at the same time. The P.O.W.E.R. framework aligns with the math learning objectives, providing instructors with a resource that has been consistently integrated throughout the text.

A **STUDY STRATEGIES** feature begins each chapter. Utilizing the P.O.W.E.R. framework, these boxes present steps for mastering the different skills students will use to succeed in their developmental math course. For example, these boxes will contain strategies on time management, taking good notes and, as seen in the sample below, taking a math test.

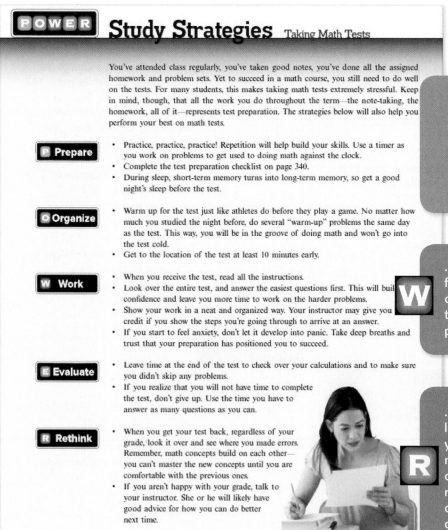

POWER Study Strategies — Taking Math Tests

You've attended class regularly, you've taken good notes, you've done all the assigned homework and problem sets. Yet to succeed in a math course, you still need to do well on the tests. For many students, this makes taking math tests extremely stressful. Keep in mind, though, that all the work you do throughout the term—the note-taking, the homework, all of it—represents test preparation. The strategies below will also help you perform your best on math tests.

P Prepare
- Practice, practice, practice! Repetition will help build your skills. Use a timer as you work on problems to get used to doing math against the clock.
- Complete the test preparation checklist on page 340.
- During sleep, short-term memory turns into long-term memory, so get a good night's sleep before the test.

O Organize
- Warm up for the test just like athletes do before they play a game. No matter how much you studied the night before, do several "warm-up" problems the same day as the test. This way, you will be in the groove of doing math and won't go into the test cold.
- Get to the location of the test at least 10 minutes early.

W Work
- When you receive the test, read all the instructions.
- Look over the entire test, and answer the easiest questions first. This will build confidence and leave you more time to work on the harder problems.
- Show your work in a neat and organized way. Your instructor may give you credit if you show the steps you're going through to arrive at an answer.
- If you start to feel anxiety, don't let it develop into panic. Take deep breaths and trust that your preparation has positioned you to succeed.

E Evaluate
- Leave time at the end of the test to check over your calculations and to make sure you didn't skip any problems.
- If you realize that you will not have time to complete the test, don't give up. Use the time you have to answer as many questions as you can.

R Rethink
- When you get your test back, regardless of your grade, look it over and see where you made errors. Remember, math concepts build on each other—you can't master the new concepts until you are comfortable with the previous ones.
- If you aren't happy with your grade, talk to your instructor. She or he will likely have good advice for how you can do better next time.

...all the work you do throughout the term—the note-taking, the homework, all of it—represents test preparation. The strategies below will help you perform your best on math tests.

W ...answer the easiest questions first. This will build your confidence and leave you more time to work on the harder problems.

R ...when you get your test back... look it over and see where you made errors. Remember, math concepts build on each other—you can't master the new concepts until you are comfortable with the previous ones.

Chapter 5 P O W E R Plan

P Prepare	**O Organize**
What are your goals for Chapter 5?	**How can you accomplish each goal?**
1 Be prepared before and during class.	• Don't stay out late the night before, and be sure to set your alarm clock! • Bring a pencil, notebook paper, and textbook to class. • Avoid distractions by turning off your cell phone during class. • Pay attention, take good notes, and ask questions. • Complete your homework on time, and ask questions on problems you do not understand.
2 Understand the homework to the point where you could do it without needing any help or hints.	• Read the directions, and show all of your steps. • Go to the professor's office for help. • Rework homework and quiz problems, and find similar problems for practice.
3 Use the P.O.W.E.R. framework to learn ways to improve the way you take math tests: *Is Anxiety the Hardest Problem on the Test?*	• Read the Study Strategy as it is outlined in the P.O.W.E.R. framework. • Decide which steps you might need to improve. • Complete the emPOWERme that appears before the Chapter Summary.
4 Write your own goal. _____ _____	• _____ _____
What are your objectives for Chapter 5?	**How can you accomplish each objective?**
1 Learn to read, write, and round decimals.	• Use place value, number lines, and writing a decimal as a fraction or mixed number to help understand what a decimal represents. • The same rounding principles you learned previously still apply to decimals.
2 Learn how to perform basic operations on decimals.	• Write the procedures for adding, subtracting, multiplying, and dividing with decimals in your own words. • Know how to multiply or divide by a power of 10. • Be able to solve applied problems using decimals.
3 Learn to write a fraction as a decimal.	• Understand the two ways to write a fraction as a decimal. One way is to use division, and the other is to write an equivalent fraction with a denominator that is a power of 10. • Know how to compare a decimal and a fraction.
4 Understand how to use measures of central tendency.	• Be able to find a mean, weighted mean, median, and mode. • Know what the different measures represent.
5 Write your own goal. _____ _____	• _____ _____

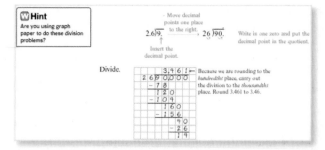

WORK HINTS provide additional explanation and point out common places where students might go wrong when solving a problem. Along with the ***Be Careful*** boxes, these tools act as a built-in tutor, helping students navigate the material and learn concepts even outside of class.

IN-CLASS EXAMPLES are available only in the Annotated Instructor Edition. These examples offer instructors additional problems to work through in class. In-class example problems align with the Guided Student Notes resource available with this package.

EXAMPLE 3

In-Class Example 3

Write each fraction as a decimal.

a) $\dfrac{3}{100}$ b) $\dfrac{859}{1000}$ c) $\dfrac{273}{10,000}$

Answer: a) 0.03 b) 0.859 c) 0.0273

Write each fraction as a decimal.

a) $\dfrac{9}{100}$ b) $\dfrac{137}{1000}$ c) $\dfrac{421}{10,000}$

Solution

a) Reading the fraction to ourselves will help us determine how to correctly write the decimal.

$\dfrac{9}{100}$ is read as "nine hundredths."

PUTTING IT ALL TOGETHER One of the challenges students struggle with is putting all of the steps they've learned together and *applying* that knowledge to a problem. **Putting It All Together** sections will help students understand the big picture and work through the toughest challenge when solving applications—*problem recognition,* or knowing *when* to use *what* method or thought process. These sections include a summary and several problems that help students reason through a problem using conversational, yet mathematically correct, language.

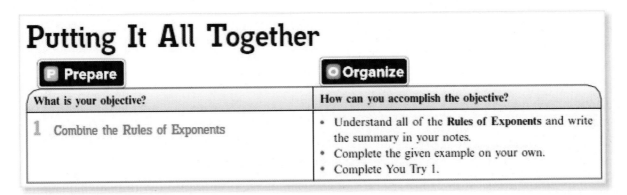

Putting It All Together

P Prepare	**O Organize**
What is your objective?	How can you accomplish the objective?
1 Combine the Rules of Exponents	• Understand all of the **Rules of Exponents** and write the summary in your notes. • Complete the given example on your own. • Complete You Try 1.

em**POWER**me boxes circle back to the opening **Study Strategies** and give students a checklist to evaluate how well they followed through on all of the positive habits recommended to successfully master a skill.

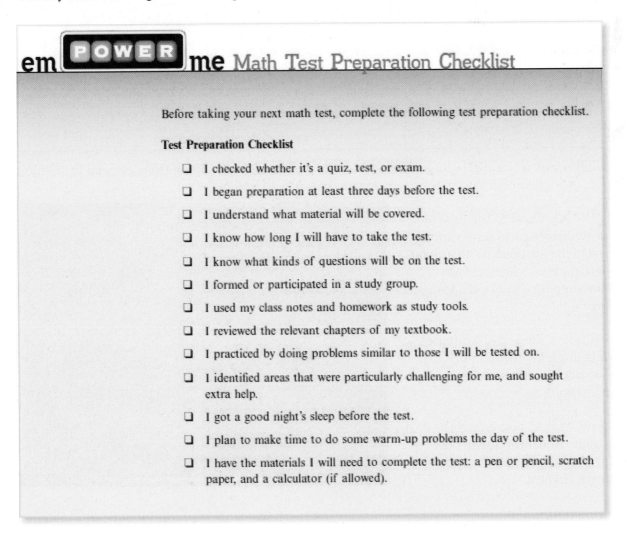

em **POWER** me Math Test Preparation Checklist

Before taking your next math test, complete the following test preparation checklist.

Test Preparation Checklist

- ❑ I checked whether it's a quiz, test, or exam.
- ❑ I began preparation at least three days before the test.
- ❑ I understand what material will be covered.
- ❑ I know how long I will have to take the test.
- ❑ I know what kinds of questions will be on the test.
- ❑ I formed or participated in a study group.
- ❑ I used my class notes and homework as study tools.
- ❑ I reviewed the relevant chapters of my textbook.
- ❑ I practiced by doing problems similar to those I will be tested on.
- ❑ I identified areas that were particularly challenging for me, and sought extra help.
- ❑ I got a good night's sleep before the test.
- ❑ I plan to make time to do some warm-up problems the day of the test.
- ❑ I have the materials I will need to complete the test: a pen or pencil, scratch paper, and a calculator (if allowed).

The Messersmith/Perez/Feldman Series offers instructors a robust resources package to help you with all of your teaching needs.

Resources in your P.O.W.E.R. tool kit include:

- Connect Hosted by ALEKS®
- ALEKS 360*
- Instructor Solutions Manual
- Student Solutions Manual
- Guided Student Notes*
- Classroom Worksheets*

- Instructor Resource Manual
- Test Bank Files
- Computerized Test Bank
- Faculty Development and Digital Training*
- PowerPoint Presentations
- Extensive Video Package*

*Details of these resources are included in the following pages!

Videos

Hundreds of videos are available to guide students through the content, offering support and instruction even outside your classroom.

Exercise Videos – These 3–5-minute clips show students how to solve various exercises from the textbook. With around thirty videos for every chapter, your students are supported even outside the classroom.

Lecture Videos – These 5–10-minute videos walk students through key learning objectives and problems from the textbook.

P.O.W.E.R. Videos – These engaging segments guide students through the P.O.W.E.R. framework and the study skills for each chapter.

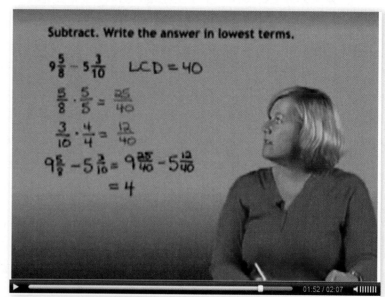

Faculty Development and Digital Training

McGraw-Hill is excited to partner with our customers to ensure success in the classroom with our course solutions.

Looking for ways to be more effective in the classroom? Interested in learning how to integrate student success skills in your developmental math courses?

Workshops are available on these topics for anyone using or considering the Messersmith/Perez/Feldman P.O.W.E.R. Math Series. Led by the authors, contributors, and McGraw-Hill P.O.W.E.R. Learning consultants, each workshop is tailored to the needs of individual campuses or programs.

New to McGraw-Hill Digital Solutions? Need help setting up your course, using reports, and building assignments?

No need to wait for that big group training session during faculty development week. The McGraw-Hill Digital Implementation Team is a select group of advisors and experts in Connect Hosted by ALEKS®. The Digital Implementation Team will work one-on-one with each instructor to make sure you are trained on the program and have everything you need to ensure a good experience for you and your students.

Are you redesigning a course or expanding your use of technology? Are you interested in getting ideas from other instructors who have used ALEKS™ or Connect Hosted by ALEKS in their courses?

Digital Faculty Consultants (DFCs) are instructors who have effectively incorporated technology such as ALEKS and Connect Hosted by ALEKS Corp. in their courses. Discuss goals and best practices and improve outcomes in your course through peer-to-peer interaction and idea sharing.

Contact your local representative for more information about any of the faculty development, training, and support opportunities through McGraw-Hill. http://catalogs.mhhe.com/mhhe/findRep.do

Need a tool to help your students take better notes?

GUIDED STUDENT NOTES

By taking advantage of Guided Student Notes, your students will have more time to learn the material and participate in solving in-class problems while, at the same time, becoming better note takers. Ample examples are included for appropriate coverage of a topic that will not overwhelm students. Use them as they are or download and edit the Guided Student Notes according to your teaching style.

Guided Student Notes
Messersmith – Introductory Algebra

Name:_____

5.1a Basic Rules of Exponents
Product Rule and Power Rule

Base   **Exponent**

Identify the base and the exponent in each expression and evaluate.

1) 3^4

2) $(-3)^4$

3) -3^4

4) -5^2

5) $(-5)^3$

6) $2(5)^2$

7) $4a(-3)^2$

8) $-(2)^4$

Product Rule   **Power Rule**

Find each product.

9) $5^2 \cdot 5$

10) $y^4 y^9$

11) $-4x^5 \cdot (-10x^8)$

12) $d \cdot d^7 \cdot d^4$

Simplify using the power

13) $\left(4^6\right)^3$

14) $\left(m^2\right)^5$

15) $\left(q^8\right)^7$

16) $\left(df^2\right)^3$

1

Guided Student Notes
Messersmith – Introductory Algebra

Name:_____

5.1b Basic Rules of Exponents
Combining the Rules

How to Solve Exponent Problems That Require Several Rules

Simplify each expression. Some problems may require a combination of both product and power rules.

1) $(4f)^2 (2f)^3$

2) $(2x)^3 (2x)^2$

3) $2x\left(4x^2t^4\right)^2$

4) $-3\left(4a^2b^3\right)^2$

5) $\left(\dfrac{3d^2}{6}\right)^3$

6) $\dfrac{\left(12y^3\right)^2}{\left(4z^{10}\right)^3}$

1

Develop your students' basic skills
with a ready-made resource.

WORKSHEETS FOR STUDENT AND INSTRUCTOR USE

Worksheets for every section are available as an instructor supplement. These author-created worksheets provide a quick, engaging way for students to work on key concepts. They save instructors from having to create their own supplemental material and address potential stumbling blocks in student understanding. Classroom tested and easy to implement, they are also a great resource for standardizing instruction across a mathematics department.

The worksheets fall into three categories: Worksheets to Improve Basic Skills; Worksheets to Help Teach New Concepts; and Worksheets to Tie Concepts Together.

The worksheets are available in an instructor edition, with answers, and in a student edition, without answers.

Worksheet 3C   Name: Answer Key
Messersmith – Beginning Algebra

Find 2 numbers that...

MULTIPLY TO	and ADD TO	ANSWER
27	6	9 and 3
72	18	12 and 6
24	11	8 and 3
4	3	1 and 4
10	7	5 and 2
121	22	11 and 11
54	3	9 and 6
54	29	27 and 2
16	10	8 and 2
30	17	15 and 2
9	6	3 and 3
8	2	4 and 2
21	10	7 and 3
60	19	15 and 4
56	15	8 and 7
28	3	4 and 7
72	6	12 and 6
100	25	20 and 5
40	6	4 and 10
11	12	11 and 1
20	12	10 and 2
35	2	7 and 5
77	18	11 and 7
108	21	12 and 9
3	2	3 and 1

Worksheet 5E   Na
Messersmith – Introductory Algebra

Perform the indicated operations and simplify.

1) $\frac{1}{3}+\frac{2}{5}$

3) $\frac{4}{7}+\left(-\frac{7}{8}\right)$

5) $-\frac{5}{9}+\left(-\frac{1}{3}\right)$

7) $\frac{3}{4}+\frac{1}{6}$

Connect Math Hosted by ALEKS® Corp.

Built By Today's Educators, For Today's Students

Fewer clicks means more time for you...

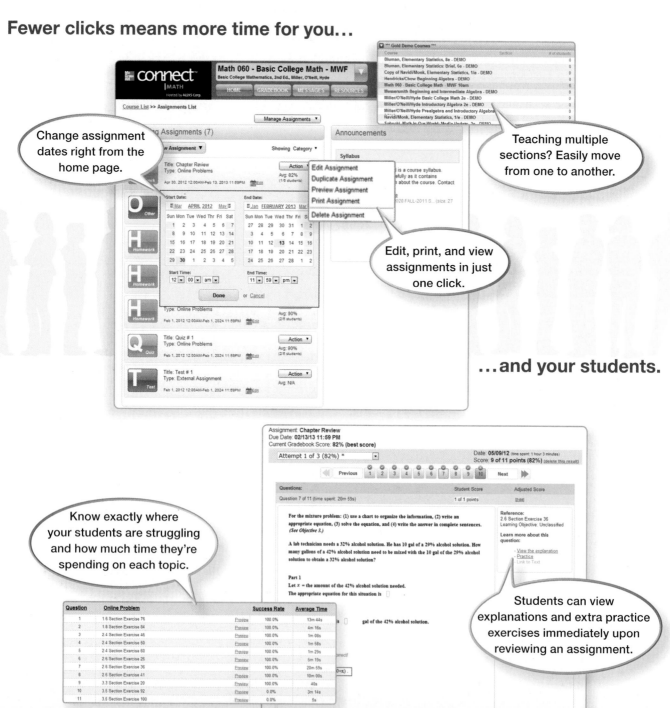

Change assignment dates right from the home page.

Teaching multiple sections? Easily move from one to another.

Edit, print, and view assignments in just one click.

...and your students.

Know exactly where your students are struggling and how much time they're spending on each topic.

Students can view explanations and extra practice exercises immediately upon reviewing an assignment.

Quality Content For Today's Online Learners

Online Exercises were carefully selected and developed to provide a seamless transition from textbook to technology.

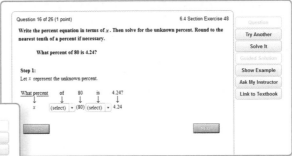

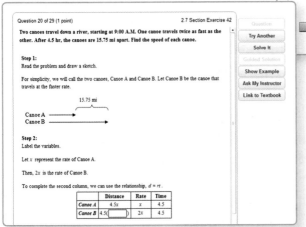

For consistency, the guided solutions match the style and voice of the original text as though the author is guiding the students through the problems.

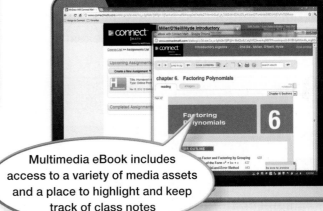

Multimedia eBook includes access to a variety of media assets and a place to highlight and keep track of class notes

ALEKS Corporation's experience with algorithm development ensures a commitment to accuracy and a meaningful experience for students to demonstrate their understanding with a focus towards online learning.

The ALEKS® Initial Assessment is an artificially intelligent (AI), diagnostic assessment that identifies precisely what a student knows. Instructors can then use this information to make more informed decisions on what topics to cover in more detail with the class.

www.successinmath.com

Hosted by **ALEKS Corp.**

ALEKS®

ALEKS is a unique, online program that significantly raises student proficiency and success rates in mathematics, while reducing faculty workload and office-hour lines. ALEKS uses artificial intelligence and adaptive questioning to assess precisely a student's knowledge, and deliver individualized learning tailored to the student's needs. With a comprehensive library of math courses, ALEKS delivers an unparalleled adaptive learning system that has helped millions of students achieve math success.

ALEKS Delivers a Unique Math Experience:

- **Research-Based, Artificial Intelligence** precisely measures each student's knowledge
- **Individualized Learning** presents the exact topics each student is most **ready to learn**
- **Adaptive, Open-Response Environment** includes comprehensive tutorials and resources
- **Detailed, Automated Reports** track student and class progress toward course mastery
- **Course Management Tools** include textbook integration, custom features, and more

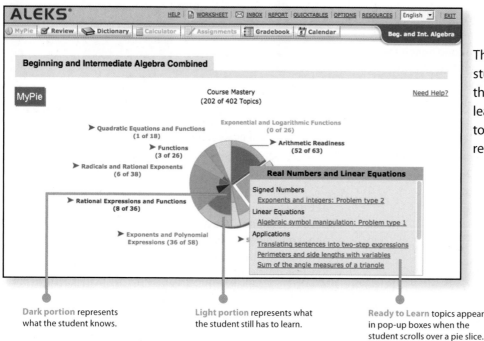

The ALEKS Pie summarizes a student's current knowledge, then delivers an individualized learning path with the exact topics the student is most ready to learn.

Dark portion represents what the student knows.

Light portion represents what the student still has to learn.

Ready to Learn topics appear in pop-up boxes when the student scrolls over a pie slice.

> "My experience with ALEKS has been effective, efficient, and eloquent. **Our students' pass rates improved from 49 percent to 82 percent with ALEKS.** We also saw student retention rates increase by 12% in the next course. Students feel empowered as they guide their own learning through ALEKS."
>
> —Professor Eden Donahou, *Seminole State College of Florida*

To learn more about ALEKS, please visit: **www.aleks.com/highered/math**

ALEKS® Prep Products

ALEKS Prep products focus on prerequisite and introductory material, and can be used during the first six weeks of the term to ensure student success in math courses ranging from Beginning Algebra through Calculus. ALEKS Prep quickly fills gaps in prerequisite knowledge by assessing precisely each student's preparedness and delivering individualized instruction on the exact topics students are most ready to learn. As a result, instructors can focus on core course concepts and see improved student performance with fewer drops.

> **"**ALEKS is wonderful. It is a professional product that takes very little time as an instructor to administer. Many of our students have taken Calculus in high school, but they have forgotten important algebra skills. ALEKS gives our students an opportunity to review these important skills.**"**
>
> —**Professor Edward E. Allen,** *Wake Forest University*

A Total Course Solution

A cost-effective total course solution: fully integrated, interactive eBook combined with the power of ALEKS adaptive learning and assessment.

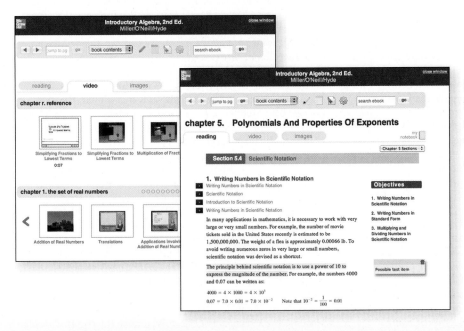

Students can easily access the full eBook content, multimedia resources, and their notes from within their ALEKS Student Accounts.

To learn more about ALEKS, please visit: **www.aleks.com/highered/math**

Acknowledgments

Manuscript Reviewers and Focus Group Participants

Thank you to all of the dedicated instructors who reviewed manuscript, participated in focus groups, and provided thoughtful feedback throughout the development of the *P.O.W.E.R.* series.

Darla Aguilar, *Pima Community College;* Scott Albert, *College of DuPage;* Bhagirathi Anand, *Long Beach City College;* Raul Arana, *Lee College;* Jan Archibald, *Ventura College;* Morgan Arnold, *Central Georgia Technical College;* Christy Babu, *Laredo Community College;* Michele Bach, *Kansas City Kansas Community College;* Kelly Bails, *Parkland College;* Vince Bander, *Pierce College, Pullallup;* Kim Banks, *Florence Darlington Technical College;* Michael Bartlett, *University of Wisconsin—Marinette;* Sarah Baxter, *Gloucester County College;* Michelle Beard, *Ventura College;* Annette Benbow, *Tarrant County College, Northwest;* Abraham Biggs, *Broward College;* Leslie Bolinger Horton, *Quinsigamond Community College;* Jessica Bosworth, *Nassau Community College;* Joseph Brenkert, *Front Range Community College;* Michelle Briles, *Gloucester County College;* Kelly Brooks, *Daytona State College (and Pierce);* Connie Buller, *Metropolitan Community College;* Rebecca Burkala, *Rose State College;* Gail Burkett, *Palm Beach State College;* Gale Burtch, *Ivy Tech Community College;* Jennifer Caldwell, *Mesa Community College;* Edie Carter, *Amarillo College;* Allison Cath, *Ivy Tech Community College of Indiana, Indianapolis;* Dawn Chapman, *Columbus Tech College;* Christopher Chappa, *Tyler Junior College;* Chris Chappa, *Tyler Junior College;* Charles Choo, *University of Pittsburgh at Titusville;* Patricia Clark, *Sinclair Community College;* Judy Kim Clark, *Wayne Community College;* Karen Cliffe, *Southwestern College;* Sherry Clune, *Front Range Community College;* Heather Coltharp, *West Kentucky Comm & Tech School;* Ela Coronado, *Front Range Community College;* Danny Cowan, *Tarrant County College, Northwest;* Susanna Crawford, *Solano College;* George Daugavietis, *Solano Community College;* Joseph De Guzman, *Norco College;* Michaelle Downey, *Ivy Tech Community College;* Dale Duke, *Oklahoma City Community College;* Rhonda Duncan, *Midlands Technical College;* Marcial Echenique, *Broward College;* Sarah Ellis, *Dona Ana Community College;* Onunwor Enyinda, *Stark State College;* Chana Epstein, *Sullivan County Community College;* Karen Ernst, *Hawkeye Community College;* Stephen Ester, *St. Petersburg College;* Rosemary Farrar, *Southern West Virginia Community & Technical College;* John Fay, *Chaffey College;* Stephanie Fernandes, *Lewis and Clark Community College;* James Fiebiger, *Front Range Community College;* Angela Fipps, *Durham Technical Community College;* Jennifer Fisher, *Caldwell Community College & Technical Institute;* Elaine Fitt, *Bucks County Community College;* Carol Fletcher, *Hinds Community College;* Claude Fortune, *Atlantic Cape Community College;* Marilyn Frydrych, *Pikes Peak Community College;* Robert Fusco, *Broward College;* Jared Ganson, *Nassau Community College;* Kristine Glasener, *Cuyahoga Community College;* Ernest Gobert, *Oklahoma City Community College;* Linda Golovin, *Caldwell College;* Suzette Goss, *Lone Star College Kingwood;* Sharon Graber, *Lee College;* Susan Grody, *Broward College;* Leonard Groeneveld, *Springfield Tech Community College;* Joseph Guiciardi, *Community College of Allegheny County;* Susanna Gunther, *Solano College;* Lucy Gurrola, *Dona Ana Community College;* Frederick Hageman, *Delaware Technical & Community College;* Tamela Hanebrink, *Southeast Missouri State University;* Deborah Hanus, *Brookhaven College;* John Hargraves, *St. John's River State College;* Michael Helinger, *Clinton Community College;* Mary Hill, *College of DuPage;* Jody Hinson, *Cape Fear Community College;* Kayana Hoagland, *South Puget Sound Community College;* Tracey Hollister, *Casper College;* Wendy Houston, *Everett Community College;* Mary Howard, *Thomas Nelson Community College;* Lisa Hugdahl, *Milwaukee Area Tech College—Milwaukee;* Larry Huntzinger, *Western Oklahoma State College;* Manoj Illickal, *Nassau Community College;* Sharon Jackson, *Brookhaven College;* Lisa Jackson, *Black River Technical College;* Christina Jacobs, *Washington State University;* Gretta Johnson, *Amarillo College;* Lisa Juliano, *El Paso Community College, Northwest Campus;* Elias M. Jureidini, *Lamar State College/Orange;* Ismail Karahouni, *Lamar University;* Cliffe Karen, *Southwestern College;* David Kater, *San Diego City College;* Joe Kemble, *Lamar University;* Esmarie Kennedy, *San Antonio College;* Ahmed Khago, *Lamar University;* Michael Kirby, *Tidewater Community College VA Beach Campus;* Corrine Kirkbride, *Solano Community College;* Mary Ann Klicka, *Bucks County Community College;* Alex Kolesnik, *Ventura College;* Tatyana Kravchuk, *Northern Virginia Community College;* Randa Kress, *Idaho State University;* Julianne Labbiento, *Lehigh Carbon Community College;* Robert Leifson, *Pierce College;* Greg Liano, *Brookdale Community College;* Charyl Link, *Kansas City Kansas Community College;* Wanda Long, *Saint Charles County Community College;* Lorraine Lopez, *San Antonio College;* Luke Mannion, *St. John's University;* Shakir Manshad, *New Mexico State University;* Robert Marinelli, *Durham Technical Community College;* Lydia Matthews-Morales, *Ventura College;* Melvin Mays, *Metropolitan Community College (Omaha NE);* Carrie McCammon, *Ivy Tech Community College;* Milisa Mcilwain, *Meridian Community College;* Valerie Melvin, *Cape Fear Community College;* Christopher Merlo, *Nassau Community College;* Leslie Meyer, *Ivy Tech Community College/ Central Indiana;* Beverly Meyers, *Jefferson College;* Laura Middaugh, *McHenry County College;* Karen Mifflin, *Palomar College;* Kris Mudunuri, *Long Beach City College;* Donald Munsey, *Louisiana Delta Community College;* Randall Nichols, *Delta College;* Joshua Niemczyk, *Saint Charles County Community College;* Katherine Ocker Stone, *Tusculum College;* Karen Orr, *Roane State;* Staci Osborn, *Cuyahoga Community College;* Steven Ottmann, *Southeast Community College, Lincoln Nebraska;* William Parker, *Greenville Technical College;* Joanne Peeples, *El Paso Community College;* Paul Peery, *Lee College;* Betty Peterson, *Mercer County Community College;* Carol Ann Poore, *Hinds Community College;* Hadley Pridgen, *Gulf Coast State College;* William Radulovich, *Florida State College @ Jacksonville;* Lakshminarayan Rajaram, *St. Petersburg College;* Kumars Ranjbaran, *Mountain View College;* Darian Ransom, *Southeastern Community College;* Nimisha Raval, *Central Georgia Technical College;* Amy Riipinen, *Hibbing Community College;* Janet Roads, *Moberly Area Community College;* Marianne Roarty, *Metropolitan Community College;* Jennifer Robb, *Scott Community College;* Marie Robison, *McHenry County College;* Daphne Anne Rossiter, *Mesa Community College;* Anna Roth, *Gloucester County College;* Daria Santerre, *Norwalk Community College;* Kala Sathappan,

College of Southern Nevada; Patricia Schubert, *Saddleback College;* William H. Shaw, *Coppin State University;* Azzam Shihabi, *Long Beach City College;* Jed Soifer, *Atlantic Cape Community College;* Lee Ann Spahr, *Durham Technical Community College;* Marie St. James, *Saint Clair County Community College;* Mike Stack, *College of DuPage;* Ann Starkey, *Stark State College of Technology;* Thomas Steinmann, *Lewis and Clark Community College;* Claudia Stewart, *Casper College;* Kim Taylor, *Florence Darlington Technical College;* Laura Taylor, *Cape Fear Community College;* Janet Teeguarden, *Ivy Tech Community College;* Janine Termine, *Bucks County Community College;* Yan Tian, *Palomar College;* Lisa Tolliver, *Brown Mackie South Bend;* David Usinski,

Erie Community College; Hien Van Eaton, *Liberty University;* Theresa Vecchiarelli, *Nassau Community College;* Val Villegas, *Southwestern College;* David Walker, *Hinds Community College;* Ursula Walsh, *Minneapolis Community & Tech College;* Dottie Walton, *Cuyahoga Community College;* LuAnn Walton, *San Juan College;* Thomas Wells, *Delta College;* Kathryn Wetzel, *Amarillo College;* Marjorie Whitmore, *North West Arkansas Community College;* Ross Wiliams, *Stark State College of Technology;* Gerald Williams, *San Juan College;* Michelle Wolcott, *Pierce College, Puyallup;* Mary Young, *Brookdale Community College;* Loris Zucca, *Lone Star College, Kingwood;* Michael Zwilling, *University of Mount Union*

Student Focus Group Participants

Thanks to the students who reviewed elements of P.O.W.E.R. and talked candidly with the editorial team about their experiences in math courses.

Eire Aatnite, *Roosevelt University;* Megan Bekker, *Northeastern Illinois University;* Hiran Crespo, *Northeastern Illinois University;* John J. Frederick, Jr., *Harold Washington College;* Omar Gonzalez, *Wright College;* Yamizaret Guzman, *Western Illinois University;* Ashley Grayson, *Northeastern Illinois University;* Nathan Hurde, *University of Illinois at Chicago;* Zainab Khomusi, *University of Illinois at Chicago;* Amanda Koger, *Roosevelt University;* Diana Kotchounian, *Roosevelt University;*

Adrana Martinez, *DePaul University;* Laurien Mosley, *Western Illinois University;* Jeffrey Moy, *University of Illinois at Chicago;* Jaimie O'Leary, *Northeastern Illinois University;* Trupti Patel, *University of Illinois at Chicago;* Pete Rodriguez, *Truman College;* Kyaw Sint Lay Wu, *University of Illinois at Chicago;* Shona L. Thomas, *Northeastern Illinois University;* Nina Turnage, *Roosevelt University;* Brittany K. Vernon, *Roosevelt University;* Kyaw Sint Lay Wu, *University of Illinois at Chicago*

Digital Contributors

Special thanks go to the faculty members who contributed their time and expertise to the digital offerings with *P.O.W.E.R.*

Jennifer Caldwell, *Mesa Community College*
Chris Chappa, *Tyler Junior College*
Kim Cozean, *Saddleback College*
Cindy Cummins, *Ozarks Technical Community College*
Rob Fusco, *Bergen Community College*
Sharon Jackson, *Brookhaven College*

Corrine Kirkbride, *Solano Community College*
Brianna Kurtz, *Daytona State College*
Christy Peterson, *College of DuPage*
Janine Termine, *Bucks County Community College*
Linda Schott, *Ozarks Technical Community College*

From the Authors

The authors would like to thank many people at McGraw-Hill. First, our editorial team: Elizabeth O'Brien, Liz Recker, and most of all, Dawn "Dawesome" Bercier, who believed in, championed, and put never-ending energy into our project from the beginning. To Ryan Blankenship, Marty Lange, Kurt Strand, and Brian Kibby: thank you for your continued support and vision that allows us to help students far beyond our own classrooms. Also, Kim Leistner, Nicole Lloyd, Peggy Selle, Peter Vanaria and Stewart Mattson have been instrumental in what they do to help bring our books and digital products to completion.

We offer sincere thanks to Vicki Garringer, Jennifer Caldwell and Sharon Bailey for their contributions to the series.

From Larry Perez: Thank you to my wife, Georgette, for your patience, support, and understanding throughout this endeavor. Thank you to Patrick Quigley and Candice Harrington for your friendship and support. Also, I must thank Dr. Harriet Edwards and Dr. Raghu Mathur for modeling inspirational and innovative pedagogy, examples which I still strive to emulate.

From Bob Feldman: I am grateful to my children, Jonathan, Joshua, and Sarah; my daughters-in-law Leigh and Julie; my smart, handsome, and talented grandsons Alex and Miles, and most of all to my wife, Katherine (who no longer can claim to be the sole mathematician in our family). I thank them all, with great love.

From Sherri Messersmith: Thank you to my daughters, Alex and Cailen, for being the smart, strong, supportive young women that you are; and to my husband, Phil, for understanding the crazy schedule I must keep that often does not complement your own. To Sue, Mary, Sheila, and Jill: everyone should have girlfriends like you. Thank you to the baristas at my hometown Starbucks for your always-smiling faces at 6 am and for letting me occupy the same seat for hours on end. Larry and Bob, thank you for agreeing to become my coauthors and for bringing your expertise to these books. Bill Mulford, we are immensely grateful for your hard work and creativity and for introducing Bob, Larry, and me in the first place. Working with our team of four has been a joy. And, finally, thank you Bill, for your friendship, your patience, and for working with me since the very beginning more than 8 years ago, without question the best student I've ever had.

Sherri Messersmith
Larry Perez
Bob Feldman

Application Index

speeds of two cars, 280–281, 284
speeds of two planes, 281, 284, 304
speeds of two trains, 284
speeds of two vehicles, 660
taxi fare per mile, 160
time of ball in air, 627
time to drive between cities, 521
time driven, 131
time parked in garage, 160
time in space, 282
time to travel at specified speed, 525
velocity of boat, 612
velocity of roller coaster, 612
visitors to Las Vegas, 185–186

EDUCATION

applications to college, 118
area of bulletin board decorations, 33
art history projects chosen, 282
cost of fund-raiser items, 283
credits earned by two students, 113
foreign students in U.S. institutions, 253
homework problems completed, 16
perimeter of bulletin board decorations, 33
public high school graduation rates, 183
public schools budget, 49–50
rate of doing homework, 523
science and engineering doctorates, 197–198
student absences, 16
students graduating in 4 years, 522
students in tutoring sessions, 169
students studying languages, 118
students taking school bus vs. driving, 522
students using tutoring service, 522
test grades to maintain average grade, 160, 168
time to type report, 523
tuition at private 4-year institution, 193–195

ENTERTAINMENT

album sales after Michael Jackson's death, 167
album sales per artist, 120
BET Award nominations, 282
cost of attraction tickets, 277–278
cost of concert tickets, 651
cost of lottery tickets, 283
cost of songs on iTunes, 197
dimensions of iPod, 282
film earnings on opening days, 282
height of fireworks, 441
height of TV screen, 550
music festivals attendance, 119
prices of concert tickets, 283
ratings for World Series, 50
revenue from music club tickets, 442
savings on video game, 147–148
songs on two iPods, 112–113
value of record albums, 238
weeks albums were on charts, 274–275

ENVIRONMENT AND NATURE

average high temperature in Paris, 134
average low temperature in Buenos Aires, 134
carbon emissions from China, 49
carbon emissions per person, 339
cats at animal shelter, 168

crude oil reserves by state, 303
difference between coldest temperatures, 47
difference in snowfall, 50
farmland in Nebraska, 343
greenhouse gas emission levels, 169
height of dropped rock, 440
highest temperature in Minnesota, 50
lowest temperature in Africa, 82
lowest temperature in Georgia, 659
quills on a porcupine, 343
rainfall in April, 118
speed of current, 285, 523, 534
speed of wind, 285, 523
temperature after storm, 118
temperature conversion, 230
visitors to national parks, 49
weight of kitten, 230
weight of Tasmanian devil joey, 217

FINANCE AND INVESTMENT

amount invested in two accounts, 660
annual salaries of social workers, 186
annual wage of mathematicians, 230
base salary, 150
down payment for house, 170
earnings deposited in savings account, 16, 145–146
earnings for hours worked, 519
exchange rate for euros, 141
exchange rate for pounds, 141
exchange rate for rupee, 217
gross pay and hours worked, 217
income donated to charity, 149
insurance policy coverage, 168
investment in two accounts, 284
monthly income plus bonus, 146
original price of house, 240, 383
percent increase in Social Security benefits, 168
salary growth over time, 239
sales tax paid in Seattle, 186–187
total cost of car, 150
weekly earnings, 525

FOOD

acres of crops planted, 118
area of lazy Susan, 33
average household spending on food, 339
avocadoes eaten during Super Bowl, 339
baking soda for cookie dough, 141
best buy on Gatorade, 136
best buy on ice cream, 136
best buy on olive oil, 168
brown sugar for cookie recipe, 16
caffeine in coffee, 118
caffeine in energy drink, 141
caffeine in soda, 141
calories in snacks, 118
calories in White Castle order, 284
cost of apples, 231
cost of candy bar and soda, 304
cost of coffee, 139
cost of hamburger and fries, 283
cost of popcorn items, 283
cost of yogurt, 141
cups of coffee sold, 118
fat in Caramel Frappuccinos, 535

flour for cookie recipe, 119
flour in partially filled bags, 16
fruit juice consumption per person, 183
gluten-free flour mixture, 522
height of coffee can, 131
juices in fruit drink, 283
kitchen scraps in yard waste, 141
lengths of subs, 119
milk for ricotta cheese, 522
milk produced per cow, 339
nut mixture, 283
orange juice for marinade, 141
percent increase in gluten-free product sales, 149
percent increase in peanut butter price, 149
perimeter of lazy Susan, 33
pounds of chocolate purchased, 376
sodium in Taco Bell order, 283–284
tea mixture, 283
types of milk sold, 304
value of squash crop, 238
volume of cooler, 32
volume of ice cream cone, 33

INTERNET AND TECHNOLOGY

cell phone data services used by teenagers, 17
cell phone minutes per month, 158
data sent from moon, 339
e-mails per day, 167
Facebook friends per person, 120
increase in Twitter visitors, 42
ink droplets per photo printed, 339
markdown on USB thumb drive, 149
percent decrease in Facebook users, 149
photos uploaded to Facebook, 658
sales tax rate on computer, 149
samples of sound read on by a CD player, 647
students with Internet access on phone, 512
text messages per month, 343
time to set up computer, 523
visits to news websites after election, 343
weight of computers, 118
width of laptop screen, 550

SCIENCE AND CHEMISTRY

acid solutions, 280, 283, 304
alcohol solution, 283
current through electric circuit, 521
distance from light source, 611
force to stretch a spring, 197, 525
frequency of piano string, 535
hydrogen peroxide solutions, 278–280
industrial cleaning solution, 522
intensity of light at a distance, 525
mass of water, 343
sound level at a distance, 525
time to fill sink, 523
time to fill tub, 523
volume of laboratory water bath, 33
weight of object above Earth, 525

SOCIOLOGY AND DEMOGRAPHICS

decrease in new housing, 42
decrease in population of Michigan, 42
immigrants from two countries, 282

The Real Number System and Geometry

Math at Work:

Landscape Architect

Jill Cavanaugh works as a landscape architect. She designs gardens, parks, and other outdoor spaces. To do her job, she uses multiplication, division, and geometry formulas on a daily basis.

For instance, when Jill is asked to create the landscape for a new house, her first task is to draw the plans. Often, the ground in front of the house will be dug out into shapes that include rectangles and circles, then shrubs and flowers will be planted, and mulch will cover the ground. To determine the volume of mulch that will be needed, Jill must use the formulas for the area of a rectangle and a circle and then multiply by the depth of the mulch. She will calculate the total cost of this landscaping job only after determining the costs of the plants, the mulch, and the labor.

In approaching a problem as complex as designing and creating the landscaping for an entire house, Jill takes a systematic, organized approach, working step by step to complete the job. This is the same technique you can use when you face a difficult math homework set or sit down for an exam. Keep your thinking and actions organized, and you will have a good chance to succeed.

In this chapter, we will review formulas from geometry as well as some concepts from arithmetic. We will also introduce P.O.W.E.R., a framework that you can use to help you meet challenges in this course and in virtually any other setting.

The **P.O.W.E.R. Framework** is based on an acronym—a word formed from the first letters of a series of steps—that will help you take in, process, and make use of new information. P.O.W.E.R. stands for **P**repare, **O**rganize, **W**ork, **E**valuate, and **R**ethink. It's a simple framework, but an effective one. As you may have noticed by now, its steps have been integrated into this book in order to help you succeed in this course.

In fact, the P.O.W.E.R. framework can help with any academic task, from finishing a problem set, to taking an exam, to managing your course schedule. The P.O.W.E.R. framework can be used outside the classroom, too, providing an effective approach to challenges as diverse as finding a new apartment and doing the weekly grocery shopping.

Whether you are familiar with the P.O.W.E.R. framework already or are encountering it for the first time, take a moment now to review each of its steps in depth.

- Think about what you are ultimately trying to accomplish: Define both your short-term and long-term goals.
- **Long-term goals** relate to major accomplishments that take some time to achieve. **Short-term goals** are relatively limited steps you would take on the road to accomplishing your long-term goals.

- Organize the tools you need to complete your task.
- Effective organization involves both gathering the *physical* tools you will need for your task (for example, a textbook, scrap paper, and so forth) and doing the *mental* work (reviewing lecture notes or major concepts in your book, say) to ensure you are ready to succeed.

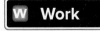

- Using the previous steps as your foundation, do the work to carry out your task.
- In doing math work, keep in mind the lessons you have learned earlier in the term or in past classes.
- Stay motivated, think positively, and focus on achieving the maximum you are capable of.

- Make an honest comparison of your work and your original goals for it.
- Based on your evaluation, revise your work if you feel it can be improved.
- In evaluating math work, try to determine the source of any errors you might have made: Did you make a careless calculation mistake? Is there a concept you are struggling with?

R Rethink

- Think critically both about the work you have done and the process you used to accomplish it.
- Take a "big-picture" look at what you accomplished. Has your work brought you closer to your long-term goals?

Chapter 1 P O W E R Plan

P Prepare

What are your goals for Chapter 1?

1 Be prepared before and during class.

2 Understand the homework to the point where you could do it without needing any help or hints.

3 Use the P.O.W.E.R. framework to help you organize your study: *Measuring Your Math Confidence.*

4 Write your own goal.

_____

_____

What are your objectives for Chapter 1?

1 Review the operations with fractions.

2 Learn to use exponents.

3 Learn to use the order of operations in all situations.

4 Review the basic concepts of geometry.

5 Understand sets of numbers and absolute value.

6 Be able to add, subtract, multiply, and divide real numbers, and translate English expressions to mathematical expressions.

7 Know the properties of real numbers, and apply them to algebraic expressions.

8 Write your own goal.

_____

_____

O Organize

How can you accomplish each goal?

- Don't stay out late the night before and be sure to set your alarm clock!
- Bring a pencil, notebook paper, and textbook to class.
- Avoid distractions by turning off your cell phone during class.
- Pay attention, take good notes, and ask questions.
- Complete your homework on time and ask questions on problems you do not understand.

- Read the directions and show all of your steps.
- Go to the professor's office for help.
- Rework homework and quiz problems, and find similar problems for practice.

- Read the Study Strategy that explains how to use P.O.W.E.R.
- What does P.O.W.E.R. stand for?
- Complete the emPOWERme that appears before the Chapter Summary.

- _____

How can you accomplish each objective?

- Review the rules for writing a fraction in lowest terms. Understand how to multiply, divide, add, and subtract fractions and mixed numbers.

- Exponents indicate repeated multiplication of the base.
- Be able to quickly evaluate and simplify commonly used powers.

- Create your own acronym like **P**lease **E**xcuse **M**y **D**ear **A**unt **S**ally to help you know how to evaluate mathematical expressions.

- Learn all classifications of angles, lines, and triangles.
- Learn all formulas for finding perimeter, circumference, area, and volume of common shapes.

- Be able to identify as well as compare and contrast natural numbers, whole numbers, integers, rational, and irrational numbers.
- Be able to compare real numbers using a number line as well as symbols.
- Find the absolute value of a number, and know what it means.

- Know the rules for adding numbers of the same sign.
- Know how to use absolute value when subtracting numbers or adding numbers with different signs.
- Know the procedures for multiplying and dividing real numbers and when the results will be positive or negative.

- Be able to use the commutative, associative, identity, inverse, and distributive properties.
- Use the properties to help you combine like terms in an expression.

- _____

E Evaluate	Complete the Chapter Review and Chapter Test. How did you do?	R Rethink	• How did you perform on the goals for the chapter? If you had the chance to do this chapter over, what would you do differently?

- Think of a job you might like to have and describe how you would need to use what you have just learned to effectively do that job.
- How has the P.O.W.E.R. framework helped you master the objectives of this chapter? Where else could you use this framework? Make it a point to use P.O.W.E.R. to complete another task this week.

1.1 Review of Fractions

P Prepare	O Organize
What are your objectives for Section 1.1?	**How can you accomplish each objective?**
1 Understand What a Fraction Represents	• Write your own definition of a **fraction,** and include the words *numerator* and *denominator*. • Complete Example 1 on your own. • Complete You Try 1.
2 Write Fractions in Lowest Terms	• Know how to write the prime factorization of a number. • In your own words, write a procedure for writing a fraction in lowest terms. • Complete the examples on your own. • Complete You Trys 2–6.
3 Multiply and Divide Fractions	• In your own words, write a procedure for multiplying fractions and mixed numbers. • Be able to find the reciprocal of a number. • In your own words, write a procedure for dividing fractions and mixed numbers. • Complete the examples on your own. • Complete You Trys 7 and 8.
4 Add and Subtract Fractions	• Understand that to add or subtract fractions, the fractions must have a common denominator. • Know how to find a least common denominator for a group of fractions. • Be able to rewrite a fraction with a different denominator. • In your own words, write a procedure for **Adding and Subtracting Fractions with Unlike Denominators.** • Complete the examples on your own. • Complete You Trys 9–11.

Read the explanations, follow the examples, take notes, and complete the You Trys.

Why review fractions and arithmetic skills? Because the manipulations done in arithmetic and with fractions are precisely the same skills needed to learn algebra.

Let's begin by defining some numbers used in arithmetic:

Natural numbers: 1, 2, 3, 4, 5, …

Whole numbers: 0, 1, 2, 3, 4, 5, …

Natural numbers are often thought of as the counting numbers. **Whole numbers** consist of the natural numbers and zero.

Natural and whole numbers are used to represent complete quantities. To represent a part of a quantity, we can use a fraction.

1 Understand What a Fraction Represents

What is a fraction?

Definition

A **fraction** is a number in the form $\frac{a}{b}$, where $b \neq 0$. a is called the **numerator,** and b is the **denominator.**

 Note

1) A fraction describes a part of a whole quantity.

2) $\frac{a}{b}$ means $a \div b$.

EXAMPLE 1

In-Class Example 1

What part of the figure is shaded?

Answer: $\frac{4}{5}$

What part of the figure is shaded?

Solution

The whole figure is divided into three equal parts. Two of the parts are shaded. Therefore, the part of the figure that is shaded is $\frac{2}{3}$.

$\dfrac{2}{3}$ $\begin{array}{l}\rightarrow \text{Number of shaded parts} \\ \rightarrow \text{Total number of equal parts in the figure}\end{array}$

[YOU TRY 1] What part of the figure is shaded?

---

2 Write Fractions in Lowest Terms

A fraction is in **lowest terms** when the numerator and denominator have no common factors except 1. Before discussing how to write a fraction in lowest terms, we need to know about factors.

Consider the number 12.

$$12 = 3 \cdot 4$$

        ↑   ↑   ↑

    Product Factor Factor

3 and 4 are **factors** of 12. (When we use the term *factors,* we mean natural numbers.) Multiplying 3 and 4 results in 12. 12 is the **product.**

Does 12 have any other factors?

EXAMPLE 2

In-Class Example 2
Find all factors of 18.

Answer: 1, 2, 3, 6, 9, 18

Find all factors of 12.

Solution

$$12 = 3 \cdot 4 \qquad \text{Factors are 3 and 4.}$$
$$12 = 2 \cdot 6 \qquad \text{Factors are 2 and 6.}$$
$$12 = 1 \cdot 12 \qquad \text{Factors are 1 and 12.}$$

These are all of the ways to write 12 as the product of two factors. The factors of 12 are 1, 2, 3, 4, 6, and 12.

[YOU TRY 2] Find all factors of 30.

We can also write 12 as a product of *prime numbers.*

> ## Definition
>
> A **prime number** is a natural number whose only two *different* factors are 1 and itself. (The factors are natural numbers.)

EXAMPLE 3

In-Class Example 3
Is 13 a prime number?

Answer: yes

Is 7 a prime number?

Solution

Yes. The only way to write 7 as a product of natural numbers is $1 \cdot 7$.

[YOU TRY 3] Is 19 a prime number?

> ## Definition
>
> A **composite number** is a natural number with factors other than 1 and itself. Therefore, if a natural number is not prime, it is composite (with the exception of 0 and 1).

[YOU TRY 4]

a) What are the first six prime numbers?

b) What are the first six composite numbers?

To perform various operations in arithmetic and algebra, it is helpful to write a number as the product of its **prime factors.** This is called finding the **prime factorization** of a number. We can use a **factor tree** to help us find the prime factorization of a number.

EXAMPLE 4

In-Class Example 4

Write 260 as the product of its prime factors.

Answer: $2 \cdot 2 \cdot 5 \cdot 13$

Hint

Write a procedure for using a factor tree to write the prime factorization of a number.

Write 120 as the product of its prime factors.

Solution

```
            120
           /   \
         10  ·  12
        / \    / \
      ②·⑤  ②· 6
                / \
              ②·③
```

Think of *any* two natural numbers that multiply to 120.

10 and 12 are not prime, so write them as the product of two factors. Circle the primes.

6 is not prime, so write it as the product of two factors. The factors are primes. Circle them.

Prime factorization: $120 = 2 \cdot 2 \cdot 2 \cdot 3 \cdot 5$.

[YOU TRY 5]

Use a factor tree to write each number as the product of its prime factors.

a) 20   b) 36   c) 90

Let's return to writing a fraction in lowest terms.

EXAMPLE 5

In-Class Example 5

Write $\dfrac{9}{27}$ in lowest terms.

Answer: $\dfrac{1}{3}$

Hint

Describe two ways to write a fraction in lowest terms.

Write $\dfrac{48}{42}$ in lowest terms.

Solution

$\dfrac{48}{42}$ is an **improper fraction.** A fraction is *improper* if its numerator is greater than or equal to its denominator. We will use two methods to express this fraction in lowest terms.

Method 1

Using a factor tree to get the prime factorizations of 48 and 42 and then dividing out common factors, we have

$$\frac{48}{42} = \frac{\overset{1}{\cancel{2}} \cdot 2 \cdot 2 \cdot 2 \cdot \overset{1}{\cancel{3}}}{\underset{1}{\cancel{2}} \cdot \underset{1}{\cancel{3}} \cdot 7} = \frac{2 \cdot 2 \cdot 2}{7} = \frac{8}{7} \text{ or } 1\frac{1}{7}$$

The answer may be expressed as an improper fraction, $\dfrac{8}{7}$, or as a **mixed number,** $1\dfrac{1}{7}$, as long as each is in lowest terms.

Method 2

48 and 42 are each divisible by 6, so we can divide each by 6.

$$\frac{48}{42} = \frac{48 \div 6}{42 \div 6} = \frac{8}{7} \text{ or } 1\frac{1}{7}$$

[**YOU TRY 6**] Write each fraction in lowest terms.

a) $\dfrac{8}{14}$ b) $\dfrac{63}{36}$

3 Multiply and Divide Fractions

> **Procedure** Multiplying Fractions
>
> To multiply fractions, $\dfrac{a}{b} \cdot \dfrac{c}{d}$, we multiply the numerators and multiply the denominators. That is,
>
> $$\frac{a}{b} \cdot \frac{c}{d} = \frac{a \cdot c}{b \cdot d} \quad \text{if } b \neq 0 \text{ and } d \neq 0.$$

EXAMPLE 6

In-Class Example 6

Multiply. Write each answer in lowest terms.

a) $\dfrac{5}{7} \cdot \dfrac{9}{4}$ b) $\dfrac{8}{5} \cdot \dfrac{25}{4}$

c) $3\dfrac{3}{4} \cdot 2\dfrac{5}{6}$

Answer:

a) $\dfrac{45}{28}$ b) 10   c) $\dfrac{85}{8}$ or $10\dfrac{5}{8}$

Multiply. Write each answer in lowest terms.

a) $\dfrac{3}{8} \cdot \dfrac{7}{4}$ b) $\dfrac{10}{21} \cdot \dfrac{21}{25}$ c) $4\dfrac{2}{5} \cdot 1\dfrac{7}{8}$

Solution

a) $\dfrac{3}{8} \cdot \dfrac{7}{4} = \dfrac{3 \cdot 7}{8 \cdot 4}$ Multiply numerators; multiply denominators.

$\phantom{\dfrac{3}{8} \cdot \dfrac{7}{4}} = \dfrac{21}{32}$ 21 and 32 contain no common factors, so $\dfrac{21}{32}$ is in lowest terms.

b) $\dfrac{10}{21} \cdot \dfrac{21}{25}$

We can take out the common factors before we multiply.

5 is the greatest common factor of 10 and 25. Divide 10 and 25 by **5**.

$$\frac{\overset{2}{\cancel{10}}}{\underset{1}{\cancel{21}}} \cdot \frac{\overset{1}{\cancel{21}}}{\underset{5}{\cancel{25}}} = \frac{2}{1} \cdot \frac{1}{5} = \frac{2 \cdot 1}{1 \cdot 5} = \boxed{\frac{2}{5}}$$

21 is the greatest common factor of 21 and 21. Divide each 21 by **21**.

Usually, it is easier to remove the common factors before multiplying rather than after finding the product.

c) $4\dfrac{2}{5} \cdot 1\dfrac{7}{8}$

Before multiplying mixed numbers, we must change them to improper fractions. Recall that $4\dfrac{2}{5}$ is the same as $4 + \dfrac{2}{5}$. Here is one way to rewrite $4\dfrac{2}{5}$ as an improper fraction:

1) Multiply the denominator and the whole number: $5 \cdot 4 = 20$.

2) Add the numerator: $20 + 2 = 22$.

3) Put the sum over the denominator: $\dfrac{22}{5}$

> **W Hint**
> Do you prefer to divide out common factors in the numerators and denominators *before* or *after* multiplying?

To summarize, $4\dfrac{2}{5} = \dfrac{(5 \cdot 4) + 2}{5} = \dfrac{20 + 2}{5} = \dfrac{22}{5}$.

Then, $1\dfrac{7}{8} = \dfrac{(8 \cdot 1) + 7}{8} = \dfrac{8 + 7}{8} = \dfrac{15}{8}$.

$$4\dfrac{2}{5} \cdot 1\dfrac{7}{8} = \dfrac{22}{5} \cdot \dfrac{15}{8}$$

5 and 15 each divide by **5**.

$$= \dfrac{\overset{11}{\cancel{22}}}{\underset{1}{\cancel{5}}} \cdot \dfrac{\overset{3}{\cancel{15}}}{\underset{4}{\cancel{8}}}$$

8 and 22 each divide by **2**.

$$= \dfrac{11}{1} \cdot \dfrac{3}{4}$$

$$= \dfrac{33}{4} \text{ or } 8\dfrac{1}{4} \quad \text{Write the result as an improper fraction or as a mixed number.}$$

[YOU TRY 7] Multiply. Write the answer in lowest terms.

a) $\dfrac{1}{5} \cdot \dfrac{4}{9}$ b) $\dfrac{8}{25} \cdot \dfrac{15}{32}$ c) $3\dfrac{3}{4} \cdot 2\dfrac{2}{3}$

To divide fractions, we must define a reciprocal.

Definition

The **reciprocal** of a number, $\dfrac{a}{b}$, is $\dfrac{b}{a}$ since $\dfrac{a}{b} \cdot \dfrac{b}{a} = 1$. That is, a nonzero number times its reciprocal equals 1. Notice that $a \neq 0$ and $b \neq 0$.

For example, the reciprocal of $\dfrac{5}{9}$ is $\dfrac{9}{5}$ since $\dfrac{\overset{1}{\cancel{5}}}{\underset{1}{\cancel{9}}} \cdot \dfrac{\overset{1}{\cancel{9}}}{\underset{1}{\cancel{5}}} = \dfrac{1}{1} = 1$.

Procedure Dividing Fractions

Division of fractions: Let a, b, c, and d represent numbers so that b, c, and d do not equal zero. Then,

$$\frac{a}{b} \div \frac{c}{d} = \frac{a}{b} \cdot \frac{d}{c}.$$

Note

To perform division involving fractions, multiply the first fraction by the reciprocal of the second.

EXAMPLE 7

Divide. Write the answer in lowest terms.

a) $\dfrac{3}{8} \div \dfrac{10}{11}$ b) $\dfrac{3}{2} \div 9$ c) $5\dfrac{1}{4} \div 1\dfrac{1}{13}$

Solution

a) $\dfrac{3}{8} \div \dfrac{10}{11} = \dfrac{3}{8} \cdot \dfrac{11}{10}$ Multiply $\dfrac{3}{8}$ by the reciprocal of $\dfrac{10}{11}$.

$\qquad\qquad = \dfrac{33}{80}$ Multiply.

b) $\dfrac{3}{2} \div 9 = \dfrac{3}{2} \cdot \dfrac{1}{9}$ The reciprocal of 9 is $\dfrac{1}{9}$.

$\qquad\quad = \dfrac{\overset{1}{3}}{2} \cdot \dfrac{1}{\underset{3}{9}}$ Divide out a common factor of 3.

$\qquad\quad = \dfrac{1}{6}$ Multiply.

c) $5\dfrac{1}{4} \div 1\dfrac{1}{13} = \dfrac{21}{4} \div \dfrac{14}{13}$ Change the mixed numbers to improper fractions.

$\qquad\qquad = \dfrac{21}{4} \cdot \dfrac{13}{14}$ Multiply $\dfrac{21}{4}$ by the reciprocal of $\dfrac{14}{13}$.

$\qquad\qquad = \dfrac{\overset{3}{21}}{4} \cdot \dfrac{13}{\underset{2}{14}}$ Divide out a common factor of 7.

$\qquad\qquad = \dfrac{39}{8} \text{ or } 4\dfrac{7}{8}$ Express the answer as an improper fraction or mixed number.

In-Class Example 7

Divide. Write each answer in lowest terms.

a) $\dfrac{5}{11} \div \dfrac{3}{10}$ b) $\dfrac{2}{3} \div 4$

c) $2\dfrac{1}{6} \div 5\dfrac{1}{5}$

Answer: a) $\dfrac{50}{33}$ or $1\dfrac{17}{33}$

b) $\dfrac{1}{6}$ c) $\dfrac{5}{12}$

W Hint

In your own words, write a procedure for dividing fractions and mixed numbers.

YOU TRY 8

Divide. Write the answer in lowest terms.

a) $\dfrac{2}{7} \div \dfrac{3}{5}$ b) $\dfrac{3}{10} \div \dfrac{9}{16}$ c) $9\dfrac{1}{6} \div 5$

---

4 Add and Subtract Fractions

The pizza at the top left is cut into eight equal slices. If you eat two pieces and your friend eats three pieces, what fraction of the pizza was eaten?

Five out of the eight pieces were eaten. As a fraction, we can say that you and your friend ate $\frac{5}{8}$ of the pizza.

Let's set up this problem as the sum of two fractions.

Fraction you ate + Fraction your friend ate = Fraction of the pizza eaten

$$\frac{2}{8} \quad + \quad \frac{3}{8} \quad = \quad \frac{5}{8}$$

To add $\frac{2}{8} + \frac{3}{8}$, we added the numerators and kept the denominator the same. Notice that these fractions have the same denominator.

Procedure Adding and Subtracting Fractions

Let a, b, and c be numbers such that $c \neq 0$.

$$\frac{a}{c} + \frac{b}{c} = \frac{a+b}{c} \quad \text{and} \quad \frac{a}{c} - \frac{b}{c} = \frac{a-b}{c}$$

To add or subtract fractions, the denominators must be the same. (This is called a **common denominator**.) Then, add (or subtract) the numerators and keep the same denominator.

EXAMPLE 8

Perform the operation and simplify.

a) $\frac{3}{11} + \frac{5}{11}$ 　　 b) $\frac{17}{30} - \frac{13}{30}$

Solution

a) $\frac{3}{11} + \frac{5}{11} = \frac{3+5}{11}$ 　　 Add the numerators, and keep the denominator the same.

$= \frac{8}{11}$

b) $\frac{17}{30} - \frac{13}{30} = \frac{17-13}{30}$ 　　 Subtract the numerators, and keep the denominator the same.

$= \frac{4}{30}$ 　　 This is not in lowest terms, so simplify it.

$= \frac{2}{15}$ 　　 Simplify.

In-Class Example 8

Perform the operation and simplify.

a) $\frac{7}{15} + \frac{2}{15}$ 　 b) $\frac{14}{27} - \frac{5}{27}$

Answer: a) $\frac{3}{5}$ 　 b) $\frac{1}{3}$

[**YOU TRY 9**]

Perform the operation and simplify.

a) $\frac{5}{9} + \frac{2}{9}$ 　　 b) $\frac{19}{20} - \frac{7}{20}$

The examples given so far contain common denominators. How do we add or subtract fractions that do not have common denominators? We find the least common denominator for the fractions and rewrite each fraction with this denominator.

The **least common denominator (LCD)** of two fractions is the least common multiple of the numbers in the denominators.

EXAMPLE 9

In-Class Example 9

Find the LCD for $\frac{1}{2}$ and $\frac{2}{3}$.

Answer: 6

Find the LCD for $\frac{3}{4}$ and $\frac{1}{6}$.

Solution

Method 1

List some multiples of 4 and 6.

4:   4, 8, $\boxed{12}$, 16, 20, *24*, ...

6:   6, $\boxed{12}$, 18, *24*, 30, ...

Although 24 is a multiple of 6 and of 4, the *least* common multiple, and therefore the least common denominator, is 12.

Method 2

We can also use the prime factorizations of 4 and 6 to find the LCD.

To find the LCD:

1) Find the prime factorization of each number.

2) The least common denominator will include each different factor appearing in the factorizations.

W Hint

In your own words, explain how to find the LCD of a group of fractions.

3) If a factor appears more than once in any prime factorization, use it in the LCD the *maximum number of times* it appears in any single factorization. Multiply the factors.

$$4 = 2 \cdot 2$$
$$6 = 2 \cdot 3$$

The least common multiple of 4 and 6 is

$$\underbrace{2 \cdot 2}_{\substack{\text{2 appears at} \\ \text{most twice in} \\ \text{any single} \\ \text{factorization.}}} \cdot \underbrace{3}_{\substack{\text{3 appears} \\ \text{once in a} \\ \text{factorization.}}} = 12$$

The LCD of $\frac{3}{4}$ and $\frac{1}{6}$ is 12.

[YOU TRY 10]   Find the LCD for $\frac{5}{6}$ and $\frac{4}{9}$.

To add or subtract fractions with unlike denominators, begin by identifying the least common denominator. Then, we must rewrite each fraction with this LCD. This will not change the value of the fraction; we will obtain an *equivalent* fraction.

> **Procedure** Adding or Subtracting Fractions with Unlike Denominators
>
> To add or subtract fractions with unlike denominators:
>
> 1) Determine, and write down, the least common denominator (LCD).
> 2) Rewrite each fraction with the LCD.
> 3) Add or subtract.
> 4) Express the answer in lowest terms.

EXAMPLE 10

Add or subtract.

a) $\dfrac{2}{9} + \dfrac{1}{6}$ b) $6\dfrac{7}{8} - 3\dfrac{1}{2}$

In-Class Example 10

Add or subtract.

a) $\dfrac{4}{9} + \dfrac{5}{12}$ b) $3\dfrac{3}{4} - 1\dfrac{5}{6}$

Answer: a) $\dfrac{31}{36}$

b) $\dfrac{23}{12}$ or $1\dfrac{11}{12}$

W Hint

Whenever you multiply, divide, add, or subtract fractions, you should *always* look at the result and ask yourself, "Is it in lowest terms?" If not, write the answer in lowest terms.

Solution

a) $\dfrac{2}{9} + \dfrac{1}{6}$ $\text{LCD} = 18$ Identify the least common denominator.

$\dfrac{2}{9} \cdot \dfrac{2}{2} = \dfrac{4}{18}$ $\dfrac{1}{6} \cdot \dfrac{3}{3} = \dfrac{3}{18}$ Rewrite each fraction with a denominator of 18.

$\dfrac{2}{9} + \dfrac{1}{6} = \dfrac{4}{18} + \dfrac{3}{18} = \dfrac{7}{18}$

b) $6\dfrac{7}{8} - 3\dfrac{1}{2}$

Method 1

Keep the numbers in mixed number form. Subtract the whole number parts and subtract the fractional parts. Get a common denominator for the fractional parts.

For $6\dfrac{7}{8}$ and $3\dfrac{1}{2}$, the LCD is 8.   Identify the least common denominator.

$6\dfrac{7}{8}$: $\dfrac{7}{8}$ has the LCD of 8.

$3\dfrac{1}{2}$: $\dfrac{1}{2} \cdot \dfrac{4}{4} = \dfrac{4}{8}$.   So, $3\dfrac{1}{2} = 3\dfrac{4}{8}$.   Rewrite $\dfrac{1}{2}$ with a denominator of 8.

$6\dfrac{7}{8} - 3\dfrac{1}{2} = 6\dfrac{7}{8} - 3\dfrac{4}{8}$

$= 3\dfrac{3}{8}$ Subtract whole number parts and subtract fractional parts.

Method 2

Rewrite each mixed number as an improper fraction, get a common denominator, then subtract.

$6\dfrac{7}{8} - 3\dfrac{1}{2} = \dfrac{55}{8} - \dfrac{7}{2}$ $\text{LCD} = 8$ $\dfrac{55}{8}$ already has a denominator of 8.

$\dfrac{7}{2} \cdot \dfrac{4}{4} = \dfrac{28}{8}$ Rewrite $\dfrac{7}{2}$ with a denominator of 8.

$6\dfrac{7}{8} - 3\dfrac{1}{2} = \dfrac{55}{8} - \dfrac{7}{2} = \dfrac{55}{8} - \dfrac{28}{8} = \dfrac{27}{8}$ or $3\dfrac{3}{8}$

E Evaluate **1.1** Exercises   Do the exercises, and check your work.

*Additional answers can be found in the Answers to Exercises appendix.

Objective 1: Understand What a Fraction Represents

1) What fraction of each figure is shaded? If the fraction is not in lowest terms, reduce it.

a)  $\dfrac{2}{5}$

b)  $\dfrac{2}{3}$

c)  1

2) What fraction of each figure is *not* shaded? If the fraction is not in lowest terms, reduce it.

a)  $\dfrac{1}{3}$ b) 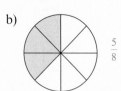 $\dfrac{5}{8}$

c) 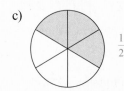 $\dfrac{1}{2}$

3) Draw a rectangle divided into 8 equal parts. Shade in $\dfrac{4}{8}$ of the rectangle. Write another fraction to represent how much of the rectangle is shaded. $\dfrac{1}{2}$

4) Draw a rectangle divided into 6 equal parts. Shade in $\dfrac{2}{6}$ of the rectangle. Write another fraction to represent how much of the rectangle is shaded. $\dfrac{1}{3}$

Objective 2: Write Fractions in Lowest Terms

5) Find all factors of each number.

a) 18   1, 2, 3, 6, 9, 18

b) 40   1, 2, 4, 5, 8, 10, 20, 40

c) 23   1, 23

6) Find all factors of each number.

a) 20   1, 2, 4, 5, 10, 20

b) 17   1, 17

c) 60   1, 2, 3, 4, 5, 6, 10, 12, 15, 20, 30, 60

7) Identify each number as prime or composite.

a) 27   composite

b) 34   composite

c) 11   prime

8) Identify each number as prime or composite.

a) 2   prime

b) 57   composite

c) 90   composite

9) Is 3072 prime or composite? Explain your answer.
Composite. It is divisible by 2 and has other factors as well.

10) Is 4185 prime or composite? Explain your answer.
Composite. It is divisible by 5 and has other factors as well.

11) Use a factor tree to find the prime factorization of each number.

   a) 18 $2 \cdot 3 \cdot 3$ b) 54 $2 \cdot 3 \cdot 3 \cdot 3$

   c) 42 $2 \cdot 3 \cdot 7$ d) 150 $2 \cdot 3 \cdot 5 \cdot 5$

12) Explain, in words, how to use a factor tree to find the prime factorization of 72.   Answers may vary.

13) Write each fraction in lowest terms.

   a) $\dfrac{9}{12}$ $\dfrac{3}{4}$ b) $\dfrac{54}{72}$ $\dfrac{3}{4}$

   c) $\dfrac{84}{35}$ $\dfrac{12}{5}$ or $2\dfrac{2}{5}$ d) $\dfrac{120}{280}$ $\dfrac{3}{7}$

14) Write each fraction in lowest terms.

   a) $\dfrac{21}{35}$ $\dfrac{3}{5}$ b) $\dfrac{48}{80}$ $\dfrac{3}{5}$

   c) $\dfrac{125}{500}$ $\dfrac{1}{4}$ d) $\dfrac{900}{450}$ 2

Objective 3: Multiply and Divide Fractions

15) Multiply. Write the answer in lowest terms.

   a) $\dfrac{2}{7} \cdot \dfrac{3}{5}$ $\dfrac{6}{35}$ b) $\dfrac{15}{26} \cdot \dfrac{4}{9}$ $\dfrac{10}{39}$

   c) $\dfrac{1}{2} \cdot \dfrac{14}{15}$ $\dfrac{7}{15}$ d) $\dfrac{42}{55} \cdot \dfrac{22}{35}$ $\dfrac{12}{25}$

   e) $4 \cdot \dfrac{1}{8}$ $\dfrac{1}{2}$ f) $6\dfrac{1}{8} \cdot \dfrac{2}{7}$ $\dfrac{7}{4}$ or $1\dfrac{3}{4}$

16) Multiply. Write the answer in lowest terms.

   a) $\dfrac{1}{6} \cdot \dfrac{5}{9}$ $\dfrac{5}{54}$ b) $\dfrac{9}{20} \cdot \dfrac{6}{7}$ $\dfrac{27}{70}$

   c) $\dfrac{12}{25} \cdot \dfrac{25}{36}$ $\dfrac{1}{3}$ d) $\dfrac{30}{49} \cdot \dfrac{21}{100}$ $\dfrac{9}{70}$

   e) $\dfrac{7}{15} \cdot 10$ $\dfrac{14}{3}$ or $4\dfrac{2}{3}$ f) $7\dfrac{5}{7} \cdot 1\dfrac{5}{9}$ 12

17) When Elizabeth multiplies $5\dfrac{1}{2} \cdot 2\dfrac{1}{3}$, she gets $10\dfrac{1}{6}$.

   What was her mistake? What is the correct answer?

18) Explain how to multiply mixed numbers.
   Convert mixed numbers to improper fractions, then multiply.

19) Divide. Write the answer in lowest terms.

   a) $\dfrac{1}{42} \div \dfrac{2}{7}$ $\dfrac{1}{12}$ b) $\dfrac{3}{11} \div \dfrac{4}{5}$ $\dfrac{15}{44}$

   c) $\dfrac{18}{35} \div \dfrac{9}{10}$ $\dfrac{4}{7}$ d) $\dfrac{14}{15} \div \dfrac{2}{15}$ 7

   e) $6\dfrac{2}{5} \div 1\dfrac{13}{15}$ $\dfrac{24}{7}$ or $3\dfrac{3}{7}$ f) $\dfrac{4}{7} \div 8$ $\dfrac{1}{14}$

20) Explain how to divide mixed numbers.

Objective 4: Add and Subtract Fractions

21) Find the least common multiple of 10 and 15.   30

22) Find the least common multiple of 12 and 9.   36

23) Find the least common denominator for each group of fractions.

   a) $\dfrac{9}{10}, \dfrac{11}{30}$ 30   b) $\dfrac{7}{8}, \dfrac{5}{12}$ 24

   c) $\dfrac{4}{9}, \dfrac{1}{6}, \dfrac{3}{4}$ 36

24) Find the least common denominator for each group of fractions.

   a) $\dfrac{3}{14}, \dfrac{2}{7}$ 14   b) $\dfrac{17}{25}, \dfrac{3}{10}$ 50

   c) $\dfrac{29}{30}, \dfrac{3}{4}, \dfrac{9}{20}$ 60

25) Add or subtract. Write the answer in lowest terms.

   a) $\dfrac{6}{11} + \dfrac{2}{11}$ $\dfrac{8}{11}$ b) $\dfrac{19}{20} - \dfrac{7}{20}$ $\dfrac{3}{5}$

   c) $\dfrac{4}{25} + \dfrac{2}{25} + \dfrac{9}{25}$ $\dfrac{3}{5}$ d) $\dfrac{2}{9} + \dfrac{1}{6}$ $\dfrac{7}{18}$

   e) $\dfrac{3}{5} + \dfrac{11}{30}$ $\dfrac{29}{30}$ f) $\dfrac{13}{18} - \dfrac{2}{3}$ $\dfrac{1}{18}$

   g) $\dfrac{4}{7} + \dfrac{5}{9}$ $\dfrac{71}{63}$ or $1\dfrac{8}{63}$ h) $\dfrac{5}{6} - \dfrac{1}{4}$ $\dfrac{7}{12}$

   i) $\dfrac{3}{10} + \dfrac{7}{20} + \dfrac{3}{4}$ $\dfrac{7}{5}$ or $1\dfrac{2}{5}$ j) $\dfrac{1}{6} + \dfrac{2}{9} + \dfrac{10}{27}$ $\dfrac{41}{54}$

26) Add or subtract. Write the answer in lowest terms.

   a) $\dfrac{8}{9} - \dfrac{5}{9}$ $\dfrac{1}{3}$ b) $\dfrac{14}{15} - \dfrac{2}{15}$ $\dfrac{4}{5}$

   c) $\dfrac{11}{36} + \dfrac{13}{36}$ $\dfrac{2}{3}$ d) $\dfrac{16}{45} + \dfrac{8}{45} + \dfrac{11}{45}$ $\dfrac{7}{9}$

   e) $\dfrac{15}{16} - \dfrac{3}{4}$ $\dfrac{3}{16}$ f) $\dfrac{1}{8} + \dfrac{1}{6}$ $\dfrac{7}{24}$

   g) $\dfrac{5}{8} - \dfrac{2}{9}$ $\dfrac{29}{72}$ h) $\dfrac{23}{30} - \dfrac{19}{90}$ $\dfrac{5}{9}$

   i) $\dfrac{1}{6} + \dfrac{1}{4} + \dfrac{2}{3}$ $\dfrac{13}{12}$ or $1\dfrac{1}{12}$ j) $\dfrac{3}{10} + \dfrac{2}{5} + \dfrac{4}{15}$ $\dfrac{29}{30}$

27) Add or subtract. Write the answer in lowest terms.

a) $8\frac{5}{11} + 6\frac{2}{11}$ $14\frac{7}{11}$ b) $2\frac{1}{10} + 9\frac{3}{10}$ $11\frac{2}{5}$

c) $7\frac{11}{12} - 1\frac{5}{12}$ $6\frac{1}{2}$ d) $3\frac{1}{5} + 2\frac{1}{4}$ $5\frac{9}{20}$

e) $5\frac{2}{3} - 4\frac{4}{15}$ $1\frac{2}{5}$ f) $9\frac{5}{8} - 5\frac{3}{10}$ $4\frac{13}{40}$

g) $4\frac{3}{7} + 6\frac{3}{4}$ $11\frac{5}{28}$ h) $7\frac{13}{20} + \frac{4}{5}$ $8\frac{9}{20}$

28) Add or subtract. Write the answer in lowest terms.

a) $3\frac{2}{7} + 1\frac{3}{7}$ $4\frac{5}{7}$ b) $8\frac{5}{16} + 7\frac{3}{16}$ $15\frac{1}{2}$

c) $5\frac{13}{20} - 3\frac{5}{20}$ $2\frac{2}{5}$ d) $10\frac{8}{9} - 2\frac{1}{3}$ $8\frac{5}{9}$

e) $1\frac{5}{12} + 2\frac{3}{8}$ $3\frac{19}{24}$ f) $4\frac{1}{9} + 7\frac{2}{5}$ $11\frac{23}{45}$

g) $1\frac{5}{6} + 4\frac{11}{18}$ $6\frac{4}{9}$ h) $3\frac{7}{8} + 4\frac{2}{5}$ $8\frac{11}{40}$

Mixed Exercises: Objectives 3 and 4

29) For Valentine's Day, Alex wants to sew teddy bears for her friends. Each bear requires $1\frac{2}{3}$ yd of fabric. If she has 7 yd of material, how many bears can Alex make? How much fabric will be left over?   Four bears: $\frac{1}{3}$ yd remaining

30) A chocolate chip cookie recipe that makes 24 cookies uses $\frac{3}{4}$ cup of brown sugar. If Raphael wants to make 48 cookies, how much brown sugar does he need? $1\frac{1}{2}$ cups

31) Nine weeks into the 2009 Major League Baseball season, Nyjer Morgan of the Pittsburgh Pirates had been up to bat 175 times. He got a hit $\frac{2}{7}$ of the time. How many hits did Nyjer have?   50

32) When all children are present, Ms. Yamoto has 30 children in her fifth-grade class. One day during flu season, $\frac{3}{5}$ of them were absent. How many children were absent on this day?   18

33) Mr. Burnett plans to have a picture measuring $18\frac{3}{8}''$ by $12\frac{1}{4}''$ custom framed. The frame he chose is $2\frac{1}{8}''$ wide. What will be the new length and width of the picture plus the frame? $16\frac{1}{2}$ in. by $22\frac{5}{8}$ in.

34) Andre is building a table in his workshop. For the legs, he bought wood that is 30 in. long. If the legs are to be $26\frac{3}{4}$ in. tall, how many inches must he cut off to get the desired height? $3\frac{1}{4}$ in.

35) When Rosa opens the kitchen cabinet, she finds three partially filled bags of flour. One contains $\frac{2}{3}$ cup, another contains $1\frac{1}{4}$ cups, and the third contains $1\frac{1}{2}$ cups. How much flour does she have all together? $3\frac{5}{12}$ cups

36) Tamika takes the same route to school every day. (See the figure.) How far does she walk to school? $\frac{9}{10}$ mi

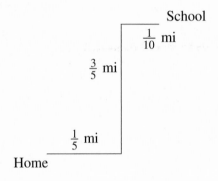

37) The gas tank of Jenny's car holds $11\frac{3}{5}$ gal, while Scott's car holds $16\frac{3}{4}$ gal. How much more gasoline does Scott's car hold? $5\frac{3}{20}$ gal

38) Mr. Johnston is building a brick wall along his driveway. He estimates that one row of brick plus mortar will be $4\frac{1}{4}$ in. high. How many rows will he need to construct a wall that is 34 in. high?   8

39) For homework, Bill's math teacher assigned 42 problems. Bill finished $\frac{5}{6}$ of them. How many problems did he do?   35

40) Clarice's parents tell her that she must deposit $\frac{1}{3}$ of the money she earns from babysitting into her savings account, but she can keep the rest. If she earns $117 in 1 week during the summer, how much does she deposit, and how much does she keep?   deposit: $39; keep: $78

41) A welder must construct a beam with a total length of $32\frac{7}{8}$ in. If he has already joined a $14\frac{1}{6}$-in. beam with a $10\frac{3}{4}$-in. beam, find the length of a third beam needed to reach the total length. $7\frac{23}{24}$ in.

42) Telephia, a market research company, surveyed 1500 teenage cell phone owners. The company learned that $\frac{2}{3}$ of them use cell phone data services.

How many teenagers surveyed use cell phone data services? (*American Demographics,* May 2004, Vol. 26, Issue 4, p. 10)   1000

R Rethink

R1) Were there any problems you could not answer or that you got wrong? If so, write them down or circle them. Look back in the section to see whether you can figure out how to do them, or ask your instructor for help.

R2) Can you think of a situation where you used fractions in the last week? What was that situation?

R3) Write your own application problem involving fractions, and give it to another student to solve.

1.2 Exponents and Order of Operations

P Prepare

O Organize

What are your objectives for Section 1.2?	How can you accomplish each objective?
1 Use Exponents	• Understand the prime factorization of a number, exponent, and exponential expression. • Memorize the table of commonly-used powers. • Complete the given examples on your own. • Complete You Trys 1 and 2.
2 Use the Order of Operations	• Write **The Order of Operations** in your own words. • Understand how the phrase **P**lease **E**xcuse **M**y **D**ear **A**unt **S**ally can help you master the order of operations. • Complete the given examples on your own. • Complete You Trys 3–5.

W Work

Read the explanations, follow the examples, take notes, and complete the You Trys.

1 Use Exponents

In Section 1.1, we discussed the prime factorization of a number. The prime factorization of 8 is $8 = 2 \cdot 2 \cdot 2$.

We can write $2 \cdot 2 \cdot 2$ another way, by using an *exponent*.

$$2 \cdot 2 \cdot 2 = 2^3 \leftarrow \text{exponent (or power)}$$
$$\uparrow$$
$$\text{base}$$

2 is the *base*. 2 is a *factor* that appears three times. 3 is the *exponent* or *power*. An **exponent** represents repeated multiplication. We read 2^3 as "2 to the third power" or "2 cubed." 2^3 is called an **exponential expression.**

Hint

Notice that the exponent tells you how many times the base is multiplied by itself.

EXAMPLE 1

In-Class Example 1

Rewrite each product in exponential form.
a) $5 \cdot 5 \cdot 5 \cdot 5$
b) $4 \cdot 4 \cdot 4 \cdot 4 \cdot 4 \cdot 4$

Answer: a) 5^4 b) 4^6

Rewrite each product in exponential form.

a) $9 \cdot 9 \cdot 9 \cdot 9$ b) $7 \cdot 7$

Solution

a) $9 \cdot 9 \cdot 9 \cdot 9 = 9^4$ 9 is the base. It appears as a factor 4 times. So, 4 is the exponent.

b) $7 \cdot 7 = 7^2$ 7 is the base. 2 is the exponent.
    This is read as "7 squared."

[YOU TRY 1] Rewrite each product in exponential form.

a) $8 \cdot 8 \cdot 8 \cdot 8 \cdot 8$ b) $\dfrac{3}{2} \cdot \dfrac{3}{2} \cdot \dfrac{3}{2} \cdot \dfrac{3}{2}$

We can also *evaluate* an exponential expression.

EXAMPLE 2

In-Class Example 2

Evaluate.
a) 4^3 b) 11^2
c) $\left(\dfrac{2}{3}\right)^4$ d) 1^5

Answer: a) 64 b) 121
c) $\dfrac{16}{81}$ d) 1

Evaluate.

a) 2^5 b) 5^3 c) $\left(\dfrac{4}{7}\right)^2$ d) 8^1 e) 1^4

Solution

a) $2^5 = 2 \cdot 2 \cdot 2 \cdot 2 \cdot 2 = 32$ 2 appears as a factor 5 times.

b) $5^3 = 5 \cdot 5 \cdot 5 = 125$ 5 appears as a factor 3 times.

c) $\left(\dfrac{4}{7}\right)^2 = \dfrac{4}{7} \cdot \dfrac{4}{7} = \dfrac{16}{49}$ $\dfrac{4}{7}$ appears as a factor 2 times.

d) $8^1 = 8$ 8 is a factor only once.

e) $1^4 = 1 \cdot 1 \cdot 1 \cdot 1 = 1$ 1 appears as a factor 4 times.

Note

1 raised to any natural number power is 1 since 1 multiplied by itself equals 1.

Evaluate.

a) 3^4 b) 8^2 c) $\left(\dfrac{3}{4}\right)^3$

It is generally agreed that there are some skills in arithmetic that everyone should have in order to be able to acquire other math skills. Knowing the basic multiplication facts, for example, is essential for learning how to add, subtract, multiply, and divide fractions as well as how to perform many other operations in arithmetic and algebra. Similarly, memorizing powers of certain bases is necessary for learning how to apply the rules of exponents (Chapter 5) and for working with radicals (Chapter 9). Therefore, the powers listed here must be memorized in order to be successful in the previously mentioned, as well as other, topics. Throughout this book, it is assumed that students know these powers:

W Hint

Write down any patterns you see in the "Powers to Memorize" table.

Powers to Memorize						
$2^1 = 2$	$3^1 = 3$	$4^1 = 4$	$5^1 = 5$	$6^1 = 6$	$8^1 = 8$	$10^1 = 10$
$2^2 = 4$	$3^2 = 9$	$4^2 = 16$	$5^2 = 25$	$6^2 = 36$	$8^2 = 64$	$10^2 = 100$
$2^3 = 8$	$3^3 = 27$	$4^3 = 64$	$5^3 = 125$			$10^3 = 1000$
$2^4 = 16$	$3^4 = 81$					
$2^5 = 32$				$7^1 = 7$	$9^1 = 9$	$11^1 = 11$
$2^6 = 64$				$7^2 = 49$	$9^2 = 81$	$11^2 = 121$
						$12^1 = 12$
						$12^2 = 144$
						$13^1 = 13$
						$13^2 = 169$

(Hint: Making flash cards might help you learn these facts.)

2   Use the Order of Operations

We will begin this topic with a problem for the student:

Evaluate $40 - 24 \div 8 + (5 - 3)^2$.

What answer did you get? 41? or 6? or 33? Or, did you get another result?

Most likely you obtained one of the three answers just given. Only one is correct, however. If we do not have rules to guide us in evaluating expressions, it is easy to get the incorrect answer.

Therefore, here are the rules we follow. This is called the **order of operations.**

Procedure The Order of Operations

Simplify expressions in the following order:

1) If parentheses or other grouping symbols appear in an expression, simplify what is in these grouping symbols first.

2) Simplify expressions with exponents.

3) Perform multiplication and division from left to right.

4) Perform addition and subtraction from left to right.

Think about the "You Try" problem. Did you evaluate it using the order of operations? Let's look at that expression.

EXAMPLE 3

Evaluate $40 - 24 \div 8 + (5 - 3)^2$.

In-Class Example 3

Evaluate.
$9 \cdot 5 - (12 \div 4) + 8 \cdot 4 - 32$.

Answer: 42

 Hint

When practicing these problems, be sure to write out every step.

Solution

$$40 - 24 \div 8 + (5 - 3)^2$$

$\qquad 40 - 24 \div 8 + 2^2$ First, perform the operation in the parentheses.

$\qquad 40 - 24 \div 8 + 4$ Exponents are done before division, addition, and subtraction.

$\qquad\quad 40 - 3 + 4$ Perform division before addition and subtraction.

$\qquad\qquad 37 + 4$ When an expression contains only addition and subtraction, perform the operations starting at the left and moving to the right.

$\qquad\qquad\quad 41$

[YOU TRY 4]

Evaluate: $12 \cdot 3 - (2 + 1)^2 \div 9$.

A good way to remember the order of operations is to remember the sentence, "**P**lease **E**xcuse **M**y **D**ear **A**unt **S**ally" (**P**arentheses, **E**xponents, **M**ultiplication and **D**ivision from left to right, **A**ddition and **S**ubtraction from left to right). Don't forget that multiplication and division are at the same "level" in the process of performing operations and that addition and subtraction are at the same "level."

EXAMPLE 4

Evaluate.

In-Class Example 4

Evaluate.
a) $2[30 - (28 \div 2)] + 5$
b) $\dfrac{3 + (5 - 2)^3}{2^2 + 24 \div 4}$

Answer: a) 37   b) 3

a) $4[3 + (10 \div 2)] - 11$ b) $\dfrac{(9 - 6)^3 \cdot 2}{26 - 4 \cdot 5}$

Solution

a) $4[3 + (10 \div 2)] - 11$

This expression contains two sets of grouping symbols: **brackets** [ ] and **parentheses** ( ). Perform the operation in the **innermost** grouping symbol first—the parentheses in this case.

$\qquad 4[3 + (10 \div 2)] - 11 = 4[3 + 5] - 11$ Innermost grouping symbol

$\qquad\qquad\qquad\qquad\qquad\quad = 4[8] - 11$ Brackets

$\qquad\qquad\qquad\qquad\qquad\quad = 32 - 11$ Perform multiplication before subtraction.

$\qquad\qquad\qquad\qquad\qquad\quad = 21$ Subtract.

b) $\dfrac{(9 - 6)^3 \cdot 2}{26 - 4 \cdot 5}$

The fraction bar in this expression acts as a grouping symbol. Therefore, simplify the numerator, simplify the denominator, then simplify the resulting fraction, if possible.

$$\frac{(9 - 6)^3 \cdot 2}{26 - 4 \cdot 5} = \frac{3^3 \cdot 2}{26 - 20} \qquad \text{Parentheses} \quad \text{Multiply.}$$

$$= \frac{27 \cdot 2}{6} \qquad \text{Exponent} \quad \text{Subtract.}$$

$$= \frac{54}{6} \qquad \text{Multiply.}$$

$$= 9 \qquad \text{Divide.}$$

Using Technology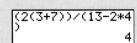

We can use a graphing calculator to check our answer when we evaluate an expression by hand. The order of operations is built into the calculator. For example, evaluate the expression $\dfrac{2(3 + 7)}{13 - 2 \cdot 4}$. To evaluate the expression using a graphing calculator, enter the following on the home screen: (2(3+7))/(13−2*4) and then press ENTER. The result is 4, as shown on the screen.

```
(2(3+7))/(13-2*4
)
                4
```

Notice that it is important to enclose the numerator and denominator in parentheses since the fraction bar acts as both a division and a grouping symbol.

Evaluate each expression by hand, and then verify your answer using a graphing calculator.

1) $45 - 3 \cdot 2 + 7$

2) $24 \div \dfrac{6}{7} - 5 \cdot 4$

3) $5 + 2(9 - 6)^2$

4) $3 + 2[37 - (4 + 1)^2 - 2 \cdot 6]$

5) $\dfrac{5(7 - 3)}{50 - 3^2 \cdot 4}$

6) $\dfrac{25 - (1 + 3)^2}{6 + 14 \div 2 - 8}$

ANSWERS TO [**YOU TRY**] **EXERCISES**

1) a) 8^5 b) $\left(\dfrac{3}{2}\right)^4$ 2) a) 81 b) 64 c) $\dfrac{27}{64}$ 3) 41   4) 35   5) a) 24 b) 33 c) 31 d) 1

ANSWERS TO TECHNOLOGY EXERCISES

1) 46   2) 8   3) 23   4) 3   5) $\dfrac{10}{7}$ or $1\dfrac{3}{7}$ 6) $\dfrac{9}{5}$ or $1\dfrac{4}{5}$

E Evaluate **1.2** Exercises   Do the exercises, and check your work.

*Additional answers can be found in the Answers to Exercises appendix.

Objective 1: Use Exponents

 1) Identify the base and the exponent.

  a) 6^4 base: 6; exponent: 4

  b) 2^3 base: 2; exponent: 3

  c) $\left(\dfrac{9}{8}\right)^5$ base: $\dfrac{9}{8}$; exponent: 5

2) Identify the base and the exponent.

  a) 5^1 base: 5; exponent: 1

  b) 1^8 base: 1; exponent: 8

  c) $\left(\dfrac{3}{7}\right)^2$ base: $\dfrac{3}{7}$; exponent: 2

3) Write in exponential form.

   a) $9 \cdot 9 \cdot 9 \cdot 9$ 9^4

   b) $2 \cdot 2 \cdot 2 \cdot 2 \cdot 2 \cdot 2 \cdot 2 \cdot 2$ 2^8

   c) $\dfrac{1}{4} \cdot \dfrac{1}{4} \cdot \dfrac{1}{4}$ $\left(\dfrac{1}{4}\right)^3$

4) Explain, in words, why $7 \cdot 7 \cdot 7 \cdot 7 \cdot 7 = 7^5$.

5) Evaluate.

   a) 8^2 64   b) 11^2 121

   c) 2^4 16   d) 5^3 125

   e) 3^4 81   f) 12^2 144

   g) 1^2 1   h) $\left(\dfrac{3}{10}\right)^2$ $\dfrac{9}{100}$

   i) $\left(\dfrac{1}{2}\right)^6$ $\dfrac{1}{64}$ j) $(0.3)^2$ 0.09

6) Evaluate.

   a) 9^2 81   b) 13^2 169

   c) 3^3 27   d) 2^5 32

   e) 4^3 64   f) 1^4 1

   g) 6^2 36   h) $\left(\dfrac{7}{5}\right)^2$ $\dfrac{49}{25}$

   i) $\left(\dfrac{2}{3}\right)^4$ $\dfrac{16}{81}$ j) $(0.02)^2$ 0.0004

7) Evaluate $(0.5)^2$ two different ways.   0.25

8) Explain why $1^{200} = 1$.   1 raised to any power equals 1.

9) Raising a positive decimal number that is less than 1 to a natural number power *always, sometimes,* or *never* has a result that is less than that number.   always

10) Raising a positive proper fraction to a natural number power *always, sometimes,* or *never* has a result that is less than that fraction.   always

11) Raising a positive improper fraction (*not equivalent to* 1) to a natural number power *always, sometimes,* or *never* has a result that is less than that fraction.   never

12) Which two numbers, when raised to a natural number power, have a result equal to themselves?   0 and 1

Objective 2: Use the Order of Operations

13) In your own words, summarize the order of operations.   Answers may vary.

Evaluate.

14) $20 + 12 - 5$ 27   15) $17 - 2 + 4$ 19

16) $51 - 18 + 2 - 11$ 24   17) $48 \div 2 + 14$ 38

18) $15 \cdot 2 - 1$ 29   19) $20 - 3 \cdot 2 + 9$ 23

20) $28 + 21 \div 7 - 4$ 27   21) $8 + 12 \cdot \dfrac{3}{4}$ 17

22) $27 \div \dfrac{9}{5} - 1$ 14   23) $\dfrac{2}{5} \cdot \dfrac{1}{8} + \dfrac{2}{3} \cdot \dfrac{9}{10}$ $\dfrac{13}{20}$

24) $\dfrac{4}{9} \cdot \dfrac{5}{6} - \dfrac{1}{6} \cdot \dfrac{2}{3}$ $\dfrac{7}{27}$ 25) $2 \cdot \dfrac{3}{4} - \left(\dfrac{2}{3}\right)^2$ $\dfrac{19}{18}$ or $1\dfrac{1}{18}$

26) $\left(\dfrac{3}{2}\right)^2 - \left(\dfrac{5}{4}\right)^2$ $\dfrac{11}{16}$ 27) $25 - 11 \cdot 2 + 1$ 4

28) $2 + 16 + 14 \div 2$ 25

29) $39 - 3(9 - 7)^3$ 15

30) $1 + 2(7 - 1)^2$ 73

31) $60 \div 15 + 5 \cdot 3$ 19

32) $27 \div (10 - 7)^2 + 8 \cdot 3$ 27

33) $7[45 \div (19 - 10)] + 2$ 37

34) $6[3 + (14 - 12)^3] - 10$ 56

35) $1 + 2[(3 + 2)^3 \div (11 - 6)^2]$ 11

36) $(4 + 7)^2 - 3[5(6 + 2) - 4^2]$ 49

37) $\dfrac{4(7 - 2)^2}{12^2 - 8 \cdot 3}$ $\dfrac{5}{6}$ 38) $\dfrac{(8 + 4)^2 - 2^6}{7 \cdot 8 - 6 \cdot 9}$ 40

39) $\dfrac{4(9 - 6)^3}{2^2 + 3 \cdot 8}$ $\dfrac{27}{7}$ or $3\dfrac{6}{7}$ 40) $\dfrac{7 + 3(10 - 8)^4}{6 + 10 \div 2 + 11}$ $\dfrac{5}{2}$

R Rethink

R1) Where have you encountered exponents other than in this math course?

R2) Where in a 12 × 12 multiplication table can you rewrite the products using an exponent?

R3) Which objective is the most difficult for you?

1.3 Geometry Review

P Prepare

O Organize

What are your objectives for Section 1.3?	How can you accomplish each objective?
1 Identify Angles and Parallel and Perpendicular Lines	• Learn and draw examples of *acute, right, obtuse, straight, complementary,* and *supplementary angles.* • Complete the given example on your own. • Complete You Try 1. • Understand the relationship between *vertical angles* along with *parallel* and *perpendicular lines.*
2 Identify Triangles	• Know how to identify a triangle by its angles: *acute, obtuse,* and *right triangles.* • Know how to identify a triangle by its sides: *equilateral, isosceles,* and *scalene triangles.* • Complete the given example on your own. • Complete You Try 2.
3 Use Area, Perimeter, and Circumference Formulas	• Understand and memorize the formulas to determine the perimeter and area of common figures noted on p. 26. • Take notes on the terms required to determine the circumference and area of a circle. • Complete the given examples on your own. • Complete You Trys 3–5.
4 Use Volume Formulas	• Understand and memorize the formulas to determine the volume of common figures noted on p. 29. • Complete the given example on your own. • Complete You Try 6.

 **W Work**

Read the explanations, follow the examples, take notes, and complete the You Trys.

We often use geometry to solve algebraic problems. We must review some basic geometric concepts that we will need beginning in Chapter 2.

Let's begin by looking at angles. An angle can be measured in **degrees.** For example, 45° is read as "45 degrees."

1 Identify Angles and Parallel and Perpendicular Lines

Definitions

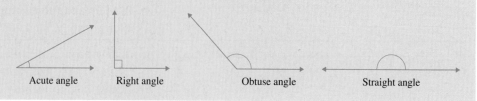

An **acute angle** is an angle whose measure is greater than 0° and less than 90°.

A **right angle** is an angle whose measure is 90°, indicated by the ⌐ symbol.

An **obtuse angle** is an angle whose measure is greater than 90° and less than 180°.

A **straight angle** is an angle whose measure is 180°.

| Acute angle | Right angle | Obtuse angle | Straight angle |

Two angles are **complementary** if their measures add to 90°.

Two angles are **supplementary** if their measures add to 180°.

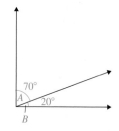

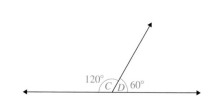

A and *B* are **complementary angles** since $m\angle A + m\angle B = 70° + 20° = 90°$.

C and *D* are **supplementary angles** since $m\angle C + m\angle D = 120° + 60° = 180°$.

Note

The measure of angle *A* is denoted by $m\angle A$.

EXAMPLE 1

$m\angle A = 41°$. Find its complement.

In-Class Example 1

$m\angle A = 34°$. Find its complement.

Answer: 56°

Solution

$$\text{Complement} = 90° - 41° = 49°$$

Since the sum of two complementary angles is 90°, if one angle measures 41°, its complement has a measure of $90° - 41° = 49°$.

[YOU TRY 1] $m\angle A = 62°$. Find its supplement.

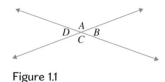

Figure 1.1

When two lines intersect, four angles are formed (see Figure 1.1). The pair of opposite angles are called **vertical angles**. Angles *A* and *C* are *vertical angles*, and angles *B* and *D* are *vertical angles*. *The measures of vertical angles are equal.* Therefore, $m\angle A = m\angle C$ and $m\angle B = m\angle D$.

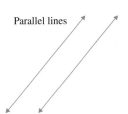
Parallel lines are lines in the same plane that do not intersect (Figure 1.2). **Perpendicular lines** are lines that intersect at right angles (Figure 1.3).

2 Identify Triangles

We can classify triangles by their angles and by their sides.

Perpendicular lines

Figure 1.3

Definitions

An **acute triangle** is one in which all three angles are acute.

An **obtuse triangle** contains one obtuse angle.

A **right triangle** contains one right angle.

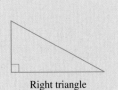

Acute triangle   Obtuse triangle   Right triangle

Property

The sum of the measures of the angles of any triangle is 180°.

W Hint

Use the acronym "EIS" or make up your own way to help you remember these three different types of triangles.

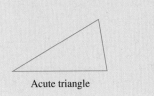

Equilateral triangle   Isosceles triangle   Scalene triangle

If a triangle has three sides of equal length, it is an **equilateral triangle.** (Each angle measure of an equilateral triangle is 60°.)

If a triangle has two sides of equal length, it is an **isosceles triangle.** (The angles opposite the equal sides have the same measure.)

If a triangle has no sides of equal length, it is a **scalene triangle.** (No angles have the same measure.)

EXAMPLE 2

Find the measures of angles A and B in this isosceles triangle.

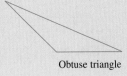

In-Class Example 2

Find the measures of angles A and B in this isosceles triangle.

Answer: $m\angle A = 27°$; $m\angle B = 126°$

Solution

The single hash marks on the two sides of the triangle mean that those sides are of equal length.

$$m\angle B = 39° \quad \text{Angle measures opposite sides of equal length are the same.}$$

$$39° + m\angle B = 39° + 39° = 78°$$

We have found that the sum of two of the angles is 78°. Since all of the angle measures add up to 180°,

$$m\angle A = 180° - 78° = 102°$$

[YOU TRY 2] Find the measures of angles A and B in this isosceles triangle.

3 Use Area, Perimeter, and Circumference Formulas

The **perimeter** of a figure is the distance around the figure, while the **area** of a figure is the number of square units enclosed within the figure. For some familiar shapes, we have the following formulas:

Figure		Perimeter	Area
Rectangle:		$P = 2l + 2w$	$A = lw$
Square:		$P = 4s$	$A = s^2$
Triangle: h = height		$P = a + b + c$	$A = \dfrac{1}{2}bh$
Parallelogram: h = height		$P = 2a + 2b$	$A = bh$
Trapezoid: h = height		$P = a + c + b_1 + b_2$	$A = \dfrac{1}{2}h(b_1 + b_2)$

The perimeter of a circle is called the **circumference.** The **radius,** r, is the distance from the center of the circle to a point on the circle. A line segment that passes through the center of the circle and has its endpoints on the circle is called a **diameter.**

Pi, π, is the ratio of the circumference of any circle to its diameter. $\pi \approx 3.14159265\ldots$, but we will use 3.14 as an approximation for π. The symbol $\approx$ is read as "is approximately equal to."

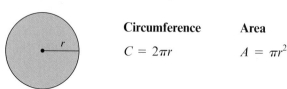

Circumference   **Area**

$C = 2\pi r$ $A = \pi r^2$

EXAMPLE 3   Find the perimeter and area of each figure.

In-Class Example 3

Find the perimeter and area of each figure.

a)

5 in.

9 in.

a)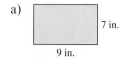

7 in.

9 in.

b)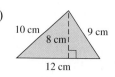

10 cm   8 cm   9 cm

12 cm

b)

9 in. 8 in.

7 in.

10 in.

Answer:
a) $P = 28$ in., $A = 45$ in^2
b) $P = 27$ in., $A = 35$ in^2

W Hint

Notice that perimeter is always expressed using one dimension, while area is expressed in two dimensions or square units.

Solution

a) This figure is a rectangle.

Perimeter: $P = 2l + 2w$

$P = 2(9$ in.$) + 2(7$ in.$)$ Substitute the values.

$P = 18$ in. $+ 14$ in.

$P = 32$ in.

Area: $A = lw$

$A = (9$ in.$)(7$ in.$)$ Substitute the values.

$A = 63$ in^2 or 63 square inches

b) This figure is a triangle.

Perimeter: $P = a + b + c$

$P = 9$ cm $+ 12$ cm $+ 10$ cm   Substitute the values.

$P = 31$ cm

Area: $A = \dfrac{1}{2}bh$

$A = \dfrac{1}{2}(12$ cm$)(8$ cm$)$ Substitute the values.

$A = 48$ cm^2 or 48 square centimeters

[YOU TRY 3] Find the perimeter and area of the figure.

8 cm

11 cm

EXAMPLE 4

In-Class Example 4

Find (a) the circumference and (b) the area of the circle. Give an exact answer for each and give an approximation using 3.14 for π.

Answer:
a) $C = 12\pi$ cm; $C \approx 37.68$ cm
b) $A = 36\pi$ cm^2; $A \approx 113.04$ cm^2

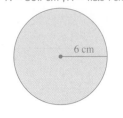

6 cm

Find (a) the circumference and (b) the area of the circle shown below right. Give an exact answer for each, and give an approximation using 3.14 for π.

Solution

a) The formula for the circumference of a circle is $C = 2\pi r$.
The radius of the given circle is 4 cm. Replace r with 4 cm.

$C = 2\pi r$

$= 2\pi(4$ cm$)$ Replace r with 4 cm.

$= 8\pi$ cm   Multiply.

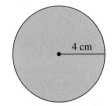

4 cm

Leaving the answer in terms of π gives us the exact circumference of the circle, 8π cm.

To find an approximation for the circumference, substitute 3.14 for π and simplify.

$C = 8\pi$ cm

$\approx 8(3.14)$ cm $= 25.12$ cm

b) The formula for the area of a circle is $A = \pi r^2$. Replace r with 4 cm.

$A = \pi r^2$

$= \pi(4$ cm$)^2$ Replace r with 4 cm.

$= 16\pi$ cm^2 $4^2 = 16$

Leaving the answer in terms of π gives us the exact area of the circle, 16π cm^2. To find an approximation for the area, substitute 3.14 for π and simplify.

$$A = 16\pi \text{ cm}^2$$
$$\approx 16(3.14) \text{ cm}^2$$
$$= 50.24 \text{ cm}^2$$

[YOU TRY 4] Find (a) the circumference and (b) the area of the circle. Give an exact answer for each, and give an approximation using 3.14 for π.

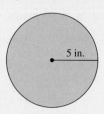

We can find the area and perimeter of some figures by combining formulas for more than one figure.

EXAMPLE 5

Find the perimeter and area of the figure shown here.

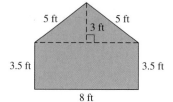

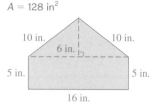
Solution

Perimeter: The perimeter is the distance around the figure.

$$P = 5 \text{ ft} + 5 \text{ ft} + 3.5 \text{ ft} + 8 \text{ ft} + 3.5 \text{ ft}$$
$$P = 25 \text{ ft}$$

Area: To find the area of this figure, think of it as two regions: a triangle and a rectangle.

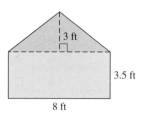

Total area = Area of triangle + Area of rectangle

$$= \frac{1}{2} bh + lw$$

$$= \frac{1}{2}(8 \text{ ft})(3 \text{ ft}) + (8 \text{ ft})(3.5 \text{ ft})$$

$$= 12 \text{ ft}^2 + 28 \text{ ft}^2$$

$$= 40 \text{ ft}^2$$

[YOU TRY 5] Find the perimeter and area of the figure.

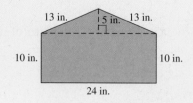

4 Use Volume Formulas

The **volume** of a three-dimensional object is the amount of space occupied by the object. Volume is measured in cubic units such as cubic inches (in^3), cubic centimeters (cm^3), cubic feet (ft^3), and so on. Volume also describes the amount of a substance that can be enclosed within a three-dimensional object. Therefore, volume can also be measured in quarts, liters, gallons, and so on. In the figures, l = length, w = width, h = height, s = length of a side, and r = radius.

Volumes of Three-Dimensional Figures		
Rectangular solid		$V = lwh$
Cube		$V = s^3$
Right circular cylinder		$V = \pi r^2 h$
Sphere		$V = \dfrac{4}{3}\pi r^3$
Right circular cone		$V = \dfrac{1}{3}\pi r^2 h$

EXAMPLE 6

Find the volume of each. In (b), give the answer in terms of π.

In-Class Example 6

Find the volume of each. In (b), give the answer in terms of π.
a)   b)

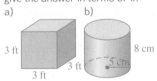

3 ft

3 ft
3 ft

8 cm

5 cm

Answer: a) $V = 27 \ ft^3$
b) $V = 200\pi \ cm^3$

a)

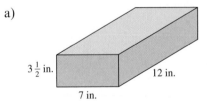

$3\dfrac{1}{2}$ in.   12 in.

7 in.

b)

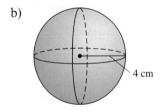

4 cm

W Hint

Notice that volume is always expressed using three dimensions or cubic units.

Solution

a) $V = lwh$ Volume of a rectangular solid

$= (12 \text{ in.})(7 \text{ in.})\left(3\dfrac{1}{2}\text{ in.}\right)$ Substitute values.

$= (12 \text{ in.})(7 \text{ in.})\left(\dfrac{7}{2}\text{ in.}\right)$ Change to an improper fraction.

$= \left(84 \cdot \dfrac{7}{2}\right)\text{in}^3$ Multiply.

$= 294 \text{ in}^3$ or 294 cubic inches

b) $V = \dfrac{4}{3}\pi r^3$ Volume of a sphere

   $= \dfrac{4}{3}\pi(4\text{ cm})^3$ Replace r with 4 cm.

   $= \dfrac{4}{3}\pi(64\text{ cm}^3)$ $4^3 = 64$

   $= \dfrac{256}{3}\pi\text{ cm}^3$ Multiply.

[**YOU TRY 6**] Find the volume of each figure. In (b), give the answer in terms of π.

a) A box with length = 3 ft, width = 2 ft, and height = 1.5 ft

b) A sphere with radius = 3 in.

ANSWERS TO [YOU TRY] EXERCISES

1) 118°   2) $m\angle A = 130°$; $m\angle B = 25°$ 3) $P = 38$ cm; $A = 88$ cm²   4) a) $C = 10\pi$ in.;
$C \approx 31.4$ in. b) $A = 25\pi$ in²; $A \approx 78.5$ in²   5) $P = 70$ in.; $A = 300$ in²   6) a) 9 ft³ b) 36π in³

E Evaluate **1.3** Exercises   Do the exercises, and check your work.

*Additional answers can be found in the Answers to Exercises appendix.

Objective 1: Identify Angles and Parallel and Perpendicular Lines

(24) 1) An angle whose measure is between 0° and 90° is a(n) __acute__ angle.

2) An angle whose measure is 90° is a(n) __right__ angle.

3) An angle whose measure is 180° is a(n) __straight__ angle.

4) An angle whose measure is between 90° and 180° is a(n) __obtuse__ angle.

5) If the sum of two angles is 180°, the angles are _____.
If the sum of two angles is 90°, the angles are _____.
supplementary; complementary

6) If two angles are supplementary, can both of them be obtuse? Explain. No. If both angles are obtuse, then their sum will be greater than 180°, so they cannot be supplementary.

Find the complement of each angle.

7) 59° 31°   8) 84° 6°

9) 12° 78°   10) 40° 50°

Find the supplement of each angle.

11) 143° 37°   12) 62° 118°

13) 38° 142°   14) 155° 25°

Find the measure of the missing angles.

15)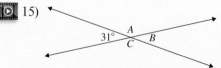
$m\angle A = m\angle C = 149°$;
$m\angle B = 31°$

16)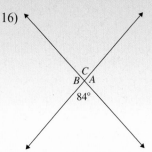
$m\angle A = m\angle B = 96°$; $m\angle C = 84°$

Objective 2: Identify Triangles

17) The sum of the angles in a triangle is __180__ degrees.

Find the missing angle and classify each triangle as acute, obtuse, or right.

18) 22°; right

19) 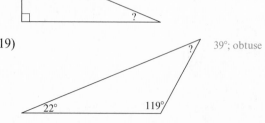 39°; obtuse

20) 62°; acute   21) 39°; right

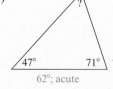

22) Can a triangle contain more than one obtuse angle?
Explain. No. The sum of the angles in a triangle is 180°. If a triangle had two obtuse angles, their sum would be greater than 180°.

Classify each triangle as equilateral, isosceles, or scalene.

23)
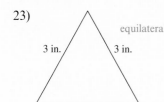
equilateral

3 in.   3 in.

3 in.

24)
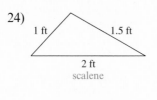
1 ft   1.5 ft

2 ft
scalene

25)
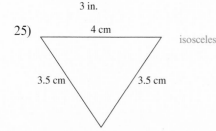
4 cm

isosceles

3.5 cm   3.5 cm

26) What can you say about the measures of the angles in an equilateral triangle? Each of them is a 60° angle.

27) True or False: A right triangle can also be isosceles. true

28) True or False: If a triangle has two sides of equal length, then the angles opposite these sides are equal. true

Objective 3: Use Area, Perimeter, and Circumference Formulas

Find the area and perimeter of each figure. Include the correct units.

29)

$A = 80 \text{ ft}^2$; $P = 36$ ft

8 ft

10 ft

30)
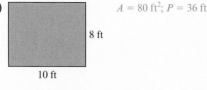
8 mm

4 mm   3.7 mm   4 mm

8 mm

$A = 29.6 \text{ mm}^2$; $P = 24$ mm

31)
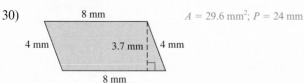
$A = 42 \text{ cm}^2$; $P = 29.25$ cm

8 cm   7.25 cm
6 cm
14 cm

32)
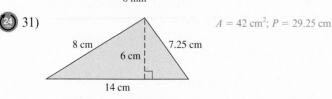
13 in.   7.8 in.
5 in.

18 in.

$A = 45 \text{ in}^2$; $P = 38.8$ in.

33)
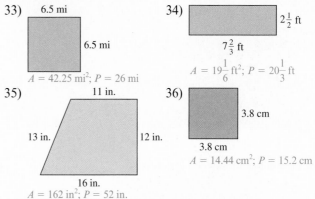
6.5 mi

6.5 mi

$A = 42.25 \text{ mi}^2$; $P = 26$ mi

34)
$2\frac{1}{2}$ ft

$7\frac{2}{3}$ ft

$A = 19\frac{1}{6} \text{ ft}^2$; $P = 20\frac{1}{3}$ ft

35)
11 in.

13 in.   12 in.

16 in.

$A = 162 \text{ in}^2$; $P = 52$ in.

36)
3.8 cm

3.8 cm

$A = 14.44 \text{ cm}^2$; $P = 15.2$ cm

Find (a) the area and (b) the circumference of the circle. Give an exact answer for each, and give an approximation using 3.14 for π. Include the correct units.

37)
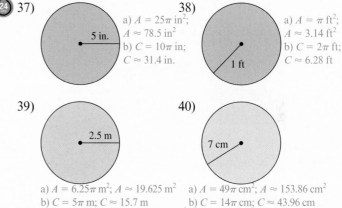
5 in.

a) $A = 25\pi \text{ in}^2$;
$A \approx 78.5 \text{ in}^2$
b) $C = 10\pi$ in;
$C \approx 31.4$ in.

38)
1 ft

a) $A = \pi \text{ ft}^2$;
$A \approx 3.14 \text{ ft}^2$
b) $C = 2\pi$ ft;
$C \approx 6.28$ ft

39)
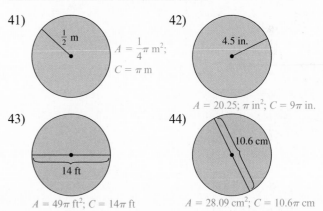
2.5 m

40)
7 cm

a) $A = 6.25\pi \text{ m}^2$; $A \approx 19.625 \text{ m}^2$
b) $C = 5\pi$ m; $C \approx 15.7$ m

a) $A = 49\pi \text{ cm}^2$; $A \approx 153.86 \text{ cm}^2$
b) $C = 14\pi$ cm; $C \approx 43.96$ cm

Find the exact area and circumference of the circle in terms of π. Include the correct units.

41)
$\frac{1}{2}$ m

$A = \frac{1}{4}\pi \text{ m}^2$;
$C = \pi$ m

42)
4.5 in.

$A = 20.25; \pi \text{ in}^2$; $C = 9\pi$ in.

43)
14 ft

$A = 49\pi \text{ ft}^2$; $C = 14\pi$ ft

44)
10.6 cm

$A = 28.09 \text{ cm}^2$; $C = 10.6\pi$ cm

Find the area and perimeter of each figure. Include the correct units.

45)
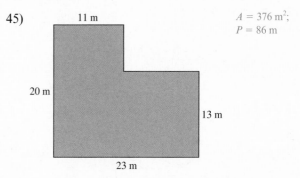
11 m

20 m

13 m

23 m

$A = 376 \text{ m}^2$;
$P = 86$ m

46)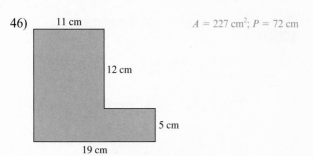
11 cm
12 cm
5 cm
19 cm
$A = 227 \text{ cm}^2; P = 72 \text{ cm}$

47)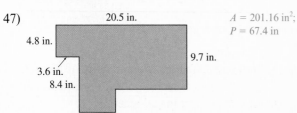
20.5 in.
4.8 in.
3.6 in.
8.4 in.
9.7 in.
5.7 in.
$A = 201.16 \text{ in}^2;$
$P = 67.4 \text{ in}$

48)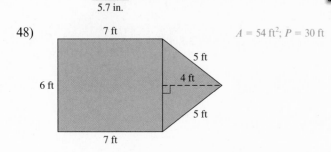
7 ft
5 ft
4 ft
6 ft
5 ft
7 ft
$A = 54 \text{ ft}^2; P = 30 \text{ ft}$

Find the area of the shaded region. Use 3.14 for π. Include the correct units.

49)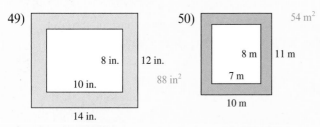
8 in.
12 in.
10 in.
14 in.
88 in²

50)
54 m²
8 m
11 m
7 m
10 m

51)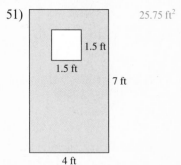
1.5 ft
1.5 ft
7 ft
4 ft
25.75 ft²

52)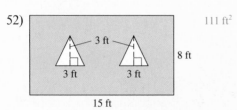
3 ft
8 ft
3 ft
3 ft
15 ft
111 ft²

Objective 4: Use Volume Formulas

Find the volume of each figure. Where appropriate, give the answer in terms of π. Include the correct units.

(24) 53)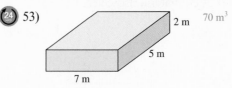
2 m
5 m
7 m
70 m³

54)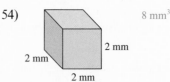
8 mm³
2 mm
2 mm
2 mm

55)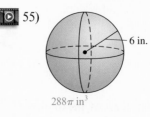
6 in.
$288\pi \text{ in}^3$

56)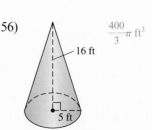
$\dfrac{400}{3}\pi \text{ ft}^3$
16 ft
5 ft

57)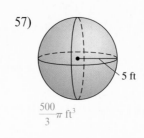
5 ft
$\dfrac{500}{3}\pi \text{ ft}^3$

58)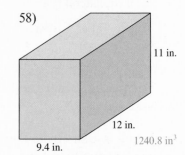
11 in.
12 in.
9.4 in.
1240.8 in³

59)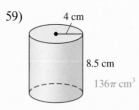
4 cm
8.5 cm
$136\pi \text{ cm}^3$

60)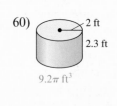
2 ft
2.3 ft
$9.2\pi \text{ ft}^3$

Mixed Exercises: Objectives 3 and 4

Applications of Perimeter, Area, and Volume: Use 3.14 for π, and include the correct units.

61) To lower her energy costs, Yun would like to replace her rectangular storefront window with low-emissivity (low-e) glass that costs $20.00/ft². The window measures 9 ft by 6.5 ft, and she can spend at most $900.

a) How much glass does she need?   58.5 ft²

b) Can she afford the low-e glass for this window?
No, it would cost $1170 to use this glass.

62) An insulated rectangular cooler is 15″ long, 10″ wide, and 13.6″ high. What is the volume of the cooler?
2040 in³

63) The lazy Susan on a table in a Chinese restaurant has a 10-inch radius. (A lazy Susan is a rotating tray used to serve food.)

a) What is the perimeter of the lazy Susan? 62.8 in.

b) What is its area? 314 in²

64) Yessenia wants a custom-made area rug measuring 5 ft by 8 ft. She has budgeted $500. She likes the Alhambra carpet sample that costs $9.80/ft² and the Sahara pattern that costs $12.20/ft². Can she afford either of these fabrics to make the area rug, or does she have to choose the cheaper one to remain within her budget? Support your answer by determining how much it would cost to have the rug made in each pattern. The Alhambra would cost $392.00 and the Sahara would cost $488.00, so she could afford either one.

65) A rectangular reflecting pool is 30 ft long, 19 ft wide, and 1.5 ft deep. How many gallons of water will this pool hold? (1 ft³ ≈ 7.48 gallons) 6395.4 gal

66) Find the perimeter of home plate given the dimensions below. 58 in.

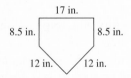

67) Ralph wants to childproof his house now that his daughter has learned to walk. The round, glass-top coffee table in his living room has a diameter of 36 inches. How much soft padding does Ralph need to cover the edges around the table? 113.04 in.

68) Nadia is remodeling her kitchen and her favorite granite countertop costs $80.00/ft², including installation. The layout of the countertop is shown below, where the counter has a uniform width of $2\frac{1}{4}$ ft. If she can spend at most $2500.00, can she afford her first-choice granite?
No. This granite countertop would cost $2970.00.

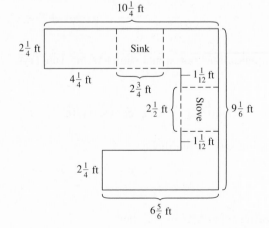

69) The chamber of a rectangular laboratory water bath measures $6'' \times 11\frac{3}{4}'' \times 5\frac{1}{2}''$.

a) How many cubic inches of water will the water bath hold? $387\frac{3}{4}$ in³ or 387.75 in³

b) How many liters of water will the water bath hold? (1 in³ ≈ 0.016 liter) 6.204 liters

70) A container of lip balm is in the shape of a right circular cylinder with a radius of 1.5 cm and a height of 2 cm. How much lip balm will the container hold? 14.13 cm³

71) A town's public works department will install a flower garden in the shape of a trapezoid. It will be enclosed by decorative fencing that costs $23.50/ft.

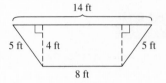

a) Find the area of the garden. 44 ft²

b) Find the cost of the fence. $752

72) Jaden is making decorations for the bulletin board in his fifth-grade classroom. Each equilateral triangle has a height of 15.6 inches and sides of length 18 inches.

a) Find the area of each triangle. 140.4 in²

b) Find the perimeter of each triangle. 54 in.

73) The dimensions of Riyad's home office are $10' \times 12'$. He plans to install laminated hardwood flooring that costs $2.69/ft². How much will the flooring cost? $322.80

74) Find the volume of the ice cream pictured below. Assume that the right circular cone is completely filled and that the scoop on top is half of a sphere. 6.28 in³

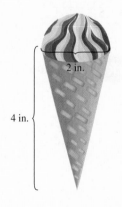

R1) Besides a math course, in what other courses have you used geometric shapes?

R2) Can you think of any jobs that would require computing an area?

R3) Think of a situation where you have to calculate the area of something to determine its cost.

1.4 Sets of Numbers and Absolute Value

P **Prepare**   **O** **Organize**

What are your objectives for Section 1.4?	How can you accomplish each objective?
1 Identify and Graph Numbers on a Number Line	• Know the definitions of *natural numbers, whole numbers,* and *integers,* and be able to compare the sets of numbers. • Be able to identify and graph any of these numbers on a number line. • Write the definitions of a *rational number,* an *irrational number,* and the *real numbers* in your own words. • Complete the given examples on your own. • Complete You Trys 1–5.
2 Compare Numbers Using Inequality Symbols	• Learn the inequality symbols by using each to compare two numbers in your notes. • Be able to use signed numbers to represent increases or decreases. • Complete the given examples on your own. • Complete You Trys 6–7.
3 Find the Additive Inverse and Absolute Value of a Number	• Write the definitions of *additive inverse* and *absolute value* in your own words. • Complete the given examples on your own. • Complete You Trys 8 and 9.

Read the explanations, follow the examples, take notes, and complete the You Trys.

1 Identify and Graph Numbers on a Number Line

In Section 1.1, we defined the following sets of numbers:

Natural numbers: {1, 2, 3, 4, …}

Whole numbers: {0, 1, 2, 3, 4, …}

We will begin this section by discussing other sets of numbers.

On a **number line,** positive numbers are to the right of zero and negative numbers are to the left of zero.

Negative numbers   Positive numbers

0 is neither positive nor negative.

The natural numbers, their negatives, and 0 form the set of numbers called *integers*.

Definition

The set of **integers** includes the set of natural numbers, their negatives, and zero. The set of *integers* is $\{\dots, -3, -2, -1, 0, 1, 2, 3, \dots\}$.

EXAMPLE 1

Graph each number on a number line.

$$4, 1, -6, 0, -3$$

Solution

4 and 1 are to the right of zero since they are positive.

-3 is three units to the left of zero, and -6 is six units to the left of zero.

[YOU TRY 1]

Graph each number on a number line. $2, -4, 5, -1, -2$

Positive and negative numbers are also called **signed numbers.**

EXAMPLE 2

Given the set of numbers $\left\{4, -7, 0, \dfrac{3}{4}, -6, 10, -3\right\}$, list the

a) whole numbers   b) natural numbers   c) integers

Solution

a) whole numbers: 0, 4, 10   b) natural numbers: 4, 10

c) integers: $-7, -6, -3, 0, 4, 10$

[YOU TRY 2]

Given the set of numbers $\left\{-1, 5, \dfrac{2}{7}, 8, -\dfrac{4}{5}, 0, -12\right\}$, list the

a) whole numbers   b) natural numbers   c) integers

Notice in Example 2 that $\dfrac{3}{4}$ did not belong to any of these sets. That is because the whole numbers, natural numbers, and integers do not contain any fractional parts. $\dfrac{3}{4}$ is a *rational number*.

Definition

A **rational number** is any number of the form $\dfrac{p}{q}$, where p and q are integers and $q \neq 0$.

That is, a rational number is any number that can be written as a fraction where the numerator and denominator are integers and the denominator does not equal zero.

Rational numbers include much more than numbers like $\dfrac{3}{4}$, which are already in fractional form.

EXAMPLE 3

Explain why each of the following numbers is rational.

In-Class Example 3

Explain why each of the following numbers is rational.
a) 6   b) 0.01   c) -9
d) $3\dfrac{1}{5}$ e) $0.\overline{3}$ f) $\sqrt{25}$

Answer:

a) 6 can be written as $\dfrac{6}{1}$.

b) 0.01 can be written as $\dfrac{1}{100}$.

c) -9 can be written as $\dfrac{-9}{1}$.

d) $3\dfrac{1}{5}$ can be written as $\dfrac{16}{5}$.

e) $0.\overline{3}$ can be written as $\dfrac{1}{3}$.

f) $\sqrt{25} = 5$ and $5 = \dfrac{5}{1}$.

a) 7   b) 0.8   c) -5 d) $6\dfrac{1}{4}$ e) $0.\overline{3}$ f) $\sqrt{4}$

Solution

Rational Number	Reason
7	7 can be written as $\dfrac{7}{1}$.
0.8	0.8 can be written as $\dfrac{8}{10}$.
-5	-5 can be written as $\dfrac{-5}{1}$.
$6\dfrac{1}{4}$	$6\dfrac{1}{4}$ can be written as $\dfrac{25}{4}$.
$0.\overline{3}$	$0.\overline{3}$ can be written as $\dfrac{1}{3}$.
$\sqrt{4}$	$\sqrt{4} = 2$ and $2 = \dfrac{2}{1}$.

$\sqrt{4}$ is read as "the square root of 4." This means, "What number times itself equals 4?" That number is 2.

YOU TRY 3

Explain why each of the following numbers is rational.

a) 12   b) 0.7   c) -8 d) $2\dfrac{3}{4}$ e) $0.\overline{6}$ f) $\sqrt{100}$

W Hint

Notice that square roots of perfect squares are always rational numbers!

To summarize, the set of rational numbers includes

1) integers, whole numbers, and natural numbers.

2) repeating decimals.

3) terminating decimals.

4) fractions and mixed numbers.

The set of rational numbers does *not* include nonrepeating, nonterminating decimals. These decimals cannot be written as the quotient of two integers. Numbers such as these are called *irrational numbers*.

Definition

The set of numbers that cannot be written as the quotient of two integers is called the set of **irrational numbers.** Written in decimal form, an *irrational number* is a nonrepeating, nonterminating decimal.

Explain why each of the following numbers is irrational.

a) 0.8271316...   b) π c) $\sqrt{3}$

Solution

Irrational Number	Reason
0.827136...	It is a nonrepeating, nonterminating decimal.
π	$\pi \approx 3.14159265...$ It is a nonrepeating, nonterminating decimal.
$\sqrt{3}$	3 is not a perfect square, and the decimal equivalent of the square root of a nonperfect square is a nonrepeating, nonterminating decimal. Here, $\sqrt{3} \approx 1.73205...$

[YOU TRY 4]

Explain why each of the following numbers is irrational.

a) 2.41895...   b) $\sqrt{2}$

If we put together the sets of numbers we have discussed up to this point, we get the *real numbers*.

Definition

The set of **real numbers** consists of the rational and irrational numbers.

We summarize the information next with examples of the different sets of numbers:

Real Numbers

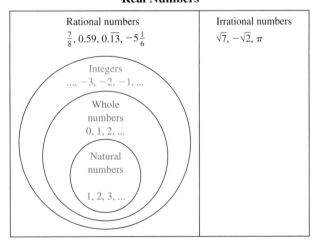

From the figure we can see, for example, that all whole numbers {0, 1, 2, 3...} are integers, but not all integers are whole numbers (−3, for example).

EXAMPLE 5

Given the set of numbers $\left\{ -16, 3.82, 0, 29, 0.\overline{7}, -\dfrac{11}{12}, \sqrt{10}, 5.302981... \right\}$, list the

a) integers   b) natural numbers   c) whole numbers

d) rational numbers   e) irrational numbers   f) real numbers

In-Class Example 5

Given the set of numbers

$$\left\{23, -6, \frac{4}{9}, 7.13, \sqrt{19}, 0, -0.\overline{41}, 2.35821..., \pi \right\}, \text{ list the}$$

a) integers b) natural numbers c) whole numbers d) rational numbers e) irrational numbers f) real numbers.

Answer:

a) $23, -6, 0$ b) 23 c) $23, 0$

d) $23, -6, \frac{4}{9}, 7.13, 0, -0.\overline{41}$

e) $\sqrt{19}, 2.35821..., \pi$

f) $23, -6, \frac{4}{9}, 7.13, \sqrt{19}, 0, -0.\overline{41}, 2.35821..., \pi$

Solution

a) integers: $-16, 0, 29$

b) natural numbers: 29

c) whole numbers: $0, 29$

d) rational numbers: $-16, 3.82, 0, 29, 0.\overline{7}, -\frac{11}{12}$. Each of these numbers can be written as the quotient of two integers.

e) irrational numbers: $\sqrt{10}, 5.302981...$

f) real numbers: All of the numbers in this set are real.

$$\left\{-16, 3.82, 0, 29, 0.\overline{7}, -\frac{11}{12}, \sqrt{10}, 5.302981...\right\}$$

[YOU TRY 5]

Given the set of numbers $\left\{\frac{9}{8}, \sqrt{14}, 34, -41, 6.5, 0.\overline{21}, 0, 7.412835... \right\}$, list the

a) whole numbers b) integers

c) rational numbers d) irrational numbers

---

2 Compare Numbers Using Inequality Symbols

Let's review the inequality symbols.

W Hint

As we move to the *left* on the number line, the numbers get smaller. As we move to the *right* on the number line, the numbers get larger.

$<$ less than	$\leq$ less than or equal to
$>$ greater than	$\geq$ greater than or equal to
$\neq$ not equal to	$\approx$ approximately equal to

We use these symbols to compare numbers as in $5 > 2$, $6 \leq 17$, $4 \neq 9$, and so on. How do we compare negative numbers?

EXAMPLE 6

Insert $>$ or $<$ to make the statement true. Look at the number line, if necessary.

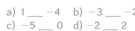
$$-5\ -4\ -3\ -2\ -1\ \ 0\ \ 1\ \ 2\ \ 3\ \ 4\ \ 5$$

a) $4 \_\_ 2$ b) $2 \_\_ 4$ c) $-3 \_\_ 1$ d) $-2 \_\_ -5$ e) $-4 \_\_ -1$

In-Class Example 6

Insert $>$ or $<$ to make the statement true. Look at the number line, if necessary.

$$-5\ -4\ -3\ -2\ -1\ \ 0\ \ 1\ \ 2\ \ 3\ \ 4\ \ 5$$

a) $1 \_\_ -4$ b) $-3 \_\_ -2$ c) $-5 \_\_ 0$ d) $-2 \_\_ 2$

Answer:

a) $1 > -4$ b) $-3 < -2$ c) $-5 < 0$ d) $-2 < 2$

Solution

a) $4 \underline{>} 2$ 4 is to the right of 2.

b) $2 \underline{<} 4$ 2 is to the left of 4.

c) $-3 \underline{<} 1$ -3 is to the left of 1.

d) $-2 \underline{>} -5$ -2 is to the right of -5.

e) $-4 \underline{<} -1$ -4 is to the left of -1.

[YOU TRY 6]

Insert $>$ or $<$ to make the statement true.

a) $7 \_\_\_ 3$ b) $-5 \_\_\_ -1$ c) $-6 \_\_\_ -14$

---

www.mhhe.com/messersmith

Note
The inequality sign always points to the number farther *left* on the number line.

Look at a) and b) in Example 6, and you will notice that the same numbers, 2 and 4, are being compared but the order and symbol are reversed. We can rewrite the statements by reversing the inequality symbol and interchanging the numbers being compared.

Signed numbers are used in many different ways.

EXAMPLE 7

In-Class Example 7

Use a signed number to represent the change in each situation.
a) After changing her diet, Jill decreased her cholesterol level by 14 points.
b) Over the last six months, mortgage rates rose 0.8%.

Answer: a) −14 b) 0.8%

Use a signed number to represent the change in each situation.

a) During the recession, the number of employees at a manufacturing company decreased by 850.

b) In February 2011, Facebook gained 23,800,000 users worldwide. (www.insidefacebook.com)

Solution

a) −850   The negative number represents a decrease in the number of employees.

b) 23,800,000   The positive number represents an increase in the number of Facebook users.

[YOU TRY 7]

Use a signed number to represent the change.
After getting off the highway, Huda decreased his car's speed by 25 mph.

---

3 Find the Additive Inverse and Absolute Value of a Number

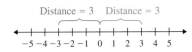

Notice that both −3 and 3 are a distance of 3 units from 0 but are on opposite sides of 0. We say that 3 and −3 are *additive inverses*.

 Hint

Notice that the additive inverse of a negative number is always positive. The additive inverse of a positive number is always negative.

Definition

Two numbers are **additive inverses** if they are the same distance from 0 on the number line but on opposite sides of 0. Therefore, if a is any real number, then $-a$ is its additive inverse.

Furthermore, $-(-a) = a$. We can see this on the number line.

EXAMPLE 8

In-Class Example 8

Find $-(-10)$.

Answer: 10

Find $-(-2)$.

Solution

So, beginning with -2, the number on the opposite side of zero and 2 units away from zero is 2.
$-(-2) = 2$

[YOU TRY 8] Find $-(-13)$.

We can explain *distance from zero* in another way: *absolute value*.

W Hint

The absolute value of a number describes its *distance* from zero and *not* what side of zero the number is on. So, the absolute value of a number is always positive or zero.

Definition

The **absolute value** of a number is the distance between that number and 0 on the number line. Furthermore, if a is any real number, then the **absolute value of a**, denoted by $|a|$, is

$$\text{i) } a \text{ if } a \geq 0$$
$$\text{ii) } -a \text{ if } a < 0$$

Remember, $|a|$ is never negative.

EXAMPLE 9

In-Class Example 9

Evaluate each.
a) $|12|$ b) $|-18|$ c) $|0|$
d) $-|9|$ e) $|13 - 10|$

Answer:
a) 12 b) 18 c) 0
d) -9 e) 3

Evaluate each.

a) $|6|$ b) $|-5|$ c) $|0|$ d) $-|12|$ e) $|14 - 5|$

Solution

a) $|6| = 6$ 6 is 6 units from 0.

b) $|-5| = 5$ -5 is 5 units from 0.

c) $|0| = 0$

d) $-|12| = -12$ First, evaluate $|12|$: $|12| = 12$. Then, apply the negative symbol to get -12.

e) $|14 - 5| = |9|$ The absolute value symbols work like parentheses. First, evaluate what is inside: $14 - 5 = 9$.

   $= 9$ Find the absolute value.

[YOU TRY 9] Evaluate each.

a) $|19|$ b) $|-8|$ c) $-|7|$ d) $|20 - 9|$

ANSWERS TO [YOU TRY] EXERCISES

1) 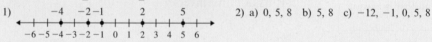   2) a) 0, 5, 8 b) 5, 8 c) $-12, -1, 0, 5, 8$

3) a) $12 = \dfrac{12}{1}$ b) $0.7 = \dfrac{7}{10}$ c) $-8 = \dfrac{-8}{1}$ d) $2\dfrac{3}{4} = \dfrac{11}{4}$ e) $0.\overline{6} = \dfrac{2}{3}$ f) $\sqrt{100} = 10$ and $10 = \dfrac{10}{1}$

4) a) It is a nonrepeating, nonterminating decimal. b) 2 is not a perfect square, so the decimal equivalent of $\sqrt{2}$ is a nonrepeating, nonterminating decimal.   5) a) 34, 0 b) 34, -41, 0
c) $\dfrac{9}{8}$, 34, $-41, 6.5, 0.\overline{21}, 0$ d) $\sqrt{14}, 7.412835\ldots$ 6) a) $>$ b) $<$ c) $>$ 7) -25 8) 13

9) a) 19 b) 8 c) -7 d) 11

*Additional answers can be found in the Answers to Exercises appendix.

Objective 1: Identify and Graph Numbers on a Number Line

1) In your own words, explain the difference between the set of rational numbers and the set of irrational numbers. Give two examples of each type of number.
Answers may vary.

2) In your own words, explain the difference between the set of whole numbers and the set of natural numbers. Give two examples of each type of number.
Answers may vary.

In Exercises 3 and 4, given each set of numbers, list the

a) natural numbers   b) whole numbers

c) integers   d) rational numbers

e) irrational numbers   f) real numbers

3) $\left\{ 17, 3.8, \dfrac{4}{5}, 0, \sqrt{10}, -25, 6.\overline{7}, -2\dfrac{1}{8}, 9.721983... \right\}$

a) 17 b) 17, 0 c) 17, 0, −25 d) $17, 3.8, \dfrac{4}{5}, 0, -25, 6.\overline{7}, -2\dfrac{1}{8}$
e) $\sqrt{10}$, 9.721983... f) all numbers in the set

4) $\left\{ -6, \sqrt{23}, 21, 5.\overline{62}, 0.4, 3\dfrac{2}{9}, 0, -\dfrac{7}{8}, 2.074816... \right\}$

a) 21 b) 21, 0 c) −6, 21, 0 d) $-6, 21, 5.\overline{62}, 0.4, 3\dfrac{2}{9}, 0, -\dfrac{7}{8}$
e) $\sqrt{23}$, 2.074816... f) all numbers in the set

Determine whether each statement is true or false.

5) Every whole number is a real number.   true

6) Every real number is an integer.   false

7) Every rational number is a whole number.   false

8) Every whole number is an integer.   true

9) Every natural number is a whole number.   true

10) Every integer is a rational number.   true

Graph the numbers on a number line. Label each.

11) $5, -2, \dfrac{3}{2}, -3\dfrac{1}{2}, 0$

12) $-4, 3, \dfrac{7}{8}, 4\dfrac{1}{3}, -2\dfrac{1}{4}$

13) $-6.8, -\dfrac{3}{8}, 0.2, 1\dfrac{8}{9}, -4\dfrac{1}{3}$

14) $-3.25, \dfrac{2}{3}, 2, -1\dfrac{3}{8}, 4.1$

Objective 2: Compare Numbers Using Inequality Symbols

Insert > or < to make the statement true.

15) 7 ___ 4   >   16) 3 ___ 11   <

17) −4 ___ −1   <   18) −2 ___ −3   >

19) 9 ___ −2   >   20) −3 ___ 5   <

21) 0 ___ −2   >   22) 8 ___ 10   <

23) $-\dfrac{1}{2}$ ___ $-\dfrac{3}{4}$ >   24) $-\dfrac{3}{5}$ ___ $-\dfrac{7}{10}$ >

Write each statement with the inequality symbol reversed.

25) 6 < 11   11 > 6   26) −5 < −1   −1 > −5

27) 4 > −8   −8 < 4   28) 3 > −12   −12 < 3

Objective 3: Find the Additive Inverse and Absolute Value of a Number

29) What does the absolute value of a number represent?
the distance of the number from zero

30) If a is a real number and if $|a|$ is not a positive number, then what is the value of a?   0

Find the additive inverse of each.

31) 8   −8   32) 6   −6

33) −15   15   34) −1   1

35) $-\dfrac{3}{4}$ $\dfrac{3}{4}$ 36) 4.7   −4.7

Evaluate.

37) $|-10|$ 10   38) $|9|$ 9

39) $\left|\dfrac{9}{4}\right|$ $\dfrac{9}{4}$ 40) $\left|-\dfrac{5}{6}\right|$ $\dfrac{5}{6}$

41) $-|-14|$ −14   42) $-|27|$ −27

43) $|17 - 4|$ 13   44) $-|10 - 6|$ −4

45) $-\left|-4\dfrac{1}{7}\right|$ $-4\dfrac{1}{7}$ 46) $|-9.6|$ 9.6

Determine whether each statement is true or false.

47) The absolute value of the difference between any two real numbers represents the distance between the two numbers on a number line. true

48) The additive inverse of the absolute value of any nonzero real number is always negative. true

Write each group of numbers from smallest to largest.

49) $7, -2, 3.8, -10, 0, \dfrac{9}{10}$ $-10, -2, 0, \dfrac{9}{10}, 3.8, 7$

50) $2.6, 2.06, -1, -5\dfrac{3}{8}, 3, \dfrac{7}{4}$ $-5\dfrac{3}{8}, -1, \dfrac{7}{4}, 2.06, 2.6, 3$

51) $7\dfrac{5}{6}, -5, -6.5, -6.51, 7\dfrac{1}{3}, 2$ $-6.51, -6.5, -5, 2, 7\dfrac{1}{3}, 7\dfrac{5}{6}$

52) $-\dfrac{3}{4}, 0, -0.5, 4, -1, \dfrac{15}{2}$ $-1, -\dfrac{3}{4}, -0.5, 0, 4, \dfrac{15}{2}$

Mixed Exercises: Objectives 2 and 3
Decide whether each statement is true or false.

53) $16 \geq -11$ true 54) $-19 < -18$ true

55) $\dfrac{7}{11} \leq \dfrac{5}{9}$ false 56) $-1.7 \geq -1.6$ false

57) $-|-28| = 28$ false 58) $-|13| = -13$ true

59) $-5\dfrac{3}{10} < -5\dfrac{3}{4}$ false 60) $\dfrac{3}{2} \leq \dfrac{3}{4}$ false

Use a signed number to represent the change in each situation.

61) In 2010, Alex Rodriguez of the New York Yankees had 125 RBIs (runs batted in) while in 2011 he had 62 RBIs. That was a decrease of 63 RBIs.
(http://newyork.yankees.mlb.com) -63

62) In 2000, the average price of a gallon of gas in Los Angeles was $1.62. In 2010, the average price was $3.11 per gallon, an increase of $1.49 per gallon.
(www.census.gov) $1.49 per gallon

63) In June 2011, an estimated 30.6 million people visited the Twitter website. In July 2011, there were about 32.8 million visitors to the site. This is an increase of 2.2 million people.
(http://news.cnet.com) 2.2 million

64) According to the *Statistical Abstract of the United States,* the population of Michigan decreased by about 121,000 from July 1, 2005 to July 1, 2009.
(www.census.gov) $-121,000$

24 65) From 2006 to 2010, the number of new housing starts decreased by about 1,214,000.
(www.census.gov) $-1,214,000$

66) The number of military officers on active duty increased by 5700 from 2009 to 2010.
(www.census.gov) 5700

R Rethink

R1) If the absolute value of a number is greater than the absolute value of another, what does that mean in terms of their locations on a number line?

R2) In a computer programming language, what do you think "abs(−24)" means?

R3) Why is the additive inverse of a negative number always positive?

1.5 Addition and Subtraction of Real Numbers

P Prepare	**O Organize**
What are your objectives for Section 1.5?	**How can you accomplish each objective?**
1 Add Integers Using a Number Line	• Complete the given examples on your own. • Complete You Try 1.
2 Add Real Numbers with the Same Sign	• Write the procedure for **Adding Numbers with the Same Sign** in your own words. • Complete the given example on your own. • Complete You Try 2.
3 Add Real Numbers with Different Signs	• Write the procedure for **Adding Numbers with Different Signs** in your own words. • Complete the given example on your own. • Complete You Try 3.
4 Subtract Real Numbers	• Understand the definition on p. 46, and write a procedure that will help you subtract numbers. • Use the procedure you developed to complete the given example on your own. • Complete You Try 4.
5 Solve Applied Problems	• Complete the given example on your own. • Complete You Try 5.
6 Apply the Order of Operations to Real Numbers	• Add any additional steps to the Order of Operations you learned in Section 1.1. • Complete the given example on your own. • Complete You Try 6.
7 Translate English Expressions into Mathematical Expressions	• Take notes on the common English expressions, and be able to identify whether an expression refers to addition or subtraction. • Complete the given example on your own. • Complete You Try 7.

W Work   **Read the explanations, follow the examples, take notes, and complete the You Trys.**

In Section 1.4, we defined real numbers. In this section, we will discuss adding and subtracting real numbers.

1 Add Integers Using a Number Line

Let's use a number line to add numbers.

EXAMPLE 1   Use a number line to add each pair of numbers.

    a) $2 + 5$ b) $-1 + (-4)$ c) $2 + (-5)$ d) $-8 + 12$

Use a number line to add
each pair of numbers.
a) $4 + 3$ b) $-2 + (-9)$
c) $-7 + 8$ d) $-5 + 3$

Answer: a) 7   b) -11
c) 1   d) -2

Solution

a) $2 + 5$: Start at 2 and move 5 units to the right.

$2 + 5 = 7$

b) $-1 + (-4)$: Start at -1 and move 4 units to the left. (Move to the left when adding a negative.)

$-1 + (-4) = -5$

c) $2 + (-5)$: Start at 2 and move 5 units to the left.

$2 + (-5) = -3$

d) $-8 + 12$: Start at -8 and move 12 units to the right.

$-8 + 12 = 4$

[**YOU TRY 1**]   Use a number line to add each pair of numbers.

a) $1 + 3$ b) $-3 + (-2)$ c) $8 + (-6)$ d) $-10 + 7$

2 Add Real Numbers with the Same Sign

We found that

$$2 + 5 = 7, \quad -1 + (-4) = -5, \quad 2 + (-5) = -3, \quad -8 + 12 = 4.$$

Notice that when we add two numbers with the same sign, the result has the same sign as the numbers being added.

> **Procedure** Adding Numbers with the Same Sign
>
> To add numbers with the same sign, find the absolute value of each number and add them. The sum will have the same sign as the numbers being added.

Apply this rule to $-1 + (-4)$.

The result will be negative
↓

$$-1 + (-4) = -(|-1| + |-4|) = -(1 + 4) = -5$$

Add the
absolute
value of
each number.

EXAMPLE 2

Add −23 + (−41).

In-Class Example 2

Add −52 + (−79).

Answer: −131

Solution

$-23 + (-41) = -(|-23| + |-41|) = -(23 + 41) = -64$

YOU TRY 2

Add.

a) −6 + (−10)   b) −38 + (−56)

---

3 Add Real Numbers with Different Signs

W Hint

Try writing this procedure using steps.

Step 1: _____

Step 2: _____

In Example 1, we found that 2 + (−5) = −3 and −8 + 12 = 4.

> **Procedure** Adding Numbers with Different Signs
>
> To add two numbers with different signs, find the absolute value of each number. Subtract the smaller absolute value from the larger. The sum will have the sign of the number with the larger absolute value.

EXAMPLE 3

In-Class Example 3

Add.

a) −43 + 27 b) 7.8 + (−5.1)

c) $\dfrac{4}{9} + \left(-\dfrac{5}{6}\right)$ d) −14 + 14

Answer: a) −16 b) 2.7

c) $-\dfrac{7}{18}$ d) 0

Add.

a) −17 + 5   b) 9.8 + (−6.3)   c) $\dfrac{1}{5} + \left(-\dfrac{2}{3}\right)$ d) −8 + 8

Solution

a) −17 + 5 = −12   The sum will be negative since the number with the larger absolute value, |−17|, is negative.

b) 9.8 + (−6.3) = 3.5   The sum will be positive since the number with the larger absolute value, |9.8|, is positive.

c) $\dfrac{1}{5} + \left(-\dfrac{2}{3}\right) = \dfrac{3}{15} + \left(-\dfrac{10}{15}\right)$ Get a common denominator.
The sum will be negative since the number with the larger

$= -\dfrac{7}{15}$ absolute value, $\left|-\dfrac{10}{15}\right|$, is negative.

d) −8 + 8 = 0

Note

The sum of a number and its additive inverse is always 0. That is, if a is a real number, then a + (−a) = 0. Notice in part d) of Example 3 that −8 and 8 are additive inverses.

YOU TRY 3

Add.

a) 20 + (−19)   b) −14 + (−2)   c) $-\dfrac{3}{7} + \dfrac{1}{4}$ d) 7.2 + (−7.2)

---

4 Subtract Real Numbers

We can use the additive inverse to subtract numbers. Let's start with a basic subtraction problem, and use a number line to find $8 - 5$.

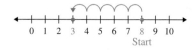

$$0 \quad 1 \quad 2 \quad 3 \quad 4 \quad 5 \quad 6 \quad 7 \quad 8 \quad 9 \quad 10$$
Start

Start at 8. Then to subtract 5, move 5 units to the left to get 3.

$$8 - 5 = 3$$

We use the same procedure to find $8 + (-5)$. This leads us to a definition of subtraction:

> ## Definition
>
> If a and b are real numbers, then $a - b = a + (-b)$.

The definition tells us that to subtract $a - b$,

1) change subtraction to addition.
2) find the additive inverse of b.
3) add a and the additive inverse of b.

EXAMPLE 4

Ⓦ Hint

Notice the use of parentheses when a negative number follows an addition or subtraction sign.

Subtract.

a) $4 - 9$ b) $-10 - 8$ c) $6 - (-25)$

Solution

a) $4 - 9 = 4 + (-9) = -5$

Change to addition.   Additive inverse of 9

b) $-10 - 8 = -10 + (-8) = -18$

Change to addition.   Additive inverse of 8

c) $6 - (-25) = 6 + 25 = 31$

Change to addition.   Additive inverse of -25

[YOU TRY 4]   Subtract.

a) $2 - 14$ b) $-9 - 13$ c) $31 - 14$ d) $23 - (-34)$

In part c) of Example 4, $6 - (-25)$ changed to $6 + 25$. This illustrates that *subtracting a negative number is equivalent to adding a positive number.* Therefore, $-7 - (-15) = -7 + 15 = 8$.

---

5 Solve Applied Problems

We can use signed numbers to solve real-life problems.

In-Class Example 5

The lowest temperature ever recorded in Chicago, Illinois, was −27°F. The highest temperature in Chicago was 105°F. What is the difference between these two temperatures?

Answer: 132°F

According to the National Weather Service, the coldest temperature ever recorded in Wyoming was −66°F on February 9, 1944. The record high was 115°F on August 8, 1983. What is the difference between these two temperatures? (www.ncdc.noaa.gov)

Solution

$$\text{Difference} = \text{Highest temperature} - \text{Lowest temperature}$$
$$= \quad\quad 115 \quad\quad - \quad\quad (-66)$$
$$= 115 + 66$$
$$= 181$$

The difference between the temperatures is 181°F.

[YOU TRY 5] The best score in a golf tournament was −16, and the worst score was +9. What is the difference between these two scores?

6 Apply the Order of Operations to Real Numbers

We discussed the order of operations in Section 1.2. Let's explore it further with the real numbers.

EXAMPLE 6

In-Class Example 6

Simplify.
a) $(4 - 12) + (5 - 2)$
b) $|-11 + 46| - 2|6 - 9|$

Answer: a) −5 b) 29

Simplify.

a) $(10 - 18) + (-4 + 6)$ b) $|-31 - 4| - 7|9 - 4|$

Solution

a) $(10 - 18) + (-4 + 6) = -8 + 2$ First, perform the operations in parentheses.
$$\qquad\qquad\qquad\qquad\quad = -6 \qquad\qquad \text{Add.}$$

b) $|-31 - 4| - 7|9 - 4| = |-31 + (-4)| - 7|9 - 4|$
$$= |-35| - 7|5| \qquad \text{Perform the operations in the absolute values.}$$
$$= 35 - 7(5) \qquad \text{Evaluate the absolute values.}$$
$$= 35 - 35$$
$$= 0$$

[YOU TRY 6] Simplify.

a) $[12 + (-5)] - [-16 + (-8)]$ b) $-\dfrac{4}{9} + \left(\dfrac{1}{6} - \dfrac{2}{3}\right)$ c) $-|7 - 15| - |4 - 2|$

7 Translate English Expressions into Mathematical Expressions

Knowing how to translate English expressions into mathematical expressions is a skill students need to learn algebra. Here, we will discuss how to "translate" from English into mathematics. Let's look at some key words and phrases you may encounter.

English Expression	Mathematical Operation
sum, more than, increased by	addition
difference between, less than, decreased by	subtraction

EXAMPLE 7

In-Class Example 7

Write a mathematical expression for each and simplify.
a) 18 more than 6
b) 5 less than 81
c) −14 decreased by 12
d) the sum of −1 and 28
e) 21 more than the sum of −8 and −7

Answer: a) 6 + 18; 24
b) 81 − 5; 76
c) −14 − 12; −26
d) −1 + 28; 27
e) [−8 + (−7)] + 21; 6

Write a mathematical expression for each and simplify.

a) 9 more than −2   b) 10 less than 41   c) −8 decreased by 17

d) the sum of 13 and −4   e) 8 less than the sum of −11 and −3

Solution

a) 9 more than −2

9 more than a quantity means we *add* 9 to the quantity, in this case, −2.

$$-2 + 9 = 7$$

b) 10 less than 41

10 less than a quantity means we *subtract 10 from* that quantity, in this case, 41.

$$41 - 10 = 31$$

c) −8 decreased by 17

If −8 is being *decreased by 17*, then we subtract 17 *from* −8.

$$-8 - 17 = -8 + (-17)$$
$$= -25$$

d) the sum of 13 and −4

Sum means add. $13 + (-4) = 9$

W Hint

Review inequalities in Section 1.4, and notice the difference between "10 less than 41" and "10 is less than 41."

e) 8 less than the sum of −11 and −3

8 less than means we are subtracting 8 *from* something. From what? From the *sum of −11 and −3*.

Sum means add, so we must find the sum of −11 and −3 and subtract 8 from it.

$$[-11 + (-3)] - 8 = -14 - 8 \qquad \text{First, perform the operation in the brackets.}$$
$$= -14 + (-8) \qquad \text{Change to addition.}$$
$$= -22 \qquad \text{Add.}$$

[YOU TRY 7]

Write a mathematical expression for each and simplify.

a) −14 increased by 6   b) 27 less than 15

c) The sum of 23 and −7 decreased by 5

ANSWERS TO [YOU TRY] EXERCISES

1) a) 4 b) −5 c) 2 d) −3   2) a) −16 b) −94   3) a) 1 b) −16 c) $-\dfrac{5}{28}$ d) 0

4) a) −12 b) −22 c) 17 d) 57   5) 25   6) a) 31 b) $-\dfrac{17}{18}$ c) −10

7) a) −14 + 6; −8 b) 15 − 27; −12 c) [23 + (−7)] − 5; 11

*Additional answers can be found in the Answers to Exercises appendix.

Mixed Exercises: Objectives 1–4 and 6

1) Explain, in your own words, how to subtract two negative numbers.   Answers may vary.

2) Explain, in your own words, how to add two negative numbers.   Answers may vary.

3) Explain, in your own words, how to add a positive and a negative number.   Answers may vary.

Use a number line to represent each sum or difference.

4) $-8 + 5$

5) $6 - 11$

6) $-1 - 5$

7) $-2 + (-7)$

8) $10 + (-6)$

Add or subtract as indicated.

9) $8 + (-15)$ -7

10) $-12 + (-6)$ -18

11) $-3 - 11$ -14

12) $-7 + 13$ 6

13) $-31 + 54$ 23

14) $19 - (-14)$ 33

15) $-26 - (-15)$ -11

16) $-20 - (-30)$ 10

17) $-352 - 498$ -850

18) $217 + (-521)$ -304

19) $-\dfrac{7}{12} + \dfrac{3}{4}$ $\dfrac{1}{6}$

20) $\dfrac{3}{10} - \dfrac{11}{15}$ $-\dfrac{13}{30}$

21) $-\dfrac{1}{6} - \dfrac{7}{8}$ $-\dfrac{25}{24}$ or $-1\dfrac{1}{24}$

22) $\dfrac{2}{9} - \left(-\dfrac{2}{5}\right)$ $\dfrac{28}{45}$

23) $-\dfrac{4}{9} - \left(-\dfrac{4}{15}\right)$ $-\dfrac{8}{45}$

24) $-\dfrac{1}{8} + \left(-\dfrac{3}{4}\right)$ $-\dfrac{7}{8}$

25) $19.4 + (-16.7)$ 2.7

26) $-31.3 - (-19.82)$ -11.48

27) $-25.8 - (-16.57)$ -9.23

28) $7.3 - 21.9$ -14.6

29) $9 - (5 - 11)$ 15

30) $-2 + (3 - 8)$ -7

31) $-1 + (-6 - 4)$ -11

32) $14 - (-10 - 2)$ 26

33) $(-3 - 1) - (-8 + 6)$ -2

34) $[14 + (-9)] + (1 - 8)$ -2

35) $-16 + 4 + 3 - 10$ -19

36) $8 - 28 + 3 - 7$ -24

37) $5 - (-30) - 14 + 2$ 23

38) $-17 - (-9) + 1 - 10$ -17

39) $\dfrac{4}{9} - \left(\dfrac{2}{3} + \dfrac{5}{6}\right)$ $-\dfrac{19}{18}$ or $-1\dfrac{1}{18}$

40) $-\dfrac{1}{2} + \left(\dfrac{3}{5} - \dfrac{3}{10}\right)$ $-\dfrac{1}{5}$

41) $\left(\dfrac{1}{8} - \dfrac{1}{2}\right) + \left(\dfrac{3}{4} - \dfrac{1}{6}\right)$ $\dfrac{5}{24}$

42) $\dfrac{11}{12} - \left(\dfrac{3}{8} - \dfrac{2}{3}\right)$ $\dfrac{29}{24}$ or $1\dfrac{5}{24}$

43) $(2.7 + 3.8) - (1.4 - 6.9)$ 12

44) $-9.7 - (-5.5 + 1.1)$ -5.3

45) $|7 - 11| + |6 + (-13)|$ 11

46) $|8 - (-1)| - |3 + 12|$ -6

47) $-|-2 - (-3)| - 2|-5 + 8|$ -7

48) $|-6 + 7| + 5|-20 - (-11)|$ 46

49) The sum of a positive number and a negative number is *always, sometimes,* or *never* negative.   sometimes

50) The sum of any two negative numbers is *always, sometimes,* or *never* positive.   never

Determine whether each statement is true or false. For any real numbers a and b,

51) $|a + b| = |a| + |b|$ false

52) $|a - b| = |b - a|$ true

53) $|a + b| = a + b$ false

54) $|a| + |b| = a + b$ false

55) $-b - (-b) = 0$ true

56) $a + (-a) = 0$ true

Objective 5: Solve Applied Problems

Write an expression for each and simplify. Answer the question with a complete sentence.

57) Tiger Woods won his first Masters championship in 1997 at age 21 with a score of -18. When he won the championship in 2005, his score was 6 strokes higher. What was Tiger's score when he won the Masters in 2005? (www.masters.com)
$-18 + 6 = -12$. His score in the 2005 Masters was -12.

58) In 1999, the U.S. National Park System recorded 287.1 million visits while in 2010 there were 281.3 million visits. What was the difference in the number of visits from 1999 to 2010? (www.nationalparkstraveler.com)
$281.3 - 287.1 = -5.8$. There were 5.8 million fewer visitors to the national parks in 2010 than in 1999.

59) In 2006, China's carbon emissions were 6,110,000 thousand metric tons and the carbon emissions of the United States totaled 5,790,000 thousand metric tons. By how much did China's carbon emissions exceed those of the United States? (www.pbl.nl)
$6,110,000 - 5,790,000 = 320,000$. The carbon emissions of China were 320,000 thousand metric tons more than those of the United States.

60) The budget of the Cincinnati Public Schools was $428,554,470 in the 2006–2007 school year. This was $22,430 less than the previous school year. What was the budget in the 2005–2006 school year?
(www.cps-k12.org) $428,554,470 + 22,430 = 428,576,900$.
The 2005–2006 budget was $428,576,900.

61) From 2009 to 2010, the number of flights going through O'Hare Airport in Chicago increased by 54,718. There were 882,617 flights in 2010. How many flights went through O'Hare in 2009?

(www.ohare.com) $882,617 - 54,718 = 827,899$. There were 827,899 flights at O'Hare in 2009.

62) The lowest temperature ever recorded in Minnesota was $-60°F$ while the highest temperature on record was $175°$ greater than that. What was the warmest temperature ever recorded in Minnesota?

(www.ncdc.noaa.gov) $-60 + 175 = 115$. The highest temperature recorded in Minnesota was $115°F$.

63) The bar graph shows the total number of daily newspapers in the United States in various years. Use a signed number to represent the change in the number of dailies over the given years.

(www.naa.org)

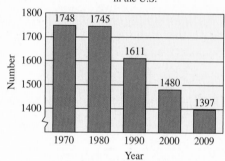

Number of Daily Newspapers in the U.S.

a) 1970–1980 -3 b) 1980–1990 -134

c) 2000–2009 -83 d) 1970–2009 -351

64) The bar graph shows the TV ratings for the World Series over a 5-year period. Each ratings number represents the percentage of people watching TV at the time of the World Series who were tuned into the games. Use a signed number to represent the change in ratings over the given years.

(www.baseball-almanac.com)

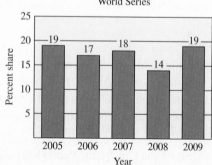

TV Ratings for the World Series

a) 2005–2006 -2 b) 2006–2007 1

c) 2007–2008 -4 d) 2008–2009 5

65) The bar graph shows the average number of days a woman was in the hospital for childbirth. Use a signed number to represent the change in hospitalization time over the given years. (www.cdc.gov)

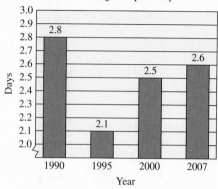

Average Hospital Stay

a) 1990–1995 -0.7 b) 1995–2000 0.4

c) 2000–2007 0.1 d) 1990–2007 -0.2

66) The bar graph shows snowfall totals for different seasons in Syracuse, NY. Use a signed number to represent the difference in snowfall totals over different years. (www.erh.noaa.gov)

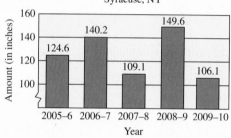

Snowfall Totals for Syracuse, NY

a) 2005–6 to 2007–8   b) 2005–6 to 2006–7
-15.5 15.6

c) 2008–9 to 2009–10   d) 2005–6 to 2009–10
-43.5 -18.5

Objective 7: Translate English Expressions into Mathematical Expressions

Write a mathematical expression for each and simplify.

67) 7 more than 5 $5 + 7; 12$

68) 3 more than 11 $11 + 3; 14$

69) 16 less than 10 $10 - 16; -6$

70) 15 less than 4 $4 - 15; -11$

71) -8 less than 9 $9 - (-8); 17$

72) -25 less than -19 $-19 - (-25); 6$

73) The sum of -21 and 13 $-21 + 13; -8$

74) The sum of -7 and 20 $-7 + 20; 13$

75) -20 increased by 30 $-20 + 30;\ 10$

76) -37 increased by 22 $-37 + 22;\ -15$

77) 23 decreased by 19 $23 - 19;\ 4$

78) 8 decreased by 18 $8 - 18;\ -10$

79) 18 less than the sum of -5 and 11 $(-5 + 11) - 18;\ -12$

80) 35 less than the sum of -17 and 3 $(-17 + 3) - 35;\ -49$

R Rethink

R1) Which exercises in this section do you find most challenging?

R2) Which is easier for you, $7 - 9$ or $7 + (-9)$? Why?

R3) Where have you used the addition of signed numbers outside of your math class?

R4) If the absolute value of a number is greater than the absolute value of another, how do you determine whether their difference is positive or negative?

1.6 Multiplication and Division of Real Numbers

P Prepare

O Organize

What are your objectives for Section 1.6?	How can you accomplish each objective?
1 Multiply Real Numbers	• Write the procedure for **Multiplying Real Numbers** in your own words. • Complete the given examples on your own. • Complete You Trys 1 and 2.
2 Evaluate Exponential Expressions	• Create a procedure to help you quickly determine whether an exponential expression will give a positive or negative result. • Complete the given examples on your own. • Complete You Trys 3 and 4.
3 Divide Real Numbers	• Write the procedure for **Dividing Real Numbers** in your own words. • Complete the given example on your own. • Complete You Try 5.
4 Apply the Order of Operations	• Add any additional steps to the Order of Operations you updated in Section 1.4. • Complete the given example on your own. • Complete You Try 6.
5 Translate English Expressions into Mathematical Expressions	• Take notes on the common English expressions, and be able to identify whether an expression refers to multiplication or division. • Complete the given example on your own. • Complete You Try 7.

Read the explanations, follow the examples, take notes, and complete the You Trys.

1 Multiply Real Numbers

What is the meaning of $4 \cdot 5$? It is repeated addition.

$$4 \cdot 5 = 4 + 4 + 4 + 4 + 4 = 20$$

So, what is the meaning of $-4 \cdot 5$? It, too, represents repeated addition.

$$-4 \cdot 5 = -4 + (-4) + (-4) + (-4) + (-4) = -20$$

Let's make a table of some products:

$\times$	5	4	③	2	1	0	-1	-2	-3	-4	-5
④	20	16	12	8	4	0	-4	-8	-12	-16	-20

$$4 \cdot 3 = 12$$

The bottom row represents the product of 4 and the number above it ($4 \cdot 3 = 12$). Notice that as the numbers in the first row decrease by 1, the numbers in the bottom row decrease by 4. Therefore, once we get to $4 \cdot (-1)$, the product is negative. From the table we can see that,

Note

The product of a positive number and a negative number is negative.

EXAMPLE 1

Multiply.

a) $-6 \cdot 9$ b) $\dfrac{3}{8} \cdot (-12)$ c) $-5 \cdot 0$

Solution

a) $-6 \cdot 9 = -54$ b) $\dfrac{3}{8} \cdot (-12) = \dfrac{3}{8} \cdot \left(-\dfrac{\overset{3}{\cancel{12}}}{1} \right) = -\dfrac{9}{2}$

c) $-5 \cdot 0 = 0$ The product of zero and any real number is zero.

[YOU TRY 1]

Multiply.

a) $-7 \cdot 3$ b) $\dfrac{8}{15} \cdot (-10)$

What is the sign of the product of two negative numbers? Again, we'll make a table.

$\times$	3	2	1	0	-1	-2	-3
-4	-12	-8	-4	0	4	8	12

As the numbers in the top row decrease by 1, the numbers in the bottom row *increase* by 4. When we reach $-4 \cdot (-1)$, our product is a positive number, 4. The table illustrates that the product of two negative numbers is positive.

We can summarize our findings this way:

> ### Procedure Multiplying Real Numbers
>
> 1) The product of two positive numbers is positive.
> 2) The product of two negative numbers is positive.
> 3) The product of a positive number and a negative number is negative.
> 4) The product of any real number and zero is zero.

EXAMPLE 2

In-Class Example 2

Multiply.
a) $-9 \cdot (-12)$ b) $-2.5 \cdot 10$
c) $-\dfrac{9}{10} \cdot \left(-\dfrac{4}{15}\right)$
d) $-5 \cdot \left(-\dfrac{1}{6}\right) \cdot (-24)$

Answer: a) 108 b) -25
c) $\dfrac{6}{25}$ d) -20

Multiply.

a) $-8 \cdot (-5)$ b) $-1.5 \cdot 6$ c) $-\dfrac{3}{8} \cdot \left(-\dfrac{4}{5}\right)$ d) $-5 \cdot (-2) \cdot (-3)$

Solution

a) $-8 \cdot (-5) = 40$ The product of two negative numbers is positive.

b) $-1.5 \cdot 6 = -9$ The product of a negative number and a positive number is negative.

c) $-\dfrac{3}{8} \cdot \left(-\dfrac{4}{5}\right) = -\dfrac{3}{8} \cdot \left(-\dfrac{\overset{1}{4}}{\underset{2}{5}}\right) = \dfrac{3}{10}$ The product of two negatives is positive.

d) $\underbrace{-5 \cdot (-2)}_{10} \cdot (-3) = 10 \cdot (-3) = -30$ Order of operations—multiply from left to right.

[YOU TRY 2] Multiply.

a) $-6 \cdot 7$ b) $-\dfrac{8}{9} \cdot \dfrac{3}{4}$ c) $-4 \cdot (-1) \cdot (-5) \cdot (-2)$

> **Note**
> It is helpful to know that
> 1) an **even number** of negative factors in a product gives a positive result.
>
> $$-3 \cdot 1 \cdot (-2) \cdot (-1) \cdot (-4) = 24 \quad \text{Four negative factors}$$
>
> 2) an **odd number** of negative factors in a product gives a negative result.
>
> $$5 \cdot (-3) \cdot (-1) \cdot (-2) \cdot (3) = -90 \quad \text{Three negative factors}$$

2 Evaluate Exponential Expressions

In Section 1.2, we discussed exponential expressions. Recall that exponential notation is a shorthand way to represent repeated multiplication:

$$2^4 = 2 \cdot 2 \cdot 2 \cdot 2 = 16$$

Now we will discuss exponents and negative numbers. Consider a base of -2 raised to different powers. (The -2 is in parentheses to indicate that it is the base.)

$$(-2)^1 = -2$$
$$(-2)^2 = -2 \cdot (-2) = 4$$
$$(-2)^3 = -2 \cdot (-2) \cdot (-2) = -8$$
$$(-2)^4 = -2 \cdot (-2) \cdot (-2) \cdot (-2) = 16$$
$$(-2)^5 = -2 \cdot (-2) \cdot (-2) \cdot (-2) \cdot (-2) = -32$$
$$(-2)^6 = -2 \cdot (-2) \cdot (-2) \cdot (-2) \cdot (-2) \cdot (-2) = 64$$

Do you notice that

1) -2 raised to an *odd* power gives a negative result?

and

2) -2 raised to an *even* power gives a positive result?

This will always be true.

> **W Hint**
>
> A negative number raised to an *odd* power will give a *negative* result. A negative number raised to an *even* power will give a *positive* result.

EXAMPLE 3

In-Class Example 3

Evaluate.
a) $(-3)^3$ b) $(-12)^2$

Answer: a) -27 b) 144

Evaluate.

a) $(-6)^2$ b) $(-10)^3$

Solution

a) $(-6)^2 = 36$ b) $(-10)^3 = -1000$

[YOU TRY 3]

Evaluate.

a) $(-9)^2$ b) $(-5)^3$

> **W Hint**
>
> The order of operations tells us to evaluate exponents before multiplying. Therefore, to evaluate -2^4, first we evaluate the base using the exponent, then we multiply by -1.

How do $(-2)^4$ and -2^4 differ? Let's identify their bases and evaluate each.

$(-2)^4$: Base $= -2$ $(-2)^4 = 16$

-2^4: Since there are no parentheses,

-2^4 is equivalent to $-1 \cdot 2^4$. Therefore, the base is 2.

$$-2^4 = -1 \cdot 2^4$$
$$= -1 \cdot 2 \cdot 2 \cdot 2 \cdot 2$$
$$= -16$$

So, $(-2)^4 = 16$ and $-2^4 = -16$.

 BE CAREFUL When working with exponential expressions, be able to identify the base.

EXAMPLE 4

Evaluate.

a) $(-5)^3$ b) -9^2 c) $\left(-\dfrac{1}{7}\right)^2$

Solution

a) $(-5)^3$: Base $= -5$ $(-5)^3 = -5 \cdot (-5) \cdot (-5) = -125$

b) -9^2: Base $= 9$ $-9^2 = -1 \cdot 9^2$
$$= -1 \cdot 9 \cdot 9$$
$$= -81$$

c) $\left(-\dfrac{1}{7}\right)^2$: Base $= -\dfrac{1}{7}$ $\left(-\dfrac{1}{7}\right)^2 = -\dfrac{1}{7} \cdot \left(-\dfrac{1}{7}\right) = \dfrac{1}{49}$

YOU TRY 4

Evaluate.

a) -3^4 b) $(-11)^2$ c) -8^2 d) $-\left(-\dfrac{2}{3}\right)^3$

3 Divide Real Numbers

Here are the rules for dividing signed numbers:

Procedure Dividing Signed Numbers

1) The quotient of two positive numbers is a positive number.
2) The quotient of two negative numbers is a positive number.
3) The quotient of a positive and a negative number is a negative number.

EXAMPLE 5

Divide.

a) $-36 \div 9$ b) $-\dfrac{1}{10} \div \left(-\dfrac{3}{5}\right)$ c) $\dfrac{-8}{-1}$ d) $\dfrac{-24}{42}$

Solution

a) $-36 \div 9 = -4$

b) $-\dfrac{1}{10} \div \left(-\dfrac{3}{5}\right) = -\dfrac{1}{10} \cdot \left(-\dfrac{5}{3}\right)$ When dividing by a fraction, multiply by the reciprocal.

$$= -\dfrac{1}{\underset{2}{10}} \cdot \left(-\dfrac{\overset{1}{5}}{3}\right) = \dfrac{1}{6}$$

W Hint

Remember, write all answers in lowest terms.

c) $\dfrac{-8}{-1} = 8$ The quotient of two negative numbers is positive, and $\dfrac{8}{1}$ simplifies to 8.

d) $\dfrac{-24}{42} = -\dfrac{24}{42}$ The quotient of a negative number and a positive number is negative, so simplify $\dfrac{24}{42}$.

$$= -\dfrac{4}{7}$$ 24 and 42 each divide by 6.

It is important to note here in part d) that there are three ways to write the answer: $-\dfrac{4}{7}, \dfrac{-4}{7},$ or $\dfrac{4}{-7}$. These are equivalent. However, we usually write the negative sign in front of the entire fraction as in $-\dfrac{4}{7}$.

[YOU TRY 5] Divide.

a) $-\dfrac{8}{5} \div \left(-\dfrac{6}{5}\right)$ b) $\dfrac{-30}{-10}$ c) $\dfrac{21}{-56}$

---

4 Apply the Order of Operations

EXAMPLE 6

In-Class Example 6

Simplify.
a) $-28 \div (-7) + 3^2$
b) $4(-11) - 4(8 - 15)$

Answer: a) 13 b) −16

Simplify.

a) $-24 \div 12 - 2^2$ b) $-5(-3) - 4(2 - 3)$

Solution

a) $-24 \div 12 - 2^2 = -24 \div 12 - 4$ Simplify exponent first.
$$= -2 - 4$$ Perform division before subtraction.
$$= -6$$

b) $-5(-3) - 4(2 - 3) = -5(-3) - 4(-1)$ Simplify the difference in parentheses.
$$= 15 - (-4)$$ Find the products.
$$= 15 + 4$$ Change subtraction to addition.
$$= 19$$

[YOU TRY 6] Simplify.

a) $-13 - 4(-5 + 2)$ b) $(-10)^2 + 2[8 - 5(4)]$

---

5 Translate English Expressions into Mathematical Expressions

Here are some words and phrases you may encounter and how they would translate into mathematical expressions:

English Expression	**Mathematical Operation**
times, product of	multiplication
divided by, quotient of	division

EXAMPLE 7

In-Class Example 7

Write a mathematical expression for each and simplify.
a) The quotient of 9 and −3
b) The product of −8 and the sum of 11 and −2

Write a mathematical expression for each and simplify.

a) The quotient of -56 and 7

b) The product of 4 and the sum of 15 and -6

c) Twice the difference of -10 and -3

d) Half of the sum of -8 and 3

---

c) Twice the difference
   of −14 and 8
d) Half of the sum of 16
   and −42

Answer: a) $\dfrac{9}{-3} = -3$

b) $-8[11 + (-2)] = -72$

c) $2[-14 - 8] = -44$

d) $\dfrac{1}{2}[16 + (-42)] = -13$

Solution

a) The quotient of −56 and 7:
   Quotient means division with −56 in the numerator and 7 in the denominator.
   The expression is $\dfrac{-56}{7} = -8$.

b) The product of 4 and the sum of 15 and −6:
   The *sum of* 15 *and* −6 means we must add the two numbers. *Product* means multiply.

$$\underbrace{4[\overbrace{15 + (-6)}^{\text{Sum of 15 and }-6}]}_{\substack{\text{Product of 4} \\ \text{and the sum}}} = 4(9) = 36$$

c) Twice the difference of −10 and −3:
   The *difference of* −10 *and* −3 will be in parentheses with −3 being subtracted from −10. *Twice* means "two times."

$$2[-10 - (-3)] = 2(-10 + 3) = 2(-7) = -14$$

d) Half of the sum of −8 and 3:
   The *sum of* −8 *and* 3 means that we will add the two numbers. They will be in parentheses. *Half of* means multiply by $\dfrac{1}{2}$.

$$\frac{1}{2}(-8 + 3) = \frac{1}{2}(-5) = -\frac{5}{2}$$

[YOU TRY 7]

Write a mathematical expression for each and simplify.

a) 12 less than the product of −7 and 4

b) Twice the sum of 19 and −11

c) The sum of −41 and −23, divided by the square of −2

ANSWERS TO [YOU TRY] EXERCISES

1) a) −21 b) $-\dfrac{16}{3}$ 2) a) −42 b) $-\dfrac{2}{3}$ c) 40   3) a) 81 b) −125

4) a) −81 b) 121 c) −64 d) $\dfrac{8}{27}$ 5) a) $\dfrac{4}{3}$ b) 3 c) $-\dfrac{3}{8}$ 6) a) −1 b) 76

7) a) $(-7) \cdot 4 - 12$; −40 b) $2[19 + (-11)]$; 16 c) $\dfrac{-41 + (-23)}{(-2)^2}$; −16

E Evaluate **1.6** **Exercises**   Do the exercises, and check your work.

*Additional answers can be found in the Answers to Exercises appendix.

Objective 1: Multiply Real Numbers

Fill in the blank with *positive* or *negative*.

1) The product of a positive number and a negative number is __negative__.

2) The product of two negative numbers is __positive__.

Multiply.

3) $-8 \cdot 7$ −56

4) $4 \cdot (-9)$ −36

5) $-15 \cdot (-3)$ 45

6) $-23 \cdot (-48)$ 1104

7) $-4 \cdot 3 \cdot (-7)$ 84

8) $-5 \cdot (-1) \cdot (-11)$ −55

9) $\frac{4}{33} \cdot \left(-\frac{11}{10}\right)$ $-\frac{2}{15}$

10) $-\frac{14}{27} \cdot \left(-\frac{15}{28}\right)$ $\frac{5}{18}$

11) $(-0.5)(-2.8)$ 1.4

12) $(-6.1)(5.7)$ -34.77

13) $-9 \cdot (-5) \cdot (-1) \cdot (-3)$ 135

14) $-1 \cdot (-6) \cdot (4) \cdot (-2) \cdot (3)$ -144

15) $\frac{3}{10} \cdot (-7) \cdot (8) \cdot (-1) \cdot (-5)$ -84

16) $-\frac{5}{6} \cdot (-4) \cdot 0 \cdot 3$ 0

Objective 2: Evaluate Exponential Expressions

17) For what values of k is k^5 a negative quantity?
when k is negative

18) For what values of k is k^5 a positive quantity?
when k is positive

19) For what values of k is $-k^2$ a negative quantity?
when $k \neq 0$

20) Explain the difference between how you would evaluate $(-8)^2$ and -8^2. Then, evaluate each.
$(-8)^2$ means $-8 \cdot (-8) = 64$. -8^2 means $-1 \cdot 8^2$. So, $-1 \cdot 8^2 = -1 \cdot 64 = -64$.

Evaluate.

21) $(-6)^2$ 36

22) -6^2 -36

23) -5^3 -125

24) $(-2)^4$ 16

25) $(-3)^2$ 9

26) $(-1)^5$ -1

27) -7^2 -49

28) -4^3 -64

29) -2^5 -32

30) $(-12)^2$ 144

Objective 3: Divide Real Numbers

Fill in the blank with *positive* or *negative*.

31) The quotient of two negative numbers is __positive__.

32) The quotient of a negative number and a positive number is __negative__.

Divide.

33) $-50 \div (-5)$ 10

34) $-84 \div 12$ -7

35) $\frac{64}{-16}$ -4

36) $\frac{-54}{-9}$ 6

37) $\frac{-2.4}{0.3}$ -8

38) $\frac{16}{-0.5}$ -32

39) $-\frac{12}{13} \div \left(-\frac{6}{5}\right)$ $\frac{10}{13}$

40) $20 \div \left(-\frac{15}{7}\right)$ $-\frac{28}{3}$ or $-9\frac{1}{3}$

41) $-\frac{0}{7}$ 0

42) $\frac{0}{-6}$ 0

43) $\frac{270}{-180}$ $-\frac{3}{2}$ or $-1\frac{1}{2}$

44) $\frac{-64}{-320}$ $\frac{1}{5}$

Objective 4: Apply the Order of Operations

Use the order of operations to simplify.

45) $7 + 8(-5)$ -33

46) $-40 \div 2 - 10$ -30

47) $(9 - 14)^2 - (-3)(6)$ 43

48) $-23 - 6^2 \div 4$ -32

49) $10 - 2(1 - 4)^3 \div 9$ 16

50) $-7(4) + (-8 + 6)^4 + 5$ -7

51) $\left(-\frac{3}{4}\right)(8) - 2[7 - (-3)(-6)]$ 16

52) $-2^5 - (-3)(4) + 5[(-9 + 30) \div 7]$ -5

53) $\frac{-46 - 3(-12)}{(-5)(-2)(-4)}$ $\frac{1}{4}$

54) $\frac{(8)(-6) + 10 - 7}{(-5 + 1)^2 - 12 + 5}$ -5

Objective 5: Translate English Expressions into Mathematical Expressions

Write a mathematical expression for each and simplify.

55) The product of -12 and 6 $-12 \cdot 6; -72$

56) The quotient of -80 and -4 $\frac{-80}{-4}; 20$

57) 9 more than the product of -7 and -5
$(-7)(-5) + 9; 44$

58) The product of -10 and 2 increased by 11
$(-10)(2) + 11; -9$

59) The quotient of 63 and -9 increased by 7 $\frac{63}{-9} + 7; 0$

60) 8 more than the quotient of 54 and -6 $\frac{54}{-6} + 8; -1$

61) 19 less than the product of -4 and -8
$(-4)(-8) - 19; 13$

62) The product of -16 and -3 decreased by 20
$(-16)(-3) - 20; 28$

63) The quotient of -100 and 4 decreased by the sum of -7 and 2 $\frac{-100}{4} - (-7 + 2); -20$

64) The quotient of -35 and 5 increased by the product of -11 and -2 $\frac{-35}{5} + (-11)(-2); 15$

65) Twice the sum of 18 and -31 $2[18 + (-31)]; -26$

66) Twice the difference of -5 and -15 $2[-5 - (-15)]; (20)$

67) Two-thirds of -27 $\frac{2}{3}(-27); -18$

68) Half of -30 $\frac{1}{2}(-30); -15$

69) The product of 12 and -5 increased by half of 36
$12(-5) + \frac{1}{2}(36); -42$

70) One-third of -18 decreased by half the sum of -21 and -5 $\frac{1}{3}(-18) - \frac{1}{2}[-21 + (-5)]; 7$

R1) What are the similarities between multiplying and dividing signed numbers?

R2) In Objective 2, why is the use of parentheses so important?

R3) In Objective 5, think about the order in which the integer values are given. How does this affect their placement into the mathematical expression?

1.7 Properties of Real Numbers

P Prepare   **O** Organize

What are your objectives for Section 1.7?	How can you accomplish each objective?
1 Evaluate Algebraic Expressions	• Write the definitions of a *variable* and an *algebraic expression* in your own words. • Complete the given examples on your own. • Complete You Trys 1–3.
2 Use the Commutative Properties	• Learn the **commutative properties** for addition and multiplication. • Complete the given example on your own. • Complete You Try 4.
3 Use the Associative Properties	• Learn the **associative properties** for addition and multiplication. • Complete the given examples on your own. • Complete You Trys 5 and 6.
4 Use the Identity and Inverse Properties	• Know that 0 is the additive identity and 1 is the multiplicative identity, and use that information to develop the **identity properties** in your notes. • Know how to find an additive inverse and a reciprocal or multiplicative inverse, and use that information to develop the **inverse properties** in your notes. • Complete the given example on your own. • Complete You Try 7.
5 Use the Distributive Property	• Learn the **distributive properties.** • Complete the given examples on your own. • Complete You Trys 8 and 9.

W Work   **Read the explanations, follow the examples, take notes, and complete the You Trys.**

1 Evaluate Algebraic Expressions

Before we learn about the properties of real numbers, we must learn about variables and expressions.

Here is an algebraic expression: $2x + 5$. We call x the *variable*.

Definition

A **variable** is a symbol, usually a letter, used to represent an unknown number.

In the expression $2x + 5$, there is no symbol written between 2 and x. This means the operation is multiplication.

$$2x \quad \text{means} \quad 2 \cdot x$$

We combine variables with numbers and operation symbols to form *algebraic expressions*.

Definition

An **algebraic expression** is a collection of numbers, variables, and grouping symbols connected by operation symbols such as $+$, $-$, $\times$, and $\div$.

Examples of expressions: $\quad 9a, \quad 8 - n, \quad 4k - 3h + 10, \quad 5c^2 + 2(c - 6)$

We can **evaluate** an algebraic expression by substituting a value for a variable and simplifying. The value of an algebraic expression changes depending on the value that is substituted.

EXAMPLE 1

Evaluate $2x + 5$ when a) $x = 7$ and b) $x = -3$.

In-Class Example 1

Evaluate $3p + 7$ when
a) $p = 10$ and b) $p = -4$.

Answer: a) 37 b) -5

Solution

a) $2x + 5$ when $x = 7$ Substitute 7 for x.
    $= 2(7) + 5$ Use parentheses when substituting a value for a variable.
    $= 14 + 5$ Multiply.
    $= 19$ Add.

W Hint

A good strategy for reading a math book is to write out the example as you are reading it.

b) $2x + 5$ when $x = -3$ Substitute -3 for x.
    $= 2(-3) + 5$ Use parentheses when substituting a value for a variable.
    $= -6 + 5$ Multiply.
    $= -1$ Add.

[**YOU TRY 1**] Evaluate $8t + 1$ when a) $t = 4$ and b) $t = -5$.

EXAMPLE 2

In-Class Example 2

Evaluate $\dfrac{4c - 5d}{c + d}$ when $c = -8$ and $d = 2$.

Answer: 7

Evaluate $\dfrac{8a - 3b}{a + 4b}$ when $a = -6$ and $b = 2$.

Solution

$$\dfrac{8a - 3b}{a + 4b} = \dfrac{8(-6) - 3(2)}{-6 + 4(2)} \qquad \text{Substitute } -6 \text{ for } a, \text{ and } 2 \text{ for } b.$$

$$= \dfrac{-48 - 6}{-6 + 8} \qquad \text{Multiply.}$$

$$= \dfrac{-54}{2} \qquad \text{Subtract in the numerator; add in the denominator.}$$

$$= -27 \qquad \text{Simplify.}$$

[YOU TRY 2]

Evaluate $\dfrac{m + 9n}{4m - n}$ when $m = 6$ and $n = -4$.

EXAMPLE 3

In-Class Example 3

Find the value of $2g^2 - gh - 7$ when $g = -3$ and $h = 5$.

Answer: 26

Evaluate $5x^2 - xy - 11$ when $x = -2$ and $y = 3$.

Solution

$$5x^2 - xy - 11 = 5(-2)^2 - (-2)(3) - 11 \qquad \text{Substitute } -2 \text{ for } x \text{ and } 3 \text{ for } y.$$

$$= 5 \cdot 4 - (-6) - 11 \qquad \text{Evaluate exponent; multiply.}$$

$$= 20 + 6 - 11 \qquad \text{Multiply.}$$

$$= 15$$

[YOU TRY 3]

Evaluate $6u^2 - 5uv + 4$ when $u = -1$ and $v = 4$.

Next, we will learn about the properties of real numbers so that we can use them to work with algebraic expressions.

Like the order of operations, the properties of real numbers guide us in our work with numbers and variables. We begin with the commutative properties of real numbers.

2 Use the Commutative Properties

True or false?

1) $7 + 3 = 3 + 7$ *True:* $7 + 3 = 10$ and $3 + 7 = 10$

2) $8 - 2 = 2 - 8$ *False:* $8 - 2 = 6$ but $2 - 8 = -6$

3) $(-6)(5) = (5)(-6)$ *True:* $(-6)(5) = -30$ and $(5)(-6) = -30$

In 1), we see that adding 7 and 3 in any order still equals 10. The third equation shows that multiplying $(-6)(5)$ and $(5)(-6)$ both equal -30. But, 2) illustrates that changing the order in which numbers are subtracted does *not* necessarily give the same result: $8 - 2 \neq 2 - 8$.

Therefore, subtraction is **not commutative,** while the addition and multiplication of real numbers **are commutative.** This gives us our first properties of real numbers.

Property Commutative Properties

If a and b are real numbers, then

1) $a + b = b + a$ Commutative property of addition

2) $ab = ba$ Commutative property of multiplication

We have already shown that subtraction is not commutative. Is division commutative? No. For example,

$$20 \div 4 \overset{?}{=} 4 \div 20$$

$$5 \neq \frac{1}{5}$$

EXAMPLE 4

Use the commutative property to rewrite each expression.

a) $12 + 5$ b) $k \cdot 3$

Solution

a) $12 + 5 = 5 + 12$ b) $k \cdot 3 = 3 \cdot k$ or $3k$

In-Class Example 4

Use the commutative property to rewrite each expression.
a) $2 + 29$ b) $m \cdot 10$

Answer: a) $2 + 29 = 29 + 2$
b) $m \cdot 10 = 10 \cdot m$ or $10m$

[YOU TRY 4]

Use the commutative property to rewrite each expression.

a) $1 + 16$ b) $n \cdot 6$

3 Use the Associative Properties

Another important property involves the use of grouping symbols. Let's determine whether these two statements are true:

$$\begin{array}{ccc} (9 + 4) + 2 \overset{?}{=} 9 + (4 + 2) & & (2 \cdot 3)4 \overset{?}{=} 2(3 \cdot 4) \\ 13 + 2 \overset{?}{=} 9 + 6 & \text{and} & (6)4 \overset{?}{=} 2(12) \\ 15 = 15 & & 24 = 24 \\ \text{TRUE} & & \text{TRUE} \end{array}$$

We can generalize and say that when adding or multiplying real numbers, the way in which we group them to evaluate them will not affect the result. Notice that the *order* in which the numbers are written does not change.

Property Associative Properties

If a, b, and c are real numbers, then

1) $(a + b) + c = a + (b + c)$ Associative property of addition

2) $(ab)c = a(bc)$ Associative property of multiplication

Sometimes, applying the associative property can simplify calculations.

EXAMPLE 5

In-Class Example 5

Apply the associative property to simplify $\left(4 \cdot \dfrac{3}{11}\right)11$.

Answer: 12

Apply the associative property to simplify $\left(7 \cdot \dfrac{2}{5}\right)5$.

Solution

By the associative property, $\left(7 \cdot \dfrac{2}{5}\right)5 = 7 \cdot \left(\dfrac{2}{\cancel{5}} \cdot \cancel{5}\right) = 7 \cdot 2 = 14$

[YOU TRY 5]

Apply the associative property to simplify $\left(9 \cdot \dfrac{4}{3}\right)3$.

EXAMPLE 6

In-Class Example 6

Use the associative property to simplify each expression.

a) $(12 + y) + 3$

b) $\left(-\dfrac{5}{8} \cdot \dfrac{2}{7}\right)\dfrac{7}{2}$

Answer: a) $15 + y$ b) $-\dfrac{5}{8}$

Use the associative property to simplify each expression.

a) $-6 + (10 + y)$ b) $\left(-\dfrac{3}{11} \cdot \dfrac{8}{5}\right)\dfrac{5}{8}$

Solution

a) $-6 + (10 + y) = (-6 + 10) + y = 4 + y$

b) $\left(-\dfrac{3}{11} \cdot \dfrac{8}{5}\right)\dfrac{5}{8} = -\dfrac{3}{11}\left(\dfrac{8}{5} \cdot \dfrac{5}{8}\right) = -\dfrac{3}{11}(1) = -\dfrac{3}{11}$ A number times its reciprocal equals 1.

[YOU TRY 6]

Use the associative property to simplify each expression.

a) $(k + 3) + 9$ b) $\left(-\dfrac{9}{7} \cdot \dfrac{8}{5}\right)\dfrac{5}{8}$

The identity properties of addition and multiplication are also ones we need to know.

4 Use the Identity and Inverse Properties

For addition we know that, for example,

$$5 + 0 = 5, \qquad 0 + \dfrac{2}{3} = \dfrac{2}{3}, \qquad -14 + 0 = -14$$

When zero is added to a number, the value of the number is unchanged. *Zero* is the **identity element for addition** (also called the **additive identity**).

What is the identity element for multiplication?

$$-4(1) = -4 \qquad 1(3.82) = 3.82 \qquad \dfrac{9}{2}(1) = \dfrac{9}{2}$$

W Hint

Notice that the identity property of addition uses the number 0, and that the identity property of multiplication uses the number 1.

When a number is multiplied by 1, the value of the number is unchanged. *One* is the **identity element for multiplication** (also called the **multiplicative identity**).

Property Identity Properties

If a is a real number, then

1) $a + 0 = 0 + a = a$ Identity property of addition
2) $a \cdot 1 = 1 \cdot a = a$ Identity property of multiplication

The next properties we will discuss give us the additive and multiplicative identities as results. In Section 1.4, we introduced an **additive inverse.**

Number	Additive Inverse
3	−3
−11	11
$-\dfrac{7}{9}$	$\dfrac{7}{9}$

Let's add each number and its additive inverse:

$$3 + (-3) = 0, \qquad -11 + 11 = 0, \qquad -\frac{7}{9} + \frac{7}{9} = 0$$

Given a number such as $\dfrac{3}{5}$, we know that its **reciprocal** (or **multiplicative inverse**) is $\dfrac{5}{3}$. We have also established the fact that the product of a number and its reciprocal is 1 as in $\dfrac{3}{5} \cdot \dfrac{5}{3} = 1$.

Therefore, multiplying a number b by its reciprocal (multiplicative inverse) $\dfrac{1}{b}$ gives us the identity element for multiplication, 1. That is,

$$b \cdot \frac{1}{b} = \frac{1}{b} \cdot b = 1$$

Property Inverse Properties

If a is any real number and b is a real number not equal to 0, then

1) $a + (-a) = -a + a = 0$ Inverse property of addition
2) $b \cdot \dfrac{1}{b} = \dfrac{1}{b} \cdot b = 1$ Inverse property of multiplication

 Hint

Notice that the inverse property for addition has a result of 0, and the inverse property for multiplication gives a result of 1.

EXAMPLE 7

In-Class Example 7

Which property is illustrated by each statement?
a) $-6 + 6 = 0$
b) $3.8 + 0 = 3.8$
c) $(-14)\left(-\dfrac{1}{14}\right) = 1$
d) $9(1) = 9$

Which property is illustrated by each statement?

a) $0 + 12 = 12$ b) $-9.4 + 9.4 = 0$

c) $\dfrac{1}{7} \cdot 7 = 1$ d) $2(1) = 2$

Solution

a) $0 + 12 = 12$ Identity property of addition

b) $-9.4 + 9.4 = 0$ Inverse property of addition

c) $\dfrac{1}{7} \cdot 7 = 1$ Inverse property of multiplication

d) $2(1) = 2$ Identity property of multiplication

[YOU TRY 7] Which property is illustrated by each statement?

a) $5 \cdot \dfrac{1}{5} = 1$ b) $-26 + 26 = 0$ c) $2.7(1) = 2.7$ d) $-4 + 0 = -4$

---

5 Use the Distributive Property

The last property we will discuss is the **distributive property.** It involves both multiplication and addition or multiplication and subtraction.

> **Property** Distributive Properties
>
> If a, b, and c are real numbers, then
>
> 1) $a(b + c) = ab + ac$ and $(b + c)a = ba + ca$
> 2) $a(b - c) = ab - ac$ and $(b - c)a = ba - ca$

EXAMPLE 8

Evaluate using the distributive property.

a) $3(2 + 8)$ b) $-(6 + 3)$

Solution

a) $3(2 + 8) = 3 \cdot 2 + 3 \cdot 8$ Apply distributive property.
$$= 6 + 24$$
$$= 30$$

> **W Hint**
>
> Write out each example as you are reading it!

Note that we would get the same result if we would apply the order of operations:

$$3(2 + 8) = 3(10) = 30$$

b) $-(6 + 3) = -1(6 + 3)$
$$= -1 \cdot 6 + (-1)(3)$$ Apply distributive property.
$$= -6 + (-3)$$
$$= -9$$

A negative sign in front of parentheses is the same as multiplying by -1.

[YOU TRY 8] Evaluate using the distributive property.

a) $2(11 - 5)$ b) $-5(3 - 7)$ c) $-(4 + 9)$

The distributive property can be applied when there are more than two terms in parentheses and when there are variables.

EXAMPLE 9

In-Class Example 9

Use the distributive property to rewrite each expression. Simplify if possible.
a) $-(x + 5 - 7)$
b) $-3(4 - 9y)$

Answer: a) $2 - x$
b) $27y - 12$

Use the distributive property to rewrite each expression. Simplify if possible.

a) $7(x + 4)$ b) $-(-5c + 4d - 6)$

Solution

a) $7(x + 4) = 7x + 7 \cdot 4$ Apply distributive property.

$\qquad = 7x + 28$

b) $-(-5c + 4d - 6) = -1(-5c + 4d - 6)$

$\qquad = -1(-5c) + (-1)(4d) - (-1)(6)$ Apply distributive property.

$\qquad = 5c + (-4d) - (-6)$ Multiply.

$\qquad = 5c - 4d + 6$

[YOU TRY 9]

Use the distributive property to rewrite each expression. Simplify if possible.

a) $6(a + 2)$ b) $5(2x - 7y - 4z)$ c) $-(-r + 4s - 9)$

The properties stated previously are summarized next.

Summary Properties of Real Numbers

If a, b, and c are real numbers, then

Commutative properties:	$a + b = b + a$ and $ab = ba$
Associative properties:	$(a + b) + c = a + (b + c)$ and $(ab)c = a(bc)$
Identity properties:	$a + 0 = 0 + a = a$
	$a \cdot 1 = 1 \cdot a = a$
Inverse properties:	$a + (-a) = -a + a = 0$
	$b \cdot \dfrac{1}{b} = \dfrac{1}{b} \cdot b = 1 \ (b \neq 0)$
Distributive properties:	$a(b + c) = ab + ac$ and $(b + c)a = ba + ca$
	$a(b - c) = ab - ac$ and $(b - c)a = ba - ca$

ANSWERS TO [YOU TRY] EXERCISES

1) a) 33   b) -39 2) $-\dfrac{15}{14}$ 3) 30   4) a) $16 + 1$ b) $6n$ 5) 36   6) a) $k + 12$

b) $-\dfrac{9}{7}$ or $-1\dfrac{2}{7}$ 7) a) inverse property of multiplication   b) inverse property of addition

c) identity property of multiplication   d) identity property of addition   8) a) 12   b) 20   c) -13

9) a) $6a + 12$ b) $10x - 35y - 20z$ c) $r - 4s + 9$

*Additional answers can be found in the Answers to Exercises appendix.

Objective 1: Evaluate Algebraic Expressions

1) What is an algebraic expression? It is a collection of numbers, variables, and grouping symbols connected by operation symbols such as $+$, $-$, $\times$, and $\div$.

2) Is the value of an algebraic expression always the same? Explain. No. The value of the expression changes as the value of the variables change.

Evaluate $4m + 3$ for each value of m.

3) $m = 5$ 23
4) $m = 9$ 39
5) $m = -6$ -21
6) $m = -7$ -25

Evaluate $-6d + 11$ for each value of d.

7) $d = 2$ -1
8) $d = 1$ 5
9) $d = -9$ 65
10) $d = -8$ 59
11) $d = \dfrac{2}{3}$ 7
12) $d = \dfrac{5}{6}$ 6
13) $d = 3.4$ -9.4
14) $d = 4.7$ -17.2

Evaluate each expression for the given values.

15) $3r - t + 4$ when $r = -7$ and $t = -4$ -13

16) $5u - v + 9$ when $u = -6$ and $v = -10$ -11

17) $\dfrac{x + 4y}{x - 2y}$ when $x = 8$ and $y = 3$ 10

18) $\dfrac{a + 9b}{a - 3b}$ when $a = 12$ and $b = 2$ 5

19) $\dfrac{2d - c}{c + 5d + 1}$ when $c = 1$ and $d = -4$ $\dfrac{1}{2}$

20) $\dfrac{4k - h}{6h + 2k - 1}$ when $h = 5$ and $k = -1$ $-\dfrac{1}{3}$

21) $m^2 + 4mn + n^2$ when $m = -3$ and $n = 2$ -11

22) $p^2 + 3pq + q^2$ when $p = -4$ and $q = 5$ -19

23) $-8g^2 - gh + 14$ when $g = -1$ and $h = -6$ 0

24) $-7a^2 - ab + 19$ when $a = -1$ and $b = -8$ 4

Evaluate each expression when $x = 2$, $y = -3$, and $z = 5$.

25) $\dfrac{z^2 - x^2}{2y + z}$ -21

26) $\dfrac{y^2 - 1}{x - 2z}$ -1

27) $\dfrac{y^3}{x - z^2 + 2}$ $\dfrac{9}{7}$

28) $\dfrac{x^4 + 2y}{y + z^2 - 10}$ $\dfrac{5}{6}$

Mixed Exercises: Objectives 2–5

29) What is the identity element for multiplication?   1

30) What is the identity element for addition?   0

31) What is the additive inverse of 5? -5

32) What is the multiplicative inverse of 8? $\dfrac{1}{8}$

Which property of real numbers is illustrated by each example? Choose from the commutative, associative, identity, inverse, or distributive property.

33) $9(2 + 8) = 9 \cdot 2 + 9 \cdot 8$ distributive

34) $(-16 + 7) + 3 = -16 + (7 + 3)$ associative

35) $14 \cdot 1 = 14$ identity

36) $\left(\dfrac{9}{2}\right)\left(\dfrac{2}{9}\right) = 1$ inverse

37) $-10 + 18 = 18 + (-10)$ commutative

38) $4 \cdot 6 - 4 \cdot 1 = 4(6 - 1)$ distributive

39) $5(2 \cdot 3) = (5 \cdot 2) \cdot 3$ associative

40) $11 \cdot 7 = 7 \cdot 11$ commutative

41) $(11 + 3)(-1) = 11(-1) + 3(-1)$ distributive

42) $4 + (2 + 3) = (4 + 2) + 3$ associative

43) $-3 + 3 = 0$ inverse

44) $-7 + 0 = -7$ identity

45) $\dfrac{4}{5} = \dfrac{4}{5} \cdot \dfrac{3}{3} = \dfrac{12}{15}$ identity

46) $8(p - q) = 8p - 8q$ distributive

47) $7(4w) - 7(w) = 7(4w - w)$ distributive

48) $9ac + 9mn = 9(ac + mn)$ distributive

Rewrite each expression using the indicated property.

49) $p + 19$; commutative $19 + p$

50) $5(m + n)$; distributive $5m + 5n$

51) $8 + (1 + 9)$; associative $(8 + 1) + 9$

52) $-2c + 0$; identity $-2c$

53) $3(k - 7)$; distributive $3k - 21$

54) $10 + 9x$; commutative $9x + 10$

55) $y + 0$; identity $\quad y$

56) $\left(4 \cdot \dfrac{2}{7}\right) \cdot 7$; associative $\quad 4 \cdot \left(\dfrac{2}{7} \cdot 7\right)$

57) $-(n + 5)$; distributive $\quad -n - 5$

58) $(2 + 1) + 3$; associative $\quad 2 + (1 + 3)$

Use the commutative property to simplify each expression.

59) $-\dfrac{5}{2} + \dfrac{4}{9} + \dfrac{11}{2} + \dfrac{5}{9}$ $\quad 4$

60) $-\dfrac{8}{3} + \dfrac{2}{5} + \dfrac{2}{3} + \dfrac{8}{5}$ $\quad 0$

61) $\dfrac{1}{6} \cdot \dfrac{11}{7} \cdot \dfrac{6}{5}$ $\quad \dfrac{11}{35}$

62) $\dfrac{3}{4} \cdot \dfrac{2}{9} \cdot \dfrac{4}{3}$ $\quad \dfrac{2}{9}$

63) Is $2a - 7$ equivalent to $7 - 2a$? Why or why not?
No. Subtraction is not commutative.

64) Is $6 + t$ equivalent to $t + 6$? Why or why not?
Yes. Addition is commutative.

Rewrite each expression using the distributive property. Simplify if possible.

65) $2(1 + 9)$ $\quad 2 \cdot 1 + 2 \cdot 9 = 2 + 18 = 20$

66) $3(9 + 4)$ $\quad 3 \cdot 9 + 3 \cdot 4 = 27 + 12 = 39$

67) $-2(5 + 7)$ $\quad -2 \cdot 5 + (-2) \cdot 7 = -10 + (-14) = -24$

68) $-5(3 + 7)$ $\quad -5 \cdot 3 + (-5) \cdot 7 = -15 + (-35) = -50$

69) $9 \cdot a + 9 \cdot b$ $\quad 9(a + b)$

70) $4 \cdot c - 4 \cdot d$ $\quad 4(c - d)$

71) $8 \cdot 3 - 8 \cdot 10$ $\quad 8(3 - 10) = 8(-7) = -56$

72) $12 \cdot 2 + 12 \cdot 5$ $\quad 12(2 + 5) = 12(7) = 84$

73) $4(8 - 3)$ $\quad 4 \cdot 8 - 4 \cdot 3 = 32 - 12 = 20$

74) $-6(5 - 11)$ $\quad -6 \cdot 5 - (-6) \cdot (11) = -30 + 66 = 36$

75) $-(10 - 4)$ $\quad -10 + 4 = -6$

76) $-(3 + 9)$ $\quad -3 - 9 = -12$

77) $8(y + 3)$ $\quad 8y + 8 \cdot 3 = 8y + 24$

78) $4(k + 11)$ $\quad 4k + 4 \cdot 11 = 4k + 44$

79) $-\dfrac{2}{3}(z + 6)$ $\quad -\dfrac{2}{3}z + \left(-\dfrac{2}{3}\right) \cdot 6 = -\dfrac{2}{3}z - 4$

80) $-\dfrac{3}{5}(m + 10)$ $\quad -\dfrac{3}{5}m + \left(-\dfrac{3}{5}\right) \cdot 10 = -\dfrac{3}{5}m - 6$

81) $-3(x - 4y - 6)$
$-3x - (-3) \cdot (4y) - (-3) \cdot (6) = -3x + 12y + 18$

82) $6(2a - 5b + 1)$ $\quad 6 \cdot (2a) - 6 \cdot (5b) + 6 \cdot (1) = 12a - 30b + 6$

83) $-(-8c + 9d - 14)$ $\quad 8c - 9d + 14$

84) $-(x - 10y - 4z)$ $\quad -x + 10y + 4z$

R Rethink

R1) Which exercises do you need help mastering?

R2) Think about ways to use the distributive property to help you multiply less common numbers. For example, you could rewrite $9 \cdot 13$ as $9(10 + 3)$. If you apply the distributive property to $9(10 + 3)$, does it make it easier? Explain. Come up with another example and multiply.

1.8 Simplifying Algebraic Expressions

P Prepare **O Organize**

What are your objectives for Section 1.8?	How can you accomplish each objective?
1 Identify the Terms and Coefficients in an Expression	• Write the definitions of a *term, constant term,* and *coefficient* in your own words. • Complete the given example on your own. • Complete You Try 1.
2 Identify Like Terms	• Write the definition of *like terms* in your own words. • Complete the given example on your own. • Complete You Try 2.
3 Combine Like Terms	• Use the **distributive property** to combine *like terms*. • Complete the given examples on your own. • Complete You Trys 3 and 4.
4 Translate English Expressions into Mathematical Expressions	• Use the tools you learned from the previous sections to create a procedure from Example 5. • Complete You Try 5.

W Work   Read the explanations, follow the examples, take notes, and complete the You Trys.

Let's learn some vocabulary associated with algebraic expressions before we learn how to combine like terms.

1 Identify the Terms and Coefficients in an Expression

Here is an algebraic expression: $8x^3 - 5x^2 + \frac{2}{7}x + 4$

This expression contains four terms. A **term** is a number or a variable or a product or quotient of numbers and variables. 4 is the **constant** or **constant term.** The value of a constant does not change. Each term has a **coefficient.**

Term	Coefficient
$8x^3$	8
$-5x^2$	-5
$\frac{2}{7}x$	$\frac{2}{7}$
4	4

EXAMPLE 1

List the terms and coefficients of $4x^2y + 7xy - x + \frac{y}{9} - 12$.

In-Class Example 1

List the terms and coefficients of

$12x^2y - xy^2 + 7x - \frac{3}{4}y + 20.$

Answer:

Term	Coeff.
$12x^2y$	12
$-xy^2$	-1
$7x$	7
$-\frac{3}{4}y$	$-\frac{3}{4}$
20	20

Solution

Term	Coefficient
$4x^2y$	4
$7xy$	7
$-x$	-1
$\frac{y}{9}$	$\frac{1}{9}$
-12	-12

The minus sign indicates a negative coefficient.

$\frac{y}{9}$ can be rewritten as $\frac{1}{9}y$.

-12 is also called the *constant*.

2 Identify Like Terms

In the expression $15a + 11a - 8a + 3a$, there are four **terms:** $15a$, $11a$, $-8a$, $3a$. In fact, they are **like terms.** *Like terms contain the same variables with the same exponents.*

EXAMPLE 2

In-Class Example 2

Determine whether the following groups of terms are like terms.

a) $2p^2, -0.1p^2, \dfrac{p^2}{5}$

b) $18f^3, -f^4, 0.4f$ c) $-7t, u$

d) $8p^3q, -14p^3q, p^3q$

Answer:

a) yes b) no c) no

d) yes

Determine whether the following groups of terms are like terms.

a) $4y^2, -9y^2, \dfrac{2}{3}y^2$ b) $-5x^6, 0.8x^9, 3x^4$

c) $6a^2b^3, a^2b^3, -\dfrac{5}{8}a^2b^3$ d) $9c, 4d$

Solution

a) $4y^2, -9y^2, \dfrac{2}{3}y^2$

Yes. Each contains the variable y with an exponent of 2. They are y^2-terms.

b) $-5x^6, 0.8x^9, 3x^4$

No. Although each contains the variable x, the exponents are not the same.

c) $6a^2b^3, a^2b^3, -\dfrac{5}{8}a^2b^3$

Yes. Each term contains the product of a^2 and b^3.

d) $9c, 4d$

No. The terms contain different variables.

[YOU TRY 2] Determine whether the following groups of terms are like terms.

a) $2k^2, -9k^2, \dfrac{1}{5}k^2$ b) $-xy^2, 8xy^2, 7xy^2$ c) $3r^3s^2, -10r^2s^3$

3 Combine Like Terms

To simplify an expression like $15a + 11a - 8a + 3a$, we combine like terms using the distributive property.

$$
\begin{aligned}
15a + 11a - 8a + 3a &= (15 + 11 - 8 + 3)a && \text{Distributive property} \\
&= (26 - 8 + 3)a && \text{Order of operations} \\
&= (18 + 3)a && \text{Order of operations} \\
&= 21a
\end{aligned}
$$

We can add and subtract only those terms that are like terms.

EXAMPLE 3

In-Class Example 3

Combine like terms.
a) $13b - 4b$
b) $d - 8 + 5d - 1$
c) $\frac{7}{10}y^2 - \frac{1}{3}y^2$
d) $9w^3 + 4w - w^3$

Answer:
a) $9b$ b) $6d - 9$
c) $\frac{11}{30}y^2$ d) $8w^3 + 4w$

Combine like terms.

a) $-9k + 2k$ b) $n + 8 - 4n + 3$ c) $\frac{3}{5}t^2 + \frac{1}{4}t^2$

d) $10x^2 + 6x - 2x^2 + 5x$

Solution

a) We can use the distributive property to combine like terms.

$$-9k + 2k = (-9 + 2)k = -7k$$

Notice that using the distributive property to combine like terms is the same as combining the coefficients of the terms and leaving the variable and its exponent the same.

b) $n + 8 - 4n + 3 = n - 4n + 8 + 3$ Rewrite like terms together.
           $= -3n + 11$ Remember, n is the same as $1n$.

c) $\frac{3}{5}t^2 + \frac{1}{4}t^2 = \frac{12}{20}t^2 + \frac{5}{20}t^2$ Get a common denominator.

           $= \frac{17}{20}t^2$

d) $10x^2 + 6x - 2x^2 + 5x = 10x^2 - 2x^2 + 6x + 5x$ Rewrite like terms together.
                          $= 8x^2 + 11x$

$8x^2 + 11x$ cannot be simplified more because the terms are *not* like terms.

[YOU TRY 3] Combine like terms.

a) $6z + 5z$ b) $q - 9 - 4q + 11$ c) $\frac{5}{6}c^2 - \frac{2}{3}c^2$

d) $2y^2 + 8y + y^2 - 3y$

If an expression contains parentheses, we use the distributive property to clear the parentheses, and then combine like terms.

EXAMPLE 4

In-Class Example 4

Combine like terms.
a) $2(8f - 1) - 4f + 7$
b) $5(3n + 2) - (-4n + 8)$
c) $\frac{1}{4}(6 - 5t) + \frac{2}{3}(3t + 9)$

Answer: a) $12f + 5$
b) $19n + 2$ c) $\frac{3}{4}t + \frac{15}{2}$

Combine like terms.

a) $5(2c + 3) - 3c + 4$ b) $3(2n + 1) - (6n - 11)$

c) $\frac{3}{8}(8 - 4p) + \frac{5}{6}(2p - 6)$

Solution

a) $5(2c + 3) - 3c + 4 = 10c + 15 - 3c + 4$ Distributive property
                      $= 10c - 3c + 15 + 4$ Rewrite like terms together.
                      $= 7c + 19$

b) $3(2n + 1) - (6n - 11) = 3(2n + 1) - 1(6n - 11)$ Remember, $-(6n - 11)$ is the same as $-1(6n - 11)$.

                       $= 6n + 3 - 6n + 11$ Distributive property
                       $= 6n - 6n + 3 + 11$ Rewrite like terms together.
                       $= 0n + 14$ $0n = 0$
                       $= 14$

c) $\frac{3}{8}(8 - 4p) + \frac{5}{6}(2p - 6) = \frac{3}{8}(8) - \frac{3}{8}(4p) + \frac{5}{6}(2p) - \frac{5}{6}(6)$ Distributive property

$$= 3 - \frac{3}{2}p + \frac{5}{3}p - 5 \qquad \text{Multiply.}$$

$$= -\frac{3}{2}p + \frac{5}{3}p + 3 - 5 \qquad \text{Rewrite like terms together.}$$

$$= -\frac{9}{6}p + \frac{10}{6}p + 3 - 5 \qquad \text{Get a common denominator.}$$

$$= \frac{1}{6}p - 2 \qquad \text{Combine like terms.}$$

[YOU TRY 4]   Combine like terms.

a) $9d^2 - 7 + 2d^2 + 3$ b) $10 - 3(2k + 5) + k - 6$

---

4   Translate English Expressions into Mathematical Expressions

Translating from English into a mathematical expression is a skill that is necessary to solve applied problems. Let's practice writing mathematical expressions.

Read the phrase carefully, choose a variable to represent the unknown quantity, then translate the phrase into a mathematical expression.

EXAMPLE 5

Write a mathematical expression for each and simplify. Define the unknown with a variable.

a) Seven more than twice a number

b) The sum of a number and four times the same number

Solution

a) Seven more than twice a number

   i) **Define the unknown.** This means that you should clearly state on your paper what the variable represents.

$$\text{Let } x = \text{the number.}$$

   ii) **Slowly, break down the phrase.** How do you write an expression for "seven more than" something?

$$+\ 7$$

   iii) **What does "twice a number" mean?** It means two times the number. Since our number is represented by x, "twice a number" is $2x$.

   iv) **Put the information together:**

Seven more than twice a number

$2x \qquad + \qquad 7$

The expression is $2x + 7$.

---

b) The sum of a number and four times the same number

   i) **Define the unknown.**

$$\text{Let } y = \text{the number.}$$

   ii) **Slowly, break down the phrase.** What does *sum* mean? **Add.** So, we have to add a number and four times the same number:

$$\text{Number} + 4(\text{Number})$$

   iii) Since y represents the number, *four times the number* is $4y$.

   iv) Therefore, to translate from English into a mathematical expression, we know that we must add the number, y, to four times the number, $4y$. Our expression is $y + 4y$. It simplifies to $5y$.

[YOU TRY 5] Write a mathematical expression for each and simplify. Let x equal the unknown number.

a) Five less than twice a number

b) The sum of a number and two times the same number

Using Technology

A graphing calculator can be used to evaluate an algebraic expression. This is especially valuable when evaluating expressions for several values of the given variables.

We will evaluate the expression $\dfrac{x^2 - 2xy}{3x + y}$ when $x = -3$ and $y = 8$.

Method 1

Substitute the values for the variables and evaluate the arithmetic expression on the home screen. Each value substituted for a variable should be enclosed in parentheses to guarantee a correct answer. For example $(-3)^2$ gives the result 9, whereas -3^2 gives the result -9. Be careful to press the negative key $\boxed{(-)}$ when entering a negative sign and the minus key $\boxed{-}$ when entering the minus operator.

```
((-3)²-2(-3)(8))
/(3(-3)+8)
                -57
```

Method 2

Store the given values in the variables and evaluate the algebraic expression on the home screen.

To store -3 in the variable x, press $\boxed{(-)}$ $\boxed{3}$ $\boxed{\text{STO▸}}$ $\boxed{\text{X, T, θ, n}}$ $\boxed{\text{ENTER}}$.

To store 8 in the variable y, press $\boxed{8}$ $\boxed{\text{STO▸}}$ $\boxed{\text{ALPHA}}$ $\boxed{1}$ $\boxed{\text{ENTER}}$.

Enter $\dfrac{x^2 - 2xy}{3x + y}$ on the home screen.

```
-3→X
               -3
8→Y
                8
(X²-2XY)/(3X+Y)
              -57
```

The advantage of Method 2 is that we can easily store two different values in x and y. For example, store 5 in x and -2 in y. It is not necessary to enter the expression again because the calculator can recall previous entries.

Press $\boxed{\text{2nd}}$ $\boxed{\text{ENTER}}$ three times; then press $\boxed{\text{ENTER}}$.

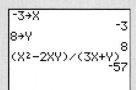

```
5→X
                5
-2→Y
               -2
(X²-2XY)/(3X+Y)
       3.461538462
```

To convert this decimal to a fraction, press MATH
ENTER ENTER.

Evaluate each expression when $x = -5$ and $y = 2$.

1. $3y - 4x$
2. $2xy - 5y$
3. $y^3 - 2x^2$
4. $\dfrac{x - y}{4x}$
5. $\dfrac{2x + 5y}{x - y}$
6. $\dfrac{x - y^2}{2x}$

```
Ans▶Frac
                45/13
```

ANSWERS TO YOU TRY **EXERCISES**

1)

Term	Coeff.
$-15r^3$	-15
r^2	1
$-4r$	-4
8	8

2) a) yes b) yes c) no   3) a) $11z$ b) $-3q + 2$ c) $\dfrac{1}{6}c^2$
d) $3y^2 + 5y$ 4) a) $11d^2 - 4$ b) $-5k - 11$ 5) a) $2x - 5$
b) $x + 2x; 3x$

ANSWERS TO TECHNOLOGY EXERCISES

1. 26   2. -30 3. -42 4. $\dfrac{7}{20}$ 5. 0   6. $\dfrac{9}{10}$

E Evaluate **1.8** Exercises   Do the exercises, and check your work.

*Additional answers can be found in the Answers to Exercises appendix.

Objective 1: Identify the Terms and Coefficients in an Expression

For each expression, list the terms and their coefficients. Also, identify the constant.

1) $7p^2 - 6p + 4$

2) $-8z + \dfrac{5}{6}$

3) $x^2y^2 + 2xy - y + 11$

4) $w^3 - w^2 + 9w - 5$

5) $-2g^5 + \dfrac{g^4}{5} + 3.8g^2 + g - 1$

6) $121c^2 - d^2$

Objective 2: Identify Like Terms

7) Are $9k$ and $9k^2$ *like* terms? Why or why not?
No. The exponents are different.

8) Are $\dfrac{3}{4}n$ and $8n$ *like* terms? Why or why not?
Yes. Both are n-terms.

9) Are a^3b and $-7a^3b$ *like* terms? Why or why not?
Yes. Both are a^3b-terms.

10) Write three *like* terms that are x^2-terms.
Answers may vary.

Objective 3: Combine Like Terms

Combine like terms and simplify.

11) $10p + 9 + 14p - 2$ $24p + 7$

12) $11 - k^2 + 12k^2 - 3 + 6k^2$ $17k^2 + 8$

13) $-18y^2 - 2y^2 + 19 + y^2 - 2 + 13$ $-19y^2 + 30$

14) $-7x - 3x - 1 + 9x + 6 - 2x$ $-3x + 5$

15) $\dfrac{4}{9} + 3r - \dfrac{2}{3} + \dfrac{1}{5}r$ $\dfrac{16}{5}r - \dfrac{2}{9}$

16) $6a - \dfrac{3}{8}a + 2 + \dfrac{1}{4} - \dfrac{3}{4}a$ $\dfrac{39}{8}a + \dfrac{9}{4}$

17) $2(3w + 5) + w$ $7w + 10$

18) $-8d^2 + 6(d^2 - 3) + 7$ $-2d^2 - 11$

19) $9 - 4(3 - x) - 4x + 3$ 0

20) $m + 11 + 3(2m - 5) + 1$ $7m - 3$

21) $3g - (8g + 3) + 5$ $-5g + 2$

22) $-6 + 4(10b - 11) - 8(5b + 2)$ -66

23) $-5(t - 2) - (10 - 2t)$ $-3t$

24) $11 + 8(3u - 4) - 2(u + 6) + 9$ $22u - 24$

25) $3[2(5x + 7) - 11] + 4(7 - x)$ $26x + 37$

26) $22 - [6 + 5(2w - 3)] - (7w + 16)$ $-17w + 15$

27) $\frac{4}{5}(2z + 10) - \frac{1}{2}(z + 3)$ $\frac{11}{10}z + \frac{13}{2}$

28) $\frac{2}{3}(6c - 7) + \frac{5}{12}(2c + 5)$ $\frac{29}{6}c - \frac{31}{12}$

29) $1 + \frac{3}{4}(10t - 3) + \frac{5}{8}\left(t + \frac{1}{10}\right)$ $\frac{65}{8}t - \frac{19}{16}$

30) $\frac{7}{15} - \frac{9}{10}(2y + 1) - \frac{2}{5}(4y - 3)$ $-\frac{17}{5}y + \frac{23}{30}$

31) $2.5(x - 4) - 1.2(3x + 8)$ $-1.1x - 19.6$

32) $9.4 - 3.8(2a + 5) + 0.6 + 1.9a$ $-5.7a - 9$

Objective 4: Translate English Expressions into Mathematical Expressions

Write a mathematical expression for each phrase, and combine like terms if possible. Let x represent the unknown quantity.

33) Eighteen more than a number $x + 18$

34) Eleven more than a number $x + 11$

35) Six subtracted from a number $x - 6$

36) Eight subtracted from a number $x - 8$

37) Three less than a number $x - 3$

38) Fourteen less than a number $x - 14$

39) The sum of twelve and twice a number $12 + 2x$

40) The sum of nine and twice a number $9 + 2x$

41) One less than half a number $\frac{1}{2}x - 1$

42) Ten less than one-third of a number $\frac{1}{3}x - 10$

43) Five more than the sum of a number and six
   $(x + 6) + 5; x + 11$

44) Eight more than the sum of a number and twelve
   $(x + 12) + 8; x + 20$

45) Seven less than the sum of three and twice a number
   $(3 + 2x) - 7; 2x - 4$

46) Two more than the sum of a number and nine
   $(x + 9) + 2; x + 11$

47) The sum of a number and fifteen decreased by five
   $(x + 15) - 5; x + 10$

48) The sum of -8 and twice a number increased by three
   $(-8 + 2x) + 3; 2x - 5$

49) The sum of a number and three times the number
   $x + 3x; 4x$

50) The sum of a number and five times the number
   $x + 5x; 6x$

R Rethink

R1) Think about the situations in your everyday life when you combine like terms.

R2) Think about how a cashier counts money in the cash register at the end of a shift. How does the cashier use like terms to count the money?

R3) Which exercises do you need help mastering?

Group Activity — The Group Activity can be found online on Connect.

Many students find math intimidating. Some believe that they just aren't "good with numbers"; others find it scary that in math there is typically only one correct answer. Regardless of the reasoning, even successful students have their confidence shaken by the prospect of doing math. Yet math is a subject like any other. It can be mastered with hard work and effective study skills. Too often, the fear of math itself prevents students from learning math.

Are you confident in your math abilities? To find out, complete the following chart:

Statement	Strongly Agree	Agree	Disagree	Strongly Disagree
1. On the whole, I am satisfied with my math abilities.				
2. I have no instinct for numbers at all.				
3. I feel that I can eventually understand most math concepts.				
4. I can do math as well as most other people.				
5. I feel I do not do calculations very well in my head.				
6. I have dreams about coming unprepared to a math test.				
7. I know that if I am patient, I can solve just about any math problem.				
8. I can't really do math without a calculator.				
9. Some people are born good at math, but for me it is always going to be a challenge.				
10. When I sit down to do my math homework, I believe that I will complete it.				

Scoring: For statements 1, 3, 4, 7, and 10, score as follows:

Strongly Agree	4 points
Agree	3 points
Disagree	2 points
Strongly Disagree	1 point

For statements 2, 5, 6, 8, and 9, score as follows:

Strongly Agree	1 point
Agree	2 points
Disagree	3 points
Strongly Disagree	4 points

If the total score for all your answers is 30 or higher, your math confidence is high. If you scored in the 20's or below, though, your lack of confidence may be negatively impacting your chances to succeed in math. Consider discussing your feelings about math with your instructor. He or she will be able to direct you to resources where you can get extra help to boost your confidence.

Chapter 1:   Summary

Definition/Procedure	Example

1.1 Review of Fractions

Reducing Fractions
A fraction is in **lowest terms** when the numerator and denominator have no common factors other than 1. **(p. 6)**

Write $\dfrac{36}{48}$ in lowest terms.

$$\frac{36}{48} = \frac{36 \div 12}{48 \div 12} = \frac{3}{4}.$$

Multiplying Fractions
To multiply fractions, multiply the numerators and multiply the denominators. Common factors can be divided out either before or after multiplying. **(p. 8)**

Multiply $\dfrac{21}{45} \cdot \dfrac{9}{14}$. $\dfrac{\overset{3}{\cancel{21}}}{\cancel{45}_{5}} \cdot \dfrac{\overset{1}{\cancel{9}}}{\cancel{14}_{2}}$ ← 9 and 45 each divide by 9.
← 21 and 14 each divide by 7.

$$= \frac{3}{5} \cdot \frac{1}{2} = \frac{3}{10}$$

Dividing Fractions
To divide fractions, multiply the first fraction by the reciprocal of the second. **(p. 10)**

Divide $\dfrac{7}{5} \div \dfrac{4}{3}$. $\dfrac{7}{5} \div \dfrac{4}{3} = \dfrac{7}{5} \cdot \dfrac{3}{4} = \dfrac{21}{20}$ or $1\dfrac{1}{20}$

Adding and Subtracting Fractions
To add or subtract fractions,
1) Identify the least common denominator (LCD).
2) Write each fraction as an equivalent fraction using the LCD.
3) Add or subtract.
4) Express the answer in lowest terms. **(p. 11)**

Add $\dfrac{5}{11} + \dfrac{2}{11}$. $\dfrac{5}{11} + \dfrac{2}{11} = \dfrac{7}{11}$

Subtract $\dfrac{8}{9} - \dfrac{3}{4}$. $\dfrac{8}{9} - \dfrac{3}{4} = \dfrac{32}{36} - \dfrac{27}{36} = \dfrac{5}{36}$

1.2 Exponents and Order of Operations

Exponents
An **exponent** represents repeated multiplication. **(p. 18)**

Write $9 \cdot 9 \cdot 9 \cdot 9 \cdot 9$ in exponential form.
$$9 \cdot 9 \cdot 9 \cdot 9 \cdot 9 = 9^5$$

Evaluate 2^4. $2^4 = 2 \cdot 2 \cdot 2 \cdot 2 = 16$

Order of Operations
Parentheses, **E**xponents, **M**ultiplication and **D**ivision, **A**ddition and **S**ubtraction. **(p. 19)**

Evaluate $8 + (5 - 1)^2 - 6 \cdot 3$.
$$8 + (5 - 1)^2 - 6 \cdot 3$$

$= 8 + 4^2 - 6 \cdot 3$	Parentheses
$= 8 + 16 - 6 \cdot 3$	Exponents
$= 8 + 16 - 18$	Multiply.
$= 24 - 18$	Add.
$= 6$	Subtract.

1.3 Geometry Review

Important Angles
The definitions for an acute angle, an obtuse angle, and a right angle can be found on **p. 24**.

Two angles are **complementary** if the sum of their measures is 90°.

Two angles are **supplementary** if the sum of their measures is 180°. **(p. 24)**

The measure of an angle is 73°. Find the measure of its complement and its supplement.

The measure of its complement is 17° since $90° - 73° = 17°$.

The measure of its supplement is 107° since $180° - 73° = 107°$.

Definition/Procedure	Example
Triangle Properties The sum of the measures of the angles of any triangle is 180°. An **equilateral triangle** has three sides of equal length. Each angle measures 60°. An **isosceles triangle** has two sides of equal length. The angles opposite the sides with equal measure have the same measure. A **scalene triangle** has no sides of equal length. No angles have the same measure. **(p. 25)**	Find the measure of $\angle C$. 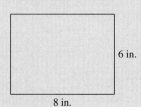 $m\angle A + m\angle B = 63° + 94° = 157°$ $m\angle C = 180° - 157° = 23°$

Perimeter and Area
The formulas for the perimeter and area of a rectangle, square, triangle, parallelogram, and trapezoid can be found on **p. 26**.
	Find the area and perimeter of this rectangle.
	
	6 in.
	
	8 in.
	
	Area = (Length)(Width) Perimeter = 2(Length) + 2(Width)
	     = (8 in.)(6 in.) = 2(8 in.) + 2(6 in.)
	     = 48 in^2 = 16 in. + 12 in. = 28 in.

Volume
The formulas for the volume of a rectangular solid, cube, right circular cylinder, sphere, and right circular cone can be found on **p. 29**.
	Find the volume of the cylinder pictured here.
	
	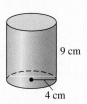
	9 cm
	
	4 cm
	
	Give an exact answer and give an approximation using 3.14 for π.
	$V = \pi r^2 h$ $V = 144\pi \text{ cm}^3$
	  $= \pi(4 \text{ cm})^2(9 \text{ cm})$ $\approx 144(3.14) \text{ cm}^3$
	  $= \pi(16 \text{ cm}^2)(9 \text{ cm})$ $\approx 452.16 \text{ cm}^3$
	  $= 144\pi \text{ cm}^3$

1.4 Sets of Numbers and Absolute Value

Natural numbers: $\{1, 2, 3, 4, \ldots\}$ **Whole numbers:** $\{0, 1, 2, 3, 4, \ldots\}$ **Integers:** $\{\ldots, -3, -2, -1, 0, 1, 2, 3, \ldots\}$ **(p. 35)**	
A **rational number** is any number of the form $\dfrac{p}{q}$, where p and q are integers and $q \neq 0$. **(p. 36)**	The following numbers are rational: $-3, 10, \dfrac{5}{8}, 7.4, 2.\overline{3}$
An **irrational number** cannot be written as the quotient of two integers. **(p. 36)**	The following numbers are irrational: $\sqrt{6}, 9.2731\ldots$
The set of **real numbers** includes the rational and irrational numbers. **(p. 37)**	Any number that can be represented on the number line is a real number.
The **additive inverse** of a is $-a$. **(p. 39)**	The additive inverse of 4 is -4.
Absolute Value $|a|$ is the distance of a from zero. **(p. 40)**	$|-6| = 6$

Definition/Procedure	Example

1.5 Addition and Subtraction of Real Numbers

Adding Real Numbers

To add numbers with the **same sign,** add the absolute values of the numbers. The sum will have the same sign as the numbers being added. **(p. 44)**

$$-3 + (-9) = -12$$

To add two numbers with **different signs,** subtract the smaller absolute value from the larger. The sum will have the sign of the number with the larger absolute value. **(p. 45)**

$$-20 + 15 = -5$$

Subtracting Real Numbers

To subtract $a - b$, change subtraction to addition, and add the additive inverse of b: $a - b = a + (-b)$. **(p. 46)**

$$2 - 11 = 2 + (-11) = -9$$
$$-17 - (-7) = -17 + 7 = -10$$

1.6 Multiplication and Division of Real Numbers

Multiplying Real Numbers

The product of two real numbers with the *same* sign is positive.

$$8 \cdot 3 = 24 \qquad -7 \cdot (-8) = 56$$

The product of a positive number and a negative number is *negative*.

$$-2 \cdot 5 = -10 \qquad 9 \cdot (-1) = -9$$

An *even number* of negative factors in a product gives a *positive* result.

$$\underbrace{(-1)(-6)(-3)(2)(-4)}_{\text{4 negative factors}} = 144$$

An *odd number* of negative factors in a product gives a *negative* result. **(p. 52)**

$$\underbrace{(5)(-2)(-3)(1)(-1)}_{\text{3 negative factors}} = -30$$

Evaluating Exponential Expressions

Exponential notation is a shorthand way to represent repeated multiplication. **(p. 53)**

Evaluate $(-3)^4$.   The base is -3.
$$(-3)^4 = (-3)(-3)(-3)(-3) = 81$$

Evaluate -3^4.   The base is 3.
$$-3^4 = -1 \cdot 3^4 = -1 \cdot 3 \cdot 3 \cdot 3 \cdot 3 = -81$$

Dividing Real Numbers

The quotient of two numbers with the *same* sign is positive. The quotient of two numbers with *different* signs is negative. **(p. 55)**

$$\frac{40}{2} = 20 \qquad -18 \div (-3) = 6$$
$$\frac{-56}{8} = -7 \qquad 48 \div (-4) = -12$$

1.7 Properties of Real Numbers

A **variable** is a symbol, usually a letter, used to represent an unknown number. **(p. 60)**

In the expression $9c + 2$, c is the variable.

An **algebraic expression** is a collection of numbers, variables, and grouping symbols connected by operation symbols such as $+$, $-$, $\times$, and $\div$. **(p. 60)**

$$4y^2 - 7y + \frac{3}{5}$$

Evaluating Expressions

We can evaluate expressions for different values of the variables. **(p. 60)**

Evaluate $2xy - 5y + 1$ when $x = -3$ and $y = 4$.

Substitute -3 for x and 4 for y and simplify.

$$\begin{aligned} 2xy - 5y + 1 &= 2(-3)(4) - 5(4) + 1 \\ &= -24 - 20 + 1 \\ &= -24 + (-20) + 1 \\ &= -43 \end{aligned}$$

Definition/Procedure	Example

Properties of Real Numbers

If a, b, and c are real numbers, then the following properties hold.

Commutative Properties:

$a + b = b + a$

$ab = ba$

$10 + 3 = 3 + 10$

$(-6)(5) = (5)(-6)$

Associative Properties:

$(a + b) + c = a + (b + c)$

$(ab)c = a(bc)$

$(9 + 4) + 2 = 9 + (4 + 2)$

$(5 \cdot 2) \cdot 8 = 5 \cdot (2 \cdot 8)$

Identity Properties:

$a + 0 = 0 + a = a$

$a \cdot 1 = 1 \cdot a = a$

$7 + 0 = 7 \qquad \dfrac{2}{3} \cdot 1 = \dfrac{2}{3}$

Inverse Properties:

$a + (-a) = -a + a = 0$

$b \cdot \dfrac{1}{b} = \dfrac{1}{b} \cdot b = 1$

$11 + (-11) = 0 \qquad 5 \cdot \dfrac{1}{5} = 1$

Distributive Properties:

$a(b + c) = ab + ac$ and $(b + c)a = ba + ca$

$a(b - c) = ab - ac$ and $(b - c)a = ba - ca$ **(p. 61)**

$6(5 + 8) = 6 \cdot 5 + 6 \cdot 8$
$= 30 + 48$
$= 78$

$9(w - 2) = 9w - 9 \cdot 2$
$= 9w - 18$

1.8 Simplifying Algebraic Expressions

A **term** in an expression is a number or a variable or a product or quotient of numbers and variables.

The value of a **constant** does not change.

The **coefficient** of a variable is the number that the variable is multiplied by. **(p. 69)**

List the terms and coefficients of $7x^3 - x^2 - \dfrac{2}{5}x + 8$.

Term	Coefficient
$7x^3$	7
$-x^2$	-1
$-\dfrac{2}{5}x$	$-\dfrac{2}{5}$
8	8

8 is the constant.

Like Terms

Like terms contain the same variables with the same exponents. **(p. 70)**

In the group of terms $5k^2$, $-8k$, $-4k^2$, $\dfrac{1}{3}k$,

$5k^2$ and $-4k^2$ are like terms and $-8k$ and $\dfrac{1}{3}k$ are like terms.

Combining Like Terms

We can simplify expressions by combining like terms. **(p. 70)**

Combine like terms and simplify.

$4n^2 - 3n + 1 - 2(6n^2 - 5n + 7)$
$= 4n^2 - 3n + 1 - 12n^2 + 10n - 14$ Distributive property
$= -8n^2 + 7n - 13$ Combine like terms.

Writing Mathematical Expressions (p. 72)

Write a mathematical expression for the following:
Sixteen more than twice a number

Let x = the number.

$\underbrace{\text{Sixteen more than}}_{+16} \qquad \underbrace{\text{twice a number}}_{2x}$

$2x + 16$

Chapter 1: Review Exercises

*Additional answers can be found in the Answers to Exercises appendix.

(1.1)

1) Find all factors of each number.

    a) 16   1, 2, 4, 8, 16   b) 37   1, 37

2) Find the prime factorization of each number.

    a) 28 $2 \cdot 2 \cdot 7$ b) 66 $2 \cdot 3 \cdot 11$

3) Write each fraction in lowest terms.

    a) $\dfrac{12}{30}$ $\dfrac{2}{5}$ b) $\dfrac{414}{702}$ $\dfrac{23}{39}$

Perform the indicated operation. Write the answer in lowest terms.

4) $\dfrac{4}{11} \cdot \dfrac{3}{5}$ $\dfrac{12}{55}$ 5) $\dfrac{45}{64} \cdot \dfrac{32}{75}$ $\dfrac{3}{10}$

6) $\dfrac{5}{8} \div \dfrac{3}{10}$ $\dfrac{25}{12}$ or $2\dfrac{1}{12}$ 7) $35 \div \dfrac{7}{8}$ 40

8) $4\dfrac{2}{3} \cdot 1\dfrac{1}{8}$ $\dfrac{21}{4}$ or $5\dfrac{1}{4}$ 9) $\dfrac{30}{49} \div 2\dfrac{6}{7}$ $\dfrac{3}{14}$

10) $\dfrac{2}{9} + \dfrac{4}{9}$ $\dfrac{2}{3}$ 11) $\dfrac{2}{3} + \dfrac{1}{4}$ $\dfrac{11}{12}$

12) $\dfrac{9}{40} + \dfrac{7}{16}$ $\dfrac{53}{80}$ 13) $\dfrac{1}{5} + \dfrac{1}{3} + \dfrac{1}{6}$ $\dfrac{7}{10}$

14) $\dfrac{21}{25} - \dfrac{11}{25}$ $\dfrac{2}{5}$ 15) $\dfrac{5}{8} - \dfrac{2}{7}$ $\dfrac{19}{56}$

16) $3\dfrac{2}{9} + 5\dfrac{3}{8}$ $8\dfrac{43}{72}$ 17) $9\dfrac{3}{8} - 2\dfrac{5}{6}$ $6\dfrac{13}{24}$

18) A pattern for a skirt calls for $1\dfrac{7}{8}$ yd of fabric. If Mary Kate wants to make one skirt for herself and one for her twin, how much fabric will she need? $3\dfrac{3}{4}$ yd

(1.2) Evaluate.

19) 3^4 81   20) 2^6 64

21) $\left(\dfrac{3}{4}\right)^3$ $\dfrac{27}{64}$ 22) $(0.6)^2$ 0.36

23) $13 - 7 + 4$ 10   24) $8 \cdot 3 + 20 \div 4$ 29

25) $\dfrac{12 - 56 \div 8}{(1 + 5)^2 - 2^4}$ $\dfrac{1}{4}$

(1.3)

26) The complement of 51° is _____.   39°

27) The supplement of 78° is _____.   102°

28) Is this triangle acute, obtuse, or right? Find the missing angle.

                      obtuse; 26°

    ?

          130°   24°

Find the area and perimeter of each figure. Include the correct units.

29)   30)

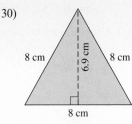

$1\dfrac{7}{8}$ miles

$3\dfrac{1}{2}$ miles

$A = 6\dfrac{9}{16}$ mi²; $P = 10\dfrac{3}{4}$ mi

8 cm   6.9 cm   8 cm

8 cm

$A = 27.6$ cm²; $P = 24$ cm

31)   11 in.   32)   5 ft

5 in.   7 ft

3in.   8 in.   6 ft

5 in.   4 ft

$A = 100$ in²; $P = 40$ in. $A = 58$ ft²; $P = 36$ ft

Find a) the area and b) the circumference of each circle. Give an exact answer for each and give an approximation using 3.14 for π. Include the correct units.

33)   34)

3 in.   10 cm

a) $A = 9\pi$ in²; $A \approx 28.26$ in²   a) $A = 100\pi$ cm²; $A \approx 314$ cm²

b) $C = 6\pi$ in.; $C \approx 18.84$ in.   b) $C = 0\pi$ cm; $C \approx 62.8$ cm

Find the area of the shaded region. Include the correct units.

35)   124 ft²

8 ft   12 ft

14 ft

18 ft

Find the volume of each figure. Where appropriate, give the answer in terms of π. Include the correct units.

36)   37)   1 ft

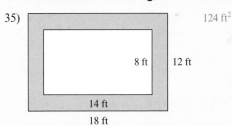

              2 m   1.3 ft

          5 m

7 m   70 m³ 1.3π ft³

38) 8π cm³   39)

6 cm $2\dfrac{1}{2}$ in.

2 cm $2\dfrac{1}{2}$ in.

$2\dfrac{1}{2}$ in.

$\dfrac{125}{8}$ in³ or $15\dfrac{5}{8}$ in³

40) The radius of a basketball is approximately 4.7 inches. Find its circumference to the nearest tenth of an inch.   29.5 in.

(1.4)

41) A rational number can *always, sometimes,* or *never* be expressed as a fraction.   always

42) An irrational number can *always, sometimes,* or *never* be expressed as a fraction.   never

43) Given this set of numbers,

$$\left\{ \frac{7}{15}, -16, 0, 3.\overline{2}, 8.5, \sqrt{31}, 4, 6.01832\ldots \right\}$$

list the

   a) integers $\{-16, 0, 4\}$

   b) rational numbers $\left\{ \frac{7}{15}, -16, 0, 3.\overline{2}, 8.5, 4 \right\}$

   c) natural numbers $\{4\}$

   d) whole numbers $\{0, 4\}$

   e) irrational numbers $\{\sqrt{31}, 6.01832\ldots\}$

44) Graph and label these numbers on a number line.

$$-3.5, 4, \frac{9}{10}, 2\frac{1}{3}, -\frac{3}{4}, -5$$

45) Evaluate.

   a) $|-18|$ 18   b) $-|7|$ -7

(1.5) Add or subtract as indicated.

46) $-38 + 13$ -25 47) $-21 - (-40)$ 19

48) $-1.9 + 2.3$ 0.4   49) $\frac{5}{12} - \frac{5}{8}$ $-\frac{5}{24}$

50) The lowest temperature on record in the country of Greenland is $-87°F$. The coldest temperature ever reached on the African continent was in Morocco and is $76°$ higher than Greenland's record low. What is the lowest temperature ever recorded in Africa? $-11°F$ (www.ncdc.noaa.gov)

(1.6) Multiply or divide as indicated.

51) $\left(-\frac{3}{2}\right)(8)$ -12 52) $(-4.9)(-3.6)$ 17.64

53) $(-4)(3)(-2)(-1)(-3)$ 72   54) $\left(-\frac{2}{3}\right)(-5)(2)(-6)$ -40

55) $-108 \div 9$ -12 56) $\frac{56}{-84}$ $-\frac{2}{3}$

57) $-3\frac{1}{8} \div \left(-\frac{5}{6}\right)$ $\frac{15}{4}$ or $3\frac{3}{4}$ 58) $-\frac{9}{10} \div 12$ $-\frac{3}{40}$

Determine whether each statement is *always, sometimes,* or *never* true.

59) The sum of a positive number and a negative number is positive.   sometimes

60) The product of a positive number and a negative number is negative.   always

61) The absolute value of a number is 0 or positive.   always

62) The absolute value of a number tells its distance from 0.   always

63) A negative number raised to an odd power is negative.   always

64) A negative number raised to an even power is negative.   never

Evaluate.

65) -6^2 -36 66) $(-6)^2$ 36

67) $(-2)^6$ 64   68) -1^{10} -1

69) 3^3 27   70) $(-5)^3$ -125

Use the order of operations to simplify.

71) $56 \div (-7) - 1$ -9

72) $15 - (2 - 5)^3$ 42

73) $-11 + 4 \cdot 3 + (-8 + 6)^5$ -31

74) $\dfrac{1 + 6(7 - 3)}{2[3 - 2(8 - 1)] - 3}$ -1

Write a mathematical expression for each and simplify.

75) The quotient of -120 and -3 $\frac{-120}{-3}$; 40

76) Twice the sum of 22 and -10 $2[22 + (-10)]$; 24

77) 15 less than the product of -4 and 7 $(-4) \cdot 7 - 15$; -43

78) 11 more than half of -18 $(-18) \cdot \frac{1}{2} + 11$; 2

(1.7)

79) Evaluate $9x - 4y$ when $x = -3$ and $y = 7$. -55

80) Evaluate $\dfrac{2a + b}{a^3 - b^2}$ when $a = -3$ and $b = 5$. $\frac{1}{52}$

Which property of real numbers is illustrated by each example? Choose from the commutative, associative, identity, inverse, or distributive property.

81) $12 + (5 + 3) = (12 + 5) + 3$ associative

82) $\left(\frac{2}{5}\right)\left(\frac{5}{2}\right) = 1$ inverse

83) $0 + 19 = 19$ identity

84) $-4(7 + 2) = -4(7) + (-4)(2)$ distributive

85) $8 \cdot 3 = 3 \cdot 8$ commutative

Rewrite each expression using the distributive property. Simplify if possible.

86) $7(3 - 9)$ $7 \cdot 3 - 7 \cdot 9 = 21 - 63 = -42$

87) $(10 + 4)5$ $10 \cdot 5 + 4 \cdot 5 = 50 + 20 = 70$

88) $-(15 - 3)$ $-15 + 3 = -12$

89) $-6(9p - 4q + 1)$ $(-6) \cdot 9p - (-6)(4q) + (-6) \cdot 1$
$= -54p + 24q - 6$

90) List the terms and coefficients of

$$5z^4 - 8z^3 + \frac{3}{5}z^2 - z + 14$$

Combine like terms and simplify.

91) $9m - 14 + 3m + 4$ $\quad 12m - 10$

92) $-5c + d - 2c + 8d$ $\quad -7c + 9d$

93) $15y^2 + 8y - 4 + 2y^2 - 11y + 1$ $\quad 17y^2 - 3y - 1$

94) $7t + 10 - 3(2t + 3)$ $\quad t + 1$

95) $\frac{3}{2}(5n - 4) + \frac{1}{4}(n + 6)$ $\quad \frac{31}{4}n - \frac{9}{2}$

96) $1.4(a + 5) - (a + 2)$ $\quad 0.4a + 5$

Chapter 1:   Test

*Additional answers can be found in the Answers to Exercises appendix.

1) Find the prime factorization of 210. $\quad 2 \cdot 3 \cdot 5 \cdot 7$

2) Write in lowest terms:

   a) $\dfrac{45}{72}$ $\quad \frac{5}{8}$ b) $\dfrac{420}{560}$ $\quad \frac{3}{4}$

Perform the indicated operations. Write all answers in lowest terms.

3) $\dfrac{7}{16} \cdot \dfrac{10}{21}$ $\quad \frac{5}{24}$

4) $\dfrac{5}{12} + \dfrac{2}{9}$ $\quad \frac{23}{36}$

5) $10\dfrac{2}{3} - 3\dfrac{1}{4}$ $\quad 7\frac{5}{12}$

6) $\dfrac{4}{9} \div 12$ $\quad \frac{1}{27}$

7) $\dfrac{3}{5} - \dfrac{17}{20}$ $\quad -\frac{1}{4}$

8) $-31 - (-14)$ $\quad -17$

9) $16 + 8 \div 2$ $\quad 20$

10) $\dfrac{1}{8} \cdot \left(-\dfrac{2}{3}\right)$ $\quad -\frac{1}{12}$

11) $-15 \cdot (-4)$ $\quad 60$

12) $-9.5 + 5.8$ $\quad -3.7$

13) $23 - 6[-4 + (9 - 11)^4]$ $\quad -49$

14) $\dfrac{7 \cdot 2 - 4}{48 \div 3 - 8^0}$ $\quad \frac{2}{3}$

15) An extreme sports athlete has reached an altitude of 14,693 ft while ice climbing and has dived to a depth of 518 ft below sea level. What is the difference between these two elevations? $\quad 15{,}211$ ft

16) Evaluate.

   a) 5^3 $\quad 125$

   b) $|-43|$ $\quad 43$

   c) $-|18 - 40| - 3|9 - 4|$ $\quad -37$

17) Write $(-2)^4$ and -2^4 without exponents. Evaluate each. $\quad (-2)^4 = -2 \cdot (-2) \cdot (-2) \cdot (-2) = 16$
$-2^4 = -1 \cdot 2^4 = -1 \cdot 2 \cdot 2 \cdot 2 \cdot 2 = -16$

18) The supplement of $31°$ is _____. $\quad 149°$

19) Find the missing angle, and classify the triangle as acute, obtuse, or right. $\quad 49°$; acute

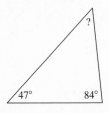

20) Find the area and perimeter of each figure. Include the correct units.

   a)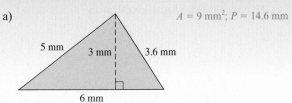
   $A = 9$ mm^2; $P = 14.6$ mm

   b)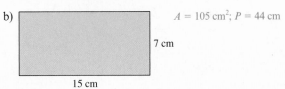
   $A = 105$ cm^2; $P = 44$ cm

   c)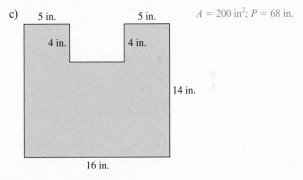
   $A = 200$ in^2; $P = 68$ in.

21) Find the volume of this figure: $\quad 9$ ft^3

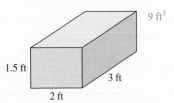

22) Given this set of numbers,
$$\left\{ 3\frac{1}{5}, 22, -7, \sqrt{43}, 0, 1.\overline{5}, 8.0934 \ldots \right\}$$ list the

   a) whole numbers $\quad 22, 0$

   b) natural numbers $\quad 22$

   c) irrational numbers $\quad \sqrt{43}, 8.0934 \ldots$

   d) integers $\quad 22, -7, 0$

   e) rational numbers $\quad 3\frac{1}{5}, 22, -7, 0, 6.2, 1.\overline{5}$

23) Graph the numbers on a number line. Label each.

$4, -5, \dfrac{2}{3}, -3\dfrac{1}{2}, -\dfrac{5}{6}, 2.2$

24) Write a mathematical expression for each and simplify.

a) The sum of -4 and 27 $-4 + 27$; 23

b) The product of 5 and -6 subtracted from 17

                                       $17 - 5(-6)$; 47

25) List the terms and coefficients of

$4p^3 - p^2 + \dfrac{1}{3}p - 10$

26) Evaluate $\dfrac{x^2 - y^2}{6y + x}$ when $x = 3$ and $y = -4$. $\dfrac{1}{3}$

27) Which property of real numbers is illustrated by each example? Choose from the commutative, associative, identity, inverse, or distributive property.

a) $9 \cdot 5 = 5 \cdot 9$ commutative

b) $16 + (4 + 7) = (16 + 4) + 7$ associative

c) $\left(\dfrac{10}{3}\right)\left(\dfrac{3}{10}\right) = 1$ inverse

d) $8(1 - 4) = 8 \cdot 1 - 8 \cdot 4$ distributive

28) Rewrite each expression using the distributive property. Simplify if possible.

a) $-4(2 + 7)$ $(-4) \cdot 2 + (-4) \cdot 7 = -8 + (-28) = -36$

b) $3(8m - 3n + 11)$ $3 \cdot 8m - 3 \cdot 3n + 3 \cdot 11 = 24m - 9n + 33$

29) Combine like terms and simplify.

a) $-8k^2 + 3k - 5 + 2k^2 + k - 9$ $-6k^2 + 4k - 14$

b) $\dfrac{4}{3}(6c - 5) - \dfrac{1}{2}(4c + 3)$ $6c - \dfrac{49}{6}$

30) Write a mathematical expression for "nine less than twice a number." Let x represent the number. $2x - 9$

Linear Equations and Inequalities

Math at Work:

Landscape Architect

A landscape architect like Jill Cavanaugh has to take many factors into account in order to do her job. She needs to understand what her client wants. Just as importantly, though, she needs to be aware of the physical, legal, and other issues that affect how landscape architecture is done.

For instance, Jill is designing the driveway, patio, and walkway for this new home. The town has a building code which states that at most 70% of the lot can be covered with an impervious surface such as the house, driveway, patio, and walkway leading up to the front door. So, her design must satisfy this restriction.

To begin, Jill must determine the area of the land and find 70% of that number to figure out how much land can be covered with hard surfaces. She must subtract the area covered by the house to determine how much land she has left for the driveway, patio, and walkway. Using her design experience and problem-solving skills, she must come up with a plan for building the driveway, patio, and walkway that will not only please her clients but will meet building codes as well.

Clearly, Jill's math skills are a big asset to her career. Her reading skills are helpful, too. She has to read and understand a wide variety of materials to find the information she needs regarding building codes, construction practices, and so forth.

In this chapter, we will learn to work with linear equations and inequalities and discuss strategies for reading math and other types of textbooks.

 **Study Strategies** Reading Math (and Other) Textbooks

Whether you are studying math or a foreign language, computer programming or art, it's likely that you will need to read material and answer questions contained in a textbook. So, no matter what you study, having effective strategies for textbook reading will help you succeed in college. The strategies below can be used when reading any textbook and should be particularly helpful when you are working with a math textbook like this one.

 **Prepare**
- When you get a new textbook, study the frontmatter, such as the preface and the introduction. It is in these sections that the author or authors introduce the goals and ideas behind the book.
- Before you start reading, review advance organizers like outlines, overviews, or section objectives. Students often skip this material, but it provides a helpful preview of what will be covered in your reading.

 **Organize**
- Gather the tools you'll need, such as a pen and paper to take notes and a highlighter to mark up the textbook itself.
- Plan ahead so you have time to read the material carefully and think about it deeply. Never try to read a textbook assignment in the 5 minutes before class starts!

 Work
- Don't try to read a whole chapter without stopping. Instead, read a section of a chapter, then give yourself a short break.
- As you read, take notes on a separate piece of paper and mark up the textbook page itself with arrows, check marks, etc. As you read the examples, work out the problems on your own paper.
- Highlight and/or underline essential information, but be careful not to mark too much. One guideline: No more than 10% of what you read should be highlighted or underlined.
- Complete all the exercises and sample problems. With math in particular, practice is the key to mastering concepts.
- Pay particular attention to formulas, marking them clearly so it is easy to find them later. Formulas are fundamental to building math skills.

Evaluate
- Try to connect major ideas to your own life and experiences. For example, think of an outside-the-classroom situation in which you might use a formula you have learned.
- Explain the material out loud, either to a classmate or just to yourself. This will help ensure that you understand everything covered in the reading.

 **Rethink**
- Within 24 hours of first reading the assignment, look it over again, along with your notes. This will help fix the material in your mind and could save you hours of study time later.

Chapter 2 $\boxed{\textbf{P O W E R}}$ Plan

What are your goals for Chapter 2?	How can you accomplish each goal?
1 Be prepared before and during class.	• Don't stay out late the night before, and be sure to set your alarm clock! • Bring a pencil, notebook paper, and textbook to class. • Avoid distractions by turning off your cell phone during class. • Pay attention, take good notes, and ask questions. • Complete your homework on time, and ask questions on problems you do not understand.
2 Understand the homework to the point where you could do it without needing any help or hints.	• Read the directions, and show all of your steps. • Go to the professor's office for help. • Rework homework and quiz problems, and find similar problems for practice.
3 Use the P.O.W.E.R. framework to help you read math textbooks: *Discover Your Reading Style.*	• Read the Study Strategy that explains how to read a math textbook. • What does P.O.W.E.R. stand for? • Complete the emPOWERme that appears before the Chapter Summary.
4 Write your own goal. _____ _____	• _____ _____

What are your objectives for Chapter 2?	How can you accomplish each objective?
1 Learn to solve linear equations in one variable.	• Know the difference between an equation and an expression. • Be able to use the addition, subtraction, multiplication, and division properties of equality to solve an equation. • Apply the procedure for solving a linear equation to equations with variables on one side or both sides of the equation. • Know how to eliminate fractions or decimals before solving an equation. • Know the three different outcomes when solving a linear equation.
2 Use the five steps for solving applied problems.	• Learn and understand the basic steps for solving any applied problem. • Understand the terminology related to lengths and consecutive integers to help solve applied problems. • Apply the procedure for solving a linear equation to solve geometry, angle measures, and percent problems. • Be able to solve a formula for a specific variable.
3 Know how to solve problems using proportions.	• Understand what a ratio represents. • Know how to set up a proportion and use cross products. • Apply the definition of proportions to solve problems involving similar triangles.
4 Learn to solve linear inequalities in one variable.	• Understand set and interval notations, and be able to graph a linear inequality. • Use the properties of inequalities to solve an inequality. • Create a procedure for solving three-part inequalities.
5 Write your own goal. _____ _____	• _____ _____

E Evaluate	Complete the Chapter Review and Chapter Test. How did you do?	**R** Rethink	• Describe how the first two properties of equality helped outline what you needed to know to master most of this chapter.
			• How often do you encounter questions or problems that could now be solved by creating an equation or inequality? How does translating that problem into an equation or inequality help?
			• How has the Study Strategy for this chapter helped you more effectively read the textbooks for the courses you are taking this term?

2.1 Solving Linear Equations Part I

P Prepare	**O** Organize
What are your objectives for Section 2.1?	**How can you accomplish each objective?**
1 Define a Linear Equation in One Variable	• Write the definition of a *linear equation in one variable* in your own words, and write an example.
	• Know the differences between expressions and equations and what solving an equation means.
2 Use the Addition and Subtraction Properties of Equality	• Follow the explanation to understand and then learn the **Addition and Subtraction Properties of Equality.**
	• Complete the given examples on your own.
	• Complete You Trys 1 and 2.
3 Use the Multiplication and Division Properties of Equality	• Learn the **Multiplication and Division Properties of Equality.**
	• Complete the given examples on your own.
	• Complete You Trys 3–6.
4 Solve Equations of the Form $ax + b = c$	• Be able to combine the properties learned in Objectives 2 and 3.
	• Complete the given examples on your own.
	• Complete You Trys 7 and 8.

 **W Work** Read the explanations, follow the examples, take notes, and complete the You Trys.

1 Define a Linear Equation in One Variable

What is an equation? It is a mathematical statement that two expressions are equal. For example, $4 + 3 = 7$ is an equation.

An equation contains an "=" sign, and an expression does not.

$$3x + 5 = 17 \text{ is an } equation.$$
$$3x + 5x \text{ is an } expression.$$

We can **solve** equations, and we can **simplify** expressions.

There are many different types of algebraic equations, and in Sections 2.1–2.3, we will learn how to solve *linear* equations. Here are some examples of linear equations in one variable:

$$p - 1 = 4 \qquad 3x + 5 = 17 \qquad 8(n + 1) - 7 = 2n + 3 \qquad -\frac{5}{6}y + \frac{1}{3} = y - 2$$

Definition

A **linear equation in one variable** is an equation that can be written in the form $ax + b = 0$, where a and b are real numbers and $a \neq 0$.

The exponent of the variable, x, in a linear equation like $ax + b = 0$ is 1. For this reason, linear equations are also known as first-degree equations. Equations like $k^2 - 13k + 36 = 0$ and $\sqrt{w - 3} = 2$ are not linear equations and are presented later in the text.

To **solve an equation** means to find the value or values of the variable that make the equation true. For example, the solution of the equation $p - 1 = 4$ is $p = 5$ since substituting 5 for the variable makes the equation true.

$$p - 1 = 4$$
$$5 - 1 = 4 \quad \text{True}$$

Usually, we use set notation to list all the solutions of an equation. The **solution set** of an equation is the set of all numbers that make the equation true. Therefore, $\{5\}$ is the solution set of the equation $p - 1 = 4$. We also say that 5 *satisfies* the equation $p - 1 = 4$.

2 Use the Addition and Subtraction Properties of Equality

Begin with the true statement $8 = 8$. What happens if we add the same number, say 2, to each side? Is the statement still true? Yes.

$$8 = 8$$
$$8 + 2 = 8 + 2$$
$$10 = 10 \quad \text{True}$$

The statement would also be true if we subtracted 2 from both sides of $8 = 8$.

$$8 = 8 \text{ and } 8 + 2 = 8 + 2 \text{ are } equivalent \ equations.$$
$$8 = 8 \text{ and } 8 - 2 = 8 - 2 \text{ are } equivalent \ equations \text{ as well.}$$

We can use these principles to solve an algebraic equation because doing so will not change the equation's solution.

> **Property** Addition and Subtraction Properties of Equality
>
> Let a, b, and c be expressions representing real numbers. Then,
>
> 1) If $a = b$, then $a + c = b + c$ Addition property of equality
> 2) If $a = b$, then $a - c = b - c$ Subtraction property of equality

EXAMPLE 1

In-Class Example 1

Solve $c - 12 = 7$. Check the solution.

Answer: {19}

Solve $x - 8 = 3$. Check the solution.

Solution

Remember, to solve the equation means to find the value of the variable that makes the statement true. To do this, we want to get the variable on a side by itself. We call this **isolating the variable.**

On the left side of the equal sign, the 8 is being **subtracted from** the x. To isolate x, we perform the "opposite" operation—that is, we **add 8** to each side.

$$x - 8 = 3$$
$$x - 8 + 8 = 3 + 8 \qquad \text{Add 8 to each side.}$$
$$x = 11 \qquad \text{Simplify.}$$

Check: Substitute 11 for x in the original equation.

$$x - 8 = 3$$
$$11 - 8 = 3$$
$$3 = 3 \checkmark$$

The solution set is {11}.

[YOU TRY 1]

Solve $b - 5 = 9$. Check the solution.

Equations can contain variables on either side of the equal sign.

EXAMPLE 2

In-Class Example 2

Solve $-8 = z + 4$. Check the solution.

Answer: {−12}

Solve $-5 = m + 4$. Check the solution.

Solution

Notice that the 4 is being **added to** the variable, m. We will **subtract 4** from each side to isolate the variable.

$$-5 = m + 4$$
$$-5 - 4 = m + 4 - 4 \qquad \text{Subtract 4 from each side.}$$
$$-9 = m \qquad \text{Simplify.}$$

Check: Substitute -9 for m in the original equation.

$$-5 = m + 4$$
$$-5 = -9 + 4$$
$$-5 = -5 \checkmark$$

The solution set is {−9}.

Solve $-2 = y + 6$. Check the solution.

3 Use the Multiplication and Division Properties of Equality

It is also true that if we multiply both sides of an equation by the same nonzero number or divide both sides of an equation by the same nonzero number, then we will obtain an equivalent equation.

W Hint

Did you notice that dividing both sides of an equation by c is the same as multiplying both sides by $1/c$?

Property Multiplication and Division Properties of Equality

Let a, b, and c be expressions representing real numbers where $c \neq 0$. Then,

1) If $a = b$, then $ac = bc$ Multiplication property of equality

2) If $a = b$, then $\dfrac{a}{c} = \dfrac{b}{c}$ Division property of equality

---

EXAMPLE 3

Solve $3k = -9.6$. Check the solution.

In-Class Example 3

Solve $2v = -8.6$. Check the solution.

Answer: $\{-4.3\}$

Solution

On the left-hand side of the equation, the k is being **multiplied** by 3. So, we will perform the "opposite" operation and **divide** each side by 3.

$$3k = -9.6$$
$$\frac{3k}{3} = \frac{-9.6}{3} \quad \text{Divide each side by 3.}$$
$$k = -3.2 \quad \text{Simplify.}$$

Check: Substitute -3.2 for k in the original equation.

$$3k = -9.6$$
$$3(-3.2) = -9.6$$
$$-9.6 = -9.6 \checkmark$$

The solution set is $\{-3.2\}$.

YOU TRY 3

Solve $-8w = 42.4$. Check the solution.

EXAMPLE 4

Solve $-m = 19$. Check the solution.

In-Class Example 4

Solve $-t = 4$. Check the solution.

Answer: $\{-4\}$

Solution

The negative sign in front of the m tells us that the coefficient of m is -1. Since m is being **multiplied** by -1, we will **divide** each side by -1.

$$-m = 19$$
$$\frac{-1m}{-1} = \frac{19}{-1} \quad \text{Rewrite } -m \text{ as } -1m; \text{ divide each side by } -1.$$
$$m = -19 \quad \text{Simplify.}$$

The check is left to the student. The solution set is $\{-19\}$.

[YOU TRY 4] Solve $-a = -25$. Check the solution.

EXAMPLE 5

In-Class Example 5

Solve $\dfrac{d}{4} = -9$. Check the solution.

Answer: $\{-36\}$

Solve $\dfrac{x}{4} = 5$. Check the solution.

Solution

The x is being **divided** by 4. Therefore, we will **multiply** each side by 4 to get the x on a side by itself.

$$\frac{x}{4} = 5$$

$$4 \cdot \frac{x}{4} = 4 \cdot 5 \qquad \text{Multiply each side by 4.}$$

$$1x = 20 \qquad \text{Simplify.}$$

$$x = 20$$

The check is left to the student. The solution set is $\{20\}$.

[YOU TRY 5] Solve $\dfrac{h}{7} = 8$. Check the solution.

EXAMPLE 6

In-Class Example 6

Solve $\dfrac{6}{11}x = 12$. Check the solution.

Answer: $\{22\}$

Ⓦ Hint

Which property from Section 1.6 are you using to complete this example?

Solve $\dfrac{3}{8}y = 12$. Check the solution.

Solution

On the left-hand side, the y is being **multiplied** by $\dfrac{3}{8}$. So, we could divide each side by $\dfrac{3}{8}$. However, recall that dividing a quantity by a fraction is the same as **multiplying by the reciprocal** of the fraction. Therefore, we will multiply each side by the reciprocal of $\dfrac{3}{8}$.

$$\frac{3}{8}y = 12$$

$$\frac{8}{3} \cdot \frac{3}{8}y = \frac{8}{3} \cdot 12 \qquad \text{The reciprocal of } \frac{3}{8} \text{ is } \frac{8}{3}. \text{ Multiply each side by } \frac{8}{3}.$$

$$1y = \frac{8}{\cancel{3}} \cdot \overset{4}{\cancel{12}} \qquad \text{Perform the multiplication.}$$

$$y = 32 \qquad \text{Simplify.}$$

The check is left to the student. The solution set is $\{32\}$.

[YOU TRY 6] Solve $-\dfrac{5}{9}c = 20$. Check the solution.

---

4 Solve Equations of the Form $ax + b = c$

So far, we have not combined the properties of addition, subtraction, multiplication, and division to solve an equation. But that is exactly what we must do to solve equations like

$$3p + 7 = 31 \quad \text{and} \quad 4x + 9 - 6x + 2 = 17$$

EXAMPLE 7

In-Class Example 7

Solve $6u - 1 = 17$.

Answer: {3}

Solve $3p + 7 = 31$.

Solution

In this equation, there is a number, 7, being **added** to the term containing the variable, and the variable is being multiplied by a number, 3. **In general, we first eliminate the number being added to or subtracted from the variable.** Then we eliminate the coefficient.

$$
\begin{aligned}
3p + 7 &= 31 \\
3p + 7 - 7 &= 31 - 7 \qquad &\text{Subtract 7 from each side.} \\
3p &= 24 \qquad &\text{Combine like terms.} \\
\frac{3p}{3} &= \frac{24}{3} \qquad &\text{Divide by 3.} \\
p &= 8 \qquad &\text{Simplify.}
\end{aligned}
$$

Check:
$$
\begin{aligned}
3p + 7 &= 31 \\
3(8) + 7 &= 31 \\
24 + 7 &= 31 \\
31 &= 31 \quad \checkmark
\end{aligned}
$$

The solution set is {8}.

[YOU TRY 7] Solve $2n + 9 = 15$.

EXAMPLE 8

In-Class Example 8

Solve $-\dfrac{8}{3}x + 5 = 41$.

Answer: $\left\{-\dfrac{27}{2}\right\}$

Solve $-\dfrac{6}{5}c - 1 = 13$.

Solution

On the left-hand side, the c is being multiplied by $-\dfrac{6}{5}$, and 1 is being subtracted from the c-term. To solve the equation, begin by eliminating the number being subtracted from the c-term.

W Hint

Write down all of the steps in the example on your paper as you are reading it.

$$
\begin{aligned}
-\frac{6}{5}c - 1 &= 13 \\[4pt]
-\frac{6}{5}c - 1 + 1 &= 13 + 1 \qquad &\text{Add 1 to each side.} \\[4pt]
-\frac{6}{5}c &= 14 \qquad &\text{Combine like terms.} \\[4pt]
-\frac{5}{6} \cdot \left(-\frac{6}{5}c\right) &= -\frac{5}{6} \cdot 14 \qquad &\text{Multiply each side by the reciprocal of } -\frac{6}{5}. \\[4pt]
1c &= -\frac{5}{\overset{}{\underset{3}{6}}} \cdot \overset{7}{14} \qquad &\text{Simplify.} \\[4pt]
c &= -\frac{35}{3}
\end{aligned}
$$

The check is left to the student. The solution set is $\left\{-\dfrac{35}{3}\right\}$.

SECTION 2.1 **Solving Linear Equations Part I**

$\left[\text{YOU TRY 8}\right]$ Solve $-\dfrac{4}{9}z + 3 = -7$.

ANSWERS TO $\left[\text{YOU TRY}\right]$ EXERCISES

1) $\{14\}$ 2) $\{-8\}$ 3) $\{-5.3\}$ 4) $\{25\}$ 5) $\{56\}$

6) $\{-36\}$ 7) $\{3\}$ 8) $\left\{\dfrac{45}{2}\right\}$

E Evaluate **2.1** Exercises   Do the exercises, and check your work.

*Additional answers can be found in the Answers to Exercises appendix.

Objective 1: Define a Linear Equation in One Variable

Identify each as an expression or an equation.

1) $9c + 4 - 2c$ expression

2) $5n - 3 = 6$ equation

3) $y + 10(8y + 1) = -13$ equation

4) $4 + 2(5p - 3)$ expression

5) Can we solve $-6x + 10x$? Why or why not?
No, it is an expression.

6) Can we solve $-6x + 10x = 28$? Why or why not?
Yes, it is an equation.

7) Which of the following are linear equations in one variable?   b, c

a) $2k + 9 - 7k + 1$

b) $-8 = \dfrac{3}{2}n - 1$

c) $4(3t - 1) + 9t = 12$

d) $w^2 + 13w + 36 = 0$

8) Explain how to check the solution of an equation.
Answers may vary.

Determine whether the given value is a solution to the equation.

9) $a - 4 = -9$; $a = 5$ no

10) $-5c = -10$; $c = 2$ yes

11) $-12y = 8$; $y = -\dfrac{2}{3}$ yes

12) $5 = a + 14$; $a = -9$ yes

13) $1.3 = 2p - 1.7$; $p = 1.5$ yes

14) $20m + 3 = 16$; $m = \dfrac{4}{5}$ no

Objective 2: Use the Addition and Subtraction Properties of Equality

Solve each equation, and check the solution.

15) $n - 5 = 12$ $\{17\}$

16) $z + 8 = -2$ $\{-10\}$

17) $b + 10 = 4$ $\{-6\}$

18) $x - 3 = 9$ $\{12\}$

19) $-16 = k - 12$ $\{-4\}$

20) $23 = r + 14$ $\{9\}$

21) $6 = 6 + y$ $\{0\}$

22) $5 = -1 + k$ $\{6\}$

23) $a - 2.9 = -3.6$ $\{-0.7\}$

24) $w + 4.7 = 9.1$ $\{4.4\}$

25) $12 = x + 7.2$ $\{4.8\}$

26) $-8.3 = y - 5.6$ $\{-2.7\}$

27) $h + \dfrac{5}{6} = \dfrac{1}{3}$ $\left\{-\dfrac{1}{2}\right\}$

28) $b - \dfrac{3}{8} = \dfrac{3}{10}$ $\left\{\dfrac{27}{40}\right\}$

29) $-\dfrac{2}{5} = -\dfrac{5}{4} + c$ $\left\{\dfrac{17}{20}\right\}$

30) $-\dfrac{8}{9} = d + 2$ $\left\{-\dfrac{26}{9}\right\}$

31) Write an equation that can be solved with the subtraction property of equality and that has a solution set of $\{-7\}$.   Answers may vary.

32) Write an equation that can be solved with the addition property of equality and that has a solution set of $\{10\}$.   Answers may vary.

Objective 3: Use the Multiplication and Division Properties of Equality

Solve each equation, and check the solution.

33) $2n = 8$ $\{4\}$

34) $9k = 72$ $\{8\}$

35) $-5z = 35$ $\{-7\}$

36) $-8b = -24$ $\{3\}$

37) $-48 = -4r$ $\{12\}$

38) $-54 = 6m$ $\{-9\}$

39) $63 = -28y$ $\left\{-\dfrac{9}{4}\right\}$

40) $75 = 50c$ $\left\{\dfrac{3}{2}\right\}$

41) $10n = 2.3$ $\{0.23\}$

42) $-4x = -28.4$ $\{7.1\}$

43) $-7 = -0.5d$ $\{14\}$

44) $-3.9 = 1.3p$ $\{-3\}$

45) $-x = 1$ $\{-1\}$

46) $-h = -3$ $\{3\}$

47) $-6.5 = -v$ $\{6.5\}$

48) $\dfrac{4}{9} = -t$ $\left\{-\dfrac{4}{9}\right\}$

49) $\dfrac{a}{4} = 12$ $\{48\}$

50) $\dfrac{w}{5} = 4$ $\{20\}$

51) $-\dfrac{m}{3} = 13$ $\{-39\}$

52) $7 = -\dfrac{n}{11}$ $\{-77\}$

53) $\dfrac{w}{6} = -\dfrac{3}{4}$ $\left\{-\dfrac{9}{2}\right\}$

54) $-\dfrac{a}{21} = -\dfrac{2}{3}$ $\{14\}$

83) $-1 = \dfrac{10}{11}c + 5$ $\left\{-\dfrac{33}{5}\right\}$

84) $2 = -\dfrac{9}{4}a - 10$ $\left\{-\dfrac{16}{3}\right\}$

55) $\dfrac{1}{5}q = -9$ $\{-45\}$

56) $\dfrac{1}{8}y = 3$ $\{24\}$

85) $2 - \dfrac{5}{6}t = -2$ $\left\{\dfrac{24}{5}\right\}$

86) $5 + \dfrac{3}{4}h = -1$ $\{-8\}$

57) $-\dfrac{1}{6}m = -14$ $\{84\}$

58) $-\dfrac{1}{3}z = 9$ $\{-27\}$

87) $\dfrac{3}{4} = \dfrac{1}{2} - \dfrac{1}{6}z$ $\left\{-\dfrac{3}{2}\right\}$

88) $1 = \dfrac{3}{5} - \dfrac{2}{3}k$ $\left\{-\dfrac{3}{5}\right\}$

59) $\dfrac{5}{12} = \dfrac{1}{4}c$ $\left\{\dfrac{5}{3}\right\}$

60) $-\dfrac{5}{9} = -\dfrac{1}{6}r$ $\left\{\dfrac{10}{3}\right\}$

89) $0.2p + 9.3 = 5.7$ $\{-18\}$

90) $0.5x - 2.6 = 4.9$ $\{15\}$

61) $\dfrac{1}{3}y = -\dfrac{11}{15}$ $\left\{-\dfrac{11}{5}\right\}$

62) $-\dfrac{1}{7}a = \dfrac{3}{2}$ $\left\{-\dfrac{21}{2}\right\}$

91) $3.8c - 7.62 = 2.64$ $\{2.7\}$

92) $4.3a + 1.98 = -14.36$ $\{-3.8\}$

63) $-\dfrac{5}{3}d = -30$ $\{18\}$

64) $35 = \dfrac{5}{3}k$ $\{21\}$

93) $14.74 = -20.6 - 5.7u$ $\{-6.2\}$

94) $10.5 - 9.2m = -36.42$ $\{5.1\}$

65) $-21 = \dfrac{3}{2}d$ $\{-14\}$

66) $-\dfrac{4}{7}w = -36$ $\{63\}$

Mixed Exercises: Objectives 1–4
Solve and check each equation.

95) $-9z = 6$ $\left\{-\dfrac{2}{3}\right\}$

96) $u - 23 = 52$ $\{75\}$

97) $3a - 11 = 16$ $\{9\}$

98) $20 = -\dfrac{1}{4}x$ $\{-80\}$

Objective 4: Solve Equations of the Form $ax + b = c$
Solve each equation, and check the solution.

67) $5z + 8 = 43$ $\{7\}$

68) $2y - 5 = 3$ $\{4\}$

99) $-\dfrac{c}{6} = -9$ $\{54\}$

100) $6w - 1 = -19$ $\{-3\}$

69) $4n - 15 = -19$ $\{-1\}$

70) $7c + 4 = 18$ $\{2\}$

101) $-34 = n + 15$ $\{-49\}$

102) $67.9 = 7y$ $\{9.7\}$

71) $8d - 15 = -15$ $\{0\}$

72) $10m + 7 = 7$ $\{0\}$

103) $-\dfrac{1}{7}p = -8$ $\{56\}$

104) $-6 = \dfrac{m}{8}$ $\{-48\}$

73) $-11 = 5t - 9$ $\left\{-\dfrac{2}{5}\right\}$

74) $7 = 4k + 13$ $\left\{-\dfrac{3}{2}\right\}$

75) $-6h + 19 = 3$ $\left\{\dfrac{8}{3}\right\}$

76) $-8q - 11 = 9$ $\left\{-\dfrac{5}{2}\right\}$

105) $8.33 - 6.35d = 17.22$ $\{-1.4\}$

106) $\dfrac{10}{7}r + 3 = 18$ $\left\{\dfrac{21}{2}\right\}$

77) $10 = 3 - 7y$ $\{-1\}$

78) $-6 = 9 - 3p$ $\{5\}$

107) $-15 = 9 + \dfrac{4}{5}c$ $\{-30\}$

79) $\dfrac{1}{2}d + 7 = 12$ $\{10\}$

80) $\dfrac{1}{3}x + 4 = 11$ $\{21\}$

108) $-7.92q + 41.95 = 22.15$ $\{2.5\}$

81) $\dfrac{4}{5}b - 9 = -13$ $\{-5\}$

82) $-\dfrac{12}{7}r + 5 = 3$ $\left\{\dfrac{7}{6}\right\}$

109) $-\dfrac{3}{4}k + \dfrac{2}{5} = -2$ $\left\{\dfrac{16}{5}\right\}$

110) $\dfrac{2}{3} = \dfrac{1}{2} - \dfrac{3}{8}t$ $\left\{-\dfrac{4}{9}\right\}$

R Rethink

R1) Why is it that we generally add numbers to or subtract numbers from both sides of an equation before we multiply or divide? (Hint: Compare to the order of operations.)

R2) What is the difference between an expression and an equation?

R3) Which exercises in this section do you need to practice more?

R4) Were there any problems that you could not do? If so, write them down or circle them and ask your instructor how to do them.

2.2 Solving Linear Equations Part II

P Prepare

What are your objectives for Section 2.2?	How can you accomplish each objective?
1 Summarize the Steps for Solving a Linear Equation	• Write the procedure for **How to Solve a Linear Equation** in your own words.
2 Solve Equations Containing Variables on One Side of the Equal Sign	• Follow the procedure for solving equations, and notice the subtle differences between the examples. • Complete the given examples on your own. • Complete You Trys 1–3.
3 Solve Equations Containing Variables on Both Sides of the Equal Sign	• Follow the procedure for solving equations, and notice the subtle differences between the examples. • Complete the given example on your own. • Complete You Try 4.

O Organize

Read the explanations, follow the examples, take notes, and complete the You Trys.

1 Summarize the Steps for Solving a Linear Equation

In Section 2.1, we learned how to solve equations such as

$$x - 8 = 3 \qquad 3k = -9.6 \qquad \frac{3}{8}y = 12 \qquad -8.85 = 2.1y - 5.49$$

Each of these equations contains only one variable term. In this section, we will discuss how to solve equations in which more than one term contains a variable and where variables appear on both sides of the equal sign. We can use the following steps.

> **Procedure** How to Solve a Linear Equation
>
> *Step 1:* **Clear parentheses** and **combine like terms** on each side of the equation.
>
> *Step 2:* **Get the variable on one side of the equal sign and the constant on the other side of the equal sign** (isolate the variable) using the addition or subtraction property of equality.
>
> *Step 3:* **Solve for the variable** using the multiplication or division property of equality.
>
> *Step 4:* **Check the solution** in the original equation.

2 Solve Equations Containing Variables on One Side of the Equal Sign

Let's start by solving equations containing variables on only one side of the equal sign.

EXAMPLE 1

In-Class Example 1

Solve $9b - 8 - 3b + 5 = 21$.

Answer: {4}

EXAMPLE 1

Solve $4x + 9 - 6x + 2 = 17$.

Solution

Step 1: Because there are two x-terms on the left side of the equal sign, begin by combining like terms.

$$4x + 9 - 6x + 2 = 17$$
$$-2x + 11 = 17 \quad \text{Combine like terms.}$$

Step 2: Isolate the variable.

$$-2x + 11 - 11 = 17 - 11 \quad \text{Subtract 11 from each side.}$$
$$-2x = 6 \quad \text{Combine like terms.}$$

Step 3: Solve for x using the division property of equality.

$$\frac{-2x}{-2} = \frac{6}{-2} \quad \text{Divide each side by } -2.$$
$$x = -3 \quad \text{Simplify.}$$

Step 4: Check:
$$4x + 9 - 6x + 2 = 17$$
$$4(-3) + 9 - 6(-3) + 2 = 17$$
$$-12 + 9 + 18 + 2 = 17$$
$$17 = 17 \quad \checkmark$$

The solution set is $\{-3\}$.

[YOU TRY 1] Solve $15 - 7u - 6 + 2u = -1$.

EXAMPLE 2

In-Class Example 2

Solve
$7(2 - r) - 2(3r + 8) = 24$.

Answer: {−2}

EXAMPLE 2

Solve $2(1 - 3h) - 5(2h + 3) = -21$.

Solution

Step 1: Clear the parentheses and combine like terms.

$$2(1 - 3h) - 5(2h + 3) = -21$$
$$2 - 6h - 10h - 15 = -21 \quad \text{Distribute.}$$
$$-16h - 13 = -21 \quad \text{Combine like terms.}$$

Step 2: Isolate the variable.

$$-16h - 13 + 13 = -21 + 13 \quad \text{Add 13 to each side.}$$
$$-16h = -8 \quad \text{Combine like terms.}$$

W Hint

Combine like terms on each side of the equation *before* isolating the variable.

Step 3: Solve for h using the division property of equality.

$$\frac{-16h}{-16} = \frac{-8}{-16} \qquad \text{Divide each side by } -16.$$

$$h = \frac{1}{2} \qquad \text{Simplify.}$$

Step 4: The check is left to the student. The solution set is $\left\{\frac{1}{2}\right\}$.

[YOU TRY 2] Solve $-3(4y - 3) + 4(y + 1) = 15$.

EXAMPLE 3

In-Class Example 3

Solve $\frac{3}{4}(c + 1) - \frac{5}{6} = \frac{1}{4}$.

Answer: $\left\{\frac{4}{9}\right\}$

Solve $\frac{1}{2}(3b + 8) + \frac{3}{4} = -\frac{1}{2}$.

Solution

Step 1: Clear the parentheses and combine like terms.

$$\frac{1}{2}(3b + 8) + \frac{3}{4} = -\frac{1}{2}$$

$$\frac{3}{2}b + 4 + \frac{3}{4} = -\frac{1}{2} \qquad \text{Distribute.}$$

$$\frac{3}{2}b + \frac{16}{4} + \frac{3}{4} = -\frac{1}{2} \qquad \text{Get a common denominator for the like terms.}$$

$$\frac{3}{2}b + \frac{19}{4} = -\frac{1}{2} \qquad \text{Combine like terms.}$$

Step 2: Isolate the variable.

$$\frac{3}{2}b + \frac{19}{4} - \frac{19}{4} = -\frac{1}{2} - \frac{19}{4} \qquad \text{Subtract } \frac{19}{4} \text{ from each side.}$$

$$\frac{3}{2}b = -\frac{2}{4} - \frac{19}{4} \qquad \text{Get a common denominator.}$$

$$\frac{3}{2}b = -\frac{21}{4} \qquad \text{Simplify.}$$

Step 3: Solve for b using the multiplication property of equality.

$$\frac{2}{3} \cdot \frac{3}{2}b = \frac{2}{3} \cdot \left(-\frac{21}{4}\right) \qquad \text{Multiply both sides by the reciprocal of } \frac{3}{2}.$$

$$b = \frac{\overset{1}{\cancel{2}}}{\underset{1}{\cancel{3}}} \cdot \left(-\frac{\overset{-7}{\cancel{21}}}{\underset{2}{\cancel{4}}}\right) \qquad \text{Perform the multiplication.}$$

$$b = -\frac{7}{2} \qquad \text{Simplify.}$$

Step 4: The check is left to the student. The solution set is $\left\{-\frac{7}{2}\right\}$.

[YOU TRY 3] Solve $\frac{1}{3}(2m - 1) + \frac{5}{9} = \frac{4}{3}$.

In the next section, we will learn another way to solve an equation containing several fractions like the one in Example 3.

3 Solve Equations Containing Variables on Both Sides of the Equal Sign

Now let's see how we use the steps to solve equations containing variables on both sides of the equal sign. Remember that we want to get the variables on one side of the equal sign and the constants on the other side so that we can combine like terms and isolate the variable.

EXAMPLE 4

W Hint

Are you working out the problems on your paper as you are reading the examples?

Solve $9t + 4 - (7t - 2) = t + 6(t + 1)$.

Solution

Step 1: Clear the parentheses and combine like terms.

$$9t + 4 - (7t - 2) = t + 6(t + 1)$$
$$9t + 4 - 7t + 2 = t + 6t + 6 \qquad \text{Distribute.}$$
$$2t + 6 = 7t + 6 \qquad \text{Combine like terms.}$$

Step 2: Isolate the variable.

$$2t - 7t + 6 = 7t - 7t + 6 \qquad \text{Subtract } 7t \text{ from each side.}$$
$$-5t + 6 = 6 \qquad \text{Combine like terms.}$$
$$-5t + 6 - 6 = 6 - 6 \qquad \text{Subtract 6 from each side.}$$
$$-5t = 0 \qquad \text{Combine like terms.}$$

Step 3: Solve for t using the division property of equality.

$$\frac{-5t}{-5} = \frac{0}{-5} \qquad \text{Divide each side by } -5.$$
$$t = 0 \qquad \text{Simplify.}$$

Step 4: Check:

$$9t + 4 - (7t - 2) = t + 6(t + 1)$$
$$9(0) + 4 - [7(0) - 2] = 0 + 6[(0) + 1]$$
$$0 + 4 - (0 - 2) = 0 + 6(1)$$
$$4 - (-2) = 0 + 6$$
$$6 = 6 \quad \checkmark$$

The solution set is $\{0\}$.

[YOU TRY 4] Solve $5 + 3(a + 4) = 7a - (9 - 10a) + 4$.

ANSWERS TO [YOU TRY] EXERCISES

1) $\{2\}$ 2) $\left\{-\dfrac{1}{4}\right\}$ 3) $\left\{\dfrac{5}{3}\right\}$ 4) $\left\{\dfrac{11}{7}\right\}$

Additional answers can be found in the Answers to Exercises appendix.

Objective 1: Summarize the Steps for Solving a Linear Equation

1) Explain, in your own words, the steps for solving a linear equation.

2) What is the first step for solving $8n + 3 + 2n - 9 = 13$? Do not solve the equation.   Combine like terms on the left side.

Objective 2: Solve Equations Containing Variables on One Side of the Equal Sign

Solve each equation.

Fill It In

Fill in the blanks with either the missing mathematical step or reason for the given step.

3) $3x + 7 + 5x + 4 = 27$

$8x + 11 = 27$ — Combine like terms.

$8x + 11 - 11 = 27 - 11$ — Subtraction property of equality

$8x = 16$ — Combine like terms.

$\dfrac{8x}{8} = \dfrac{16}{8}$ — Division property of equality

$x = 2$ — Simplify.

The solution set is $\underline{\{2\}}$.

4) $5 - 2(3k + 1) + 2k = 23$

$5 - 6k - 2 + 2k = 23$ — Distribute.

$-4k + 3 = 23$ — Combine like terms.

$-4k + 3 - 3 = 23 - 3$ — Subtraction property of equality

$-4k = 20$ — Combine like terms.

$\dfrac{-4k}{-4} = \dfrac{20}{-4}$ — Division property of equality

$k = -5$ — Simplify.

The solution set is $\underline{\{-5\}}$.

For Exercises 5 and 6, choose from *always, sometimes,* or *never*.

5) An equation containing fractions will *always, sometimes,* or *never* have a solution that is a fraction.   sometimes

6) An equation that does *not* contain fractions will *always, sometimes,* or *never* have a solution that is a fraction.   sometimes

Solve each equation, and check the solution.

7) $6a - 10 + 4a + 9 = 39$ $\{4\}$

8) $7m + 11 + 2m - 5 = 33$ $\{3\}$

9) $-15 + 8y - 10y + 1 = 8$ $\{-11\}$

10) $3 - 3p - 2p + 9 = 2$ $\{2\}$

11) $30 = 5c + 14 - 11c + 1$ $\left\{-\dfrac{5}{2}\right\}$

12) $-42 = 4x - 17 + 5x + 8$ $\left\{-\dfrac{11}{3}\right\}$

13) $5 - 3m + 9m + 10 - 7m = -4$ $\{19\}$

14) $1 + 10z - 14 - 2z - 9z = -5$ $\{-8\}$

15) $5 = -12p + 7 + 4p - 12$ $\left\{-\dfrac{5}{4}\right\}$

16) $-40 = 13t + 2 - 4t - 11 - 5t + 3$ $\left\{-\dfrac{17}{2}\right\}$

17) $\dfrac{1}{4}n + 2 + \dfrac{1}{2}n - \dfrac{3}{2} = \dfrac{11}{4}$ $\{3\}$

18) $\dfrac{1}{6} + \dfrac{1}{2}w - \dfrac{4}{3} + \dfrac{1}{3}w = \dfrac{1}{2}$ $\{2\}$

19) $4.2d - 1.7 - 2.2d + 4.3 = -1.4$ $\{-2\}$

20) $5.9h + 2.8 - 3.7 - 3.9 = 1.1$ $\{1\}$

21) $2(5x + 3) - 3x + 4 = -11$ $\{-3\}$

22) $6(2c - 1) + 3 - 7c = 42$ $\{9\}$

23) $7(b - 5) + 5(b + 4) = 45$ $\{5\}$

24) $4(z - 2) + 3(z + 8) = -12$ $\{-4\}$

25) $8 - 3(2k + 9) + 2(7 + k) = 2$ $\left\{-\dfrac{7}{4}\right\}$

26) $1 - 5(3y + 2) + 3(4 + y) = -12$ $\left\{\dfrac{5}{4}\right\}$

27) $-23 = 4(3x - 7) - (8x - 5)$ $\{0\}$

28) $38 = 9(2a + 3) - (10a - 7)$ $\left\{\dfrac{1}{2}\right\}$

29) $8 = 5(4n + 3) - 3(2n - 7) - 20$ $\left\{-\dfrac{4}{7}\right\}$

30) $-4 = 3(2z - 5) - 2(5z - 1) + 9$ $\{0\}$

31) $2(7u - 3) - (u + 9) - 3(2u + 1) = 24$ $\{6\}$

32) $6(4h + 7) + 2(h - 5) - (h - 11) = 18$ $\{-1\}$

33) $\dfrac{1}{3}(3w + 4) - \dfrac{2}{3} = -\dfrac{1}{3}$ $\{-1\}$

34) $\dfrac{3}{4}(2r - 5) + \dfrac{1}{2} = \dfrac{5}{4}$ $\{3\}$

35) $\dfrac{1}{2}(c - 2) + \dfrac{1}{4}(2c + 1) = \dfrac{5}{4}$ $\{2\}$

36) $\frac{2}{3}(m + 3) - \frac{4}{15}(3m + 7) = \frac{4}{5}$ $\{-5\}$

37) $\frac{4}{3}(t + 1) - \frac{1}{6}(4t - 3) = 2$ $\left\{\frac{1}{4}\right\}$

38) $\frac{1}{4}(3x - 2) - \frac{1}{2}(x - 1) = -\frac{1}{7}$ $\left\{-\frac{4}{7}\right\}$

Objective 3: Solve Equations Containing Variables on Both Sides of the Equal Sign

Solve each equation, and check the solution.

39) $9y + 8 = 4y - 17$ $\{-5\}$

40) $12b - 5 = 8b + 11$ $\{4\}$

24 41) $5k - 6 = 7k - 8$ $\{1\}$

42) $3v + 14 = 9v - 22$ $\{6\}$

43) $-15w + 4 = 24 - 7w$ $\left\{-\frac{5}{2}\right\}$

44) $-7x + 13 = 3 - 13x$ $\left\{-\frac{5}{3}\right\}$

45) $1.8z - 1.1 = 1.4z + 1.7$ $\{7\}$

46) $2.2q + 1.9 = 2.8q + 7.3$ $\{-9\}$

47) $18 - h + 5h - 11 = 9h + 19 - 3h$ $\{-6\}$

48) $4m - 1 - 6m + 7 = 11m + 3 - 10m$ $\{1\}$

49) $2t + 7 - 6t + 12 = 4t + 5 - 7t - 1$ $\{15\}$

50) $4 + 8a - 17 + 3a = 7a + 1 + 5a - 10$ $\{-4\}$

51) $6.1r + 1.6 - 3.7r - 0.3 = r - 1.7 + 0.2r - 0.6$ $\{-3\}$

52) $-7.5k + 3.2 + 3.8k + 0.9 = 0.1k - 2.1 - 3.4k + 7$ $\{-2\}$

53) $1 + 5(4n - 7) = 4(7n - 3) - 30$ $\{1\}$

54) $10 + 2(z - 9) = 3(z + 1) - 6$ $\{-5\}$

55) $2(1 - 8c) = 5 - 3(6c + 1) + 4c$ $\{0\}$

56) $13u + 6 - 5(2u - 3) = 1 + 4(u + 5)$ $\{0\}$

24 57) $9 - (8p - 5) + 4p = 6(2p + 1)$ $\left\{\frac{1}{2}\right\}$

58) $2(6d + 5) = 16 - (7d - 4) + 11d$ $\left\{\frac{5}{4}\right\}$

59) $-3(4r + 9) + 2(3r + 8) = r - (9r - 5)$ $\{8\}$

60) $2(3t - 4) - 6(t + 1) = -t + 4(t + 10)$ $\{-18\}$

61) $m + \frac{1}{2}(3m + 4) - 5 = \frac{2}{3}(2m - 1) + \frac{5}{6}$ $\left\{\frac{19}{7}\right\}$

62) $2x - \frac{2}{3}(5x - 6) - 10 = \frac{1}{4}(3x + 1) + \frac{1}{2}$ $\left\{-\frac{81}{25}\right\}$

R Rethink

R1) When we isolate the variable on one side of an equation, does it matter on which side we put the variable?

R2) Does $x = 5$ mean the same thing as $5 = x$?

R3) Write a paragraph that describes how to solve linear equations and include the properties learned so far.

2.3 Solving Linear Equations Part III

P Prepare

O Organize

What are your objectives for Section 2.3?	How can you accomplish each objective?
1 Solve Equations Containing Fractions	• Write the procedure for **Eliminating Fractions from an Equation** in your own words, and add it to the procedure developed for solving linear equations in one variable. • Complete the given example on your own. • Complete You Try 1.
2 Solve Equations Containing Decimals	• Write the procedure for **Eliminating Decimals from an Equation** in your own words, and add it to the procedure developed for solving linear equations in one variable. • Complete the given example on your own. • Complete You Try 2.
3 Solve Equations with No Solution or an Infinite Number of Solutions	• Complete and understand the given examples on your own. • Learn the **Outcomes When Solving Linear Equations,** and know how to interpret the results when solving an equation. • Complete You Try 3.
4 Use the Five Steps for Solving Applied Problems	• Summarize the **Five Steps for Solving Applied Problems** in your own words. • Complete the given examples on your own. • Complete You Trys 4 and 5.

W Work

Read the explanations, follow the examples, take notes, and complete the You Trys.

When equations contain fractions or decimals, they may appear difficult to solve. We begin this section by learning how to eliminate the fractions or decimals so that it will be easier to solve such equations.

1 Solve Equations Containing Fractions

To solve $\frac{1}{2}(3b + 8) + \frac{3}{4} = -\frac{1}{2}$ in Section 2.2, we began by using the distributive property to clear the parentheses, and we worked with the fractions throughout the solving process. But, there is another way we can solve equations containing several fractions. Before applying the steps for solving a linear equation, we can eliminate the fractions from the equation.

Procedure Eliminating Fractions from an Equation

To eliminate the fractions, determine the least common denominator for all the fractions in the equation. Then multiply both sides of the equation by the least common denominator (LCD).

Let's solve the equation we solved in Section 2.2 using this new approach.

EXAMPLE 1

In-Class Example 1

Solve
$$\frac{1}{9}m + \frac{1}{2} = \frac{2}{9}(3m - 2).$$

Answer: $\left\{\frac{17}{10}\right\}$

Solve $\frac{1}{2}(3b + 8) + \frac{3}{4} = -\frac{1}{2}$.

Solution

The least common denominator of all the fractions in the equation is 4. Multiply both sides of the equation by 4 to eliminate the fractions.

$$4\left[\frac{1}{2}(3b + 8) + \frac{3}{4}\right] = 4\left(-\frac{1}{2}\right)$$

Step 1: Distribute the 4, clear the parentheses, and combine like terms.

$$4 \cdot \frac{1}{2}(3b + 8) + 4 \cdot \frac{3}{4} = -2 \qquad \text{Distribute.}$$
$$2(3b + 8) + 3 = -2 \qquad \text{Multiply.}$$
$$6b + 16 + 3 = -2 \qquad \text{Distribute.}$$
$$6b + 19 = -2 \qquad \text{Combine like terms.}$$

Step 2: Isolate the variable.

$$6b + 19 - 19 = -2 - 19 \qquad \text{Subtract 19 from each side.}$$
$$6b = -21 \qquad \text{Combine like terms.}$$

Step 3: Solve for b using the division property of equality.

$$\frac{6b}{6} = \frac{-21}{6} \qquad \text{Divide each side by 6.}$$
$$b = -\frac{7}{2} \qquad \text{Simplify.}$$

Step 4: The check is left to the student. The solution set is $\left\{-\frac{7}{2}\right\}$. This is the same as the result we obtained in Section 2.2, Example 3.

[YOU TRY 1] Solve $\frac{1}{6}x + \frac{5}{4} = \frac{1}{2}x - \frac{5}{12}$.

2 Solve Equations Containing Decimals

Just as we can eliminate the fractions from an equation to make it easier to solve, we can eliminate decimals from an equation before applying the four-step equation-solving process.

Procedure Eliminating Decimals from an Equation

To eliminate the decimals from an equation, multiply both sides of the equation by the smallest power of 10 that will eliminate all decimals from the problem.

EXAMPLE 2

In-Class Example 2

Solve
$0.2n - 0.15(n + 4) = 0.25$

Answer: {17}

EXAMPLE 2

Solve $0.05a + 0.2(a + 3) = 0.1$

Solution

We want to eliminate the decimals. The number containing a decimal place farthest to the right is 0.05. The 5 is in the *hundredths* place. Therefore, multiply both sides of the equation by 100 to eliminate all decimals in the equation.

$$100[0.05a + 0.2(a + 3)] = 100(0.1)$$

Step 1: Distribute the 100, clear the parentheses, and combine like terms.

$$
\begin{aligned}
100 \cdot (0.05a) + 100[0.2(a + 3)] &= 10 &&\text{Distribute.} \\
5a + 20(a + 3) &= 10 &&\text{Multiply.} \\
5a + 20a + 60 &= 10 &&\text{Distribute.} \\
25a + 60 &= 10 &&\text{Combine like terms.}
\end{aligned}
$$

Step 2: Isolate the variable.

$$
\begin{aligned}
25a + 60 - 60 &= 10 - 60 &&\text{Subtract 60 from each side.} \\
25a &= -50 &&\text{Combine like terms.}
\end{aligned}
$$

Step 3: Solve for a using the division property of equality.

$$
\begin{aligned}
\frac{25a}{25} &= \frac{-50}{25} &&\text{Divide each side by 25.} \\
a &= -2 &&\text{Simplify.}
\end{aligned}
$$

Step 4: The check is left to the student. The solution set is $\{-2\}$.

[YOU TRY 2]   Solve $0.08k - 0.2(k + 5) = -0.1$

3 Solve Equations with No Solution or an Infinite Number of Solutions

Does every equation have a solution? Consider the next example.

EXAMPLE 3

In-Class Example 3

Solve
$9p + 4 = p + 4(2p - 7)$

Answer: ∅

EXAMPLE 3

Solve $11w - 9 = 5w + 2(3w + 1)$

Solution

$$
\begin{aligned}
11w - 9 &= 5w + 2(3w + 1) \\
11w - 9 &= 5w + 6w + 2 &&\text{Distribute.} \\
11w - 9 &= 11w + 2 &&\text{Combine like terms.} \\
11w - 11w - 9 &= 11w - 11w + 2 &&\text{Subtract } 11w \text{ from each side.} \\
-9 &= 2 &&\text{False}
\end{aligned}
$$

Notice that the variable has "dropped out." Is $-9 = 2$ a true statement? No! This means that there is no value for w that will make the statement true. This means that the equation has *no solution*. We say that the solution set is the **empty set,** or **null set,** and it is denoted by $\varnothing$.

We have seen that a linear equation may have one solution or no solution. There is a third possibility—a linear equation may have an infinite number of solutions.

EXAMPLE 4

EXAMPLE 4 Solve $10r - 3r + 15 = 7r + 15$.

In-Class Example 4

Solve
$3y + 16 = 5y - 2(y - 8)$.

Answer: {all real numbers}

Solution

$$10r - 3r + 15 = 7r + 15$$
$$7r + 15 = 7r + 15 \qquad \text{Combine like terms.}$$
$$7r - 7r + 15 = 7r - 7r + 15 \qquad \text{Subtract } 7r \text{ from each side.}$$
$$15 = 15 \qquad \text{True}$$

W Hint

Check to see if $r = -2$, $r = 0$, and $r = 6$ are solutions to the equation.

Here, the variable has "dropped out," and we are left with an equation, $15 = 15$, that is true. This means that any real number we substitute for r will make the original equation true. Therefore, this equation has an *infinite number of solutions*. The solution set is **{all real numbers}.**

[**YOU TRY 3**] Solve.

a) $9 - 4(3c + 1) = 15 - 12c - 10$ b) $12z + 7 - 10z = 2z + 5$

Summary Outcomes When Solving Linear Equations

There are three possible outcomes when solving a linear equation. The equation may have

1) **one solution.** Solution set: {a real number}. An equation that is true for some values and not for others is called a **conditional equation.**

or

2) **no solution.** In this case, the variable will drop out, and there will be a false statement such as $-9 = 2$. Solution set: $\varnothing$. An equation that has no solution is called a **contradiction.**

or

3) **an infinite number of solutions.** In this case, the variable will drop out, and there will be a true statement such as $15 = 15$. Solution set: {all real numbers}. An equation that has all real numbers as its solution set is called an **identity.**

4   Use the Five Steps for Solving Applied Problems

Mathematical equations can be used to describe many situations in the real world. To do this, we must learn how to translate information presented in English into an algebraic equation. We will begin slowly and work our way up to more challenging problems. Yes, it may be difficult at first, but with patience and persistence, you can do it!

Although no single method will work for solving all applied problems, the following approach is suggested to help in the problem-solving process.

Hint

Rewrite the steps for solving applied problems in your own words.

Procedure Steps for Solving Applied Problems

Step 1: Read the problem carefully, more than once if necessary, until you understand it. Draw a picture, if applicable. Identify what you are being asked to find.

Step 2: Choose a variable to represent an unknown quantity. If there are any other unknowns, define them in terms of the variable.

Step 3: Translate the problem from English into an equation using the chosen variable. Here are some suggestions for doing so:

- Restate the problem in your own words.
- Read and think of the problem in "small parts."
- Make a chart to separate these "small parts" of the problem to help you translate into mathematical terms.
- Write an equation in English, then translate it into an algebraic equation.

Step 4: Solve the equation.

Step 5: Check the answer in the original problem, and **interpret** the solution as it relates to the problem. Be sure your answer makes sense in the context of the problem.

EXAMPLE 5

In-Class Example 5

Write the following statement as an equation, and find the number.

Seven more than twice a number is nineteen. Find the number.

Answer: $7 + 2x = 19$; 6

Write the following statement as an equation, and find the number.

Nine more than twice a number is fifteen. Find the number.

Solution

Step 1: Read the problem carefully. We must find an unknown number.

Step 2: Choose a variable to represent the unknown.

$$\text{Let } x = \text{the number.}$$

Step 3: Translate the information that appears in English into an algebraic equation by rereading the problem slowly and "in parts."

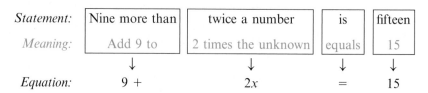

Statement:	Nine more than	twice a number	is	fifteen
Meaning:	Add 9 to	2 times the unknown	equals	15
	↓	↓	↓	↓
Equation:	9 +	2x	=	15

The equation is $9 + 2x = 15$.

Step 4: Solve the equation.

$$9 + 2x = 15$$
$$9 - 9 + 2x = 15 - 9 \qquad \text{Subtract 9 from each side.}$$
$$2x = 6 \qquad \text{Combine like terms.}$$
$$x = 3 \qquad \text{Divide each side by 2.}$$

Step 5: Check the answer. Does the answer make sense? Nine more than twice three is $9 + 2(3) = 15$. The answer is correct. The number is 3.

[YOU TRY 4] Write the following statement as an equation, and find the number. *Three more than twice a number is twenty-nine.*

Sometimes, dealing with subtraction in an application can be confusing. So let's look at an arithmetic problem first.

EXAMPLE 6

In-Class Example 6

What is three less than five?

Answer: 2

What is two less than seven?

Solution

To solve this problem, do we subtract $7 - 2$ or $2 - 7$? "Two less than seven" is written as $7 - 2$, and $7 - 2 = 5$. Five is two less than seven. To get the correct answer, the 2 is *subtracted from* the 7.

Keep this in mind as you read the next problem.

EXAMPLE 7

In-Class Example 7

Write the following statement as an equation, and find the number.
  Seven less than four times a number is the same as that number increased by twenty. Find the number.

Answer: $4x - 7 = x + 20$; 9

Write the following statement as an equation, and find the number.
  Five less than three times a number is the same as the number increased by seven. Find the number.

Solution

Step 1: Read the problem carefully. We must find an unknown number.

Step 2: Choose a variable to represent the unknown.

$$\text{Let } x = \text{the number.}$$

Step 3: Translate the information that appears in English into an algebraic equation by rereading the problem slowly and "in parts."

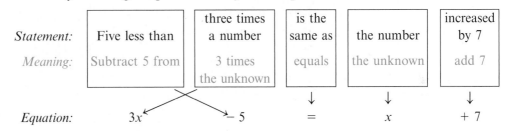

The equation is $3x - 5 = x + 7$.

Step 4: Solve the equation.

$$
\begin{aligned}
3x - 5 &= x + 7 \\
3x - x - 5 &= x - x + 7 \quad &&\text{Subtract } x \text{ from each side.} \\
2x - 5 &= 7 \quad &&\text{Combine like terms.} \\
2x - 5 + 5 &= 7 + 5 \quad &&\text{Add 5 to each side.} \\
2x &= 12 \quad &&\text{Combine like terms.} \\
x &= 6 \quad &&\text{Divide each side by 2.}
\end{aligned}
$$

Step 5: Check the answer. Does it make sense? Five less than three times 6 is $3(6) - 5 = 13$. The number increased by seven is $6 + 7 = 13$. The answer is correct. The number is 6.

[**YOU TRY 5**]

Write the following statement as an equation, and find the number.
  Three less than five times a number is the same as the number increased by thirteen.

We can use a graphing calculator to solve a linear equation in one variable. First, enter the left side of the equation in Y_1 and the right side of the equation in Y_2. Then graph the equations. The x-coordinate of the point of intersection is the solution to the equation.

We will solve $x + 2 = -3x + 7$ algebraically and by using a graphing calculator, and then compare the results. First, use algebra to solve $x + 2 = -3x + 7$.

You should get $\dfrac{5}{4}$.

Next, use a graphing calculator to solve $x + 2 = -3x + 7$.

1) Enter $x + 2$ in Y_1 by pressing $\boxed{Y=}$ and entering $x + 2$ to the right of $\backslash Y_1=$. Then press $\boxed{\text{ENTER}}$.

2) Enter $-3x + 7$ in Y_2 by pressing the $\boxed{Y=}$ key and entering $-3x + 7$ to the right of $\backslash Y_2=$. Press $\boxed{\text{ENTER}}$.

3) Press $\boxed{\text{ZOOM}}$ and select 6:ZStandard to graph the equations.

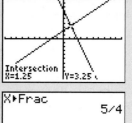

4) To find the intersection point, press $\boxed{\text{2nd}}$ $\boxed{\text{TRACE}}$ and select 5:intersect. Press $\boxed{\text{ENTER}}$ three times. The x-coordinate of the intersection point is shown on the left side of the screen and is stored in the variable x.

5) Return to the home screen by pressing $\boxed{\text{2nd}}$ $\boxed{\text{MODE}}$. Press $\boxed{\text{X,T,}\theta\text{,n}}$ $\boxed{\text{ENTER}}$ to display the solution. Since the result in this case is a decimal value, we can convert it to a fraction by pressing $\boxed{\text{X,T,}\theta\text{,n}}$ $\boxed{\text{MATH}}$, selecting Frac, then pressing $\boxed{\text{ENTER}}$.

The calculator then gives us the solution set $\left\{\dfrac{5}{4}\right\}$.

Solve each equation algebraically, then verify your answer using a graphing calculator.

1) $x + 6 = -2x - 3$

2) $2x + 3 = -x - 4$

3) $\dfrac{5}{6}x + \dfrac{1}{2} = \dfrac{1}{6}x - \dfrac{3}{4}$

4) $0.3x - 1 = -0.2x - 5$

5) $3x - 7 = -x + 5$

6) $6x - 7 = 5$

ANSWERS TO $\boxed{\text{YOU TRY}}$ **EXERCISES**

1) $\{5\}$　　2) $\{-7.5\}$　　3) a) $\{$all real numbers$\}$　b) $\varnothing$　　4) $2x + 3 = 29$; 13
5) $5x - 3 = x + 13$; 4

ANSWERS TO TECHNOLOGY EXERCISES

1) $\{-3\}$　　2) $\left\{-\dfrac{7}{3}\right\}$　　3) $\left\{-\dfrac{15}{8}\right\}$　　4) $\{-8\}$　　5) $\{3\}$　　6) $\{2\}$

---

*Additional answers can be found in the Answers to Exercises appendix.

Mixed Exercises: Objectives 1 and 2

1) If an equation contains fractions, what is the first step you can perform to make it easier to solve?

2) If an equation contains decimals, what is the first step you can perform to make it easier to solve?

3) How can you eliminate the fractions from the equation $\frac{3}{8}x - \frac{1}{2} = \frac{1}{8}x + \frac{3}{4}$?
   Multiply both sides of the equation by 8.

4) How can you eliminate the decimals from the equation $0.02n + 0.1(n - 3) = 0.06$?
   Multiply both sides of the equation by 100.

Solve each equation by first clearing the fractions or decimals.

5) $\frac{3}{8}x - \frac{1}{2} = \frac{1}{8}x + \frac{3}{4}$ $\{5\}$

6) $\frac{1}{2}c + \frac{7}{4} = \frac{5}{4}c - \frac{1}{2}$ $\{3\}$

7) $\frac{4}{7}t + \frac{1}{14} = \frac{3}{14}t + \frac{3}{2}$ $\{4\}$

8) $\frac{1}{2} - \frac{7}{12}k = \frac{1}{6}k + \frac{11}{4}$ $\{-3\}$

9) $\frac{1}{3} - \frac{1}{2}m = \frac{1}{6}m + \frac{7}{9}$ $\left\{-\frac{2}{3}\right\}$

10) $\frac{1}{15}p - \frac{1}{2} = \frac{1}{5}p - \frac{3}{10}$ $\left\{-\frac{3}{2}\right\}$

11) $\frac{1}{3} + \frac{1}{9}(k + 5) - \frac{k}{4} = 2$ $\{-8\}$

12) $\frac{5}{8}(2w + 3) + \frac{5}{4}w = \frac{3}{4}(4w + 1)$ $\left\{\frac{9}{4}\right\}$

13) $\frac{3}{4}(y + 7) + \frac{1}{2}(3y - 5) = \frac{9}{4}(2y - 1)$ $\left\{\frac{20}{9}\right\}$

14) $\frac{2}{3}(5z - 2) - \frac{4}{9}(3z - 2) = \frac{1}{3}(2z + 1)$ $\left\{\frac{7}{12}\right\}$

15) $\frac{1}{2}(4r + 1) - r = \frac{2}{5}(2r - 3) + \frac{3}{2}$ $\{-1\}$

16) $\frac{2}{3}(3h - 5) + 1 = \frac{3}{2}(h - 2) + \frac{1}{6}h$ $\{-2\}$

17) $0.06d + 0.13 = 0.31$ $\{3\}$

18) $0.09x - 0.14 = 0.4$ $\{6\}$

19) $0.04n - 0.05(n + 2) = 0.1$ $\{-20\}$

20) $0.07t + 0.02(3t + 8) = -0.1$ $\{-2\}$

21) $0.2(c - 4) + 1 = 0.15(c + 2)$ $\{2\}$

22) $0.12(5q - 1) - q = 0.15(7 - 2q)$ $\{-11.7\}$

23) $0.35a - a = 0.03(5a + 4)$ $\{-0.15\}$

24) $0.3(x - 2) + 1 = 0.25(x + 9)$ $\{37\}$

25) $0.06w + 0.1(20 - w) = 0.08(20)$ $\{10\}$

26) $0.17m + 0.05(16 - m) = 0.11(16)$ $\{8\}$

27) $0.07k + 0.15(200) = 0.09(k + 200)$ $\{600\}$

28) $0.2p + 0.08(120) = 0.16(p + 120)$ $\{240\}$

Objective 3: Solve Equations with No Solution or an Infinite Number of Solutions

For Exercises 29 and 30, choose from *always, sometimes,* or *never*.

29) An equation will *always, sometimes,* or *never* have one solution. sometimes

30) When the variable is eliminated from an equation, the solution set is *always, sometimes,* or *never* $\varnothing$.
   sometimes

31) How do you know that an equation has no solution?
   The variable is eliminated, and you get a false statement like $5 = -12$.

32) How do you know that the solution set of an equation is {all real numbers}?
   The variable is eliminated, and you get a true statement like $8 = 8$.

Determine whether each of the following equations has a solution set of {all real numbers} or has no solution, $\varnothing$.

33) $9(c + 6) - 2c = 4c + 1 + 3c$ $\varnothing$

34) $-21n + 22 = 3(4 - 7n) + 10$ {all real numbers}

35) $5t + 2(t + 3) - 4t = 4(t + 1) - (t - 2)$ {all real numbers}

36) $8z + 11 + 5z - 9 = 16 - 6(3 - 2z) + z$ $\varnothing$

37) $\frac{5}{6}k - \frac{2}{3} = \frac{1}{6}(5k - 4) + \frac{1}{2}$ $\varnothing$

38) $0.4y + 0.3(20 - y) = 0.1y + 6$ {all real numbers}

Mixed Exercises: Sections 2.1–2.3

The following set of exercises contains equations from Sections 2.1–2.3. Solve each equation.

39) $\frac{n}{5} = 20$ $\{100\}$ 40) $z + 18 = -5$ $\{-23\}$

41) $-19 = 6 - p$ $\{25\}$ 42) $-a = 34$ $\{-34\}$

43) $-5.4 = -0.9m$ $\{6\}$ 44) $\frac{15}{7}h = 25$ $\left\{\frac{35}{3}\right\}$

45) $51 = 4y - 13$ {16}

46) $3c + 8 = 5c + 11$ $\left\{-\dfrac{3}{2}\right\}$

47) $9 - (7k - 2) + 2k = 4(k + 3) + 5$ $\left\{\dfrac{2}{3}\right\}$

48) $0.3t + 0.18(5000 - t) = 0.21(5000)$ {1250}

49) $-\dfrac{5}{4}r + 17 = 7$ {8}

50) $-2.3 = 2.4z + 1.3$ {−1.5}

51) $8(3t + 4) = 10t - 3 + 7(2t + 5)$ {all real numbers}

52) $-6 - (a + 9) + 7 = 3a + 2(4a - 1)$ $\left\{-\dfrac{1}{2}\right\}$

53) $\dfrac{5}{3}w + \dfrac{2}{5} = w - \dfrac{7}{3}$ $\left\{-\dfrac{41}{10}\right\}$

54) $2d + 7 = -4d + 3(2d - 5)$ $\varnothing$

55) $7(2q + 3) - 3(q + 5) = 6$ {0}

56) $-11 = \dfrac{4}{5}k - 17$ $\left\{\dfrac{15}{2}\right\}$

57) $0.16h + 0.4(2000) = 0.22(2000 + h)$ {6000}

58) $\dfrac{4}{9} + \dfrac{2}{3}(c - 1) + \dfrac{5}{9}c = \dfrac{2}{9}(5c + 3)$ {8}

59) $-9r + 4r - 11 + 2 = 3r + 7 - 8r + 9$ $\varnothing$

60) $2u - 4.6 = -4.6$ {0}

61) $\dfrac{1}{2}(2r + 9) - \dfrac{1}{3}(r + 12) = 1$ $\left\{\dfrac{3}{4}\right\}$

62) $t + 18 = 3(5 - t) + 4t + 3$ {all real numbers}

Objective 4: Use the Five Steps for Solving Applied Problems

63) What are the five steps for solving applied problems?

64) If you are solving an applied problem in which you have to find the length of a side of a rectangle, would a solution of −18 be reasonable? Explain your answer.
No. The length of a side of a rectangle cannot be negative.

Write each statement as an equation, and find the number.

65) Twelve more than a number is five. $x + 12 = 5$; −7

66) Fifteen more than a number is nineteen. $x + 15 = 19$; 4

67) Nine less than a number is twelve. $x - 9 = 12$; 21

68) Fourteen less than a number is three. $x - 14 = 3$; 17

69) Five more than twice a number is seventeen.
$2x + 5 = 17$; 6

70) Seven more than twice a number is twenty-three.
$2x + 7 = 23$; 8

71) Eleven more than twice a number is thirteen.
$2x + 11 = 13$; 1

72) Eighteen more than twice a number is eight.
$2x + 18 = 8$; −5

73) Three times a number decreased by eight is forty.
$3x - 8 = 40$; 16

74) Five less than four times a number is forty-three.
$4x - 5 = 43$; 12

75) Three-fourths of a number is thirty-three. $\dfrac{3}{4}x = 33$; 44

76) Two-thirds of a number is twenty-six. $\dfrac{2}{3}x = 26$; 39

77) Nine less than half a number is three. $\dfrac{1}{2}x - 9 = 3$; 24

78) Two less than one-fourth of a number is three. $\dfrac{1}{4}x - 2 = 3$; 20

79) Three less than twice a number is the same as the number increased by eight. $2x - 3 = x + 8$; 11

80) Twelve less than five times a number is the same as the number increased by sixteen. $5x - 12 = x + 16$; 7

81) Ten more than one-third of a number is the same as the number decreased by two. $\dfrac{1}{3}x + 10 = x - 2$; 18

82) A number decreased by nine is the same as seven more than half the number. $x - 9 = \dfrac{1}{2}x + 7$; 32

83) If twenty-four is subtracted from a number, the result is the same as the number divided by nine. $x - 24 = \dfrac{x}{9}$; 27

84) If forty-five is subtracted from a number, the result is the same as the number divided by four. $x - 45 = \dfrac{x}{4}$; 60

85) If two-thirds of a number is added to the number, the result is twenty-five. $x + \dfrac{2}{3}x = 25$; 15

86) If three-eighths of a number is added to twice the number, the result is thirty-eight. $2x + \dfrac{3}{8}x = 38$; 16

87) When a number is decreased by twice the number, the result is thirteen. $x - 2x = 13$; −13

88) When three times a number is subtracted from the number, the result is ten. $x - 3x = 10$; −5

R Rethink

R1) How do you know what number to use to clear the decimals or fractions from an equation?

R2) Do you find it more difficult to solve equations containing decimals or fractions? Why do you think this is true?

R3) Are there any problems you could not do? If so, write them on your paper or circle them, and ask your instructor for help.

2.4 Applications of Linear Equations

P Prepare

O Organize

What are your objectives for Section 2.4?	How can you accomplish each objective?
1 Solve Problems Involving General Quantities	• Use the **Five Steps for Solving Applied Problems** to complete the given examples on your own. • Complete You Trys 1 and 2.
2 Solve Problems Involving Lengths	• Use the **Five Steps for Solving Applied Problems** to complete the given example on your own. • Complete You Try 3.
3 Solve Consecutive Integer Problems	• Understand how to represent *consecutive integers* and *consecutive odd/even integers* using a variable and expressions. • Use the **Five Steps for Solving Applied Problems** to complete the given examples on your own. • Complete You Trys 4 and 5.

 **Work**

Read the explanations, follow the examples, take notes, and complete the You Trys.

In the previous section, we learned the Five Steps for Solving Applied Problems and used this procedure to solve problems involving unknown numbers. Now we will apply this problem-solving technique to other types of applications.

1 Solve Problems Involving General Quantities

EXAMPLE 1

In-Class Example 1

Write an equation, and solve. At Roosevelt High School, there are six more students on the newspaper staff than on the boys' varsity tennis team and a total of 42 students in these two clubs. How many students work on the newspaper, and how many are on the boys' tennis team if none of the students are involved in both organizations?

Answer: tennis team: 18; newspaper: 24

Write an equation, and solve.

Swimmers Michael Phelps and Natalie Coughlin both competed in the 2004 Olympics in Athens and in the 2008 Olympics in Beijing, where they won a total of 27 medals. Phelps won five more medals than Coughlin. How many Olympic medals did each athlete win? (http://swimming.teamusa.org)

Solution

Step 1: **Read** the problem carefully, and identify what we are being asked to find.

We must find the number of medals each Olympian won.

Step 2: **Choose a variable** to represent an unknown, and define the other unknown in terms of this variable.

In the statement "Phelps won five more medals than Coughlin," the number of medals that Michael Phelps won is expressed *in terms of* the number of medals won by Natalie Coughlin. Therefore, let

x = the number of medals Coughlin won

Define the other unknown (the number of medals that Michael Phelps won) in terms of x. The statement "Phelps won five more medals than Coughlin" means

number of Coughlin's medals + 5 = number of Phelps' medals

$x + 5$ = number of Phelps' medals

Step 3: **Translate** the information that appears in English into an algebraic equation. One approach is to restate the problem in your own words.

Since these two athletes won a total of 27 medals, we can think of the situation in this problem as:

The number of medals Coughlin won plus the number of medals Phelps won was 27.

Let's write this as an equation.

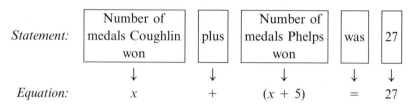

The equation is $x + (x + 5) = 27$.

W Hint

Notice that we use only one variable and then use an expression $(x + 5)$ to represent the other quantity.

Step 4: **Solve** the equation.

$$x + (x + 5) = 27$$
$$2x + 5 = 27$$
$$2x + 5 - 5 = 27 - 5 \qquad \text{Subtract 5 from each side.}$$
$$2x = 22 \qquad \text{Combine like terms.}$$
$$\frac{2x}{2} = \frac{22}{2} \qquad \text{Divide each side by 2.}$$
$$x = 11 \qquad \text{Simplify.}$$

Step 5: **Check** the answer, and **interpret** the solution as it relates to the problem.

Since x represents the number of medals that Natalie Coughlin won, she won 11 medals.

The expression $x + 5$ represents the number of medals Michael Phelps won, so he won $x + 5 = 11 + 5 = 16$ medals.

The answer makes sense because the total number of medals they won was $11 + 16 = 27$.

[YOU TRY 1] Write an equation, and solve.

An employee at a cellular phone store is doing inventory. The store has 23 more conventional cell phones in stock than smart phones. If the store has a total of 73 phones, how many of each type of phone is in stock?

EXAMPLE 2

In-Class Example 2

Write an equation, and solve. A tee-ball team for 5-year-olds has half as many girls as boys. If there are 15 children on the team, how many are boys and how many are girls?

Answer: 10 boys, 5 girls

Write an equation, and solve.

Nick has half as many songs on his iPod as Mariah. Together they have a total of 4887 songs. How many songs does each of them have?

Solution

Step 1: **Read** the problem carefully, and identify what we are being asked to find.

We must find the number of songs on Nick's iPod and the number on Mariah's iPod.

Step 2: **Choose a variable** to represent an unknown, and define the other unknown in terms of this variable.

www.mhhe.com/messersmith

In the sentence "Nick has half as many songs on his iPod as Mariah," the number of songs Nick has is expressed *in terms of* the number of songs Mariah has. Therefore, let

$$x = \text{the number of songs on Mariah's iPod}$$

Define the other unknown in terms of x.

$$\frac{1}{2}x = \text{the number of songs on Nick's iPod}$$

Step 3: **Translate** the information that appears in English into an algebraic equation. One approach is to restate the problem in your own words.

Since Mariah and Nick have a total of 4887 songs, we can think of the situation in this problem as:

The number of Mariah's songs *plus* the number of Nick's songs *equals* 4887.

Let's write this as an equation.

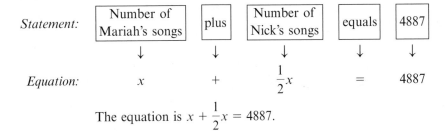

Statement:	Number of Mariah's songs	plus	Number of Nick's songs	equals	4887
	↓	↓	↓	↓	↓
Equation:	x	$+$	$\frac{1}{2}x$	$=$	4887

The equation is $x + \frac{1}{2}x = 4887$.

Step 4: **Solve** the equation.

$$x + \frac{1}{2}x = 4887$$

$$\frac{3}{2}x = 4887 \qquad \text{Combine like terms.}$$

$$\frac{2}{3} \cdot \frac{3}{2}x = \frac{2}{3} \cdot 4887 \qquad \text{Multiply by the reciprocal of } \frac{3}{2}.$$

$$x = 3258 \qquad \text{Multiply.}$$

Step 5: **Check** the answer, and **interpret** the solution as it relates to the problem.

Mariah has 3258 songs on her iPod.

The expression $\frac{1}{2}x$ represents the number of songs on Nick's iPod, so there are $\frac{1}{2}(3258) = 1629$ songs on Nick's iPod.

The answer makes sense because the total number of songs on their iPods is $3258 + 1629 = 4887$ songs.

[YOU TRY 2] Write an equation, and solve.

Terrance and Janay are in college. Terrance has earned twice as many credits as Janay. How many credits does each student have if together they have earned 51 semester hours?

---

2 Solve Problems Involving Lengths

EXAMPLE 3

In-Class Example 3

Write an equation, and solve. A carpenter has a board that is 15 ft long. He needs to cut it into two pieces so that one piece is 3 ft shorter than the other. Find the length of each piece.

Answer: 9 ft and 6 ft

Write an equation, and solve.

A plumber has a section of PVC pipe that is 12 ft long. He needs to cut it into two pieces so that one piece is 2 ft shorter than the other. How long will each piece be?

Solution

Step 1: Read the problem carefully, and identify what we are being asked to find.

We must find the length of each of two pieces of pipe.

A picture will be very helpful in this problem.

Step 2: Choose a variable to represent an unknown, and define the other unknown in terms of this variable.

One piece of pipe must be 2 ft shorter than the other piece. Therefore, let

$$x = \text{the length of one piece}$$

Define the other unknown in terms of x.

$$x - 2 = \text{the length of the second piece}$$

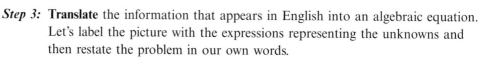

Step 3: Translate the information that appears in English into an algebraic equation. Let's label the picture with the expressions representing the unknowns and then restate the problem in our own words.

From the picture we can see that the

length of one piece plus the length of the second piece equals 12 ft

Let's write this as an equation.

Statement:	Length of one piece	plus	Length of second piece	equals	12 ft
	↓	↓	↓	↓	↓
Equation:	x	$+$	$(x - 2)$	$=$	12

The equation is $x + (x - 2) = 12$.

W Hint

How does this type of problem differ from the problem in the previous objective?

Step 4: Solve the equation.

$$x + (x - 2) = 12$$
$$2x - 2 = 12$$
$$2x - 2 + 2 = 12 + 2 \qquad \text{Add 2 to each side.}$$
$$2x = 14 \qquad \text{Combine like terms.}$$
$$\frac{2x}{2} = \frac{14}{2} \qquad \text{Divide each side by 2.}$$
$$x = 7 \qquad \text{Simplify.}$$

Step 5: Check the answer, and **interpret** the solution as it relates to the problem.

One piece of pipe is 7 ft long.

The expression $x - 2$ represents the length of the other piece of pipe, so the length of the other piece is $x - 2 = 7 - 2 = 5$ ft.

The answer makes sense because the length of the original pipe was 7 ft + 5 ft = 12 ft.

www.mhhe.com/messersmith

Write an equation, and solve.

An electrician has a 20-ft wire. He needs to cut the wire so that one piece is 4 ft shorter than the other. What will be the length of each piece?

3 Solve Consecutive Integer Problems

Consecutive means one after the other, in order. In this section, we will look at consecutive integers, consecutive even integers, and consecutive odd integers.

Consecutive integers differ by 1. Look at the consecutive integers 5, 6, 7, and 8. If $x = 5$, then $x + 1 = 6$, $x + 2 = 7$, and $x + 3 = 8$. Therefore, to define the unknowns for consecutive integers, let

$$x = \text{first integer}$$
$$x + 1 = \text{second integer}$$
$$x + 2 = \text{third integer}$$
$$x + 3 = \text{fourth integer}$$

and so on.

EXAMPLE 4

The sum of three consecutive integers is 87. Find the integers.

In-Class Example 4

The sum of three consecutive integers is 57. Find the integers.

Answer: 18, 19, 20

Solution

Step 1: **Read** the problem carefully, and identify what we are being asked to find.

We must find three consecutive integers with a sum of 87.

Step 2: **Choose a variable** to represent an unknown, and define the other unknowns in terms of this variable.

There are three unknowns. We will let x represent the first consecutive integer and then define the other unknowns in terms of x.

$$x = \text{the first integer}$$

Define the other unknowns in terms of x.

$$x + 1 = \text{the second integer} \qquad x + 2 = \text{the third integer}$$

Step 3: **Translate** the information that appears in English into an algebraic equation. What does the original statement mean?

"The sum of three consecutive integers is 87" means that when the three numbers are *added,* the sum is 87.

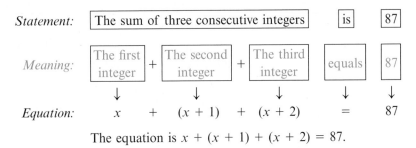

The equation is $x + (x + 1) + (x + 2) = 87$.

Step 4: **Solve** the equation.

$$x + (x + 1) + (x + 2) = 87$$
$$3x + 3 = 87$$
$$3x + 3 - 3 = 87 - 3 \qquad \text{Subtract 3 from each side.}$$
$$3x = 84 \qquad \text{Combine like terms.}$$
$$\frac{3x}{3} = \frac{84}{3} \qquad \text{Divide each side by 3.}$$
$$x = 28 \qquad \text{Simplify.}$$

Step 5: **Check** the answer, and **interpret** the solution as it relates to the problem.

The first integer is 28. The second integer is 29 since $x + 1 = 28 + 1 = 29$, and the third integer is 30 since $x + 2 = 28 + 2 = 30$.

The answer makes sense because their sum is $28 + 29 + 30 = 87$.

[YOU TRY 4] The sum of three consecutive integers is 162. Find the integers.

Next, let's look at **consecutive even integers,** which are even numbers that differ by 2, such as -10, -8, -6, and -4. If x is the first even integer, we have

$$
\begin{array}{cccc}
-10 & -8 & -6 & -4 \\
x & x + 2 & x + 4 & x + 6
\end{array}
$$

Therefore, to define the unknowns for consecutive even integers, let

$$
\begin{aligned}
x &= \text{the first even integer} \\
x + 2 &= \text{the second even integer} \\
x + 4 &= \text{the third even integer} \\
x + 6 &= \text{the fourth even integer}
\end{aligned}
$$

W Hint

Both consecutive even and consecutive odd integers are two units apart.

and so on.

We also use the same reasoning to define the unknowns for **consecutive odd integers** such as 9, 11, 13, and 15.

EXAMPLE 5

In-Class Example 5

The sum of two consecutive odd integers is 11 more than three times the larger number. Find the integers.

Answer: -15, -13

The sum of two consecutive odd integers is 19 more than five times the larger integer. Find the integers.

Solution

Step 1: **Read** the problem carefully, and identify what we are being asked to find.

We must find two consecutive odd integers.

Step 2: **Choose a variable** to represent an unknown, and define the other unknown in terms of this variable.

There are two unknowns. We will let x represent the first consecutive odd integer and then define the other unknown in terms of x.

$$
\begin{aligned}
x &= \text{the first odd integer} \\
x + 2 &= \text{the second odd integer}
\end{aligned}
$$

Step 3: **Translate** the information that appears in English into an algebraic equation. Read the problem slowly and carefully, breaking it into small parts.

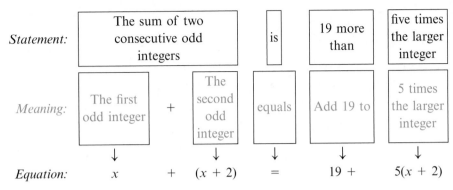

Statement:	The sum of two consecutive odd integers		is	19 more than	five times the larger integer
Meaning:	The first odd integer	+ The second odd integer	equals	Add 19 to	5 times the larger integer
	↓	↓	↓	↓	↓
Equation:	x	+ $(x + 2)$	=	19 +	$5(x + 2)$

The equation is $x + (x + 2) = 19 + 5(x + 2)$.

Step 4: **Solve** the equation.

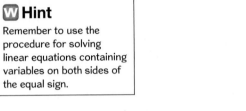

Hint

Remember to use the procedure for solving linear equations containing variables on both sides of the equal sign.

$$x + (x + 2) = 19 + 5(x + 2)$$

$2x + 2 = 19 + 5x + 10$	Combine like terms; distribute.
$2x + 2 = 5x + 29$	Combine like terms.
$2x + 2 - 2 = 5x + 29 - 2$	Subtract 2 from each side.
$2x = 5x + 27$	Combine like terms.
$2x - 5x = 5x - 5x + 27$	Subtract $5x$ from each side.
$-3x = 27$	Combine like terms.
$\dfrac{-3x}{-3} = \dfrac{27}{-3}$	Divide each side by -3.
$x = -9$	Simplify.

Step 5: **Check** the answer, and **interpret** the solution as it relates to the problem.

The first odd integer is -9. The second integer is -7 since $x + 2 = -9 + 2 = -7$.

Check these numbers in the original statement of the problem. The sum of -9 and -7 is -16. Then, 19 more than five times the larger integer is $19 + 5(-7) = 19 + (-35) = -16$. The numbers are -9 and -7.

[YOU TRY 5] The sum of two consecutive even integers is 16 less than three times the larger number. Find the integers.

ANSWERS TO [YOU TRY] **EXERCISES**

1) smart phones: 25; conventional phones: 48   2) Janay: 17 hours; Terrance: 34 hours
3) 8 ft and 12 ft   4) 53, 54, 55   5) 12 and 14

*Additional answers can be found in the Answers to Exercises appendix.

Objective 1: Solve Problems Involving General Quantities

1) During the month of June, a car dealership sold 14 more compact cars than SUVs. Write an expression for the number of compact cars sold if c SUVs were sold. *c + 14*

2) During a Little League game, the Tigers scored 3 more runs than the Eagles. Write an expression for the number of runs the Tigers scored if the Eagles scored r runs. *r + 3*

3) A restaurant had 37 fewer customers on a Wednesday night than on a Thursday night. If there were c customers on Thursday, write an expression for the number of customers on Wednesday. *c − 37*

4) After a storm rolled through Omaha, the temperature dropped 15 degrees. If the temperature before the storm was t degrees, write an expression for the temperature after the storm. *t − 15*

5) Due to the increased use of e-mail to send documents, the shipping expenses of a small business in 2012 were half of what they were in 2000. Write an expression for the cost of shipping in 2012 if the cost in 2000 was s dollars. $\frac{1}{2}s$

6) A coffee shop serves three times as many cups of regular coffee as decaffeinated coffee. If the shop serves d cups of decaffeinated coffee, write an expression for the number of cups of regular coffee it sells. *3d*

7) An electrician cuts a 14-foot wire into two pieces. If one is x feet long, how long is the other piece? *14 − x*

14 ft

x

8) Ralph worked for a total of 8.5 hours one day, some at his office and some at home. If he worked h hours in his office, write an expression for the number of hours he worked at home. *8.5 − h*

9) If you are asked to find the number of children in a class, why would 26.5 not be a reasonable answer? The number of children must be a whole number.

10) If you are asked to find the length of a piece of wire, why would −7 not be a reasonable answer? The length of a wire cannot be a negative number.

11) If you are asked to find consecutive odd integers, why would −10 not be a reasonable answer? It is an even number.

12) If you are asked to find the number of workers at an ice cream shop, why would $5\frac{1}{4}$ not be a reasonable answer? There can't be a fractional number of people.

Solve using the five-step method. See Examples 1 and 2.

13) The wettest April (greatest rainfall amount) for Albuquerque, NM, was recorded in 1905. The amount was 1.2 inches more than the amount recorded for the second-wettest April, in 2004. If the total rainfall for these two months was 7.2 inches, how much rain fell in April of each year? (www.srh.noaa.gov) 1905: 4.2 in.; 2004: 3.0 in.

14) Bo-Lin applied to three more colleges than his sister Liling. Together they applied to 13 schools. To how many colleges did each apply? Bo-Lin: 8; Liling: 5

15) Miguel Indurain of Spain won the Tour de France two fewer times than Lance Armstrong. They won a total of 12 titles. How many times did each cyclist win this race? (www.letour.fr) Lance: 7; Miguel: 5

16) In 2011, an Apple MacBook Air weighed about 13 lb less than the Apple Macintosh Portable did in 1989. Find the weight of each computer if they weighed a total of 19 lb. (http://oldcomputers.net; www.apple.com) MacBook Air: 3 lb; Macintosh Portable: 16 lb

17) A 12-oz cup of regular coffee at Starbucks has 13 times the amount of caffeine found in the same-sized serving of decaffeinated coffee. Together they contain 280 mg of caffeine. How much caffeine is in each type of coffee? (www.starbucks.com) regular: 260 mg; decaf: 20 mg

18) A farmer plants soybeans and corn on his 540 acres of land. He plants twice as many acres with soybeans as with corn. How many acres are planted with each crop? corn: 180 acres; soybeans: 360 acres

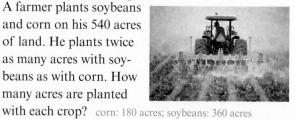

19) In the sophomore class at Dixon High School, the number of students taking French is two-thirds of the number taking Spanish. How many students are studying each language if the total number of students in French and Spanish is 310? Spanish: 186; French: 124

20) A serving of salsa contains one-sixth of the number of calories of the same-sized serving of guacamole. Find the number of calories in each snack if they contain a total of 175 calories. salsa: 25 calories; guacamole: 150 calories

Objective 2: Solve Problems Involving Lengths

Solve using the five-step method. See Example 3.

⓴ 21) A plumber has a 36-in. pipe. He must cut it into two pieces so that one piece is 14 inches longer than the other. How long is each piece?   11 in., 25 in.

22) A 40-in. board is to be cut into two pieces so that one piece is 8 inches shorter than the other. Find the length of each piece.   16 in., 24 in.

23) Trisha has a 28.5-inch piece of wire to make a necklace and a bracelet. She has to cut the wire so that the piece for the necklace will be twice as long as the piece for the bracelet. Find the length of each piece.   bracelet: 9.5 in.; necklace: 19 in.

24) Ethan has a 20-ft piece of rope that he will cut into two pieces. One piece will be one-fourth the length of the other piece. Find the length of each piece of rope.   16 ft, 4 ft

▶ 25) Derek orders a 6-ft sub sandwich for himself and two friends. Cory wants his piece to be 2 feet longer than Tamara's piece, and Tamara wants half as much as Derek. Find the length of each person's sub.   Derek: 2 ft; Cory: 3 ft; Tamara: 1 ft

26) A 24-ft pipe must be cut into three pieces. The longest piece will be twice as long as the shortest piece, and the medium-sized piece will be 4 feet longer than the shortest piece. Find the length of each piece of pipe.   5 ft, 9 ft, 10 ft

Objective 3: Solve Consecutive Integer Problems

Solve using the five-step method. See Examples 4 and 5.

▶⓴ 27) The sum of three consecutive integers is 126. Find the integers.   41, 42, 43

28) The sum of two consecutive integers is 171. Find the integers.   85, 86

29) Find two consecutive even integers such that twice the smaller is 16 more than the larger.   18, 20

30) Find two consecutive odd integers such that the smaller one is 12 more than one-third the larger.   19, 21

31) Find three consecutive odd integers such that their sum is five more than four times the largest integer.   −15, −13, −11

32) Find three consecutive even integers such that their sum is 12 less than twice the smallest.   −18, −16, −14

33) Two consecutive page numbers in a book add up to 215. Find the page numbers.   107, 108

34) The addresses on the west side of Hampton Street are consecutive even numbers. Two consecutive house numbers add up to 7446. Find the addresses of these two houses.   3722 Hampton St., 3724 Hampton St.

Mixed Exercises: Objectives 1–3

Solve using the five-step method.

▶ 35) In a fishing derby, Jimmy caught six more trout than his sister Kelly. How many fish did each person catch if they caught a total of 20 fish?
Jimmy: 13; Kelly: 7

36) Five times the sum of two consecutive integers is two more than three times the larger integer. Find the integers.   0, 1

37) A 16-ft steel beam is to be cut into two pieces so that one piece is 1 foot longer than twice the other. Find the length of each piece.   5 ft, 11 ft

38) A plumber has a 9-ft piece of copper pipe that has to be cut into three pieces. The longest piece will be 4 ft longer than the shortest piece. The medium-sized piece will be three times the length of the shortest. Find the length of each piece of pipe.   1 ft, 3 ft, 5 ft

39) The attendance at the 2011 Lollapalooza Festival was about 30,000 more than three times the attendance at Bonnaroo that year. The total number of people attending those festivals was about 350,000. How many people went to each event?
(www.chicagotribune.com, www.mtv.com)
Bonnaroo: 80,000; Lollapalooza: 270,000

40) A cookie recipe uses twice as much flour as sugar. If the total amount of these ingredients is $2\frac{1}{4}$ cups, how much sugar and how much flour are in these cookies?   sugar: $\frac{3}{4}$ cup; flour: $1\frac{1}{2}$ cups

41) The sum of three consecutive page numbers in a book is 174. What are the page numbers?   57, 58, 59

42) At a ribbon-cutting ceremony, the mayor cuts a 12-ft ribbon into two pieces so that the length of one piece is 2 ft shorter than the other. Find the length of each piece.   7 ft, 5 ft

43) Charlie has 218 fewer Facebook friends than Ileana. The number of Deepa's Facebook friends is 28 more than half of Ileana's. Together, Charlie, Deepa, and Ileana have 2175 friends on Facebook. How many Facebook friends does each of them have?   Ileana: 946, Charlie: 728, Deepa: 501

44) Find three consecutive odd integers such that three times the middle number is 23 more than the sum of the other two.   21, 23, 25

45) A builder is installing hardwood floors. He has to cut a 72-in piece into three separate pieces so that the smallest piece is one-third the length of the longest piece, and the third piece is 12 inches shorter than the longest. How long is each piece?   12 in., 24 in., 36 in.

46) In 2008, there were 1368 more cases of tuberculosis in the United States than in 2009. In 2010, the number of cases was 355 less than in 2009. If the total number of TB cases in those three years was 35,624, how many people had TB each year? (www.cdc.gov)   2008: 12,905, 2009: 11,537, 2010: 11,182

47) Three of the best-selling albums of 2010 were Eminem's *Recovery*, Justin Bieber's *My World 2.0*, and Lady Gaga's *The Fame*. Justin Bieber's album sold 0.7 million more copies than Lady Gaga's. The number of Eminem albums sold was 0.2 million more than twice the number of Lady Gaga's sales. The three artists sold a total of about 7.3 million albums. How many albums did each artist sell? (www.billboard.com)   Eminem: 3.4 million, Justin Bieber: 2.3 million, Lady Gaga: 1.6 million

48) Workers cutting down a large tree have a rope that is 33 ft long. They need to cut it into two pieces so that one piece is half the length of the other piece. How long is each piece of rope?   11 ft, 22 ft

49) One-sixth of the smallest of three consecutive even integers is three less than one-tenth the sum of the other even integers. Find the integers.   72, 74, 76

50) Caedon's mom is a math teacher, and when he asks her on which pages he can find the magazine article on LeBron James, she says, "The article is on three consecutive pages so that 62 less than four times the last page number is the same as the sum of all the page numbers." On what page does the LeBron James article begin?   57

R Rethink

R1) When solving an application problem, how do you determine which unknown to let the variable *x* represent?

R1) Where have you recently encountered a problem similar to those you just solved? Write an application problem and have a classmate solve it.

2.5 Geometry Applications and Solving Formulas

What are your objectives for Section 2.5?	How can you accomplish each objective?
1 Substitute Values into a Formula, and Find the Unknown Variable	• Understand and be able to identify a *formula*. • Complete the given example on your own, and create a procedure. • Complete You Try 1.
2 Solve Problems Using Formulas from Geometry	• Review the geometry formulas in Section 1.3. • Use the **Five Steps for Solving Applied Problems** to complete the given examples on your own. • Complete You Trys 2 and 3.
3 Solve Problems Involving Angle Measures	• Review the special relationships between angles in Section 1.3. • Use the **Five Steps for Solving Applied Problems** to complete the given examples on your own. • Complete You Trys 4–6.
4 Solve a Formula for a Specific Variable	• Understand the similarities between solving for a variable in an equation that contains numbers and one that contains only other variables. • Complete the given example on your own. • Complete You Trys 7 and 8.

W Work

Read the explanations, follow the examples, take notes, and complete the You Trys.

A **formula** is a rule containing variables and mathematical symbols to state relationships between certain quantities.

Some examples of formulas we have used already are

$$P = 2l + 2w \qquad A = \frac{1}{2}bh \qquad C = 2\pi r$$

In this section we will solve problems using *formulas*, and then we will learn how to solve a formula for a specific variable.

1 Substitute Values into a Formula, and Find the Unknown Variable

EXAMPLE 1

In-Class Example 1

The volume of a pyramid is $V = \frac{1}{3}Ah$. If $V = 60$ when $h = 9$, find A.

Answer: 20

The formula for the area of a triangle is $A = \frac{1}{2}bh$. If $A = 30$ when $b = 8$, find h.

Solution

The only unknown variable is h since we are given the values of A and b. Substitute $A = 30$ and $b = 8$ into the formula, and solve for h.

$$A = \frac{1}{2}bh$$

$$30 = \frac{1}{2}(8)h \qquad \text{Substitute the given values.}$$

Since h is the only remaining variable in the equation, we can solve for it.

$$30 = 4h \qquad \text{Multiply.}$$
$$\frac{30}{4} = \frac{4h}{4} \qquad \text{Divide by 4.}$$
$$\frac{15}{2} = h \qquad \text{Simplify.}$$

[**YOU TRY 1**] The area of a trapezoid is $A = \frac{1}{2}h(b_1 + b_2)$. If $A = 21$ when $b_1 = 10$ and $b_2 = 4$, find h.

2 Solve Problems Using Formulas from Geometry

Next we will solve applied problems using concepts and formulas from geometry. Unlike in Example 1, you will not be given a formula. You will need to know the geometry formulas that we reviewed in Section 1.3.

EXAMPLE 2

In-Class Example 2

The area of a rectangular garden is 105 ft². Find the length if the width is 7 ft.

Answer: 15 ft

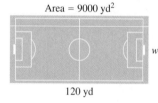
Area = 9000 yd²

120 yd

W Hint

On a sheet of paper, make a list of all the geometric formulas that are used in this section.

A soccer field is in the shape of a rectangle and has an area of 9000 yd². Its length is 120 yd. What is the width of the field?

Solution

Step 1: **Read** the problem carefully, and identify what we are being asked to find.

We must find the length of the soccer field.

A picture will be very helpful in this problem.

Step 2: **Choose a variable** to represent the unknown.

$w =$ the width of the soccer field

Label the picture with the length, 120 yd, and the width, w.

Step 3: **Translate** the information that appears in English into an algebraic equation. We will use a known geometry formula. How do we know which formula to use? List the information we are given and what we want to find:

The field is in the shape of a rectangle; its area = 9000 yd² and its length = 120 yd. We must find the width. Which formula involves the area, length, and width of a rectangle?

$$A = lw$$

Substitute the known values into the formula for the area of a rectangle, and solve for w.

$$A = lw$$
$$9000 = 120w \qquad \text{Substitute the known values.}$$

Step 4: **Solve** the equation.

$$9000 = 120w$$
$$\frac{9000}{120} = \frac{120w}{120} \qquad \text{Divide by 120.}$$
$$75 = w \qquad \text{Simplify.}$$

Step 5: **Check** the answer and **interpret** the solution as it relates to the problem.

If $w = 75$ yd, then $l \cdot w = 120$ yd $\cdot$ 75 yd $= 9000$ yd^2. Therefore, the width of the soccer field is 75 yd.

Note

Remember to include the correct units in your answer!

[**YOU TRY 2**] The area of a rectangular room is 270 ft^2. Find the length of the room if the width is 15 ft.

EXAMPLE 3

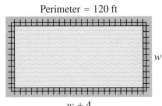

Perimeter = 120 ft

$w + 4$

Stewart wants to put a rectangular safety fence around his backyard pool. He calculates that he will need 120 ft of fencing and that the length will be 4 ft longer than the width. Find the dimensions of the safety fence.

Solution

Step 1: **Read** the problem carefully, and identify what we are being asked to find.

We must find the length and width of the safety fence.

Draw a picture.

Step 2: **Choose a variable** to represent an unknown, and define the other unknown in terms of this variable.

The length is 4 ft longer than the width. Therefore, let

$w =$ the width of the safety fence

Define the other unknown in terms of w.

$w + 4 =$ the length of the safety fence

Label the picture with the expressions for the width and length.

Step 3: **Translate** the information that appears in English into an algebraic equation.

Use a known geometry formula. What does the 120 ft of fencing represent? *Since the fencing will go around the pool, the 120 ft represents the perimeter of the rectangular safety fence.* We need to use a formula that involves the length, width, and perimeter of a rectangle. The formula we will use is

$$P = 2l + 2w$$

Substitute the known values and expressions into the formula.

$$P = 2l + 2w$$
$$120 = 2(w + 4) + 2w \qquad \text{Substitute.}$$

Step 4: **Solve** the equation.

$$120 = 2(w + 4) + 2w$$
$$120 = 2w + 8 + 2w \qquad \text{Distribute.}$$
$$120 = 4w + 8 \qquad \text{Combine like terms.}$$
$$120 - 8 = 4w + 8 - 8 \qquad \text{Subtract 8 from each side.}$$
$$112 = 4w \qquad \text{Combine like terms.}$$
$$\frac{112}{4} = \frac{4w}{4} \qquad \text{Divide each side by 4.}$$
$$28 = w \qquad \text{Simplify.}$$

Step 5: **Check** the answer, and **interpret** the solution as it relates to the problem.

The width of the safety fence is 28 ft. The length is $w + 4 = 28 + 4 = 32$ ft.

The answer makes sense because the perimeter of the fence is
$2(32 \text{ ft}) + 2(28 \text{ ft}) = 64 \text{ ft} + 56 \text{ ft} = 120 \text{ ft}.$

[YOU TRY 3] Marina wants to make a rectangular dog run in her backyard. It will take 46 feet of fencing to enclose it, and the length will be 1 foot less than three times the width. Find the dimensions of the dog run.

---

3 Solve Problems Involving Angle Measures

Recall from Section 1.3 that the sum of the angle measures in a triangle is 180°. We will use this fact in our next example.

EXAMPLE 4

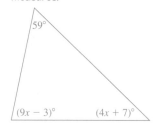
Find the missing angle measures.

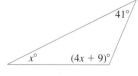

Solution

Step 1: **Read** the problem carefully, and identify what we are being asked to find.

Find the missing angle measures.

Step 2: The unknowns are already defined. We must find x, the measure of one angle, and then $4x + 9$, the measure of the other angle.

Step 3: **Translate** the information into an algebraic equation. Since the sum of the angles in a triangle is 180°, we can write

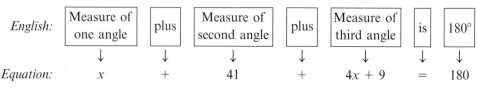

The equation is $x + 41 + (4x + 9) = 180$.

Step 4: **Solve** the equation.

$$x + 41 + (4x + 9) = 180$$
$$5x + 50 = 180 \qquad \text{Combine like terms.}$$
$$5x + 50 - 50 = 180 - 50 \qquad \text{Subtract 50 from each side.}$$
$$5x = 130 \qquad \text{Combine like terms.}$$
$$\frac{5x}{5} = \frac{130}{5} \qquad \text{Divide each side by 5.}$$
$$x = 26 \qquad \text{Simplify.}$$

Step 5: **Check** the answer, and **interpret** the solution as it relates to the problem.

One angle, x, has a measure of $26°$. The other unknown angle measure is $4x + 9 = 4(26) + 9 = 113°$.

The answer makes sense because the sum of the angle measures is $26° + 41° + 113° = 180°$.

$\begin{bmatrix} \text{YOU TRY 4} \end{bmatrix}$ Find the missing angle measures.

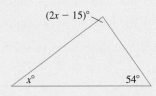

Let's look at another type of problem involving angle measures.

Find the measure of each indicated angle.

$(6x - 9)°$ $(5x + 1)°$

Solution

The indicated angles are *vertical angles,* and **vertical angles** have the same measure. (See Section 1.3.) Since their measures are the same, set $6x - 9 = 5x + 1$ and solve for x.

$$6x - 9 = 5x + 1$$
$$6x - 9 + 9 = 5x + 1 + 9 \qquad \text{Add 9 to each side.}$$
$$6x = 5x + 10 \qquad \text{Combine like terms.}$$
$$6x - 5x = 5x - 5x + 10 \qquad \text{Subtract } 5x \text{ from each side.}$$
$$x = 10 \qquad \text{Combine like terms.}$$

Be careful! Although $x = 10$, the angle measure is *not* 10. To find the angle measures, substitute $x = 10$ into the expressions for the angles.

The measure of the angle on the left is $6x - 9 = 6(10) - 9 = 51°$. The other angle measure is also $51°$ since these are vertical angles. We can verify this by substituting 10 into the expression for the other angle, $5x + 1$: $5x + 1 = 5(10) + 1 = 51°$.

Find the measure of each indicated angle.

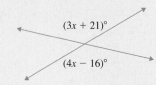

$(3x + 21)°$

$(4x - 16)°$

In Section 1.3, we learned that two angles are **complementary** if the sum of their angles is 90°, and two angles are **supplementary** if the sum of their angles is 180°.

For example, if the measure of $\angle A$ is 71°, then

    a) the measure of its complement is $90° - 71° = 19°$.

    b) the measure of its supplement is $180° - 71° = 109°$.

Now let's say the measure of an angle is x. Using the same reasoning as above,

    a) the measure of its complement is $90 - x$.

    b) the measure of its supplement is $180 - x$.

We will use these ideas to solve the problem in Example 6.

EXAMPLE 6

In-Class Example 6

Three times the complement of an angle is 68° more than the supplement of the angle. Find the angle.

Answer: 11°

The supplement of an angle is 34° more than twice the complement of the angle. Find the measure of the angle.

Solution

Step 1: **Read** the problem carefully, and identify what we are being asked to find.

        We must find the measure of the angle.

Step 2: **Choose a variable** to represent an unknown, and define the other unknowns in terms of this variable.

This problem has three unknowns: the measures of the angle, its complement, and its supplement. Choose a variable to represent the *original angle,* then define the other unknowns in terms of this variable.

$$x = \text{the measure of the angle}$$

Define the other unknowns in terms of x.

$$90 - x = \text{the measure of the complement}$$
$$180 - x = \text{the measure of the supplement}$$

> **W Hint**
>
> On a sheet of paper, make a list of all the cases where angle measurements are used in this section.

Step 3: **Translate** the information that appears in English into an algebraic equation.

	The supplement of an angle	is	34° more than	twice the complement of the angle
Statement:	↓	↓	↓	↓
Equation:	$180 - x$	$=$	$34 +$	$2(90 - x)$

The equation is $180 - x = 34 + 2(90 - x)$.

Step 4: **Solve** the equation.

$$180 - x = 34 + 2(90 - x)$$
$$180 - x = 34 + 180 - 2x \qquad \text{Distribute.}$$
$$180 - x = 214 - 2x \qquad \text{Combine like terms.}$$
$$180 - 180 - x = 214 - 180 - 2x \qquad \text{Subtract 180 from each side.}$$
$$-x = 34 - 2x \qquad \text{Combine like terms.}$$
$$-x + 2x = 34 - 2x + 2x \qquad \text{Add } 2x \text{ to each side.}$$
$$x = 34 \qquad \text{Simplify.}$$

Step 5: **Check** the answer, and **interpret** the solution as it relates to the problem.

The measure of the angle is 34°.

To check the answer, we first need to find its complement and supplement. The complement is 90° − 34° = 56°, and its supplement is 180° − 34° = 146°. Now we can check these values in the original statement: The supplement is 146°. Thirty-four degrees more than twice the complement is 34° + 2(56°) = 34° + 112° = 146°.

[YOU TRY 6] Twice the complement of an angle is 18° less than the supplement of the angle. Find the measure of the angle.

---

4 Solve a Formula for a Specific Variable

The formula $P = 2l + 2w$ allows us to find the perimeter of a rectangle when we know its length (l) and width (w). But what if we were solving problems where we repeatedly needed to find the value of w? Then, we could rewrite $P = 2l + 2w$ so that it is solved for w:

$$w = \frac{P - 2l}{2}$$

Doing this means that we have *solved the formula $P = 2l + 2w$ for the specific variable w*.

Solving a formula for a specific variable may seem confusing at first because the formula contains more than one letter. Keep in mind that we will solve for a specific variable the same way we have been solving equations up to this point.

We'll start by solving $3x + 4 = 19$ step-by-step for x and then applying the same procedure to solving $ax + b = c$ for x.

EXAMPLE 7 Solve $3x + 4 = 19$ and $ax + b = c$ for x.

In-Class Example 7

Solve
$2x + 9 = 17$ and $kx + c = m$
for x.

Answer: {4}; $x = \dfrac{m - c}{k}$

Solution

Look at these equations carefully, and notice that they have the same form. Read the parts of the solution in numerical order.

Part 1 Solve $3x + 4 = 19$.

Don't quickly run through the solution of this equation. **The emphasis here is on the steps used to solve the equation and why we use those steps!**

$$3\boxed{x} + 4 = 19$$

We are solving for x. We'll put a box around it. What is the first step? "Get rid of" what is being added to the $3x$; that is, "get rid of" the 4 on the left. Subtract 4 from each side.

$$3\boxed{x} + 4 - 4 = 19 - 4$$

Combine like terms.

$$3\boxed{x} = 15$$

Part 3 We need to solve $3\boxed{x} = 15$ for x. We need to eliminate the 3 on the left. Since x is being multiplied by 3, we will **divide** each side by 3.

$$\frac{3\boxed{x}}{3} = \frac{15}{3}$$

Simplify.

$$x = 5$$

The solution set is {5}.

Part 2 Solve $ax + b = c$ for x.

Since we are solving for x, we'll put a box around it.

$$a\boxed{x} + b = c$$

The goal is to get the x on a side by itself. What do we do first? As in Part 1, "get rid of" what is being added to the ax term; that is, "get rid of" the b on the left. Since b is being added to ax, we will subtract it from each side. (We are performing the same steps as in Part 1!)

$$a\boxed{x} + b - b = c - b$$

Combine like terms.

$$a\boxed{x} = c - b$$

We cannot combine the terms on the right, so it remains $c - b$.

Part 4 Now, we have to solve $a\boxed{x} = c - b$ for x. We need to eliminate the a on the left. Since x is being multiplied by a, we will **divide** each side by a.

$$\frac{a\boxed{x}}{a} = \frac{c - b}{a}$$

These are the same steps used in Part 3!

Simplify.

$$\frac{a\boxed{x}}{a} = \frac{c - b}{a}$$

$$x = \frac{c - b}{a} \text{ or } \frac{c}{a} - \frac{b}{a}$$

Hint

Why can the solution be written two different ways?

Note

To obtain the result $x = \dfrac{c}{a} - \dfrac{b}{a}$, we distributed the a in the denominator to each term in the numerator. Either form of the answer is correct.

When you are solving a formula for a specific variable, think about the steps you use to solve an equation in one variable.

[**YOU TRY 7**] Solve $rt - n = k$ for t.

EXAMPLE 8

$U = \frac{1}{2}LI^2$ is a formula used in physics. Solve this equation for L.

Solution

$$U = \frac{1}{2}\boxed{L}I^2 \qquad \text{Solve for } L. \text{ Put it in a box.}$$

$$2U = 2 \cdot \frac{1}{2}\boxed{L}I^2 \qquad \text{Multiply by 2 to eliminate the fraction.}$$

$$\frac{2U}{I^2} = \frac{\boxed{L}I^2}{I^2} \qquad \text{Divide each side by } I^2.$$

$$\frac{2U}{I^2} = L \qquad \text{Simplify.}$$

EXAMPLE 9

$A = \frac{1}{2}h(b_1 + b_2)$ is the formula for the area of a trapezoid. Solve it for b_1.

Solution

There are two ways to solve this for b_1.

Method 1: We will put b_1 in a box to remind us that this is what we must solve for. In Method 1, we will start by eliminating the fraction.

$$2A = 2 \cdot \frac{1}{2}h(\boxed{b_1} + b_2) \qquad \text{Multiply each side by 2.}$$

$$2A = h(\boxed{b_1} + b_2) \qquad \text{Simplify.}$$

$$\frac{2A}{h} = \frac{h(\boxed{b_1} + b_2)}{h} \qquad \text{Divide each side by } h.$$

$$\frac{2A}{h} = \boxed{b_1} + b_2$$

$$\frac{2A}{h} - b_2 = \boxed{b_1} + b_2 - b_2 \qquad \text{Subtract } b_2 \text{ from each side.}$$

$$\frac{2A}{h} - b_2 = b_1 \qquad \text{Simplify.}$$

Method 2: Another way to solve $A = \frac{1}{2}h(b_1 + b_2)$ for b_1 is to begin by distributing $\frac{1}{2}h$ on the right.

$$A = \frac{1}{2}h\boxed{b_1} + \frac{1}{2}hb_2 \qquad \text{Distribute.}$$

$$2A = 2\left(\frac{1}{2}h\boxed{b_1} + \frac{1}{2}hb_2\right) \qquad \text{Multiply by 2 to eliminate the fractions.}$$

$$2A = h\boxed{b_1} + hb_2 \qquad \text{Distribute.}$$

$$2A - hb_2 = h\boxed{b_1} + hb_2 - hb_2 \qquad \text{Subtract } hb_2 \text{ from each side.}$$

$$2A - hb_2 = h\boxed{b_1} \qquad \text{Simplify.}$$

$$\frac{2A - hb_2}{h} = \frac{h\boxed{b_1}}{h} \qquad \text{Divide by } h.$$

$$\frac{2A - hb_2}{h} = b_1 \qquad \text{Simplify.}$$

Therefore, b_1 can be written as $b_1 = \dfrac{2A}{h} - \dfrac{hb_2}{h}$ or $b_1 = \dfrac{2A}{h} - b_2$. These two forms are equivalent.

E Evaluate **2.5** Exercises   Do the exercises, and check your work.

*Additional answers can be found in the Answers to Exercises appendix.

Objective 1: Substitute Values into a Formula, and Find the Unknown Variable

1) If you are using the formula $A = \dfrac{1}{2}bh$, is it reasonable to get an answer of $h = -6$? Explain your answer.
No. The height of a triangle cannot be a negative number.

2) If you are finding the area of a rectangle and the lengths of the sides are given in inches, the area of the rectangle would be expressed in which unit?   square inches

3) If you are asked to find the volume of a sphere and the radius is given in centimeters, the volume would be expressed in which unit?   cubic centimeters

4) If you are asked to find the perimeter of a football field and the length and width are given in yards, the perimeter of the field would be expressed in which unit?   yards

Determine whether each statement is true or false.

5) To find the area of a triangle, the units of the base and height must be the same before they are substituted into the formula.   true

6) For any formula involving the irrational number π, the value of at least one of the variables must be an irrational number.   false

Substitute the given values into the formula and solve for the remaining variable.

7) $A = lw$; If $A = 44$ when $l = 16$, find w. $\dfrac{11}{4}$

8) $A = \dfrac{1}{2}bh$; If $A = 21$ when $h = 14$, find b.   3

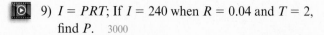 9) $I = PRT$; If $I = 240$ when $R = 0.04$ and $T = 2$, find P.   3000

10) $I = PRT$; If $I = 600$ when $P = 2500$ and $T = 4$, find R.   0.06

11) $d = rt$
(Distance formula: *distance = rate · time*);
If $d = 150$ when $r = 60$, find t.   2.5

12) $d = rt$
(Distance formula: *distance = rate · time*);
If $r = 36$ and $t = 0.75$, find d.   27

13) $C = 2\pi r$; If $r = 4.6$, find C. 9.2π

14) $C = 2\pi r$; If $C = 15\pi$, find r. $\dfrac{15}{2}$

15) $P = 2l + 2w$; If $P = 11$ when $w = \dfrac{3}{2}$, find l.   4

16) $P = s_1 + s_2 + s_3$ (Perimeter of a triangle);
If $P = 11.6$ when $s_2 = 2.7$ and $s_3 = 3.8$, find s_1.   5.1

17) $V = lwh$; If $V = 52$ when $l = 6.5$ and $h = 2$, find w.   4

18) $V = \dfrac{1}{3}Ah$ (Volume of a pyramid); If $V = 16$ when $A = 24$, find h.   2

19) $V = \dfrac{1}{3}\pi r^2 h$; If $V = 48\pi$ when $r = 4$, find h.   9

20) $V = \dfrac{1}{3}\pi r^2 h$; If $V = 50\pi$ when $r = 5$, find h.   6

21) $S = 2\pi r^2 + 2\pi rh$ (Surface area of a right circular cylinder); If $S = 154\pi$ when $r = 7$, find h.   4

22) $S = 2\pi r^2 + 2\pi rh$; If $S = 132\pi$ when $r = 6$, find h.   5

23) $A = \dfrac{1}{2}h(b_1 + b_2)$; If $A = 136$ when $b_1 = 7$ and $h = 16$, find b_2.   10

24) $A = \dfrac{1}{2}h(b_1 + b_2)$; If $A = 1.5$ when $b_1 = 3$ and $b_2 = 1$, find h.   0.75

Objective 2: Solve Problems Using Formulas from Geometry

Use a known formula to solve. See Example 2.

25) The area of a tennis court is 2808 ft². Find the length of the court if it is 36 ft wide.   78 ft

26) A rectangular tabletop has an area of 13.5 ft². What is the width of the table if it is 4.5 ft long?   3 ft

27) A rectangular flower box holds 1232 in³ of soil. Find the height of the box if it is 22 in. long and 7 in. wide.   8 in.

28) A rectangular storage box is 2.5 ft wide, 4 ft long, and 1.5 ft high. What is the storage capacity of the box?   15 ft³

29) The center circle on a soccer field has a radius of 10 yd. What is the area of the center circle? Use 3.14 for π.   314 yd²

30) The face of the clock on Big Ben in London has a radius of 11.5 feet. What is the area of this circular clock face? Use 3.14 for π. (www.bigben.freeservers.com)   415.265 ft²

31) Abbas drove 134 miles on the highway in 2 hours. What was his average speed?   67 mph

32) If Reza drove 108 miles at 72 mph, without stopping, for how long did she drive?   1.5 hr

33) A stainless steel garbage can is in the shape of a right circular cylinder. If its radius is 6 inches and its volume is 864π in³, what is the height of the can?   24 in.

34) A coffee can in the shape of a right circular cylinder has a volume of 50π in³. Find the height of the can if its diameter is 5 inches.   8 in.

35) A flag is in the shape of a triangle and has an area of 6 ft². Find the length of the base if its height is 4 ft.   3 ft

36) A championship banner hanging from the rafters of a stadium is in the shape of a triangle and has an area of 20 ft². How long is the banner if its base is 5 ft?   8 ft

Use a known formula to solve. See Example 3.

37) Vivian is making a rectangular wooden picture frame that will have a width that is 10 in. shorter than its length. If she will use 92 in. of wood, what are the dimensions of the frame?   18 in. × 28 in.

38) A construction crew is making repairs next to a school, so they have to enclose the rectangular area with a fence. They determine that they will need 176 ft of fencing for the work area, which is 22 ft longer than it is wide. Find the dimensions of the fenced area.   33 ft × 55 ft

39) The "lane" on a basketball court is a rectangle that has a perimeter of 62 ft. Find the dimensions of the "lane" given that its length is 5 ft less than twice the width.   12 ft × 19 ft

40) A rectangular white-board in a classroom is twice as long as it is high. Its perimeter is 24 ft. What are the dimensions of the whiteboard?   4 ft × 8 ft

41) One base of a trapezoid is 2 in. longer than three times the other base. Find the lengths of the bases if the trapezoid is 5 in. high and has an area of 25 in².   2 in., 8 in.

42) A caution flag on the side of a road is shaped like a trapezoid. One base of the trapezoid is 1 ft shorter than the other base. Find the lengths of the bases if the trapezoid is 4 ft high and has an area of 10 ft².   2 ft, 3 ft

43) A triangular sign in a store window has a perimeter of 5.5 ft. Two of the sides of the triangle are the same length while the third side is 1 foot longer than those sides. Find the lengths of the sides of the sign.   1.5 ft, 1.5 ft, 2.5 ft

44) A triangle has a perimeter of 31 in. The longest side is 1 in. less than twice the shortest side, and the third side is 4 in. longer than the shortest side. Find the lengths of the sides.   7 in., 11 in., 13 in.

Objective 3: Solve Problems Involving Angle Measures

Find the missing angle measures.

45) $m\angle A = 35°, m\angle C = 62°$

Triangle with vertex B at top (angle $83°$), vertex A at bottom left (angle $(x - 27)°$), and vertex C at bottom right (angle $x°$).

46) 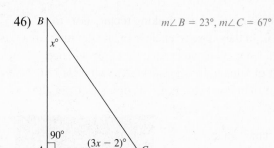 $m\angle B = 23°, m\angle C = 67°$

54) 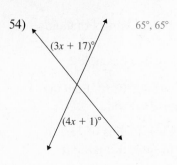 65°, 65°

$(3x + 17)°$

$(4x + 1)°$

47) 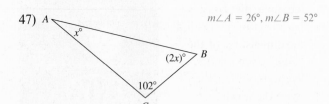 $m\angle A = 26°, m\angle B = 52°$

55) 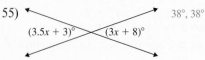 38°, 38°

$(3.5x + 3)°$ $(3x + 8)°$

48) 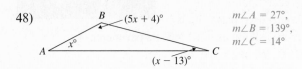 $m\angle A = 27°,$
$m\angle B = 139°,$
$m\angle C = 14°$

56) 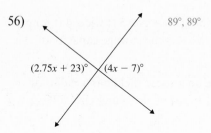 89°, 89°

$(2.75x + 23)°$ $(4x - 7)°$

49) 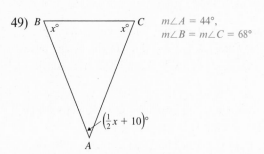 $m\angle A = 44°,$
$m\angle B = m\angle C = 68°$

57) 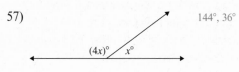 144°, 36°

$(4x)°$ $x°$

58) 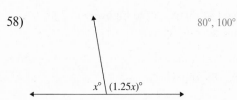 80°, 100°

$x°$ $(1.25x)°$

50) 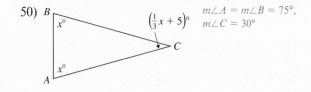 $m\angle A = m\angle B = 75°,$
$m\angle C = 30°$

59) 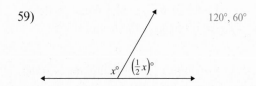 120°, 60°

$x°$ $\left(\frac{1}{2}x\right)°$

Find the measure of each indicated angle.

60) 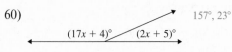 157°, 23°

$(17x + 4)°$ $(2x + 5)°$

51) 43°, 43°

$(x + 28)°$ $(3x - 2)°$

61) 73°, 107°

$(x + 30)°$ $(2x + 21)°$

52) 150°, 150°

$(5x)°$
$\left(\frac{8}{3}x + 70\right)°$

53) 172°, 172°

$(4x - 12)°$
$\left(\frac{5}{2}x + 57\right)°$

62) 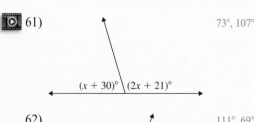 111°, 69°

$\left(\frac{13}{3}x + 20\right)°$ $(4x - 15)°$

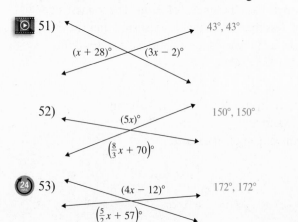

63) If x = the measure of an angle, write an expression for its supplement. $180 - x$

64) If x = the measure of an angle, write an expression for its complement. $90 - x$

Write an equation and solve.

(24 HRS) 65) The supplement of an angle is 63° more than twice the measure of its complement. Find the measure of the angle.   63°

66) Twice the complement of an angle is 49° less than its supplement. Find the measure of the angle.   49°

67) Six times an angle is 12° less than its supplement. Find the measure of the angle.   24°

68) An angle is 1° less than 12 times its complement. Find the measure of the angle.   83°

69) Four times the complement of an angle is 40° less than twice the angle's supplement. Find the angle, its complement, and its supplement.
angle: 20°; comp: 70°; supp: 160°

70) Twice the supplement of an angle is 30° more than eight times its complement. Find the angle, its complement, and its supplement.
angle: 65°; comp: 25°; supp: 115°

71) The sum of an angle and half its supplement is seven times its complement. Find the measure of the angle.   72°

72) The sum of an angle and three times its complement is 62° more than its supplement. Find the measure of the angle.   28°

73) The sum of four times an angle and twice its complement is 270°. Find the angle.   45°

74) The sum of twice an angle and half its supplement is 192°. Find the angle.   68°

Objective 4: Solve a Formula for a Specific Variable

75) Solve for x.
   a) $x + 16 = 37$ $x = 21$ b) $x + h = y$ $x = y - h$
   c) $x + r = c$ $x = c - r$

76) Solve for t.
   a) $t - 8 = 17$ $t = 25$ b) $t - p = z$ $t = z + p$
   c) $t - k = n$ $t = n + k$

77) Solve for c.
   a) $8c = 56$ $c = 7$ b) $ac = d$ $c = \dfrac{d}{a}$
   c) $mc = v$ $c = \dfrac{v}{m}$

78) Solve for k.
   a) $9k = 54$ $k = 6$ b) $nk = t$ $k = \dfrac{t}{n}$
   c) $wk = h$ $k = \dfrac{h}{w}$

79) Solve for a.
   a) $\dfrac{a}{4} = 11$ $a = 44$ b) $\dfrac{a}{y} = r$ $a = ry$
   c) $\dfrac{a}{w} = d$ $a = dw$

80) Solve for d.
   a) $\dfrac{d}{6} = 3$ $d = 18$ b) $\dfrac{d}{t} = q$ $d = qt$
   c) $\dfrac{d}{x} = a$ $d = ax$

(24 HRS) 81) Solve for d.
   a) $8d - 7 = 17$ $d = 3$
   b) $kd - a = z$ $d = \dfrac{z + a}{k}$

82) Solve for w.
   a) $5w + 18 = 3$ $w = -3$
   b) $pw + r = \pi$ $w = \dfrac{\pi - r}{p}$

83) Solve for h.
   a) $9h + 23 = 17$ $h = -\dfrac{2}{3}$
   b) $qh + v = n$ $h = \dfrac{n - v}{q}$

84) Solve for b.
   a) $12b - 5 = 17$ $b = \dfrac{11}{6}$
   b) $mb - c = a$ $b = \dfrac{a + c}{m}$

Solve each formula for the indicated variable.

85) $F = ma$ for m (Physics) $m = \dfrac{F}{a}$

86) $C = 2\pi r$ for r $r = \dfrac{C}{2\pi}$

87) $n = \dfrac{c}{v}$ for c (Physics) $c = nv$

88) $f = \dfrac{R}{2}$ for R (Physics) $R = 2f$

89) $E = \sigma T^4$ for σ (Meteorology) $\quad \sigma = \dfrac{E}{T^4}$

90) $p = \rho gy$ for ρ (Geology) $\quad \rho = \dfrac{p}{gy}$

91) $V = \dfrac{1}{3}\pi r^2 h$ for h $\quad h = \dfrac{3V}{\pi r^2}$

92) $d = rt$ for r $\quad r = \dfrac{d}{t}$

93) $R = \dfrac{E}{I}$ for E (Electricity) $\quad E = IR$

94) $A = \dfrac{1}{2}bh$ for b $\quad b = \dfrac{2A}{h}$

95) $I = PRT$ for R $\quad R = \dfrac{I}{PT}$

96) $I = PRT$ for P $\quad P = \dfrac{I}{RT}$

97) $P = 2l + 2w$ for l $\quad l = \dfrac{P - 2w}{2}$ or $l = \dfrac{P}{2} - w$

98) $A = P + PRT$ for T (Finance) $\quad T = \dfrac{A - P}{PR}$ or $T = \dfrac{A}{PR} - \dfrac{1}{R}$

99) $H = \dfrac{D^2 N}{2.5}$ for N (Auto mechanics) $\quad N = \dfrac{2.5H}{D^2}$

100) $V = \dfrac{AH}{3}$ for A (Geometry) $\quad A = \dfrac{3V}{H}$

101) $A = \dfrac{1}{2}h(b_1 + b_2)$ for b_2 $\quad b_2 = \dfrac{2A}{h} - b_1$ or $b_2 = \dfrac{2A - hb_1}{h}$

102) $A = \pi(R^2 - r^2)$ for r^2 (Geometry) $\quad r^2 = R^2 - \dfrac{A}{\pi}$ or $r^2 = \dfrac{\pi R^2 - A}{\pi}$

103) The perimeter, P, of a rectangle is $P = 2l + 2w$, where l = length and w = width.

a) Solve $P = 2l + 2w$ for w. $\quad w = \dfrac{P - 2l}{2}$ or $w = \dfrac{P}{2} - l$

b) Find the width of the rectangle with perimeter 28 cm and length 11 cm. $\quad$ 3 cm

104) The area, A, of a triangle is $A = \dfrac{1}{2}bh$, where b = length of the base and h = height.

a) Solve $A = \dfrac{1}{2}bh$ for h. $\quad h = \dfrac{2A}{b}$

b) Find the height of the triangle that has an area of 39 cm^2 and a base of length 13 cm. $\quad$ 6 cm

105) The formula $C = \dfrac{5}{9}(F - 32)$ can be used to convert from degrees Fahrenheit, F, to degrees Celsius, C.

a) Solve this formula for F. $\quad F = \dfrac{9}{5}C + 32$

b) The average high temperature in Paris, France, in May is 20°C. Use the result in Part a) to find the equivalent temperature in degrees Fahrenheit. (www.bbc.co.uk) $\quad$ 68°F

106) The average low temperature in Buenos Aires, Argentina, in June is 5°C. Use the result in Exercise 107 a) to find the equivalent temperature in degrees Fahrenheit. (www.bbc.co.uk) $\quad$ 41°F

R Rethink

R1) What kinds of landscape features are often rectangular in shape?

R2) What kinds of jobs use angle measurements?

R3) What kinds of sports require that a player project a ball using some type of strategic angle measurement?

2.6 Ratios and Proportions

P Prepare

O Organize

What are your objectives for Section 2.6?	How can you accomplish each objective?
1 Use Ratios	• Write the definition of *ratio* in your own words. • Complete the given examples on your own. • Complete You Trys 1 and 2.
2 Solve a Proportion	• Write the definition of *proportion* in your own words. • Understand what *cross products* are and how to use them. • Complete the given examples on your own. • Complete You Trys 3–5.
3 Solve Problems Involving Similar Triangles	• Review similar triangles in Section 1.3. • Follow the explanation to complete the given example on your own. • Complete You Try 6.

W Work

Read the explanations, follow the examples, take notes, and complete the You Trys.

1 Use Ratios

We hear about *ratios* and use them in many ways in everyday life. For example, if a survey on cell phone use revealed that 80 teenagers prefer texting their friends while 25 prefer calling their friends, we could write the ratio of teens who prefer texting to teens who prefer calling as

$$\frac{\text{Number who prefer texting}}{\text{Number who prefer calling}} = \frac{80}{25} = \frac{16}{5}$$

Here is a formal definition of a ratio:

Definition

A **ratio** is a quotient of two quantities. The ratio of the number x to the number y, where $y \neq 0$, can be written as $\frac{x}{y}$, x to y, or $x:y$.

EXAMPLE 1

In-Class Example 1

Write the ratio of 6 feet to 4 yards.

Answer: $\frac{1}{2}$

Write the ratio of 4 feet to 2 yards.

Solution

Write each quantity with the same units. Let's change yards to feet. Since there are 3 feet in 1 yard,

$$2 \text{ yards} = 2 \cdot 3 \text{ feet} = 6 \text{ feet}$$

Then the ratio of 4 feet to 2 yards is

$$\frac{4 \text{ feet}}{2 \text{ yd}} = \frac{4 \text{ feet}}{6 \text{ feet}} = \frac{4}{6} = \frac{2}{3}$$

[YOU TRY 1] Write the ratio of 3 feet to 24 inches.

We can use ratios to help us figure out which item in a store gives us the most value for our money. To do this, we will determine the *unit price* of each item. The **unit price** is the ratio of the price of the item to the amount of the item.

EXAMPLE 2

In-Class Example 2

A store sells toothpaste in three different sizes. A 4-oz tube sells for $2.79, a 6-oz tube costs $3.19, and an 8-oz tube sells for $3.99. Which size is the best buy, and what is the unit price?

Answer: The 8-oz tube is the best buy. Its unit price is $0.499/oz. The unit price of the 4-oz tube is $0.698/oz, and the unit price of the 6-oz tube is $0.532/oz.

A store sells Haagen-Dazs vanilla ice cream in three different sizes. The sizes and prices are listed here. Which size is the best buy?

Size	Price
4 oz	$1.00
14 oz	$3.49
28 oz	$7.39

Solution

For each carton of ice cream, we must find the unit price, or how much the ice cream costs per ounce. We will find the unit price by dividing.

$$\text{Unit price} = \frac{\text{Price of ice cream}}{\text{Number of ounces in the container}} = \text{Cost per ounce}$$

Size	Unit Price
4 oz	$\frac{\$1.00}{4 \text{ oz}} = \$0.250 \text{ per oz}$
14 oz	$\frac{\$3.49}{14 \text{ oz}} = \$0.249 \text{ per oz}$
28 oz	$\frac{\$7.39}{28 \text{ oz}} = \$0.264 \text{ per oz}$

We round the answers to the thousandths place because, as you can see, there is not much difference in the unit price. Since the 14-oz carton of ice cream has the smallest unit price, it is the best buy.

[YOU TRY 2] A store sells Gatorade fruit punch in three different sizes. A 20-oz bottle costs $1.00, a 32-oz bottle sells for $1.89, and the price of a 128-oz bottle is $5.49. Which size is the best buy, and what is its unit price?

2 Solve a Proportion

We have learned that a ratio is a way to compare two quantities. If two ratios are equivalent, like $\frac{4}{6}$ and $\frac{2}{3}$, we can set them equal to make a *proportion*.

Definition

A **proportion** is a statement that two ratios are equal.

How can we be certain that a proportion is true? We can find the **cross products.** If the cross products are equal, then the proportion is true. If the cross products are not equal, then the proportion is false.

> ## Property
>
> **Cross Products**   If $\dfrac{a}{b} = \dfrac{c}{d}$, then $ad = bc$ provided that $b \neq 0$ and $d \neq 0$.

We will see later in the book that finding the cross products is the same as multiplying both sides of the equation by a common denominator of the fractions.

EXAMPLE 3

 Hint

Notice that cross products are used only across an equal sign!

Determine whether each proportion is true or false.

a) $\dfrac{5}{7} = \dfrac{15}{21}$ b) $\dfrac{2}{9} = \dfrac{7}{36}$

Solution

a) Find the cross products.

$$\frac{5}{7} \diagup\!\!\!\!\diagdown \frac{15}{21} \quad \text{Multiply.}$$

$5 \cdot 21 = 7 \cdot 15$ Set the cross products equal.
$105 = 105$ True

The cross products are equal, so the proportion is true.

b) Find the cross products.

$$\frac{2}{9} \diagup\!\!\!\!\diagdown \frac{7}{36} \quad \text{Multiply.}$$

$2 \cdot 36 = 9 \cdot 7$ Set the cross products equal.
$72 = 63$ False

The cross products are not equal, so the proportion is false.

[YOU TRY 3]   Determine whether each proportion is true or false.

a) $\dfrac{4}{9} = \dfrac{24}{56}$ b) $\dfrac{3}{8} = \dfrac{12}{32}$

We can use cross products to solve equations.

EXAMPLE 4

 Hint

Be sure to use parentheses when the numerator or denominator contain more than one term.

Solve each proportion.

a) $\dfrac{16}{24} = \dfrac{x}{3}$ b) $\dfrac{k + 2}{2} = \dfrac{k - 4}{5}$

Solution

Find the cross products.

a) $\dfrac{16}{24} \diagup\!\!\!\!\diagdown \dfrac{x}{3}$ Multiply.

$16 \cdot 3 = 24 \cdot x$ Set the cross products equal.
$48 = 24x$ Multiply.
$2 = x$ Divide by 24.

The solution set is {2}.

b) $\dfrac{k + 2}{2} \diagup\!\!\!\!\diagdown \dfrac{k - 4}{5}$ Multiply.

$5(k + 2) = 2(k - 4)$ Set the cross products equal.
$5k + 10 = 2k - 8$ Distribute.
$3k + 10 = -8$ Subtract $2k$.
$3k = -18$ Subtract 10.
$k = -6$ Divide by 3.

The solution set is {−6}.

Solve each proportion.

a) $\dfrac{2}{3} = \dfrac{w}{27}$ b) $\dfrac{b-6}{12} = \dfrac{b+2}{20}$

Proportions are often used to solve real-world problems. When we solve problems by setting up a proportion, we must be sure that the numerators contain the same quantities and the denominators contain the same quantities.

EXAMPLE 5

Hint

Again, use the same five-step method for solving applied problems.

Write an equation and solve.

Cailen is an artist, and she wants to make turquoise paint by mixing the green and blue paints that she already has. To make turquoise, she will have to mix 4 parts of green with 3 parts of blue. If she uses 6 oz of green paint, how much blue paint should she use?

Solution

Step 1: **Read** the problem carefully, and identify what we are being asked to find.

We must find the amount of blue paint needed.

Step 2: **Choose a variable** to represent the unknown.

x = the number of ounces of blue paint

Step 3: **Translate** the information that appears in English into an algebraic equation. Write a proportion. We will write our ratios in the form of

$\dfrac{\text{Amount of green paint}}{\text{Amount of blue paint}}$ so that the numerators contain the same quantities and the denominators contain the same quantities.

$$\text{Amount of green paint} \rightarrow \dfrac{4}{3} = \dfrac{6}{x} \leftarrow \text{Amount of green paint}$$
$$\text{Amount of blue paint} \rightarrow \quad\quad\quad \leftarrow \text{Amount of blue paint}$$

The equation is $\dfrac{4}{3} = \dfrac{6}{x}$.

Step 4: **Solve** the equation.

$$\dfrac{4}{3} \diagdown\!\!\!\diagup \dfrac{6}{x} \quad \text{Multiply.}$$

$4x = 6 \cdot 3$ Set the cross products equal.
$4x = 18$ Multiply.
$x = 4.5$ Divide by 4.

Step 5: **Check** the answer, and **interpret** the solution as it relates to the problem.

Cailen should mix 4.5 oz of blue paint with the 6 oz of green paint to make the turquoise paint she needs. The check is left to the student.

[YOU TRY 5]

Write an equation and solve.

If 3 lb of coffee costs \$21.60, how much would 5 lb of the same coffee cost?

3 Solve Problems Involving Similar Triangles

Another application of proportions is for solving similar triangles.

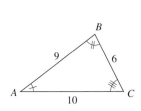 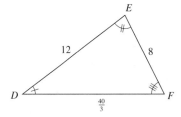

$$m\angle A = m\angle D, \quad m\angle B = m\angle E, \quad \text{and} \quad m\angle C = m\angle F$$

We say that $\triangle ABC$ and $\triangle DEF$ are *similar triangles*. Two triangles are **similar** if they have the same shape, the corresponding angles have the same measure, and the corresponding sides are proportional.

The ratio of each of the corresponding sides is $\frac{3}{4}$:

$$\frac{9}{12} = \frac{3}{4}; \quad \frac{6}{8} = \frac{3}{4}; \quad \frac{10}{\frac{40}{3}} = 10 \cdot \frac{3}{40} = \frac{3}{4}$$

We can use a proportion to find the length of an unknown side in two similar triangles.

EXAMPLE 6

Given the following similar triangles, find x.

In-Class Example 6

Given the following similar triangles, find x.

Answer: 16

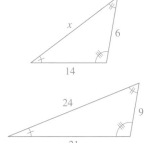

Solution

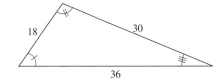

$$\frac{12}{18} = \frac{x}{30} \qquad \text{Set the ratios of two corresponding sides equal to each other. (Set up a proportion.)}$$

$$12 \cdot 30 = 18 \cdot x \qquad \text{Solve the proportion.}$$

$$360 = 18x \qquad \text{Multiply.}$$

$$20 = x \qquad \text{Divide by 18.}$$

 Hint

Note that there is more than one way to set up the proportion to solve for x.

[YOU TRY 6] Given the following similar triangles, find x.

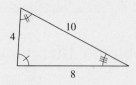

W Hint

Did you check your answers by hand before checking them here?

ANSWERS TO [YOU TRY] EXERCISES

1) $\frac{3}{2}$ 2) 128-oz bottle; $0.043/oz   3) a) false   b) true   4) a) {18}   b) {18}

5) $36.00   6) 15

E Evaluate **2.6** Exercises   Do the exercises, and check your work.

*Additional answers can be found in the Answers to Exercises appendix.

Objective 1: Use Ratios

1) Write three ratios that are equivalent to $\frac{3}{4}$.

Answers may vary, but some possible answers are $\frac{6}{8}$, $\frac{9}{12}$, and $\frac{12}{16}$.

2) Is 0.65 equivalent to the ratio 13 to 20? Explain.

Yes, the ratio "13 to 20" can be written as $\frac{13}{20}$. This reduces to 0.65.

3) Is a percent a type of ratio? Explain.

4) Write 57% as a ratio. $57\% = \frac{57}{100}$

Write as a ratio in lowest terms.

5) 16 girls to 12 boys $\frac{4}{3}$

6) 9 managers to 90 employees $\frac{1}{10}$

7) 4 coaches to 50 team members $\frac{2}{25}$

8) 30 blue marbles to 18 red marbles $\frac{5}{3}$

9) 20 feet to 80 feet $\frac{1}{4}$

10) 7 minutes to 4 minutes $\frac{7}{4}$

11) 2 feet to 36 inches $\frac{2}{3}$

12) 30 minutes to 3 hours $\frac{1}{6}$

13) 18 hours to 2 days $\frac{3}{8}$

14) 20 inches to 3 yards $\frac{5}{27}$

A store sells the same product in different sizes. Determine which size is the best buy based on the unit price of each item.

15) Batteries

Number	Price
8	$ 6.29
16	$12.99

package of 8: $0.786 per battery

16) Cat litter

Size	Price
30 lb	$ 8.48
50 lb	$12.98

50-lb bag: $0.260 per lb

17) Mayonnaise

Size	Price
8 oz	$2.69
15 oz	$3.59
48 oz	$8.49

48-oz jar: $0.177 per oz

18) Applesauce

Size	Price
16 oz	$1.69
24 oz	$2.29
48 oz	$3.39

48-oz jar: $0.071 per oz

19) Cereal

Size	Price
11 oz	$4.49
16 oz	$5.15
24 oz	$6.29

24-oz box: $0.262 per oz

20) Shampoo

Size	Price
14 oz	$3.19
25 oz	$5.29
32 oz	$6.99

25-oz bottle: $0.212 per oz

Objective 2: Solve a Proportion

🖊 21) What is the difference between a ratio and a proportion? A ratio is a quotient of two quantities. A proportion is a statement that two ratios are equal.

🖊 22) In the proportion $\dfrac{a}{b} = \dfrac{c}{d}$, can $b = 0$? Explain.
No. The denominator of a fraction cannot equal zero.

Determine whether each proportion is true or false.

(24) 23) $\dfrac{4}{7} = \dfrac{20}{35}$ true

24) $\dfrac{54}{64} = \dfrac{7}{8}$ false

25) $\dfrac{72}{54} = \dfrac{8}{7}$ false

26) $\dfrac{120}{140} = \dfrac{30}{35}$ true

🎬 27) $\dfrac{8}{10} = \dfrac{2}{\frac{5}{2}}$ true

28) $\dfrac{3}{4} = \dfrac{\frac{1}{2}}{\frac{2}{3}}$ true

Solve each proportion.

29) $\dfrac{8}{36} = \dfrac{c}{9}$ {2}

30) $\dfrac{n}{3} = \dfrac{20}{15}$ {4}

(24) 31) $\dfrac{w}{15} = \dfrac{32}{12}$ {40}

32) $\dfrac{8}{14} = \dfrac{d}{21}$ {12}

33) $\dfrac{40}{24} = \dfrac{30}{a}$ {18}

34) $\dfrac{10}{x} = \dfrac{12}{54}$ {45}

35) $\dfrac{2}{k} = \dfrac{9}{12}$ $\left\{\dfrac{8}{3}\right\}$

36) $\dfrac{15}{27} = \dfrac{m}{6}$ $\left\{\dfrac{10}{3}\right\}$

37) $\dfrac{3z + 10}{14} = \dfrac{2}{7}$ {−2}

38) $\dfrac{8t − 9}{20} = \dfrac{3}{4}$ {3}

39) $\dfrac{r + 7}{9} = \dfrac{r − 5}{3}$ {11}

40) $\dfrac{b + 6}{5} = \dfrac{b + 10}{15}$ {−4}

41) $\dfrac{3h + 15}{16} = \dfrac{2h + 5}{4}$ {−1}

42) $\dfrac{a + 7}{8} = \dfrac{4a − 11}{6}$ {5}

43) $\dfrac{4m − 1}{6} = \dfrac{6m}{10}$ $\left\{\dfrac{5}{2}\right\}$

44) $\dfrac{9w + 8}{10} = \dfrac{5 − 3w}{12}$ $\left\{−\dfrac{1}{3}\right\}$

Set up a proportion and solve.

(24) 45) If 4 containers of yogurt cost $2.36, find the cost of 6 containers of yogurt. $3.54

46) Find the cost of 3 scarves if 2 scarves cost $29.00. $43.50

47) A marinade for chicken uses 2 parts of lime juice for every 3 parts of orange juice. If the marinade uses $\dfrac{1}{3}$ cup of lime juice, how much orange juice should be used? $\dfrac{1}{2}$ cup

48) The ratio of salt to baking soda in a cookie recipe is 0.75 to 1. If a recipe calls for $1\dfrac{1}{2}$ teaspoons of salt, how much baking soda is in the cookie dough?   2 teaspoons

49) A 12-oz serving of Mountain Dew contains 55 mg of caffeine. How much caffeine is in an 18-oz serving of Mountain Dew? (www.energyfiend.com)   82.5 mg

50) An 8-oz serving of Red Bull energy drink contains about 80 mg of caffeine. Approximately how much caffeine is in 12 oz of Red Bull? (www.energyfiend.com)   120 mg

51) Approximately 9 out of 10 smokers began smoking before the age of 21. In a group of 400 smokers, about how many of them started before they reached their 21st birthday? (www.lungusa.org)   360

52) Ridgemont High School administrators estimate that 2 out of 3 members of its student body attended the homecoming football game. If there are 1941 students in the school, how many went to the game?   1294

53) At the end of a week, Ernest put 20 lb of yard waste and some kitchen scraps on the compost pile. If the ratio of yard waste to kitchen scraps was 5 to 2, how many pounds of kitchen scraps did he put on the pile?   8 lb

54) On a map of the United States, 1 inch represents 120 miles. If two cities are 3.5 inches apart on the map, what is the actual distance between the two cities?   420 miles

55) On July 4, 2009, the exchange rate was such that $20.00 (American) was worth 14.30 Euros. How many Euros could you get for $50.00? (www.xe.com)   35.75 Euros

56) On July 4, 2009, the exchange rate was such that 100 British pounds were worth $163.29 (American). How many dollars could you get for 280 British pounds? (www.xe.com) $457.21

Objective 3: Solve Problems Involving Similar Triangles

Given the following similar triangles, find x.

57) $x = 10$

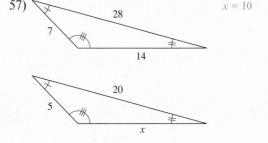

58) 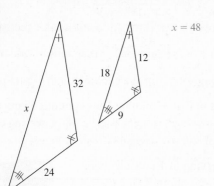 $x = 48$

59) 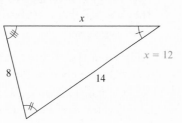 $x = 13$

60) 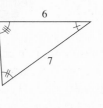 $x = 12$

61) 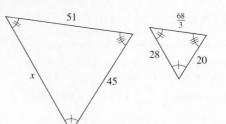 $x = 63$

62) 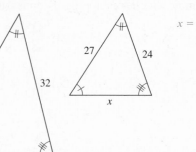 $x = \frac{39}{2}$

R **Rethink**

R1) When setting up a ratio, does it matter which quantity is placed in the numerator and which quantity is placed in the denominator?

R2) When setting up a proportion, does it matter which quantities are placed in the numerators and which quantities are placed in the denominators?

R3) Where can you use proportions to help you make better financial decisions now?

2.7 Applications of Percents

P Prepare

O Organize

What are your objectives for Section 2.7?	How can you accomplish each objective?
1 Use an Equation to Solve a Percent Problem	• Compare the difference between using arithmetic and using an equation to find a percentage. • Complete the given examples on your own. • Complete You Trys 1 and 2.
2 Solve Applications Involving Percents	• Adapt the **Five Steps for Solving Applied Problems** to help solve problems that contain percents. • Complete the given examples on your own. • Complete You Trys 3–5.

W Work   Read the explanations, follow the examples, take notes, and complete the You Trys.

Problems involving percents are everywhere—in stores, in banks, in laboratories, and in many more places. In this section, we will learn how to solve some algebraic problems involving percents. Before doing this, let's look at a problem involving only arithmetic so that you can see the relationship between an arithmetic problem and an algebraic problem.

1   Use an Equation to Solve a Percent Problem

EXAMPLE 1

In-Class Example 1

a) Use multiplication to find 30% of 150.
b) Use an equation to find 30% of 150.

Answer: a) 45   b) 45

Hint

How are these two approaches different?

a)   Use multiplication to find 20% of 120.

b)   Use an equation to find 20% of 120.

Solution

a)   The *of* in 20% *of* 120 means multiply. Change 20% to a decimal, then multiply.

$$20\% \text{ of } 120 = 0.20 \cdot 120 = 24$$

b)   Let's think of finding 20% of 120 as the question, "What is 20% of 120?" Let $x =$ the unknown quantity.

$$x = \text{the unknown quantity, } 20\% \text{ of } 120$$

Write the question in English, then write an equation.

English:	What	is	20%	of	120?
	↓	↓	↓	↓	↓
Equation:	x	$=$	0.20	$\cdot$	120

Change the percent to a decimal.

The equation is $x = 0.20 \cdot 120$. Solve.

$$x = 24 \quad \text{Multiply.}$$

Therefore, 24 is 20% of 120.

a)   Use multiplication to find 15% of 300.

b)   Use an equation to find 15% of 300.

Using an equation to solve Example 1 may be longer than just using multiplication, but sometimes we use an equation like this to solve an applied problem. Also, using an equation is a more efficient way to find other quantities in a percent problem.

EXAMPLE 2

In-Class Example 2

Use an equation to solve each problem.
a) 49 is 28% of what number?
b) 59.5 is what percent of 70?

Answer: a) 175   b) 85%

Use an equation to solve each problem.

a)   21 is 14% of what number?   b)   58.5 is what percent of 90?

Solution

a)   Let x represent the unknown value.

$$x = \text{the number}$$

Write the question in English, and write an equation.

Statement: 21   is   14%   of   what number?

Equation: 21   =   0.14   ·   x
Change the
percent to
a decimal.

The equation is $21 = 0.14 \cdot x$. Solve it.

$$21 = 0.14x \quad \text{Write the equation without the multiplication symbol.}$$

$$\frac{21}{0.14} = \frac{0.14}{0.14}x \quad \text{Divide both sides by 0.14.}$$

$$150 = x \quad \text{Perform the division.}$$

Therefore, 21 is 14% of 150. We can check the answer by finding 14% of 150.

$$14\% \text{ of } 150 = 0.14 \cdot 150 = 21$$

The answer, 150, is correct.

b)   What is the unknown quantity? It is a percent. Let x represent the percent.

$$x = \text{the percent}$$

Write the question in English, understand its meaning in terms of math, and write an equation.

Statement: 58.5   is   what percent   of   90?

Equation: 58.5   =   x   ·   90

The equation is $58.5 = x \cdot 90$. The commutative property says that we can write $x \cdot 90$ as $90 \cdot x$. So, we can think of the equation as $58.5 = 90 \cdot x$ or $58.5 = 90x$. Solve this equation.

$$58.5 = 90x$$
$$\frac{58.5}{90} = \frac{90}{90}x \qquad \text{Divide both sides by 90.}$$
$$0.65 = x \qquad \text{Perform the division.}$$

The final answer is not 0.65 because x represents a percent. The last step is to change 0.65 to a percent.

$$0.65 = 65\% \qquad \text{Change 0.65 to a percent.}$$

Therefore, 58.5 is 65% of 90. We can check the answer using multiplication.

$$65\% \text{ of } 90 = 0.65 \cdot 90 = 58.5$$

The answer, 65%, is correct.

 BE CAREFUL When you are asked to find a percent, the value that you get for x is *not* the final answer. You must change that number to a percent.

[YOU TRY 2]

Use an equation to solve each problem.

a) 33% of what number is 132?   b) 43.2 is what percent of 60?

We can use equations to solve applied problems.

2 Solve Applications Involving Percents

EXAMPLE 3

Rukshad earns $2600 per month, and she puts 5% in her bank account to save for college expenses. How much does she save each month? How much of her check remains?

Solution

Step 1: **Read** the problem carefully, and identify what you are being asked to find.

We must find the amount Rukshad saves each month. Then, determine how much of her check remains.

Step 2: **Choose a variable** to represent the unknown.

$x =$ the amount Rukshad saves each month

Step 3: **Translate** the information that appears in English into an equation. We can think of this problem as, "The amount saved each month is 5% of $2600."

Statement: | Amount saved each month | is | 5% | of | $2600 |

$$\downarrow \qquad \downarrow \quad \downarrow \quad \downarrow \qquad \downarrow$$

Equation: $\qquad x \qquad = \quad 0.05 \quad \cdot \qquad 2600$

The equation is $x = 0.05 \cdot 2600$.

Step 4: **Solve** the equation.

$$x = 0.05 \cdot 2600 = 130$$

Step 5: **Check** the answer, and **interpret** the solution as it relates to the problem.

Rukshad saves $130 each month. Double-check to see that $0.05 \cdot 2600 = 130$. The amount of her check that remains is $2600 - \$130 = \$2470$.

> **[YOU TRY 3]** Ciu Han earns $3800 per month plus a bonus, based on performance, that is a percent of her monthly income. In May, she earned a 7% bonus. Find the amount of the bonus and Ciu Han's total income in May.

EXAMPLE 4

W Hint

Did you notice that this problem could be solved using the statement "75% of the distance from home to Disneyland is 270 miles"?

The Montez family is driving to Disneyland for their family vacation, and they have already driven 220 mi. Mr. Montez has calculated that they have completed approximately 75% of their trip. Find the total distance from their home to Disneyland. (Round the answer to the nearest mile.)

Solution

Step 1: **Read** the problem carefully, and identify what you are being asked to find.

We must find the distance from the Montez family's home to Disneyland.

Step 2: **Choose a variable** to represent the unknown.

$$x = \text{the distance from home to Disneyland}$$

Step 3: **Translate** the information that appears in English into an equation. Since the family has already driven 220 mi, we can think of this problem as

220 miles is 75% of the distance from home to Disneyland.

Statement: | 220 miles | is | 75% | of | the distance from home to Disneyland. |

$$\downarrow \qquad \downarrow \quad \downarrow \quad \downarrow \qquad \downarrow$$

Equation: $\qquad 220 \qquad = \quad 0.75 \quad \cdot \qquad x$

The equation is $220 = 0.75 \cdot x$.

Step 4: **Solve** the equation.

$$220 = 0.75x$$
$$\frac{220}{0.75} = \frac{0.75}{0.75}x \qquad \text{Divide both sides by 0.75.}$$
$$293.\overline{3} = x \qquad \text{Perform the division.}$$

Step 5: **Check** the answer, and **interpret** the solution as it relates to the problem.

Rounding to the nearest mile, the total distance between their home and Disneyland is 293 mi. To check the answer, find 75% of 293: $0.75 \cdot 293 = 219.75$. This is close to 220 mi, but not exact, because we rounded the distance to the nearest mile.

[**YOU TRY 4**]

At noon on election day, approximately 32% of registered voters in a particular district had already voted. If 501 people voted before noon, how many registered voters live in that district? (Round to the nearest person.)

Note

We could also solve Example 4 using a proportion as in the previous section.

In Example 4, $x =$ the distance from home to Disneyland. Because the family has driven 220 mi out of x total miles of the trip, we can write a ratio,

$$\frac{220}{x} \quad \begin{array}{l} \leftarrow \text{Number of miles completed} \\ \leftarrow \text{Total miles} \end{array}$$

They have completed 75% of their trip, so we can think of 75% as

$$\frac{75}{100} \quad \begin{array}{l} \leftarrow \text{Number of miles completed} \\ \leftarrow \text{Total miles} \end{array}$$

Write a proportion and solve.

$$\begin{array}{c} \text{Number of miles completed} \rightarrow \dfrac{220}{x} = \dfrac{75}{100} \leftarrow \text{Number of miles completed} \\ \text{Total miles} \rightarrow \phantom{\dfrac{220}{x}} \phantom{\dfrac{75}{100}} \leftarrow \text{Total miles} \end{array}$$

$$220 \cdot 100 = 75x \qquad \text{Set the cross products equal.}$$
$$22{,}000 = 75x \qquad \text{Multiply.}$$
$$293.\overline{3} = x \qquad \text{Divide by 75.}$$

Rounding to the nearest mile, the distance from their home to Disneyland is 293 mi.

Sometimes, we are asked to find a percentage in an applied problem.

EXAMPLE 5

In-Class Example 5

Jim bought a digital camera on sale for $104.00, and the original price was $130.00. What percent of the original price did Jim save?

Answer: 20%

Lara bought a sweater on sale for $36.00 that normally sells for $60.00. What percent of the original price did Lara save?

Solution

Step 1: **Read** the problem carefully, and identify what we are being asked to find.

We must find the percent of the original price that was saved.

Step 2: **Choose a variable** to represent the unknown.

$$x = \text{the percent savings}$$

Step 3: **Translate** the information that appears in English into an equation. Since we must find the *percent* saved, we must first find the *amount* of money Lara saved.

Amount of money saved = Original price − Sale price
Amount of money saved = $60.00   − $36.00 = $24.00

Now, we can think of the problem this way:

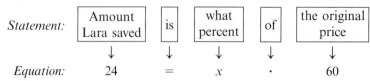

The amount Lara saved is what percent of the original price?

Statement:	Amount Lara saved	is	what percent	of	the original price
	↓	↓	↓	↓	↓
Equation:	24	=	x	·	60

The equation is $24 = x \cdot 60$ or $24 = 60x$.

Step 4: **Solve** the equation.

$$24 = 60x$$
$$\frac{24}{60} = \frac{60}{60}x \qquad \text{Divide both sides by 60.}$$
$$0.4 = x \qquad \text{Perform the division.}$$

Step 5: **Check** the answer, and **interpret** the solution as it relates to the problem.

Change $x = 0.4$ to a percent: 40%. Lara saved 40% on the sweater.

Check: The amount Lara saved was 40% of $60.00 or $0.40 \cdot 60 = \$24.00$. The sale price of the sweater was $60.00 − $24.00 = $36.00.

[YOU TRY 5]

Shane bought a video game on sale for $48.75; it normally costs $65.00. What percent did he save on the original price?

ANSWERS TO [YOU TRY] EXERCISES

1) a) 45 b) 45   2) a) 400 b) 72%   3) amount of bonus: $266; total May income: $4066
4) 1566 registered voters   5) 25%

E Evaluate **2.7** Exercises   Do the exercises, and check your work.

*Additional answers can be found in the Answers to Exercises appendix.

Objective 1: Use an Equation to Solve a Percent Problem
Write an equation and solve.

1) Find 70% of 90.   63

2) Find 60% of 40.   24

3) What is 14% of 52?   7.28

4) What is 31% of 65?   20.15

 5) 92 is 23% of what number?   400

6) 30 is 12% of what number?   250

7) 6.5% of what number is 9.75?   150

8) 8.4% of what number is 15.12?   180

9) 75 is what percent of 125?   60%

10) 90 is what percent of 225?   40%

11) What percent of 32 is 48?   150%

12) What percent of 76 is 95?   125%

Objective 2: Solve Applications Involving Percents

Solve using the Five Steps for Solving Applied Problems.

13) A baby stroller with a regular price of $69.00 is on sale at 20% off.

   a) Find the amount of the discount. $13.80

   b) Find the sale price. $55.20

14) The Johnson family has decided to give 5% of their annual income to charity. If their annual income is $140,000,

   a) how much do they donate to charity? $7000

   b) how much of their income remains? $133,000

15) Korina buys a new phone for $148.00 and also has to pay 6.5% sales tax. Find the total cost of the phone. $157.62

16) Antoine buys a jacket for $86.00 plus 5.5% sales tax. Find the total cost of the jacket. $90.73

17) In March, Bruno paid $4.40 for a calendar, and this was 25% of the original price. What was the original price of the calendar? $17.60

18) After the holidays, Pilar paid $1.95 for a roll of wrapping paper. This was 30% of the original price. What was the original price of the wrapping paper? $6.50

19) Corey saved $105.00 on a surfboard with a regular price of $525.00. What percent of the original price did he save? 20%

20) A USB thumb drive that regularly sells for $23.00 is marked down by $3.45. What was the percent markdown off the original price? 15%

21) Before the recession, an Internet company had 1100 employees. After the recession, the company had 737 employees. Find the percent decrease in the number of employees. 33%

22) In 2011, the Wellington Police Department issued 450 speeding tickets. In 2012, that number fell to 414. Find the percent decrease in the number of tickets issued. 8%

23) Sameet buys a computer for $460.00 plus $22.08 in sales tax. Find the sales tax rate. 4.8%

24) Alhaji buys a bike for $170.00 plus $10.54 in sales tax. Find the sales tax rate. 6.2%

25) Retail sales of gluten-free products in the United States in 2010 were $2.3 billion. This amount is expected to increase to $5.5 billion in 2015. What is the expected percent increase in sales? Round to the nearest percent. (www.celiac.com) 139%

26) Before a drought in the summer of 2011, a jar of peanut butter sold for $2.99. After damage to the peanut crop, the same jar cost $3.79. Find the percent increase in the price. Round to the nearest percent. 27%

27) At the beginning of May 2011, Facebook had 155.2 million users in the United States. At the end of May, that number fell to 149.4 million. Find the percent decrease. Round to the nearest tenth of a percent. (www.insidefacebook.com) 3.7%

28) At the beginning of May 2011, Facebook had 23.7 million users in Mexico. That number rose to 25.6 million by the end of the month. Find the percent increase. Round to the nearest tenth of a percent. (www.insidefacebook.com) 8.0%

29) A white female born in 1995 has a life expectancy of 79.6 years. For a white female born in 2020, the life expectancy is expected to be 82.4 years. What is the percent increase in life expectancy? Round to the nearest tenth of a percent. (www.census.gov) 3.5%

30) In 1982, the number of 18–20-year-olds in the United States who died while driving drunk was 1458. In 2009, that number fell to 537. Find the percent decrease in the number of deaths due to drunk driving in this age group. Round to the nearest tenth of a percent. (www.iihs.com) 63.2%

Mixed Exercises: Sections 2.4–2.7
This section contains applications from Sections 2.4–2.7.

Write an equation and solve.

31) The width of a kitchen table is 2.5 ft less than the length. The perimeter of the table is 17 ft. Find the dimensions of the table. 3 ft × 5.5 ft

32) The sum of three consecutive odd integers is 231. What are the numbers? 75, 77, and 79

33) Approximately 83 out of every 1000 Americans have diabetes. The population of the United States is about 311,000,000. How many Americans have diabetes? (http://diabetes.niddk.nih.gov) 25,813,000

34) Find the measure of each angle in the triangle.
$m\angle A = 62°, m\angle B = 28°, m\angle C = 90°$

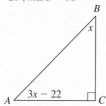

35) At the end of 2000, Walmart operated 3989 stores worldwide. As of September 2011, this number rose to 9759 stores. What is the percent increase in the number of stores? Round to the nearest percent. (http://walmartstores.com) 145%

36) Randy's bonus last year was $6720, and this was 12% of his base salary. What is his base salary? $56,000

37) A pipe is 72 in. long and must be cut into two pieces so that one piece is half as long as the other. Find the length of each piece. 48 in. and 24 in.

38) In Texas in 2011, approximately 17 out of every 250 child care centers were nationally accredited. If 637 child care centers had national accreditation, find the total number of child care centers in Texas. Round to the nearest whole number. (www.naccrra.org) 9368

39) Camilla will be participating in a triathlon. The swimming portion will be 3 mi long, and the cycling leg is 9 mi more than four times the length of the running leg. The length of the entire race is 182 mi. Find the lengths of the running and cycling portions of the race. running: 34 mi, cycling: 145 mi

40) The cost of a round-trip train ticket from Washington D.C. to New York City is $5 more than twice the cost of a bus ticket. Together, a bus ticket and a train ticket cost $191. Find the cost of each ticket. bus: $62, train: $129

41) The sum of the complement of an angle and five times the angle is five degrees less than the supplement of the angle. What is the measure of the angle? 17°

42) Azad buys a new car for $14,600 plus 6.25% sales tax. What is the total cost of the car? $15,512.50

R Rethink

R1) In which case do you want a bank to give you a low percentage rate: when you save money or when you borrow money?

R2) Can you think of any jobs that use percents often?

R3) When was the last time you saw a percent, and what did it describe?

2.8 Solving Linear Inequalities in One Variable

P Prepare

O Organize

What are your objectives for Section 2.8?	How can you accomplish each objective?
1 Use Graphs and Set and Interval Notations	• Write the definition of a *linear inequality in one variable* in your notes. • Complete the given example on your own, and create a procedure to help reproduce these results for future exercises. • Complete You Try 1.
2 Solve Inequalities Using the Addition and Subtraction Properties of Inequality	• Learn the **Addition and Subtraction Properties of Inequality,** and write examples that include numbers to help you understand the properties. • Complete the given example on your own. • Complete You Try 2.
3 Solve Inequalities Using the Multiplication Property of Inequality	• Learn the **Multiplication and Division Properties of Inequality,** and write examples that include numbers to help you understand the property. • Complete the given example on your own. • Complete You Try 3.
4 Solve Inequalities Using a Combination of the Properties	• Write a procedure similar to the one used in Objective 3 of Section 2.2, but make it specific to the properties of inequality. • Complete the given example on your own. • Complete You Try 4.
5 Solve Three-Part Inequalities	• Understand the definitions of *three-part inequality* (*compound inequality*), *lower bound,* and *upper bound.* • Know how to write the solution in interval notation and set notation. • Complete the given example on your own. • Complete You Try 5.
6 Solve Applications Involving Linear Inequalities	• Know key phrases to look for when dealing with inequalities, and use the **Five Steps for Solving Applied Problems.** • Complete the given example on your own. • Complete You Try 6.

 **W Work**   **Read the explanations, follow the examples, take notes, and complete the You Trys.**

Recall the inequality symbols

$<$ "is less than" $\leq$ "is less than or equal to"

$>$ "is greater than" $\geq$ "is greater than or equal to"

We will use the symbols to form *linear inequalities in one variable.*

Definition

A **linear inequality in one variable** can be written in the form $ax + b < c$, $ax + b \le c$, $ax + b > c$, or $ax + b \ge c$, where a, b, and c are real numbers and $a \ne 0$.

Some examples of linear inequalities in one variable are $2x + 11 \le 19$ and $y > -4$. The solution to a linear inequality is a set of numbers that can be represented in one of three ways:

1) On a graph

2) In *set notation*

3) In *interval notation*

In this section, we will learn how to solve linear inequalities in one variable and learn how to represent the solution in each of these three ways.

1 Use Graphs and Set and Interval Notations

EXAMPLE 1

In-Class Example 1

Graph each inequality, and express the solution in set notation and interval notation.
a) $x < 5$ b) $w \ge -4$

Answer: a)

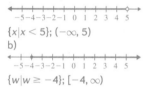

$\{x \mid x < 5\}$; $(-\infty, 5)$
b)

$\{w \mid w \ge -4\}$; $[-4, \infty)$

W Hint

$x \le 2$ means that x is to the left of, or equal to, 2 on the number line.

W Hint

The variable does not appear anywhere in interval notation.

Graph each inequality, and express the solution in set notation and interval notation.

a) $x \le 2$ b) $k > -3$

Solution

a) $x \le 2$

When we graph $x \le 2$, we are finding the solution set of $x \le 2$. What value(s) of x will make the inequality true? The largest solution is 2. Also, any number *less than* 2 will make $x \le 2$ true. We represent this **on the number line** as follows:

The graph illustrates that the solution is the set of all numbers less than and including 2.

Notice that the dot on 2 is shaded. This tells us that 2 is included in the solution set. The shading to the left of 2 indicates that *any* real number (not just integers) in this region is a solution.

We can write the solution set in **set notation** this way: $\{x \mid x \le 2\}$. This means

$$\{x \qquad\qquad | \qquad\qquad x \le 2\}$$

↑	↑	↑
The set of all values of x	such that	x is less than or equal to 2.

In **interval notation,** we write $(-\infty, 2]$

$-\infty$ is not a number. x gets infinitely more negative without bound. Use a "(" instead of a bracket.

The bracket indicates the 2 is included in the interval.

b) $k > -3$

We will plot -3 as an *open circle* on the number line because the symbol is ">" and *not* "≥." The inequality $k > -3$ means that we must find the set of all numbers, k, *greater than* (but *not* equal to) -3. Shade to the right of -3.

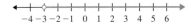

The graph illustrates that the solution is the set of all numbers greater than -3 but not including -3.

We can write the solution set in *set notation* this way: $\{k| k > -3\}$

In *interval notation,* we write

$$(-3, \infty)$$

The "(" indicates that -3 is the lower bound of the interval but that it is not included.

∞ is not a number. k gets increasingly bigger without bound. Use a ")" instead of a bracket.

Hint

Note that the infinity symbol never gets a bracket in interval notation.

Hints for using interval notation:

1) The variable never appears in interval notation.
2) A number *included* in the solution set gets a bracket: $x \le -2 \;\rightarrow\; (-\infty, -2]$
3) A number *not included* in the solution set gets a parenthesis: $k > -3 \;\rightarrow\; (-3, \infty)$
4) The symbols $-\infty$ and ∞ *always* get parentheses.
5) The smaller number is always placed to the left. The larger number is placed to the right.
6) Even if we are not asked to graph the solution set, the graph may be helpful in writing the interval notation correctly.

[YOU TRY 1]

Graph each inequality, and express the solution in interval notation.

a) $z \ge -1$ b) $n < 4$

Note

We can also use a bracket, [, for the closed circle and a parenthesis,), for an open circle on the number line.

Examples: $(-\infty, 2]$

$(-3, \infty)$

2 Solve Inequalities Using the Addition and Subtraction Properties of Inequality

The addition and subtraction properties of equality help us to solve equations. Similar properties hold for inequalities as well.

W Hint

Notice that whenever you add or subtract numbers on both sides of an inequality, the direction of the inequality is unchanged.

Property Addition and Subtraction Properties of Inequality

Let a, b, and c be real numbers. Then,

1) $a < b$ and $a + c < b + c$ are equivalent

and

2) $a < b$ and $a - c < b - c$ are equivalent.

Adding the same number to both sides of an inequality or subtracting the same number from both sides of an inequality will not change the solution.

Note

The above properties hold for any of the inequality symbols.

EXAMPLE 2

In-Class Example 2

Solve $c - 6 < -3$. Graph the solution set, and write the answer in interval and set notations.

Answer:

$(-\infty, 3)$; $\{c | c < 3\}$

Solve $n - 9 \geq -8$. Graph the solution set, and write the answer in interval and set notations.

Solution

$$n - 9 \geq -8$$
$$n - 9 + 9 \geq -8 + 9 \qquad \text{Add 9 to each side.}$$
$$n \geq 1$$

The solution set in interval notation is $[1, \infty)$. In set notation, we write $\{n | n \geq 1\}$.

$\Big[$ **YOU TRY 2** $\Big]$ Solve $q - 5 \geq -3$. Graph the solution set, and write the answer in interval and set notations.

---

3   Solve Inequalities Using the Multiplication Property of Inequality

Let's see how multiplication works in inequalities.

Below, we begin with an inequality we know is true and multiply both sides by a *positive* number on the left and by a *negative* number on the right.

$2 < 5$	True		$2 < 5$	True
$3(2) < 3(5)$	Multiply by 3.		$-3(2) < -3(5)$	Multiply by -3.
$6 < 15$	True		$-6 < -15$	False

In order to make $-6 < -15$ into a *true* statement, we must *reverse the direction of the inequality symbol.*

$$-6 > -15 \qquad \text{True}$$

If you begin with a true inequality and *divide* by a positive number or by a negative number, the results will be the same as above since division can be defined in terms of multiplication. This leads us to the multiplication and division properties of inequality.

Property Multiplication and Division Properties of Inequality

Let a, b, and c be real numbers.

1) If c is a *positive* number, then $a < b$ and $ac < bc$ are equivalent inequalities and have the same solutions.

2) If c is a *negative* number, then $a < b$ and $ac > bc$ are equivalent inequalities and have the same solutions.

3) If c is a *positive* number, then $a < b$ and $\dfrac{a}{c} < \dfrac{b}{c}$ are equivalent inequalities and have the same solutions.

2) If c is a *negative* number, then $a < b$ and $\dfrac{a}{c} > \dfrac{b}{c}$ are equivalent inequalities and have the same solutions.

For the most part, the procedures used to solve linear inequalities are the same as those for solving linear equations **except** *when you multiply or divide an inequality by a negative number, you must reverse the direction of the inequality symbol.*

EXAMPLE 3

In-Class Example 3

Solve each inequality. Graph the solution set, and write the answer in interval and set notations.
a) $4y > -16$
b) $-4y > 16$

Answer: a)

$(-4, \infty)$; $\{y|y > -4\}$
b)

$(-\infty, -4)$; $\{y|y < -4\}$

Solve each inequality. Graph the solution set, and write the answer in interval and set notations.

a) $-6t \le 12$ b) $6t \le -12$

Solution

a) $-6t \le 12$

First, divide each side by -6. *Since we are dividing by a negative number, we must remember to reverse the direction of the inequality symbol.*

$$-6t \le 12$$
$$\frac{-6t}{-6} \ge \frac{12}{-6} \qquad \text{Divide by } -6, \text{ so reverse the inequality symbol.}$$
$$t \ge -2$$

Interval notation: $[-2, \infty)$ Set notation: $\{t|t \ge -2\}$

b) $6t \le -12$

First, divide by 6. Since we are dividing by a *positive* number, the inequality symbol remains the same.

$$6t \le -12$$
$$\frac{6t}{6} \le \frac{-12}{6} \qquad \text{Divide by 6. Do } not \text{ reverse the inequality symbol}$$
$$t \le -2$$

Interval notation: $(-\infty, -2]$ Set notation: $\{t|t \le -2\}$

[YOU TRY 3]

Solve $-\dfrac{1}{2}t < 3$. Graph the solution set, and write the answer in interval and set notations.

---

4 Solve Inequalities Using a Combination of the Properties

Often it is necessary to combine the properties to solve an inequality.

EXAMPLE 4

In-Class Example 4

Solve
$5(2 - 3h) + 9h \leq 2(h + 1)$.
Graph the solution set, and write the answer in interval and set notations.

Answer:

$[1, \infty); \{h \,|\, h \geq 1\}$

W Hint

How does solving the inequality in Example 4 compare to solving the equation $3(1 - 4a) + 15 = 2(2a + 5)$?

Solve $3(1 - 4a) + 15 < 2(2a + 5)$. Graph the solution set, and write the answer in interval and set notations.

Solution

$$3(1 - 4a) + 15 < 2(2a + 5)$$

$3 - 12a + 15 < 4a + 10$	Distribute.
$18 - 12a < 4a + 10$	Combine like terms.
$18 - 12a - 4a < 4a - 4a + 10$	Subtract $4a$ from each side.
$18 - 16a < 10$	
$18 - 18 - 16a < 10 - 18$	Subtract 18 from each side.
$-16a < -8$	
$\dfrac{-16a}{-16} > \dfrac{-8}{-16}$	Divide both sides by -16. Reverse the inequality symbol.
$a > \dfrac{1}{2}$	Simplify.

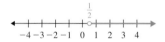

The solution set in interval notation is $\left(\dfrac{1}{2}, \infty\right)$.

In set notation, we write $\left\{ a \,\middle|\, a > \dfrac{1}{2} \right\}$

[YOU TRY 4] Solve $5(b + 2) - 3 > 4(2b - 1)$. Graph the solution set, and write the answer in interval and set notations.

---

5 Solve Three-Part Inequalities

A **three-part inequality** states that one number is between two other numbers. Some examples are $5 < 8 < 12$, $-4 \leq x \leq 1$, and $0 < r + 2 < 5$. They are also called **compound inequalities** because they contain more than one inequality symbol.

The inequality $-4 \leq x \leq 1$ means that x is *between* -4 and 1, and -4 and 1 are included in the interval.

On a number line, the inequality would be represented as

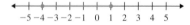

Notice that the **lower bound** of the interval on the number line is -4 (including -4), and the **upper bound** is 1 (including 1). Therefore, we can write the interval notation as

$$[-4, 1]$$

The endpoint, -4, is included in the interval, so use a bracket.

The endpoint, 1, is included in the interval, so use a bracket.

The set notation to represent $-4 \leq x \leq 1$ is $\{x \,|\, -4 \leq x \leq 1\}$.

To solve a three-part inequality, you must remember that *whatever operation you perform on one part of the inequality must be performed on all three parts.* All properties of inequalities apply.

EXAMPLE 5

In-Class Example 5

Solve $-11 \le 3x + 4 < 7$. Graph the solution set, and write the solution in interval notation.

Answer: $[-5, 1)$

$-5\,-4\,-3\,-2\,-1\ \ 0\ \ 1\ \ 2\ \ 3\ \ 4\ \ 5$

Solve $3 \le 2m + 7 < 13$. Graph the solution set, and write the answer in interval notation.

Solution

$$3 \le 2m + 7 < 13$$

$$3 - 7 \le 2m + 7 - 7 < 13 - 7 \qquad \text{Subtract 7 from each part.}$$

$$-4 \le 2m < 6$$

$$\frac{-4}{2} \le \frac{2m}{2} < \frac{6}{2} \qquad \text{Divide each part of the inequality by 2.}$$

$$-2 \le m < 3 \qquad \text{Simplify.}$$

The graph of the solution set is

$$-7\,-6\,-5\,-4\,-3\,-2\,-1\ \ 0\ \ 1\ \ 2\ \ 3$$

In interval notation, we write the solution as $[-2, 3)$.

[**YOU TRY 5**] Solve $-1 < 4p + 11 \le 15$. Graph the solution set, and write the answer in interval notation.

---

6 Solve Applications Involving Linear Inequalities

Certain phrases in applied problems indicate the use of inequality symbols:

at least: $\ge$	no less than: $\ge$
at most: $\le$	no more than: $\le$

There are others. Next, we will look at an example of a problem involving the use of an inequality symbol. We will use the same steps that we used to solve applications involving equations.

EXAMPLE 6

In-Class Example 6

Refer to Example 6. If the cost of the shower is $350 for the first 20 people plus $14 for each additional guest, how many people could attend the baby shower if Keisha can spend at most $650?

Answer: 41

Keisha is planning a baby shower for her sister. The restaurant charges $450 for the first 25 people plus $15 for each additional guest. If Keisha can spend at most $700, find the greatest number of people who can attend the shower.

Solution

Step 1: **Read** the problem carefully. We must find the greatest number of people who can attend the shower.

Step 2: **Choose a variable** to represent the unknown quantity. We know that the first 25 people will cost $450, but we do not know how many *additional* guests Keisha can afford to invite.

$x = $ number of people **over** the first 25 who attend the shower

Step 3: Translate from English to an algebraic inequality.

	Cost of first 25 people	+	Cost of additional guests	is at most	$700

English:

Inequality: 450   + $15x$ $\leq$ 700

The inequality is $450 + 15x \leq 700$.

Step 4: Solve the inequality.

$$450 + 15x \leq 700$$
$$15x \leq 250 \qquad \text{Subtract 450.}$$
$$x \leq 16.\overline{6} \qquad \text{Divide by 15.}$$

Step 5: Check the answer, and **interpret** the solution as it relates to the problem.

The result was $x \leq 16.\overline{6}$, where x represents the number of additional people who can attend the baby shower. Since it is not possible to have $16.\overline{6}$ people, in order to stay within budget Keisha can afford to pay for at most 16 additional guests *over* the initial 25.

Therefore, the greatest number of people who can attend the shower is

The first 25 + Additional = Total

25   +   16   = 41

At most, 41 people can attend the baby shower.

Does the answer make sense?

Total cost of shower = $450 + $15(16) = $450 + $240 = $690

We can see that one more guest (at a cost of $15) would put Keisha over budget.

[YOU TRY 6]

Tristan's basic mobile phone plan gives him 500 minutes of calling per month for $40.00. Each additional minute costs $0.25. If he can spend at most $55.00 per month on his phone bill, find the greatest number of minutes Tristan can talk each month.

ANSWERS TO [YOU TRY] EXERCISES

1) a) $[-1, \infty)$

b) $(-\infty, 4)$

2) interval: $[2, \infty)$, set: $\{q | q \geq 2\}$

3) interval: $[-6, \infty)$, set: $\{t | t > -6\}$

4) interval: $\left(-\infty, \dfrac{11}{3}\right)$, set: $\left\{b \left| b < \dfrac{11}{3}\right.\right\}$

5) $(-3, 1]$ 6) 560

*Additional answers can be found in the Answers to Exercises appendix.

Objective 1: Use Graphs and Set and Interval Notations

1) When do you use brackets when writing a solution set in interval notation?
Use brackets when there is a $\leq$ or $\geq$ symbol.

2) When do you use parentheses when writing a solution set in interval notation?   Use parentheses when there is a $<$ or $>$ symbol or when you use ∞ or $-\infty$.

Write each set of numbers in interval notation.

3)  $(-\infty, 4)$

4) $[-1, \infty)$

5) $[-3, \infty)$

6) $(-\infty, 0)$

Graph the inequality. Express the solution in a) set notation and b) interval notation.

7) $k \leq 2$
a) $\{k|k \leq 2\}$ b) $(-\infty, 2]$

8) $y \geq 3$
a) $\{y|y \geq 3\}$ b) $[3, \infty)$

9) $c < \dfrac{5}{2}$
a) $\left\{c \middle| c < \dfrac{5}{2}\right\}$ b) $\left(-\infty, \dfrac{5}{2}\right)$

10) $n > -\dfrac{11}{3}$
a) $\left\{n \middle| n > -\dfrac{11}{3}\right\}$ b) $\left(-\dfrac{11}{3}, \infty\right)$

11) $a \geq -4$
a) $\{a|a \geq -4\}$ b) $[-4, \infty)$

12) $x < -1$
a) $\{x|x < -1\}$ b) $(-\infty, -1)$

Mixed Exercises: Objectives 2 and 3

13) When solving an inequality, when do you change the direction of the inequality symbol?
when you multiply or divide the inequality by a negative number

14) What is the solution set of $-4x \leq 12$?   b

a) $(-\infty, -3]$ b) $[-3, \infty)$

c) $(-\infty, 3]$ d) $[3, \infty)$

Solve each inequality. Graph the solution set, and write the answer in a) set notation and b) interval notation.

15) $k + 9 \geq 7$
a) $\{k|k \geq -2\}$ b) $[-2, \infty)$

16) $t - 3 \leq 2$
a) $\{t|t \leq 5\}$ b) $(-\infty, 5]$

17) $c - 10 \leq -6$
a) $\{c|c \leq 4\}$ b) $(-\infty, 4]$

18) $x + 12 \geq 8$
a) $\{x|x \geq -4\}$ b) $[-4, \infty)$

19) $-3 + d < -4$
a) $\{d|d < -1\}$ b) $(-\infty, -1)$

20) $1 + k > 1$
a) $\{k|k > 0\}$ b) $(0, \infty)$

21) $16 < z + 11$
a) $\{z|z > 5\}$ b) $(5, \infty)$

22) $-5 > p - 7$
a) $\{p|p < 2\}$ b) $(-\infty, 2)$

23) $5m > 15$
a) $\{m|m > 3\}$ b) $(3, \infty)$

24) $10r > 40$
a) $\{r|r > 4\}$ b) $(4, \infty)$

25) $12x < -21$
a) $\left\{x \middle| x < -\dfrac{7}{4}\right\}$ b) $\left(-\infty, -\dfrac{7}{4}\right)$

26) $6y < -22$
a) $\left\{y \middle| y < -\dfrac{11}{3}\right\}$ b) $\left(-\infty, -\dfrac{11}{3}\right)$

27) $-4b \leq 32$
a) $\{b|b \geq -8\}$ b) $[-8, \infty)$

28) $-7b \geq 21$
a) $\{b|b \leq -3\}$ b) $(-\infty, -3]$

29) $-24a < -40$
a) $\left\{a \middle| a > \dfrac{5}{3}\right\}$ b) $\left(\dfrac{5}{3}, \infty\right)$

30) $-12n > -36$
a) $\{n|n < 3\}$ b) $(-\infty, 3)$

31) $\dfrac{1}{3}k \geq -5$
a) $\{k|k \geq -15\}$ b) $[-15, \infty)$

32) $\dfrac{1}{2}w < -3$
a) $\{w|w < -6\}$ b) $(-\infty, -6)$

33) $-\dfrac{3}{8}c < -3$
a) $\{c|c > 8\}$ b) $(8, \infty)$

34) $-\dfrac{7}{2}d \geq 35$
a) $\{d|d \leq -10\}$ b) $(-\infty, -10]$

Objective 4: Solve Inequalities Using a Combination of the Properties

Solve each inequality. Graph the solution set, and write the answer in interval notation.

35) $4p - 11 \leq 17$ $(-\infty, 7]$

36) $6y + 5 > -13$ $(-3, \infty)$

37) $9 - 2w \leq 11$ $[-1, \infty)$

38) $17 - 7x \geq 20$ $\left(-\infty, -\dfrac{3}{7}\right]$

39) $-\dfrac{3}{4}m + 10 > 1$ $(-\infty, 12)$

40) $\dfrac{1}{2}k - 3 < -2$ $(-\infty, 2)$

41) $3c + 10 > 5c + 13$ $\left(-\infty, -\dfrac{3}{2}\right)$

42) $a + 2 < 2a + 3$ $(-1, \infty)$

43) $3(n + 1) - 16 \leq 2(6 - n)$ $(-\infty, 5]$

44) $-6 - (t + 8) \leq 2(11 - 3t) + 4$ $(-\infty, 8]$

45) $\dfrac{8}{3}(2k + 1) > \dfrac{1}{6}k + \dfrac{8}{3}$ $(0, \infty)$

46) $\dfrac{11}{6} + \dfrac{3}{2}(d - 2) > \dfrac{2}{3}(d + 5) + \dfrac{1}{2}d$ $\left(\dfrac{27}{2}, \infty\right)$

47) $0.05x + 0.09(40 - x) > 0.07(40)$ $(-\infty, 20)$

48) $0.02c + 0.1(30) < 0.08(30 + c)$ $(10, \infty)$

Objective 5: Solve Three-Part Inequalities

Write each set of numbers in interval notation.

49) $[-5, 3]$

50) $(1, 4)$

51) $(-3, 0]$

52) $[-2, 5)$

Graph the inequality. Express the solution in a) set notation and b) interval notation.

53) $-4 < y < 0$
a) $\{y|-4 < y < 0\}$ b) $(-4, 0)$

54) $1 \leq t \leq 4$
a) $\{t|1 \leq t \leq 4\}$ b) $[1, 4]$

55) $-3 \leq k \leq 2$
a) $\{k|-3 \leq k \leq 2\}$ b) $[-3, 2]$

56) $-2 < p < 1$
a) $\{p|-2 < p < 1\}$ b) $(-2, 1)$

57) $\dfrac{1}{2} < n \leq 3$
a) $\left\{n \middle| \dfrac{1}{2} < n \leq 3\right\}$ b) $\left(\dfrac{1}{2}, 3\right]$

58) $-2 \leq a < 3$
a) $\{a|-2 \leq a < 3\}$ b) $[-2, 3)$

Solve each inequality. Graph the solution set, and write the answer in interval notation.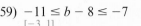

59) $-11 \leq b - 8 \leq -7$
   [-3, 1]

60) $4 < k + 9 < 10$ $(-5, 1)$

61) $-10 < 2a < 7$ $\left(-5, \frac{7}{2}\right)$

62) $-5 \leq 5m \leq -2$ $\left[-1, -\frac{2}{5}\right]$

63) $-5 \leq 4x - 13 \leq 7$
   [2, 5]

64) $-4 < 2y - 7 < -1$ $\left(\frac{3}{2}, 3\right)$

(24) 65) $-17 < \frac{3}{2}c - 5 < 1$
   $(-8, 4)$

66) $2 \leq \frac{1}{2}n + 3 \leq 5$ $[-2, 4]$

67) $-6 \leq 4c - 13 < -1$
   $\left[\frac{7}{4}, 3\right)$

68) $-4 < 3w - 1 \leq 3$ $\left(-1, \frac{4}{3}\right]$

69) $5 < \frac{1}{3}y + 4 \leq 6$
   (3, 6]

70) $1 \leq \frac{5}{2}n - 4 < 6$
   [2, 4)

71) $4 \leq \frac{k + 11}{4} \leq 5$ $[5, 9]$

72) $0 < \frac{5t + 2}{3} < \frac{7}{3}$ $\left(-\frac{2}{5}, 1\right)$

Objective 6: Solve Applications Involving Linear Inequalities

Write an inequality for each problem, and solve.

73) Leslie is planning a party for her daughter at Princess Party Palace. The cost of a party is $180 for the first 10 children plus $16.00 for each additional child. If Leslie can spend at most $300, find the greatest number of children who can attend the party.   17

74) Big-City Parking Garage charges $36.00 for the first 4 hours plus $3.00 for each additional half-hour. Eduardo has $50.00 for parking. For how long can Eduardo park his car in this garage?   at most 6 hours

 75) Heinrich is planning an Oktoberfest party at the House of Bratwurst. It costs $150.00 to rent a tent plus $11.50 per person for food. If Heinrich can spend at most $450.00, find the greatest number of people he can invite to the party.   26

76) A marketing company plans to hold a meeting in a large conference room at a hotel. The cost of renting the room is $500, and the hotel will provide snacks and beverages for an additional $8.00 per person. If the company has budgeted $1000.00 for the room and refreshments, find the greatest number of people who can attend the meeting.   62

77) A taxi in a large city charges $2.50 plus $0.40 for every $\frac{1}{5}$ of a mile. How many miles can you go if you have $14.50?   at most 6 mi

78) A taxi in a small city charges $2.00 plus $0.30 for every $\frac{1}{4}$ of a mile. How many miles can you go if you have $14.00?   at most 10 mi

79) Melinda's first two test grades in Psychology were 87 and 94. What does she need to make on the third test to maintain an average of at least 90?   89 or higher

80) Eliana's first three test scores in Algebra were 92, 85, and 96. What does she need to make on the fourth test to maintain an average of at least 90?   87 or higher

R Rethink

R1) Which exercises in this section did you find most challenging?

R2) Were there any problems you could not do? If so, write them down or circle them and ask your instructor how to do them.

R3) What kind of paid services do you use that place a limit on the number of times you use the service? If there is a cost for going over the limit, how much is it?

Group Activity — The Group Activity can be found online on Connect.

Not all people read the same way. Some people take a big-picture approach when they are reading, trying to understand the material as a whole. Other people read with great attention to detail, analyzing each point as they come to it. Use the following questions to learn how *you* read—that is, your characteristic reading style. Rate how well each statement below describes you. Use this rating scale:

1 = Doesn't describe me at all

2 = Describes me only slightly

3 = Describes me fairly well

4 = Describes me very well

		1	2	3	4
1.	I often reread passages in books that I particularly like.				
2.	I often read good passages aloud to whomever is around.				
3.	I often stop while reading to check that I understood what I just read.				
4.	If I come across a long, unfamiliar name, I try to sound it out and pronounce it correctly.				
5.	If there's a word I don't understand, I look it up in a dictionary right away or mark it to look it up later.				
6.	Before I start reading a textbook or other serious book or article, I look for clues about how it is organized.				
7.	I often question what I'm reading and "argue" with the author.				
8.	I often try to guess what the chapter I'm about to read will cover.				
9.	I often write comments or make notes in books that I own.				
10.	I'm always finding typographical errors in books and articles I read.				

Reading styles range from a big-picture, noncritical style to a very analytic, critical style. Add up the points you assigned yourself. Use this informal scale to find your reading style:

10–12 = Very broad reading style

13–20 = Mostly broad reading style

21–28 = Mostly analytic reading style

29–40 = Very analytic reading style

How do you think your reading style affects the way you learn material in textbooks? Is your style related to the kinds of subjects you prefer? Do you think your reading style affects your leisure reading?

Chapter 2: Summary

Definition/Procedure	Example

2.1 Solving Linear Equations Part I

The Addition and Subtraction Properties of Equality

1) If $a = b$, then $a + c = b + c$.
2) If $a = b$, then $a - c = b - c$. **(p. 89)**

Solve $\quad 3 + b = 20$

$\qquad 3 - 3 + b = 20 - 3 \quad$ Subtract 3 from each side.

$\qquad\qquad\quad b = 17$

The solution set is $\{17\}$.

The Multiplication and Division Properties of Equality

1) If $a = b$, then $ac = bc$.
2) If $a = b$, then $\dfrac{a}{c} = \dfrac{b}{c}\,(c \neq 0)$. **(p. 91)**

Solve $\quad \dfrac{3}{5}m = 6$

$\qquad \dfrac{5}{3} \cdot \dfrac{3}{5}m = \dfrac{5}{3} \cdot 6 \quad$ Multiply each side by $\dfrac{5}{3}$.

$\qquad\qquad m = 10$

The solution set is $\{10\}$.

2.2 Solving Linear Equations Part II

How to Solve a Linear Equation

Step 1: **Clear parentheses** and **combine like terms** on each side of the equation.

Step 2: **Get the variable on one side of the equal sign and the constant on the other side of the equal sign** (isolate the variable) using the addition or subtraction property of equality.

Step 3: **Solve for the variable** using the multiplication or division property of equality.

Step 4: **Check the solution** in the original equation. **(p. 96)**

Solve $2(c + 2) + 11 = 5c + 9$.

$2c + 4 + 11 = 5c + 9$	Distribute.
$2c + 15 = 5c + 9$	Combine like terms.
$2c - 5c + 15 = 5c - 5c + 9$	Get variable terms on one side.
$-3c + 15 = 9$	
$-3c = -6$	Get constants on one side.
$\dfrac{-3c}{-3} = \dfrac{-6}{-3}$	Division property of equality
$c = 2$	

The solution set is $\{2\}$.

2.3 Solving Linear Equations Part III

Solve Equations Containing Fractions or Decimals

To eliminate the fractions, determine the least common denominator (LCD) for all of the fractions in the equation. Then, multiply both sides of the equation by the LCD. **(p. 102)**

To eliminate the decimals from an equation, multiply both sides of the equation by the smallest power of 10 that will eliminate all decimals from the problem. **(p. 104)**

Solve $\dfrac{3}{4}y - 3 = \dfrac{1}{4}y - \dfrac{2}{3}$

$12\left(\dfrac{3}{4}y - 3\right) = 12\left(\dfrac{1}{4}y - \dfrac{2}{3}\right)$	Multiply each side of the equation by 12.
$9y - 36 = 3y - 8$	Distribute.
$9y - 3y - 36 = 3y - 3y - 8$	Get the y-terms on one side.
$6y - 36 = -8$	
$6y - 36 + 36 = -8 + 36$	Get the constants on the other side.
$6y = 28$	
$\dfrac{6y}{6} = \dfrac{28}{6}$	Divide each side by 6.
$y = \dfrac{28}{6} = \dfrac{14}{3}$	Simplify.

The solution set is $\left\{\dfrac{14}{3}\right\}$.

Definition/Procedure	Example

Steps for Solving Applied Problems

Step 1: **Read** and reread the problem. Draw a picture, if applicable.

Step 2: **Choose a variable** to represent an unknown. Define other unknown quantities in terms of the variable.

Step 3: **Translate** from English into math.

Step 4: **Solve** the equation.

Step 5: **Check** the answer in the original problem, and **interpret** the solution as it relates to the problem. **(p. 106)**

Nine less than twice a number is the same as the number plus thirteen. Find the number.

Step 1: **Read** the problem carefully, then read it again.

Step 2: **Choose a variable** to represent the unknown.
$$x = \text{the number}$$

Step 3: "Nine less than twice a number is the same as the number plus thirteen" means $2x - 9 = x + 13$.

Step 4: **Solve** the equation.
$$2x - 9 = x + 13$$
$$2x - 9 + 9 = x + 13 + 9$$
$$2x = x + 22$$
$$x = 22$$

Step 5: The number is 22. The **check** is left to the student.

2.4 Applications of Linear Equations

The **Five Steps for Solving Applied Problems** can be used to solve problems involving general quantities, lengths, and consecutive integers. **(p. 111)**

The sum of three consecutive even integers is 72. Find the integers.

Step 1: **Read** the problem carefully, then read it again.

Step 2: **Choose** a variable, and define the unknowns.
$$x = \text{the first even integer}$$
$$x + 2 = \text{the second even integer}$$
$$x + 4 = \text{the third even integer}$$

Step 3: "The sum of three consecutive even integers is 72" means

First even	+	Second even	+	Third even	=	72
x	+	$(x + 2)$	+	$(x + 4)$	=	72

Equation: $x + (x + 2) + (x + 4) = 72$

Step 4: **Solve** $x + (x + 2) + (x + 4) = 72$
$$3x + 6 = 72$$
$$3x + 6 - 6 = 72 - 6$$
$$3x = 66$$
$$\frac{3x}{3} = \frac{66}{3}$$
$$x = 22$$

Step 5: Find the values of all the unknowns.
$$x = 22, \quad x + 2 = 24, \quad x + 4 = 26$$

The numbers are 22, 24, and 26. **Check** to verify the sum is 72: $22 + 24 + 26 = 72$.

2.5 Geometry Applications and Solving Formulas

Formulas from geometry can be used to solve applications. **(p. 122)**

A rectangular bulletin board has an area of 180 in^2. It is 12 in. wide. Find its length.

Use $A = lw$.　　Formula for the area of a rectangle

$A = 180 \text{ in}^2, \quad w = 12 \text{ in}.$　　Find l.

$$A = lw$$
$$180 = l(12) \quad \text{Substitute values into } A = lw.$$
$$\frac{180}{12} = \frac{l(12)}{12}$$
$$15 = l$$

The length is 15 in.

Definition/Procedure	Example
To solve a formula for a specific variable, think about the steps involved in solving a linear equation in one variable. **(p. 127)**	Solve $C = kr - w$ for k. $C + w = \boxed{k}r - w + w$ Add w to each side. $C + w = \boxed{k}r$ $\dfrac{C + w}{r} = \dfrac{\boxed{k}r}{r}$ Divide each side by r. $\dfrac{C + w}{r} = k$

2.6 Ratios and Proportions

A **ratio** is a quotient of two quantities. The ratio of x to y, where $y \neq 0$, can be written as $\dfrac{x}{y}$, x to y, or $x : y$. **(p. 135)** A **proportion** is a statement that two ratios are equal. To solve a proportion, set the cross products equal to each other and solve the equation. **(p. 136)**	The ratio of 12 in. to 16 in. is $\dfrac{12 \text{ in.}}{16 \text{ in.}} = \dfrac{3}{4}$. Write in lowest terms. Solve $\dfrac{n-3}{4} = \dfrac{n+1}{5}$. $\dfrac{n-3}{4} \diagdown \dfrac{n+1}{5}$ Multiply. Multiply. $5(n-3) = 4(n+1)$ Set the cross products equal. $5n - 15 = 4n + 4$ Distribute. $n - 15 = 4$ Subtract $4n$. $n = 19$ Add 15. The solution set is $\{19\}$.

We can use the Five-Step Method to solve problems involving proportions. **(p. 138)**	If Geri can watch 4 movies in 3 weeks, how long will it take her to watch 7 movies?
	
	Step 1: **Read** the problem carefully twice.
	Step 2: **Choose** a variable to represent the unknown.
	     x = number of weeks to watch 7 movies
	
	Step 3: Set up a proportion. $\dfrac{4 \text{ movies}}{3 \text{ weeks}} = \dfrac{7 \text{ movies}}{x \text{ weeks}}$
	
	Equation: $\dfrac{4}{3} = \dfrac{7}{x}$
	
	Step 4: **Solve** $\dfrac{4}{3} \diagdown \dfrac{7}{x}$. Multiply.
	         Multiply.
	$4x = 3(7)$ Set cross products equal.
	$\dfrac{4x}{4} = \dfrac{21}{4}$ Divide by 4
	
	$x = \dfrac{21}{4} = 5\dfrac{1}{4}$
	
	Step 5: It will take Geri $5\dfrac{1}{4}$ weeks to watch 7 movies. The check is left to the student.

Definition/Procedure	Example

2.7 Applications of Percents

We can use an equation to solve a percent problem. **(p. 143)**

Use an equation to solve: 98 is 28% of what number?

Solution

$$\text{Let } x = \text{the number.}$$

Write the question in English, then write the equation.

Statement: $\boxed{98}$ $\boxed{\text{is}}$ $\boxed{28\%}$ $\boxed{\text{of}}$ $\boxed{\text{what number?}}$

Equation: 98 = 0.28 · x

The equation is $98 = 0.28x$. Solve it.

$$98 = 0.28x$$

$$\frac{98}{0.28} = \frac{0.28x}{0.28} \qquad \text{Divide by 0.28.}$$

$$350 = x \qquad \text{Perform the division.}$$

98 is 28% of 350.

Solving Applications Involving Percents

We can use an equation to solve many different types of applications involving percents. **(p. 145)**

Use the Five-Step Process to solve the problem.

In 2000, the population of Lansing, Michigan, was 119,128. In 2010, the population was 114,297. Find the percent decrease in the population. Round to the nearest tenth of a percent. (http://quickfacts.census.gov)

Solution

Step 1: **Read** the problem carefully, and identify what we are being asked to find.

We must find the percent decrease in the population.

Step 2: **Choose a variable** to represent the unknown.

$$\text{Let } x = \text{the percent decrease.}$$

Step 3: **Translate** the information that appears in English into an equation. Since we must find the *percent* decrease, we must first find the *amount* of the decrease.

Amount of decrease = Population in 2000 − Population in 2010

Amount of decrease = 119,128 − 114,297 = 4831

Now, we can think of the problem this way:

The amount of the decrease is what percent of the population in 2000?

Statement: $\boxed{\begin{array}{c}\text{Amount of}\\\text{the decrease}\end{array}}$ $\boxed{\text{is}}$ $\boxed{\begin{array}{c}\text{what}\\\text{percent}\end{array}}$ $\boxed{\text{of}}$ $\boxed{\begin{array}{c}\text{the population}\\\text{in 2000?}\end{array}}$

Equation: 4831 = x · 119,128

The equation is $4831 = x \cdot 119{,}128$ or $4831 = 119{,}128x$.

Definition/Procedure	Example
	Step 4: **Solve** the equation.

$$4831 = 119{,}128x$$
$$\frac{4831}{119{,}128} = \frac{119{,}128x}{119{,}128} \qquad \text{Divide both sides by 119,128.}$$
$$0.04055\ldots = x \qquad \text{Perform the division.}$$

Step 5: **Check** the answer, and **interpret** the solution as it relates to the problem.

Change $x = 0.04055\ldots$ to a percent, and round to the nearest tenth: 4.1%. The population decreased by approximately 4.1% from 2000 to 2010.

Check: The amount of the decrease was approximately 4.1% of 119,128 or $0.041 \cdot 119{,}128 \approx 4884$. Then, the population in 2010 was $119{,}128 - 4884 = 114{,}244$. This is not exactly the population given in the problem because of rounding the percentage, but it is close. The answer is correct. |

2.8 Solving Linear Inequalities in One Variable

A **linear inequality in one variable** can be written in the form $ax + b < c, ax + b \le c, ax + b > c,$ or $ax + b \ge c.$

We solve linear inequalities in very much the same way we solve linear equations *except that when we multiply or divide by a negative number, we must reverse the direction of the inequality symbol.*

We can graph the solution set, write the solution in set notation, or write the solution in interval notation. **(p. 151)** | Solve $x - 9 \le -7$. Graph the solution set, and write the answer in both set notation and interval notation.

$$x - 9 \le -7$$
$$x - 9 + 9 \le -7 + 9 \qquad \text{Add 9 to each side.}$$
$$x \le 2$$

$$-3\ -2\ -1\ \ 0\ \ 1\ \ 2\ \ 3$$

$\{x \mid x \le 2\}$ Set notation
$(-\infty, 2]$ Interval notation |
| A **three-part inequality** states that one number is between two other numbers.

To **solve a three-part inequality,** remember that each operation you perform on one part of the inequality must be performed on all three parts. **(p. 156)** | Solve $1 \le 3k + 7 < 19$.

$$1 \le 3k + 7 < 19$$
$$1 - 7 \le 3k + 7 - 7 < 19 - 7 \qquad \text{Subtract 7.}$$
$$-6 \le 3k < 12$$
$$\frac{-6}{3} \le \frac{3k}{3} < \frac{12}{3} \qquad \text{Divide by 3.}$$
$$-2 \le k < 4$$

The graph of the solution set is

$$-5\ -4\ -3\ -2\ -1\ \ 0\ \ 1\ \ 2\ \ 3\ \ 4\ \ 5$$

In interval notation, the solution is $[-2, 4)$. |

Chapter 2: Review Exercises

(2.1–2.3) Determine whether the given value is a solution to the equation.

1) $\frac{3}{2}k - 5 = 1$; $k = -4$ no

2) $5 - 2(3p + 1) = 9p - 2$; $p = \frac{1}{3}$ yes

3) How do you know that an equation has no solution?
The variables are eliminated, and you get a false statement like $5 = 13$.

4) What can you do to make it easier to solve an equation with fractions? Multiply both sides in the equation by the LCD of all the fractions to eliminate the fractions.

Solve each equation.

5) $h + 14 = -5$ $\{-19\}$ 6) $w - 9 = 16$ $\{25\}$

7) $-7g = 56$ $\{-8\}$ 8) $-0.78 = -0.6t$ $\{1.3\}$

9) $4 = \frac{c}{9}$ $\{36\}$ 10) $-\frac{10}{3}y = 16$ $\left\{-\frac{24}{5}\right\}$

11) $23 = 4m - 7$ $\left\{\frac{15}{2}\right\}$ 12) $\frac{1}{6}v - 7 = -3$ $\{24\}$

13) $4c + 9 + 2(c - 12) = 15$ $\{5\}$

14) $\frac{5}{9}x + \frac{1}{6} = -\frac{3}{2}$ $\{-3\}$ 15) $2z + 11 = 8z + 15$ $\left\{-\frac{2}{3}\right\}$

16) $8 - 5(2y - 3) = 14 - 9y$ $\{9\}$

17) $k + 3(2k - 5) = 4(k - 2) - 7$ $\{0\}$

18) $10 - 7b = 4 - 5(2b + 9) + 3b$ $\varnothing$

19) $0.18a + 0.1(20 - a) = 0.14(20)$ $\{10\}$

20) $16 = -\frac{12}{5}d$ $\left\{-\frac{20}{3}\right\}$ 21) $3(r + 4) - r = 2(r + 6)$
$\{$all real numbers$\}$

22) $\frac{1}{2}(n - 5) - 1 = \frac{2}{3}(n - 6)$ $\{3\}$

Write each statement as an equation, and find the number.

23) Nine less than twice a number is twenty-five.
$2x - 9 = 25$; 17

24) One more than two-thirds of a number is the same as the number decreased by three. $\frac{2}{3}x + 1 = x - 3$; 12

(2.4) Solve using the five-step method.

25) Kendrick received 24 fewer e-mails on Friday than he did on Thursday. If he received a total of 126 e-mails on those two days, how many did he get on each day?
Thursday: 75; Friday: 51

26) The number of Michael Jackson solo albums sold the week after his death was 42.2 times the number sold the previous week. If a total of 432,000 albums were sold during those two weeks, how many albums were sold the week after his death? (http://abcnews.go.com) 422,000

27) A plumber has a 36-in. pipe that he has to cut into two pieces so that one piece is 8 in. longer than the other. Find the length of each piece. 14 in., 22 in.

28) The sum of three consecutive integers is 249. Find the integers. 82, 83, 84

(2.5) Substitute the given values into the formula, and solve for the remaining variable.

29) $P = 2l + 2w$; If $P = 32$ when $l = 9$, find w. 7

30) $V = \frac{1}{3}\pi r^2 h$; If $V = 60\pi$ when $r = 6$, find h. 5

Use a known formula to solve.

31) The base of a triangle measures 12 in. If the area of the triangle is 42 in^2, find the height. 7 in.

32) The Statue of Liberty holds a tablet in her left hand that is inscribed with the date, in Roman numerals, that the Declaration of Independence was signed. The length of this rectangular tablet is 120 in. more than the width, and the perimeter of the tablet is 892 in. What are the dimensions of the tablet? (www.nps.gov) 163 in. × 283 in.

33) Find the missing angle measures.
$m\angle A = 55°, m\angle B = 55°, m\angle C = 70°$

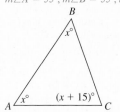

Find the measure of each indicated angle.

34)   32°, 148°

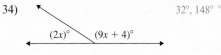
$(2x)°$ $(9x + 4)°$

35)   61°, 61°

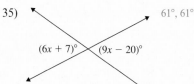
$(6x + 7)°$ $(9x - 20)°$

Solve using the five-step method.

36) The sum of the supplement of an angle and twice the angle is 10° more than four times the measure of its complement. Find the measure of the angle. 38°

Solve for the indicated variable.

37) $p - n = z$ for p $p = z + n$

38) $r = ct + a$ for t $t = \frac{r - a}{c}$

39) $A = \frac{1}{2}bh$ for b $b = \frac{2A}{h}$

40) $M = \frac{1}{4}k(d + D)$ for D $D = \frac{4M}{k} - d$ or $D = \frac{4M - dk}{k}$

41) Can 15% be written as a ratio? Explain. Yes. It can be written as $\frac{15}{100}$ or $\frac{3}{20}$.

42) What is the difference between a ratio and a proportion? A ratio is a quotient of two quantities. A proportion is a statement that two ratios are equal.

43) Write the ratio of 12 girls to 15 boys in lowest terms. $\frac{4}{5}$

44) A store sells olive oil in three different sizes. Which size is the best buy, and what is its unit price?

Size	Price
17 oz	$ 8.69
25 oz	$11.79
101 oz	$46.99

101-oz container: $0.465 per oz

Solve each proportion.

45) $\dfrac{x}{15} = \dfrac{8}{10}$ {12}   46) $\dfrac{2c + 3}{6} = \dfrac{c - 4}{2}$ {15}

Set up a proportion and solve.

47) The 2009 Youth Risk Behavior Survey found that about 5 out of 12 high school students drank some amount of alcohol in the 30 days preceding the survey. If a high school has 2400 students, how many would be expected to have used alcohol within a 30-day period? (www.cdc.gov) 1000

48) Given these two similar triangles, find x.

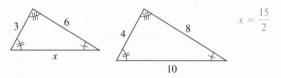

$x = \dfrac{15}{2}$

(2.7) Write an equation and solve.

49) What is 30% of 150? 45

50) 23% of what number is 36.87? 160

51) 13 is what percent of 25? 52%

52) The local humane society currently has 60 cats. If 300 animals are in the shelter, what percent are cats? 20%

53) In January 2012, the Social Security Administration increased benefits to its recipients for the first time since 2009. Before the increase, James received $1200/month. After the increase, he received $1243.20/month. Find the percent increase in Social Security benefits. (http://quickfacts.census.gov) 3.6%

54) Rusty is an insurance agent and is writing a summary of an insurance policy for the Scheel family. Their *dwelling coverage* is for replacement of their house, while their *personal property coverage* is for replacement of the items in their house. According to the insurance company's guidelines, personal property coverage is 75% of dwelling coverage. If the Scheel family lost their home in a fire and their dwelling coverage is $187,000, how much would they receive from the insurance agency to cover the items lost inside their house? $140,250

(2.8) Solve each inequality. Graph the solution set, and write the answer in interval notation.

55) $w + 8 > 5$ $(-3, \infty)$

56) $-6k \le 15$ $\left[-\dfrac{5}{2}, \infty\right)$

57) $5x - 2 \le 18$ $(-\infty, 4]$

58) $3(3c + 8) - 7 > 2(7c + 1) - 5$ $(-\infty, 4)$

59) $-19 \le 7p + 9 \le 2$ $[-4, -1]$

60) $-3 < \dfrac{3}{4}a - 6 \le 0$ $(4, 8]$

61) $1 \le \dfrac{1}{2}t + 1 < 2$ $[0, 2)$

62) *Write an inequality, and solve.* Gia's scores on her first three History tests were 94, 88, and 91. What does she need to make on her fourth test to have an average of at least 90? 87 or higher

Mixed Exercises: Solving Equations, Inequalities, and Applications

Solve each equation or inequality. Graph the solution set of an inequality, and also write it in interval notation.

63) $-8k + 13 = -7$ $\left\{\dfrac{5}{2}\right\}$

64) $-7 - 4(3w - 2) = 1 - 9w$ {0}

65) $29 = -\dfrac{4}{7}m + 5$ {-42}

66) $\dfrac{c}{20} = \dfrac{18}{12}$ {30}

67) $-8m > 4$ $\left(-\infty, -\dfrac{1}{2}\right)$

68) $-3 \le 5t + 2 < 7$ $[-1, 1)$

69) $10p + 11 = 5(2p + 3) - 1$ $\varnothing$

70) $0.14a + 0.06(36 - a) = 0.12(36)$ {27}

71) $3n - 17 > 5 + 2(5n + 3)$ $(-\infty, -4)$

72) $6 < 2v + 9 < 13$ $\left(-\dfrac{3}{2}, 2\right)$

73) $\dfrac{2x + 9}{5} = \dfrac{x + 1}{2}$ {13}

74) $14 = 8 - h$ {-6}

75) $-4 \le \dfrac{5}{3}y - 9 \le 1$ $[3, 6]$

76) $10 - z \le 8$ $[2, \infty)$

77) $\dfrac{5}{6} - \dfrac{3}{4}(r + 2) = \dfrac{1}{2}r + \dfrac{7}{12}$ {−1}

78) $\dfrac{1}{4}d + \dfrac{9}{4} = 1 + \dfrac{1}{4}(d + 5)$ {all real numbers}

Solve using the five-step procedure.

79) The sum of an angle and three times its complement is 33° more than its supplement. Find the measure of the angle.   57°

80) A library offers free tutoring after school for children in grades 1–5. The number of students who attended on Friday was half the number who attended on Thursday. How many students came for tutoring each day if the total number of students served on both days was 42?
Thursday: 28; Friday: 14

81) The sum of two consecutive odd integers is 21 less than three times the larger integer. Find the numbers.   17, 19

82) The perimeter of a triangle is 35 cm. One side is 3 cm longer than the shortest side, and the longest side is twice as long as the shortest. How long is each side of the triangle?
8 cm, 11 cm, 16 cm

83) A 2008 poll revealed that 9 out of 25 residents of Quebec, Canada, wanted to secede from the rest of the country. If 1000 people were surveyed, how many said they would like to see Quebec separate from the rest of Canada? (www.bloomberg.com)   360

84) If a certain environmental bill is passed by Congress, the United States would have to reduce greenhouse gas emissions by 17% from 2005 levels by the year 2020. The University of New Hampshire has been at the forefront of reducing emissions, and if this bill is passed they would be required to have a greenhouse gas emission level of about 56,440 MTCDE (metric tons carbon dioxide equivalents) by 2020. Find their approximate emission level in 2005. (www.sustainableunh.unh.edu, www.knoxnews.com)
68,000 MTCDE

Chapter 2:   Test

*Additional answers can be found in the Answers to Exercises appendix.

Solve each equation.

1) $-18y = 14$ $\left\{-\dfrac{7}{9}\right\}$

2) $16 = 7 - a$ {−9}

3) $\dfrac{8}{3}n - 11 = 5$ {6}

4) $3c - 2 = 8c + 13$ {−3}

5) $\dfrac{1}{2} - \dfrac{1}{6}(x - 5) = \dfrac{1}{3}(x + 1) + \dfrac{2}{3}$ $\left\{\dfrac{2}{3}\right\}$

6) $7(3k + 4) = 11k + 8 + 10k - 20$ ∅

7) $\dfrac{9 - w}{4} = \dfrac{3w + 1}{2}$ {1}

8) What is the difference between a ratio and a proportion?
A ratio is a quotient of two quantities. A proportion is a statement that two ratios are equal.

Solve using the five-step method.

9) Ramon's tuition bill for Spring semester of 2013 was $174 more than his bill for the Fall 2012 semester. He paid a total of $2784 for both semesters. Find the amount of Ramon's tuition bill each semester.
Fall: $1305, Spring: $1479

10) The sum of three consecutive even integers is 114. Find the numbers.   36, 38, 40

11) Ray buys 14 gallons of gas and pays $40.60. His wife, Debra, goes to the same gas station later that day and buys 11 gallons of the same gasoline. How much did she spend?
$31.90

12) The tray table on the back of an airplane seat is in the shape of a rectangle. It is 5 in. longer than it is wide and has a perimeter of 50 in. Find the dimensions.   10 in. by 15 in.

13) Given the following similar triangles, find x.   4.5

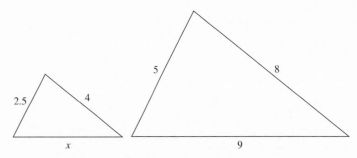

14) Nina bought a television on sale for $240 that usually sells for $300. What percent of the original price did she save?   20%

Solve for the indicated variable.

15) $B = \dfrac{an}{4}$ for a $a = \dfrac{4B}{n}$

16) $S = 2\pi r^2 + 2\pi rh$ for h $h = \dfrac{S - 2\pi r^2}{2\pi r}$ or $h = \dfrac{S}{2\pi r} - r$

17) Find the measure of each indicated angle.

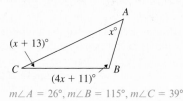

$(x + 13)°$ $x°$ A
C $(4x + 11)°$ B

$m\angle A = 26°, m\angle B = 115°, m\angle C = 39°$

Solve. Graph the solution set, and write the answer in interval notation.

18) $6m + 19 \le 7$ $(-\infty, -2]$

19) $1 - 2(3x - 5) < 2x + 5$ $\left(\dfrac{3}{4}, \infty\right)$

20) $-5 \le \dfrac{3}{2}a - 5 < 4$ $[0, 6)$ ⟵|—|—|—|—|—|○—|⟶
  0 1 2 3 4 5 6 7

Chapter 2:   Cumulative Review for Chapters 1 and 2

*Additional answers can be found in the Answers to Exercises appendix.

Perform the operations and simplify.

1) $\dfrac{3}{8} - \dfrac{5}{6}$ $-\dfrac{11}{24}$

2) $\dfrac{5}{8} \cdot 12$ $\dfrac{15}{2}$

3) $26 - 14 \div 2 + 5 \cdot 7$ 54

4) $-82 + 15 + 10(1 - 3)$ -87

5) $-39 - |7 - 15|$ -47

6) Find the area of a triangle with a base of length 9 cm and height of 6 cm. $27\ \text{cm}^2$

Given the set of numbers $\left\{\dfrac{3}{4}, -5, \sqrt{11}, 2.5, 0, 9, 0.\overline{4}\right\}$, identify

7) the integers. $\{-5, 0, 9\}$

8) the rational numbers. $\left\{\dfrac{3}{4}, -5, 2.5, 0, 0.\overline{4}, 9\right\}$

9) the whole numbers. $\{0, 9\}$

10) Which property is illustrated by $6(5 + 2) = 6 \cdot 5 + 6 \cdot 2$?
  distributive

11) Does the commutative property apply to the subtraction of real numbers? Explain. No. For example, $10 - 3 \ne 3 - 10$.

12) Combine like terms.
  $11y^2 - 14y + 6y^2 + y - 5y$ $17y^2 - 18y$

Solve.

13) $8t - 17 = 10t + 6$ $\left\{-\dfrac{23}{2}\right\}$

14) $\dfrac{3}{2}n + 14 = 20$ $\{4\}$

15) $3(7w - 5) - w = -7 + 4(5w - 2)$ {all real numbers}

16) $\dfrac{x + 3}{10} = \dfrac{2x - 1}{4}$ $\left\{\dfrac{11}{8}\right\}$

17) $-\dfrac{1}{2}c + \dfrac{1}{5}(2c - 3) = \dfrac{3}{10}(2c + 1) - \dfrac{3}{4}c$ $\{18\}$

18) Solve for k. $M = \dfrac{1}{4}k(d + D)$ $k = \dfrac{4M}{d + D}$

19) Solve $\dfrac{x + 2}{2} = \dfrac{2x - 5}{7}$ $\{-8\}$

20) The sum of three consecutive odd integers is 69. Find the integers. $21, 23, 25$

21) Find the measure of each indicated angle. $145°, 35°$

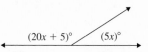

$(20x + 5)°$ $(5x)°$

22) What is 112% of 63? 70.56

23) Nancy and Sean put down $40,000 on a new house. If this represents a 20% down payment, how much did the house cost? $200,000

Solve. Write the answer in interval notation.

24) $7k + 4 \ge 9k + 16$ $(-\infty, -6]$

25) $-17 < 6b - 11 < 4$ $\left(-1, \dfrac{5}{2}\right)$

170 CHAPTER 2 **Linear Equations and Inequalities** www.mhhe.com/messersmith

Linear Equations in Two Variables

Math at Work:

Landscape Architect

Let's take one last look at Jill Cavanaugh's use of math in her work as a landscape architect.

Jill uses slope in many different ways. She explains that one important application of slope is in designing driveways after a new house has been built. Towns often have building codes that restrict the slope or steepness of a driveway. In this case, the rise of the land is the difference in height between the top and the bottom of the driveway. The run is the linear horizontal distance between those two points. By computing *rise/run,* a landscape architect knows how to design the driveway so that it meets the town's building code. This is especially important in cold-weather climates, where if a driveway is too steep, a car will too easily slide into the street. If the driveway doesn't meet the code, it may have to be removed and rebuilt, or coils that radiate heat might have to be installed under the driveway to melt the snow in the wintertime. Either way, a mistake in calculating slope could cost Jill's company or her client a lot of extra money.

It is also essential for Jill to make her calculations and carry out her work in a timely manner. Delays cost her—and her clients—money! So Jill has learned to be an effective time manager, figuring out how long projects will take and creating a schedule to ensure that everything gets done on time.

In Chapter 3, we will learn about slope, its meaning, and different ways to use it. We will also introduce some strategies you can use to help you manage your time, in the classroom or in your career.

 **Study Strategies** Time Management

Do you know people who say things like, "Oh, I'll find the time to do that later"? Or are you yourself someone who makes such comments? While we might like to think that our schedule will arrange itself and we'll "find the time" to accomplish our tasks, the simple truth is that time management is rarely so easy. For college students in particular, it is essential to plan carefully how you will use your time. The strategies below will help ensure that you *make* the time for your most important obligations.

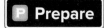

- Start the time management process by creating a **time log:** a record of how you actually spend your time, including interruptions, noting blocks of time in increments as short as 15 minutes.
- In order to build your schedule around the activities and goals that matter most to you, create a ranked list of your priorities.

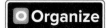

- Make a **master calendar** that shows every week of the term on a single page. Write in all your major class assignments and any significant events in your personal life.
- Create a **weekly timetable** with the days of the week across the top and the hours of the day along the side. Write in all your regularly scheduled activities, such as your classes, as well as one-time events.
- Finally, make a habit of creating a **daily to-do list** that you carry with you every day.

W Work

- Use the schedules you've made as tools to help you manage your time and stay focused on your priorities.
- Avoid "time traps" like procrastination and situations that you know will distract you from your priorities.

E Evaluate

- Review your daily to-do list to determine whether you are accomplishing your short-term tasks.
- Use your weekly timetable and master calendar to see whether you are on track to meet your long-term obligations.

R Rethink

- Consider how you feel about time during the day. If you find yourself rushed and stressed, think about ways to reduce your obligations.
  Remember: One person can only do so much in a 24-hour day!
- Reassess your priorities. Are you making time for what really matters to you?

Chapter 3 **POWER** Plan

P Prepare

What are your goals for Chapter 3?

1 Be prepared before and during class.

2 Understand the homework to the point where you could do it without needing any help or hints.

3 Use the P.O.W.E.R. framework to help manage your time: *Find Your Time Style.*

4 Write your own goal.
_____
_____

O Organize

How can you accomplish each goal?

- Don't stay out late the night before, and be sure to set your alarm clock!
- Bring a pencil, notebook paper, and textbook to class.
- Avoid distractions by turning off your cell phone during class.
- Pay attention, take good notes, and ask questions.
- Complete your homework on time, and ask questions on problems you do not understand.

- Read the directions, and show all of your steps.
- Go to the professor's office for help.
- Rework homework and quiz problems, and find similar problems for practice.

- Read the Study Strategy that explains how to manage your time.
- What does P.O.W.E.R. stand for?
- Complete the emPOWERme that appears before the Chapter Summary.

- _____
  _____

What are your objectives for Chapter 3?

1 Understand the basic concepts of linear equations.

2 Know how to graph a linear equation in two variables using various methods.

3 Understand slope, the slope-intercept form of a line, and how the slopes of parallel lines and perpendicular lines are related.

4 Know how to write an equation of a line.

5 Write your own goal.
_____
_____

How can you accomplish each objective?

- Know how to identify a linear equation in two variables and use the Cartesian coordinate system.
- Be able to work with ordered pairs; determine whether they are a solution, complete an ordered pair, and plot them.

- Know the properties for solutions of linear equations in two variables and the graphs of linear equations in two variables.
- Graph by finding ordered pairs, by finding intercepts, and plotting points.
- Know how to quickly graph linear equations of the form $Ax + By = 0$, $x = a$, and $y = b$.

- Know three different ways to represent the slope of a line.
- Know how to rewrite an equation and graph a line expressed in slope-intercept form.
- Understand parallel and perpendicular lines, and be able to determine whether lines are parallel or perpendicular.

- Write an equation in standard form.
- Write an equation of a line given its slope and y-intercept.
- Use the point-slope formula to write an equation of a line.
- Write equations of horizontal and vertical lines.
- Write equations of lines that are parallel or perpendicular to another line.

- _____
  _____

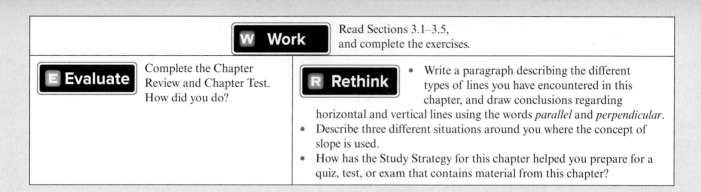

3.1 Introduction to Linear Equations in Two Variables

P Prepare

O Organize

What are your objectives for Section 3.1?	How can you accomplish each objective?
1 Define a Linear Equation in Two Variables	• Write the definition of *linear equation in two variables* in your own words, and write an example. • Understand the *Cartesian coordinate system* or *rectangular coordinate system*.
2 Decide Whether an Ordered Pair Is a Solution of a Given Equation	• Know how to substitute values into an equation. • Complete the given example on your own. • Complete You Try 1.
3 Complete Ordered Pairs for a Given Equation	• Be able to substitute either the *x*-value or the *y*-value, and solve for the missing variable. • Complete the given examples on your own. • Complete You Trys 2 and 3.
4 Plot Ordered Pairs	• Draw the *Cartesian coordinate system* in your notes with the quadrants, *x*-axis, *y*-axis, and origin labeled. • Follow the examples to create a procedure for plotting points. • Complete You Trys 4 and 5.
5 Solve Applied Problems Involving Ordered Pairs	• Follow the example to know what questions to ask when solving. • Complete the given example on your own.

W Work   **Read the explanations, follow the examples, take notes, and complete the You Trys.**

Graphs are everywhere—online, in newspapers, and in books. The accompanying graph shows how many billions of dollars consumers spent shopping online for clothing during the years 2003–2009.

We can get different types of information from this graph. For example, in the year 2003, consumers spent about $0.7 billion on clothes, while in 2009, they spent about $3.0 billion on clothes. The graph also illustrates a general trend in online shopping: More and more people are buying their clothing online.

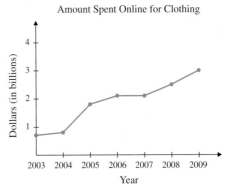

Amount Spent Online for Clothing

Source: www.census.gov

1 Define a Linear Equation in Two Variables

Later in this section, we will see that graphs like the one above are based on the *Cartesian coordinate system,* also known as the *rectangular coordinate system,* which gives us a way to graphically represent the relationship between two quantities. We will also learn about different ways to represent relationships between two quantities, like year and online spending, when we learn about *linear equations in two variables.* Let's begin with a definition.

Definition

A **linear equation in two variables** can be written in the form $Ax + By = C$ where A, B, and C are real numbers and where both A and B do not equal zero.

Some examples of linear equations in two variables are

$$5x - 2y = 11 \qquad y = \frac{3}{4}x + 1 \qquad -3a + b = 2 \qquad y = x \qquad x = -3$$

(We can write $x = -3$ as $x + 0y = -3$; therefore it is a linear equation in two variables.)

2 Decide Whether an Ordered Pair Is a Solution of a Given Equation

A solution to a linear equation in two variables is written as an *ordered pair* so that when the values are substituted for the appropriate variables, we obtain a true statement.

EXAMPLE 1

Determine whether each ordered pair is a solution of $5x - 2y = 11$.

a) $(1, -3)$ b) $\left(\frac{3}{5}, 4\right)$

Solution

a) Solutions to the equation $5x - 2y = 11$ are written in the form (x, y), where (x, y) is called an **ordered pair.** Therefore, the ordered pair $(1, -3)$ means that $x = 1$ and $y = -3$.

$$(1, -3)$$
x-coordinate   y-coordinate

To determine whether $(1, -3)$ is a solution of $5x - 2y = 11$, we substitute 1 for x and -3 for y. Remember to put these values in parentheses.

$$5x - 2y = 11$$
$$5(1) - 2(-3) = 11 \qquad \text{Substitute 1 for } x \text{ and } -3 \text{ for } y.$$
$$5 + 6 = 11 \qquad \text{Multiply.}$$
$$11 = 11 \qquad \text{True}$$

Since substituting $x = 1$ and $y = -3$ into the equation gives the true statement $11 = 11$, $(1, -3)$ *is a solution* of $5x - 2y = 11$. We say that $(1, -3)$ *satisfies* $5x - 2y = 11$.

b) The ordered pair $\left(\dfrac{3}{5}, 4\right)$ tells us that $x = \dfrac{3}{5}$ and $y = 4$.

$$5x - 2y = 11$$
$$5\left(\frac{3}{5}\right) - 2(4) = 11 \qquad \text{Substitute } \frac{3}{5} \text{ for } x \text{ and 4 for } y.$$
$$3 - 8 = 11 \qquad \text{Multiply.}$$
$$-5 = 11 \qquad \text{False}$$

Since substituting $\left(\dfrac{3}{5}, 4\right)$ into the equation gives the false statement $-5 = 11$, the ordered pair is *not* a solution to the equation.

[YOU TRY 1]

Determine whether each ordered pair is a solution of the equation $y = -\dfrac{3}{4}x + 5$.

a) $(12, -4)$ b) $(0, 7)$ c) $(-8, 11)$

If the variables in the equation are not x and y, then the variables in the ordered pairs are written in alphabetical order. For example, solutions to $-3a + b = 2$ are ordered pairs of the form (a, b).

3   Complete Ordered Pairs for a Given Equation

Often, we are given the value of one variable in an equation and we can find the value of the other variable that makes the equation true.

EXAMPLE 2

Complete the ordered pair $(-3, \quad)$ for $y = 2x + 10$.

Solution

To complete the ordered pair $(-3, \quad)$, we must find the value of y from $y = 2x + 10$ when $x = -3$.

$$y = 2x + 10$$
$$y = 2(-3) + 10 \qquad \text{Substitute } -3 \text{ for } x.$$
$$y = -6 + 10$$
$$y = 4$$

When $x = -3$, $y = 4$. The ordered pair is $(-3, 4)$.

[YOU TRY 2]

Complete the ordered pair $(5, \quad)$ for $y = 3x - 7$.

If we want to complete more than one ordered pair for a particular equation, we can organize the information in a **table of values.**

EXAMPLE 3

In-Class Example 3

Complete the table of values for each equation, and write the information as ordered pairs.
a) $5x - y = 9$

x	y
3	
	-4
0	
	-19

b) $y = 9$

x	y
4	
2	
-1	

Answer:
a)

x	y	
3	6	(3, 6)
1	-4	(1, -4)
0	-9	(0, -9)
-2	-19	(-2, -19)

b)

x	y	
4	9	(4, 9)
2	9	(2, 9)
-1	9	(-1, 9)

Complete the table of values for each equation, and write the information as ordered pairs.

a) $-x + 3y = 8$

x	y
1	
	-4
	$\frac{2}{3}$

b) $y = 2$

x	y
7	
-5	
0	

Solution

a) $-x + 3y = 8$

x	y
1	
	-4
	$\frac{2}{3}$

The first ordered pair is (1,   ), and we must find y.

$$-x + 3y = 8$$
$$-(1) + 3y = 8 \qquad \text{Substitute 1 for } x.$$
$$-1 + 3y = 8$$
$$3y = 9 \qquad \text{Add 1 to each side.}$$
$$y = 3 \qquad \text{Divide by 3.}$$

The ordered pair is (1, 3).

The second ordered pair is
(, -4), and we must find x.

The ordered pair is (-20, -4).

$$-x + 3y = 8$$
$$-x + 3(-4) = 8 \qquad \text{Substitute } -4 \text{ for } y.$$
$$-x + (-12) = 8 \qquad \text{Multiply.}$$
$$-x = 20 \qquad \text{Add 12 to each side.}$$
$$x = -20 \qquad \text{Divide by } -1.$$

The third ordered pair is $\left(\ \ , \frac{2}{3}\right)$,
and we must find x.

The ordered pair is $\left(-6, \frac{2}{3}\right)$.

$$-x + 3y = 8$$
$$-x + 3\left(\frac{2}{3}\right) = 8 \qquad \text{Substitute } \frac{2}{3} \text{ for } y.$$
$$-x + 2 = 8 \qquad \text{Multiply.}$$
$$-x = 6 \qquad \text{Subtract 2 from each side.}$$
$$x = -6 \qquad \text{Divide by } -1.$$

As you complete each ordered pair, fill in the table of values. The completed table will look like this:

x	y
1	3
-20	-4
-6	$\frac{2}{3}$

b) $y = 2$

x	y
7	
-5	
0	

The first ordered pair is (7,   ), and we must find y. The equation $y = 2$ means that *no matter the value of x, y always equals 2.* Therefore, when $x = 7$, $y = 2$.

The ordered pair is (7, 2).

Since $y = 2$ for every value of x, we can complete the table of values as follows:

The ordered pairs are (7, 2), (-5, 2), and (0, 2).

x	y
7	2
-5	2
0	2

W Hint

For the equation $y = 2$, you will see that the y-values never change and are always equal to 2.

Complete the table of values for each equation, and write the information as ordered pairs.

a) $x - 2y = 9$

x	y
5	
12	
	−7
	$\frac{5}{2}$

b) $x = -3$

x	y
	1
	3
	−8

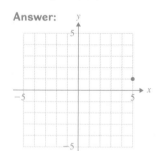

4 Plot Ordered Pairs

When we completed the table of values for the last two equations, we were finding solutions to each linear equation in two variables.

How can we represent the solutions graphically? We will use the **Cartesian coordinate system,** also known as the **rectangular coordinate system,** to graph the ordered pairs, (x, y).

In the Cartesian coordinate system, we have a horizontal number line, called the **x-axis,** and a vertical number line, called the **y-axis.**

The x-axis and y-axis in the Cartesian coordinate system determine a flat surface called a **plane.** The axes divide this plane into four **quadrants,** as shown in the figure. The point at which the x-axis and y-axis intersect is called the **origin.** The arrow at one end of the x-axis and one end of the y-axis indicates the positive direction on each axis.

Ordered pairs can be represented by **points** in the plane. Therefore, to graph the ordered pair $(4, 2)$ we *plot the point* $(4, 2)$. We will do this in Example 4.

W Hint

You move counterclockwise from Quadrant I to move through the quadrants in order.

EXAMPLE 4

In-Class Example 4

Plot the point $(5, 1)$.

Answer:

Plot the point $(4, 2)$.

Solution

Since $x = 4$, we say that the *x-coordinate* of the point is 4. Likewise, the *y-coordinate* is 2.

The *origin* has coordinates $(0, 0)$. The **coordinates** of a point tell us how far from the origin, in the x-direction and y-direction, the point is located. So, the coordinates of the point $(4, 2)$ tell us that to locate the point we do the following:

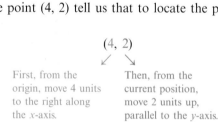

$(4, 2)$

First, from the origin, move 4 units to the right along the x-axis.

Then, from the current position, move 2 units up, parallel to the y-axis.

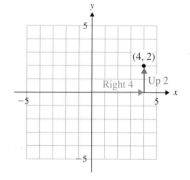

W Hint

Always start at the origin to first move horizontally and then move vertically.

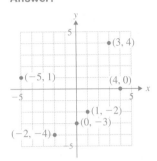
Plot the points.

a) (−2, 5)   b) (1, −4)   c) $\left(\dfrac{5}{2}, 3\right)$

d) (−5, −2)   e) (0, 1)   f) (−4, 0)

Solution

The points are plotted on the graph below.

a)   (−2, 5)   b)   (1, −4)

First   Then   First   Then
From the origin, From the   From the origin, From the
move left 2 units current position,   move right 1 unit current position,
on the x-axis.   move 5 units   on the x-axis.   move 4 units
                   up, parallel to   down, parallel
                   the y-axis.   to the y-axis.

c) $\left(\dfrac{5}{2}, 3\right)$

Think of $\dfrac{5}{2}$ as $2\dfrac{1}{2}$. From the origin, move right $2\dfrac{1}{2}$
units, then up 3 units.

d) (−5, −2)   From the origin, move left 5 units, then
down 2 units.

e) (0, 1)   The x-coordinate of 0 means that we
don't move in the x-direction
(horizontally). From the origin, move
up 1 on the y-axis.

f) (−4, 0)   From the origin, move left 4 units. Since the y-coordinate is zero, we
do not move in the y-direction (vertically).

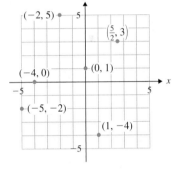

[YOU TRY 4] Plot the points.

a) (3, 1)   b) (−2, 4)   c) (0, −5)   d) (2, 0)   e) (−4, −3)   f) $\left(1, \dfrac{7}{2}\right)$

Note

The coordinate system should always be labeled to indicate how many units
each mark represents.

We can graph sets of ordered pairs for a linear equation in two variables.

EXAMPLE 6

Complete the table of values for $2x - y = 5$, then plot the points.

x	y
0	
1	
	3

In-Class Example 6

Complete the table of values for $-3x + y = 1$, then plot the points.

x	y
−1	
0	
	7

Answer: $(-1, -2), (0, 1), (2, 7)$

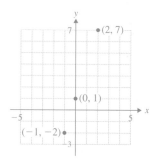

Solution

The first ordered pair is $(0, \ \ )$, and we must find y.

$$2x - y = 5$$
$$2(0) - y = 5 \qquad \text{Substitute 0 for } x.$$
$$0 - y = 5$$
$$-y = 5$$
$$y = -5 \qquad \text{Divide by } -1.$$

The ordered pair is $(0, -5)$.

The second ordered pair is $(1, \ \ )$, and we must find y.

$$2x - y = 5$$
$$2(1) - y = 5 \qquad \text{Substitute 1 for } x.$$
$$2 - y = 5$$
$$-y = 3 \qquad \text{Subtract 2 from each side.}$$
$$y = -3 \qquad \text{Divide by } -1.$$

The ordered pair is $(1, -3)$.

The third ordered pair is $(\ \ , 3)$, and we must find x.

$$2x - y = 5$$
$$2x - (3) = 5 \qquad \text{Substitute 3 for } y.$$
$$2x = 8 \qquad \text{Add 3 to each side.}$$
$$x = 4 \qquad \text{Divide by 2.}$$

The ordered pair is $(4, 3)$.

Each of the points $(0, -5)$, $(1, -3)$, and $(4, 3)$ satisfies the equation $2x - y = 5$.

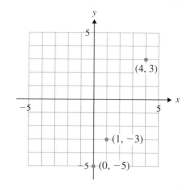

[**YOU TRY 5**] Complete the table of values for $3x + y = 1$, then plot the points.

x	y
0	
−1	
	−5

5 Solve Applied Problems Involving Ordered Pairs

Next, we will look at an application of ordered pairs.

EXAMPLE 7

The length of an 18-year-old female's hair is measured to be 250 millimeters (mm) (almost 10 in.). The length of her hair after x days can be approximated by

$$y = 0.30x + 250$$

where y is the length of her hair in millimeters.

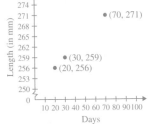
a) Find the length of her hair (i) 10 days, (ii) 60 days, and (iii) 90 days after the initial measurement, and write the results as ordered pairs.

b) Graph the ordered pairs.

c) How long would it take for her hair to reach a length of 274 mm (almost 11 in.)?

Solution

a) The problem states that in the equation $y = 0.30x + 250$,

x = number of days after the hair was measured
y = length of the hair (in millimeters)

x	y
10	
60	
90	

We must determine the length of her hair after 10 days, 60 days, and 90 days. We can organize the information in a table of values.

i) $x = 10$:

$y = 0.30x + 250$
$y = 0.30(10) + 250$ Substitute 10 for x.
$y = 3 + 250$ Multiply.
$y = 253$

After 10 days, her hair is 253 mm long. We can write this as the ordered pair (10, 253).

ii) $x = 60$:

$y = 0.30x + 250$
$y = 0.30(60) + 250$ Substitute 60 for x.
$y = 18 + 250$ Multiply.
$y = 268$

After 60 days, her hair is 268 mm long. We can write this as the ordered pair (60, 268).

iii) $x = 90$:

$y = 0.30x + 250$
$y = 0.30(90) + 250$ Substitute 90 for x.
$y = 27 + 250$ Multiply.
$y = 277$

After 90 days, her hair is 277 mm long. We can write this as the ordered pair (90, 277).

We can complete the table of values:

x	y
10	253
60	268
90	277

The ordered pairs are (10, 253), (60, 268), and (90, 277).

b) Graph the ordered pairs.

The x-axis represents the number of days after the hair was measured. Since it does not make sense to talk about a negative number of days, we will not continue the x-axis in the negative direction.

The y-axis represents the length of the female's hair. Likewise, a negative number does not make sense in this situation, so we will not continue the y-axis in the negative direction.

The scales on the x-axis and y-axis are different. This is because the size of the numbers they represent are quite different.

Here are the ordered pairs we must graph: (10, 253), (60, 268), and (90, 277).

The *x*-values are 10, 60, and 90, so we will let each mark in the *x*-direction represent 10 units.

The *y*-values are 253, 268, and 277. While the numbers are rather large, they do not actually differ by much. We will begin labeling the *y*-axis at 250, but each mark in the *y*-direction will represent 3 units. Because there is a large jump in values from 0 to 250 on the *y*-axis, we indicate this with "⌇" on the axis between the 0 and 250.

Notice also that we have labeled both axes. The ordered pairs are plotted on the following graph.

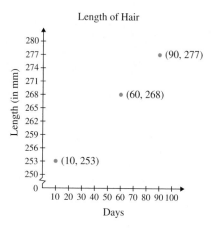

c) We must determine how many days it would take for the hair to grow to a length of 274 mm.

The length, 274 mm, is the *y*-value. We must find the value of *x* that corresponds to *y* = 274 since *x* represents the number of days.

The equation relating *x* and *y* is $y = 0.30x + 250$. We will substitute 274 for *y* and solve for *x*.

$$y = 0.30x + 250$$
$$274 = 0.30x + 250$$
$$24 = 0.30x$$
$$80 = x$$

It will take 80 days for her hair to grow to a length of 274 mm.

ANSWERS TO [YOU TRY] EXERCISES

1) a) yes b) no c) yes   2) (5, 8)   3) a) $(5, -2)$, $(12, \frac{3}{2})$, $(-5, -7)$, $(14, \frac{5}{2})$
b) $(-3, 1)$, $(-3, 3)$, $(-3, -8)$
4)   5) (0, 1), (−1, 4), (2, −5)

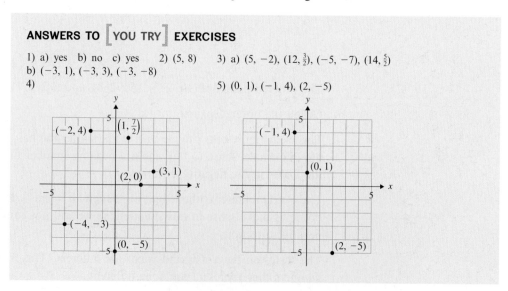

Do the exercises, and check your work.

*Additional answers can be found in the Answers to Exercises appendix.

Mixed Exercises: Objectives 1 and 2

The graph shows the number of gallons of fruit juice consumed per person for the years 2005–2009.

(U.S. Dept of Agriculture)

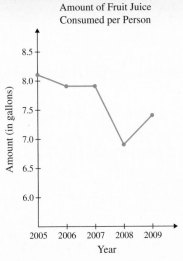

Amount of Fruit Juice
Consumed per Person

1) How many gallons of fruit juice were consumed per person in 2005?   8.1 gallons

2) During which year was the consumption level about 7.4 gallons per person?   2009

3) During which two years was consumption the same, and how much juice was consumed each of these years?   2006 and 2007; 7.9 gallons

4) During which year did people drink the least amount of fruit juice?   2008

5) What was the general consumption trend from 2007 to 2008?   Consumption was decreasing.

6) Compare the consumption level in 2006 with that in 2008.   In 2006, people consumed about 1.0 more gallons of fruit juice than in 2008.

The bar graph shows the public high school graduation rate in certain states in 2008. (www.higheredinfo.org)

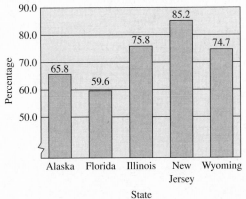

Public High School
Graduation Rate, 2008

7) Which state had the highest graduation rate, and what percentage of its public high school students graduated?   New Jersey; 85.2%

8) Which states graduated between 70% and 80% of its students?   Illinois and Wyoming

9) How does the graduation rate of Florida compare with that of New Jersey?   Florida's graduation rate is about 25.6% less than New Jersey's.

10) Which state had a graduation rate of about 65.8%?   Alaska

11) Explain the difference between a linear equation in one variable and a linear equation in two variables. Give an example of each.   Answers may vary.

12) True or False: $3x + 6y^2 = -1$ is a linear equation in two variables.   false

Determine whether each ordered pair is a solution of the given equation.

13) $2x + 5y = 1$; $(-2, 1)$ yes

14) $2x + 7y = -4$; $(2, -5)$ no

15) $-3x - 2y = -15$; $(7, -3)$ yes

16) $y = 5x - 6$; $(3, 9)$ yes

17) $y = -\dfrac{3}{2}x - 7$; $(8, 5)$ no

18) $5y = \dfrac{2}{3}x + 1$; $(6, 1)$ yes

19) $y = -7$; $(9, -7)$ yes

20) $x = 8$; $(-10, 8)$ no

Objective 3: Complete Ordered Pairs for a Given Equation

Complete the ordered pair for each equation.

21) $y = 3x - 7$; $(4, \ \ )$ 5

22) $y = -2x + 3$; $(6, \ \ )$ -9

23) $2x - 15y = 13$; $\left(\ , -\dfrac{4}{3} \right)$ $-\dfrac{7}{2}$

24) $-x + 10y = 8$; $\left(\ , \dfrac{2}{5} \right)$ -4

25) $x = 5$; $(\ , -200)$ 5

26) $y = -10$; $(12, \ \ )$ -10

Complete the table of values for each equation.

27) $y = 2x - 4$

x	y
0	-4
1	-2
-1	-6
-2	-8

28) $y = -5x + 1$

x	y
0	1
1	-4
2	-9
-1	6

29) $y = 4x$

x	y
0	0
$\frac{1}{2}$	2
3	12
−5	−20

30) $y = 9x - 8$

x	y
0	−8
$-\frac{1}{3}$	−11
−1	−17
1	1

31) $5x + 4y = -8$

x	y
0	−2
$-\frac{8}{5}$	0
1	$-\frac{13}{4}$
$\frac{12}{5}$	1

32) $2x - y = 12$

x	y
6	0
0	−12
5	−2
$\frac{5}{2}$	−7

33) $y = -2$

x	y
0	−2
−3	−2
8	−2
17	−2

34) $x = 20$

x	y
20	0
20	3
20	−4
20	−9

35) Explain, in words, how to complete the table of values for $x = -13$. Answers may vary.

x	y
	0
	2
	−1

36) Explain, in words, how to complete the ordered pair (, −3) for $y = -x - 2$. Answers may vary.

Objective 4: Plot Ordered Pairs

Name each point with an ordered pair, and identify the quadrant in which each point lies.

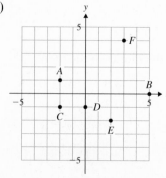

37)

A: (−2, 1); quadrant II
B: (5, 0); no quadrant
C: (−2, −1); quadrant III
D: (0, −1); no quadrant
E: (2, −2); quadrant IV
F: (3, 4); quadrant I

38)

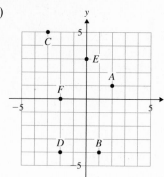

A: (2, 1); quadrant I
B: (1, −4); quadrant IV
C: (−3, 5); quadrant II
D: (−2, −4); quadrant III
E: (0, 3); no quadrant
F: (−2, 0); no quadrant

Graph each ordered pair, and explain how you plotted the points.

39) (2, 4)   40) (4, 1)

41) (−3, −5)   42) (−2, 1)

Graph each ordered pair.

43) (−6, 1)   44) (−2, −3)

45) (0, −1)   46) (4, −5)

47) (0, 4)   48) (−5, 0)

49) (−2, 0)   50) (0, −1)

51) $\left(-2, \frac{3}{2}\right)$ 52) $\left(\frac{4}{3}, 3\right)$

53) $\left(3, -\frac{1}{4}\right)$ 54) $\left(-2, -\frac{9}{4}\right)$

55) $\left(0, -\frac{11}{5}\right)$ 56) $\left(-\frac{9}{2}, -\frac{2}{3}\right)$

Answer *always, sometimes,* or *never.*

57) When writing an ordered pair for a linear equation of the form $Ax + By = C$, the y-coordinate is *always, sometimes,* or *never* placed before the x-coordinate as in (y, x).   never

58) A linear equation will *always, sometimes,* or *never* contain a squared variable term.   never

59) A linear equation will *always, sometimes,* or *never* contain two different variable terms.   sometimes

60) A linear equation will *always, sometimes,* or *never* contain one variable term.   sometimes

Mixed Exercises: Objectives 3 and 4

Complete the table of values for each equation, and plot the points.

61) $y = -4x + 3$

x	y
0	3
$\frac{3}{4}$	0
2	-5
-1	7

62) $y = -3x + 4$

x	y
0	4
$\frac{4}{3}$	0
2	-2
-1	7

63) $y = x$

x	y
0	0
-1	-1
3	3
-5	-5

64) $y = -2x$

x	y
0	0
-2	4
2	-4
-3	6

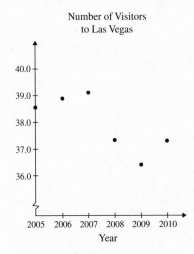 65) $3x + 4y = 12$

x	y
0	3
4	0
1	$\frac{9}{4}$
-4	6

66) $2x - 3y = 6$

x	y
0	-2
3	0
6	2
2	$-\frac{2}{3}$

67) $y + 1 = 0$

x	y
0	-1
1	-1
-3	-1
-1	-1

68) $x = 3$

x	y
3	0
3	-2
3	3
3	1

69) $y = \frac{1}{4}x + 2$

x	y
0	2
-2	$\frac{3}{2}$
4	3
-1	$\frac{7}{4}$

70) $y = -\frac{5}{2}x + 3$

x	y
0	3
4	-7
2	-2
1	$\frac{1}{2}$

71) For $y = \frac{2}{3}x - 7$,

a) find y when $x = 3$, $x = 6$, and $x = -3$. Write the results as ordered pairs. $(3, -5), (6, -3), (-3, -9)$

b) find y when $x = 1$, $x = 5$, and $x = -2$. Write the results as ordered pairs.

c) why is it easier to find the y-values in part a) than in part b)?

72) Which ordered pair is a solution to every linear equation of the form $y = mx$, where m is a real number? $(0, 0)$

Fill in the blank with *positive, negative,* or *zero*.

73) The x-coordinate of every point in quadrant III is __negative__.

74) The y-coordinate of every point in quadrant I is __positive__.

75) The x-coordinate of every point in quadrant II is __negative__.

76) The y-coordinate of every point in quadrant II is __positive__.

77) The x-coordinate of every point in quadrant I is __positive__.

78) The y-coordinate of every point in quadrant IV is __negative__.

79) The x-coordinate of every point on the y-axis is __zero__.

80) The y-coordinate of every point on the x-axis is __zero__.

Objective 5: Solve Applied Problems Involving Ordered Pairs

81) The graph shows the number of people who visited Las Vegas from 2005 to 2010. (www.lvcva.com)

a) If a point on the graph is represented by the ordered pair (x, y), then what do x and y represent? x represents the year; y represents the number of visitors in millions

b) What does the ordered pair (2010, 37.3) represent in the context of this problem? In 2010, there were 37.3 million visitors to Las Vegas.

c) Approximately how many people went to Las Vegas in 2006? 38.9 million

d) In which year were there approximately 38.6 million visitors?   2005

e) Approximately how many more people visited Las Vegas in 2008 than in 2009?   1 million

f) Represent the following with an ordered pair: During which year did Las Vegas have the most visitors, and how many visitors were there?
(2007, 39.2)

82) The graph shows the average amount of time people spent commuting to work in the Los Angeles metropolitan area from 2006 to 2010.
(American Community Survey, U.S. Census)

Average Commute Time in
Los Angeles Area

a) If a point on the graph is represented by the ordered pair (x, y), then what do x and y represent?
x represents the year; y represents the average commute time

b) What does the ordered pair (2007, 28.5) represent in the context of this problem?

c) Which year during this time period had the shortest commute? What was the approximate commute time?   2006; 27.9 minutes

d) When was the average commute 29.5 minutes?
2008 and 2009

e) Write an ordered pair to represent when the average commute time was 28.1 minutes.
(2010, 28.1)

83) The percentage of deadly crashes involving alcohol is given in the table. (www.nhtsa.gov)

Year	Percentage
2005	40.0
2006	41.0
2007	31.7
2008	31.0
2009	32.0

a) Write the information as ordered pairs (x, y), where x represents the year and y represents the percentage of accidents involving alcohol.
(2005, 40.0), (2006, 41.0), (2007, 31.7), (2008, 31.0), (2009, 32.0)

b) Label a coordinate system, choose an appropriate scale, and graph the ordered pairs.

c) Explain the meaning of the ordered pair (2007, 31.7) in the context of the problem.
In the year 2007, 31.7% of all fatal accidents involved alcohol.

84) The average annual salary of a social worker is given in the table. (www.bls.gov)

Year	Salary
2007	$47,170
2008	$48,180
2009	$50,470
2010	$52,270

a) Write the information as ordered pairs (x, y) where x represents the year and y represents the average annual salary.
(2007, $47,170), (2008, $48,180), (2009, $50,470), (2010, $52,270)

b) Label a coordinate system, choose an appropriate scale, and graph the ordered pairs.

c) Explain the meaning of the ordered pair (2009, $50,470) in the context of the problem.
In 2009, the average salary of a social worker was $50,470.

85) The amount of sales tax paid by consumers in Seattle in 2011 is given by $y = 0.095x$, where x is the price of an item in dollars and y is the amount of tax to be paid.

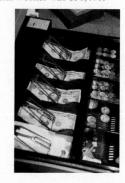

a) Complete the table of values, and write the information as ordered pairs.

x	y
100.00	9.50
140.00	13.30
210.72	20.0184
250.00	23.75

b) Label a coordinate system, choose an appropriate scale, and graph the ordered pairs.

c) Explain the meaning of the ordered pair (140.00, 13.30) in the context of the problem.
If a bill totals $140.00, the sales tax will be $13.30.

d) How much tax would a customer pay if the cost of an item was $210.72? $20.02

e) Look at the graph. Is there a pattern indicated by the points? They lie on a straight line.

f) If a customer paid $19.00 in sales tax, what was the cost of the item purchased? $200.00

86) Kyle is driving from Atlanta to Oklahoma City. His distance from Atlanta, y (in miles), is given by $y = 66x$, where x represents the number of hours driven.

a) Complete the table of values, and write the information as ordered pairs.

x	y
1	66
1.5	99
2	132
4.5	297

b) Label a coordinate system, choose an appropriate scale, and graph the ordered pairs.

c) Explain the meaning of the ordered pair (4.5, 297) in the context of the problem.
After 4.5 hours, Kyle is 297 miles from Atlanta.

d) Look at the graph. Is there a pattern indicated by the points? They lie on a straight line.

e) What does the 66 in $y = 66x$ represent?
Kyle's speed

f) How many hours of driving time will it take for Kyle to get to Oklahoma City if the distance between Atlanta and Oklahoma City is about 860 miles? about 13 hr

R Rethink

R1) If the x-coordinate of an ordered pair is 0, on which axis does the point lie?

R2) If the y-coordinate of an ordered pair is 0, on which axis does the point lie?

R3) Why do we need two number lines when plotting ordered pairs?

R4) Which objective is the most difficult for you? Which exercises do you need to work again to fully master?

3.2 Graphing by Plotting Points and Finding Intercepts

P Prepare

O Organize

What are your objectives for Section 3.2?	How can you accomplish each objective?
1 Graph a Linear Equation by Plotting Points	• Learn the properties for **Solutions of Linear Equations in Two Variables** and **The Graph of a Linear Equation in Two Variables.** • Create a procedure after following the examples for graphing linear equations by plotting points. • Complete the given example on your own. • Complete You Try 1.
2 Graph a Linear Equation in Two Variables by Finding the Intercepts	• Write the definitions of x- and y-intercept in your own words. • Learn the procedure for **Finding Intercepts,** and write it in your notes. • Complete the given example on your own. • Complete You Try 2.
3 Graph a Linear Equation of the Form $Ax + By = 0$	• Write the property **The Graph of $Ax + By = 0$** in your own words, and learn it. • Complete the given example on your own. • Complete You Try 3.
4 Graph Linear Equations of the Forms $x = a$ and $y = b$	• Write the properties **The Graph of $x = a$** and **The Graph of $y = b$** in your own words, and learn them. • Complete the given examples on your own. • Complete You Trys 4 and 5.
5 Model Data with a Linear Equation	• Do not let "bigger numbers" scare you. Follow the example to learn how to scale the Cartesian coordinate system. • Follow the example, and then complete the example in your notes without looking at the solution.

W Work

Read the explanations, follow the examples, take notes, and complete the You Trys.

1 Graph a Linear Equation by Plotting Points

In Example 3 of Section 3.1, we found that the ordered pairs (1, 3), (−20, −4), and $\left(-6, \frac{2}{3}\right)$ are three solutions to the equation $-x + 3y = 8$. But how many solutions does the equation have? It has an infinite number of solutions. Every linear equation in two variables has an infinite number of solutions because we can choose any real number for one of the variables and we will get another real number for the other variable.

Property Solutions of Linear Equations in Two Variables

Every linear equation in two variables has an infinite number of solutions, and the solutions are ordered pairs.

How can we represent all of the solutions to a linear equation in two variables? We can represent them with a graph, and that graph is a *line*.

Property The Graph of a Linear Equation in Two Variables

The graph of a linear equation in two variables, $Ax + By = C$, is a straight **line.** Each point on the line is a solution to the equation.

EXAMPLE 1

In-Class Example 1

Graph $x - 3y = 4$.

Answer:

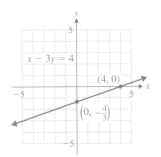

Graph $-x + 2y = 4$.

Solution

We will find three ordered pairs that satisfy the equation. Let's complete a table of values for $x = 0$, $x = 2$, and $x = -4$.

$x = 0$:

$$-x + 2y = 4$$
$$-(0) + 2y = 4$$
$$2y = 4$$
$$y = 2$$

$x = 2$:

$$-x + 2y = 4$$
$$-(2) + 2y = 4$$
$$-2 + 2y = 4$$
$$2y = 6$$
$$y = 3$$

$x = -4$:

$$-x + 2y = 4$$
$$-(-4) + 2y = 4$$
$$4 + 2y = 4$$
$$2y = 0$$
$$y = 0$$

We get the table of values

x	y
0	2
2	3
-4	0

Plot the points $(0, 2)$, $(2, 3)$, and $(-4, 0)$, and draw the line through them.

W Hint

If possible, use graph paper when graphing lines. Don't forget to label the x-axis and y-axis and to draw arrowheads on both ends of your graphed line!

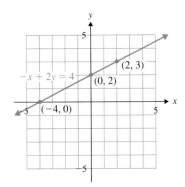

The line represents all solutions to the equation $-x + 2y = 4$. Every point on the line is a solution to the equation. The arrows on the ends of the line indicate that the line extends indefinitely in each direction. Although it is true that we need to find only two points to graph the a line, it is best to plot at least three as a check.

[YOU TRY 1]

Graph each line.

a) $3x + 2y = 6$ b) $y = 4x - 8$

---

2 Graph a Linear Equation in Two Variables by Finding the Intercepts

In Example 1, the line crosses the *x*-axis at −4 and crosses the *y*-axis at 2. These points are called **intercepts**.

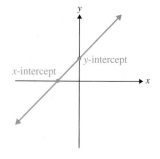

Definitions

The *x*-**intercept** of the graph of an equation is the point where the graph intersects the *x*-axis.

The *y*-**intercept** of the graph of an equation is the point where the graph intersects the *y*-axis.

What is the *y*-coordinate of any point on the *x*-axis? It is zero. Likewise, the *x*-coordinate of any point on the *y*-axis is zero. (See the graph at the right for some points on the axes.)

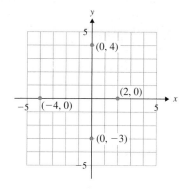

We use these facts to find the intercepts of the graph of an equation.

Procedure Finding Intercepts

To find the *x-intercept* of the graph of an equation, let $y = 0$ and solve for *x*.

To find the *y-intercept* of the graph of an equation, let $x = 0$ and solve for *y*.

Finding intercepts is very helpful for graphing linear equations in two variables.

EXAMPLE 2

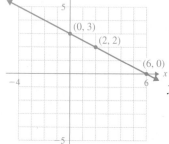

Graph $y = -\frac{1}{3}x + 1$ by finding the intercepts and one other point.

Solution

We will begin by finding the intercepts.

x-intercept: Let $y = 0$, and solve for *x*. $0 = -\frac{1}{3}x + 1$

$$-1 = -\frac{1}{3}x$$

$$3 = x \qquad \text{Multiply both sides by } -3 \text{ to solve for } x.$$

The *x*-intercept is (3, 0).

y-intercept: Let $x = 0$, and solve for *y*. $y = -\frac{1}{3}(0) + 1$

$$y = 0 + 1$$

$$y = 1$$

The *y*-intercept is (0, 1).

We must find another point. Let's look closely at the equation $y = -\dfrac{1}{3}x + 1$. The coefficient of x is $-\dfrac{1}{3}$. If we choose a value for x that is a multiple of 3 (the denominator of the fraction), then $-\dfrac{1}{3}x$ will not be a fraction.

Let $x = -3$.
$$y = -\frac{1}{3}x + 1$$
$$y = -\frac{1}{3}(-3) + 1$$
$$y = 1 + 1$$
$$y = 2$$

The third point is $(-3, 2)$.

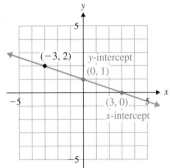

Plot the points, and draw the line through them.
See the graph above.

[YOU TRY 2] Graph $y = \dfrac{4}{3}x - 2$ by finding the intercepts and one other point.

3 Graph a Linear Equation of the Form $Ax + By = 0$

Sometimes the x- and y-intercepts are the same point.

EXAMPLE 3

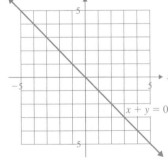

Graph $-2x + y = 0$.

Solution

If we begin by finding the x-intercept, let $y = 0$ and solve for x.
$$-2x + y = 0$$
$$-2x + (0) = 0$$
$$-2x = 0$$
$$x = 0$$

The x-intercept is $(0, 0)$. But this is the same as the y-intercept since we find the y-intercept by substituting 0 for x and solving for y. Therefore, *the x- and y-intercepts are the same point.*

Instead of the intercepts giving us two points on the graph of $-2x + y = 0$, we have only one. We will find two other points on the line.

$x = 2$:	$-2x + y = 0$	$x = -2$:	$-2x + y = 0$
	$-2(2) + y = 0$		$-2(-2) + y = 0$
	$-4 + y = 0$		$4 + y = 0$
	$y = 4$		$y = -4$

The ordered pairs (2, 4) and (−2, −4) are also
solutions to the equation. Plot all three points on
the graph, and draw the line through them.

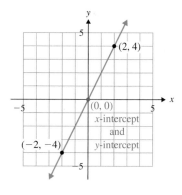

Property The Graph of $Ax + By = 0$

If A and B are nonzero real numbers, then the graph of $Ax + By = 0$ is a line
passing through the origin, (0, 0).

[**YOU TRY 3**]   Graph $x − y = 0$.

4 Graph Linear Equations of the Forms $x = a$ and $y = b$

An equation like $x = −2$ is a linear equation in two variables since it can be written in
the form $x + 0y = −2$. The same is true for $y = 3$. It can be written as $0x + y = 3$.
Let's see how we can graph these equations.

EXAMPLE 4

In-Class Example 4

Graph $x = 3$.

Answer:

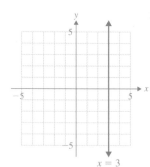

Graph $x = −2$.

Solution

The equation $x = −2$ means that *no matter the value of
y, x always equals* −2. We can make a table of values
where we choose any value for y, but x is always −2.

Plot the points, and draw a line through them. The
graph of $x = −2$ is a **vertical line.**

x	y
−2	0
−2	1
−2	−2

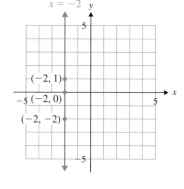

We can generalize the result as follows:

Property The Graph of $x = a$

If a is a constant, then the graph of $x = a$ is a **vertical line** going through the
point $(a, 0)$.

YOU TRY 4   Graph $x = 2$.

EXAMPLE 5

In-Class Example 5

Graph $y = -1$.

Answer:

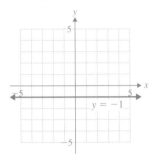

Hint

If y is constant, then the graphed line is horizontal, perpendicular to the y-axis.

Graph $y = 3$.

Solution

The equation $y = 3$ means that *no matter the value of x, y always equals* 3. Make a table of values where we choose any value for x, but y is always 3.

Plot the points, and draw a line through them. The graph of $y = 3$ is a **horizontal line**.

x	y
0	3
2	3
-2	3

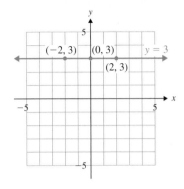

We can generalize the result as follows:

Property The Graph of $y = b$

If b is a constant, then the graph of $y = b$ is a **horizontal line** going through the point $(0, b)$.

YOU TRY 5   Graph $y = -4$.

5 Model Data with a Linear Equation

Linear equations are often used to model (or describe mathematically) real-world data. We can use these equations to learn what has happened in the past or predict what will happen in the future.

EXAMPLE 6

In-Class Example 6

Refer to Example 6. Find the approximate cost of tuition and fees in 2008 and 2009. Write the information as ordered pairs.

Answer: (2, 21,254); (3, 22,426.50)

The average annual cost of college tuition and fees at private, 4-year institutions can be modeled by

$$y = 1172.50x + 18{,}909$$

where x is the number of years after 2006 and y is the average tuition and fees, in dollars. (www.census.gov)

a) Find the y-intercept of the graph of this equation and explain its meaning.

b) Find the approximate cost of tuition and fees in 2007 and 2010. Write the information as ordered pairs.

c) Graph $y = 1172.50x + 18{,}909$.

d) Use the graph to approximate the average cost of tuition and fees in 2011. Is this the same result as when you use the equation to estimate the average cost?

Solution

a) To find the *y*-intercept, let $x = 0$.

$$y = 1172.50(0) + 18{,}909$$
$$y = 18{,}909$$

The *y*-intercept is (0, 18,909). What does this represent?

The problem states that x is the number of years *after* 2006. Therefore, $x = 0$ represents zero years after 2006, which is the year 2006.

The *y*-intercept (0, 18,909) tells us that in 2006, the average cost of tuition and fees at a private 4-year institution was $18,909.

b) The approximate cost of tuition and fees in

2007: First, realize that $x \neq 2007$. x is the number of years *after* 2006. Since 2007 is 1 year after 2006, $x = 1$. Let $x = 1$ in $y = 1172.50x + 18{,}909$ and find *y*.

$$y = 1172.50(1) + 18{,}909$$
$$y = 1172.50 + 18{,}909$$
$$y = 20{,}081.50$$

In 2007, the approximate cost of college tuition and fees at these schools was $20,081.50. We can write this information as the ordered pair (1, 20,081.50).

2010: Begin by finding x. 2010 is 4 years after 2006, so $x = 4$.

$$y = 1172.50(4) + 18{,}909$$
$$y = 4690 + 18{,}909$$
$$y = 23{,}599$$

In 2010, the approximate cost of college tuition and fees at private 4-year schools was $23,599.

The ordered pair (4, 23,599) can be written from this information.

c) We will plot the points (0, 18,909), (1, 20,081.50), and (4, 23,599). Label the axes, and choose an appropriate scale for each.

The *x*-coordinates of the ordered pairs range from 0 to 4, so we will let each mark in the *x*-direction represent 1 unit. (Sometimes, we let each tick mark represent more than one unit.) The *y*-coordinates of the ordered pairs range from 18,909 to 23,599. We will let each mark in the *y*-direction represent 2000 units.

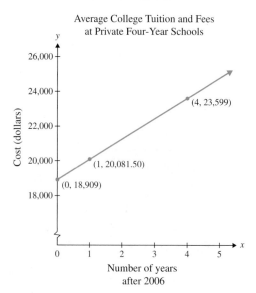

Average College Tuition and Fees at Private Four-Year Schools

www.mhhe.com/messersmith

d) Using the graph to estimate the cost of tuition and fees in 2011, we locate $x = 5$ on the x-axis since 2011 is 5 years after 2006. When $x = 5$, we move straight up the graph to $y \approx 24{,}700$. Our approximation from the graph is $24,700.

If we use the equation and let $x = 5$, we get

$$y = 1172.50x + 18{,}909$$
$$y = 1172.50(5) + 18{,}909$$
$$y = 24{,}771.50$$

From the equation, we find that the cost of college tuition and fees at private 4-year schools was about $24,771.50. The numbers are not exactly the same, but they are close.

Using Technology

A graphing calculator can be used to graph an equation and to verify information that we find using algebra. We will graph the equation $y = -\dfrac{1}{2}x + 2$ and then find the intercepts both algebraically and using the calculator.

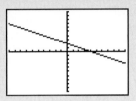

First, enter the equation into the calculator. Press ZOOM and select 6:Zstandard to graph the equation.

1) Find, algebraically, the y-intercept of the graph of $y = -\dfrac{1}{2}x + 2$. Is it consistent with the graph of the equation?   (0, 2); yes, consistent

2) Find, algebraically, the x-intercept of the graph of $y = -\dfrac{1}{2}x + 2$. Is it consistent with the graph of the equation?   (4, 0); yes, consistent

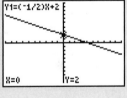

Now let's verify the intercepts using the graphing calculator. To find the y-intercept, press TRACE after displaying the graph. The cursor is automatically placed at the center x-value on the screen, which is at the point (0, 2) as shown next on the left. To find the x-intercept, press TRACE, type 4, and press ENTER. The calculator displays (4, 0) as shown next on the right. This is consistent with the intercepts found in 1 and 2, using algebra.

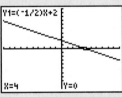

Use algebra to find the x- and y-intercepts of the graph of each equation. Then, use the graphing calculator to verify your results.

1) $y = 2x - 4$ 2) $y = x + 3$ 3) $y = -x + 5$
4) $2x - 5y = 10$ 5) $3x + 4y = 24$ 6) $3x - 7y = 21$

1) a)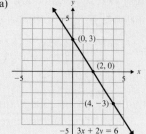

b)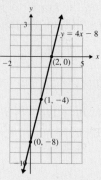

2)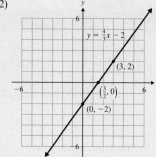

3)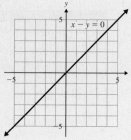

4)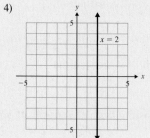

5)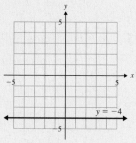

ANSWERS TO TECHNOLOGY EXERCISES

1) (2, 0), (0, −4)	2) (−3, 0), (0, 3)	3) (5, 0), (0, 5)
4) (5, 0), (0, −2)	5) (8, 0), (0, 6)	6) (7, 0), (0, −3)

E Evaluate **3.2** Exercises

Do the exercises, and check your work.

*Additional answers can be found in the Answers to Exercises appendix.

Objective 1: Graph a Linear Equation by Plotting Points

1) The graph of a linear equation in two variables is a ___line___.

2) Every linear equation in two variables has how many solutions? _an infinite number_

Complete the table of values, and graph each equation.

3) $y = -2x + 4$

x	y
0	4
−1	6
2	0
3	−2

4) $y = 3x - 2$

x	y
0	−2
1	1
2	4
−1	−5

5) $y = \dfrac{3}{2}x + 7$

x	y
0	7
2	10
−2	4
−4	1

6) $y = -\dfrac{5}{3}x + 3$

x	y
0	3
−3	8
3	−2
6	−7

7) $2x = 3 - y$

x	y
$\frac{3}{2}$	0
0	3
$\frac{1}{2}$	2
−1	5

8) $-x + 5y = 10$

x	y
0	2
−10	0
10	4
−3	$\frac{7}{5}$

9) $x = -\dfrac{4}{9}$

x	y
$-\frac{4}{9}$	5
$-\frac{4}{9}$	0
$-\frac{4}{9}$	−1
$-\frac{4}{9}$	−2

10) $y + 5 = 0$

x	y
0	−5
−3	−5
−1	−5
2	−5

Mixed Exercises: Objectives 1–4

11) What is the *y*-intercept of the graph of an equation? How do you find it? It is the point where the graph intersects the *y*-axis. Let $x = 0$ in the equation, and solve for *y*.

12) What is the *x*-intercept of the graph of an equation? How do you find it? It is the point where the graph intersects the *x*-axis. Let $y = 0$ in the equation, and solve for *x*.

Graph each equation by finding the intercepts and at least one other point.

13) $y = x - 1$

14) $y = -x + 3$

15) $3x - 4y = 12$

16) $2x - 7y = 14$

17) $x = -\dfrac{4}{3}y - 2$

18) $x = \dfrac{5}{4}y - 5$

19) $2x - y = 8$

20) $3x + y = -6$

21) $y = -x$

22) $y = 3x$

23) $4x - 3y = 0$

24) $6y - 5x = 0$

25) $x = 5$

26) $y = -4$

27) $y = 0$

28) $x = 0$

29) $x - \dfrac{4}{3} = 0$

30) $y + 1 = 0$

31) $4x - y = 9$

32) $x + 3y = -5$

For Exercises 33–36, choose from *always, sometimes,* or *never*.

33) The *x*-intercept of the graph of an equation will *always, sometimes,* or *never* have a *y*-coordinate of 0. always

34) The *y*-intercept of the graph of an equation will *always, sometimes,* or *never* have a *y*-coordinate of 0. sometimes

35) The graph of a linear equation will *always, sometimes,* or *never* pass through all four quadrants of the Cartesian coordinate system. never

36) If a linear equation contains only one variable, its graph will *always, sometimes,* or *never* pass through the origin. sometimes

37) Which ordered pair is a solution to every linear equation of the form $Ax + By = 0$? (0, 0)

38) True or False: The graph of $Ax + By = 0$ will always pass through the origin. true

Objective 5: Model Data with a Linear Equation

39) The cost of downloading popular songs from iTunes is given by $y = 1.29x$, where *x* represents the number of songs downloaded and *y* represents the cost, in dollars.

a) Make a table of values using $x = 0, 4, 7,$ and 12, and write the information as ordered pairs.

b) Explain the meaning of each ordered pair in the context of the problem.

c) Graph the equation. Use an appropriate scale.

d) How many songs could you download for $11.61? 9

40) The force, *y*, measured in newtons (N), required to stretch a particular spring *x* meters is given by $y = 100x$.

a) Make a table of values using $x = 0, 0.5, 1.0,$ and 1.5, and write the information as ordered pairs.

b) Explain the meaning of each ordered pair in the context of the problem.

c) Graph the equation. Use an appropriate scale.

d) If the spring was pulled with a force of 80 N, how far did it stretch? 0.8 m

41) The number of doctorate degrees awarded in science and engineering in the United States from 2005 to 2009 can be modeled by $y = 1394x + 28{,}405$, where x represents the number of years after 2005, and y represents the number of doctorate degrees awarded. The actual data are graphed here. (www.nsf.gov)

Number of Science and Engineering
Doctorates Awarded in the U.S.

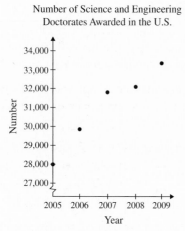

a) From the graph, estimate the number of science and engineering doctorates awarded in 2006 and 2009.
2006: 29,800; 2009: 33,500

b) Determine the number of degrees awarded during the same years using the equation. Are the numbers close?
2006: 29,799; 2009: 33,981; yes, they are close.

c) Graph the line that models the data given on the original graph.

d) What is the y-intercept of the graph of this equation, and what does it represent? How close is it to the actual point plotted on the given graph?

e) If the trend continues, how many science and engineering doctorates will be awarded in 2018? Use the equation.   46,527

42) The amount of money Americans spent on skin and scuba diving equipment from 2004 to 2007 can be modeled by $y = 8.6x + 350.6$, where x represents the number of years after 2004, and y represents the amount spent on equipment in millions of dollars. The actual data are graphed here. (www.census.gov)

Amount Spent on
Skin and Scuba Equipment

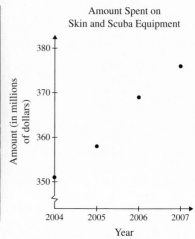

a) From the graph, estimate the amount spent in 2005 and 2006.   2005: $358 mil; 2006: $369 mil

b) Determine the amount of money spent during the same years using the equation. Are the numbers close?
2005: $359.2 mil; 2006: $367.8 mil; yes, they are close.

c) Graph the line that models the data given on the original graph.

d) What is the y-intercept of the graph of this equation, and what does it represent? How close is it to the actual point plotted on the given graph?

e) If the trend continues, how much will be spent on skin and scuba gear in 2014? Use the equation.
$436.6 mil

R Rethink

R1) In which other courses, besides math, have you had to use an equation of a line?

R2) When using the intercepts to graph a line, why should you always find a third point?

R3) How can you tell whether a line is going to be drawn vertically?

R4) How can you tell whether a line is going to be drawn horizontally?

R5) Explain how graph paper has helped you to draw your graphs neatly.

3.3 The Slope of a Line

P Prepare

What are your objectives for Section 3.3?	How can you accomplish each objective?
1 Understand the Concept of Slope	• Write the property for **Slope of a Line** in your own words. • Complete the given example on your own. • Complete You Try 1.
2 Find the Slope of a Line Given Two Points on the Line	• Learn the formula for finding **The Slope of a Line,** and write it in your notes. • Understand and learn the property that explains **Positive and Negative Slopes.** • Complete the given example on your own. • Complete You Try 2.
3 Use Slope to Solve Applied Problems	• Follow the example to know what questions to ask when solving applied problems. • Complete the given example on your own.
4 Find the Slope of Horizontal and Vertical Lines	• Write the property for **Slopes of Horizontal and Vertical Lines** in your own words. • Complete the given example on your own. • Complete You Try 3.
5 Use Slope and One Point on a Line to Graph the Line	• Follow Example 5 to create a procedure for using the slope and one point to graph the line. • Complete You Try 4.

O Organize

W Work

Read the explanations, follow the examples, take notes, and complete the You Trys.

1 Understand the Concept of Slope

In Section 3.2, we learned to graph lines by plotting points. You may have noticed that some lines are steeper than others. Their "slants" are different, too.

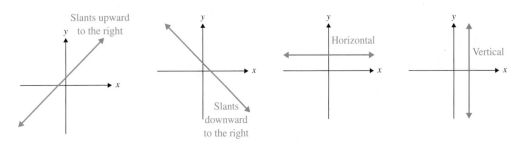

We can describe the steepness of a line with its *slope*.

Property Slope of a Line

The **slope** of a line measures its steepness. It is the ratio of the vertical change in y to the horizontal change in x. Slope is denoted by m.

We can also think of slope as a rate of change. *Slope* is the rate of change between two points. More specifically, it describes the rate of change in y to the change in x.

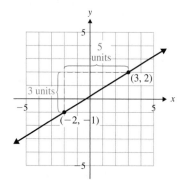

Hint

Use two clearly defined points on the line to find the slope.

$$\text{Slope} = \frac{3}{5} \begin{array}{l} \leftarrow \text{Vertical change} \\ \leftarrow \text{Horizontal change} \end{array} \qquad \text{Slope} = 4 \text{ or } \frac{4}{1} \begin{array}{l} \leftarrow \text{Vertical change} \\ \leftarrow \text{Horizontal change} \end{array}$$

Hint

As the magnitude of the slope gets larger, the line gets steeper.

For example, in the graph on the left, the line changes 3 units vertically for every 5 units it changes horizontally. Its slope is $\frac{3}{5}$. The line on the right changes 4 units vertically for every 1 unit of horizontal change. It has a slope of $\frac{4}{1}$ or 4.

Notice that the line with slope 4 is steeper than the line that has a slope of $\frac{3}{5}$.

EXAMPLE 1

In-Class Example 1

Use Example 1.

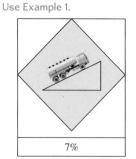

7%

A sign along a highway through the Rocky Mountains is shown on the left. What does it mean?

Solution

Percent means "out of 100." Therefore, we can write 7% as $\frac{7}{100}$. We can interpret $\frac{7}{100}$ as the ratio of the vertical change in the road to horizontal change in the road.

$$\text{The slope of the road is } \frac{7}{100}. \begin{array}{l} \leftarrow \text{Vertical change} \\ \leftarrow \text{Horizontal change} \end{array}$$

The highway rises 7 ft for every 100 horizontal feet.

[YOU TRY 1]

The slope of a conveyer belt is $\frac{5}{12}$, where the dimensions of the ramp are in inches. What does this mean?

---

2 Find the Slope of a Line Given Two Points on the Line

Here is line L. The points (x_1, y_1) and (x_2, y_2) are two points on line L. *We will find the ratio of the vertical change in y to the horizontal change in x between the points (x_1, y_1) and (x_2, y_2).*

To get from (x_1, y_1) to (x_2, y_2), we move *vertically* to point P then *horizontally* to (x_2, y_2). The x-coordinate of point P is x_1, and the y-coordinate of P is y_2.

When we moved *vertically* from (x_1, y_1) to point $P(x_1, y_2)$, how far did we go? We moved a vertical distance $y_2 - y_1$.

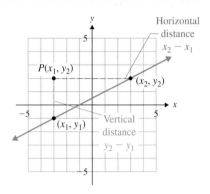

> **Note**
> The vertical change is $y_2 - y_1$ and is called the **rise**.

Then we moved *horizontally* from point $P(x_1, y_2)$ to (x_2, y_2). How far did we go? We moved a horizontal distance $x_2 - x_1$.

> **Note**
> The horizontal change is $x_2 - x_1$ and is called the **run**.

We said that the slope of a line is the ratio of the vertical change (rise) to the horizontal change (run). Therefore,

> **Formula** The Slope of a Line
>
> The **slope**, m, of a line containing the points (x_1, y_1) and (x_2, y_2) is given by
> $$m = \frac{\text{Vertical change}}{\text{Horizontal change}} = \frac{y_2 - y_1}{x_2 - x_1}$$

W Hint

Notice that the slope of a line is a ratio. It is the ratio of $(y_2 - y_1)$ to $(x_2 - x_1)$.

We can also think of slope as:

$$\frac{\text{Rise}}{\text{Run}} \quad \text{or} \quad \frac{\text{Change in } y}{\text{Change in } x}.$$

Let's look at some different ways to determine the slope of a line.

EXAMPLE 2   Determine the slope of each line.

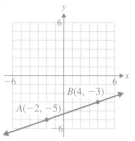
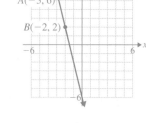
a)

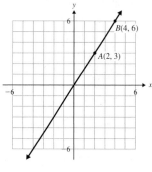

b)

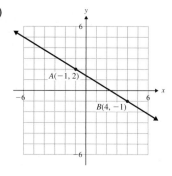

Solution

a) We will find the slope in two ways.

i) First, we will find the vertical change and the horizontal change by counting these changes as we go from A to B.

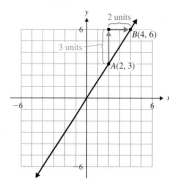

Vertical change (change in y) from A to B: 3 units

Horizontal change (change in x) from A to B: 2 units

Slope $= \dfrac{\text{Change in } y}{\text{Change in } x} = \dfrac{3}{2}$ or $m = \dfrac{3}{2}$

W Hint

If a line slopes *upward* from left to right, the slope is positive. If the line slopes *downward* from left to right, the slope is negative.

---

ii) We can also find the slope using the formula.
Let $(x_1, y_1) = (2, 3)$ and $(x_2, y_2) = (4, 6)$.

$$m = \frac{y_2 - y_1}{x_2 - x_1} = \frac{6 - 3}{4 - 2} = \frac{3}{2}$$

You can see that we get the same result either way we find the slope.

b) i) First, find the slope by counting the vertical change and horizontal change as we go from A to B.

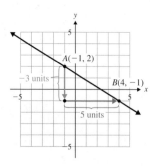

Vertical change (change in y) from A to B: -3 units

Horizontal change (change in x) from A to B: 5 units

Slope $= \dfrac{\text{Change in } y}{\text{Change in } x} = \dfrac{-3}{5} = -\dfrac{3}{5}$

or $m = -\dfrac{3}{5}$

ii) We can also find the slope using the formula.

Let $(x_1, y_1) = (-1, 2)$ and $(x_2, y_2) = (4, -1)$.

$$m = \frac{y_2 - y_1}{x_2 - x_1} = \frac{-1 - 2}{4 - (-1)} = \frac{-3}{5} = -\frac{3}{5}$$

Again, we obtain the same result using either method for finding the slope.

Note

The slope of $-\dfrac{3}{5}$ can be thought of as $\dfrac{-3}{5}, \dfrac{3}{-5}$, or $-\dfrac{3}{5}$.

[YOU TRY 2] Determine the slope of each line by

a) counting the vertical change and horizontal change.   b) using the formula.

a)

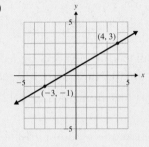

b)

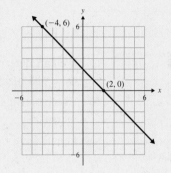

Notice that in Example 2a), the line has a positive slope and slants upward from left to right. As the value of x increases, the value of y increases as well. The line in Example 2b) has a negative slope and slants downward from left to right. Notice, in this case, that as the line goes from left to right, the value of x increases while the value of y decreases. We can summarize these results with the following general statements.

Property Positive and Negative Slopes

A line with a **positive slope** slants upward from left to right. As the value of x increases, the value of y increases as well.

A line with a **negative slope** slants downward from left to right. As the value of x increases, the value of y decreases.

3 Use Slope to Solve Applied Problems

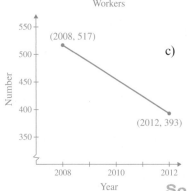
The graph models the number of members of a certain health club from 2008 to 2012.

a) How many members did the club have in 2008? in 2012?

b) What does the sign of the slope of the line segment mean in the context of the problem?

c) Find the slope of the line segment, and explain what it means in the context of the problem.

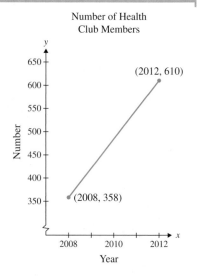

Number of Health Club Members

(2012, 610)

(2008, 358)

Solution

a) We can determine the number of members by reading the graph. In 2008, there were 358 members, and in 2012 there were 610 members.

b) The positive slope tells us that from 2008 to 2012 the number of members was increasing.

c) Let $(x_1, y_1) = (2008, 358)$ and $(x_2, y_2) = (2012, 610)$.

$$\text{Slope} = \frac{y_2 - y_1}{x_2 - x_1} = \frac{610 - 358}{2012 - 2008} = \frac{252}{4} = 63$$

The slope of the line is 63. Therefore, the number of members of the health club increased by 63 per year between 2008 and 2012.

4 Find the Slope of Horizontal and Vertical Lines

Find the slope of the line containing each pair of points.

a) (−4, 1) and (2, 1)   b) (2, 4) and (2, −3)

Solution

a) Let $(x_1, y_1) = (-4, 1)$ and $(x_2, y_2) = (2, 1)$.

$$m = \frac{y_2 - y_1}{x_2 - x_1} = \frac{1 - 1}{2 - (-4)} = \frac{0}{6} = 0$$

If we plot the points, we see that they lie on a horizontal line. Each point on the line has a y-coordinate of 1, so $y_2 - y_1$ *always* equals zero. **The slope of every horizontal line is zero.**

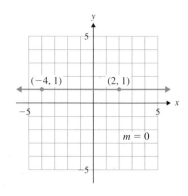

b) Let $(x_1, y_1) = (2, 4)$ and $(x_2, y_2) = (2, -3)$.

$$m = \frac{y_2 - y_1}{x_2 - x_1} = \frac{-3 - 4}{2 - 2} = \frac{-7}{0} \quad \text{undefined}$$

We say that the slope is undefined. Plotting these points gives us a vertical line. Each point on the line has an x-coordinate of 2, so $x_2 - x_1$ *always* equals zero.

The slope of every vertical line is undefined.

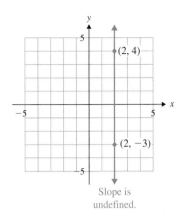

Slope is undefined.

[YOU TRY 3] Find the slope of the line containing each pair of points.

a) (4, 9) and (−3, 9)   b) (−7, 2) and (−7, 0)

Property Slopes of Horizontal and Vertical Lines

The slope of a horizontal line, $y = b$, is **zero.** The slope of a vertical line, $x = a$, is **undefined.** We can also say that a vertical line has **no slope.** (a and b are constants.)

---

5 Use Slope and One Point on a Line to Graph the Line

We have seen how we can find the slope of a line given two points on the line. Now, we will see how we can use the slope and *one* point on the line to graph the line.

EXAMPLE 5

Graph the line containing the point

a) $(-1, -2)$ with a slope of $\frac{3}{2}$.   b) (0, 1) with a slope of -3.

In-Class Example 5

Graph the line containing the point

a) (1, −3) with a slope of $\frac{1}{2}$.

b) (0, 2) with a slope of −3.

Answer:

a)

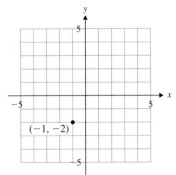

Solution

a) Plot the point.

Use the slope to find another point on the line.

$$m = \frac{3}{2} = \frac{\text{Change in } y}{\text{Change in } x}$$

To get from the point $(-1, -2)$ to another point on the line, move up 3 units in the y-direction and right 2 units in the x-direction.

Plot this second point, and draw a line through the two points.

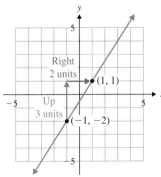

b)

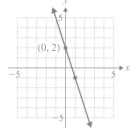

b) Plot the point (0, 1).

What does the slope, $m = -3$, mean?

$$m = -3 = \frac{-3}{1} = \frac{\text{Change in } y}{\text{Change in } x}$$

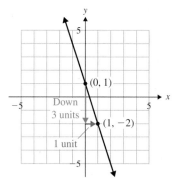

To get from (0, 1) to another point on the line, we will move *down* 3 units in the y-direction and *right* 1 unit in the x-direction. We end up at (1, −2).

Plot this point, and draw a line through (0, 1) and (1, −2).

In part b), we could have written $m = -3$ as $m = \frac{3}{-1}$. This would have given us a different point on the same line.

YOU TRY 4   Graph the line containing the point

a) (−2, 1) with a slope of $-\frac{3}{2}$.   b) (0, −3) with a slope of 2.

c) (3, 2) with an undefined slope.

Using Technology

When we look at the graph of a linear equation, we should be able to estimate its slope. Use the equation $y = x$ as a guideline.

Step 1: Graph the equation $y = x$.

We can make the graph a thick line (so we can tell it apart from the others) by moving the arrow all the way to the left and pressing ENTER:

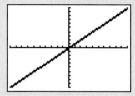

Step 2: Keeping this equation, graph the equation $y = 2x$:

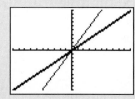

a. Is the new graph steeper or flatter than the graph of $y = x$?

b. Make a guess as to whether $y = 3x$ will be steeper or flatter than $y = x$. Test your guess by graphing $y = 3x$.

Step 3: Clear the equation $y = 2x$, and graph the equation $y = 0.5x$:

 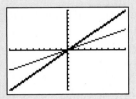

a. Is the new graph steeper or flatter than the graph of $y = x$?

b. Make a guess as to whether $y = 0.65x$ will be steeper or flatter than $y = x$. Test your guess by graphing $y = 0.65x$.

Step 4: Test similar situations, except with negative slopes: $y = -x$

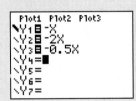 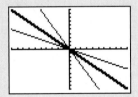

Did you notice that we have the same relationship, except in the opposite direction? That is, $y = 2x$ is steeper than $y = x$ in the positive direction, and $y = -2x$ is steeper than $y = -x$, but in the negative direction. And $y = 0.5x$ is flatter than $y = x$ in the positive direction, and $y = -0.5x$ is flatter than $y = -x$, but in the negative direction.

ANSWERS TO $\boxed{\text{YOU TRY}}$ **EXERCISES**

1) The belt rises 5 in. for every 12 horizontal inches.   2) a) $m = \dfrac{4}{7}$ b) $m = -1$

3) a) $m = 0$ b) undefined

4) a)   b)   c)

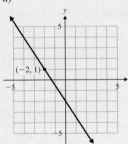 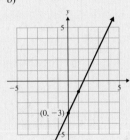 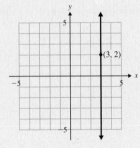

E Evaluate **3.3** Exercises   Do the exercises, and check your work.

*Additional answers can be found in the Answers to Exercises appendix.

Objective 1: Understand the Concept of Slope

1) Explain the meaning of slope.

2) Describe the slant of a line with a negative slope.
   It slants downward from left to right.

3) Describe the slant of a line with a positive slope.
   It slants upward from left to right.

4) The slope of a horizontal line is ___zero___.

5) The slope of a vertical line is _undefined_.

6) If a line contains the points (x_1, y_1) and (x_2, y_2), write the formula for the slope of the line. $m = \dfrac{y_2 - y_1}{x_2 - x_1}$

Mixed Exercises: Objectives 2 and 4

Determine the slope of each line by

a) counting the vertical change and the horizontal change as you move from one point to the other on the line;

and

b) using the slope formula. (See Example 2.)

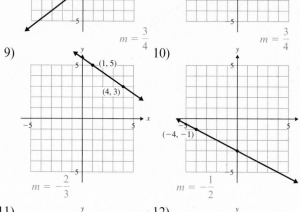

7)

$m = \dfrac{3}{4}$

8)

(4, 5)

$m = \dfrac{3}{4}$

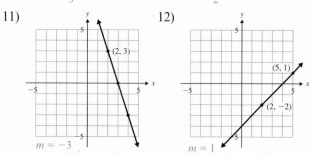

9)

(1, 5)

(4, 3)

$m = -\dfrac{2}{3}$

10)

(−4, −1)

$m = -\dfrac{1}{2}$

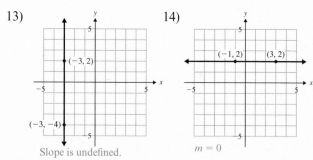

11)

(2, 3)

$m = -3$

12)

(5, 1)

(2, −2)

$m = 1$

13)

(−3, 2)

(−3, −4)

Slope is undefined.

14)

(−1, 2)   (3, 2)

$m = 0$

15) Graph a line with a positive slope and a negative y-intercept.

16) Graph a line with a negative slope and a positive x-intercept.

Use the slope formula to find the slope of the line containing each pair of points.

17) (2, 1) and (0, −3)   2

18) (0, 3) and (9, 6) $\dfrac{1}{3}$

19) (2, −6) and (−1, 6)   −4

20) (−3, 9) and (2, 4)   −1

21) (−4, 3) and (1, −8) $-\dfrac{11}{5}$

22) (2, 0) and (−5, 4) $-\dfrac{4}{7}$

23) (−2, −2) and (−2, 7)   undefined

24) (0, −6) and (−9, −6)   0

25) (3, 5) and (−1, 5)   0

26) (1, 3) and (1, −1)   undefined

27) $\left(\dfrac{2}{3}, \dfrac{5}{2}\right)$ and $\left(-\dfrac{1}{2}, 2\right)$ $\dfrac{3}{7}$

28) $\left(-\dfrac{1}{5}, \dfrac{3}{4}\right)$ and $\left(\dfrac{1}{3}, -\dfrac{3}{5}\right)$ $-\dfrac{81}{32}$

29) (3.5, −1.4) and (7.5, 1.6)   0.75

30) (−1.7, 10.2) and (0.8, −0.8)   −4.4

Objective 3: Use Slope to Solve Applied Problems

31) The longest run at Ski Dubai, an indoor ski resort in the Middle East, has a vertical drop of about 60 m with a horizontal distance of about 395 m. What is the slope of this ski run? (www.skidxb.com) $\dfrac{79}{12}$

32) The federal government requires that all wheelchair ramps in new buildings have a maximum slope of $\dfrac{1}{12}$. Does the following ramp meet this requirement? Give a reason for your answer. (www.access-board.gov)

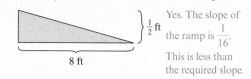

$\dfrac{1}{2}$ ft

8 ft

Yes. The slope of the ramp is $\dfrac{1}{16}$. This is less than the required slope.

Use the following information for Exercises 33 and 34.

To minimize accidents, the Park District Risk Management Agency recommends that playground slides and sledding hills have a maximum slope of about 0.577. (Illinois Parks and Recreation)

33) Does this slide meet the agency's recommendations?

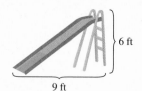

6 ft

9 ft

No. The slope of the slide is 0.6̄. This is more than the recommended slope.

34) Does this sledding hill meet the agency's recommendations?

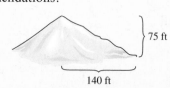

75 ft

140 ft

Yes. The slope of the sledding hill is about 0.536. This is less than the recommended slope.

35) In Granby, Colorado, the first 50 ft of a driveway cannot have a slope of more than 5%. If the first 50 ft of a driveway rises 0.75 ft for every 20 ft of horizontal distance, does this driveway meet the building code? (http://co.grand.co.us)   Yes. The slope of the driveway is 0.0375. This is less than the maximum slope allowed.

Use the following information for Exercises 36–38.

The steepness (slope) of a roof on a house in a certain town cannot exceed $\frac{7}{12}$, also known as a 7:12 *pitch*.

The first number refers to the rise of the roof. The second number refers to how far over you must go (the run) to attain that rise.

36) Find the slope of a roof with a 12:20 pitch. $\frac{3}{5}$

37) Find the slope of a roof with a 12:26 pitch. $\frac{6}{13}$

38) Does the slope in Exercise 36 meet the town's building code? Give a reason for your answer. No. The slope of the roof is $\frac{3}{5}$ or 0.6. This is greater than the maximum slope allowed.

39) The graph shows the approximate number of people in the United States injured in a motor vehicle accident from 2003 to 2007. (www.bts.gov)

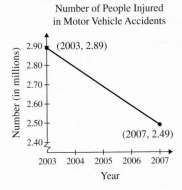

Number of People Injured in Motor Vehicle Accidents

a) Approximately how many people were injured in 2003? in 2005? 2.89 mil; 2.70 mil

b) Without computing the slope, determine whether it is positive or negative. negative

c) What does the sign of the slope mean in the context of the problem? The number of injuries is decreasing.

d) Find the slope of the line segment, and explain what it means in the context of the problem.

40) The graph shows the approximate number of prescriptions filled by mail order from 2004 to 2007. (www.census.gov)

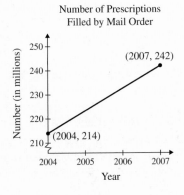

Number of Prescriptions Filled by Mail Order

a) Approximately how many prescriptions were filled by mail order in 2004? in 2007? 214 mil; 242 mil

b) Without computing the slope, determine whether it is positive or negative. positive

c) What does the sign of the slope mean in the context of the problem? The number of prescriptions being obtained by mail order is increasing.

d) Find the slope of the line segment, and explain what it means in the context of the problem. $m = 9\frac{1}{3}$; the number of prescriptions being obtained by mail order is increasing by $9\frac{1}{3}$ million per year.

Objective 5: Use Slope and One Point on a Line to Graph the Line

Graph the line containing the given point and with the given slope.

41) $(2, 1)$; $m = \frac{3}{4}$

42) $(1, 2)$; $m = \frac{1}{3}$

43) $(-2, -3)$; $m = \frac{1}{4}$

44) $(-5, 0)$; $m = \frac{2}{5}$

45) $(1, 2)$; $m = -\frac{3}{4}$

46) $(1, -3)$; $m = -\frac{2}{5}$

47) $(-1, -3)$; $m = 3$

48) $(0, -2)$; $m = -2$

49) $(6, 2)$; $m = -4$

50) $(4, 3)$; $m = -5$

51) $(3, -4)$; $m = -1$

52) $(-1, -2)$; $m = 0$

53) $(-2, 3)$; $m = 0$

54) $(2, 0)$; slope is undefined.

55) $(-1, -4)$; slope is undefined.

56) $(0, 0)$; $m = 1$

57) $(0, 0)$; $m = -1$

R1) After completing the exercises, what steps could you take to graph a line if you knew only two points of the line?

R2) How would two lines with the same slope look on a graph?

R3) Are there any problems you could not do? If so, write them down or circle them and ask your instructor for help.

3.4 The Slope-Intercept Form of a Line

P Prepare

O Organize

What are your objectives for Section 3.4?	How can you accomplish each objective?
1 Define the Slope-Intercept Form of a Line	• Understand how the slope-intercept form of a line is derived. • Write the definition of the *slope-intercept form of a line* in your notes, and draw an example.
2 Graph a Line Expressed in Slope-Intercept Form	• Use Objective 5 from Section 3.3 to help graph the line. • Complete the given example on your own. • Complete You Try 1.
3 Rewrite an Equation in Slope-Intercept Form, and Graph the Line	• Solve for y to write equations in slope-intercept form. • Complete the given example on your own. • Complete You Try 2. • Review the different methods for graphing a line given its equation.
4 Use Slope to Determine Whether Two Lines Are Parallel or Perpendicular	• Understand the properties of **Parallel Lines** and **Perpendicular Lines.** • Complete the given examples on your own. • Develop steps to determine whether lines are parallel or perpendicular, and compare the steps. • Complete You Trys 3 and 4.

W Work

Read the explanations, follow the examples, take notes, and complete the You Trys.

In Section 3.1, we learned that a linear equation in two variables can be written in the form $Ax + By = C$ (this is called **standard form**), where A, B, and C are real numbers and where both A and B do not equal zero. Equations of lines can take other forms, too, and we will look at one of those forms in this section.

1 Define the Slope-Intercept Form of a Line

We know that if (x_1, y_1) and (x_2, y_2) are points on a line, then the slope of the line is

$$m = \frac{y_2 - y_1}{x_2 - x_1}$$

Recall that to find the y-intercept of a line, we let $x = 0$ and solve for y. Let one of the points on a line be the y-intercept $(0, b)$, where b is a number. Let another point on the line be (x, y). See the graph on the left.

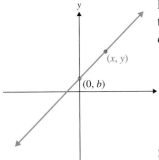

Substitute the points $(0, b)$ and (x, y) into the slope formula:

Subtract y-coordinates.
$$m = \frac{y_2 - y_1}{x_2 - x_1} = \frac{y - b}{x - 0} = \frac{y - b}{x}$$
Subtract x-coordinates.

Solve $m = \dfrac{y - b}{x}$ for y.

$$mx = \frac{y - b}{x} \cdot x \qquad \text{Multiply by } x \text{ to eliminate the fraction.}$$

$$mx = y - b$$
$$mx + b = y - b + b \qquad \text{Add } b \text{ to each side to solve for } y.$$
$$mx + b = y$$

OR

$$y = mx + b \qquad \text{Slope-intercept form}$$

W Hint

Notice that in $y = mx + b$, the letter b represents the y-coordinate of the y-intercept.

Definition

The **slope-intercept form of a line** is $y = mx + b$, where m is the slope and $(0, b)$ is the y-intercept.

2 Graph a Line Expressed in Slope-Intercept Form

EXAMPLE 1

Graph each equation.

a) $y = 4x - 3$ b) $y = \dfrac{1}{2}x$

In-Class Example 1

Graph each equation.

a) $y = -\dfrac{2}{3}x + 4$

b) $y = x$

Answer:

a)

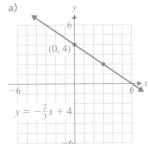

$y = -\dfrac{2}{3}x + 4$

Solution

Notice that each equation is in slope-intercept form, $y = mx + b$, where m is the slope and $(0, b)$ is the y-intercept.

a) Graph $y = 4x - 3$.

Slope $= 4$, y-intercept is $(0, -3)$.

Plot the y-intercept first, then use the slope to locate another point on the line. Since the slope is 4, think of it as $\dfrac{4}{1}$.   ← Change in y
                  ← Change in x

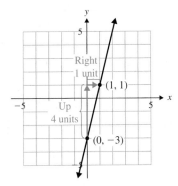

b)

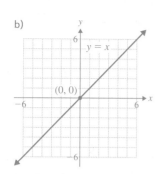

b) The equation $y = \frac{1}{2}x$ is the same as $y = \frac{1}{2}x + 0$. Identify the slope and y-intercept.

$$\text{Slope} = \frac{1}{2}, \quad y\text{-intercept is } (0, 0).$$

Plot the y-intercept, then use the slope to locate another point on the line.

$\frac{1}{2}$ is equivalent to $\frac{-1}{-2}$, so we can use $\frac{-1}{-2}$ as the slope to locate yet another point on the line.

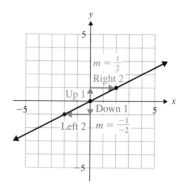

[**YOU TRY 1**] Graph each line using the slope and y-intercept.

a) $y = \frac{1}{4}x + 1$ b) $y = x - 3$ c) $y = -2x$

---

3 Rewrite an Equation in Slope-Intercept Form, and Graph the Line

Lines are not always written in slope-intercept form. They may be written in *standard form* (like $7x + 4y = 12$) or in another form such as $2x = 2y + 10$. We can put equations like these into slope-intercept form by solving the equation for y.

EXAMPLE 2

In-Class Example 2

Write $3x = 3y + 3$ in slope-intercept form, and graph.

Answer: $y = x - 1$

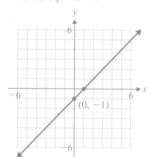

Write $7x + 4y = 12$ in slope-intercept form, and graph.

Solution

The slope-intercept form of a line is $y = mx + b$. We must solve the equation for y.

$$7x + 4y = 12$$
$$4y = -7x + 12 \qquad \text{Add } -7x \text{ to each side.}$$
$$y = -\frac{7}{4}x + 3 \qquad \text{Divide each side by 4.}$$

$$\text{Slope} = -\frac{7}{4} \text{ or } \frac{-7}{4}; \quad y\text{-intercept is } (0, 3).$$

$\left(\text{We could also have thought of the slope as } \frac{7}{-4}. \right)$

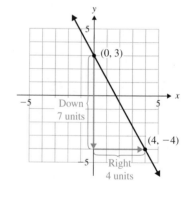

[**YOU TRY 2**] Put each equation into slope-intercept form, and graph.

a) $10x - 5y = -5$ b) $2x = -3 - 3y$

Summary Different Methods for Graphing a Line Given Its Equation

We have learned that we can use different methods for graphing lines. Given the equation of a line we can:

1) Make a table of values, plot the points, and draw the line through the points.

2) Find the x-intercept by letting $y = 0$ and solving for x, and find the y-intercept by letting $x = 0$ and solving for y. Plot the points, then draw the line through the points.

3) Put the equation into slope-intercept form, $y = mx + b$, identify the slope and y-intercept, then graph the line.

4 Use Slope to Determine Whether Two Lines Are Parallel or Perpendicular

Recall that two lines in a plane are **parallel** if they do not intersect. If we are given the equations of two lines, how can we determine whether they are parallel?

Here are the equations of two lines:

$$2x - 3y = -3 \qquad y = \frac{2}{3}x - 5$$

We will graph each line. To graph the first line, we write it in slope-intercept form.

$$-3y = -2x - 3 \qquad \text{Add } -2x \text{ to each side.}$$
$$y = \frac{-2}{-3}x - \frac{3}{-3} \qquad \text{Divide by } -3.$$
$$y = \frac{2}{3}x + 1 \qquad \text{Simplify.}$$

The slope-intercept form of the first line is $y = \frac{2}{3}x + 1$, and the second line is already in slope-intercept form, $y = \frac{2}{3}x - 5$. Now, graph each line.

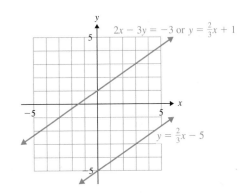

These lines are parallel. Their slopes are the same, but they have different y-intercepts. (If the y-intercepts were the same, they would be the same line.) This is how we determine whether two (nonvertical) lines are parallel. They have the same slope, but different y-intercepts.

Property Parallel Lines

Parallel lines have the same slope. (If two lines are vertical, they are parallel. However, their slopes are undefined.)

EXAMPLE 3

In-Class Example 3

Determine whether each pair of lines is parallel.
a) $-3x + 8y = 24$
   $3x - 8y = 7$
b) $y = 6x - 2$
   $12x - 3y = -5$

Answer: a) parallel
b) not parallel

W Hint

If the equation of a line is given in $Ax + By = C$ form, the slope of the line is equivalent to $-A/B$ in lowest terms.

Determine whether each pair of lines is parallel.

a) $2x + 8y = 12$
   $x + 4y = -20$

b) $y = -5x + 2$
   $5x - y = 7$

Solution

a) To determine whether the lines are parallel, we must find the slope of each line. If the slopes are the same, but the y-intercepts are different, the lines are parallel.

Write each equation in slope-intercept form.

$$2x + 8y = 12$$
$$8y = -2x + 12$$
$$y = -\frac{2}{8}x + \frac{12}{8}$$
$$y = -\frac{1}{4}x + \frac{3}{2}$$

$$x + 4y = -20$$
$$4y = -x - 20$$
$$y = -\frac{x}{4} - \frac{20}{4}$$
$$y = -\frac{1}{4}x - 5$$

Each line has a slope of $-\frac{1}{4}$. Their y-intercepts are different. Therefore, $2x + 8y = 12$ and $x + 4y = -20$ are parallel lines.

b) Again, we must find the slope of each line. $y = -5x + 2$ is already in slope-intercept form. Its slope is -5.

Write $5x - y = 7$ in slope-intercept form.

$$-y = -5x + 7 \qquad \text{Add } -5x \text{ to each side.}$$
$$y = 5x - 7 \qquad \text{Divide each side by } -1.$$

The slope of $y = -5x + 2$ is -5. The slope of $5x - y = 7$ is 5. The slopes are different; therefore, the lines are not parallel.

The slopes of two lines can tell us about another relationship between the lines. The slopes can tell us whether two lines are *perpendicular*.

Recall that two lines are **perpendicular** if they intersect at 90° angles.

The graphs of two perpendicular lines and their equations are on the left. We will see how their slopes are related.

Find the slope of each line by writing them in slope-intercept form.

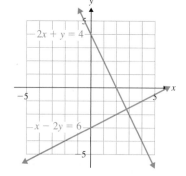

Line 1: $2x + y = 4$

$$y = -2x + 4$$

$$m = -2$$

Line 2: $x - 2y = 6$
$$-2y = -x + 6$$
$$y = \frac{-x}{-2} + \frac{6}{-2}$$
$$y = \frac{1}{2}x - 3$$

$$m = \frac{1}{2}$$

How are the slopes related? They are **negative reciprocals**. That is, if the slope of one line is a, then the slope of a line perpendicular to it is $-\frac{1}{a}$. This is how we determine whether two lines are perpendicular (where neither one is vertical).

Property Perpendicular Lines

Perpendicular lines have slopes that are negative reciprocals of each other.

EXAMPLE 4

In-Class Example 4

Determine whether each pair
of lines is perpendicular.
a) $4x - 3y = 3$
   $3x + 4y = 36$
b) $-9x + y = 2$
   $x = 9y - 5$

Answer: a) perpendicular
b) not perpendicular

Determine whether each pair of lines is perpendicular.

a) $15x - 12y = -4$ b) $2x - 9y = -9$
     $4x - 5y = 10$ $9x + 2y = 8$

Solution

a) To determine whether the lines are perpendicular, we must find the slope of each
    line. If the slopes are negative reciprocals, then the lines are perpendicular.

    Write each equation in slope-intercept form.

$$15x - 12y = -4 \qquad\qquad 4x - 5y = 10$$
$$-12y = -15x - 4 \qquad\qquad -5y = -4x + 10$$
$$y = \frac{-15}{-12}x - \frac{4}{-12} \qquad\qquad y = \frac{-4}{-5}x + \frac{10}{-5}$$
$$y = \frac{5}{4}x + \frac{1}{3} \qquad\qquad y = \frac{4}{5}x - 2$$
$$m = \frac{5}{4} \qquad\qquad m = \frac{4}{5}$$

The slopes are reciprocals, but they are not *negative* reciprocals. Therefore, the lines
are *not* perpendicular.

b) Begin by writing each equation in slope-intercept form so that we can find
    their slopes.

W Hint

Notice that if two lines are
perpendicular, the product
of their slopes will always
equal −1!

$$2x - 9y = -9 \qquad\qquad 9x + 2y = 8$$
$$-9y = -2x - 9 \qquad\qquad 2y = -9x + 8$$
$$y = \frac{-2}{-9}x - \frac{9}{-9} \qquad\qquad y = -\frac{9}{2}x + \frac{8}{2}$$
$$y = \frac{2}{9}x + 1 \qquad\qquad y = -\frac{9}{2}x + 4$$
$$m = \frac{2}{9} \qquad\qquad m = -\frac{9}{2}$$

The slopes are negative reciprocals; therefore, the lines are perpendicular.

[YOU TRY 3] Determine whether each pair of lines is parallel, perpendicular, or neither.

a) $5x - y = -2$ b) $y = \frac{8}{3}x + 9$ c) $x + 2y = 8$ d) $x = 7$
    $3x + 15y = -20$ $-32x + 12y = 15$ $2x = 4y + 3$ $y = -4$

ANSWERS TO [YOU TRY] EXERCISES

1) a)

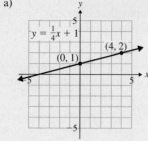

b)

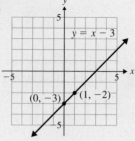

c)

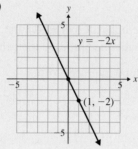

2) a)

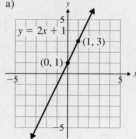

b)

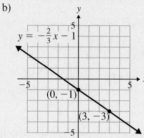

3) a) perpendicular
   b) parallel
   c) neither
   d) perpendicular

E Evaluate **3.4** Exercises

Do the exercises, and check your work.

*Additional answers can be found in the Answers to Exercises appendix.

Mixed Exercises: Objectives 1 and 2

1) The slope-intercept form of a line is $y = mx + b$.
   What is the slope? What is the y-intercept?
   The slope is m, and the y-intercept is $(0, b)$.

2) How do you put an equation that is in standard
   form, $Ax + By = C$, into slope-intercept form?
   Solve the equation for y.

Each of the following equations is in slope-intercept
form. Identify the slope and the y-intercept, then graph
each line using this information.

3) $y = \dfrac{2}{5}x - 6$ $m = \dfrac{2}{5}$, y-int: $(0, -6)$

4) $y = \dfrac{7}{5}x - 1$ $m = \dfrac{7}{5}$, y-int: $(0, -1)$

5) $y = -\dfrac{3}{2}x + 3$
   $m = -\dfrac{3}{2}$, y-int: $(0, 3)$

6) $y = -\dfrac{1}{3}x + 2$
   $m = -\dfrac{1}{3}$, y-int: $(0, 2)$

7) $y = \dfrac{3}{4}x + 2$
   $m = \dfrac{3}{4}$, y-int: $(0, 2)$

8) $y = \dfrac{2}{3}x + 5$
   $m = \dfrac{2}{3}$, y-int: $(0, 5)$

9) $y = -2x - 3$
   $m = -2$, y-int: $(0, -3)$

10) $y = 3x - 1$
    $m = 3$, y-int: $(0, -1)$

11) $y = 5x$ $m = 5$, y-int: $(0, 0)$

12) $y = -2x + 5$
    $m = -2$, y-int: $(0, 5)$

13) $y = -\dfrac{3}{2}x - \dfrac{7}{2}$ $m = -\dfrac{3}{2}$, y-int: $\left(0, -\dfrac{7}{2}\right)$

14) $y = \dfrac{3}{5}x + \dfrac{3}{4}$ $m = \dfrac{3}{5}$, y-int: $\left(0, \dfrac{3}{4}\right)$

15) $y = 6$ $m = 0$, y-int: $(0, 6)$

16) $y = -4$ $m = 0$, y-int: $(0, -4)$

Objective 3: Rewrite an Equation in Slope-Intercept Form, and Graph the Line

Put each equation into slope-intercept form, if possible, and graph.

17) $x + 3y = -6$ $y = -\frac{1}{3}x - 2$

18) $x + 2y = -8$ $y = -\frac{1}{2}x - 4$

19) $4x + 3y = 21$ $y = -\frac{4}{3}x + 7$

20) $2x - 5y = 5$ $y = \frac{2}{5}x - 1$

21) $2 = x + 3$

22) $x + 12 = 4$

23) $2x = 18 - 3y$ $y = -\frac{2}{3}x + 6$

24) $98 = 49y - 28x$ $y = \frac{4}{7}x + 2$

25) $y + 2 = -3$ $y = -5$

26) $y + 3 = 3$ $y = 0$

27) Kolya has a part-time job, and his gross pay can be described by the equation $P = 8.50h$, where P is his gross pay, in dollars, and h is the number of hours worked.

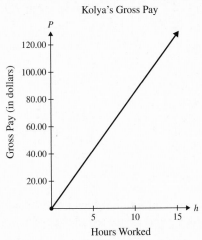

Kolya's Gross Pay

a) What is the P-intercept? What does it mean in the context of the problem?
(0, 0); if Kolya works 0 hr, he earns $0.

b) What is the slope? What does it mean in the context of the problem?
$m = 8.50$; Kolya earns $8.50 per hour.

c) Use the graph to find Kolya's gross pay when he works 12 hours. Confirm your answer using the equation. $102.00

28) The number of people, y, leaving on cruises from Florida from 2005 to 2009 can be approximated by $y = 88,000x + 4,866,600$, where x is the number of years after 2005. (www.census.gov)

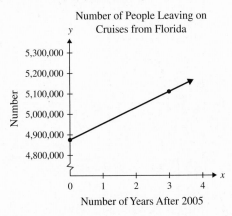

Number of People Leaving on Cruises from Florida

Number of Years After 2005

a) What is the y-intercept? What does it mean in the context of the problem?   (0, 4,875,000); in 2005, approximately 4,875,000 people left on their cruises from Florida.

b) What is the slope? What does it mean in the context of the problem? $m = 88,000$; the number of people leaving on a cruise from Florida is increasing by 88,000 per year.

c) Use the graph to determine how many people left on cruises from Florida in 2010. Confirm your answer using the equation.   5,306,600

29) A Tasmanian devil is a marsupial that lives in Australia. Once a joey leaves its mother's pouch, its weight for the first 8 weeks can be approximated by $y = 2x + 18$, where x represents the number of weeks it has been out of the pouch and y represents its weight, in ounces. (Wikipedia and Animal Planet)

a) What is the y-intercept, and what does it represent?
(0, 18); when the joey comes out of the pouch, it weighs 18 oz.

b) How much does a Tasmanian devil weigh 3 weeks after emerging from the pouch?   24 oz

c) Explain the meaning of the slope in the context of this problem.   A joey gains 2 oz per week after coming out of its mother's pouch.

d) How long would it take for a joey to weigh 32 oz?   7 weeks

30) The number of active physicians in Idaho, y, from 2005 to 2009 can be approximated by $y = 51.8x + 2424.4$, where x represents the number of years after 2005. (www.census.gov)

a) What is the y-intercept and what does it represent?   (0, 2424.4); in 2005, there were approximately 2424 active physicians in Idaho.

b) How many doctors were practicing in 2009?   about 2632

c) Explain the meaning of the slope in the context of this problem.   The number of active physicians in Idaho is increasing by 51.8 per year.

d) If the current trend continues, how many practicing doctors would Idaho have in 2018?   3098

31) On a certain day in 2011, the exchange rate between the American dollar and the Indian rupee was given by $r = 45.2d$, where d represents the number of dollars and r represents the number of rupees.

a) What is the r-intercept and what does it represent?   (0, 0); $0 = 0 rupees

b) What is the slope? What does it mean in the context of the problem?   45.2; each American dollar is worth 45.2 rupees.

c) If Madhura is going to India to visit her family, how many rupees could she get for $70.00?   3164 rupees

d) How many dollars could be exchanged for 2260 rupees? $50.00

32) The value of a car, v, in dollars, t years after it is purchased is given by $v = -1800t + 20,000$.

   a) What is the v-intercept and what does it represent? (0, 20,000); the initial value of the car is $20,000.

   b) What is the slope? What does it mean in the context of the problem? -1800; the value of the car is decreasing by $1800 per year.

   c) What is the car worth after 3 years? $14,600

   d) When will the car be worth $11,000? after 5 yr

Write the slope-intercept form for the equation of a line with the given slope and y-intercept.

33) $m = -4$; y-int: $(0, 7)$ $y = -4x + 7$

34) $m = -7$; y-int: $(0, 4)$ $y = -7x + 4$

35) $m = \dfrac{9}{5}$; y-int: $(0, -3)$ $y = \dfrac{9}{5}x - 3$

36) $m = \dfrac{7}{4}$; y-int: $(0, -2)$ $y = \dfrac{7}{4}x - 2$

37) $m = -\dfrac{5}{2}$; y-int: $(0, -1)$ $y = -\dfrac{5}{2}x - 1$

38) $m = \dfrac{1}{4}$; y-int: $(0, 7)$ $y = \dfrac{1}{4}x + 7$

39) $m = 1$; y-int: $(0, 2)$ $y = x + 2$

40) $m = -1$; y-int: $(0, 0)$ $y = -x$

41) $m = 0$; y-int: $(0, 0)$ $y = 0$

42) $m = 0$; y-int: $(0, -8)$ $y = -8$

Objective 4: Use Slope to Determine Whether Two Lines Are Parallel or Perpendicular

43) How do you know whether two lines are perpendicular? Their slopes are negative reciprocals, or one line is vertical and one is horizontal.

44) How do you know whether two lines are parallel? They have the same slopes and different y-intercepts, or both lines are vertical.

Determine whether each statement is true or false.

45) If two nonvertical lines are parallel, then the lines have different y-intercepts. true

46) If two lines are perpendicular, then both lines must cross the x-axis. false

47) If two lines are perpendicular, then one of the lines must have a positive slope and the other must have a negative slope. false

48) If two lines are parallel and neither intersects the x-axis, then both lines have a slope of 0. true

Determine whether each pair of lines is parallel, perpendicular, or neither.

49) $y = -x - 5$
   $y = x + 8$ perpendicular

50) $y = \dfrac{3}{4}x + 2$
   $y = \dfrac{3}{4}x - 1$ parallel

51) $y = \dfrac{2}{9}x + 4$
   $4x - 18y = 9$ parallel

52) $y = \dfrac{4}{5}x + 2$
   $5x + 4y = 12$ perpendicular

53) $3x - y = 4$
   $2x - 5y = -9$ neither

54) $-4x + 3y = -5$
   $4x - 6y = -3$ neither

55) $-x + y = -21$
   $y = 2x + 5$ neither

56) $x + 3y = 7$
   $y = 3x$ perpendicular

57) $x + 7y = 4$
   $y - 7x = 4$ perpendicular

58) $5y - 3x = 1$
   $3x - 5y = -8$ parallel

59) $y = -\dfrac{1}{2}x$
   $x + 2y = 4$ parallel

60) $-4x + 6y = 5$
   $2x - 3y = -12$ parallel

61) $x = -1$
   $y = 6$ perpendicular

62) $y = 12$
   $y = 4$ parallel

63) $x = -4.3$
   $x = 0$ parallel

64) $x = 7$
   $y = 0$ perpendicular

Lines L_1 and L_2 contain the given points. Determine whether lines L_1 and L_2 are parallel, perpendicular, or neither.

65) L_1: $(-1, -7), (2, 8)$
   L_2: $(10, 2), (0, 4)$ perpendicular

66) L_1: $(0, -3), (-4, -11)$
   L_2: $(-2, 0), (3, 10)$ parallel

67) L_1: $(1, 10), (3, 8)$
   L_2: $(2, 4), (-5, -17)$ neither

68) L_1: $(-1, 4), (2, -8)$
   L_2: $(8, 5), (0, 3)$ perpendicular

69) L_1: $(-3, 6), (4, -1)$
   L_2: $(-6, -5), (-10, -1)$ parallel

70) L_1: $(5, -5), (7, 11)$
   L_2: $(-3, 0), (6, 3)$ neither

71) L_1: $(-6, 2), (-6, 1)$
   L_2: $(4, 0), (4, -5)$ parallel

72) L_1: $(8, 1), (7, 1)$
   L_2: $(12, -1), (-2, -1)$ parallel

73) L_1: $(7, 2), (7, 5)$
   L_2: $(-2, 0), (1, 0)$ perpendicular

74) L_1: $(-6, 4), (-6, -1)$
   L_2: $(-1, 10), (-3, 10)$ perpendicular

R1) Which equation from this chapter is the easiest for you to remember?

R2) Which equation from this chapter is the most difficult for you to remember?

R3) Using the standard form $Ax + By = C$, can you determine the slope and y-intercept of a line by inspection? Hint: Solve the equation $Ax + By = C$ for y.

R4) Which exercises in this section do you find most challenging?

3.5 Writing an Equation of a Line

P **Prepare**

O **Organize**

What are your objectives for Section 3.5?	How can you accomplish each objective?
1 Rewrite an Equation in Standard Form	• Complete the given example on your own, and create a procedure. • Complete You Try 1.
2 Write an Equation of a Line Given Its Slope and y-intercept	• Learn the procedure for **Writing an Equation of a Line Given Its Slope and y-intercept,** and write it in your notes. • Complete the given example on your own. • Complete You Try 2.
3 Use the Point-Slope Formula to Write an Equation of a Line Given Its Slope and a Point on the Line	• Learn the formula for **Point-Slope Form of a Line,** and write it in your notes. • Complete the given example on your own, and notice the different scenarios you might see. • Complete You Try 3.
4 Use the Point-Slope Formula to Write an Equation of a Line Given Two Points on the Line	• Learn the steps for writing an equation of a line given two points on the line. • Complete the given example on your own. • Complete You Try 4.
5 Write Equations of Horizontal and Vertical Lines	• Learn the formulas for **Equations of Horizontal and Vertical Lines,** and write them in your notes. • Complete the given example on your own. • Complete You Try 5. • Write the summary for *Writing Equations of Lines* in your own words.
6 Write an Equation of a Line That Is Parallel or Perpendicular to a Given Line	• Follow the examples, and create a step-by-step procedure for finding and writing the equations of parallel and perpendicular lines. • Complete You Try 7.
7 Write a Linear Equation to Model Real-World Data	• Follow the example, and then complete the example in your notes without looking at the solution.

So far in this chapter, we have been graphing lines given their equations. Now we will write an equation of a line when we are given information about it.

1 Rewrite an Equation in Standard Form

In Section 3.4, we practiced writing equations of lines in slope-intercept form. Here we will discuss how to write a line in **standard form**, $Ax + By = C$, **with the additional conditions that A, B, and C are integers and A is positive.**

EXAMPLE 1

In-Class Example 1

Rewrite each linear equation in standard form.
a) $-5x - 11 = 2y$
b) $y = \dfrac{1}{3}x - \dfrac{3}{4}$

Answer: a) $5x + 2y = -11$
b) $4x - 12y = 9$

Rewrite each linear equation in standard form.

a) $3x + 8 = -2y$ b) $y = -\dfrac{3}{4}x + \dfrac{1}{6}$

Solution

a) In standard form, the x- and y-terms are on the same side of the equation.

$$3x + 8 = -2y$$
$$3x = -2y - 8 \quad \text{Subtract 8 from each side.}$$
$$3x + 2y = -8 \quad \text{Add } 2y \text{ to each side; the equation is now in standard form.}$$

W Hint

Review how to eliminate fractions from an equation as demonstrated in Section 2.3.

b) Since an equation $Ax + By = C$ is considered to be in standard form when A, B, and C are integers, the first step in writing $y = -\dfrac{3}{4}x + \dfrac{1}{6}$ in standard form is to eliminate the fractions.

$$y = -\frac{3}{4}x + \frac{1}{6}$$
$$12 \cdot y = 12\left(-\frac{3}{4}x + \frac{1}{6}\right) \quad \text{Multiply both sides of the equation by 12.}$$
$$12y = -9x + 2$$
$$9x + 12y = 2 \quad \text{Add } 9x \text{ to each side.}$$

The standard form is $9x + 12y = 2$.

[YOU TRY 1]

Rewrite each equation in standard form.

a) $5x = 3 + 11y$ b) $y = \dfrac{1}{3}x - 7$

In the rest of this section, we will learn how to write equations of lines given information about their graphs.

2 Write an Equation of a Line Given Its Slope and *y*-intercept

Procedure Write an Equation of a Line Given Its Slope and *y*-intercept

If we are given the slope and *y*-intercept of a line, use $y = mx + b$ and substitute those values into the equation.

EXAMPLE 2

In-Class Example 2

Find an equation of the line with slope $= -7$ and *y*-intercept (0, 4).

Answer: $y = -7x + 4$

Find an equation of the line with slope $= -6$ and *y*-intercept (0, 5).

Solution

Since we are told the slope and *y*-intercept, use $y = mx + b$.

$$m = -6 \quad \text{and} \quad b = 5$$

Substitute these values into $y = mx + b$ to get $y = -6x + 5$.

$$\boxed{\text{YOU TRY 2}}$$ Find an equation of the line with slope $= \dfrac{5}{8}$ and *y*-intercept (0, −9).

W Hint

Remember, a *y*-intercept will always have an *x*-coordinate of 0.

3 Use the Point-Slope Formula to Write an Equation of a Line Given Its Slope and a Point on the Line

When we are given the slope of a line and a point on that line, we can use another method to find its equation. This method comes from the formula for the slope of a line.

Let (x_1, y_1) be a given point on a line, and let (x, y) be any other point on the same line, as shown in the figure. The slope of that line is

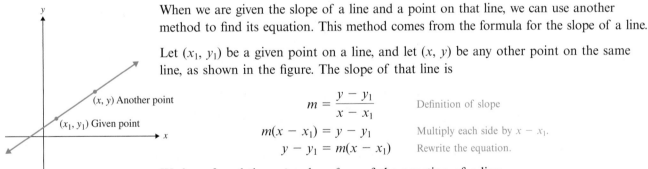

$$m = \frac{y - y_1}{x - x_1} \qquad \text{Definition of slope}$$
$$m(x - x_1) = y - y_1 \qquad \text{Multiply each side by } x - x_1.$$
$$y - y_1 = m(x - x_1) \qquad \text{Rewrite the equation.}$$

We have found the *point-slope form* of the equation of a line.

W Hint

The point-slope formula will help us write an equation of a line. We will not express our final answer in this form. We will write our answer in either slope-intercept form or in standard form.

Formula Point-Slope Form of a Line

The **point-slope form of a line** is

$$y - y_1 = m(x - x_1)$$

where (x_1, y_1) is a point on the line and m is its slope.

EXAMPLE 3

In-Class Example 3

A line has slope −3 and contains the point (2, 5). Find the standard form for the equation of the line.

Answer: 3x + y = 11

W Hint

Notice that when you write the equation of a line using the point-slope formula, you will always substitute the coordinates of the point for x_1 and y_1.

A line has slope −4 and contains the point (1, 5). Find the standard form for the equation of the line.

Solution

Although we are told to find the *standard form* for the equation of the line, we do not try to immediately "jump" to standard form. First, ask yourself, *"What information am I given?"*

We are given the *slope* and a *point on the line.* Therefore, we will begin by using the point-slope formula. Our *last* step will be to put the equation in standard form.

Use $y - y_1 = m(x - x_1)$. Substitute −4 for m. Substitute (1, 5) for (x_1, y_1).

$$y - y_1 = m(x - x_1)$$
$$y - 5 = -4(x - 1) \qquad \text{Substitute 1 for } x_1 \text{ and 5 for } y_1; \text{ let } m = -4.$$
$$y - 5 = -4x + 4 \qquad \text{Distribute.}$$

To write the answer in standard form, we must get the *x*- and *y*-terms on the same side of the equation so that the coefficient of *x* is positive.

$$4x + y - 5 = 4 \qquad \text{Add } 4x \text{ to each side.}$$
$$4x + y = 9 \qquad \text{Add 5 to each side.}$$

The standard form of the equation is $4x + y = 9$.

[YOU TRY 3]

a)   A line has slope −8 and contains the point (−4, 5). Find the standard form for the equation of the line.

b)   Find an equation of the line containing the point (5, 3) with slope = 2. Express the answer in slope-intercept form.

4 Use the Point-Slope Formula to Write an Equation of a Line Given Two Points on the Line

We are now ready to discuss how to write an equation of a line when we are given two points on a line.

To write an equation of a line given two points on the line,

a) use the points to find the slope of the line
   then
b) use the slope and *either one* of the points in the point-slope formula.

EXAMPLE 4

In-Class Example 4

Write an equation of the line containing the points (−5, 7) and (2, 1). Express the answer in slope-intercept form.

Answer: $y = \frac{6}{7}x + \frac{19}{7}$

Write an equation of the line containing the points (4, 9) and (2, 6). Express the answer in slope-intercept form.

Solution

We are given two points on the line, so first, we will find the slope.

$$m = \frac{6 - 9}{2 - 4} = \frac{-3}{-2} = \frac{3}{2}$$

We will use the slope and *either one* of the points in the point-slope formula. (Each point will give the same result.) We will use (4, 9).

Substitute $\dfrac{3}{2}$ for m. Substitute $(4, 9)$ for (x_1, y_1).

$$y - y_1 = m(x - x_1)$$

$$y - 9 = \frac{3}{2}(x - 4) \qquad \text{Substitute 4 for } x_1 \text{ and 9 for } y_1; \text{ let } m = \frac{3}{2}.$$

$$y - 9 = \frac{3}{2}x - 6 \qquad \text{Distribute.}$$

We must write our answer in slope-intercept form, $y = mx + b$, so solve the equation for y.

$$y = \frac{3}{2}x + 3 \qquad \text{Add 9 to each side to solve for } y.$$

The equation is $y = \dfrac{3}{2}x + 3$.

[**YOU TRY 4**] Find the slope-intercept form for the equation of the line containing the points $(4, 2)$ and $(1, -5)$.

---

5 Write Equations of Horizontal and Vertical Lines

Earlier we learned that the slope of a horizontal line is zero and that it has equation $y = b$, where b is a constant. The slope of a vertical line is undefined, and its equation is $x = a$, where a is a constant.

> **Formula** Equations of Horizontal and Vertical Lines
>
> **Equation of a Horizontal Line:** The equation of a horizontal line containing the point (a, b) is $y = b$.
>
> **Equation of a Vertical Line:** The equation of a vertical line containing the point (a, b) is $x = a$.

EXAMPLE 5

Write an equation of the horizontal line containing the point $(7, -1)$.

In-Class Example 5

Write an equation of the horizontal line containing the point $(-5, 11)$.

Answer: $y = 11$

Solution

The equation of a horizontal line has the form $y = b$, where b is the y-coordinate of the point. The equation of the line is $y = -1$.

[**YOU TRY 5**] Write an equation of the horizontal line containing the point $(3, -8)$.

> ## Summary Writing Equations of Lines
>
> If you are given
>
> 1) **the slope and *y*-intercept of the line,** use $y = mx + b$ and substitute those values into the equation.
>
> 2) **the slope of the line and a point on the line,** use the point-slope formula:
>
> $$y - y_1 = m(x - x_1)$$
>
> Substitute the slope for *m* and the point you are given for (x_1, y_1). Write your answer in slope-intercept or standard form.
>
> 3) **two points on the line,** find the slope of the line and then use the slope and *either one* of the points in the point-slope formula. Write your answer in slope-intercept or standard form.
>
> The equation of a **horizontal line** containing the point (a, b) is **$y = b$.**
>
> The equation of a **vertical line** containing the point (a, b) is **$x = a$.**

6 Write an Equation of a Line That Is Parallel or Perpendicular to a Given Line

In Section 3.4, we learned that parallel lines have the same slope, and perpendicular lines have slopes that are negative reciprocals of each other. We will use this information to write the equation of a line that is parallel or perpendicular to a given line.

EXAMPLE 6

In-Class Example 6

A line contains the point $(4, -2)$ and is parallel to the line $y = \frac{3}{4}x + 9$. Write the equation of the line in slope-intercept form.

Answer: $y = \frac{3}{4}x - 5$

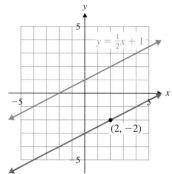

W Hint

When working with parallel lines, be sure to verify that your final equation represents the same slope as indicated in the given line.

A line contains the point $(2, -2)$ and is parallel to the line $y = \frac{1}{2}x + 1$. Write the equation of the line in slope-intercept form.

Solution

Let's look at the graph on the left to help us understand what is happening in this example.

We must find the equation of the line in red. It is the line containing the point $(2, -2)$ that is parallel to $y = \frac{1}{2}x + 1$.

The line $y = \frac{1}{2}x + 1$ has $m = \frac{1}{2}$. Therefore, the red line will have $m = \frac{1}{2}$ as well.

We know the slope, $\frac{1}{2}$, and a point on the line, $(2, -2)$, so we use the point-slope formula to find its equation.

Substitute $\frac{1}{2}$ for *m*. Substitute $(2, -2)$ for (x_1, y_1).

$$y - y_1 = m(x - x_1)$$
$$y - (-2) = \frac{1}{2}(x - 2) \quad \text{Substitute 2 for } x_1 \text{ and } -2 \text{ for } y_1; \text{ let } m = \frac{1}{2}.$$
$$y + 2 = \frac{1}{2}x - 1 \quad \text{Distribute.}$$
$$y = \frac{1}{2}x - 3 \quad \text{Subtract 2 from each side.}$$

The equation is $y = \frac{1}{2}x - 3$.

[YOU TRY 6]

A line contains the point $(-6, 2)$ and is parallel to the line $y = -\dfrac{3}{2}x + \dfrac{1}{4}$. Write the equation of the line in slope-intercept form.

EXAMPLE 7

In-Class Example 7

Find the standard form for the equation of the line that contains the point $(-5, -12)$ and that is perpendicular to $2x + 3y = 12$.

Answer: $3x - 2y = 9$

W Hint

Are you writing out the example as you are reading it?

Find the standard form for the equation of the line that contains the point $(-4, 3)$ and that is perpendicular to $3x - 4y = 8$.

Solution

Begin by finding the slope of $3x - 4y = 8$ by putting it into slope-intercept form.

$$3x - 4y = 8$$
$$-4y = -3x + 8 \qquad \text{Add } -3x \text{ to each side.}$$
$$y = \frac{-3}{-4}x + \frac{8}{-4} \qquad \text{Divide by } -4.$$
$$y = \frac{3}{4}x - 2 \qquad \text{Simplify.}$$
$$m = \frac{3}{4}$$

Then, determine the slope of the line containing $(-4, 3)$ by finding the *negative reciprocal* of the slope of the given line.

$$m_{\text{perpendicular}} = -\frac{4}{3}$$

The line we want has $m = -\dfrac{4}{3}$ and contains the point $(-4, 3)$. Use the point-slope formula to find an equation of the line.

Substitute $-\dfrac{4}{3}$ for m.

Substitute $(-4, 3)$ for (x_1, y_1).

$$y - y_1 = m(x - x_1)$$
$$y - 3 = -\frac{4}{3}(x - (-4)) \qquad \text{Substitute } -4 \text{ for } x_1 \text{ and } 3 \text{ for } y_1; \text{ let } m = -\frac{4}{3}.$$
$$y - 3 = -\frac{4}{3}(x + 4)$$
$$y - 3 = -\frac{4}{3}x - \frac{16}{3} \qquad \text{Distribute.}$$

Since we are asked to write the equation in standard form, eliminate the fractions by multiplying each side by 3.

$$3(y - 3) = 3\left(-\frac{4}{3}x - \frac{16}{3}\right)$$
$$3y - 9 = -4x - 16 \qquad \text{Distribute.}$$
$$3y = -4x - 7 \qquad \text{Add 9 to each side.}$$
$$4x + 3y = -7 \qquad \text{Add } 4x \text{ to each side.}$$

The equation is $4x + 3y = -7$.

[YOU TRY 7]

Find the equation of the line perpendicular to $5x - y = -6$ containing the point $(10, 0)$. Write the equation in standard form.

7 Write a Linear Equation to Model Real-World Data

Equations of lines are often used to describe real-world situations. We will look at an example in which we must find the equation of a line when we are given some data.

Since 2005, vehicle consumption of E85 fuel (ethanol, 85%) has increased by about 8260.7 thousand gallons per year. In 2009, approximately 70,916.6 thousand gallons were used. (Statistical Abstract of the United States)

a) Write a linear equation to model these data. Let x represent the number of years after 2005, and let y represent the amount of E85 fuel (in thousands of gallons) consumed.

b) How much E85 fuel did vehicles use in 2005? in 2007?

Solution

a) The statement "vehicle consumption of E85 fuel . . . has increased by about 8260.7 thousand gallons per year" tells us the rate of change of fuel use with respect to time. Therefore, this is the *slope*. It will be *positive* since consumption is increasing.

$$m = 8260.7$$

The statement "In 2009, approximately 70,916.6 thousand gallons were used" gives us a point on the line.

If x = the number of years after 2005, then the year 2009 corresponds to $x = 4$.

If y = number of gallons (in thousands) of E85 consumed, then 70,916.6 thousand gallons corresponds to $y = 70{,}916.6$.

A point on the line is **(4, 70,916.6)**.

Now that we know the slope and a point on the line, we can write an equation of the line using the point-slope formula.

Substitute 8260.7 for m. Substitute (4, 70,916.6) for (x_1, y_1).

$$
\begin{aligned}
y - y_1 &= m(x - x_1) \\
y - 70{,}916.6 &= 8260.7(x - 4) &&\text{Substitute 4 for } x_1 \text{ and 70,916.6 for } y_1. \\
y - 70{,}916.6 &= 8260.7x - 33{,}042.8 &&\text{Distribute.} \\
y &= 8260.7x + 37{,}873.8 &&\text{Add 70,916.6 to each side.}
\end{aligned}
$$

The equation is $y = 8260.7x + 37{,}873.8$.

b) To determine the amount of E85 used in 2005, let $x = 0$ since x = the number of years *after* 2005.

$$
\begin{aligned}
y &= 8260.7(0) + 37{,}873.8 &&\text{Substitute } x = 0. \\
y &= 37{,}837.3
\end{aligned}
$$

In 2005, vehicles used about 37,837.3 thousand gallons of E85 fuel. Notice that the equation is in slope-intercept form, $y = mx + b$, and our result is b. That is because when we find the y-intercept, we let $x = 0$.

To determine how much E85 fuel was used in 2007, let $x = 2$ since 2007 is 2 years after 2005.

$$
\begin{aligned}
y &= 8260.7(2) + 37{,}873.8 &&\text{Substitute } x = 2. \\
y &= 16{,}521.4 + 37{,}873.8 &&\text{Multiply.} \\
y &= 54{,}395.2
\end{aligned}
$$

In 2007, vehicles used approximately 54,395.2 thousand gallons of E85.

We can use a graphing calculator to explore what we have learned about perpendicular lines.

1. Graph the line $= -2x + 4$. What is its slope?

2. Find the slope of the line perpendicular to the graph of $y = -2x + 4$.

3. Find the equation of the line perpendicular to $y = -2x + 4$ that passes through the point (6, 0). Express the equation in slope-intercept form.

4. Graph both the original equation and the equation of the perpendicular line:

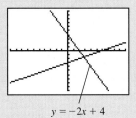

$y = -2x + 4$

5. Do the lines above appear to be perpendicular?

$y = \frac{1}{2}x - 3$

6. Press ZOOM and choose 5:Zsquare.

7. Do the graphs look perpendicular now? Because the viewing window on a graphing calculator is a rectangle, *squaring* the window will give a more accurate picture of the graphs of the equations.

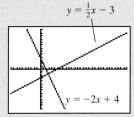

$y = -2x + 4$

ANSWERS TO [YOU TRY] EXERCISES

1) a) $5x - 11y = 3$ b) $x - 3y = 21$ 2) $y = \frac{5}{8}x - 9$ 3) a) $8x + y = -27$ b) $y = 2x - 7$

4) $y = \frac{7}{3}x - \frac{22}{3}$ 5) $y = -8$ 6) $y = -\frac{3}{2}x - 7$ 7) $x + 5y = 10$

ANSWERS TO TECHNOLOGY EXERCISES

1) -2 2) $\frac{1}{2}$ 3) $y = \frac{1}{2}x - 3$ 5) No, because they do not meet at 90° angles.

7) Yes, because they meet at 90° angles.

E Evaluate **3.5** Exercises   Do the exercises, and check your work.

*Additional answers can be found in the Answers to Exercises appendix.

Objective 1: Rewrite an Equation in Standard Form
Rewrite each equation in standard form.

1) $y = -2x - 4$ $2x + y = -4$ 2) $y = 3x + 5$
$3x - y = -5$

3) $x = y + 1$ $x - y = 1$ 4) $x = -4y - 9$
$x + 4y = -9$

5) $y = \frac{4}{5}x + 1$ $4x - 5y = -5$ 6) $y = \frac{2}{3}x - 6$
$2x - 3y = 18$

7) $y = -\frac{1}{3}x - \frac{5}{4}$ 8) $y = -\frac{1}{4}x + \frac{2}{5}$
$4x + 12y = -15$ $5x + 20y = 8$

Objective 2: Write an Equation of a Line Given Its Slope and y-intercept

9) Explain how to find an equation of a line when you are given the slope and y-intercept of the line.
Substitute the slope and y-intercept into $y = mx + b$.

Find an equation of the line with the given slope and y-intercept. Express your answer in the indicated form.

10) $m = -3$, y-int: (0, 3); slope-intercept form $y = -3x + 3$

11) $m = -7$, y-int: (0, 2); slope-intercept form $y = -7x + 2$

12) $m = 1$, y-int: $(0, -4)$; standard form $x - y = 4$

24) 13) $m = -4$, y-int: $(0, 6)$; standard form $4x + y = 6$

14) $m = -\dfrac{2}{5}$, y-int: $(0, -4)$; standard form
$$2x + 5y = -20$$

15) $m = \dfrac{2}{7}$, y-int: $(0, -3)$; standard form $2x - 7y = 21$

16) $m = 1$, y-int: $(0, 0)$; slope-intercept form $y = x$

17) $m = -1$, y-int: $(0, 0)$; slope-intercept form $y = -x$

18) $m = \dfrac{5}{9}$, y-int: $\left(0, -\dfrac{1}{3}\right)$; standard form $5x - 9y = 3$

Objective 3: Use the Point-Slope Formula to Write an Equation of a Line Given Its Slope and a Point on the Line

19) a) If (x_1, y_1) is a point on a line with slope m, then the point-slope formula is _____.
$y - y_1 = m(x - x_1)$
   b) Explain how to find an equation of a line when you are given the slope and a point on the line.
   Substitute the slope and point into the point-slope formula.

Find an equation of the line containing the given point with the given slope. Express your answer in the indicated form.

20) $(2, 3)$, $m = 4$; slope-intercept form $y = 4x - 5$

24) 21) $(5, 7)$, $m = 1$; slope-intercept form $y = x + 2$

22) $(-2, 5)$, $m = -3$; slope-intercept form $y = -3x - 1$

23) $(4, -1)$, $m = -5$; slope-intercept form $y = -5x + 19$

24) $(-1, -2)$, $m = 2$; standard form $2x - y = 0$

25) $(-2, -1)$, $m = 4$; standard form $4x - y = -7$

26) $(9, 3)$, $m = -\dfrac{1}{3}$; standard form $x + 3y = 18$

27) $(-5, 8)$, $m = \dfrac{2}{5}$; standard form $2x - 5y = -50$

28) $(-2, -3)$, $m = \dfrac{1}{8}$; slope-intercept form $y = \dfrac{1}{8}x - \dfrac{11}{4}$

29) $(5, 1)$, $m = -\dfrac{5}{4}$; slope-intercept form $y = -\dfrac{5}{4}x + \dfrac{29}{4}$

30) $(4, 0)$, $m = -\dfrac{3}{16}$; standard form $3x + 16y = 12$

31) $(-3, 0)$, $m = \dfrac{5}{6}$; standard form $5x - 6y = -15$

32) $\left(\dfrac{1}{4}, -1\right)$, $m = 3$; slope-intercept form $y = 3x - \dfrac{7}{4}$

Objective 4: Use the Point-Slope Formula to Write an Equation of a Line Given Two Points on the Line

24) 33) Explain how to find an equation of a line when you are given two points on the line.
Find the slope and use it and one of the points in the point-slope formula.

Find an equation of the line containing the two given points. Express your answer in the indicated form.

34) $(-2, 1)$ and $(8, 11)$; slope-intercept form $y = x + 3$

35) $(-1, 7)$ and $(3, -5)$; slope-intercept form $y = -3x + 4$

36) $(6, 8)$ and $(-1, -4)$; slope-intercept form $y = \dfrac{12}{7}x - \dfrac{16}{7}$

37) $(4, 5)$ and $(7, 11)$; slope-intercept form $y = 2x - 3$

38) $(2, -1)$ and $(5, 1)$; standard form $2x - 3y = 7$

39) $(-2, 4)$ and $(1, 3)$; slope-intercept form $y = -\dfrac{1}{3}x + \dfrac{10}{3}$

40) $(-1, 10)$ and $(3, -2)$; standard form $3x + y = 7$

24) 41) $(-5, 1)$ and $(4, -2)$; standard form $x + 3y = -2$

42) $(4.2, 1.3)$ and $(-3.4, -17.7)$; slope-intercept form
$y = 2.5x - 9.2$

43) $(-3, -11)$ and $(3, -1)$; standard form $5x - 3y = 18$

44) $(-6, 0)$ and $(3, -1)$; standard form $x + 9y = -6$

45) $(-2.3, 8.3)$ and $(5.1, -13.9)$; slope-intercept form
$y = -3.0x + 1.4$

46) $(-7, -4)$ and $(14, 2)$; standard form $2x - 7y = 14$

Write the slope-intercept form of the equation of each line, if possible.

47)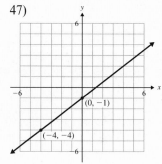

$y = \dfrac{3}{4}x - 1$

48)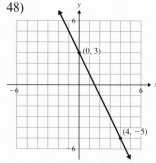

$y = -2x + 3$

49)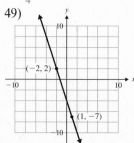

$y = -3x - 4$

50)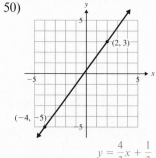

$y = \dfrac{4}{3}x + \dfrac{1}{3}$

51)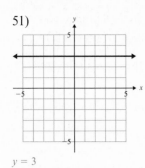

$y = 3$

52)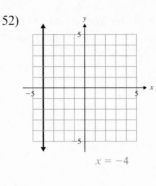

$x = -4$

Mixed Exercises: Objectives 2–5

Write the slope-intercept form of the equation of the line, if possible, given the following information.

53) contains $(-4, 7)$ and $(2, -1)$ $y = -\dfrac{4}{3}x + \dfrac{5}{3}$

54) $m = 2$ and contains $(-3, -2)$ $y = 2x + 4$

55) $m = 1$ and contains $(3, 5)$ $y = x + 2$

56) $m = \dfrac{7}{5}$ and y-intercept $(0, -4)$ $y = \dfrac{7}{5}x - 4$

57) y-intercept $(0, 6)$ and $m = 7$ $y = 7x + 6$

58) contains $(-3, -3)$ and $(1, -7)$ $y = -x - 6$

59) vertical line containing $(3, 5)$ $x = 3$

60) vertical line containing $\left(\dfrac{1}{2}, 6\right)$ $x = \dfrac{1}{2}$

61) horizontal line containing $(2, 3)$ $y = 3$

62) horizontal line containing $(5, -4)$ $y = -4$

63) $m = -4$ and y-intercept $(0, -4)$ $y = -4x - 4$

64) $m = -\dfrac{2}{3}$ and contains $(3, -1)$ $y = -\dfrac{2}{3}x + 1$

65) $m = -3$ and contains $(10, -10)$ $y = -3x + 20$

66) contains $(0, 3)$ and $(5, 0)$ $y = -\dfrac{3}{5}x + 3$

67) contains $(-4, -4)$ and $(2, -1)$ $y = \dfrac{1}{2}x - 2$

68) $m = -1$ and y-intercept $(0, 0)$ $y = -x$

Objective 6: Write an Equation of a Line That Is Parallel or Perpendicular to a Given Line

69) What can you say about the equations of two parallel lines?   They have the same slopes and different y-intercepts.

70) What can you say about the equations of two perpendicular lines?
Their slopes are negative reciprocals of each other.

Write an equation of the line *parallel* to the given line and containing the given point. Write the answer in slope-intercept form or in standard form, as indicated.

71) $y = 4x + 9$; $(0, 2)$; slope-intercept form $y = 4x + 2$

72) $y = 8x + 3$; $(0, -3)$; slope-intercept form $y = 8x - 3$

73) $y = 4x + 2$; $(-1, -4)$; standard form $4x - y = 0$

74) $y = \dfrac{2}{3}x - 6$; $(6, 6)$; standard form $2x - 3y = -6$

75) $x + 2y = 22$; $(-4, 7)$; standard form $x + 2y = 10$

76) $3x + 5y = -6$; $(-5, 8)$; standard form $3x + 5y = 25$

77) $15x - 3y = 1$; $(-2, -12)$; slope-intercept form
$y = 5x - 2$

78) $x + 6y = 12$; $(-6, 8)$; slope-intercept form
$y = -\dfrac{1}{6}x + 7$

Write an equation of the line *perpendicular* to the given line and containing the given point. Write the answer in slope-intercept form or in standard form, as indicated.

79) $y = -\dfrac{2}{3}x + 4$; $(4, 2)$; slope-intercept form
$y = \dfrac{3}{2}x - 4$

80) $y = -\dfrac{5}{3}x + 10$; $(10, 5)$; slope-intercept form
$y = \dfrac{3}{5}x - 1$

81) $y = -5x + 1$; $(10, 0)$; standard form $x - 5y = 10$

82) $y = \dfrac{1}{4}x - 9$; $(-1, 7)$; standard form $4x + y = 3$

83) $y = x$; $(4, -9)$; slope-intercept form $y = -x - 5$

84) $x + y = 9$; $(4, 4)$; slope-intercept form $y = x$

85) $x + 3y = 18$; $(4, 2)$; standard form $3x - y = 10$

86) $12x - 15y = 10$; $(16, -25)$; standard form
$5x + 4y = -20$

Write the slope-intercept form (if possible) of the equation of the line meeting the given conditions.

87) parallel to $3x + y = 8$ containing $(-4, 0)$
$y = -3x - 12$

88) perpendicular to $x - 5y = -4$ containing $(3, 5)$
$y = -5x + 20$

89) perpendicular to $y = x - 2$ containing $(2, 9)$
$y = -x + 11$

90) parallel to $y = 4x - 1$ containing $(-3, -8)$
$y = 4x + 4$

91) parallel to $y = 1$ containing $(-3, 4)$ $y = 4$

92) parallel to $x = -3$ containing $(-7, -5)$ $x = -7$

93) perpendicular to $x = 0$ containing $(9, 2)$ $y = 2$

94) perpendicular to $y = 4$ containing $(-4, -5)$ $x = -4$

95) perpendicular to $21x - 6y = 2$ containing $(4, -1)$ $y = -\frac{2}{7}x + \frac{1}{7}$

96) parallel to $-3x + 4y = 8$ containing $(9, 4)$ $y = \frac{3}{4}x - \frac{11}{4}$

97) parallel to $y = 0$ containing $\left(4, -\frac{3}{2}\right)$ $y = -\frac{3}{2}$

98) perpendicular to $y = \frac{7}{3}$ containing $(-7, 9)$ $x = -7$

Objective 7: Write a Linear Equation to Model Real-World Data

99) The graph shows the average annual wage of a mathematician in the United States from 2005 to 2008. x represents the number of years after 2005 so that $x = 0$ represents 2005, $x = 1$ represents 2006, and so on. Let y represent the average annual wage of a mathematician. (www.bls.gov)

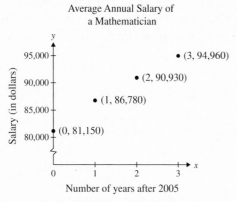

Average Annual Salary of a Mathematician

- $(3, 94{,}960)$
- $(2, 90{,}930)$
- $(1, 86{,}780)$
- $(0, 81{,}150)$

Salary (in dollars)

Number of years after 2005

a) Write a linear equation to model these data. Use the data points for 2005 and 2008, and round the slope to the nearest tenth.
$y = 4603.3x + 81{,}150$

b) Explain the meaning of the slope in the context of this problem.   The average salary of a mathematician is increasing by $4603.30 per year.

c) If the current trend continues, find the average salary of a mathematician in 2014. $122,579.70

100) The graph shows a dieter's weight over a 12-week period. Let y represent his weight x weeks after beginning his diet.

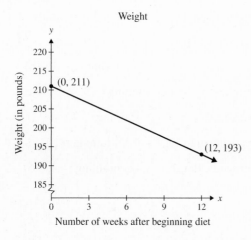

Weight

- $(0, 211)$
- $(12, 193)$

Weight (in pounds)

Number of weeks after beginning diet

a) Write a linear equation to model these data. Use the data points for week 0 and week 12.
$y = -1.5x + 211$

b) What is the meaning of the slope in the context of this problem?
The dieter is losing 1.5 pounds per week.

c) If he keeps losing weight at the same rate, what will he weigh 13 weeks after he started his diet?
191.5 lb

101) In 2011, a grocery store chain had an advertising budget of $500,000 per year. Every year since then its budget has been cut by $15,000 per year. Let y represent the advertising budget, in dollars, x years after 2011.

a) Write a linear equation to model these data.
$y = -15{,}000x + 500{,}000$

b) Explain the meaning of the slope in the context of the problem.
The budget is being cut by $15,000 per year.

c) What will the advertising budget be in 2014?
$455,000

d) If the current trend continues, in what year will the advertising budget be $365,000?   2020

102) A temperature of $-10°C$ is equivalent to $14°F$, while $15°C$ is the same as $59°F$. Let F represent the temperature on the Fahrenheit scale and C represent the temperature on the Celsius scale.

a) Write a linear equation to convert from degrees Celsius to degrees Fahrenheit. That is, write an equation for F in terms of C. $F = \frac{9}{5}C + 32$

b) Explain the meaning of the slope in the context of the problem.

c) Convert $24°C$ to degrees Fahrenheit. $75.2°F$

d) Change $95°F$ to degrees Celsius. $35°C$

103) A kitten weighs an average of 100 g at birth and should gain about 8 g per day for the first few weeks of life. Let y represent the weight of a kitten, in grams, x days after birth. (http://veterinarymedicine.dvm360.com)

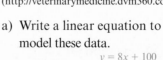

a) Write a linear equation to model these data.
$y = 8x + 100$

b) Explain the meaning of the slope in the context of the problem.   A kitten gains about 8 g per day.

c) How much would an average kitten weigh 5 days after birth? 2 weeks after birth?   140 g; 212 g

d) How long would it take for a kitten to reach a weight of 284 g?   23 days

104) In 2000, Red Delicious apples cost an average of $0.82 per lb, and in 2008 they cost $1.18 per lb. Let y represent the cost of a pound of Red Delicious apples x years after 2000.
(www.census.gov)

a) Write a linear equation to model these data. Round the slope to the nearest hundredth.
$y = 0.045x + 0.82$

b) Explain the meaning of the slope in the context of the problem. The cost of a pound of Red Delicious apples increased by about 4.5 cents per year.

c) Find the cost of a pound of apples in 2003. approximately $0.96

d) When was the average cost about $1.09/lb? 2006

105) If a woman wears a size 6 on the U.S. shoe size scale, her European size is 38. A U.S. women's size 8.5 corresponds to a European size 42. Let A represent the U.S. women's shoe size, and let E represent that size on the European scale.

a) Write a linear equation that models the European shoe size in terms of the U.S. shoe size.
$E = 1.6A + 28.4$

b) If a woman's U.S. shoe size is 7.5, what is her European shoe size? (Round to the nearest unit.) 40

106) If a man's foot is 11.5 inches long, his U.S. shoe size is 12.5. A man wears a size 8 if his foot is 10 inches long. Let L represent the length of a man's foot, and let S represent his shoe size.

a) Write a linear equation that describes the relationship between shoe size in terms of the length of a man's foot. $S = 3L - 22$

b) If a man's foot is 10.5 inches long, what is his shoe size? 9.5

R Rethink

R1) Which concepts from the previous sections did you have to learn before completing this exercise set?

R2) What is the least amount of information you need about a line in order to write its equation or graph it?

R3) Why do equations of vertical or horizontal lines involve only one variable term?

Group Activity — The Group Activity can be found online on Connect.

If you are like most college students, you have a lot of demands on your time. In addition to course obligations both inside and out of class, you may have a job, a family to care for, plus an active social life. Therefore, it's important that you make the most of your time.

To discover your personal time style, rate how well each of the statements below describes you. Use this rating scale:

1 = Doesn't describe me at all

2 = Describes me only slightly

3 = Describes me fairly well

4 = Describes me very well

		1	2	3	4
1.	I often wake up later than I should.				
2.	I am usually late for classes and appointments.				
3.	I am always in a rush getting places.				
4.	I put off big tasks and assignments until the last minute.				
5.	My friends often comment on my lateness.				
6.	I am easily interrupted, putting aside what I'm doing for something new.				
7.	When I look at a clock, I'm often surprised at how late it is.				
8.	I often forget appointments and have to reschedule them.				
9.	When faced with a big task, I feel overwhelmed and turn my mind away from it until later.				
10.	At the end of the day, I have no idea where the time went.				

Rate yourself by adding up the points you assigned yourself. Use this scale to assess your time style:

10–15 = Very efficient time user

16–20 = Efficient time user

21–30 = Time use needs work

31–40 = Victim of time

If you scored a 20 or higher, you are likely making your schedule even more hectic than it needs to be. If you make a point of paying attention to the clock and sticking to a schedule, you will likely feel less stressed and find it easier to accomplish your daily tasks.

Chapter 3:   Summary

Definition/Procedure	Example

3.1 Introduction to Linear Equations in Two Variables

A **linear equation in two variables** can be written in the form $Ax + By = C$, where A, B, and C are real numbers and where both A and B do not equal zero.

To determine whether an ordered pair is a solution of an equation, substitute the values for the variables. **(p. 175)**

Is $(4, -1)$ a solution of $3x - 5y = 17$?

$$3x - 5y = 17$$
$$3(4) - 5(-1) = 17 \quad \text{Substitute 4 for } x \text{ and } -1 \text{ for } y.$$
$$12 + 5 = 17$$
$$17 = 17 \checkmark$$

Yes, $(4, -1)$ is a solution.

3.2 Graphing by Plotting Points and Finding Intercepts

The graph of a linear equation in two variables, $Ax + By = C$, is a straight line. Each point on the line is a solution to the equation.

We can graph the line by plotting the points and drawing the line through them. **(p. 188)**

Graph $y = \dfrac{1}{3}x + 2$ by plotting points.

Make a table of values. Plot the points, and draw a line through them.

x	y
0	2
3	3
-3	1

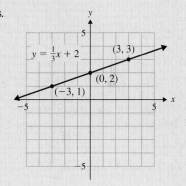

The **x-intercept** of an equation is the point where the graph intersects the *x*-axis. To find the *x-intercept* of the graph of an equation, let $y = 0$ and solve for x.

The **y-intercept** of an equation is the point where the graph intersects the *y*-axis. To find the *y-intercept* of the graph of an equation, let $x = 0$ and solve for y. **(p. 190)**

Graph $2x + 5y = -10$ by finding the intercepts and another point on the line.

x-intercept: Let $y = 0$, and solve for x.

$$2x + 5(0) = -10$$
$$2x = -10$$
$$x = -5$$

The *x*-intercept is $(-5, 0)$.

y-intercept: Let $x = 0$, and solve for y.

$$2(0) + 5y = -10$$
$$5y = -10$$
$$y = -2$$

The *y*-intercept is $(0, -2)$. Another point on the line is $(5, -4)$. Plot the points, and draw the line through them.

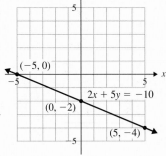

Definition/Procedure	Example

If a is a constant, then the graph of $x = a$ is a **vertical line** going through the point $(a, 0)$.

If b is a constant, then the graph of $y = b$ is a **horizontal line** going through the point $(0, b)$. **(p. 192)**

Graph $x = -2$.   Graph $y = 4$.

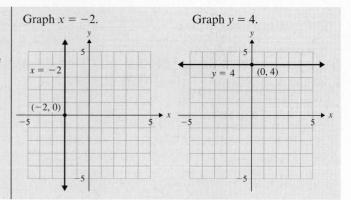

3.3 The Slope of a Line

The **slope** of a line is the ratio of the vertical change in y to the horizontal change in x. Slope is denoted by m.

The slope of a line containing the points (x_1, y_1) and (x_2, y_2) is

$$m = \frac{y_2 - y_1}{x_2 - x_1}.$$

The slope of a horizontal line is zero.
The slope of a vertical line is undefined. **(p. 200)**

Find the slope of the line containing the points $(4, -3)$ and $(-1, 5)$.

$$m = \frac{y_2 - y_1}{x_2 - x_1}$$
$$= \frac{5 - (-3)}{-1 - 4} = \frac{8}{-5} = -\frac{8}{5}$$

The slope of the line is $-\dfrac{8}{5}$.

If we know the slope of a line and a point on the line, we can graph the line. **(p. 205)**

Graph the line containing the point $(-2, 3)$ with a slope of $-\dfrac{5}{6}$.

Start with the point $(-2, 3)$, and use the slope to plot another point on the line.

$$m = \frac{-5}{6} = \frac{\text{Change in } y}{\text{Change in } x}$$

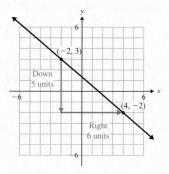

Definition/Procedure	Example

3.4 The Slope-Intercept Form of a Line

The **slope-intercept form of a line** is $y = mx + b$, where m is the slope and $(0, b)$ is the y-intercept.

If a line is written in slope-intercept form, we can use the y-intercept and the slope to graph the line. **(p. 211)**

Write the equation in slope-intercept form and graph it.

$$8x - 3y = 6$$
$$-3y = -8x + 6$$
$$y = \frac{-8}{-3}x + \frac{6}{-3}$$
$$y = \frac{8}{3}x - 2 \qquad \text{Slope-intercept form}$$

$m = \dfrac{8}{3}$, y-intercept $(0, -2)$

Plot $(0, -2)$, then use the slope to locate another point on the line. We will think of the slope as

$$m = \frac{8}{3} = \frac{\text{Change in } y}{\text{Change in } x}.$$

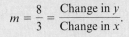

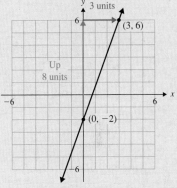

Parallel lines have the same slope.

Perpendicular lines have slopes that are negative reciprocals of each other. **(p. 213)**

Determine whether the lines $2x + y = 18$ and $x - 2y = 7$ are parallel, perpendicular, or neither.

Put each line into slope-intercept form to find their slopes.

$$2x + y = 18 \qquad\qquad x - 2y = 7$$
$$y = -2x + 18 \qquad\quad -2y = -x + 7$$
$$y = \frac{1}{2}x - \frac{7}{2}$$
$$m = -2 \qquad\qquad\qquad m = \frac{1}{2}$$

The lines are *perpendicular* since their slopes are negative reciprocals of each other.

3.5 Writing an Equation of a Line

To write the equation of a line given its slope and y-intercept, use $y = mx + b$ and substitute those values into the equation. **(p. 221)**

Find an equation of the line with slope $= 5$ and y-intercept $(0, -3)$.

Use $y = mx + b$.
$$y = 5x - 3 \qquad \text{Substitute 5 for } m \text{ and } -3 \text{ for } b.$$

If (x_1, y_1) is a point on a line and m is the slope of the line, then the equation of the line is given by $y - y_1 = m(x - x_1)$. This is the **point-slope formula.**

If we are given the slope of the line and a point on the line, we can use the point-slope formula to find an equation of the line. **(p. 221)**

Find an equation of the line containing the point $(7, -2)$ with slope $= 3$. Express the answer in standard form.

$$\text{Use } y - y_1 = m(x - x_1).$$

Substitute 3 for m. Substitute $(7, -2)$ for (x_1, y_1).

$$y - (-2) = 3(x - 7) \qquad \text{Substitute the values.}$$
$$y + 2 = 3x - 21 \qquad \text{Distribute.}$$
$$-3x + y = -23$$
$$3x - y = 23 \qquad \text{Standard form}$$

Definition/Procedure	Example

To write an equation of a line given two points on the line,

a) use the points to find the slope of the line

then

b) use the slope and *either one* of the points in the point-slope formula. **(p. 222)**

Find an equation of the line containing the points $(4, 1)$ and $(-4, 5)$. Express the answer in slope-intercept form.

$$m = \frac{5 - 1}{-4 - 4} = \frac{4}{-8} = -\frac{1}{2}$$

We will use $m = -\frac{1}{2}$ and the point $(4, 1)$ in the point-slope formula.

$$y - y_1 = m(x - x_1)$$

Substitute $-\frac{1}{2}$ for m. Substitute $(4, 1)$ for (x_1, y_1).

$$y - 1 = -\frac{1}{2}(x - 4) \qquad \text{Substitute.}$$

$$y - 1 = -\frac{1}{2}x + 2 \qquad \text{Distribute.}$$

$$y = -\frac{1}{2}x + 3 \qquad \text{Slope-intercept form}$$

The equation of a **horizontal line** containing the point (a, b) is $y = b$.

The equation of a **vertical line** containing the point (a, b) is $x = a$. **(p. 223)**

The equation of a horizontal line containing the point $(3, -2)$ is $y = -2$.

The equation of a vertical line containing the point $(6, 4)$ is $x = 6$.

To write an equation of the line parallel or perpendicular to a given line, we must first find the slope of the given line. **(p. 224)**

Write an equation of the line parallel to $4x - 5y = 20$ containing the point $(4, -3)$. Express the answer in slope-intercept form.

Find the slope of $4x - 5y = 20$.

$$-5y = -4x + 20$$

$$y = \frac{4}{5}x - 4 \qquad m = \frac{4}{5}$$

The slope of the parallel line is also $\frac{4}{5}$. Since this line contains $(4, -3)$, use the point-slope formula to write its equation.

$$y - y_1 = m(x - x_1)$$

$$y - (-3) = \frac{4}{5}(x - 4) \qquad \text{Substitute values.}$$

$$y + 3 = \frac{4}{5}x - \frac{16}{5} \qquad \text{Distribute.}$$

$$y = \frac{4}{5}x - \frac{31}{5} \qquad \text{Slope-intercept form}$$

Chapter 3: Review Exercises

*Additional answers can be found in the Answers to Exercises appendix.

(3.1) Determine whether each ordered pair is a solution of the given equation.

1) $5x - y = 13; (2, -3)$
   yes

2) $2x + 3y = 8; (-1, 5)$
   no

3) $y = -\dfrac{4}{3}x + \dfrac{7}{3}; (4, -3)$
   yes

4) $x = 6; (6, 2)$
   yes

Complete the ordered pair for each equation.

5) $y = -2x + 4; (-5, \ )$ 14

6) $y = \dfrac{5}{2}x - 3; (6, \ )$ 12

7) $y = -9; (7, \ )$ -9

8) $8x - 7y = -10; (\ , 4)$ $\dfrac{9}{4}$

Complete the table of values for each equation.

9) $y = x - 14$

x	y
0	-14
6	-8
-3	-17
-8	-22

10) $3x - 2y = 9$

x	y
3	0
0	$-\frac{9}{2}$
2	$-\frac{3}{2}$
$\frac{7}{3}$	-1

Plot the ordered pairs on the same coordinate system.

11) a) $(4, 0)$ b) $(-2, 3)$

    c) $(5, 1)$ d) $(-1, -4)$

12) a) $(0, -3)$ b) $(-4, 4)$

    c) $(1, \frac{3}{2})$ d) $(-\frac{1}{3}, -2)$

13) The cost of renting a pick-up for one day is given by $y = 0.5x + 45.00$, where x represents the number of miles driven and y represents the cost, in dollars.

    a) Complete the table of values, and write the information as ordered pairs.

x	y
10	50
18	54
29	59.50
36	63

    b) Label a coordinate system, choose an appropriate scale, and graph the ordered pairs.

    c) Explain the meaning of the ordered pair (58, 74) in the context of the problem.   The cost of renting the pick-up is $74.00 if it is driven 58 miles.

14) Fill in the blank with positive, negative, or zero.

    a) The x-coordinate of every point in quadrant III is __negative__.

    b) The y-coordinate of every point in quadrant II is __positive__.

(3.2) Complete the table of values and graph each equation.

15) $y = -2x + 4$

x	y
0	4
1	2
2	0
3	-2

16) $2x + 3y = 6$

x	y
0	2
3	0
-2	$\frac{10}{3}$
-3	4

Graph each equation by finding the intercepts and at least one other point.

17) $x - 2y = 2$

18) $3x - y = -3$

19) $y = -\dfrac{1}{2}x + 1$

20) $2x + y = 0$

21) $y = 4$

22) $x = -1$

(3.3) Determine the slope of each line.

23)

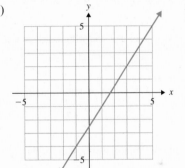

$\dfrac{3}{2}$

24)

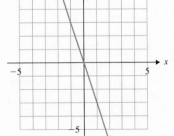

-3

Use the slope formula to find the slope of the line containing each pair of points.

25) $(5, 8)$ and $(1, -12)$ 5

26) $(-3, 4)$ and $(1, -1)$ $-\dfrac{5}{4}$

27) $(-7, -2)$ and $(2, 4)$ $\dfrac{2}{3}$

28) $(7, 3)$ and $(15, 1)$ $-\dfrac{1}{4}$

29) $\left(-\dfrac{1}{4}, 1\right)$ and $\left(\dfrac{3}{4}, -6\right)$ -7

30) $(3.7, 2.3)$ and $(5.8, 6.5)$ 2

31) $(-2, 5)$ and $(4, 5)$ 0

32) $(-9, 3)$ and $(-9, 2)$ undefined

33) Christine collects old record albums. The graph shows the value of an original, autographed copy of one of her albums from 1980.

Value of an Album

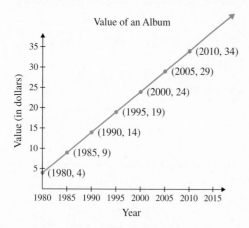

a) How much did she pay for the album in 1980? $4.00

b) Is the slope of the line positive or negative? What does the sign of the slope mean in the context of the problem?
   The slope is positive, so the value of the album is increasing over time.

c) Find the slope. What does it mean in the context of the problem?
   $m = 1$; the value of the album is increasing by $1.00 per year.

Graph the line containing the given point and with the given slope.

34) $(3, -4)$; $m = 2$

35) $(-2, 2)$; $m = -3$

36) $(1, 3)$; $m = -\dfrac{1}{2}$

37) $(-4, 1)$; slope undefined

38) $(-2, -3)$; $m = 0$

(3.4) Identify the slope and y-intercept, then graph the line.

39) $y = -x + 5$

40) $y = 4x - 2$

41) $y = \dfrac{2}{5}x - 6$

42) $y = -\dfrac{1}{2}x + 5$

43) $x + 3y = -6$

44) $18 = 6y - 15x$

45) $x + y = 0$

46) $y + 6 = 1$

47) The value of the squash crop in the United States since 2003 can be modeled by $y = 7.9x + 197.6$, where x represents the number of years after 2003, and y represents the value of the crop in millions of dollars. (U.S. Dept. of Agriculture)

Value of the Squash Crop in the U.S.

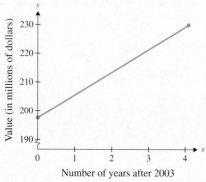

Number of years after 2003

a) What is the y-intercept? What does it mean in the context of the problem?   (0, 197.6); in 2003, the squash crop was worth about $197.6 million.

b) Has the value of the squash crop been increasing or decreasing since 2003? By how much per year?
   It has been increasing by $7.9 million per year.

c) Use the graph to estimate the value of the squash crop in the United States in 2005. Then use the equation to determine this number. $213 million; $213.4 million

Determine whether each pair of lines is parallel, perpendicular, or neither.

48) $y = \dfrac{3}{5}x - 8$

$5x + 3y = 3$ perpendicular

49) $x - 4y = 20$
   $-x + 4y = 6$ parallel

50) $5x + y = 4$ neither
   $2x + 10y = 1$

51) $x = 7$ perpendicular
   $y = -3$

52) Write the point-slope formula for the equation of a line with slope m and which contains the point (x_1, y_1).
   $y - y_1 = m(x - x_1)$

(3.5) Write the slope-intercept form of the equation of the line, if possible, given the following information.

53) $m = 6$ and contains $(-1, 4)$ $y = 6x + 10$

54) $m = -5$ and y-intercept $(0, -3)$ $y = -5x - 3$

55) $m = -\dfrac{3}{4}$ and y-intercept $(0, 7)$ $y = -\dfrac{3}{4}x + 7$

56) contains $(-4, 2)$ and $(-2, 5)$ $y = \dfrac{3}{2}x + 8$

57) contains $(4, 1)$ and $(6, -3)$ $y = -2x + 9$

58) $m = \dfrac{2}{3}$ and contains $(5, -2)$ $y = \dfrac{2}{3}x - \dfrac{16}{3}$

59) horizontal line containing $(3, 7)$ $y = 7$

60) vertical line containing $(-5, 1)$ $x = -5$

Write the standard form of the equation of the line given the following information.

61) contains $(4, 5)$ and $(-1, -10)$ $3x - y = 7$

62) $m = -\dfrac{1}{2}$ and contains $(3, 0)$ $x + 2y = 3$

63) $m = \dfrac{5}{2}$ and contains $\left(1, -\dfrac{3}{2}\right)$ $5x - 2y = 8$

64) contains $(-4, 1)$ and $(4, 3)$ $x - 4y = -8$

65) $m = -4$ and y-intercept $(0, 0)$ $4x + y = 0$

66) $m = -\dfrac{3}{7}$ and y-intercept $(0, 1)$ $3x + 7y = 7$

67) contains $(6, 1)$ and $(2, 5)$ $x + y = 7$

68) $m = \dfrac{3}{4}$ and contains $\left(-2, \dfrac{7}{2}\right)$ $3x - 4y = -20$

69) Mr. Romanski works as an advertising consultant, and his salary has been growing linearly. In 2005, he earned \$62,000, and in 2012, he earned \$86,500. Let y represent Mr. Romanski's salary, in dollars, x years after 2005.

   a) Write a linear equation to model these data. $y = 3500x + 62{,}000$

   b) Explain the meaning of the slope in the context of the problem.   Mr. Romanski's salary is increasing by \$3500 per year.

   c) How much did he earn in 2008?   \$72,500

   d) If the trend continues, in what year could he expect to earn \$100,500?   2016

Write an equation of the line *parallel* to the given line and containing the given point. Write the answer in slope-intercept form or in standard form, as indicated.

70) $y = 2x + 10$; $(2, -5)$; slope-intercept form $y = 2x - 9$

71) $y = -8x + 8$; $(-1, 14)$; slope-intercept form $y = -8x + 6$

72) $3x + y = 5$; $(-3, 5)$; standard form $3x + y = -4$

73) $x - 2y = 6$; $(4, 11)$; standard form $x - 2y = -18$

74) $3x + 4y = 1$; $(-1, 2)$; slope-intercept form $y = -\dfrac{3}{4}x + \dfrac{5}{4}$

75) $x + 5y = 10$; $(15, 7)$; slope-intercept form $y = -\dfrac{1}{5}x + 10$

Write an equation of the line *perpendicular* to the given line and containing the given point. Write the answer in slope-intercept form or in standard form, as indicated.

76) $y = -\dfrac{1}{5}x + 7$; $(1, 7)$; slope-intercept form $y = 5x + 2$

77) $y = -x + 9$; $(3, -9)$; slope-intercept form $y = x - 12$

78) $4x - 3y = 6$; $(8, -5)$; slope-intercept form $y = -\dfrac{3}{4}x + 1$

79) $2x + 3y = -3$; $(-4, -4)$; slope-intercept form $y = \dfrac{3}{2}x + 2$

80) $x + 8y = 8$; $(-2, -7)$; standard form $8x - y = -9$

81) Write an equation of the line parallel to $y = 5$ containing $(8, 4)$. $y = 4$

82) Write an equation of the line perpendicular to $x = -2$ containing $(4, -3)$. $y = -3$

Chapter 3:   Test

*Additional answers can be found in the Answers to Exercises appendix.

1) Is $(-3, -2)$ a solution of $2x - 7y = 8$?   yes

2) Complete the table of values and graph $y = \dfrac{3}{2}x - 2$.

x	y
0	-2
-2	-5
4	4
2	1

3) Fill in the blanks with *positive* or *negative*. In quadrant IV, the x-coordinate of every point is _____ and the y-coordinate is _____.   positive; negative

4) For $3x - 4y = 6$,

   a) find the x-intercept. $(2, 0)$

   b) find the y-intercept. $\left(0, -\dfrac{3}{2}\right)$

   c) find one other point on the line.   Answers may vary.

   d) graph the line.

5) Graph $y = -3$.

6) Graph $x + y = 0$.

Find the slope of the line containing the given points.

7) $(3, -1)$ and $(-5, 9)$ $-\dfrac{5}{4}$

8) $(8, 6)$ and $(11, 6)$ 0

9) Graph the line containing the point $(-1, 4)$ with slope $-\dfrac{3}{2}$.

10) Graph the line containing the point $(2, 3)$ with an undefined slope.

11) Put $3x - 2y = 10$ into slope-intercept form. Then, graph the line.

12) Write the slope-intercept form for the equation of the line with slope 7 and y-intercept $(0, -10)$. $y = 7x - 10$

13) Write the standard form for the equation of a line with slope $-\dfrac{1}{3}$ containing the point $(-3, 5)$. $x + 3y = 12$

14) Write the slope-intercept form for the equation of the line containing the points $(-5, -11)$ and $(1, 4)$. $y = \dfrac{5}{2}x + \dfrac{3}{2}$

15) Determine whether $4x + 18y = 9$ and $9x - 2y = -6$ are parallel, perpendicular, or neither.   perpendicular

16) Find the slope-intercept form of the equation of the line

   a) perpendicular to $y = 2x - 9$ containing $(-6, 10)$. $y = -\dfrac{1}{2}x + 7$

   b) parallel to $3x - 4y = -4$ containing $(11, 8)$. $y = \dfrac{3}{4}x - \dfrac{1}{4}$

Use the graph for Exercises 17–22.

The graph shows the number of children attending a neighborhood school from 2007 to 2012. Let y represent the number of children attending the school x years after 2007.

School Population

Number of students

- (0, 419)
- (1, 409)
- (2, 399)
- (3, 392)
- (4, 385)
- (5, 374)

Number of years after 2007

17) According to the graph, how many children attended this school in 2009?   399

18) Write a linear equation (in slope-intercept form) to model these data. Use the data points for 2007 and 2012. $y = -9x + 419$

19) Use the equation in part b) to determine the number of students attending the school in 2009. How does your answer in part a) compare to the number predicted by the equation?   According to the equation, 401 students attended the school in 2009. The actual number was 399.

20) Explain the meaning of the slope in the context of the problem.   The school is losing 9 students per year.

21) What is the y-intercept? What does it mean in the context of the problem?   (0, 419); in 2007, 419 students attended this school.

22) If the current trend continues, how many children can be expected to attend this school in 2015?   347

Chapter 3:   Cumulative Review for Chapters 1–3

*Additional answers can be found in the Answers to Exercises appendix.

1) Find all factors of 90.   1, 2, 3, 5, 6, 9, 10, 15, 18, 30, 45, 90

2) Write $\dfrac{336}{792}$ in lowest terms. $\dfrac{14}{33}$

3) A rectangular picture frame measures 7 in. by 12.5 in. Find the perimeter of the frame.   39 in.

Evaluate.

4) -3^4 -81

5) $\dfrac{24}{35} \cdot \dfrac{49}{60}$ $\dfrac{14}{25}$

6) $\dfrac{3}{8} - 2$ $-\dfrac{13}{8}$

7) $4 + 2^6 \div |5 - 13|$ 12

8) Write an expression for "9 less than twice 17" and simplify. $2(17) - 9;\ 25$

9) The complement of $29°$ is _____. $61°$

10) Evaluate $2x + 11y$ when $x = 4$ and $y = -2$. -14

11) Which property of real numbers does $16 + 2 = 2 + 16$ illustrate?   commutative

12) Combine like terms and simplify: $33b + 18 - 9b + 3 - 10b$ $14b + 21$

Solve.

13) $-\dfrac{2}{5}y + 9 = 15$ $\{-15\}$

14) $\dfrac{3}{2}(7c - 5) - 1 = \dfrac{2}{3}(2c + 1)$ $\{1\}$

15) $7 + 2(p - 6) = 8(p + 3) - 6p + 1$ $\varnothing$

16) Solve. Write the solution in interval notation.
$3x + 14 \le 7x + 4$ $\left[\dfrac{5}{2}, \infty\right)$

17) The Chase family put their house on the market for $306,000. This is 10% less than what they paid for it 3 years ago. What did they pay for the house? $340,000

18) Find the missing angle measures.

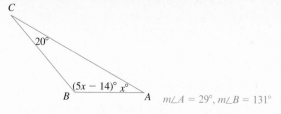

$m\angle A = 29°,\ m\angle B = 131°$

19) A 24-ft rope is cut into two pieces so that one piece is 4 ft longer than the other. Find the length of each piece.   10 ft and 14 ft

20) Determine whether $(5, -10)$ is a solution to $3x + 2y = -5$.   yes

21) Fill in the blank with *positive*, *negative*, or *zero*.

a) The y-coordinate of every point in quadrant III is _____.   negative

b) The x-coordinate of every point on the y-axis is _____.   zero

22) Find the slope of the line containing the points $(4, -11)$ and $(10, 5)$. $m = \dfrac{8}{3}$

23) Graph $3x + y = 4$.

24) Write an equation of the line with slope $\dfrac{3}{8}$ containing the point $(16, 5)$. Express the answer in standard form. $3x - 8y = 8$

25) Write an equation of the line perpendicular to $4x + 3y = -6$ containing the point $(-8, -6)$. Express the answer in slope-intercept form. $y = \dfrac{3}{4}x$

Linear Equations and Inequalities in Two Variables

Math at Work:

Motorcycle Customization

It takes both mechanical skill and artistic flair to do what Frank Esposito does at work: customize motorcycles. It also takes knowledge of mathematical expressions and exponents. Indeed, the people who build and repair motor vehicles use a lot of mathematics on the job.

For instance, suppose Frank takes apart a transmission to make repairs when he realizes that he has mixed up the gears. He is able to replace the shafts onto the bearings, but he cannot remember which gear goes on which shaft. Frank measures the distance (in inches) between the shafts, sketches the layout on a piece of paper, and comes up with a system of equations to determine which gear goes on which shaft.

If x = the radius of the gear on the left and y = the radius of the gear on the right, then the system of equations Frank must solve to determine where to put each gear is

$$x + y = 2.650$$
$$x - y = 0.746$$

Solving this system, Frank determines that $x = 1.698$ in. and $y = 0.952$ in. Now he knows on which shaft to place each gear.

Frank doesn't make mistakes like mixing up gears often. However, given how much technical information he has to remember and how many mechanical parts he has to deal with to do his job, some mistakes are inevitable. To minimize them, Frank makes a habit of writing down key pieces of information in a clear, organized way, allowing him to easily reference essential details later. You can use the same sorts of strategies to ensure the notes you take in class are effective.

In this chapter, we will learn how to write and solve systems of two equations, and learn about taking good class notes.

 **Study Strategies** Taking Notes in Class

What your instructor says in class almost always represents the most important material in any course. It is this material that you will need to master in order to succeed on homework and exams and in order to apply the course lessons to your other academic work and to your life beyond college. Therefore, taking effective class notes is essential. Use the strategies below to help you create class notes that will be useful days, weeks, and months later.

- Take a seat in the classroom where you can see and hear the instructor clearly.
- Review any assigned materials and your notes from the previous class. This will help warm up your mind for the lessons to come.
- Turn off your cell phone!

- Bring to class a writing tool that works for you. For math classes, you will usually want to use a pencil. Remember to bring more than one, or at least a pencil sharpener!
- Plan to write your notes in a loose-leaf notepad, taking notes on *only one side of the page*. This will allow you to spread your notes out in front of you when you study.

W Work

- Listen actively. Concentrate on what your instructor is saying, and try to make sense of it.
- Don't attempt to write down everything. Instead, focus on identifying and writing down the key concepts, formulas, and insights.
- Ask questions! Asking questions also helps ensure that you are engaged with what you're hearing.

- As the class is wrapping up, look over your notes. Be sure that you can read them and that they reflect what has been covered. (Use the emPOWERme on page 298 to evaluate your notes in more detail.)
- If there is a concept you didn't understand, feel free to talk to your instructor about it after class. If he or she doesn't have time to discuss it then, visit during office hours.

- As soon as the class is finished, *read over your notes*. This critical step helps transfer the information you've just learned into long-term memory.
- If you were "lost" during class, consider getting extra help in your school's math lab or from a math tutor.

Chapter 4 **POWER** Plan

P Prepare

O Organize

What are your goals for Chapter 4?	How can you accomplish each goal?
1 Be prepared before and during class.	• Don't stay out late the night before, and be sure to set your alarm clock! • Bring a pencil, notebook paper, and textbook to class. • Avoid distractions by turning off your cell phone during class. • Pay attention, take good notes, and ask questions. • Complete your homework on time, and ask questions on problems you do not understand.
2 Understand the homework to the point where you could do it without needing any help or hints.	• Read the directions, and show all of your steps. • Go to the professor's office for help. • Rework homework and quiz problems, and find similar problems for practice.
3 Use the P.O.W.E.R. framework to help you take notes in class: *Evaluate Your Class Notes*.	• Read the Study Strategy that explains how to take notes. • What does P.O.W.E.R. stand for? • Complete the emPOWERme that appears before the Chapter Summary.
4 Write your own goal.	• _____

What are your objectives for Chapter 4?	How can you accomplish each objective?
1 Know how to solve systems by graphing, the substitution method, and the elimination method.	• Understand the three different methods for solving linear systems. • Write a procedure for each method of solving a system of equations. • Be able to quickly decide the best way to solve a linear system.
2 Know how to solve application problems that involve systems of two equations.	• Learn the procedure for **Solving an Applied Problem Using a System of Equations.** • Be able to apply the procedure to problems involving general quantities, geometry, cost, mixture, and distance, rate, and time problems.
3 Know how to solve and graph linear inequalities in two variables.	• Define a *linear inequality in two variables, half plane,* and *test point*. • Learn the procedure for **Graphing a Linear Inequality in Two Variables** using a test point as well as using a slope-intercept form. • Learn the procedure for **Solving a System of Linear Inequalities by Graphing.**
4 Write your own goal.	• _____

E Evaluate Complete the Chapter Review and Chapter Test. How did you do?	**R Rethink**	• Write a paragraph describing the different types of linear systems you have encountered in the chapter. Be sure to discuss their consistency, dependence, and graphical representation. • Describe a job that you would like to have in the future that will require you to be able to follow procedures correctly in order to be successful. • How has the Study Strategy for this chapter helped you take better notes in all of your classes?

4.1 Solving Systems by Graphing

P Prepare

O Organize

What are your objectives for Section 4.1?	**How can you accomplish each objective?**
1 Determine Whether an Ordered Pair Is a Solution of a System	• Write the definition of a *solution of a system* in your own words, and write an example. • Know how to substitute values into an equation. • Complete the given example on your own. • Complete You Try 1.
2 Solve a Linear System by Graphing	• Understand that a system with one solution is considered a *consistent system* with *independent equations*. • Complete the given example on your own. • Complete You Try 2.
3 Solve a Linear System by Graphing: Special Cases	• Know that the graphs of some equations in a system might not intersect. • Understand that a system with no solution is considered an *inconsistent system* with *independent equations*. • Understand that a system with an infinite number of solutions is considered a *consistent system* with *dependent equations*. • Complete the given examples on your own. • Complete You Try 3. • Summarize how to solve systems by graphing by writing the procedure for **Solving a System by Graphing** in your own words and drawing examples for the three different cases.
4 Determine the Number of Solutions of a System Without Graphing	• Follow the explanation to use slopes and *y*-intercepts to determine the number of solutions to a system of equations. • Complete the given examples on your own. • Complete You Try 4.

What is a system of linear equations? A **system of linear equations** consists of two or more linear equations with the same variables. In Sections 4.1–4.3, we will learn how to solve systems of two equations in two variables. Some examples of such systems are

$$2x + 5y = 5 \qquad\qquad y = \frac{1}{3}x - 8 \qquad\qquad -3x + y = 1$$
$$x + 4y = -1 \qquad\qquad 5x - 6y = 10 \qquad\qquad x = -2$$

In the third system, we see that $x = -2$ is written with only one variable. However, we can think of it as an equation in two variables by writing it as $x + 0y = -2$.

It is also possible to solve systems of inequalities. In Section 4.5, we will learn how to solve linear inequalities in two variables.

1 Determine Whether an Ordered Pair Is a Solution of a System

We will begin our work with systems of equations by determining whether an ordered pair is a solution of the system.

Definition

A **solution of a system** of two equations in two variables is an ordered pair that is a solution of each equation in the system.

EXAMPLE 1

Determine whether (2, 3) is a solution of each system of equations.

a) $y = x + 1$
    $x + 2y = 8$

b) $4x - 5y = -7$
    $3x + y = 4$

Solution

a) If (2, 3) is a solution of $\begin{array}{l} y = x + 1 \\ x + 2y = 8 \end{array}$ then when we substitute 2 for x and 3 for y, the ordered pair will make each equation true.

$y = x + 1$		$x + 2y = 8$	
$3 \overset{?}{=} 2 + 1$ Substitute.		$2 + 2(3) \overset{?}{=} 8$ Substitute.	
		$2 + 6 \overset{?}{=} 8$	
$3 = 3$ True		$8 = 8$ True	

Since (2, 3) is a solution of each equation, it is a solution of the system.

b) We will substitute 2 for x and 3 for y to see whether (2, 3) satisfies (is a solution of) each equation.

$4x - 5y = -7$	$3x + y = 4$
$4(2) - 5(3) \overset{?}{=} -7$ Substitute.	$3(2) + 3 \overset{?}{=} 4$ Substitute.
$8 - 15 \overset{?}{=} -7$	$6 + 3 \overset{?}{=} 4$
$-7 = -7$ True	$9 = 4$ False

Although (2, 3) is a solution of the first equation, it does *not* satisfy $3x + y = 4$. Therefore, (2, 3) is *not* a solution of the system.

Determine whether $(-4, 3)$ is a solution of each system of equations.

a) $3x + 5y = 3$

$-2x - y = -5$

b) $y = \dfrac{1}{2}x + 5$

$-x + 3y = 13$

Let's begin solving systems of equations by graphing.

2 Solve a Linear System by Graphing

To **solve a system of equations in two variables** means to find the ordered pair (or pairs) that satisfies each equation in the system.

Recall from Chapter 3 that the graph of a linear equation is a line. This line represents all solutions of the equation.

> **W Hint**
>
> Notice that when you are solving a system of two linear equations, you are trying to find the intersection of the two lines.

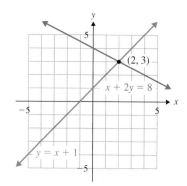

If two lines intersect at one point, that point of intersection is a solution of each equation.

For example, the graph shows the lines representing the two equations in Example 1a). The solution to that system is their point of intersection, $(2, 3)$.

> ## Definition
>
> *When solving a system of equations by graphing,* the point of intersection is the solution of the system. If a system has at least one solution, we say that the system is **consistent.** The equations are **independent** if the system has one solution.

EXAMPLE 2

In-Class Example 2

Solve the system by graphing.
$y = -x + 4$
$2x - y = -1$

Solve the system by graphing.

$y = \dfrac{1}{3}x - 2$

$2x + 3y = 3$

Solution

Graph each line on the same axes. The first equation is in slope-intercept form, and we see that $m = \dfrac{1}{3}$ and $b = -2$. Its graph is in blue.

Answer: (1, 3)

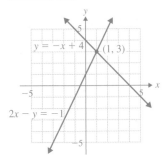

Let's graph $2x + 3y = 3$ by plotting points.

x	y
0	1
−3	3
3	−1

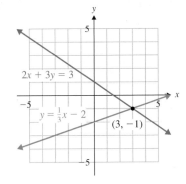

The point of intersection is $(3, -1)$, so the solution is $(3, -1)$.

The solution of the system is $(3, -1)$. The system is consistent.

W Hint

When drawing lines, be sure to extend them beyond the boundaries of your grid! Using a straightedge to graph the lines is a good idea.

Note

It is important that you use a straightedge to graph the lines. If the graph is not precise, it will be difficult to correctly locate the point of intersection. Furthermore, if the solution of a system contains numbers that are not integers, it may be impossible to accurately read the point of intersection. This is one reason why solving a system by graphing is not always the best way to find the solution. But it can be a useful method, and it is one that is used to solve problems not only in mathematics, but also in areas such as business, economics, and chemistry.

[**YOU TRY 2**] Solve the system by graphing. $\quad 3x + 2y = 2$
$$y = \frac{1}{2}x - 3$$

3 Solve a Linear System by Graphing: Special Cases

Do two lines *always* intersect? No! Then if we are trying to solve a system of two linear equations by graphing and the graphs do not intersect, what does this tell us about the solution to the system?

EXAMPLE 3

In-Class Example 3

Solve the system by graphing.
$-3x + 2y = 4$
$6x - 4y = 8$

Answer: ∅

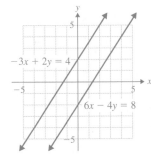

Solve the system by graphing. $\quad -2x - y = 1$
$$2x + y = 3$$

Solution

Graph each line on the same axes.

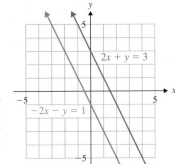

The lines are parallel; they will never intersect. Therefore, there is *no solution* to the system. We write the solution set as ∅.

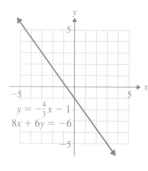

W Hint

When the graphs of the lines in a system are parallel, there is no solution.

Definition

When solving a system of equations by graphing, if the lines are parallel, then the system has **no solution.** We write this as $\varnothing$. Furthermore, a system that has no solution is **inconsistent,** and the equations are **independent.**

What if the graphs of the equations in a system are the same line?

EXAMPLE 4

In-Class Example 4

Solve the system by graphing.

$y = -\dfrac{4}{3}x - 1$

$8x + 6y = -6$

Answer: infinite number of solutions of the form

$\left\{ (x, y) \middle| y = -\dfrac{4}{3}x - 1 \right\}$

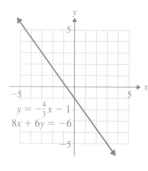

Solve the system by graphing.

$$y = \dfrac{2}{3}x + 2$$
$$12y - 8x = 24$$

Solution

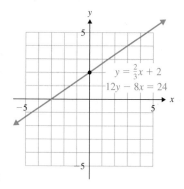

If we write the second equation in slope-intercept form, we see that it is the same as the first equation. This means that the graph of each equation is the same line. Therefore, each point on the line satisfies each equation. The system has an *infinite number of solutions* of the form $y = \dfrac{2}{3}x + 2$.

The solution set is $\left\{ (x, y) \middle| y = \dfrac{2}{3}x + 2 \right\}$.

We read this as "the set of all ordered pairs (x, y) such that $y = \dfrac{2}{3}x + 2$."

We could have used either equation to write the solution set in Example 4. However, we will use either the equation that is written in slope-intercept form or the equation written in standard form with integer coefficients that have no common factor other than 1.

W Hint

When writing the solution set for a system with an *infinite number of solutions,* always include one of the equations as part of the solution set.

Definition

When solving a system of equations by graphing, if the graph of each equation is the same line, then the system has an **infinite number of solutions.** The system is **consistent,** and the equations are **dependent.**

We will summarize what we have learned so far about solving a system of linear equations by graphing.

Procedure Solving a System by Graphing

To solve a system by graphing, graph each line on the same axes.

1) If the lines intersect at a **single point,** then the point of intersection is the solution of the system. The system is *consistent,* and the equations are *independent.* (See Figure 4.1a)

2) If the lines are **parallel,** then the system has *no solution.* We write the solution set as $\varnothing$. The system is *inconsistent.* The equations are *independent.* (See Figure 4.1b)

3) If the graphs are the **same line,** then the system has an *infinite number of solutions.* We say that the system is *consistent,* and the equations are *dependent.* (See Figure 4.1c)

Figure 4.1

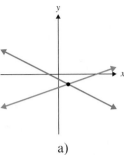

a)
One solution—the point
of intersection
Consistent system
Independent equations

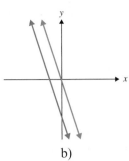

b)
No solution
Inconsistent system
Independent
equations

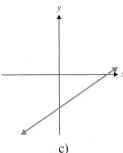

c)
Infinite number
of solutions
Consistent system
Dependent equations

Hint

Use an acronym to help you remember the three cases. For example CI for one, II for none, CD for infinite.

[YOU TRY 3] Solve each system by graphing.

a) $-x = y + 4$
    $x + y = 1$

b) $2x - 6y = 9$
    $-4x + 12y = -18$

---

4 Determine the Number of Solutions of a System Without Graphing

The graphs of lines can lead us to the solution of a system. We can also determine the number of solutions a system has without graphing.

We saw in Example 4 that if a system has lines with the same slope and the same *y*-intercept (they are the same line), then the system has an *infinite number of solutions*.

Example 3 shows that if a system contains lines with the same slope and different *y*-intercepts, then the lines are parallel and the system has *no solution*.

Finally, we learned in Example 2 that if the lines in a system have different slopes, then they will intersect and the system has *one solution*.

EXAMPLE 5

In-Class Example 5

Without graphing, determine whether each system has no solution, one solution, or an infinite number of solutions.
a) $10x - 8y = 12$
   $-15x + 12y = -18$
b) $4x - 3y = 8$
   $x + 2y = -5$
c) $-2x + 3y = 15$
   $8x - 12y = 3$

Answer: a) infinite number of solutions   b) one solution c) no solution

Without graphing, determine whether each system has no solution, one solution, or an infinite number of solutions.

a) $y = \dfrac{3}{4}x + 7$

    $5x + 8y = -8$

b) $4x - 8y = 10$

    $-6x + 12y = -15$

c) $9x + 6y = -13$

    $3x + 2y = 4$

Solution

a) The first equation is already in slope-intercept form, so write the second equation in slope-intercept form.

$$5x + 8y = -8$$
$$8y = -5x - 8$$
$$y = -\frac{5}{8}x - 1$$

The slopes, $\dfrac{3}{4}$ and $-\dfrac{5}{8}$, are different; therefore, this system has *one solution*.

---

b) Write each equation in slope-intercept form.

$$4x - 8y = 10$$
$$-8y = -4x + 10$$
$$y = \frac{-4}{-8}x + \frac{10}{-8}$$
$$y = \frac{1}{2}x - \frac{5}{4}$$

$$-6x + 12y = -15$$
$$12y = 6x - 15$$
$$y = \frac{6}{12}x - \frac{15}{12}$$
$$y = \frac{1}{2}x - \frac{5}{4}$$

The equations are the same: they have the same slope and y-intercept. Therefore, this system has an *infinite number of solutions.*

c) Write each equation in slope-intercept form.

$$9x + 6y = -13$$
$$6y = -9x - 13$$
$$y = \frac{-9}{6}x - \frac{13}{6}$$
$$y = -\frac{3}{2}x - \frac{13}{6}$$

$$3x + 2y = 4$$
$$2y = -3x + 4$$
$$y = \frac{-3}{2}x + \frac{4}{2}$$
$$y = -\frac{3}{2}x + 2$$

The equations have the same slope but different y-intercepts. If we graphed them, the lines would be parallel. Therefore, this system has *no solution.*

[YOU TRY 4]

Without graphing, determine whether each system has no solution, one solution, or an infinite number of solutions.

a) $-2x = 4y - 8$
    $x + 2y = -6$

b) $y = -\frac{5}{6}x + 1$
    $10x + 12y = 12$

c) $-5x + 3y = 12$
    $3x - y = 2$

Using Technology

In this section, we have learned that the solution of a system of equations is the point at which their graphs intersect. We can solve a system by graphing using a graphing calculator. On the calculator, we will solve the following system by graphing:

$$x + y = 5$$
$$y = 2x - 3$$

Begin by entering each equation using the $\boxed{Y=}$ key. Before entering the first equation, we must solve for y.

$$x + y = 5$$
$$y = -x + 5$$

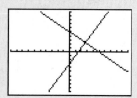

Enter $-x + 5$ in Y1 and $2x - 3$ in Y2, press $\boxed{\text{ZOOM}}$, and select 6: ZStandard to graph the equations.

Since the lines intersect, the system has a solution. How can we find that solution? Once you see from the graph that the lines intersect, press $\boxed{\text{2nd}}$ $\boxed{\text{TRACE}}$. Select 5: intersect and then press $\boxed{\text{ENTER}}$ three times. The screen will move the cursor to the point of intersection and display the solution to the system of equations on the bottom of the screen.

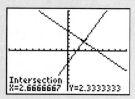

To obtain the exact solution to the system of equations, first return to the home screen by pressing 2nd MODE. To display the x-coordinate of the solution, press X,T,Θ,n MATH ENTER ENTER, and to display the y-coordinate of the solution, press ALPHA 1 MATH ENTER ENTER. The solution to the system is $\left(\dfrac{8}{3}, \dfrac{7}{3}\right)$.

```
X▸Frac
                8/3
Y▸Frac
                7/3
```

Use a graphing calculator to solve each system.

1) $y = x + 4$
$\quad y = -x + 2$

2) $y = -3x + 7$
$\quad y = x - 5$

3) $y = -4x - 2$
$\quad y = x + 5$

4) $5x + y = -1$
$\quad 4x - y = 2$

5) $5x + 2y = 7$
$\quad 2x + 4y = 3$

6) $3x + 2y = -2$
$\quad -x - 3y = -5$

ANSWERS TO [YOU TRY] EXERCISES

1) a) no   b) yes

2) $(2, -2)$

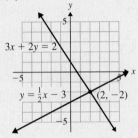

3) a) $\varnothing$

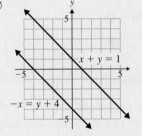

b) infinite number of solutions of the form $\{(x, y)\,|\,2x - 6y = 9\}$

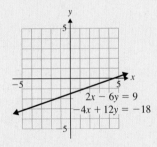

4) a) no solution   b) infinite number of solutions   c) one solution

ANSWERS TO TECHNOLOGY EXERCISES

1) $(-1, 3)$

2) $(3, -2)$

3) $\left(-\dfrac{7}{5}, \dfrac{18}{5}\right)$

4) $\left(\dfrac{1}{9}, -\dfrac{14}{9}\right)$

5) $\left(\dfrac{11}{8}, \dfrac{1}{16}\right)$

6) $\left(-\dfrac{16}{7}, \dfrac{17}{7}\right)$

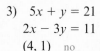

Additional answers can be found in the Answers to Exercises appendix.

Objective 1: Determine Whether an Ordered Pair Is a Solution of a System

Determine whether the ordered pair is a solution of the system of equations.

1) $x + 2y = -6$
  $-x - 3y = 13$
  $(8, -7)$ yes

2) $y - x = 4$
  $x + 3y = 8$
  $(-1, 3)$ yes

 3) $5x + y = 21$
  $2x - 3y = 11$
  $(4, 1)$ no

4) $7x + 2y = 14$
  $-5x + 6y = -12$
  $(2, 0)$ no

5) $5y - 4x = -5$
  $6x + 2y = -21$
  $\left(-\dfrac{5}{2}, -3\right)$ yes

6) $x = 9y - 7$
  $18y = 7x + 4$
  $\left(-1, \dfrac{2}{3}\right)$ no

7) $y = -x + 11$
  $x = 5y - 2$
  $(0, 9)$ no

8) $x = -y$
  $y = \dfrac{5}{8}x - 13$
  $(8, -8)$ yes

Mixed Exercises: Objectives 2 and 3

9) If you are solving a system of equations by graphing, how do you know whether the system has no solution?   The lines are parallel.

10) If you are solving a system of equations by graphing, how do you know whether the system has an infinite number of solutions?   The graphs are the same line.

Solve each system of equations by graphing. If the system is inconsistent or the equations are dependent, identify this.

 11) $y = -\dfrac{2}{3}x + 3$
  $y = x - 2$

12) $y = \dfrac{1}{2}x + 2$
  $y = 2x - 1$

13) $y = x + 1$
  $y = -\dfrac{1}{2}x + 4$

14) $y = -2x + 3$
  $y = x - 3$

15) $x + y = -1$
  $x - 2y = 14$

16) $2x - 3y = 6$
  $x + y = -7$

17) $x - 2y = 7$
  $-3x + y = -1$

18) $-x + 2y = 4$
  $3x + 4y = -12$

19) $\dfrac{3}{4}x - y = 0$
  $3x - 4y = 20$

20) $y = -x$
  $4x + 4y = 2$

21) $y = \dfrac{1}{3}x - 2$
  $4x - 12y = 24$

22) $5x + 5y = 5$
  $x + y = 1$

23) $x = 8 - 4y$
  $3x + 2y = 4$

24) $x - y = 0$
  $7x - 3y = 12$

 25) $y = -3x + 1$
  $12x + 4y = 4$

26) $2x - y = 1$
  $-2x + y = -3$

27) $x + y = 0$
  $y = \dfrac{1}{2}x + 3$

28) $x = -2$
  $y = -\dfrac{5}{2}x - 1$

29) $-3x + y = -4$
  $y = -1$

30) $5x + 2y = 6$
  $-15x - 6y = -18$

31) $y = \dfrac{3}{5}x - 6$
  $-3x + 5y = 10$

32) $y - x = -2$
  $2x + y = -5$

Write a system of equations so that the given ordered pair is a solution of the system.   For 33–38, answers may vary.

33) $(2, 5)$

34) $(3, 1)$

35) $(-4, -3)$

36) $(6, -1)$

37) $\left(-\dfrac{1}{3}, 4\right)$

38) $\left(0, \dfrac{3}{2}\right)$

For Exercises 39–42, determine which ordered pair could be a solution to the system of equations that is graphed. Explain why you chose that ordered pair.

39)

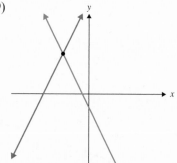

A. $(2, -6)$ B. $(3, 4)$
C. $(-3, 4)$ D. $(-2, -3)$ C; $(-3, 4)$ is in quadrant II.

40)

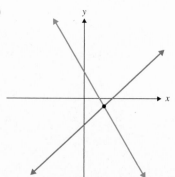

A. $\left(\dfrac{7}{2}, -\dfrac{1}{2}\right)$ B. $(-4, -1)$ A; $\left(\dfrac{7}{2}, -\dfrac{1}{2}\right)$ is in quadrant IV.

C. $\left(\dfrac{9}{4}, \dfrac{3}{4}\right)$ D. $\left(-\dfrac{10}{3}, -\dfrac{2}{3}\right)$

41)

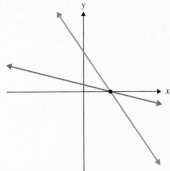

A. $(0, 3.8)$ B. $(4.1, 0)$
C. $(-2.1, 0)$ D. $(0, 5)$ B; $(4.1, 0)$ is the only point on the positive x-axis.

42)

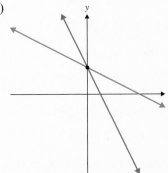

A. $(4, 0)$ B. $\left(\dfrac{1}{3}, 0\right)$
C. $(0, -3)$ D. $(0, 2)$ D; $(0, 2)$ is the only point on the positive y-axis.

Objective 4: Determine the Number of Solutions of a System Without Graphing

 43) How do you determine, *without graphing,* that a system of equations has exactly one solution?
The slopes are different.

 44) How do you determine, *without graphing,* that a system of equations has no solution?
The slopes are the same, but the y-intercepts are different.

Without graphing, determine whether each system has no solution, one solution, or an infinite number of solutions.

45) $y = 5x - 4$
$y = -3x + 7$
one solution

46) $y = \dfrac{2}{3}x + 9$
$y = \dfrac{2}{3}x + 1$
no solution

47) $y = -\dfrac{3}{8}x + 1$
$6x + 16y = -9$
no solution

48) $y = -\dfrac{1}{4}x + 3$
$2x + 8y = 24$
infinite number of solutions

49) $-15x + 9y = 27$
$10x - 6y = -18$
infinite number of solutions

50) $7x - y = 6$
$x + y = 13$
one solution

51) $3x + 12y = 9$
$x - 4y = 3$
one solution

52) $6x - 4y = -10$
$-21x + 14y = 35$
infinite number of solutions

53) $x = 5$
$x = -1$ no solution

54) $y = x$
$y = 2$ one solution

55) The graph shows the percentage of foreign students in U.S. institutions of higher learning from Saudi Arabia and the United Kingdom from 2004 to 2008. (http://nces.ed.gov)

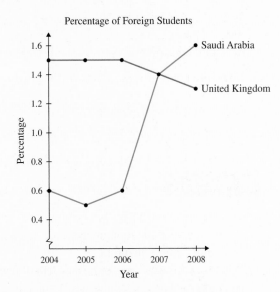

a) When was there a greater percentage of students from the United Kingdom? 2004–2006

b) Write the point of intersection of the graphs as an ordered pair in the form (year, percentage) and explain its meaning.

c) During which years did the percentage of students from the United Kingdom remain the same? 2004–2006

d) During which years did the percentage of students from Saudi Arabia increase the most? How can this be related to the slope of this line segment? 2006–2007; this line segment has the most positive slope.

56) The graph shows the approximate number of veterans living in Connecticut and Nevada from 2006 to 2010. (www.census.gov)

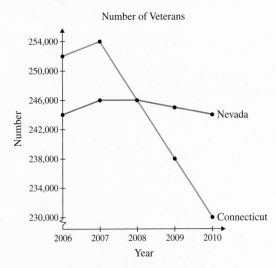

a) In which year were there approximately 7,000 fewer veterans living in Connecticut than in Nevada? Approximately how many were living in each state?   2009; CT: 238,000, NV: 245,000

b) Write the data point for Nevada in 2006 as an ordered pair of the form (year, number) and explain its meaning.   (2006, 244,000); in 2006, there were about 244,000 veterans living in Nevada.

c) Write the point of intersection of the graphs for the year 2008 as an ordered pair in the form (year, number) and explain its meaning.
(2008, 246,000); in 2008, Connecticut and Nevada each had a veteran population of about 246,000.

d) Which line segment on the Connecticut graph has a positive slope? How can this be explained in the context of this problem?   2006–2007; the number of veterans living in Connecticut increased from 2006 to 2007.

 Solve each system using a graphing calculator.

57) $y = -2x + 2$
$y = x - 7$ $(3, -4)$

58) $y = x + 1$
$y = 3x + 3$ $(-1, 0)$

59) $x - y = 3$
$x + 4y = 8$ $(4, 1)$

60) $2x + 3y = 3$
$y - x = -4$ $(3, -1)$

61) $4x + 5y = -17$
$3x - 7y = 4.45$
$(-2.25, -1.6)$

62) $-5x + 6y = 22.8$
$3x - 2y = -5.2$
$(1.8, 5.3)$

R Rethink

R1) What are you really looking for when you solve a system of linear equations?

R2) What happens when you graph a system of linear equations that has an infinite number of solutions?

R3) What is the solution to a system of linear equations when the two lines are parallel?

R4) Which objective is the most difficult for you?

4.2 Solving Systems by the Substitution Method

P Prepare

O Organize

What are your objectives for Section 4.2?	How can you accomplish each objective?
1 Solve a Linear System by Substitution	• Write the procedure for **Solving a System by the Substitution Method** in your own words. • Complete the given examples on your own. • Complete You Try 1.
2 Solve a System Containing Fractions or Decimals	• Add steps for eliminating fractions or decimals to the procedure you wrote for Objective 1. • Complete the given example on your own. • Complete You Try 2.
3 Solve a System by Substitution: Special Cases	• Know that not every system will have one solution, and review Section 4.1 to see the different types of systems. • Complete the given examples on your own. • Complete You Try 3.

Read the explanations, follow the examples, take notes, and complete the You Trys.

In Section 4.1, we learned to solve a system of equations by graphing. This method, however, is not always the *best* way to solve a system. If your graphs are not precise, you may read the solution incorrectly. And, if a solution consists of numbers that are not integers, like $\left(\frac{2}{3}, -\frac{1}{4}\right)$, it may not be possible to accurately identify the point of intersection of the graphs.

1 Solve a Linear System by Substitution

Another way to solve a system of equations is to use the *substitution method.* When we use the **substitution method,** we solve one of the equations for one of the variables in terms of the other. Then we substitute that expression into the other equation. We can do this because solving a system means finding the ordered pair, or pairs, that satisfy *both* equations. *The substitution method is especially good when one of the variables has a coefficient of 1 or −1.*

EXAMPLE 1

In-Class Example 1

Solve the system using substitution.
$y = 4x - 1$
$6x - 5y = -2$

Answer: $\left(\frac{1}{2}, 1\right)$

W Hint

Write out all of the steps in the example as you are reading it!

W Hint

Try graphing both of these equations. You will see that (1, −1) is the point of intersection.

Solve the system using substitution.

$$2x + 3y = -1$$
$$y = 2x - 3$$

Solution

The second equation, $y = 2x - 3$, is already solved for y; it tells us that y *equals* $2x - 3$. Therefore, we can substitute $2x - 3$ for y in the first equation, then solve for x.

$2x + 3y = -1$	First equation
$2x + 3(2x - 3) = -1$	Substitute.
$2x + 6x - 9 = -1$	Distribute.
$8x - 9 = -1$	Combine like terms.
$8x = 8$	Add 9 to each side.
$x = 1$	

We have found that $x = 1$, but we still need to find y. Substitute $x = 1$ into *either* equation, and solve for y. In this case, we will substitute $x = 1$ into the second equation since it is already solved for y.

$y = 2x - 3$	Second equation
$y = 2(1) - 3$	Substitute.
$y = 2 - 3$	
$y = -1$	

Check $x = 1$, $y = -1$ in *both* equations.

$2x + 3y = -1$		$y = 2x - 3$	
$2(1) + 3(-1) \stackrel{?}{=} -1$	Substitute.	$-1 \stackrel{?}{=} 2(1) - 3$	Substitute.
$2 - 3 \stackrel{?}{=} -1$		$-1 \stackrel{?}{=} 2 - 3$	
$-1 = -1$	True	$-1 = -1$	True

We write the solution of the system as an ordered pair, $(1, -1)$.

Let's summarize the steps we use to solve a system by the substitution method.

> **Procedure** Solving a System by the Substitution Method
>
> **Step 1:** Solve one of the equations for one of the variables. If possible, solve for a variable that has a coefficient of 1 or -1.
>
> **Step 2:** Substitute the expression found in *Step 1* into the *other* equation. The equation you obtain should contain only one variable.
>
> **Step 3:** Solve the equation you obtained in *Step 2*.
>
> **Step 4:** Substitute the value found in *Step 3* into either of the equations to obtain the value of the other variable.
>
> **Step 5:** Check the values in each of the original equations, and write the solution as an ordered pair.

EXAMPLE 2

Solve the system by the substitution method.

$$x - 2y = 7 \qquad (1)$$
$$2x + 3y = -21 \qquad (2)$$

Solution

We will follow the steps listed in the Procedure box.

Step 1: For which variable should we solve? The x in the first equation is the only variable with a coefficient of 1 or -1. Therefore, we will solve the first equation for x.

$$\begin{aligned} x - 2y &= 7 &&\text{First equation (1)} \\ x &= 2y + 7 &&\text{Add } 2y. \end{aligned}$$

Step 2: Substitute $2y + 7$ for the x in equation (2).

$$\begin{aligned} 2x + 3y &= -21 &&\text{Second equation (2)} \\ 2(2y + 7) + 3y &= -21 &&\text{Substitute.} \end{aligned}$$

W Hint

Notice that, in the first equation in Example 2, the coefficient of x is 1 so that it is easiest to begin by solving for this variable.

Step 3: Solve this new equation for y.

$$\begin{aligned} 2(2y + 7) + 3y &= -21 \\ 4y + 14 + 3y &= -21 &&\text{Distribute.} \\ 7y + 14 &= -21 &&\text{Combine like terms.} \\ 7y &= -35 &&\text{Subtract 14 from each side.} \\ y &= -5 \end{aligned}$$

Step 4: To determine the value of x, we can substitute -5 for y in either equation. We will use equation (1).

$$\begin{aligned} x - 2y &= 7 &&\text{Equation (1)} \\ x - 2(-5) &= 7 &&\text{Substitute.} \\ x + 10 &= 7 \\ x &= -3 \end{aligned}$$

Step 5: The check is left to the reader. The solution of the system is $(-3, -5)$.

[YOU TRY 1] Solve the system by the substitution method.

$$-3x + 4y = -2$$
$$6x - y = -3$$

If no variable in the system has a coefficient of 1 or -1, solve for any variable.

2 Solve a System Containing Fractions or Decimals

If a system contains an equation with fractions, first multiply the equation by the least common denominator to eliminate the fractions. Likewise, if an equation in the system contains decimals, begin by multiplying the equation by the lowest power of 10 that will eliminate the decimals.

EXAMPLE 3

In-Class Example 3

Solve the system by the substitution method.
$$\frac{1}{12}x + \frac{1}{4}y = \frac{3}{2}$$
$$\frac{2}{5}x - \frac{1}{2}y = -3$$

Answer: (0, 6)

Solve the system by the substitution method.

$$\frac{3}{10}x - \frac{1}{5}y = 1 \qquad (1)$$
$$-\frac{1}{12}x + \frac{1}{3}y = \frac{5}{6} \qquad (2)$$

Solution

Before applying the steps for solving the system, eliminate the fractions in each equation.

$$\frac{3}{10}x - \frac{1}{5}y = 1 \qquad\qquad\qquad -\frac{1}{12}x + \frac{1}{3}y = \frac{5}{6}$$

$$10\left(\frac{3}{10}x - \frac{1}{5}y\right) = 10 \cdot 1 \qquad \text{Multiply by the LCD: 10.} \qquad 12\left(-\frac{1}{12}x + \frac{1}{3}y\right) = 12 \cdot \frac{5}{6} \qquad \text{Multiply by the LCD: 12.}$$

$$3x - 2y = 10 \quad (3) \quad \text{Distribute.} \qquad\qquad -x + 4y = 10 \quad (4) \quad \text{Distribute.}$$

From the original equations, we obtain an equivalent system of equations.

$$3x - 2y = 10 \qquad (3)$$
$$-x + 4y = 10 \qquad (4)$$

Now, we will work with equations (3) and (4).

Apply the steps for solving the system:

Step 1: The x in equation (4) has a coefficient of -1. Solve this equation for x.

$$-x + 4y = 10 \qquad \text{Equation (4)}$$
$$-x = 10 - 4y \qquad \text{Subtract } 4y.$$
$$x = -10 + 4y \qquad \text{Divide by } -1.$$

Step 2: Substitute $-10 + 4y$ for x in equation (3).

$$3x - 2y = 10 \qquad (3)$$
$$3(-10 + 4y) - 2y = 10 \qquad \text{Substitute.}$$

Step 3: Solve the equation above for y.

$$3(-10 + 4y) - 2y = 10$$
$$-30 + 12y - 2y = 10 \qquad \text{Distribute.}$$
$$-30 + 10y = 10 \qquad \text{Combine like terms.}$$
$$10y = 40 \qquad \text{Add 30 to each side.}$$
$$y = 4 \qquad \text{Divide by 10.}$$

W Hint

Review the *eliminating fractions from an equation* procedure that you learned in Section 2.3.

Step 4: Find x by substituting 4 for y in either equation (3) or (4). Let's use equation (4) since it has smaller coefficients.

$$-x + 4y = 10 \qquad (4)$$
$$-x + 4(4) = 10 \qquad \text{Substitute.}$$
$$-x + 16 = 10 \qquad \text{Multiply.}$$
$$-x = -6 \qquad \text{Subtract 16 from each side.}$$
$$x = 6 \qquad \text{Divide by } -1.$$

Step 5: Check $x = 6$ and $y = 4$ in the original equations. The solution of the system is (6, 4).

> [**YOU TRY 2**]
>
> Solve each system by the substitution method.
>
> a) $-\dfrac{1}{6}x + \dfrac{1}{3}y = \dfrac{2}{3}$ b) $0.1x + 0.03y = -0.05$
>
> $\dfrac{3}{2}x - \dfrac{5}{2}y = -7$ $0.1x - 0.1y = 0.6$

3 Solve a System by Substitution: Special Cases

We saw in Section 4.1 that a system may have no solution or an infinite number of solutions. If we are solving a system by graphing, we know that a system has no solution if the lines are parallel, and a system has an infinite number of solutions if the graphs are the same line.

When we solve a system by *substitution,* how do we know whether the system is inconsistent or dependent? Read Examples 4 and 5 to find out.

EXAMPLE 4

In-Class Example 4

Solve the system by substitution.
 $2x + y = -3$
 $6x + 3y = 0$

Answer: ∅

W Hint

Remember that a system of linear equations has *no solution* when the two lines are parallel.

Solve the system by substitution. $3x + y = 5 \qquad (1)$
$12x + 4y = -7 \qquad (2)$

Solution

Step 1: $\qquad\qquad y = -3x + 5$ Solve equation (1) for y.

Step 2: $\qquad 12x + 4y = -7$ Substitute $-3x + 5$ for y in equation (2).
$12x + 4(-3x + 5) = -7$

Step 3: $12x + 4(-3x + 5) = -7$ Solve the resulting equation for x.
$12x - 12x + 20 = -7$ Distribute.
$20 = -7$ False

Because the variables drop out and we get a false statement, there is no solution to the system. The system is inconsistent, so the solution set is $\varnothing$.

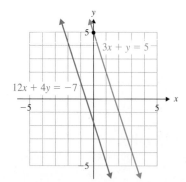

The graph of the equations in the system supports our work. The lines are parallel; therefore, the system has no solution.

EXAMPLE 5

Solve the system by substitution.

$$2x - 6y = 10 \qquad (1)$$
$$x = 3y + 5 \qquad (2)$$

Solution

Step 1: Equation (2) is already solved for x.

Step 2:
$$2x - 6y = 10$$
Substitute $3y + 5$ for x in equation (1).
$$2(3y + 5) - 6y = 10$$

Step 3:
$$2(3y + 5) - 6y = 10$$
Solve the resulting equation for y.
$$6y + 10 - 6y = 10$$
Distribute.
$$10 = 10$$
True

Because the variables drop out and we get a true statement, the system has an infinite number of solutions. The equations are dependent, and the solution set is $\{(x, y)|x - 3y = 5\}$.

W Hint

Remember that a system of linear equations has *an infinite number of solutions* when the two equations represent the same line.

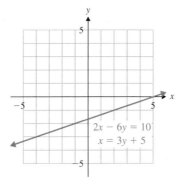

The graph shows that the equations in the system are the same line; therefore, the system has an infinite number of solutions.

Note

When you are solving a system of equations and the variables drop out:

1) If you get a *false statement*, like $20 = -7$, then the system has *no solution* and is *inconsistent*.

2) If you get a *true statement*, like $10 = 10$, then the system has an *infinite number of solutions*. The equations are *dependent*.

YOU TRY 3

Solve each system by substitution.

a) $\quad -20x + 5y = 3$
$\qquad\; 4x - y = -1$

b) $\qquad x - 4y = -7$
$\quad -2x + 8y = 14$

ANSWERS TO YOU TRY EXERCISES

1) $\left(-\dfrac{2}{3}, -1\right)$ 2) a) $(-8, -2)$ b) $(1, -5)$ 3) a) $\varnothing$ b) $\{(x, y)|x - 4y = -7\}$

Do the exercises, and check your work.

*Additional answers can be found in the Answers to Exercises appendix.

Mixed Exercises: Objectives 1 and 3

1) If you were asked to solve this system by substitution, why would it be easiest to begin by solving for y in the first equation?

$$7x + y = 1$$
$$-2x + 5y = 9$$

It is the only variable with a coefficient of 1.

2) When is the best time to use substitution to solve a system?
when one of the variables has a coefficient of 1 or -1

3) When solving a system of linear equations, how do you know whether the system has no solution?
The variables are eliminated, and you get a false statement.

4) When solving a system of linear equations, how do you know whether the system has an infinite number of solutions?
The variables are eliminated, and you get a true statement.

Solve each system by substitution.

5) $y = 4x - 3$
   $5x + y = 15$ $(2, 5)$

6) $y = 3x + 10$
   $-5x + 2y = 14$ $(-6, -8)$

7) $x = 7y + 11$
   $4x - 5y = -2$ $(-3, -2)$

8) $x = 9 - y$
   $-3x + 4y = 8$ $(4, 5)$

9) $x + 2y = -3$
   $4x + 5y = -6$ $(1, -2)$

10) $x + 4y = 1$
    $5x + 3y = 5$ $(1, 0)$

11) $2y - 7x = -14$
    $4x - y = 7$ $(0, -7)$

12) $-2x - y = 3$
    $3x + 2y = -3$ $(-3, 3)$

13) $9y - 18x = 5$
    $2x - y = 3$ $\varnothing$

14) $2x + 30y = 9$
    $x = 6 - 15y$ $\varnothing$

15) $x - 2y = 10$
    $3x - 6y = 30$

16) $6x + y = -6$
    $-12x - 2y = 12$

17) $10x + y = -5$ $\left(-\frac{4}{5}, 3\right)$
    $-5x + 2y = 10$

18) $2y - x = 4$ $\left(-1, \frac{3}{2}\right)$
    $x + 6y = 8$

19) $x = -\frac{3}{5}y + 7$
    $x + 4y = 24$ $(4, 5)$

20) $y = \frac{3}{2}x - 5$
    $2x - y = 5$ $(0, -5)$

21) $4y = 2x + 4$
    $2y - x = 2$

22) $3x + y = -12$
    $6x = 10 - 2y$ $\varnothing$

23) $2x + 3y = 6$
    $5x + 2y = -7$ $(-3, 4)$

24) $2x - 5y = -4$
    $8x - 9y = 6$ $(3, 2)$

25) $6x - 7y = -4$ $\left(\frac{5}{3}, 2\right)$
    $9x - 2y = 11$

26) $4x + 6y = -13$ $\left(-1, -\frac{3}{2}\right)$
    $7x - 4y = -1$

27) $18x + 6y = -66$
    $12x + 4y = -19$ $\varnothing$

28) $6y - 15x = -12$
    $5x - 2y = 4$

Objective 2: Solve a System Containing Fractions or Decimals

29) If an equation in a system contains fractions, what should you do first to make the system easier to solve? Multiply the equation by the LCD of the fractions to eliminate the fractions.

30) If an equation in a system contains decimals, what should you do first to make the system easier to solve? Multiply the equation by the power of 10 that will eliminate the decimals.

Solve each system by substitution.

31) $\frac{1}{4}x - \frac{1}{2}y = 1$
    $\frac{2}{3}x + \frac{1}{6}y = \frac{25}{6}$ $(6, 1)$

32) $\frac{2}{9}x + \frac{2}{9}y = 2$
    $\frac{7}{4}x - \frac{1}{8}y = \frac{3}{4}$ $(1, 8)$

33) $\frac{1}{6}x + \frac{4}{3}y = \frac{13}{3}$
    $\frac{2}{5}x + \frac{3}{2}y = \frac{18}{5}$ $(-6, 4)$

34) $\frac{1}{10}x + \frac{1}{2}y = \frac{1}{5}$
    $-\frac{1}{3}x + \frac{1}{2}y = \frac{3}{2}$ $(-3, 1)$

35) $\frac{x}{10} - \frac{y}{2} = \frac{13}{10}$
    $\frac{x}{3} + \frac{5}{4}y = -\frac{3}{2}$ $(3, -2)$

36) $-\frac{x}{3} + \frac{y}{2} = \frac{5}{3}$
    $\frac{x}{5} - \frac{4}{5}y = -1$ $(-5, 0)$

37) $y - \frac{5}{2}x = -2$
    $\frac{3}{4}x - \frac{3}{10}y = \frac{3}{5}$

38) $-\frac{2}{15}x - \frac{1}{3}y = \frac{2}{3}$
    $\frac{2}{3}x + \frac{5}{3}y = \frac{1}{2}$ $\varnothing$

39) $\frac{3}{4}x + \frac{1}{2}y = 6$
    $x = 3y + 8$ $(8, 0)$

40) $\frac{5}{3}x - \frac{4}{3}y = -\frac{4}{3}$
    $y = 2x + 4$ $(-4, -4)$

41) $0.2x - 0.1y = 0.1$
    $0.01x + 0.04y = 0.23$
    $(3, 5)$

42) $0.01x - 0.09y = -0.5$
    $0.02x + 0.05y = 0.38$
    $(4, 6)$

43) $0.6x - 0.1y = 1$
    $-0.4x + 0.5y = -1.1$
    $(1.5, -1)$

44) $0.8x + 0.7y = -1.7$
    $0.6x - 0.1y = 0.6$
    $(0.5, -3)$

45) $0.02x + 0.01y = -0.44$
    $-0.1x - 0.2y = 4$
    $(-16, -12)$

46) $0.3x + 0.1y = 3$
    $0.01x - 0.05y = -0.06$
    $(9, 3)$

47) $2.8x + 0.7y = 0.1$
    $0.04x + 0.01y = -0.06$
    $\varnothing$

48) $0.1x - 0.3y = -1.2$
    $1.5y - 0.5x = 6$

Solve by substitution. Begin by combining like terms.

49) $8 + 2(3x - 5) - 7x + 6y = 16$
    $9(y - 2) + 5x - 13y = -4$ (6, 4)

50) $3 + 4(2y - 9) + 5x - 2y = 8$
    $3(x + 3) - 4(2y + 1) - 2x = 4$ (7, 1)

51) $10(x + 3) - 7(y + 4) = 2(4x - 3y) + 3$
    $10 - 3(2x - 1) + 5y = 3y - 7x - 9$ (−4, −9)

52) $7x + 3(y - 2) = 7y + 6x - 1$
    $18 + 2(x - y) = 4(x + 2) - 5y$ (5, 0)

53) $-(y + 3) = 5(2x + 1) - 7x$
    $x + 12 - 8(y + 2) = 6(2 - y)$ (0, −8)

54) $9y - 4(2y + 3) = -2(4x + 1)$
    $16 - 5(2x + 3) = 2(4 - y)$ $\left(\frac{1}{2}, 6\right)$

55) Jamari wants to rent a cargo trailer to move his son into an apartment when he returns to college. A+ Rental charges $0.60 per mile while Rock Bottom Rental charges $70 plus $0.25 per mile. Let x = the number of miles driven, and let y = the cost of the rental. The cost of renting a cargo trailer from each company can be expressed with the following equations:

A+ Rental: $y = 0.60x$

Rock Bottom Rental: $y = 0.25x + 70$

a) How much would it cost Jamari to rent a cargo trailer from each company if he will drive a total of 160 miles?   A+: $96.00; Rock Bottom: $110.00

b) How much would it cost Jamari to rent a trailer from each company if he planned to drive 300 miles?   A+: $180.00; Rock Bottom: $145.00

c) Solve the system of equations using the substitution method, and explain the meaning of the solution.

d) Graph the system of equations, and explain when it is cheaper to rent a cargo trailer from A+ and when it is cheaper to rent it from Rock Bottom Rental. When is the cost the same?

56) To rent a pressure washer, Walsh Rentals charges $16.00 per hour while Discount Company charges $24.00 plus $12.00 per hour. Let x = the number of hours, and let y = the cost of the rental. The cost of renting a pressure washer from each company can be expressed with the following equations:

Walsh Rentals: $y = 16.00x$

Discount Company: $y = 12.00x + 24$

a) How much would it cost to rent a pressure washer from each company if it would be used for 4 hours?   Walsh: $64.00; Discount: $72.00

b) How much would it cost to rent a pressure washer from each company if it would be rented for 9 hours?   Walsh: $144.00; Discount: $132.00

c) Solve the system of equations using the substitution method, and explain the meaning of the solution.

d) Graph the system of equations, and explain when it is cheaper to rent a pressure washer from Walsh and when it is cheaper to rent it from Discount. When is the cost the same?

R Rethink

R1) So far, this chapter has demonstrated two methods for finding the intersection of two lines. Which method is easier for you?

R2) Suppose you are using the substitution method and the variables drop out. If your final statement is true, what does this mean? What does it mean if your final statement is false?

R3) Which exercises do you need to ask for help on in class?

4.3 Solving Systems by the Elimination Method

What are your objectives for Section 4.3?	How can you accomplish each objective?
1 Solve a Linear System Using the Elimination Method	• Follow Examples 1–3 in order to fully understand how the elimination method works. • Write the procedure for **Solving a System of Two Linear Equations by the Elimination Method** in your own words. • Complete the given examples on your own. • Complete You Trys 1–3.
2 Solve a Linear System Using the Elimination Method: Special Cases	• Understand that not all systems will have one solution, and review Section 4.1 to see the different types of systems. • Complete the given examples on your own. • Complete You Trys 4 and 5.
3 Use the Elimination Method Twice to Solve a Linear System	• Complete the given example on your own, and recognize when using elimination twice will be an easier way to solve a system. • Complete You Try 6.

W Work   Read the explanations, follow the examples, take notes, and complete the You Trys.

1 Solve a Linear System Using the Elimination Method

The next technique we will learn for solving a system of equations is the **elimination method.** (This is also called the **addition method.**) It is based on the addition property of equality that says that we can add the *same* quantity to each side of an equation and preserve the equality.

$$\text{If } a = b, \text{ then } a + c = b + c.$$

We can extend this idea by saying that we can add *equal* quantities to each side of an equation and still preserve the equality.

$$\text{If } a = b \text{ and } c = d, \text{ then } a + c = b + d.$$

The object of the elimination method is to add the equations (or multiples of one or both of the equations) so that one variable is eliminated. Then, we can solve for the remaining variable.

Solve the system using the elimination method.

$$x + y = 11 \quad (1)$$
$$x - y = -5 \quad (2)$$

Solution

The left side of each equation is equal to the right side of each equation. Therefore, if we add the left sides together and add the right sides together, we can set them equal. We will add these equations vertically. The y-terms are eliminated, enabling us to solve for x.

$$
\begin{array}{ll}
 x + y = 11 & (1) \\
+ \underline{ x - y = -5} & (2) \\
 2x + 0y = 6 & \text{Add equations (1) and (2).} \\
 2x = 6 & \text{Simplify.} \\
 x = 3 & \text{Divide by 2.}
\end{array}
$$

Now we substitute $x = 3$ into either equation to find the value of y. Here, we will use equation (1).

$$
\begin{array}{ll}
x + y = 11 & \text{Equation (1)} \\
3 + y = 11 & \text{Substitute 3 for } x. \\
y = 8 & \text{Subtract 3.}
\end{array}
$$

Check $x = 3$ and $y = 8$ in *both* equations.

$$
\begin{array}{ll}
x + y = 11 & \qquad x - y = -5 \\
3 + 8 \overset{?}{=} 11 \quad \text{Substitute.} & \qquad 3 - 8 \overset{?}{=} -5 \quad \text{Substitute.} \\
11 = 11 \quad \text{True} & \qquad -5 = -5 \quad \text{True}
\end{array}
$$

The solution of the system is (3, 8).

[YOU TRY 1] Solve the system using the elimination method.

$$3x + y = 10$$
$$x - y = 6$$

In Example 1, simply adding the equations eliminated a variable. But what can we do if we *cannot* eliminate a variable just by adding the equations together?

EXAMPLE 2

Solve the system using the elimination method.

$$2x + 5y = 5 \quad (1)$$
$$x + 4y = 7 \quad (2)$$

Solution

Just adding these equations will *not* eliminate a variable. The multiplication property of equality tells us that multiplying both sides of an equation by the same quantity results in an equivalent equation. If we multiply equation (2) by -2, the coefficient of x will be -2.

$$
\begin{array}{ll}
-2(x + 4y) = -2(7) & \text{Multiply equation (2) by } -2. \\
-2x - 8y = -14 & \text{New, equivalent equation}
\end{array}
$$

Original System		**Rewrite the System**
$2x + 5y = 5$	$\longrightarrow$	$2x + 5y = 5$
$x + 4y = 7$		$-2x - 8y = -14$

Add the equations in the rewritten system. The x is eliminated.

$$2x + 5y = 5$$
$$+\ \underline{-2x - 8y = -14}$$
$$0x - 3y = -9 \qquad \text{Add the equations.}$$
$$-3y = -9 \qquad \text{Simplify.}$$
$$y = 3 \qquad \text{Solve for } y.$$

Substitute $y = 3$ into (1) or (2) to find x. We will use equation (2).

$$x + 4y = 7 \qquad \text{Equation (2)}$$
$$x + 4(3) = 7 \qquad \text{Substitute 3 for } y.$$
$$x + 12 = 7$$
$$x = -5$$

The solution is $(-5, 3)$. Verify that $(-5, 3)$ satisfies equations (1) and (2).

[YOU TRY 2] Solve the system using the elimination method.

$$8x - y = -5$$
$$-6x + 2y = 15$$

Next we summarize the steps for solving a system using the elimination method.

> **Procedure** Solving a System of Two Linear Equations by the Elimination Method
>
> *Step 1:* Write each equation in the form $Ax + By = C$.
>
> *Step 2:* Determine which variable to eliminate. If necessary, multiply one or both of the equations by a number so that the coefficients of the variable to be eliminated are negatives of one another.
>
> *Step 3:* Add the equations, and solve for the remaining variable.
>
> *Step 4:* Substitute the value found in *Step 3* into either of the original equations to find the value of the other variable.
>
> *Step 5:* Check the solution in each of the original equations.

EXAMPLE 3

In-Class Example 3

Solve the system using the elimination method.
$7y = 4x - 1$
$6x + 2y = -11$

Answer: $\left(-\dfrac{3}{2}, -1\right)$

Solve the system using the elimination method.

$$2x = 9y + 4 \qquad (1)$$
$$3x - 7 = 12y \qquad (2)$$

Solution

Step 1: **Write each equation in the form $Ax + By = C$.**

$2x = 9y + 4 \qquad (1)$	$3x - 7 = 12y \qquad (2)$
$2x - 9y = 4 \qquad \text{Subtract } 9y.$	$3x - 12y = 7 \qquad \text{Subtract } 12y \text{ and add 7.}$

When we rewrite the equations in the form $Ax + By = C$, we get

$$2x - 9y = 4 \quad \text{(3)}$$
$$3x - 12y = 7 \quad \text{(4)}$$

Step 2: Determine which variable to eliminate from equations (3) and (4). Often, it is easier to eliminate the variable with the smaller coefficients. Therefore, *we will eliminate x.*

The least common multiple of 2 and 3 (the x-coefficients) is 6. Before we add the equations, one x-coefficient should be 6, and the other should be -6. Multiply equation (3) by 3 and equation (4) by -2.

Rewrite the System

$$3(2x - 9y) = 3(4) \qquad \text{3 times (3)}$$
$$-2(3x - 12y) = -2(7) \qquad \text{-2 times (4)}$$
$$\longrightarrow$$
$$6x - 27y = 12$$
$$-6x + 24y = -14$$

> **W Hint**
>
> Once again, notice that the coefficients of x are made to be opposite in sign so that they can be eliminated when the equations are added.

Step 3: Add the resulting equations to eliminate x. Solve for y.

$$\begin{aligned} 6x - 27y &= 12 \\ + \quad -6x + 24y &= -14 \\ \hline -3y &= -2 \\ y &= \frac{2}{3} \end{aligned}$$

Step 4: Substitute $y = \dfrac{2}{3}$ **into equation (1) and solve for x.**

$$2x = 9y + 4 \qquad \text{Equation (1)}$$
$$2x = 9\left(\frac{2}{3}\right) + 4 \qquad \text{Substitute.}$$
$$2x = 6 + 4 \qquad \text{Multiply.}$$
$$2x = 10 \qquad \text{Add.}$$
$$x = 5$$

Step 5: Check to verify that $\left(5, \dfrac{2}{3}\right)$ satisfies each of the original equations. The solution is $\left(5, \dfrac{2}{3}\right)$.

$\begin{bmatrix} \text{YOU TRY 3} \end{bmatrix}$ Solve the system using the elimination method.

$$5x = 2y - 14$$
$$4x + 3y = 21$$

---

2 Solve a Linear System Using the Elimination Method: Special Cases

We have seen in Sections 4.1 and 4.2 that some systems have no solution, and some have an infinite number of solutions. How does the elimination method illustrate these results?

EXAMPLE 4

In-Class Example 4

Solve the system using the
elimination method.

$4x - 2y = 7$

$3y = 6x - 1$

Answer: ∅

Solve the system using the elimination method.

$$4y = 10x + 3 \qquad (1)$$
$$6y - 15x = -8 \qquad (2)$$

Solution

Step 1: **Write each equation in the form $Ax + By = C$.**

$$
\begin{array}{ll}
4y = 10x + 3 & \qquad -10x + 4y = 3 \qquad (3) \\
6y - 15x = -8 & \qquad -15x + 6y = -8 \qquad (4)
\end{array}
$$

Step 2: **Determine which variable to eliminate from equations (3) and (4).** Eliminate y. The least common multiple of 4 and 6, the y-coefficients, is 12. One y-coefficient must be 12, and the other must be -12.

Rewrite the System

$$
\begin{array}{ll}
-3(-10x + 4y) = -3(3) & \qquad 30x - 12y = -9 \\
2(-15x + 6y) = 2(-8) & \qquad -30x + 12y = -16
\end{array}
$$

Step 3: **Add the equations.**

$$
\begin{array}{r}
30x - 12y = -9 \\
+ \; -30x + 12y = -16 \\
\hline
0 = -25 \quad \text{False}
\end{array}
$$

W Hint

Notice that, just as in Section 4.2, if the variables drop out and you end up with a false statement, there is *no solution.*

The variables drop out, and we get a false statement. Therefore, the system is inconsistent, and the solution set is ∅.

[YOU TRY 4] Solve the system using the elimination method.

$$24x + 6y = -7$$
$$4y + 3 = -16x$$

EXAMPLE 5

In-Class Example 5

Solve the system using the
elimination method.

$12x + 10y = -6$

$y = -\dfrac{6}{5}x - \dfrac{3}{5}$

Answer: infinite number
of solutions of the form

$\left\{ (x, y) \,\middle|\, y = -\dfrac{6}{5}x - \dfrac{3}{5} \right\}$

Solve the system using the elimination method.

$$12x - 18y = 9 \qquad (1)$$
$$y = \frac{2}{3}x - \frac{1}{2} \qquad (2)$$

Solution

Step 1: **Write equation (2) in the form $Ax + By = C$.**

$$y = \frac{2}{3}x - \frac{1}{2} \qquad \text{Equation (2)}$$

$$6y = 6\left(\frac{2}{3}x - \frac{1}{2}\right) \qquad \text{Multiply by 6 to eliminate fractions.}$$

$$6y = 4x - 3$$

$$-4x + 6y = -3 \qquad (3) \qquad \text{Rewrite as } Ax + By = C.$$

We can rewrite $y = \dfrac{2}{3}x - \dfrac{1}{2}$ as $-4x + 6y = -3$, equation (3).

Step 2: **Determine which variable to eliminate from equations (1) and (3).**

$$12x - 18y = 9 \qquad \text{(1)}$$
$$-4x + 6y = -3 \qquad \text{(3)}$$

Eliminate x. Multiply equation (3) by 3.

$$12x - 18y = 9 \qquad \text{(1)}$$
$$-12x + 18y = -9 \qquad \text{3 times (3)}$$

Step 3: **Add the equations.**

$$\begin{array}{r} 12x - 18y = 9 \\ + \ -12x + 18y = -9 \\ \hline 0 = 0 \qquad \text{True} \end{array}$$

The variables drop out, and we get a true statement. The equations are dependent, so there are an infinite number of solutions. The solution set is $\left\{ (x, y) \,\middle|\, y = \dfrac{2}{3}x - \dfrac{1}{2} \right\}$.

Hint

Notice that, just as in Section 4.2, if the variables drop out and you end up with a true statement, there are an *infinite number of solutions.*

[**YOU TRY 5**] Solve the system using the elimination method.

$$-6x + 8y = 4$$
$$3x - 4y = -2$$

3 Use the Elimination Method Twice to Solve a Linear System

Sometimes, applying the elimination method *twice* is the best strategy.

EXAMPLE 6

Solve using the elimination method.

$$5x - 6y = 2 \qquad \text{(1)}$$
$$9x + 4y = -3 \qquad \text{(2)}$$

In-Class Example 6

Solve using the elimination method.
$5x - 3y = -6$
$2x + 7y = 1$

Answer: $\left(-\dfrac{39}{41}, \dfrac{17}{41} \right)$

Solution

Each equation is written in the form $Ax + By = C$, so we begin with *Step 2.*

Step 2: We will eliminate y from equations (1) and (2).

Rewrite the System

$$2(5x - 6y) = 2(2) \qquad\longrightarrow\qquad 10x - 12y = 4$$
$$3(9x + 4y) = 3(-3) \qquad\qquad\qquad 27x + 12y = -9$$

Step 3: Add the resulting equations to eliminate y. Solve for x.

$$\begin{array}{r} 10x - 12y = \ \ 4 \\ + \ 27x + 12y = -9 \\ \hline 37x = -5 \end{array}$$

$$x = -\frac{5}{37} \qquad \text{Solve for } x.$$

Normally, we would substitute $x = -\dfrac{5}{37}$ into equation (1) or equation (2) and solve for y.

SECTION 4.3 **Solving Systems by the Elimination Method** **267**

This time, however, working with a number like $-\dfrac{5}{37}$ would be difficult, so *we will use the elimination method a second time.*

Go back to the original equations, (1) and (2), and use the elimination method again but eliminate the other variable, x. Then, solve for y.

Eliminate x from the original system; $5x - 6y = 2$ (1)
$\qquad\qquad\qquad\qquad\qquad\qquad\qquad\quad 9x + 4y = -3$ (2)

Rewrite the System

$$-9(5x - 6y) = -9(2) \qquad\qquad -45x + 54y = -18$$
$$5(9x + 4y) = 5(-3) \qquad\longrightarrow\qquad 45x + 20y = -15$$

Add the equations.

$$\begin{array}{r} -45x + 54y = -18 \\ + \quad 45x + 20y = -15 \\ \hline 74y = -33 \end{array}$$

$$y = -\dfrac{33}{74} \qquad \text{Solve for } y.$$

Check to verify that the solution is $\left(-\dfrac{5}{37}, -\dfrac{33}{74}\right)$.

[YOU TRY 6]   Solve using the elimination method.

$$-9x + 2y = -3$$
$$2x - 5y = 4$$

ANSWERS TO [YOU TRY] EXERCISES

1) $(4, -2)$ 2) $\left(\dfrac{1}{2}, 9\right)$ 3) $(0, 7)$ 4) $\varnothing$

5) infinite number of solutions of the form $\{(x, y)\,|\,3x - 4y = -2\}$ 6) $\left(\dfrac{7}{41}, -\dfrac{30}{41}\right)$

[E Evaluate] **4.3** Exercises   Do the exercises, and check your work.

*Additional answers can be found in the Answers to Exercises appendix.

Mixed Exercises: Objectives 1 and 2

1) What is the first step you would use to solve this system by elimination if you wanted to eliminate y?
   Add the equations.
   $$5x + y = 2$$
   $$3x - y = 6$$

2) What is the first step you would use to solve this system by elimination if you wanted to eliminate x?
   $$4x - 3y = 14$$
   $$8x - 11y = 18$$
   Multiply the first equation by -2. Then add the equations.

For Exercises 3 and 4, choose *always, sometimes,* or *never.*

3) A system of equations with two different lines written in the form $Ax + By = 0$ will *always, sometimes,* or *never* have $(0, 0)$ as the solution.   always

4) If both variables are eliminated while solving a system of equations using elimination, the system will *always, sometimes,* or *never* have an infinite number of solutions.   sometimes

Solve each system using the elimination method.

5) $x - y = -3$ $(5, 8)$
   $2x + y = 18$

6) $x + 3y = 1$ $(4, -1)$
   $-x + 5y = -9$

7) $-x + 2y = 2$ $(-6, -2)$
   $x - 7y = 8$

8) $4x - y = -15$ $(-3, 3)$
   $3x + y = -6$

9) $x + 4y = 1$
   $3x - 4y = -29$ $(-7, 2)$

10) $5x - 4y = -10$
    $-5x + 7y = 25$ $(2, 5)$

11) $-8x + 5y = -16$
    $4x - 7y = 8$ $(2, 0)$

12) $7x + 6y = 3$
    $3x + 2y = -1$ $(-3, 4)$

13) $4x + 15y = 13$
    $3x + 5y = 16$ $(7, -1)$

14) $12x + 7y = 7$
    $-3x + 8y = 8$ $(0, 1)$

15) $9x - 7y = -14$
    $4x + 3y = 6$ $(0, 2)$

16) $5x - 2y = -6$
    $4x + 5y = -18$ $(-2, -2)$

17) $-9x + 2y = -4$
    $6x - 3y = 11$ $\left(-\frac{2}{3}, -5\right)$

18) $12x - 2y = 3$
    $8x - 5y = -9$ $\left(\frac{3}{4}, 3\right)$

19) $9x - y = 2$
    $18x - 2y = 4$

20) $-4x + 7y = 13$
    $12x - 21y = -5$ $\varnothing$

21) $x = 12 - 4y$
    $2x - 7 = 9y$ $(8, 1)$

22) $5x + 3y = -11$
    $y = 6x + 4$ $(-1, -2)$

23) $4y = 9 - 3x$
    $5x - 16 = -6y$ $\left(5, -\frac{3}{2}\right)$

24) $8x = 6y - 1$
    $10y - 6 = -4x$ $\left(\frac{1}{4}, \frac{1}{2}\right)$

25) $2x - 9 = 8y$
    $20y - 5x = 6$ $\varnothing$

26) $3x + 2y = 6$
    $4y = 12 - 6x$

27) $6x - 11y = -1$
    $-7x + 13y = 2$ $(9, 5)$

28) $10x - 4y = 7$
    $12x - 3y = -15$ $\left(-\frac{9}{2}, -13\right)$

29) $9x + 6y = -2$
    $-6x - 4y = 11$ $\varnothing$

30) $4x - 9y = -3$
    $36y - 16x = 12$

31) What is the first step in solving this system by the elimination method? DO NOT SOLVE.

$$\frac{x}{4} + \frac{y}{2} = -1$$

$$\frac{3}{8}x + \frac{5}{3}y = -\frac{7}{12}$$

Eliminate the fractions. Multiply the first equation by 4, and multiply the second equation by 24.

32) What is the first step in solving this system by the elimination method? DO NOT SOLVE.

$$0.1x + 2y = -0.8$$
$$0.03x + 0.10y = 0.26$$

Eliminate the decimals. Multiply the first equation by 10, and multiply the second equation by 100.

Solve each system by elimination.

33) $\frac{4}{5}x - \frac{1}{2}y = -\frac{3}{2}$
    $2x - \frac{1}{4}y = \frac{1}{4}$ $\left(\frac{5}{8}, 4\right)$

34) $\frac{1}{3}x - \frac{4}{5}y = \frac{13}{15}$
    $\frac{1}{6}x - \frac{3}{4}y = -\frac{1}{2}$ $\left(9, \frac{8}{3}\right)$

35) $\frac{5}{4}x - \frac{1}{2}y = \frac{7}{8}$ $\left(-\frac{9}{2}, -13\right)$
    $\frac{2}{5}x - \frac{1}{10}y = -\frac{1}{2}$

36) $\frac{1}{2}x - \frac{11}{8}y = -1$ $(-2, 0)$
    $-\frac{2}{5}x + \frac{3}{10}y = \frac{4}{5}$

37) $\frac{x}{4} + \frac{y}{2} = -1$ $(-6, 1)$
    $\frac{3}{8}x + \frac{5}{3}y = -\frac{7}{12}$

38) $\frac{x}{12} - \frac{y}{6} = \frac{2}{3}$ $(8, 0)$
    $\frac{x}{4} + \frac{y}{3} = 2$

39) $\frac{x}{12} - \frac{y}{8} = \frac{7}{8}$
    $y = \frac{2}{3}x - 7$

40) $\frac{5}{3}x + \frac{1}{3}y = \frac{2}{3}$
    $\frac{3}{4}x + \frac{3}{20}y = -\frac{5}{4}$ $\varnothing$

41) $-\frac{1}{2}x + \frac{5}{4}y = \frac{3}{4}$
    $\frac{2}{5}x - \frac{1}{2}y = -\frac{1}{10}$ $(1, 1)$

42) $y = 2 - 4x$
    $\frac{1}{3}x - \frac{3}{8}y = \frac{5}{8}$ $\left(\frac{3}{4}, -1\right)$

43) $0.08x + 0.07y = -0.84$
    $0.32x - 0.06y = -2$ $(-7, -4)$

44) $0.06x + 0.05y = 0.58$
    $0.18x - 0.13y = 1.18$ $(8, 2)$

45) $0.1x + 2y = -0.8$
    $0.03x + 0.10y = 0.26$ $(12, -1)$

46) $0.6x - 0.1y = 0.5$
    $0.1x - 0.03y = -0.01$ $(2, 7)$

47) $-0.4x + 0.2y = 0.1$
    $0.6x - 0.3y = 1.5$ $\varnothing$

48) $x - 0.5y = 0.2$
    $-0.3x + 0.15y = -0.06$ infinite number of solutions of the form $\{(x, y) \mid x - 0.5y = 0.2\}$

49) $0.04x + 0.03y = 0.16$
    $0.6x + 0.2y = 1.15$ $(0.25, 5)$

50) $-0.5x + 0.8y = 0.3$
    $0.03x + 0.1y = -0.24$ $(-3, -1.5)$

51) $17x - 16(y + 1) = 4(x - y)$
    $19 - 10(x + 2) = -4(x + 6) - y + 2$ $(4, 3)$

52) $28 - 4(y + 1) = 3(x - y) + 4$
    $-5(x + 4) - y + 3 = 28 - 5(2x + 5)$ $(5, 5)$

53) $5 - 3y = 6(3x + 4) - 8(x + 2)$
    $6x - 2(5y + 2) = -7(2y - 1) - 4$ $\left(-\frac{3}{2}, 4\right)$

54) $5(y + 3) = 6(x + 1) + 6x$
    $7 - 3(2 - 3x) - y = 2(3y + 8) - 5$ $\left(\frac{1}{3}, -1\right)$

55) $6(x - 3) + x - 4y = 1 + 2(x - 9)$
    $4(2y - 3) + 10x = 5(x + 1) - 4$ $(1, 1)$

56) $8y + 2(4x + 5) - 5x = 7y - 11$
    $11y - 3(x + 2) = 16 + 2(3y - 4) - x$ $(-7, 0)$

Objective 3: Use the Elimination Method Twice to Solve a Linear System

Solve each system using the elimination method twice.

57) $4x + 5y = -6$
$3x + 8y = 15$ $\left(-\dfrac{123}{17}, \dfrac{78}{17}\right)$

58) $8x - 4y = -21$
$-5x + 6y = 12$ $\left(-\dfrac{39}{14}, -\dfrac{9}{28}\right)$

59) $4x + 9y = 7$
$6x + 11y = -14$ $\left(-\dfrac{203}{10}, \dfrac{49}{5}\right)$

60) $10x + 3y = 18$
$9x - 4y = 5$ $\left(\dfrac{87}{67}, \dfrac{112}{67}\right)$

Find k so that the given ordered pair is a solution of the given system.

61) $x + ky = 17$; $(5, 4)$ 3
$2x - 3y = -2$

62) $kx + y = -13$; $(-1, -8)$ 5
$9x - 2y = 7$

63) $3x + 4y = -9$; $(-7, 3)$ -8
$kx - 5y = 41$

64) $4x + 3y = -7$; $(2, -5)$ -2
$3x + ky = 16$

65) Given the following system of equations,

$$x - y = 5$$
$$x - y = c$$

find c so that the system has

a) an infinite number of solutions.   5

b) no solution. c can be any real number except 5.

66) Given the following system of equations,

$$2x + y = 9$$
$$2x + y = c$$

find c so that the system has

a) an infinite number of solutions.   9

b) no solution. c can be any real number except 9.

67) Given the following system of equations,

$$9x + 12y = -15$$
$$ax + 4y = -5$$

find a so that the system has

a) an infinite number of solutions.   3

b) exactly one solution. a can be any real number except 3.

68) Given the following system of equations,

$$-2x + 7y = 3$$
$$4x + by = -6$$

find b so that the system has

a) an infinite number of solutions. -14

b) exactly one solution. b can be any real number except -14.

Extension

Let a, b, and c represent nonzero constants. Solve each system for x and y.

69) $-5x + 4by = 6$
$5x + 3by = 8$ $\left(\dfrac{2}{5}, \dfrac{2}{b}\right)$

70) $ax - 6y = 4$
$-ax + 9y = 2$ $\left(\dfrac{16}{a}, 2\right)$

71) $3ax + by = 4$
$ax - by = -5$ $\left(-\dfrac{1}{4a}, \dfrac{19}{4b}\right)$

72) $2ax + by = c$
$ax + 3by = 4c$ $\left(-\dfrac{c}{5a}, \dfrac{7c}{5b}\right)$

R Rethink

R1) In this chapter, you have learned three methods for finding the intersection of two lines. Which of the three methods do you prefer and why?

R2) Suppose you want to eliminate the y-values in a system of linear equations. What do you need to do before you add the two equations together?

R3) Are there any problems you could not do? If so, write them down or circle them and ask your instructor for help.

Putting It All Together

P Prepare

O Organize

What are your objectives?	How can you accomplish each objective?
1 Review the concepts of Sections 4.1–4.3	• Understand the summary for **Choosing Between Substitution and the Elimination Method to Solve a System,** and write it in your own words. • Complete the You Try.

W Work

Read the explanations, follow the examples, take notes, and complete the You Try.

1 Review the Concepts of Sections 4.1–4.3

We have learned three methods for solving systems of linear equations:

1) Graphing   2) Substitution   3) Elimination

How do we know which method is best for a particular system? We will answer this question by summarizing what we have learned so far.

First, solving a system by graphing is the least desirable of the methods. The point of intersection can be difficult to read, especially if one of the numbers is a fraction. But, the graphing method is important in certain situations and is one you should know.

In-Class Example

Decide which method to use to solve each system, substitution or elimination, and explain why this method was chosen. Then, solve the system.
a) $5x + 2y = -2$
$y = -3x - 3$
b) $x - 5y = -10$
$7x + 3y = 6$
c) $2x + 6y = 1$
$3x + 4y = 9$

Answer: a) substitution; $(-4, 9)$ b) substitution or elimination; $(0, 2)$

c) elimination: $\left(5, -\dfrac{3}{2}\right)$

Summary Choosing Between Substitution and the Elimination Method to Solve a System

1) If at least one equation is solved for a variable and contains no fractions, **use substitution.**

$$-5x + 2y = -8$$
$$x = 4y + 16$$

2) If a variable has a coefficient of 1 or -1, you can solve for that variable and **use substitution.**

$$4x + y = 10$$
$$-3x - 8y = 7$$

Or, leave each equation in the form $Ax + By = C$ and **use elimination.** Either approach is good and is a matter of personal preference.

3) If no variable has a coefficient of 1 or -1, **use elimination.**

$$4x - 5y = -3$$
$$6x + 8y = 11$$

Remember, if an equation contains fractions or decimals, begin by eliminating them. Then, decide which method to use following the guidelines listed here.

[YOU TRY] Decide which method to use to solve each system, substitution or elimination, and explain why this method was chosen. Then, solve the system.

a) $9x - 7y = -9$
$2x + 9y = -2$

b) $9x - 2y = 0$
$x = y - 7$

c) $4x + y = 13$
$-3x - 2y = 4$

ANSWERS TO [YOU TRY] EXERCISES

a) elimination; $(-1, 0)$ b) substitution; $(2, 9)$ c) substitution or elimination; $(6, -11)$

Putting It All Together Exercises

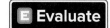 **E Evaluate**   Do the exercises, and check your work.

Additional answers can be found in the Answers to Exercises appendix.

Objective 1: Review the Concepts of Sections 4.1–4.3

Decide which method to use to solve each system, substitution or elimination, and explain why this method was chosen. Then solve the system.

1) $8x - 5y = 10$
$2x - 3y = -8$

2) $x = 2y - 7$
$8x - 3y = 9$

3) $12x - 5y = 18$
$8x + y = -1$

4) $11x + 10y = -4$
$9x - 5y = 2$

5) $y - 4x = -11$
$x = y + 8$

6) $4x - 5y = 4$
$y = \dfrac{3}{4}x - \dfrac{1}{2}$

Solve each system using either the substitution or elimination method.

7) $4x + 5y = 24$ $(6, 0)$
$x - 3y = 6$

8) $6y - 5x = 22$ $(-2, 2)$
$-9x - 8y = 2$

9) $6x + 15y = -1$
$9x = 10y - 8$ $\left(-\dfrac{2}{3}, \dfrac{1}{5}\right)$

10) $x + 2y = 9$ $(1, 4)$
$7x - y = 3$

11) $10x + 4y = 7$ $\varnothing$
$15x + 6y = -2$

12) $y = -6x + 5$
$12x + 2y = 10$

13) $10x + 9y = 4$
$x = -\dfrac{1}{2}$ $\left(-\dfrac{1}{2}, 1\right)$

14) $6x - 4y = 11$
$\dfrac{3}{2}x + \dfrac{1}{4}y = \dfrac{7}{8}$ $\left(\dfrac{5}{6}, -\dfrac{3}{2}\right)$

15) $7y - 2x = 13$ $(4, 3)$
$3x - 2y = 6$

16) $y = 6$
$12x + y = 8$ $\left(\dfrac{1}{6}, 6\right)$

17) $\dfrac{2}{5}x + \dfrac{4}{5}y = -2$ $(9, -7)$
$\dfrac{1}{6}x + \dfrac{1}{6}y = \dfrac{1}{3}$

18) $5x + 4y = 14$ $(10, -9)$
$y = -\dfrac{8}{5}x + 7$

19) $-0.3x + 0.1y = 0.4$ $(0, 4)$
$0.01x + 0.05y = 0.2$

20) $0.01x - 0.06y = 0.03$ $(-3, -1)$
$0.4x + 0.3y = -1.5$

21) $-6x + 2y = -10$
$21x - 7y = 35$

22) $\dfrac{5}{3}x + \dfrac{4}{3}y = \dfrac{2}{3}$ $\varnothing$
$10x + 8y = -5$

23) $2 = 5y - 8x$ $(-9, -14)$
$y = \dfrac{3}{2}x - \dfrac{1}{2}$

24) $\dfrac{5}{6}x - \dfrac{3}{4}y = \dfrac{2}{3}$ $\left(2, \dfrac{4}{3}\right)$
$\dfrac{1}{3}x + 2y = \dfrac{10}{3}$

25) $2x - 3y = -8$
$7x + 10y = 4$ $\left(-\dfrac{68}{41}, \dfrac{64}{41}\right)$

26) $6x = 9 - 13y$
$4x + 3y = -2$ $\left(-\dfrac{53}{34}, \dfrac{24}{17}\right)$

27) $6(2x - 3) = y + 4(x - 3)$
$5(3x + 4) + 4y = 11 - 3y + 27x$ $\left(\dfrac{3}{4}, 0\right)$

28) $3 - 5(x - 4) = 2(1 - 4y) + 2$
$2(x + 10) + y + 1 = 3x + 5(y + 6) - 17$ $\left(5, \dfrac{3}{4}\right)$

29) $2y - 2(3x + 4) = -5(y - 2) - 17$ $(1, 1)$
$4(2x + 3) = 10 + 5(y + 1)$

30) $x - y + 23 = 2y + 3(2x + 7)$ $(-2, 4)$
$9y - 8 + 4(x + 2) = 2(4x - 1) - 3x + 10y$

31) $y = -4x$
$10x + 2y = -5$ $\left(-\dfrac{5}{2}, 10\right)$

32) $x = \dfrac{2}{3}y$ $(-4, -6)$
$9x - 5y = -6$

Solve each system by graphing.

33) $y = \frac{1}{2}x + 1$ (2, 2)

   $x + y = 4$

34) $x + y = -3$ (−1, −2)

   $y = 3x + 1$

35) $x + y = 0$ (−4, 4)

   $x - 2y = -12$

36) $2y - x = 6$ (2, 4)

   $y = 2x$

37) $2x - 3y = 3$ ∅

   $y = \frac{2}{3}x + 1$

38) $y = -\frac{5}{2}x - 3$

   $10x + 4y = -12$

Solve each system using a graphing calculator.

39) $8x - 6y = -7$

   $4x - 16y = 3$

   (−1.25, −0.5)

40) $4x + 3y = -9$

   $2x + y = 2$

   (7.5, −13)

R Rethink

R1) What are the advantages and disadvantages of each method for solving a system of equations?

R2) Which characteristics of the equations in a system will help you determine whether you will solve a system by substitution or by elimination?

R3) Which method of solving systems do you prefer? Why?

R4) How will choosing the most efficient way to solve a system help you on a test?

4.4 Applications of Systems of Two Equations

P Prepare

O Organize

What are your objectives for Section 4.4?	How can you accomplish each objective?
1 Solve Problems Involving General Quantities	• Understand the procedure for **Solving an Applied Problem Using a System of Equations,** and write it in your own words. • Complete the given example on your own. • Complete You Try 1.
2 Solve Geometry Problems	• Review geometry formulas, if necessary, given in Section 1.3. • Complete the given example on your own. • Complete You Try 2.
3 Solve Problems Involving Cost	• Complete the given example on your own. • Complete You Try 3.
4 Solve Mixture Problems	• Complete the given example on your own. • Complete You Try 4.
5 Solve Distance, Rate, and Time Problems	• Complete the given example on your own. • Complete You Try 5.

 **Read the explanations, follow the examples, take notes, and complete the You Trys.**

In Section 2.3, we introduced the five-step method for solving applied problems. Here, we modify the method for problems with *two* unknowns and *two* equations.

Procedure Solving an Applied Problem Using a System of Equations

Step 1: **Read** the problem carefully, more than once if necessary. Draw a picture, if applicable. Identify what you are being asked to find.

Step 2: **Choose variables** to represent the unknown quantities. Label any pictures with the variables.

Step 3: **Write a system of equations using two variables.** It may be helpful to begin by writing the equations in words.

Step 4: **Solve** the system.

Step 5: **Check** the answer in the original problem, and **interpret** the solution as it relates to the problem. Be sure your answer makes sense in the context of the problem.

---

1 Solve Problems Involving General Quantities

EXAMPLE 1

In-Class Example 1

As of the end of 2011, the Beatles had more of their singles hit number one on the U.S. charts than any other artist. In second place was Mariah Carey. The Beatles had two more songs hit number one, and together the artists had a total of 38 number one singles. How many of each of their singles topped the U.S. charts? (www.billboard.com)

Answer: The Beatles: 20; Mariah Carey: 18

Write a system of equations, and solve.

Pink Floyd's album, *The Dark Side of the Moon,* spent more weeks on the Billboard 200 chart for top-selling albums than any other album in history. It was on the chart 251 more weeks than the second-place album, *Johnny's Greatest Hits,* by Johnny Mathis. If they were on the charts for a total of 1231 weeks, how many weeks did each album spend on the Billboard 200 chart? (www.billboard.com)

Solution

Step 1: **Read** the problem carefully, and identify what we are being asked to find.

We must find the number of weeks each album was on the chart.

Step 2: **Choose variables** to represent the unknown quantities.

x = the number of weeks *The Dark Side of the Moon* was on the Billboard 200 chart

y = the number of weeks *Johnny's Greatest Hits* was on the Billboard 200 chart

Step 3: **Write a system of equations using two variables.** First, let's think of the equations in English. Then we will translate them into algebraic equations.

To get one equation, use the information that says these two albums were on the Billboard 200 chart for a total of 1231 weeks. Write an equation in words, then translate it into an algebraic equation.

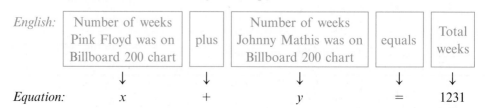

English:	Number of weeks Pink Floyd was on Billboard 200 chart	plus	Number of weeks Johnny Mathis was on Billboard 200 chart	equals	Total weeks
	↓	↓	↓	↓	↓
Equation:	x	$+$	y	$=$	1231

The first equation is $x + y = 1231$.

To get the second equation, use the information that says the Pink Floyd album was on the chart 251 weeks more than the Johnny Mathis album.

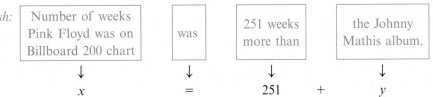

English:

Number of weeks Pink Floyd was on Billboard 200 chart	was	251 weeks more than	the Johnny Mathis album.
↓	↓	↓	↓
x	$=$	251 $+$	y

The second equation is $x = 251 + y$.

The system of equations is $x + y = 1231$
$$x = 251 + y.$$

Step 4: **Solve** the system.

$$x + y = 1231$$
$$(251 + y) + y = 1231 \qquad \text{Substitute.}$$
$$251 + 2y = 1231 \qquad \text{Combine like terms.}$$
$$2y = 980 \qquad \text{Subtract 251 from each side.}$$
$$\frac{2y}{2} = \frac{980}{2} \qquad \text{Divide each side by 2.}$$
$$y = 490 \qquad \text{Simplify.}$$

Find x by substituting $y = 490$ into $x = 251 + y$.
$$x = 251 + 490 = 741$$

The solution to the system is (741, 490).

Step 5: **Check** the answer, and **interpret** the solution as it relates to the problem.

The Dark Side of the Moon was on the Billboard 200 for 741 weeks, and *Johnny's Greatest Hits* was on the chart for 490 weeks.

They were on the chart for a total of $741 + 490 = 1231$ weeks, and the Pink Floyd album was on there $741 - 490 = 251$ weeks longer than the other album.

[YOU TRY 1]

Write a system of equations, and solve.
    In 2010, Carson City, NV, had about 26,100 more citizens than Elmira, NY. Find the population of each city if together they had approximately 84,500 residents. (www.census.gov)

Next we will see how we can use two variables and a system of equations to solve geometry problems.

> **W Hint**
> Try using the elimination method to solve this system of linear equations. Next, decide which method is easier for you!

2 Solve Geometry Problems

Write a system of equations, and solve.

A builder installed a rectangular window in a new house and needs 182 in. of trim to go around it on the inside of the house. Find the dimensions of the window if the width is 23 in. less than the length.

Solution

Step 1: **Read** the problem carefully, and identify what we are being asked to find. Draw a picture.

We must find the length and width of the window.

Step 2: **Choose variables** to represent the unknown quantities.

$$w = \text{the width of the window}$$
$$l = \text{the length of the window}$$

Label the picture with the variables.

Step 3: **Write a system of equations using two variables.**

To get one equation, we know that the width is 23 in. less than the length. We can write the equation $w = l - 23$.

If it takes 182 in. of trim to go around the window, this is the *perimeter* of the rectangular window. Use the equation for the perimeter of a rectangle.

$$2l + 2w = 182$$

The system of equations is
$$w = l - 23$$
$$2l + 2w = 182.$$

Step 4: **Solve** the system.

$$2l + 2w = 182$$
$$2l + 2(l - 23) = 182 \qquad \text{Substitute.}$$
$$2l + 2l - 46 = 182 \qquad \text{Distribute.}$$
$$4l - 46 = 182 \qquad \text{Combine like terms.}$$
$$4l = 228 \qquad \text{Add 46 to each side.}$$
$$\frac{4l}{4} = \frac{228}{4} \qquad \text{Divide each side by 4.}$$
$$l = 57 \qquad \text{Simplify.}$$

Find w by substituting $l = 57$ into $w = l - 23$.

$$w = 57 - 23 = 34$$

The solution to the system is (57, 34). (The ordered pair is written as (l, w), in alphabetical order.)

Step 5: **Check** the answer, and **interpret** the solution as it relates to the problem.

The length of the window is 57 in., and the width is 34 in. The check is left to the student.

[YOU TRY 2]

Write a system of equations and solve.

The top of a desk is twice as long as it is wide. If the perimeter of the desk is 162 in., find its dimensions.

---

3 Solve Problems Involving Cost

Write a system of equations, and solve.

Ari buys two mezzanine tickets to a Broadway play and four tickets to the top of the Empire State Building for $352. Lloyd spends $609 on four mezzanine tickets and three tickets to the top of the Empire State Building. Find the cost of a ticket to each attraction.

Solution

Step 1: **Read** the problem carefully, and identify what we are being asked to find.

We must find the cost of a ticket to a Broadway play and to the top of the Empire State Building.

Step 2: **Choose variables** to represent the unknown quantities.

x = the cost of a ticket to a Broadway play

y = the cost of a ticket to the Empire State Building

Step 3: **Write a system of equations using two variables.** First, let's think of the equations in English. Then we will translate them into algebraic equations.

First, use the information about Ari's purchase.

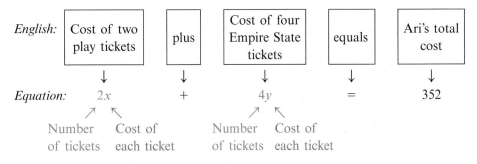

One equation is $2x + 4y = 352$.

Next, use the information about Lloyd's purchase.

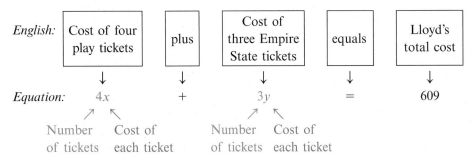

The other equation is $4x + 3y = 609$.

The system of equations is $2x + 4y = 352$
$4x + 3y = 609$.

Step 4: Solve the system.

Use the elimination method. Multiply the first equation by -2 to eliminate x.

$$\begin{array}{r} -4x - 8y = -704 \\ +\ \underline{4x + 3y = \quad 609} \\ -5y = -95 \qquad \text{Add the equations.} \\ y = 19 \end{array}$$

Find x. We will substitute $y = 19$ into $2x + 4y = 352$.

$$\begin{array}{rl} 2x + 4(19) = 352 & \text{Substitute.} \\ 2x + 76 = 352 & \text{Multiply.} \\ 2x = 276 & \text{Subtract 76 from each side.} \\ x = 138 & \end{array}$$

The solution to the system is (138, 19).

Step 5: Check the answer, and **interpret** the solution as it relates to the problem.

A Broadway play ticket costs $138.00, and a ticket to the top of the Empire State Building costs $19.00.

The check is left to the student.

[YOU TRY 3] Write a system of equations, and solve.

Torie bought three scarves and a belt for $105 while Liz bought two scarves and two belts for $98. Find the cost of a scarf and a belt.

Now we will learn how to solve mixture problems using two variables and a system of two equations.

4 Solve Mixture Problems

EXAMPLE 4

In-Class Example 4

How many ounces of a 6% alcohol solution and how many ounces of an 18% alcohol solution should be mixed to make 50 oz of a 10% alcohol solution?

Answer: 6%: $33\frac{1}{3}$ oz;

18%: $16\frac{2}{3}$ oz

EXAMPLE 4

A pharmacist needs to make 200 mL of a 10% hydrogen peroxide solution. She will make it from some 8% hydrogen peroxide solution and some 16% hydrogen peroxide solution that are in the storeroom. How much of the 8% solution and 16% solution should she use?

Solution

Step 1: Read the problem carefully, and identify what we are being asked to find. Draw a picture.

We must find the amount of 8% solution and 16% solution she should use.

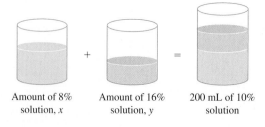

Amount of 8% solution, x +   Amount of 16% solution, y =   200 mL of 10% solution

---

W Hint

Remember that a hydrogen peroxide *solution* contains both pure hydrogen peroxide and pure water.

Step 2: **Choose variables** to represent the unknown quantities.

x = amount of 8% solution needed

y = amount of 16% solution needed

Step 3: **Write a system of equations using two variables.**

Let's begin by arranging the information in a table. To obtain the expression in the last column, multiply the percent of hydrogen peroxide in the solution by the amount of solution to get the amount of pure hydrogen peroxide in the solution.

	Percent of Hydrogen Peroxide in Solution (as a decimal)	Amount of Solution	Amount of Pure Hydrogen Peroxide in Solution
Mix these	0.08	x	$0.08x$
	0.16	y	$0.16y$
to make →	0.10	200	$0.10(200)$

To get one equation, use the information in the second column. It tells us that

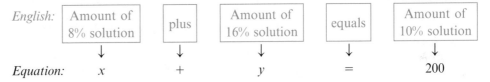

English: Amount of 8% solution **plus** Amount of 16% solution **equals** Amount of 10% solution

Equation: x $+$ y $=$ 200

The equation is $x + y = 200$.

To get the second equation, use the information in the third column. It tells us that

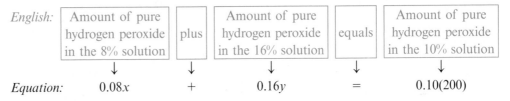

English: Amount of pure hydrogen peroxide in the 8% solution **plus** Amount of pure hydrogen peroxide in the 16% solution **equals** Amount of pure hydrogen peroxide in the 10% solution

Equation: $0.08x$ $+$ $0.16y$ $=$ $0.10(200)$

The equation is $0.08x + 0.16y = 0.10(200)$.

The system of equations is
$$x + y = 200$$
$$0.08x + 0.16y = 0.10(200).$$

Step 4: **Solve the system.**

Multiply the second equation by 100 to eliminate the decimals. Our system becomes

$$x + y = 200$$
$$8x + 16y = 2000$$

Use the elimination method. Multiply the first equation by -8 to eliminate x.

$$
\begin{aligned}
-8x - 8y &= -1600 \\
+ \quad 8x + 16y &= 2000 \\
\hline
8y &= 400 \\
y &= 50
\end{aligned}
$$

Find x. Substitute $y = 50$ into $x + y = 200$.

$$x + 50 = 200$$
$$x = 150$$

The solution to the system is (150, 50).

W Hint

Remember that x and y must have a sum of 200, and neither should be negative.

Step 5: **Check** the answer, and **interpret** the solution as it relates to the problem.

The pharmacist needs 150 mL of the 8% solution and 50 mL of the 16% solution. Check the answers in the original problem to verify that they are correct.

[**YOU TRY 4**]
Write an equation, and solve.

How many milliliters of a 5% acid solution and how many milliliters of a 17% acid solution must be mixed to obtain 60 mL of a 13% acid solution?

5 Solve Distance, Rate, and Time Problems

EXAMPLE 5

In-Class Example 5

Bart and Homer leave home on their bikes going in opposite directions. Bart pedals 2 mph faster than Homer, and after 1.5 hours they are 21 miles apart. Find each of their speeds.

Answer: Bart: 8 mph; Homer: 6 mph

Write an equation, and solve.

Two cars leave Kearney, Nebraska, one driving east and the other heading west. The eastbound car travels 4 mph faster than the westbound car, and after 2.5 hours they are 330 miles apart. Find the speed of each car.

Solution

Step 1: **Read** the problem carefully, and identify what we are being asked to find.

We must find the speed of the eastbound and westbound cars. We will draw a picture to help us see what is happening in this problem. After 2.5 hours, the positions of the cars look like this:

W Hint

Sketch a picture to visualize distance, rate, and time problems.

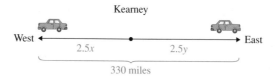

Step 2: **Choose variables** to represent the unknown quantities.

$$x = \text{the speed of the westbound car}$$
$$y = \text{the speed of the eastbound car}$$

Step 3: **Write a system of equations using two variables.**

Let's make a table using the equation $d = rt$. Fill in the time, 2.5 hr, and the rates first, then multiply those together to fill in the values for the distance.

	d	r	t
Westbound	$2.5x$	x	2.5
Eastbound	$2.5y$	y	2.5

Label the picture with the expressions for the distances.

To get one equation, look at the picture and think about the distance between the cars after 2.5 hr.

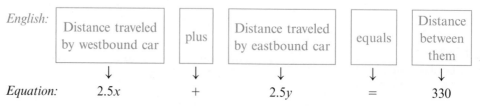

The equation is $2.5x + 2.5y = 330$.

To get the second equation, use the information that says the eastbound car goes 4 mph faster than the westbound car.

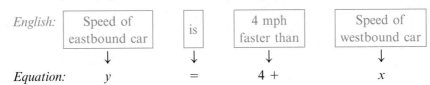

English: [Speed of eastbound car] [is] [4 mph faster than] [Speed of westbound car]

↓   ↓   ↓   ↓

Equation: y $=$ $4 +$ x

The equation is $y = 4 + x$.

The system of equations is $2.5x + 2.5y = 330$
$y = 4 + x.$

Step 4: Solve the system.

Use substitution.

$$2.5x + 2.5y = 330$$
$$2.5x + 2.5(4 + x) = 330 \qquad \text{Substitute } 4 + x \text{ for } y.$$
$$2.5x + 10 + 2.5x = 330 \qquad \text{Distribute.}$$
$$5x + 10 = 330 \qquad \text{Combine like terms.}$$
$$5x = 320 \qquad \text{Subtract 10 from each side.}$$
$$x = 64$$

Find y by substituting $x = 64$ into $y = 4 + x$.
$$y = 4 + 64 = 68$$

The solution to the system is (64, 68).

Step 5: Check the answer, and **interpret** the solution as it relates to the problem.

The speed of the westbound car is 64 mph, and the speed of the eastbound car is 68 mph.

Check.

Distance of westbound car

Distance of eastbound car

↓   ↓

$2.5(64) \quad + \quad 2.5(68) = 160 + 170 = 330 \text{ mi}$

> **W Hint**
> Remember that both x and y should be positive values when working with distance, rate, and time problems.

[YOU TRY 5]

Write an equation, and solve.

Two planes leave the same airport, one headed north and the other headed south. The northbound plane goes 100 mph slower than the southbound plane. Find each of their speeds if they are 1240 miles apart after 2 hours.

ANSWERS TO [YOU TRY] EXERCISES

1) Carson City: 55,300; Elmira: 29,200   2) width: 27 in.; length: 54 in.
3) scarf: $28; belt: $21   4) 20 mL of 5% solution; 40 mL of 17% solution
5) northbound plane: 260 mph; southbound plane: 360 mph

*Additional answers can be found in the Answers to Exercises appendix.

Objective 1: Solve Problems Involving General Quantities

Write a system of equations, and solve.

1) The sum of two numbers is 87, and one number is eleven more than the other. Find the numbers.
38 and 49

2) One number is half another number. The sum of the two numbers is 141. Find the numbers.   94 and 47

3) Through 2011, *Harry Potter and the Deathly Hallows: Part 2* earned more money on its opening weekend in the United States than any other movie. It earned $100.8 million more than *The Twilight Saga: Eclipse* on its opening weekend. Together, they earned $237.6 million. How much did each film earn on opening day? (http://boxofficemojo.com)
Harry Potter: $169.2 million; *Twilight:* $68.4 million

4) In the 1976–1977 season, Kareem Abdul-Jabbar led all players in blocked shots. He blocked 50 more shots than Bill Walton, who finished in second place. How many shots did each man block if they rejected a total of 472 shots? (www.nba.com)
Kareem: 261; Bill: 211

5) Through 2011, Beyonce had been nominated for four more BET Awards than Jay-Z. Determine how many nominations each performer received if they got a total of 54 nominations. (www.bet.com)
Beyonce: 29; Jay-Z: 25

6) From 1965 to 2000, twice as many people immigrated to the United States from the Philippines as from Vietnam. The total number of immigrants from these two countries was 2,100,000. How many people came to the United States from each country?
(www.ellisisland.org)   Vietnam: 700,000; Philippines: 1,400,000

7) According to a U.S. Census Bureau survey in 2009, about half as many people in the United States spoke Vietnamese at home as spoke Chinese. If a total of about 3,900,000 people spoke these languages in their homes, how many spoke Vietnamese and how many spoke Chinese? (www.census.gov)
Vietnamese: 1,300,000; Chinese: 2,600,000

8) During one week, a hardware store sold 27 fewer "regular" incandescent lightbulbs than energy-efficient compact fluorescent light (CFL) bulbs. How many of each type of bulb was sold if the store sold a total of 79 of these two types of lightbulbs?
incandescent: 26; CFL: 53

9) On April 12, 1961, Yuri Gagarin of the Soviet Union became the first person in space when he piloted the Vostok 1 mission. The next month, Alan B. Shepard became the first American in space in the Freedom 7 space capsule. The two of them spent a total of about 123 minutes in space, with Gagarin logging 93 more minutes than Shepard. How long did each man spend in space?
(www-pao.ksc.nasa.gov, www.enchantedlearning.com)
Gagarin: 108 min; Shepard: 15 min

10) Mr. Monet has 85 students in his Art History lecture. For their assignment on impressionists, one-fourth as many students chose to recreate an impressionist painting as chose to write a paper. How many students will be painting, and how many will be writing papers?   painting: 17; papers: 68

Objective 2: Solve Geometry Problems

11) Find the dimensions of a rectangular door that has a perimeter of 220 in. if the width is 50 in. less than the height of the door.   width: 30 in.; height: 80 in.

12) The length of a rectangle is 3.5 in. more than its width. If the perimeter is 23 in., what are the dimensions of the rectangle?   length: 7.5 in.; width: 4 in.

13) An iPod Touch is rectangular in shape and has a perimeter of 343.6 mm. Find its length and width given that it is 48.2 mm longer than it is wide.
length: 110 mm; width: 61.8 mm

14) Eliza needs 332 in. of a decorative border to sew around a rectangular quilt she just made. Its width is 26 in. less than its length. Find the dimensions of the quilt.   length: 96 in.; width: 70 in.

15) A rectangular horse corral is bordered on one side by a barn as pictured here. The length of the corral is 1.5 times the width. If 119 ft of fencing was used to make the corral, what are its dimensions?
width: 34 ft; length: 51 ft

16) The length of a rectangular mirror is twice its width. Find the dimensions of the mirror if its perimeter is 246 cm.   length: 82 cm; width: 41 cm

17) Find the measures of angles x and y if the measure of angle x is three-fifths the measure of angle y and if the angles are related according to the figure.
$m\angle x = 67.5°$; $m\angle y = 112.5°$

18) Find the measures of angles x and y if the measure of angle y is two-thirds the measure of angle x and if the angles are related according to the figure.
$m\angle x = 63°$; $m\angle y = 42°$

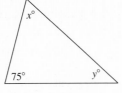

Objective 3: Solve Problems Involving Cost

19) Kenny and Kyle are huge Colorado Avalanche fans. Kenny buys a T-shirt and two souvenir hockey pucks for $36.00, and Kyle spends $64.00 on two T-shirts and three pucks. Find the price of a T-shirt and the price of a puck. T-shirt: $20.00; hockey puck: $8.00

20) Bruce Springsteen and Jimmy Buffett each played in Chicago in 2009. Four Springsteen tickets and four Buffett tickets cost $908.00, while three Springsteen tickets and two Buffett tickets cost $552.00. Find the cost of a ticket to each concert. (www.ticketmaster.com) Springsteen: $98.00; Buffett: $129.00

21) Angela and Andy watch *The Office* every week with their friends and decide to buy them some gifts. Angela buys three Dwight bobbleheads and four star mugs for $105.00, while Andy spends $74.00 on two bobbleheads and three mugs. Find the cost of each item. (www.nbcuniversalstore.com)
bobblehead: $19.00; mug: $12.00

22) Manny and Hiroki buy tickets in advance to some Los Angeles Dodgers games. Manny buys three left-field pavilion seats and six club seats for $291. Hiroki spends $292 on eight left-field pavilion seats and five club seats. Find the cost of each type of ticket. (www.dodgers.com) left-field pavilion: $9; club: $44

23) Carol orders five White Castle hamburgers and a medium order of french fries for $4.64, and Momar orders four hamburgers and two medium fries for $5.38. Find the cost of a hamburger and the cost of a medium order of french fries at White Castle. (White Castle menu)
hamburger: $0.65; medium fries: $1.39

24) Phuong buys New Jersey lottery tickets every Friday. One day she spent $17.00 on four Gold Rush tickets and three Crossword tickets. The next Friday, she bought three Gold Rush tickets and six Crossword tickets for $24.00. How much did she pay for each type of lottery ticket?
Gold Rush: $2.00; Crossword: $3.00

25) Lakeisha is selling wrapping paper products for a school fund-raiser. Her mom buys four rolls of wrapping paper and three packages of gift bags for $52.00. Her grandmother spends $29.00 on three rolls of wrapping paper and one package of gift bags. Find the cost of a roll of wrapping paper and a package of gift bags.
wrapping paper: $7.00; gift bags: $8.00

26) Alberto is selling popcorn to raise money for his Cub Scout den. His dad spends $86.00 on two tins of popcorn and three tins of caramel corn. His neighbor buys two tins of popcorn and one tin of caramel corn for $48.00. How much does each type of treat cost? popcorn: $14.50; caramel corn: $19.00

Objective 4: Solve Mixture Problems

27) How many ounces of a 9% alcohol solution and how many ounces of a 17% alcohol solution must be mixed to obtain 12 oz of a 15% alcohol solution?
9%: 3 oz; 17%: 9 oz

28) How many milliliters of a 15% acid solution and how many milliliters of a 3% acid solution must be mixed to get 45 mL of a 7% alcohol solution?
15%: 15 mL; 3%: 30 mL

29) How many liters of pure acid and how many liters of a 25% acid solution should be mixed to get 10 L of a 40% acid solution? pure acid: 2 L; 25%: 8 L

30) How many ounces of pure cranberry juice and how many ounces of a citrus fruit drink containing 10% fruit juice should be mixed to get 120 oz of a fruit drink that is 25% fruit juice?
cranberry juice: 20 oz; fruit drink: 100 oz

31) How many ounces of Asian Treasure tea that sells for $7.50/oz should be mixed with Pearadise tea that sells for $5.00/oz so that a 60-oz mixture is obtained that will sell for $6.00/oz?
Asian Treasure: 24 oz; Pearadise: 36 oz

32) How many pounds of peanuts that sell for $1.80 per pound should be mixed with cashews that sell for $4.50 per pound so that a 10-pound mixture is obtained that will sell for $2.61 per pound?
peanuts: 7 lb; cashews: 3 lb

33) During a late-night visit to Taco Bell, Giovanni orders three Crunchy Tacos and a chicken chalupa supreme. His order contains 1390 mg of sodium. Jurgis orders two Crunchy Tacos and two chicken

chalupa supremes, and his order contains 1620 mg of sodium. How much sodium is in each item? (www.tacobell.com)   taco: 290 mg; chalupa: 520 mg

34) Five White Castle hamburgers and one medium order of french fries contain 1070 calories. Four hamburgers and two orders of fries contain 1300 calories. Determine how many calories are in a White Castle hamburger and in a medium order of french fries. (www.whitecastle.com)
hamburger: 140; medium fries: 370

35) Mahmud invested $6000 in two accounts, some of it at 2% simple interest, the rest in an account earning 4% simple interest. How much did he invest in each account if he earned $190 in interest after 1 year?
2%: $2500; 4%: $3500

36) Marijke inherited $15,000 and puts some of it into an account earning 5% simple interest and the rest in an account earning 4% simple interest. She earns a total of $660 in interest after 1 year. How much did she deposit into each account?
4%: $9000; 5%: $6000

37) Oscar purchased 16 stamps. He bought some $0.44 stamps and some $0.28 stamps and spent $6.40. How many of each type of stamp did he buy?
$0.44: 12; $0.28: 4

38) Kelly saves all of her dimes and nickels in a jar on her desk. When she counts her money, she finds that she has 133 coins worth a total of $10.45. How many dimes and how many nickels does she have?   76 dimes, 57 nickels

Objective 5: Solve Distance, Rate, and Time Problems

39) Michael and Jan leave the same location but head in opposite directions on their bikes. Michael rides 1 mph faster than Jan, and after 3 hr they are 51 miles apart. How fast was each of them riding?
Michael: 9 mph; Jan: 8 mph

40) A passenger train and a freight train leave cities 400 miles apart and travel toward each other. The passenger train is traveling 16 mph faster than the freight train. Find the speed of each train if they pass each other after 5 hours.
passenger train: 48 mph; freight train: 32 mph

41) A small plane leaves an airport and heads south, while a jet takes off at the same time heading north. The speed of the small plane is 160 mph less than the speed of the jet. If they are 1280 miles apart after 2 hours, find the speeds of both planes.   small plane: 240 mph; jet: 400 mph

42) Tyreese and Justine start jogging toward each other from opposite ends of a trail 6.5 miles apart. They meet after 30 minutes. Find their speeds if Tyreese jogs 3 mph faster than Justine.
Tyreese: 8 mph; Justine: 5 mph

43) Pam and Jim leave from opposite ends of a bike trail 9 miles apart and travel toward each other. Pam is traveling 2 mph slower than Jim. Find each of their speeds if they meet after 30 minutes.
Pam: 8 mph; Jim: 10 mph

44) Stanley and Phyllis leave the office and travel in opposite directions. Stanley drives 6 mph slower than Phyllis, and after 1 hr they are 76 miles apart. How fast was each person driving?
Stanley: 35 mph; Phyllis: 41 mph

Other types of distance, rate, and time problems involve a boat traveling upstream and downstream, and a plane traveling with and against the wind. To solve problems like these, we will still use a table to help us organize our information, but we must understand what is happening in such problems.

Let's think about the case of a boat traveling upstream and downstream.

Let x = the speed of the boat in still water and let y = the speed of the current.

When the boat is going *downstream* (*with* the current), the boat is being pushed along by the current so that

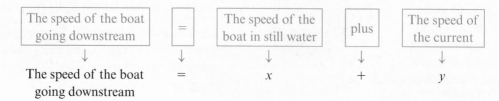

When the boat is going *upstream* (*against* the current), the boat is being slowed down by the current so that

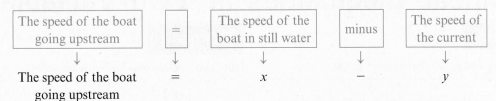

The speed of the boat going upstream	=	The speed of the boat in still water	minus	The speed of the current
↓	↓	↓	↓	↓
The speed of the boat going upstream	=	x	−	y

Use this idea to solve Exercises 45–50.

45) It takes 2 hours for a boat to travel 14 miles downstream. The boat can travel 10 miles upstream in the same amount of time. Find the speed of the boat in still water and the speed of the current. (Hint: Use the information in the following table, and write a system of equations.)
speed of boat in still water: 6 mph; speed of the current: 1 mph

	d	r	t
Downstream	14	$x + y$	2
Upstream	10	$x - y$	2

46) A boat can travel 15 miles downstream in 0.75 hours. It takes the same amount of time for the boat to travel 9 miles upstream. Find the speed of the boat in still water and the speed of the current. (Hint: Use the information in the following table, and write a system of equations.)
speed of boat in still water: 16 mph; speed of the current: 4 mph

	d	r	t
Downstream	15	$x + y$	0.75
Upstream	9	$x - y$	0.75

47) It takes 5 hours for a boat to travel 80 miles downstream. The boat can travel the same distance back upstream in 8 hours. Find the speed of the boat in still water and the speed of the current.
speed of boat in still water: 13 mph; speed of the current: 3 mph

48) A boat can travel 12 miles downstream in 1.5 hours. It takes 3 hours for the boat to travel back to the same spot going upstream. Find the speed of the boat in still water and the speed of the current.
speed of boat in still water: 6 mph; speed of the current: 2 mph

49) A jet can travel 1000 miles against the wind in 2.5 hours. Going with the wind, the jet could travel 1250 miles in the same amount of time. Find the speed of the jet in still air and speed of the wind.
speed of jet in still air: 450 mph; speed of the wind: 50 mph

50) It takes 2 hours for a small plane to travel 390 miles with the wind. Going against the wind, the plane can travel 330 miles in the same amount of time. Find the speed of the plane in still air and the speed of the wind.
speed of plane in still air: 180 mph; speed of the wind: 15 mph

R Rethink

R1) If you have two unknowns in an application problem, how many equations do you need to solve the problem?

R2) In which of your future courses do you think you will need to solve a system of linear equations?

R3) Did you check your answers by hand *before* looking at the answers in the back of the book?

R4) Choose six problems and redo them without looking back at the book or your notes for help. Were you able to do them on your own?

4.5 Linear Inequalities in Two Variables

 Prepare   **Organize**

What are your objectives for Section 4.5?	How can you accomplish each objective?
1 Define a Linear Inequality in Two Variables	• Understand the definition of a *linear inequality in two variables,* and write it in your notes. • Complete the given example on your own. • Complete You Try 1.
2 Graph a Linear Inequality in Two Variables	• Understand the definitions of a *half plane, boundary line,* and *test point.* • Learn the procedure for **Graphing a Linear Inequality in Two Variables Using the Test Point Method,** and write it in your notes. • Learn the procedure for **Using Slope-Intercept Form to Graph a Linear Inequality in Two Variables,** and write it in your notes. • Complete the given examples on your own. • Complete You Trys 2 and 3.
3 Solve a System of Linear Inequalities by Graphing	• Learn the procedure for **Solving a System of Linear Inequalities by Graphing,** and write it in your notes. • Complete the given examples on your own. • Complete You Trys 4–6.

W   **Work**   **Read the explanations, follow the examples, take notes, and complete the You Trys.**

In Chapter 2, we first learned how to solve linear inequalities in *one variable* such as $4x + 9 \geq 7$. In this section, we will first learn how to graph the solution set of a linear inequality in *two variables*. Then we will learn how to graph the solution set of *systems* of linear inequalities in two variables.

1   Define a Linear Inequality in Two Variables

Definition

A **linear inequality in two variables** is an inequality that can be written in the form $Ax + By \geq C$ or $Ax + By \leq C$, where A, B, and C are real numbers and where A and B are not both zero. ($>$ and $<$ may be substituted for $\geq$ and $\leq$.)

Here are some examples of linear inequalities in two variables.

$$2x - 5y \geq 3, \qquad y < \frac{1}{5}x + 2, \qquad x \leq 6, \qquad y > -1$$

Note

$x \leq 6$ can be considered a linear inequality in two variables because it can be written as $x + 0y \leq 6$. Likewise, $y > -1$ can be written as $0x + y > -1$.

The solutions to linear inequalities in two variables, such as $x + y \geq 1$, are *ordered pairs* of the form (x, y) that make the inequality true. We graph a linear inequality in two variables on a rectangular coordinate system.

EXAMPLE 1

In-Class Example 1

Shown here is the graph of $x - y \leq 1$. Find three points that solve $x - y \leq 1$ and three points that are not in the solution set.

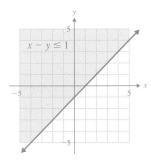

Answer: in solution set $(0, 2)$, $(0, -1)$, $(2, 4)$; not in solution set $(0, -2)$, $(4, 1)$, $(-2, -5)$

Shown here is the graph of $x + y \geq 1$. Find three points that solve $x + y \geq 1$, and find three points that are not in the solution set.

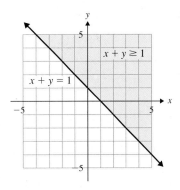

W Hint

For the inequality in Example 1, notice that if you choose *any* ordered pair in the shaded region or on the line and substitute its x- and y-values into the inequality, you will get a true statement.

Solution

The solution set of $x + y \geq 1$ consists of all points either on the line or in the shaded region. *Any* point on the line or in the shaded region will make $x + y \geq 1$ true.

Solutions of $x + y \geq 1$	Check by Substituting into $x + y \geq 1$
$(-1, 3)$	$-1 + 3 \geq 1$ True
$(4, 1)$	$4 + 1 \geq 1$ True
$(2, -1)$ (on the line)	$2 + (-1) \geq 1$ True

$(-1, 3)$, $(4, 1)$, and $(2, -1)$ are just some points that satisfy $x + y \geq 1$. There are infinitely many solutions.

Not in the Solution Set of $x + y \geq 1$	Verify by Substituting into $x + y \geq 1$
$(0, 0)$	$0 + 0 \geq 1$ False
$(-4, 1)$	$-4 + 1 \geq 1$ False
$(2, -3)$	$2 + (-3) \geq 1$ False

$(0, 0)$, $(-4, 1)$, and $(2, -3)$ are just three points that do not satisfy $x + y \geq 1$. There are infinitely many such points.

W Hint

If the inequality in Example 1 had been $x + y > 1$, then the line would have been drawn as a *dotted line,* and all points on the line would *not* be part of the solution set.

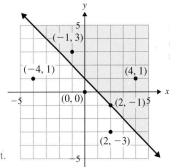

Points in the shaded region and on the line are in the solution set.

The points in the unshaded region are *not* in the solution set.

[YOU TRY 1] Shown here is the graph of $2x - 3y \geq 6$. Find three points that solve $2x - 3y \geq 6$, and find three points that are not in the solution set.

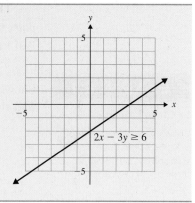

$2x - 3y \geq 6$

2 Graph a Linear Inequality in Two Variables

As you saw in the graph in Example 1, the line divides the *plane* into two regions or **half planes.** The line $x + y = 1$ is the **boundary line** between the two half planes. We can use this boundary and two different methods to graph a linear inequality in two variables. The first method we will discuss is the **test point** method.

> **Procedure** Graphing a Linear Inequality in Two Variables Using the Test Point Method
>
> **Step 1:** **Graph the boundary line.** If the inequality contains $\geq$ or $\leq$, make it a *solid line*. If the inequality contains $>$ or $<$, make it a *dotted line*.
>
> **Step 2:** **Choose a test point not on the line, and shade the appropriate region.** Substitute the test point into the inequality. If $(0, 0)$ is not on the line, it is an easy point to test in the inequality.
>
>   a) If it *makes the inequality true,* shade the side of the line *containing* the test point. All points in the shaded region are part of the solution set.
>
>   b) If the test point *does not satisfy the inequality,* shade the *other* side of the line. All points in the shaded region are part of the solution set.

EXAMPLE 2

Graph $3x + 2y \leq -6$.

In-Class Example 2

Graph $-x + y \leq 2$.

Answer:

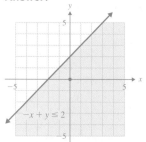

$-x + y \leq 2$

Solution

Step 1: Graph the boundary line $3x + 2y = -6$ as a solid line.

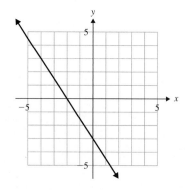

Step 2: Choose a test point not on the line, and substitute it into the inequality to determine whether or not it makes the inequality true.

Test Point	Substituting into $3x + 2y \leq -6$
(0, 0)	$3(0) + 2(0) \leq -6$
	$0 + 0 \leq -6$
	$0 \leq -6$ False

Since the test point (0, 0) does *not* satisfy the inequality, we will shade the side of the line that does *not* contain the point (0, 0).

W Hint

Don't forget that you need to choose a test point to determine which region to shade!

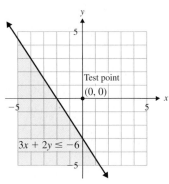

All points on the line and in the shaded region satisfy the inequality $3x + 2y \leq -6$.

EXAMPLE 3

In-Class Example 3

Graph $-2x + y > -4$.

Answer:

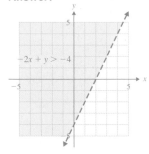

Graph $-x + 3y > -3$.

Solution

Step 1: Since the inequality symbol is $>$ and not $\geq$, graph the boundary line $-x + 3y = -3$ as a dotted line. (This means that the points *on* the line are not part of the solution set.)

Step 2: Choose a test point not on the line, and substitute it into the inequality to determine whether or not it makes the inequality true.

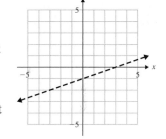

Test Point	Substitute into $-x + 3y > -3$
(0, 0)	$-(0) + 3(0) > -3$
	$0 > -3$ True

Since the test point (0, 0) satisfies the inequality, shade the side of the line containing the point (0, 0).

W Hint

Remember that drawing a boundary line as a *dashed* line means that any point *on* the boundary line will make the inequality false!

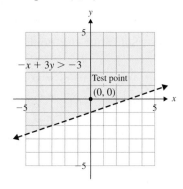

All points in the shaded region satisfy the inequality $-x + 3y > -3$.

Graph each inequality.

a) $-x + 3y \geq 1$ b) $x + 2y < 4$

Another method we can use to graph linear inequalities in two variables involves writing the boundary line in *slope-intercept form.*

> **Procedure** Using Slope-Intercept Form to Graph a Linear Inequality in Two Variables
>
> **Step 1:** Write the inequality in the form $y \geq mx + b$ ($y > mx + b$) or $y \leq mx + b$ ($y < mx + b$).
>
> **Step 2:** Graph the boundary line $y = mx + b$.
>
>   a) If the inequality contains $\geq$ or $\leq$, make it a *solid line.*
>
>   b) If the inequality contains $>$ or $<$, make it a *dotted line.*
>
> **Step 3:** Shade the appropriate side of the line.
>
>   a) If the inequality is in the form $y \geq mx + b$ or $y > mx + b$, shade *above* the line.
>
>   b) If the inequality is in the form $y \leq mx + b$ or $y < mx + b$, shade *below* the line.

EXAMPLE 4

In-Class Example 4

Graph the inequality

$y > \dfrac{5}{2}x - 1$ using the

slope-intercept method.

Answer:

a)

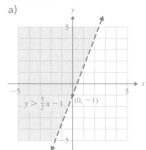

$y > \frac{5}{2}x - 1$ $(0, -1)$

Graph the inequality $y < -\dfrac{1}{3}x + 2$ using the slope-intercept method.

Solution

a) **Step 1:** $y < -\dfrac{1}{3}x + 2$ is already in the correct form.

     Step 2: Graph the boundary line $y = -\dfrac{1}{3}x + 2$ as a *dotted line.*

     Step 3: Since $y < -\dfrac{1}{3}x + 2$ has a *less than* symbol, shade *below* the line.

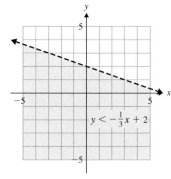

$y < -\frac{1}{3}x + 2$

All points in the shaded region satisfy $y < -\dfrac{1}{3}x + 2$. We can choose a point, such as $(0, 0)$, in the shaded region as a check. Substitute this point into the inequality.

$$0 < -\frac{1}{3}(0) + 2$$
$$0 < 0 + 2$$
$$0 < 2 \qquad \text{True}$$

[**YOU TRY 3**] Graph each inequality using the slope-intercept method.

a) $y \leq -\dfrac{3}{2}x + 2$ b) $3x - y < -1$

3 Solve a System of Linear Inequalities by Graphing

A **system of linear inequalities** contains two or more linear inequalities. The **solution set of a system of linear inequalities** consists of all the ordered pairs that make all the inequalities in the system true. To solve such a system, we will graph each inequality on the same axes. The region where the solutions intersect, or overlap, is the solution to the system. We use the following steps to solve a system of linear inequalities:

> **Procedure** Solving a System of Linear Inequalities by Graphing
>
> 1) Graph each inequality separately on the same axes. Shade lightly.
> 2) The solution set is the *intersection* (overlap) of the shaded regions. Heavily shade this region to indicate this is the solution set.

EXAMPLE 5

In-Class Example 5

Graph the solution set of the system.

$$x \geq -1$$
$$-2x + y \geq 1$$

Answer:

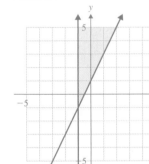

[W] **Hint**

The final solution is the set of ordered pairs that lie in the *overlapping* shaded region.

Graph the solution set of the system.

$$x \leq 1$$
$$2x + 3y \geq -3$$

Solution

To graph $x \leq 1$, graph the boundary line $x = 1$ as a solid line. The x-values are *less than* 1 to the *left* of 1, so shade to the left of the line $x = 1$.

Graph $2x + 3y \geq -3$ as shown in the middle graph. **The region shaded purple in the third graph is the *intersection* of the shaded regions and the *solution set* of the system.**

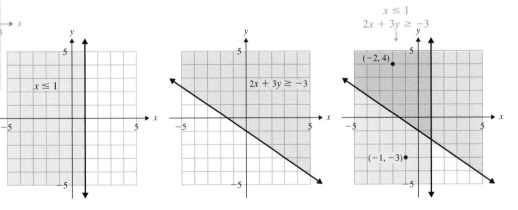

Any point inside the purple area will satisfy *both* inequalities. For example, the point $(-2, 4)$ is in this region (see the graph).

Any point outside this region of intersection will not satisfy *both* inequalities and is *not* part of the system's solution set. One such point is $(-1, -3)$.

Although we show three separate graphs in this example, it is customary to graph everything on the same axes, shading lightly at first, then to heavily shade the region that is the graph of the solution set.

[YOU TRY 4] Graph the solution set of the system.

$$y \geq 2$$
$$3x + 2y \leq -2$$

EXAMPLE 6 Graph the solution set of the system.

$$y < x + 2$$
$$-2x - y > -1$$

In-Class Example 6

Graph the solution set of the system.
$$y > x - 4$$
$$2x + y < -2$$

Answer:

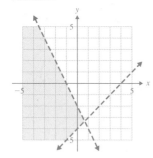

Solution

Graph each inequality separately, and lightly shade the solution set of each of them. (Notice that we use dotted lines for the boundary lines because of the inequality symbols.) The solution set of the system is the intersection (overlap) of the shaded regions. This is the purple region. The points on the boundary lines are not included in the solution set.

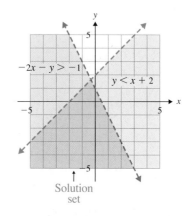

[YOU TRY 5] Graph the solution set of the system.

$$y > -\frac{1}{2}x + 2$$
$$x - y > -3$$

Next we will solve a system that contains more than two inequalities.

EXAMPLE 7 Graph the solution set of the system.

$$x \geq 0$$
$$y \geq 0$$
$$y \leq -\frac{4}{3}x + 4$$

In-Class Example 7

Graph the solution set of the system.
$$x \geq 0$$
$$y \geq 0$$
$$y \leq -\frac{3}{2}x + 3$$

Answer:

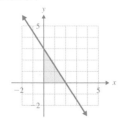

Solution

The inequalities $x \geq 0$ and $y \geq 0$ tell us that the solution set will be in the first quadrant since this is where both x and y are positive. (The axes may be included since $x = 0$ is the y-axis and $y = 0$ is the x-axis.)

Graph the third inequality. The solution set of the system is the region shaded in purple. The points on the boundary lines are included in the solution set.

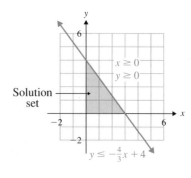

Graph the solution set of the system.

$$x \geq 0$$
$$y \geq 0$$
$$y \leq -\frac{1}{2}x + 1$$

Using Technology

A graphing calculator can be used to graph one or more linear inequalities in two variables.

To graph a linear inequality in two variables using a graphing calculator, first solve the inequality for y. Then graph the boundary line found by replacing the inequality symbol with an $=$ symbol. For example, to graph the inequality $2x - y \leq 5$, solve the inequality for y giving $y \geq 2x - 5$. Next, graph the boundary equation $y = 2x - 5$ using a solid line since the inequality symbol is $\leq$. Enter $2x - 5$ in Y1, press $\boxed{\text{ZOOM}}$, and select 6:ZStandard to graph the equation.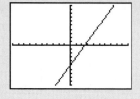

To shade above the line, press $\boxed{\text{Y=}}$ and move the cursor to the left of Y1 using the left arrow key. Press $\boxed{\text{ENTER}}$ twice, and then move the cursor to the next line as shown next on the left. Press $\boxed{\text{GRAPH}}$ to see the inequality as shown next on the right.

   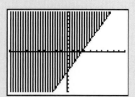

To shade below the line, press $\boxed{\text{Y=}}$ and move the cursor to the left of Y1 using the left arrow key. Press $\boxed{\text{ENTER}}$ three times, and then move the cursor to the next line as shown next on the left. Press $\boxed{\text{GRAPH}}$ to see the inequality $y \leq 2x - 5$, as shown next on the right.

   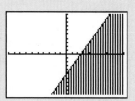

Graph the linear inequalities in two variables.

1) $y \leq 3x - 1$ 2) $y \geq x + 2$ 3) $2x - y \leq 4$

4) $x - y \geq 3$ 5) $y \leq -5x - 3$ 6) $4 - y \leq 3x$

ANSWERS TO [YOU TRY] EXERCISES

1) Answers may vary; in solution set: $(-2, -4)$, $(0, -2)$, $(4, -1)$;
   not in solution set: $(0, 0)$, $(-3, 1)$, $(2, 3)$

2) a)

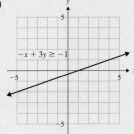

$-x + 3y \geq -1$

b)

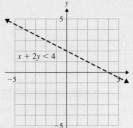

$x + 2y < 4$

3) a)

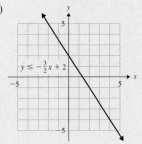

$y \leq -\frac{3}{2}x + 2$

b)

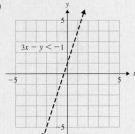

$3x - y < -1$

4)

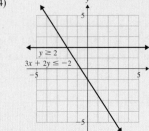

$y \geq 2$
$3x + 2y \leq -2$

5)

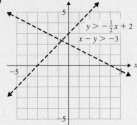

$y > -\frac{1}{2}x + 2$
$x - y > -3$

6)

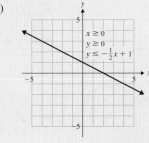

$x \geq 0$
$y \geq 0$
$y \leq -\frac{1}{2}x + 1$

ANSWERS TO TECHNOLOGY EXERCISES

1)

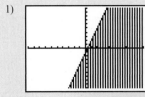

2)

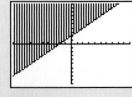

3)

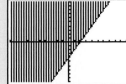

4)

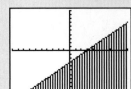

5)

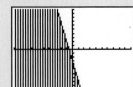

6)

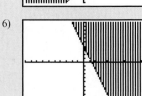

www.mhhe.com/messersmith

Do the exercises, and check your work.

*Additional answers can be found in the Answers to Exercises appendix.

Objective 1: Define a Linear Inequality in Two Variables

The graphs of linear inequalities are given in Exercises 1–6. For each, find three points that satisfy the inequality and three that are not in the solution set.

1–6, answers may vary.

1) $3x - y \geq -1$

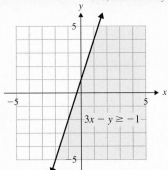

2) $4x + 3y \geq -24$

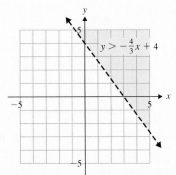

 3) $y > -\dfrac{4}{3}x + 4$

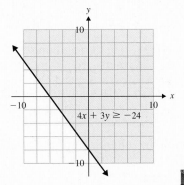

4) $y < \dfrac{1}{4}x - 2$

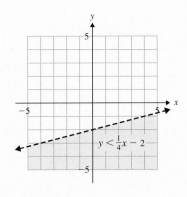

5) $y < x$

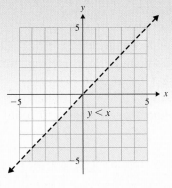

6) $x \geq 3$

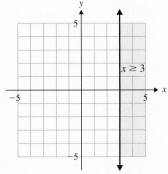

Objective 2: Graph a Linear Inequality in Two Variables

Graph using the test point method.

7) $2x + 3y \geq 6$

8) $x + 2y \leq 2$

9) $y < -x + 4$

10) $y > \dfrac{1}{2}x + 3$

11) $2x - 7y \leq 14$

12) $4x + 3y < 15$

13) $y < x$

14) $y \geq -\dfrac{2}{3}x$

15) $x \leq 4$

16) $y > -3$

Use the slope-intercept form to graph each inequality.

17) $y \leq x + 3$

18) $y \geq -2x + 1$

19) $y > \dfrac{2}{5}x - 4$

20) $y < \dfrac{1}{4}x + 1$

21) $4x + y < 2$

22) $-3x + y > -4$

23) $9x - 3y \leq -21$

24) $3x + 5y < -20$

25) $x \leq \dfrac{2}{3}y$

26) $x - 2y \geq 0$

27) To graph an inequality like $y \leq \frac{3}{4}x - 5$, which method, test point or slope-intercept, would you prefer? Why? Answers may vary.

28) To graph an inequality like $5x + 8y > 12$, which method, test point or slope-intercept, would you prefer? Why? Answers may vary.

Graph using either the test point or slope-intercept method.

29) $y > \frac{1}{3}x + 1$ 30) $y \leq -\frac{3}{4}x - 2$

31) $5x + 2y < -8$ 32) $4x + y < 7$

33) $9x - 3y \leq 21$ 34) $5x - 3y \geq -9$

35) $y > \frac{5}{2}$ 36) $x \leq -\frac{1}{2}$

37) $x - 2y \geq -5$ 38) $3x + 2y > -8$

Objective 3: Solve a System of Linear Inequalities by Graphing

The graphs of systems of linear inequalities are given next. For each, find three points that are in the solution set and three that are not. 39–42, answers may vary.

39) $y \geq \frac{4}{5}x + 2$

$y < 5$

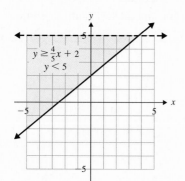

40) $x > 4$

$y \leq -\frac{2}{3}x + 2$

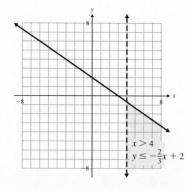

41) $x + 3y \leq 2$

$y - x \leq 0$

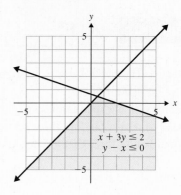

42) $3x - 2y \leq 6$

$3x + y \geq 1$

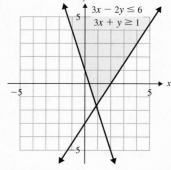

43) Is $(9, -2)$ in the solution set of the system $4x + 3y \geq 27$ and $x - 4y \geq -4$? Why or why not? Yes; it satisfies both inequalities.

44) Is $(3, 7)$ in the solution set of the system $4x + 3y \geq 27$ and $x - 4y \geq -4$? Why or why not? No; it satisfies only the first inequality.

Graph the solution set of each system.

45) $x \leq 4$

$y \geq -\frac{3}{2}x + 3$

46) $y \leq \frac{1}{4}x + 2$

$y \geq -1$

47) $y < x + 3$

$y \geq -1$

48) $x < 2$

$y > \frac{3}{2}x - 4$

49) $y < 4$

$2x - y < -2$

50) $x - 3y \geq 0$

$x \leq -3$

51) $y \leq 4x - 1$

$y \leq -3x + 3$

52) $y > -x + 2$

$y > 2x + 1$

53) $y > -\frac{3}{2}x + 2$

$y < 2x - 3$

54) $y \leq \frac{1}{4}x + 3$

$y \geq -\frac{1}{3}x + 4$

55) $y \geq \frac{2}{3}x - 4$

$4x + y \leq 3$

56) $y - 2x \leq 1$

$y \geq -\frac{1}{5}x - 2$

57) $5x - 3y > 9$
   $2x + 3y \le 12$

58) $2x - 3y < -9$
   $x + 6y < 12$

24) 63) $x \ge 0$
   $y \ge 0$
   $y \le -3x + 4$

64) $x \ge 0$
   $y \ge 0$
   $y \le -x + 1$

59) $x \le 6$
   $y \ge 1$

60) $x \ge 5$
   $y \le -3$

65) $x \ge 0$
   $y \ge 0$
   $2x - y \ge 3$

66) $x \ge 0$
   $y \ge 0$
   $2x - 3y \le -3$

61) $x \ge 0$
   $y \ge 0$
   $y \le -x + 2$

62) $x \ge 0$
   $y \ge 0$
   $y \le -\dfrac{2}{3}x + 3$

R Rethink

R1) When solving a linear inequality in two variables, how do you represent the solution set?

R2) When solving a system of linear inequalities in two variables, how do you represent the solution set?

R3) When do you draw a *solid* boundary line? When do you draw a *dashed* boundary line?

R4) Which objective do you need to spend more time on? Which exercises should you review and rework?

Group Activity — The Group Activity can be found online on Connect.

"What does that say?" "What calculation was I doing there?" "Why do I have ten pages of notes for a 40-minute class?"

For too many students, these are the questions that arise when they review their class notes. It is essential to develop your note-taking abilities so that your notes are clear and organized and so that they reflect the most important material that was covered in class. To get a sense of your current note-taking abilities, take a set of notes from a recent class and evaluate it based on the following criteria.

Statement	Not Even Slightly	Slightly	Moderately	Pretty Well	Very Well
1. I can read my notes (i.e., they are legible).					
2. Someone else can read my notes.					
3. My notes are complete; I missed nothing important.					
4. My notes represent the key points that were covered in class.					
5. My notes reflect what the instructor emphasizes.					
6. The instructor's key points are clear and understandable.					
7. The notes contain only important points, with no extraneous material.					
8. I understand not only the notes but the class content they reflect.					
9. Using only the notes, I will be able to reconstruct the essential content of the class in 3 months.					

- What do your answers tell you about the effectiveness of your note-taking skills?

- What might you do differently the next time you take notes?

Chapter 4:   Summary

Definition/Procedure	Example

4.1 Solving Systems by Graphing

A **system of linear equations** consists of two or more linear equations with the same variables. A **solution of a system** of two equations in two variables is an ordered pair that is a solution of each equation in the system. **(p. 245)**

Determine whether $(4, 2)$ is a solution of the system

$$x + 2y = 8$$
$$-3x + 4y = -4$$

$x + 2y = 8$		$-3x + 4y = -4$	
$4 + 2(2) \stackrel{?}{=} 8$	Substitute.	$-3(4) + 4(2) \stackrel{?}{=} -4$	Substitute.
$4 + 4 \stackrel{?}{=} 8$		$-12 + 8 \stackrel{?}{=} -4$	
$8 = 8$	True	$-4 = -4$	True

Since $(4, 2)$ is a solution of each equation in the system, **yes**, it is a solution of the system.

To **solve a system by graphing**, graph each line on the same axes.

a) If the lines intersect at a **single point,** then this point is the solution of the system. The system is **consistent.**

b) If the lines are **parallel,** then the system has **no solution.** We write the solution set as $\varnothing$. The system is **inconsistent.**

c) If the graphs are the **same line,** then the system has an **infinite number of solutions.** The equations are **dependent.** **(p. 246)**

Solve by graphing. $\qquad y = -\dfrac{1}{2}x + 2$

$$5x + 3y = -1$$

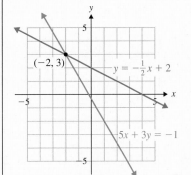

$(-2, 3)$
$y = -\dfrac{1}{2}x + 2$
$5x + 3y = -1$

The solution of the system is $(-2, 3)$.

The system is **consistent.**

4.2 Solving Systems by the Substitution Method

Steps for Solving a System by the Substitution Method

Step 1: Solve one of the equations for one of the variables. If possible, solve for a variable that has a coefficient of 1 or −1.

Step 2: Substitute the expression in *Step 1* into the *other* equation. The equation you obtain should contain only one variable.

Step 3: Solve the equation in *Step 2.*

Step 4: Substitute the value found in *Step 3* into either of the equations to obtain the value of the other variable.

Step 5: Check the values in the original equations. **(p. 256)**

Solve by the substitution method. $\qquad 7x - 3y = 8$
$$x + 2y = -11$$

Step 1: Solve for x in the second equation since its coefficient is 1.

$$x = -2y - 11$$

Step 2: Substitute $-2y - 11$ for the x in the first equation.

$$7(-2y - 11) - 3y = 8$$

Step 3: Solve the equation above for y.

$7(-2y - 11) - 3y = 8$	
$-14y - 77 - 3y = 8$	Distribute.
$-17y - 77 = 8$	Combine like terms.
$-17y = 85$	Add 77 to each side.
$y = -5$	Divide by −17.

Step 4: Substitute $y = -5$ into the equation in *Step 1* to find x.

$x = -2(-5) - 11$	Substitute −5 for y.
$x = 10 - 11$	Multiply.
$x = -1$	

Step 5: The solution is $(-1, -5)$. Verify this by substituting $(-1, -5)$ into each of the original equations.

Definition/Procedure	Example
If the variables drop out and a false equation is obtained, the system has **no solution. The system is inconsistent, and the solution set is** ∅. **(p. 258)**	Solve by the substitution method. $\quad 2x - 8y = 9$ $\qquad\qquad\qquad\qquad\qquad x = 4y + 2$ ***Step 1:*** The second equation is solved for x. ***Step 2:*** Substitute $4y + 2$ for x in the first equation. $$2(4y + 2) - 8y = 9$$ ***Step 3:*** Solve the equation above for y. $$\begin{aligned}2(4y + 2) - 8y &= 9\\ 8y + 4 - 8y &= 9 \quad \text{Distribute.}\\ 4 &= 9 \quad \text{False}\end{aligned}$$ The system has no solution. The solution set is ∅.
If the variables drop out and a true equation is obtained, the system has an **infinite number of solutions. The equations are dependent. (p. 259)**	Solve by the substitution method. $\quad y = x - 3$ $\qquad\qquad\qquad\qquad\qquad 3x - 3y = 9$ ***Step 1:*** The first equation is solved for y. ***Step 2:*** Substitute $x - 3$ for y in the second equation. $$3x - 3(x - 3) = 9$$ ***Step 3:*** Solve the equation above for x. $$\begin{aligned}3x - 3(x - 3) &= 9\\ 3x - 3x + 9 &= 9 \quad \text{Distribute.}\\ 9 &= 9 \quad \text{True}\end{aligned}$$ The system has an infinite number of solutions of the form $\{(x, y)\,|\,y = x - 3\}$.

4.3 Solving Systems by the Elimination Method

Steps for Solving a System of Two Linear Equations by the Elimination Method

Step 1: Write each equation in the form $Ax + By = C$.

Step 2: Determine which variable to eliminate. If necessary, multiply one or both of the equations by a number so that the coefficients of the variable to be eliminated are negatives of one another.

Step 3: Add the equations, and solve for the remaining variable.

Step 4: Substitute the value found in *Step 3* into either of the original equations to find the value of the other variable.

Step 5: Check the solution in each of the original equations. **(p. 264)**

Solve using the elimination method. $\quad 4x + 5y = -7$
$\qquad\qquad\qquad\qquad\qquad\qquad -5x - 6y = 8$

Eliminate x. Multiply the first equation by 5, and multiply the second equation by 4 to rewrite the system with equivalent equations.

Rewrite the system
$$5(4x + 5y) = 5(-7) \quad \rightarrow \quad 20x + 25y = -35$$
$$4(-5x - 6y) = 4(8) \quad \rightarrow \quad -20x - 24y = 32$$

Add the equations:
$$\begin{aligned}20x + 25y &= -35\\ + \underline{-20x - 24y} &= \underline{32}\\ y &= -3\end{aligned}$$

Substitute $y = -3$ into either of the original equations, and solve for x.
$$\begin{aligned}4x + 5y &= -7\\ 4x + 5(-3) &= -7\\ 4x - 15 &= -7\\ 4x &= 8\\ x &= 2\end{aligned}$$

The solution is $(2, -3)$. Verify this by substituting $(2, -3)$ into each of the original equations.

Definition/Procedure	Example

4.4 Applications of Systems of Two Equations

Use the **Five Steps for Solving Applied Problems** outlined in the section to solve an applied problem.

Step 1: **Read** the problem carefully. Draw a picture, if applicable. Identify what you are being asked to find.

Step 2: **Choose variables** to represent the unknown quantities. If applicable, label the picture with the variables.

Step 3: **Write a system of equations using two variables.** It may be helpful to begin by writing an equation in words.

Step 4: **Solve the system.**

Step 5: **Check** the answer in the original problem, and **interpret** the solution as it relates to the problem. **(p. 274)**

Amana spent $40.20 at a second-hand movie and music store when she purchased some DVDs and CDs. Each DVD cost $6.30, and each CD cost $2.50. How many DVDs and CDs did she buy if she purchased 10 items all together?

Step 1: Read the problem carefully.
Step 2: Choose variables.

$$x = \text{number of DVDs she bought}$$
$$y = \text{number of CDs she bought}$$

Step 3: One equation involves the *cost* of the items:

$$\text{Cost DVDs} + \text{Cost CDs} = \text{Total cost}$$
$$6.30x \quad + \quad 2.50y \quad = \quad 40.20$$

The second equation involves the number of items:

$$\frac{\text{Number of}}{\text{DVDs}} + \frac{\text{Number of}}{\text{CDs}} = \frac{\text{Total number}}{\text{of items}}$$
$$x \quad + \quad y \quad = \quad 10$$

The system is $6.30x + 2.50y = 40.20$.
$$x + y = 10$$

Step 4: Multiply by 10 to eliminate the decimals in the first equation, and then solve the system using substitution.

$$10(6.30x + 2.50y) = 10(40.20) \quad \text{Eliminate decimals.}$$
$$63x + 25y = 402$$

Solve the system $63x + 25y = 402$ to determine that the
$$x + y = 10$$
solution is (4, 6).

Step 5: Amana bought 4 DVDs and 6 CDs. Verify the solution.

4.5 Linear Inequalities in Two Variables

A **linear inequality in two variables** is an inequality that can be written in the form $Ax + By \geq C$ or $Ax + By \leq C$, where A, B, and C are real numbers and where A and B are not both zero. ($>$ and $<$ may be substituted for $\geq$ and $\leq$.) **(p. 286)**

Some examples of linear inequalities in two variables are

$$x + 2y \leq 3, \quad y > -\frac{1}{2}x + 4, \quad y \geq -5, \quad x < 7$$

Graphing a Linear Inequality in Two Variables Using the Test Point Method

Step 1: **Graph the boundary line.**
   a) If the inequality contains $\geq$ or $\leq$, make it a *solid line*.
   b) If the inequality contains $>$ or $<$, make it a *dotted line*.

Step 2: **Choose a test point not on the line, and shade the appropriate region.** Substitute the test point into the inequality.
   a) If it *makes the inequality true*, shade the side of the line *containing* the test point. All points in the shaded region are part of the solution set.
   b) If the test point *does not satisfy the inequality*, shade the *other* side of the line. All points in the shaded region are part of the solution set.

Graph using the test point method.
$$2x - y > -3$$

Step 1: Graph the boundary line as a *dotted* line.

Step 2: Choose a test point not on the line, and substitute it into the inequality to determine whether or not it makes the inequality true.

Test Point	Substitute into $2x - y > -3$
(0, 0)	$2(0) - (0) > -3$
	$0 - 0 > -3$
	$0 > -3$ True

Since the test point satisfies the inequality, shade the side of the line containing (0, 0).

Definition/Procedure	Example

Note: If $(0, 0)$ is not on the line, it is an easy point to test in the inequality. **(p. 288)**

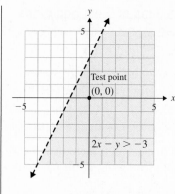

All points in the shaded region satisfy $2x - y > -3$.

Use the Slope-Intercept Form to Graph a Linear Inequality in Two Variables

Step 1: Write the inequality in the form
$y \geq mx + b \ (y > mx + b)$ or
$y \leq mx + b \ (y < mx + b)$

Step 2: Graph the boundary line $y = mx + b$.
a) If the inequality contains $\geq$ or $\leq$, make it a *solid line.*
b) If the inequality contains $>$ or $<$, make it a *dotted line.*

Step 3: Shade the appropriate side of the line.
a) If the inequality is in the form $y \geq mx + b$ or $y > mx + b$, shade *above* the line.
b) If the inequality is in the form $y \leq mx + b$ or $y < mx + b$, shade *below* the line. **(p. 290)**

Graph using the slope-intercept method.
$$-x + 4y \leq 2$$

Step 1: Solve $-x + 4y \leq 2$ for y. $\quad 4y \leq x + 2$
$$y \leq \frac{1}{4}x + \frac{1}{2}$$

Step 2: Graph $y = \frac{1}{4}x + \frac{1}{2}$ as a *solid* line.

Step 3: Since $y \leq \frac{1}{4}x + \frac{1}{2}$ has a *less than or equal to* symbol, shade *below* the line.

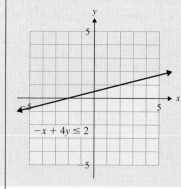

All points on the line and in the shaded region satisfy $-x + 4y \leq 2$.

Solve a System of Linear Inequalities by Graphing

1) Graph each inequality separately on the same axes. Shade lightly.
2) The solution set is the *intersection* (overlap) of the shaded regions. Heavily shade this region to indicate this is the solution set. **(p. 291)**

Graph the solution set of the system. $\quad y \geq -2x + 1$
$\qquad\qquad\qquad\qquad\qquad\qquad\qquad\qquad\quad y \geq -2$

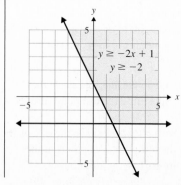

Any point in the shaded region will satisfy *both* inequalities.

Chapter 4: Review Exercises

*Additional answers can be found in the Answers to Exercises appendix.

(4.1) Determine whether the ordered pair is a solution of the system of equations.

1) $x - 5y = 13$ no
$2x + 7y = 20$
$(-4, -5)$

2) $8x + 3y = 16$ yes
$10x - 6y = 7$
$\left(\dfrac{3}{2}, \dfrac{4}{3}\right)$

3) If you are solving a system of equations by graphing, how do you know whether the system has no solution?
The lines are parallel.

Solve each system by graphing.

4) $y = \dfrac{1}{2}x + 1$ $(2, 2)$
$x + y = 4$

5) $x - 3y = 9$ $\varnothing$
$-x + 3y = 6$

6) $6x - 3y = 12$
$-2x + y = -4$

7) $-x + 2y = 1$ $(-3, -1)$
$2x + 3y = -9$

Without graphing, determine whether each system has no solution, one solution, or an infinite number of solutions.

8) $8x + 9y = -2$
$x - 4y = 1$
one solution

9) $y = -\dfrac{5}{2}x + 3$
$5x + 2y = 6$
infinite number of solutions

10) The graph shows the number of millions of barrels of crude oil reserves in Alabama and Ohio from 2005 to 2009. (www.census.gov)

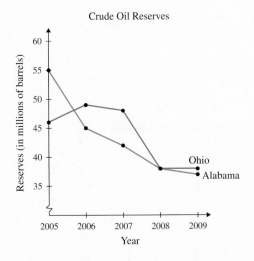

Crude Oil Reserves

a) In 2006, approximately how much more crude oil did Ohio have in reserve than Alabama?
about 4 million more barrels

b) Write the point of intersection as an ordered pair in the form (year, reserves) and explain its meaning.

c) Which line segment for Alabama has the most negative slope? How can this be explained in the context of the problem?

(4.2) Solve each system by the substitution method.

11) $9x - 2y = 8$ $(2, 5)$
$y = 2x + 1$

12) $y = -6x + 5$
$12x + 2y = 10$

13) $-x + 8y = 19$ $(5, 3)$
$4x - 3y = 11$

14) $-12x + 7y = 9$ $\left(-\dfrac{3}{4}, 0\right)$
$8x - y = -6$

(4.3) Solve each system using the elimination method.

15) $x - 7y = 3$ $(-4, -1)$
$-x + 5y = -1$

16) $5x + 4y = -23$ $\left(-5, \dfrac{1}{2}\right)$
$3x - 8y = -19$

17) $-10x + 4y = -8$ infinite number of solutions of the form
$5x - 2y = 4$ $\{(x, y)\,|\,5x - 2y = 4\}$

18) $7x - 4y = 13$ $(3, 2)$
$6x - 5y = 8$

Solve each system using the elimination method twice.

19) $2x + 9y = -6$ $\left(\dfrac{33}{43}, -\dfrac{36}{43}\right)$
$5x + y = 3$

20) $7x - 4y = 10$ $\left(\dfrac{62}{45}, -\dfrac{4}{45}\right)$
$6x + 3y = 8$

(4.2–4.3)

21) When is the best time to use substitution to solve a system?
when one of the variables has a coefficient of 1 or −1

22) If an equation in a system contains fractions, what should you do first to make the system easier to solve?   Multiply both sides of the equation by the LCD of the fractions to eliminate the fractions.

Solve each system by either the substitution or elimination method.

23) $6x + y = -8$ $\left(-\dfrac{5}{3}, 2\right)$
$9x + 7y = -1$

24) $4y - 5x = -23$ $(-1, -7)$
$2x + 3y = -23$

25) $\dfrac{1}{3}x - \dfrac{2}{9}y = -\dfrac{2}{3}$ $(0, 3)$
$\dfrac{5}{12}x + \dfrac{1}{3}y = 1$

26) $0.02x - 0.01y = 0.13$ $(10, 7)$
$-0.1x + 0.4y = 1.8$

27) $6(2x - 3) = y + 4(x - 3)$ $\left(\dfrac{3}{4}, 0\right)$
$5(3x + 4) + 4y = 11 - 3y + 27x$

28) $x - 3y = 36$ $(-9, -15)$
$y = \dfrac{5}{3}x$

29) $\dfrac{3}{4}x - \dfrac{5}{4}y = \dfrac{7}{8}$ $\varnothing$
$4 - 2(x + 5) - y = 3(1 - 2y) + x$

30) $y = -\dfrac{9}{7}x + \dfrac{6}{7}$ infinite number of solutions of the form
$18x + 14y = 12$ $\left\{(x, y)\,\middle|\, y = -\dfrac{9}{7}x + \dfrac{6}{7}\right\}$

(4.4) Write a system of equations, and solve.

31) One day in the school cafeteria, the number of children who bought white milk was twice the number who bought chocolate milk. How many cartons of each type of milk were sold if the cafeteria sold a total of 141 cartons of milk?
white: 94; chocolate: 47

32) How many ounces of a 7% acid solution and how many ounces of a 23% acid solution must be mixed to obtain 20 oz of a 17% acid solution?   7%: 7.5 oz; 23%: 12.5 oz

33) Edwin and Camille leave from opposite ends of a jogging trail 7 miles apart and travel toward each other. Edwin jogs 2 mph faster than Camille, and they meet after half an hour. How fast does each of them jog?   Edwin: 8 mph; Camille: 6 mph

34) At a movie theater concession stand, three candy bars and two small sodas cost $14.00. Four candy bars and three small sodas cost $19.50. Find the cost of a candy bar and the cost of a small soda.   candy bar: $3.00; soda: $2.50

35) The width of a rectangle is 5 cm less than the length. Find the dimensions of the rectangle if the perimeter is 38 cm.   length: 12 cm; width: 7 cm

36) Two planes leave the same airport and travel in opposite directions. The northbound plane flies 40 mph slower than the southbound plane. After 1.5 hours they are 1320 miles apart. Find the speed of each plane.   northbound: 420 mph; southbound: 460 mph

37) Shawanna saves her quarters and dimes in a piggy bank. When she opens it, she has 63 coins worth a total of $11.55. How many of each type of coin does she have?   quarters: 35; dimes: 28

38) Find the measure of angles x and y if the measure of angle x is half the measure of angle y.
$m\angle x = 29°; m\angle y = 58°$

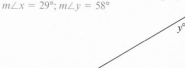

39) At a ski shop, two packs of hand warmers and one pair of socks cost $27.50. Five packs of hand warmers and three pairs of socks cost $78.00. Find the cost of a pack of hand warmers and a pair of socks.
hand warmers: $4.50; socks: $18.50

40) A 7 P.M. spinning class has 9 more members than a 10 A.M. spinning class. The two classes have a total of 71 students. How many are in each class?
7 P.M.: 40; 10 A.M.: 31

(4.5)

41) Find three points that satisfy the inequality and three that are not in the solution set.   Answers may vary.

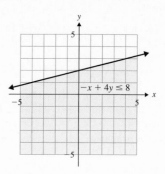

Graph each inequality.

42) $y \leq -x + 4$

43) $y > \dfrac{1}{3}x + 1$

44) $x > 2$

45) $2x + 3y \leq 9$

46) $3x - y \leq 2$

47) $y > 0$

48) Find three points that are in the solution set of the system and three that are not.   Answers may vary.

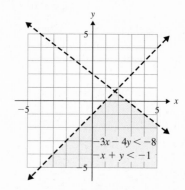

Graph the solution set of each system.

49) $x \leq 1$
    $y \geq -x + 1$

50) $y > 2x - 4$
    $2x + y < -1$

51) $-3x + y < 0$
    $3x + 2y > 4$

52) $x + 3y \leq 3$
    $x - y \geq 2$

53) $x \geq 0$
    $y \geq 0$
    $y \leq -2x + 4$

54) $x \geq 0$
    $y \geq 0$
    $2x + 5y \leq 5$

Chapter 4:   Test

*Additional answers can be found in the Answers to Exercises appendix.

1) Determine whether $\left(-\frac{2}{3}, 4\right)$ is a solution of the
system $9x + 5y = 14$ yes
$-6x - y = 0.$

Solve each system by graphing.

2) $y = -x + 2$
$3x - 4y = 20$ $(4, -2)$

3) $3y - 6x = 6$
$2x - y = 1$ $\varnothing$

4) The graph shows the unemployment rate in the civilian labor force in Vermont and Virginia in the month of September for 2008 to 2011. (www.bls.gov)

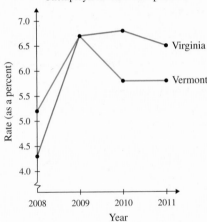

Unemployment Rate in September

a) When were more people unemployed in Vermont? Approximately what percent of the state's population was unemployed at that time?
2008; Approximately 5.2% of the population was unemployed.

b) Write the point of intersection of the graphs as an ordered pair in the form (year, percentage) and explain its meaning.   (2009, 6.7); In 2009, 6.7% of the population of Vermont and Virginia were unemployed.

c) Which line segment has the most positive slope? How can this be explained in the context of the problem?

Solve each system by the substitution method.

5) $3x - 10y = -10$
$x + 8y = -9$ $\left(-5, -\frac{1}{2}\right)$

6) $y = \frac{1}{2}x - 3$ infinite number of solutions of the form
$4x - 8y = 24$ $\left\{(x, y) \,\middle|\, y = \frac{1}{2}x - 3\right\}$

Solve each system by the elimination method.

7) $2x + 5y = 11$ $(3, 1)$
$7x - 5y = 16$

8) $3x + 4y = 24$ $(0, 6)$
$7x - 3y = -18$

9) $-6x + 9y = 14$ $\varnothing$
$4x - 6y = 5$

Solve each system using any method.

10) $11x - y = -14$ $(-2, -8)$
$-9x + 7y = -38$

11) $\frac{5}{8}x + \frac{1}{4}y = \frac{1}{4}$
$\frac{1}{3}x + \frac{1}{2}y = -\frac{4}{3}$ $(2, -4)$

12) $7 - 4(2x + 3) = x + 7 - y$
$5(x - y) + 20 = 8(2 - x) + x - 12$ $\left(-\frac{4}{3}, 0\right)$

13) Write a system of equations in two variables that has $(5, -1)$ as its only solution.   Answers may vary.

Write a system of equations, and solve.

14) The area of Yellowstone National Park is about 1.1 million fewer acres than the area of Death Valley National Park. If they cover a total of 5.5 million acres, how big is each park? (www.nps.gov)
Yellowstone: 2.2 mil acres; Death Valley: 3.3 mil acres

15) The Mahmood and Kuchar families take their kids to an amusement park. The Mahmoods buy one adult ticket and two children's tickets for $85.00. The Kuchars spend $150.00 on two adult and three children's tickets. How much did they pay for each type of ticket?   adult: $45.00; child: $20.00

16) The width of a rectangle is half its length. Find the dimensions of the rectangle if the perimeter is 114 cm.
length: 38 cm; width: 19 cm

17) How many milliliters of a 12% alcohol solution and how many milliliters of a 30% alcohol solution must be mixed to obtain 72 mL of a 20% alcohol solution?
12%: 40 mL; 30%: 32 mL

18) Rory and Lorelai leave Stars Hollow, Connecticut, and travel in opposite directions. Rory drives 4 mph faster than Lorelai, and after 1.5 hr they are 120 miles apart. How fast was each driving?   Rory: 42 mph; Lorelai: 38 mph

Graph each inequality.

19) $y > -3x - 1$

20) $x - 4y \geq -12$

Graph the solution set of each system of inequalities.

21) $x \geq -3$
$y \leq \frac{1}{2}x + 1$

22) $2x + y < 1$
$x - y < 0$

*Additional answers can be found in the Answers to Exercises appendix.

Perform the operations and simplify.

1) $\dfrac{7}{15} + \dfrac{9}{10}$ $\dfrac{41}{30}$

2) $4\dfrac{1}{5} \div \dfrac{9}{20}$ $9\dfrac{1}{3}$

3) $3(5 - 7)^3 + 18 \div 6 - 8$ -29

4) Find the area of the triangle. 30 in^2

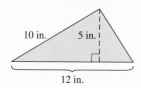
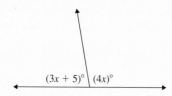

10 in.   5 in.

12 in.

5) Simplify $-3(4x^2 + 5x - 1)$. $-12x^2 - 15x + 3$

6) Find the measure of each indicated angle. $80°, 100°$

$(3x + 5)°$ $(4x)°$

Solve each equation.

7) $y + 9 = 34$ $\{25\}$

8) $-8 = -\dfrac{w}{6}$ $\{48\}$

9) $16 = -6m + 3$ $\left\{-\dfrac{13}{6}\right\}$

10) $0.04(3p - 2) - 0.02p = 0.1(p + 3)$ $\varnothing$

11) $11 - 3(2k - 1) = 2(6 - k)$ $\left\{\dfrac{1}{2}\right\}$

12) Solve. Write the answer in interval notation.

$$-5 < 4v - 9 < 15$$ $(1, 6)$

13) Write an equation and solve.

During the first week of the "Cash for Clunkers" program, the average increase in gas mileage for the new car purchased versus the car traded in was 61%. If the average gas mileage of the new cars was 25.4 miles per gallon, what was the average gas mileage of the cars traded in? Round the answer to the nearest tenth. (www.yahoo.com)   15.8 mpg

14) The area, A of a trapezoid is $A = \dfrac{1}{2}h(b_1 + b_2)$, where h = height of the trapezoid, b_1 = length of one base of the trapezoid, and b_2 = length of the second base of the trapezoid.

a) Solve the equation for h. $h = \dfrac{2A}{b_1 + b_2}$

b) Find the height of the trapezoid that has an area of 39 cm^2 and bases of length 8 cm and 5 cm.   6 cm

15) Graph $2x - 3y = 9$.

16) Find the x- and y-intercepts of the graph of $x - 8y = 16$.
   x-int: $(16, 0)$; y-int: $(0, -2)$

17) Write the slope-intercept form of the equation of the line containing $(3, 2)$ and $(-9, -1)$. $y = \dfrac{1}{4}x + \dfrac{5}{4}$

18) Determine whether the lines are parallel, perpendicular, or neither. $10x + 18y = 9$
   $9x - 5y = 17$ perpendicular

Solve each system of equations.

19) $9x - 3y = 6$
    $3x - 2y = -8$ $(4, 10)$

20) $3(2x - 1) - (y + 10) = 2(2x - 3) - 2y$
    $3x + 13 = 4x - 5(y - 3)$ $(3, 1)$

21) $x + 2y = 4$
    $-3x - 6y = 6$ $\varnothing$

22) $-\dfrac{1}{4}x - \dfrac{3}{4}y = \dfrac{1}{6}$ infinite number of solutions of the form $\{(x, y)|3x + 9y = -2\}$
    $\dfrac{1}{2}x + \dfrac{3}{2}y = -\dfrac{1}{3}$

23) $y = 4x + 1$ $(-2, -7)$
    $2x - y = 3$

Write a system of equations, and solve.

24) Dhaval used twice as many 6-foot boards as 4-foot boards when he made a treehouse for his children. If he used a total of 48 boards, how many of each size did he use?
   4-foot boards: 16; 6-foot boards: 32

25) Graph the solution set of the system:

$$y \le -2x + 1$$
$$y < 4$$

Rules of Exponents

Math at Work:

Motorcycle Customization

Let's take another look at how math is used in customizing motorcycles. One of Frank Esposito's coworkers at the motorcycle shop is Tanya Sorello, who is responsible for the frames of the motorcycles. Like Frank, Tanya uses mathematical equations to ensure that she does her work correctly.

For example, Tanya is building a chopper frame and needs to make the supports for the axle. To do this, she has to punch 1-in.-diameter holes in $\frac{3}{8}$-in.-thick mild steel plates that will be welded to the frame. The press punches two holes at a time. To determine how much power is needed to do this job, she uses a formula containing an exponent, $P = \dfrac{t^2 dN}{3.78}$. After substituting the numbers into the expression, she calculates that the power needed to punch these holes is 0.07 hp.

Tanya is used to working under tight deadlines. She has found ways to manage stress and perform at her best when it matters most. Tanya says she learned these skills when she was in school, during exam time. Mastering the high-pressure situation of sitting down to take a final exam has helped her succeed in critical moments throughout her career.

In this chapter, we will learn about working with exponents and introduce strategies you can use when you take a math test.

You've attended class regularly, you've taken good notes, you've done all the assigned homework and problem sets. Yet to succeed in a math course, you still need to do well on the tests. For many students, this makes taking math tests extremely stressful. Keep in mind, though, that all the work you do throughout the term—the note-taking, the homework, all of it—represents test preparation. The strategies below will also help you perform your best on math tests.

- Practice, practice, practice! Repetition will help build your skills. Use a timer as you work on problems to get used to doing math against the clock.
- Complete the test preparation checklist on page 340.
- During sleep, short-term memory turns into long-term memory, so get a good night's sleep before the test.

- Warm up for the test just like athletes do before they play a game. No matter how much you studied the night before, do several "warm-up" problems the same day as the test. This way, you will be in the groove of doing math and won't go into the test cold.
- Bring all supplies you will need to take the test—pencils, scratch paper, etc.
- Get to the location of the test at least 10 minutes early.

- When you receive the test, read all the instructions.
- Look over the entire test, and answer the easiest questions first. This will build your confidence and leave you more time to work on the harder problems.
- Show your work in a neat and organized way. Your instructor may give you partial credit if you show the steps you're going through to arrive at an answer.
- If you start to feel anxiety, don't let it develop into panic. Take deep breaths and trust that your preparation has positioned you to succeed.

E Evaluate

- Leave time at the end of the test to check over your calculations and to make sure you didn't skip any problems.
- If you realize that you will not have time to complete the test, don't give up. Use the time you have to answer as many questions as you can.

- When you get your test back, regardless of your grade, look it over and see where you made errors. Remember, math concepts build on each other—you can't master the new concepts until you are comfortable with the previous ones.
- If you aren't happy with your grade, talk to your instructor. She or he will likely have good advice for how you can do better next time.

Chapter 5 ⬛P⬛O⬛W⬛E⬛R Plan

⬛P Prepare   ⬛O Organize

What are your goals for Chapter 5?	How can you accomplish each goal?
1 Be prepared before and during class.	• Don't stay out late the night before, and be sure to set your alarm clock! • Bring a pencil, notebook paper, and textbook to class. • Avoid distractions by turning off your cell phone during class. • Pay attention, take good notes, and ask questions. • Complete your homework on time, and ask questions on problems you do not understand.
2 Understand the homework to the point where you could do it without needing any help or hints.	• Read the directions, and show all your steps. • Go to the professor's office for help. • Rework homework and quiz problems, and find similar problems for practice.
3 Use the P.O.W.E.R. framework to help you take math tests: *Math Test Preparation Checklist.*	• Read the Study Strategy that explains how to better take math tests. • Where can you begin using these techniques? • Complete the emPOWERme that appears before the Chapter Summary.
4 Write your own goal. _____ _____	• _____ _____

What are your objectives for Chapter 5?	How can you accomplish each objective?
1 Know how to use the product rule, power rules, and the quotient rule for exponents.	• Learn how to use each rule by itself first. • Product rule • Basic power rule • Power rule for a product • Power rule for quotient • Quotient rule for exponents • Be able to simplify an expression using any combination of the rules.
2 Understand how to work with negative exponents and zero as an exponent.	• Know that any nonzero number or variable raised to the 0 power is equal to 1. • Learn the definition of a *negative exponent,* and be able to explain it in your own words. • Know how to write an expression containing negative exponents with positive exponents.
3 Use, write, and evaluate numbers in scientific notation.	• Know how to multiply by a positive or negative power of 10, and compare it to using scientific notation. • Understand the definition of scientific notation. • Learn to write any number in scientific notation. • Be able to multiply, divide, add, and subtract numbers in scientific notation.
4 Write your own goal. _____ _____	• _____ _____

| E Evaluate | Complete the Chapter Review and Chapter Test. How did you do? | R Rethink | • Which objective was the hardest for you to master? What steps did you take to overcome difficulties in becoming an expert at that objective? |
| | | | • How has the Study Strategy for this chapter helped you take a more proactive approach to studying for your math tests? |

5.1A Basic Rules of Exponents: The Product Rule and Power Rules

P Prepare

O Organize

What are your objectives for Section 5.1A?	How can you accomplish each objective?
1 Evaluate Exponential Expressions	• Write the definition of *exponential expression* in your own words, using the words *base* and *exponent*. • Complete the given example on your own. • Complete You Try 1.
2 Use the Product Rule for Exponents	• Learn the property for the **Product Rule,** and write an example in your notes. • Complete the given example on your own. • Complete You Try 2.
3 Use the Power Rule $(a^m)^n = a^{mn}$	• Learn the property for the **Basic Power Rule,** and write an example in your notes. • Complete the given example on your own. • Complete You Try 3.
4 Use the Power Rule $(ab)^n = a^n b^n$	• Learn the property for the **Power Rule for a Product,** and write an example in your notes. • Complete the given example on your own. • Complete You Try 4.
5 Use the Power Rule $\left(\dfrac{a}{b}\right)^n = \dfrac{a^n}{b^n}$ where $b \neq 0$	• Learn the property for the **Power Rule for a Quotient,** and write an example in your notes. • Complete the given example on your own. • Complete You Try 5. • Write the summary for the product and power rules of exponents in your notes, and be able to apply them.

Read the explanations, follow the examples, take notes, and complete the You Trys.

1 Evaluate Exponential Expressions

Recall from Chapter 1 that exponential notation is used as a shorthand way to represent a multiplication problem. For example, $3 \cdot 3 \cdot 3 \cdot 3 \cdot 3$ can be written as 3^5.

Definition

An **exponential expression** of the form a^n, where a is any real number and n is a positive integer, is equivalent to $\underbrace{a \cdot a \cdot a \cdot \cdots \cdot a}_{n \text{ factors of } a}$. We say that a is the **base** and n is the **exponent.**

We can also evaluate an exponential expression.

EXAMPLE 1

Identify the base and the exponent in each expression and evaluate.

a) 2^4 b) $(-2)^4$ c) -2^4

In-Class Example 1

Identify the base and the exponent in each expression and evaluate.
a) 3^4 b) $(-3)^4$ c) -3^4

Answer:
a) base: 3; exponent: 4; 81
b) base: −3; exponent: 4; 81
c) base: 3; exponent: 4; −81

Solution

a) 2^4 2 is the base, 4 is the exponent. Therefore, $2^4 = 2 \cdot 2 \cdot 2 \cdot 2 = 16$.

b) $(-2)^4$ -2 is the base, 4 is the exponent. Therefore,
$(-2)^4 = (-2) \cdot (-2) \cdot (-2) \cdot (-2) = 16$.

c) -2^4 It may be very tempting to say that the base is -2. However, there are no parentheses in this expression. Therefore, 2 is the base, and 4 is the exponent. To evaluate,

$$-2^4 = -1 \cdot 2^4 = -1 \cdot 2 \cdot 2 \cdot 2 \cdot 2$$
$$= -16$$

 Hint

In part c), you must follow the *order of operations* and evaluate the exponential expression before multiplying by -1.

BE CAREFUL The expressions $(-a)^n$ and $-a^n$ are not always equivalent:

$$(-a)^n = \underbrace{(-a) \cdot (-a) \cdot (-a) \cdot \cdots \cdot (-a)}_{n \text{ factors of } -a}$$

$$-a^n = -1 \cdot \underbrace{a \cdot a \cdot a \cdot \cdots \cdot a}_{n \text{ factors of } a}$$

[YOU TRY 1]

Identify the base and exponent in each expression and evaluate.

a) 5^3 b) -8^2 c) $\left(-\dfrac{2}{3}\right)^3$

2 Use the Product Rule for Exponents

Is there a rule to help us *multiply* exponential expressions? Let's rewrite each of the following products as a single power of the base using what we already know:

1) $\underbrace{3 \text{ factors}}_{} \overbrace{\text{of 2}}^{2}$ $\underbrace{\text{factors}}_{}$ $\overbrace{\text{of 2}}^{}$

1) $2^3 \cdot 2^2 = \underbrace{2 \cdot 2 \cdot 2}_{3 \text{ factors of } 2} \cdot \overbrace{2 \cdot 2}^{2 \text{ factors of } 2} = 2^5$ 2) $5^4 \cdot 5^3 = \underbrace{5 \cdot 5 \cdot 5 \cdot 5}_{4 \text{ factors of } 5} \cdot \overbrace{5 \cdot 5 \cdot 5}^{3 \text{ factors of } 5} = 5^7$

Let's summarize: $2^3 \cdot 2^2 = 2^5$, $5^4 \cdot 5^3 = 5^7$

Do you notice a pattern? *When you multiply expressions with the same base, keep the same base and add the exponents.* This is called the **product rule** for exponents.

> ### Property Product Rule
>
> Let a be any real number, and let m and n be positive integers. Then,
> $$a^m \cdot a^n = a^{m+n}$$

EXAMPLE 2

In-Class Example 2

Find each product.
a) $5^2 \cdot 5$ b) $y^4 \cdot y^9$
c) $-4x^5 \cdot (-10x^8)$ d) $d \cdot d^7 \cdot d^4$

Answer: a) 125 b) y^{13}
c) $40x^{13}$ d) d^{12}

 Hint

If you do not see an exponent on the base, it is assumed that the exponent is 1. Look closely at part d).

Find each product.

a) $2^2 \cdot 2^4$ b) $x^9 \cdot x^6$ c) $5c^3 \cdot 7c^9$ d) $(-k)^8 \cdot (-k) \cdot (-k)^{11}$

Solution

a) $2^2 \cdot 2^4 = 2^{2+4} = 2^6 = 64$ Since the bases are the same, add the exponents.

b) $x^9 \cdot x^6 = x^{9+6} = x^{15}$

c) $5c^3 \cdot 7c^9 = (5 \cdot 7)(c^3 \cdot c^9)$ Use the associative and commutative properties.
     $= 35c^{12}$

d) $(-k)^8 \cdot (-k) \cdot (-k)^{11} = (-k)^{8+1+11} = (-k)^{20}$ Product rule

---

[YOU TRY 2] Find each product.

a) $3 \cdot 3^2$ b) $y^{10} \cdot y^4$ c) $-6m^5 \cdot 9m^{11}$ d) $h^4 \cdot h^6 \cdot h^4$ e) $(-3)^2 \cdot (-3)^2$

BE CAREFUL   Can the product rule be applied to $4^3 \cdot 5^2$? **No!** The bases are not the same, so we cannot add the exponents. To evaluate $4^3 \cdot 5^2$, we would evaluate $4^3 = 64$ and $5^2 = 25$, then multiply:

$$4^3 \cdot 5^2 = 64 \cdot 25 = 1600$$

---

3 Use the Power Rule $(a^m)^n = a^{mn}$

What does $(2^2)^3$ mean? We can rewrite $(2^2)^3$ first as $2^2 \cdot 2^2 \cdot 2^2$.

$$2^2 \cdot 2^2 \cdot 2^2 = 2^{2+2+2} = 2^6 = 64$$

Notice that $(2^2)^3 = 2^{2+2+2}$, or $2^{2 \cdot 3}$. This leads us to the basic power rule for exponents: *When you raise a power to another power, keep the base and multiply the exponents.*

Property Basic Power Rule

Let a be any real number, and let m and n be positive integers. Then,

$$(a^m)^n = a^{mn}$$

EXAMPLE 3

In-Class Example 3

Simplify using the power rule.
a) $(4^6)^3$ b) $(m^2)^5$ c) $(q^8)^7$

Answer: a) 4^{18} b) m^{10}
c) q^{56}

Simplify using the power rule.

a) $(3^8)^4$ b) $(n^3)^7$ c) $((-f)^4)^3$

Solution

a) $(3^8)^4 = 3^{8 \cdot 4} = 3^{32}$ b) $(n^3)^7 = n^{3 \cdot 7} = n^{21}$ c) $((-f)^4)^3 = (-f)^{4 \cdot 3} = (-f)^{12}$

[YOU TRY 3] Simplify using the power rule.

a) $(5^4)^3$ b) $(j^6)^5$ c) $((-2)^3)^2$

---

4 Use the Power Rule $(ab)^n = a^n b^n$

Hint

This property demonstrates how you can distribute an exponent to the bases if the bases are being multiplied.

We can use another power rule to simplify an expression such as $(5c)^3$. We can rewrite and simplify $(5c)^3$ as $5c \cdot 5c \cdot 5c = 5 \cdot 5 \cdot 5 \cdot c \cdot c \cdot c = 5^3 c^3 = 125c^3$. *To raise a product to a power, raise each factor to that power.*

Property Power Rule for a Product

Let a and b be real numbers, and let n be a positive integer. Then,

$$(ab)^n = a^n b^n$$

BE CAREFUL Notice that $(ab)^n = a^n b^n$ is different from $(a + b)^n$. $(a + b)^n \neq a^n + b^n$. We will study this in Chapter 6.

EXAMPLE 4

In-Class Example 4

Simplify each expression.
a) $(4x)^3$ b) $\left(\dfrac{1}{2}u\right)^4$
c) $(-3g)^3$ d) $(2w^4)^5$
e) $2(8mn)^2$

Answer: a) $64x^3$ b) $\dfrac{1}{16}u^4$
c) $-27g^3$ d) $32w^{20}$
e) $128m^2n^2$

Simplify each expression.

a) $(9y)^2$ b) $\left(\dfrac{1}{4}t\right)^3$ c) $(5c^2)^3$ d) $3(6ab)^2$

Solution

a) $(9y)^2 = 9^2 y^2 = 81y^2$ b) $\left(\dfrac{1}{4}t\right)^3 = \left(\dfrac{1}{4}\right)^3 \cdot t^3 = \dfrac{1}{64}t^3$

c) $(5c^2)^3 = 5^3 \cdot (c^2)^3 = 125c^{2 \cdot 3} = 125c^6$

d) $3(6ab)^2 = 3[6^2 \cdot (a)^2 \cdot (b)^2]$ The 3 is not in parentheses; therefore, it will not be squared.
$= 3(36a^2 b^2)$
$= 108a^2 b^2$

---

5 Use the Power Rule $\left(\dfrac{a}{b}\right)^n = \dfrac{a^n}{b^n}$ where $b \neq 0$

Another power rule allows us to simplify an expression like $\left(\dfrac{2}{x}\right)^4$. We can rewrite and simplify $\left(\dfrac{2}{x}\right)^4$ as $\dfrac{2}{x} \cdot \dfrac{2}{x} \cdot \dfrac{2}{x} \cdot \dfrac{2}{x} = \dfrac{2 \cdot 2 \cdot 2 \cdot 2}{x \cdot x \cdot x \cdot x} = \dfrac{2^4}{x^4} = \dfrac{16}{x^4}$. *To raise a quotient to a power, raise both the numerator and denominator to that power.*

Property Power Rule for a Quotient

Let a and b be real numbers, and let n be a positive integer. Then,

$$\left(\frac{a}{b}\right)^n = \frac{a^n}{b^n}, \text{ where } b \neq 0$$

EXAMPLE 5

Simplify using the power rule for quotients.

a) $\left(\dfrac{3}{8}\right)^2$ b) $\left(\dfrac{5}{x}\right)^3$ c) $\left(\dfrac{t}{u}\right)^9$

Solution

a) $\left(\dfrac{3}{8}\right)^2 = \dfrac{3^2}{8^2} = \dfrac{9}{64}$ b) $\left(\dfrac{5}{x}\right)^3 = \dfrac{5^3}{x^3} = \dfrac{125}{x^3}$ c) $\left(\dfrac{t}{u}\right)^9 = \dfrac{t^9}{u^9}$

Let's summarize the rules of exponents we have learned in this section:

Summary The Product and Power Rules of Exponents

In the rules below, a and b are any real numbers, and m and n are positive integers.

Rule		Example
Product Rule	$a^m \cdot a^n = a^{m+n}$	$p^4 \cdot p^{11} = p^{4+11} = p^{15}$
Basic Power Rule	$(a^m)^n = a^{mn}$	$(c^8)^3 = c^{8 \cdot 3} = c^{24}$
Power Rule for a Product	$(ab)^n = a^n b^n$	$(3z)^4 = 3^4 \cdot z^4 = 81z^4$
Power Rule for a Quotient	$\left(\dfrac{a}{b}\right)^n = \dfrac{a^n}{b^n}, (b \neq 0)$	$\left(\dfrac{w}{2}\right)^4 = \dfrac{w^4}{2^4} = \dfrac{w^4}{16}$

E Evaluate 5.1A Exercises

Do the exercises, and check your work.

*Additional answers can be found in the Answers to Exercises appendix.

Objective 1: Evaluate Exponential Expressions

Rewrite each expression using exponents.

1) $9 \cdot 9 \cdot 9 \cdot 9 \cdot 9 \cdot 9$ 9^6

2) $4 \cdot 4 \cdot 4 \cdot 4 \cdot 4 \cdot 4 \cdot 4$ 4^7

3) $\left(\dfrac{1}{7}\right)\left(\dfrac{1}{7}\right)\left(\dfrac{1}{7}\right)\left(\dfrac{1}{7}\right)$ $\left(\dfrac{1}{7}\right)^4$

4) $(0.8)(0.8)(0.8)$ $(0.8)^3$

5) $(-5)(-5)(-5)(-5)(-5)(-5)(-5)$ $(-5)^7$

6) $(-c)(-c)(-c)(-c)(-c)$ $(-c)^5$

7) $(-3y)(-3y)(-3y)(-3y)(-3y)(-3y)(-3y)(-3y)$
   $(-3y)^8$

8) $\left(-\dfrac{5}{4}t\right)\left(-\dfrac{5}{4}t\right)\left(-\dfrac{5}{4}t\right)\left(-\dfrac{5}{4}t\right)$ $\left(-\dfrac{5}{4}t\right)^4$

Identify the base and the exponent in each.

9) 6^8 base: 6; exponent: 8

10) 9^4 base: 9; exponent: 4

11) $(0.05)^7$ base: 0.05; exponent: 7

12) $(0.3)^{10}$ base: 0.3; exponent: 10

13) $(-8)^5$ base: -8; exponent: 5

14) $(-7)^6$ base: -7; exponent: 6

15) $(9x)^8$ base: $9x$; exponent: 8

16) $(13k)^3$ base: $13k$; exponent: 3

17) $(-11a)^2$ base: $-11a$; exponent: 2

18) $(-2w)^9$ base: $-2w$; exponent: 9

19) $5p^4$ base: p; exponent: 4

20) $-3m^5$ base: m; exponent: 5

21) $-\dfrac{3}{8}y^2$ base: y; exponent: 2

22) $\dfrac{5}{9}t^7$ base: t; exponent: 7

23) Evaluate $(3 + 4)^2$ and $3^2 + 4^2$. Are they equivalent? Why or why not?

24) Evaluate $(7 - 3)^2$ and $7^2 - 3^2$. Are they equivalent? Why or why not?

25) For any values of a and b, does $(a + b)^2 = a^2 + b^2$? Why or why not?   Answers may vary.

26) Does $-2^4 = (-2)^4$? Why or why not?

27) Are $3t^4$ and $(3t)^4$ equivalent? Why or why not?
   No, $3t^4 = 3 \cdot t^4$; $(3t)^4 = 3^4 \cdot t^4 = 81t^4$

28) Is there any value of a for which $(-a)^2 = -a^2$? Support your answer with an example.
   Yes. If $a = 0$, then $(-0)^2 = 0$ and $-0^2 = 0$.

Evaluate.

29) 2^5 32

30) 9^2 81

31) $(11)^2$ 121

32) 4^3 64

33) $(-2)^4$ 16

34) $(-5)^3$ -125

35) -3^4 -81

36) -6^2 -36

37) -2^3 -8

38) -8^2 -64

39) $\left(\dfrac{1}{5}\right)^3$ $\dfrac{1}{125}$

40) $\left(\dfrac{3}{2}\right)^4$ $\dfrac{81}{16}$

For Exercises 41–44, answer *always, sometimes,* or *never.*

41) Raising a negative base to an even exponent power will *always, sometimes,* or *never* give a negative result.   never

42) If the base of an exponential expression is 1, the result will *always, sometimes,* or *never* be 1.   always

43) If b is any integer value except zero, then the exponential expression $(-b)^3$ will *always, sometimes,* or *never* give a negative result.   sometimes

44) If a is any integer value except zero, then the exponential expression $-a^4$ will *always, sometimes,* or *never* give a positive result.   never

Objective 2: Use the Product Rule for Exponents

Evaluate the expression using the product rule, where applicable.

45) $2^2 \cdot 2^3$ 32

46) $5^2 \cdot 5$ 125

47) $3^2 \cdot 3^2$ 81

48) $2^3 \cdot 2^3$ 64

49) $5^2 \cdot 2^3$ 200

50) $4^3 \cdot 3^2$ 576

51) $\left(\dfrac{1}{2}\right)^4 \cdot \left(\dfrac{1}{2}\right)^2$ $\dfrac{1}{64}$

52) $\left(\dfrac{4}{3}\right) \cdot \left(\dfrac{4}{3}\right)^2$ $\dfrac{64}{27}$

Simplify the expression using the product rule. Leave your answer in exponential form.

53) $8^3 \cdot 8^9$ 8^{12}

54) $6^4 \cdot 6^3$ 6^7

55) $5^2 \cdot 5^4 \cdot 5^5$ 5^{11}

56) $12^4 \cdot 12 \cdot 12^2$ 12^7

57) $(-7)^2 \cdot (-7)^3 \cdot (-7)^3$ $(-7)^8$

58) $(-3)^5 \cdot (-3) \cdot (-3)^6$ $(-3)^{12}$

59) $b^2 \cdot b^4$ b^6

60) $x^4 \cdot x^3$ x^7

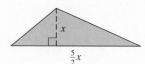 61) $k \cdot k^2 \cdot k^3$ k^6

62) $n^6 \cdot n^5 \cdot n^2$ n^{13}

63) $8y^3 \cdot y^2$ $8y^5$

64) $10c^8 \cdot c^2 \cdot c$ $10c^{11}$

(24) 65) $(9m^4)(6m^{11})$ $54m^{15}$

66) $(-10p^8)(-3p)$ $30p^9$

67) $(-6r)(7r^4)$ $-42r^5$

68) $(8h^5)(-5h^2)$ $-40h^7$

69) $(-7t^6)(t^3)(-4t^7)$ $28t^{16}$

70) $(3k^2)(-4k^5)(2k^4)$ $-24k^{11}$

71) $\left(\dfrac{5}{3}x^2\right)(12x)(-2x^3)$ $-40x^6$

72) $\left(\dfrac{7}{10}y^9\right)(-2y^4)(3y^2)$ $-\dfrac{21}{5}y^{15}$

73) $\left(\dfrac{8}{21}b\right)(-6b^8)\left(-\dfrac{7}{2}b^6\right)$ $8b^{15}$

74) $(12c^3)\left(\dfrac{14}{15}c^2\right)\left(\dfrac{5}{7}c^6\right)$ $8c^{11}$

Mixed Exercises: Objectives 3–5
Simplify the expression using one of the power rules.

75) $(y^3)^4$ y^{12}

76) $(x^5)^8$ x^{40}

77) $(w^{11})^7$ w^{77}

78) $(a^3)^2$ a^6

79) $(3^3)^2$ 729

80) $(2^2)^2$ 16

81) $((-5)^3)^2$ $(-5)^6$

82) $((-4)^5)^3$ $(-4)^{15}$

83) $\left(\dfrac{1}{3}\right)^4$ $\dfrac{1}{81}$

84) $\left(\dfrac{5}{2}\right)^3$ $\dfrac{125}{8}$

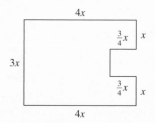 85) $\left(\dfrac{6}{a}\right)^2$ $\dfrac{36}{a^2}$

86) $\left(\dfrac{v}{4}\right)^3$ $\dfrac{v^3}{64}$

87) $\left(\dfrac{m}{n}\right)^5$ $\dfrac{m^5}{n^5}$

88) $\left(\dfrac{t}{u}\right)^{12}$ $\dfrac{t^{12}}{u^{12}}$

89) $(10y)^4$ $10{,}000y^4$

90) $(7w)^2$ $49w^2$

91) $(-3p)^4$ $81p^4$

92) $(2m)^5$ $32m^5$

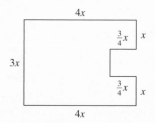 93) $(-4ab)^3$ $-64a^3b^3$

94) $(-2cd)^4$ $16c^4d^4$

95) $6(xy)^3$ $6x^3y^3$

96) $-8(mn)^5$ $-8m^5n^5$

97) $-9(tu)^4$ $-9t^4u^4$

98) $2(ab)^6$ $2a^6b^6$

Mixed Exercises: Objectives 2–5

99) Find the area and perimeter of each rectangle.

a)

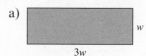

w $A = 3w^2$ sq units;
$P = 8w$ units

$3w$

b)

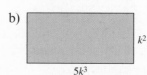

k^2 $A = 5k^5$ sq units;
$P = 10k^3 + 2k^2$ units

$5k^3$

100) Find the area.

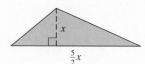

x

$\dfrac{5}{2}x$

$\dfrac{5}{4}x^2$ sq units

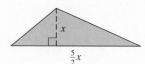 101) Find the area.

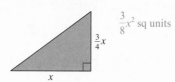

$\dfrac{3}{4}x$

x

$\dfrac{3}{8}x^2$ sq units

102) Here are the shape and dimensions of the Millers' family room. They will have wall-to-wall carpeting installed, and the carpet they have chosen costs \$2.50/ft².

$4x$

$\dfrac{3}{4}x$ x

$3x$

$\dfrac{3}{4}x$ x

$4x$

a) Write an expression for the amount of carpet they will need. (Include the correct units.) $\dfrac{45}{4}x^2$ ft²

b) Write an expression for the cost of carpeting the family room. (Include the correct units.)

$\dfrac{225}{8}x^2$ dollars

R Rethink

R1) When can you "distribute" an exponent into parentheses?

R2) When can you not "distribute" an exponent into parentheses?

R3) When do you add the exponents?

R4) When do you multiply the exponents?

R5) Which exercises in this section do you find most challenging? Where could you go for help?

5.1B Basic Rules of Exponents: Combining the Rules

P Prepare

O Organize

What is your objective for Section 5.1B?	How can you accomplish the objective?
1 Combine the Product Rule and Power Rules of Exponents	• Be able to follow the order of operations correctly, and complete the given example on your own. • Complete You Try 1.

W Work   Read the explanations, follow the examples, take notes, and complete the You Try.

1 Combine the Product Rule and Power Rules of Exponents

When we combine the rules of exponents, we follow the order of operations.

EXAMPLE 1

In-Class Example 1

Simplify.
a) $(4f)^2(2f)^3$ b) $-3(4a^2b^3)^2$
c) $\dfrac{(12y^3)^2}{(4z^{10})^3}$

Answer: a) $128f^5$
b) $-48a^4b^6$ c) $\dfrac{9y^6}{4z^{30}}$

W Hint

Notice how the parentheses are used in these examples.

Simplify.

a) $(2c)^3(3c^8)^2$ b) $2(5k^4m^3)^3$ c) $\dfrac{(6t^5)^2}{(2u^4)^3}$

Solution

a) $(2c)^3(3c^8)^2$

Because evaluating exponents comes before multiplying in the order of operations, *evaluate the exponents first.*

$$\begin{aligned}(2c)^3(3c^8)^2 &= (2^3c^3)(3^2)(c^8)^2 &&\text{Use the power rule.}\\ &= (8c^3)(9c^{16}) &&\text{Use the power rule, and evaluate exponents.}\\ &= 72c^{19} &&\text{Product rule}\end{aligned}$$

b) $2(5k^4m^3)^3$

Which operation should be performed first, multiplying $2 \cdot 5$ or simplifying $(5k^4m^3)^3$? In the order of operations, we evaluate exponents before multiplying, so *we will begin by simplifying $(5k^4m^3)^3$.*

$$\begin{aligned}2(5k^4m^3)^3 &= 2 \cdot (5)^3(k^4)^3(m^3)^3 &&\text{Order of operations and power rule}\\ &= 2 \cdot 125k^{12}m^9 &&\text{Power rule}\\ &= 250k^{12}m^9 &&\text{Multiply.}\end{aligned}$$

c) $\dfrac{(6t^5)^2}{(2u^4)^3}$

What comes first in the order of operations, dividing or evaluating exponents? *Evaluating exponents.*

$$\begin{aligned}\frac{(6t^5)^2}{(2u^4)^3} &= \frac{36t^{10}}{8u^{12}} &&\text{Power rule}\\[2mm] &= \frac{\overset{9}{36}t^{10}}{\underset{2}{8}u^{12}} &&\text{Divide out the common factor of 4.}\\[2mm] &= \frac{9t^{10}}{2u^{12}}\end{aligned}$$

When simplifying the expression in Example 1c, $\dfrac{(6t^5)^2}{(2u^4)^3}$, it may be tempting to simplify before applying the product rule, like this:

$$\dfrac{(\overset{3}{\cancel{6}}t^5)^2}{(2u^4)^3}_{\scriptstyle 1} \ne \dfrac{(3t^5)^2}{(u^4)^3} = \dfrac{9t^{10}}{u^{12}} \quad \textbf{Wrong!}$$

You can see, however, that because we did not follow the rules for the order of operations, we did *not* get the correct answer.

$\left[\text{ YOU TRY 1 }\right]$ Simplify.

a) $-4(2a^9b^6)^4$

b) $(7x^{10}y)^2(-x^4y^5)^4$

c) $\dfrac{10(m^2n^3)^5}{(5p^4)^2}$

d) $\left(\dfrac{1}{6}w^7\right)^2(3w^{11})^3$

ANSWERS TO $\left[\text{ YOU TRY }\right]$ **EXERCISE**

1) a) $-64a^{36}b^{24}$ b) $49x^{36}y^{22}$ c) $\dfrac{2m^{10}n^{15}}{5p^8}$ d) $\dfrac{3}{4}w^{47}$

E Evaluate **5.1B** Exercises   Do the exercises, and check your work.

*Additional answers can be found in the Answers to Exercises appendix.

Objective 1: Combine the Product Rule and Power Rules of Exponents

1) When evaluating expressions involving exponents, always keep in mind the order of _____. operations

2) The first step in evaluating $(9 - 3)^2$ is _____. subtraction

Simplify.

3) $(k^9)^2(k^3)^2$ k^{24}

4) $(d^5)^3(d^2)^4$ d^{23}

5) $(5z^4)^2(2z^6)^3$ $200z^{26}$

6) $(3r)^2(6r^8)^2$ $324r^{18}$

7) $6ab(-a^{10}b^2)^3$ $-6a^{31}b^7$

8) $-5pq^4(-p^4q)^4$ $-5p^{17}q^8$

9) $(9 + 2)^2$ 121

10) $(8 - 5)^3$ 27

11) $(-4t^6u^2)^3(u^4)^5$ $-64t^{18}u^{26}$

12) $(-m^2)^6(-2m^9)^4$ $16m^{48}$

13) $8(6k^7l^2)^2$ $288k^{14}l^4$

14) $5(-7c^4d)^2$ $245c^8d^2$

15) $\left(\dfrac{3}{g^5}\right)^3\left(\dfrac{1}{6}\right)^2$ $\dfrac{3}{4g^{15}}$

16) $\left(-\dfrac{2}{5}z^5\right)^3(10z)^2$ $-\dfrac{32}{5}z^{17}$

17) $\left(\dfrac{7}{8}n^2\right)^2(-4n^9)^2$ $\dfrac{49}{4}n^{22}$

18) $\left(\dfrac{2}{3}d^8\right)^4\left(\dfrac{9}{2}d^3\right)^2$ $4d^{38}$

19) $h^4(10h^3)^2(-3h^9)^2$ $900h^{28}$

20) $-v^6(-2v^5)^5(-v^4)^3$ $-32v^{43}$

21) $3w^{11}(7w^2)^2(-w^6)^5$ $-147w^{45}$

22) $5z^3(-4z)^2(2z^3)^2$ $320z^{11}$

23) $\dfrac{(12x^3)^2}{(10y^5)^2}$ $\dfrac{36x^6}{25y^{10}}$

24) $\dfrac{(-3a^4)^3}{(6b)^2}$ $-\dfrac{3a^{12}}{4b^2}$

25) $\dfrac{(4d^9)^2}{(-2c^5)^6}$ $\dfrac{d^{18}}{4c^{30}}$

26) $\dfrac{(-5m^7)^3}{(5n^{12})^2}$ $-\dfrac{5m^{21}}{n^{24}}$

27) $\dfrac{8(a^4b^7)^9}{(6c)^2}$ $\dfrac{2a^{36}b^{63}}{9c^2}$

28) $\dfrac{(3x^5)^3}{21(yz^2)^6}$ $\dfrac{9x^{15}}{7y^6z^{12}}$

29) $\dfrac{r^4(r^5)^7}{2t(11t^2)^2}$ $\dfrac{r^{39}}{242t^5}$

30) $\dfrac{k^5(k^2)^3}{7m^{10}(2m^3)^2}$ $\dfrac{k^{11}}{28m^{16}}$

31) $\left(\dfrac{4}{9}x^3y\right)^2\left(\dfrac{3}{2}x^6y^4\right)^3$ $\dfrac{2}{3}x^{24}y^{14}$

32) $(6s^8t^3)^2\left(-\dfrac{10}{3}st^4\right)^2$ $400s^{18}t^{14}$

33) $\left(-\dfrac{2}{5}c^9d^2\right)^3\left(\dfrac{5}{4}cd^6\right)^2$ $-\dfrac{1}{10}c^{29}d^{18}$

34) $-\dfrac{11}{12}\left(\dfrac{3}{2}m^3n^{10}\right)^2$ $-\dfrac{33}{16}m^6n^{20}$

35) $\left(\dfrac{5x^5y^2}{z^4}\right)^3$ $\dfrac{125x^{15}y^6}{z^{12}}$

36) $\left(-\dfrac{7a^4b}{8c^6}\right)^2$ $\dfrac{49a^8b^2}{64c^{12}}$

37) $\left(-\dfrac{3t^4u^9}{2v^7}\right)^4$ $\dfrac{81t^{16}u^{36}}{16v^{28}}$

38) $\left(\dfrac{2pr^8}{q^{11}}\right)^5$ $\dfrac{32p^5r^{40}}{q^{55}}$

39) $\left(\dfrac{12w^5}{4x^3y^6}\right)^2$ $\dfrac{9w^{10}}{x^6y^{12}}$

40) $\left(\dfrac{10b^3c^5}{15a}\right)^2$ $\dfrac{4b^6c^{10}}{9a^2}$

Determine whether the equation is *true* or *false*.

41) $(2k^2 + 2k^2) = (2k^2)^2$ false

42) $(4c^3 - 2c^2) = 2c$ false

For Exercises 43 and 44, answer *always, sometimes,* or *never*.

43) If a and b are any integer values except zero, then the exponential expression $-b^2(-a)^4$ will *always, sometimes,* or *never* give a positive result. never

44) If a and b are any integer values except zero, then the exponential expression $\left(\dfrac{-a}{b}\right)^3$ will *always, sometimes,* or *never* give a positive result. sometimes

45) The length of a side of a square is $5l^2$ units.

a) Write an expression for its perimeter. $20l^2$ units

b) Write an expression for its area. $25l^4$ sq units

46) The width of a rectangle is $2w$ units, and the length of the rectangle is $7w$ units.

a) Write an expression for its area. $14w^2$ sq units

b) Write an expression for its perimeter. $18w$ units

47) The length of a rectangle is x units, and the width of the rectangle is $\dfrac{3}{8}x$ units.

a) Write an expression for its area. $\dfrac{3}{8}x^2$ sq units

b) Write an expression for its perimeter. $\dfrac{11}{4}x$ units

48) The width of a rectangle is $4y^3$ units, and the length of the rectangle is $\dfrac{13}{2}y^3$ units.

a) Write an expression for its perimeter. $21y^3$ units

b) Write an expression for its area. $26y^6$ sq units

R Rethink

R1) How are the parentheses being used in these problems?

R2) Which exercises do you need to come back to and try again? How will this help you prepare for an exam?

5.2A Integer Exponents: Real-Number Bases

P Prepare

O Organize

What are your objectives for Section 5.2A?	How can you accomplish each objective?
1 Use 0 as an Exponent	• Understand the definition of *zero as an exponent,* and write it in your notes. • Complete the given example on your own. • Complete You Try 1.
2 Use Negative Integers as Exponents	• Understand the definition of *negative exponent,* and write it in your notes. • Complete the given example on your own. • Complete You Try 2.

W Work   Read the explanations, follow the examples, take notes, and complete the You Trys.

So far, we have defined an exponential expression such as 2^3. The exponent of 3 indicates that $2^3 = 2 \cdot 2 \cdot 2$ (3 factors of 2) so that $2^3 = 2 \cdot 2 \cdot 2 = 8$. Is it possible to have an exponent of zero or a negative exponent? If so, what do they mean?

1 Use 0 as an Exponent

Definition

Zero as an Exponent: If $a \neq 0$, then $a^0 = 1$.

How can this be possible?

Let's evaluate $2^0 \cdot 2^3$. Using the product rule, we get:

$$2^0 \cdot 2^3 = 2^{0+3} = 2^3 = 8$$

But we know that $2^3 = 8$. Therefore, if $2^0 \cdot 2^3 = 8$, then $2^0 = 1$. This is one way to understand that $a^0 = 1$.

EXAMPLE 1

In-Class Example 1

Evaluate each expression.

a) 5^0 b) -8^0 c) $(-7)^0$ d) $-3(2^0)$

Evaluate.
a) 3^0 b) -6^0 c) $(-4)^0$
d) $-2^0(4)$

Answer: a) 1 b) -1 c) 1
d) -4

Solution

a) $5^0 = 1$ b) $-8^0 = -1 \cdot 8^0 = -1 \cdot 1 = -1$

c) $(-7)^0 = 1$ d) $-3(2^0) = -3(1) = -3$

[YOU TRY 1] Evaluate.

a) 9^0 b) -2^0 c) $(-5)^0$ d) $3^0(-2)$

2 Use Negative Integers as Exponents

So far, we have worked with exponents that are zero or positive. What does a negative exponent mean?

Let's use the product rule to find $2^3 \cdot 2^{-3}$: $2^3 \cdot 2^{-3} = 2^{3+(-3)} = 2^0 = 1$

Remember that a number multiplied by its reciprocal is 1, and here we have that a quantity, 2^3, times another quantity, 2^{-3}, is 1. Therefore, 2^3 and 2^{-3} are reciprocals! This leads to the definition of a negative exponent.

W Hint

Did you notice that the signs of the bases do NOT change?

Definition

Negative Exponent: If n is any integer and a and b are not equal to zero, then

$$a^{-n} = \left(\frac{1}{a}\right)^n = \frac{1}{a^n} \quad \text{and} \quad \left(\frac{a}{b}\right)^{-n} = \left(\frac{b}{a}\right)^n.$$

Therefore, to rewrite an expression of the form a^{-n} with a positive exponent, *take the reciprocal of the base and make the exponent positive.*

EXAMPLE 2

Evaluate each expression.

a) 2^{-3} b) $\left(\dfrac{3}{2}\right)^{-4}$ c) $\left(\dfrac{1}{5}\right)^{-3}$ d) $(-7)^{-2}$

In-Class Example 2

Evaluate.

a) 3^{-3} b) $\left(\dfrac{4}{5}\right)^{-3}$ c) $\left(\dfrac{1}{3}\right)^{-2}$

d) $(-2)^{-2}$

Answer: a) $\dfrac{1}{27}$ b) $\dfrac{125}{64}$

c) 9 d) $\dfrac{1}{4}$

Solution

a) 2^{-3}: The reciprocal of 2 is $\dfrac{1}{2}$, so $2^{-3} = \left(\dfrac{1}{2}\right)^{3} = \dfrac{1^3}{2^3} = \dfrac{1}{8}$.

b) $\left(\dfrac{3}{2}\right)^{-4}$: The reciprocal of $\dfrac{3}{2}$ is $\dfrac{2}{3}$, so $\left(\dfrac{3}{2}\right)^{-4} = \left(\dfrac{2}{3}\right)^{4} = \dfrac{2^4}{3^4} = \dfrac{16}{81}$.

BE CAREFUL Notice that a negative exponent does not make the answer negative!

W Hint

Before working out these examples, try to determine by inspection whether the answer is going to be positive or negative.

c) $\left(\dfrac{1}{5}\right)^{-3}$: The reciprocal of $\dfrac{1}{5}$ is 5, so $\left(\dfrac{1}{5}\right)^{-3} = 5^3 = 125$.

d) $(-7)^{-2}$: The reciprocal of -7 is $-\dfrac{1}{7}$, so

$$(-7)^{-2} = \left(-\dfrac{1}{7}\right)^{2} = \left(-1 \cdot \dfrac{1}{7}\right)^{2} = (-1)^{2}\left(\dfrac{1}{7}\right)^{2} = 1 \cdot \dfrac{1^2}{7^2} = \dfrac{1}{49}$$

[YOU TRY 2] Evaluate.

a) $(10)^{-2}$ b) $\left(\dfrac{1}{4}\right)^{-2}$ c) $\left(\dfrac{2}{3}\right)^{-3}$ d) -5^{-3}

ANSWERS TO [YOU TRY] EXERCISES

1) a) 1 b) -1 c) 1 d) -2 2) a) $\dfrac{1}{100}$ b) 16 c) $\dfrac{27}{8}$ d) $-\dfrac{1}{125}$

E Evaluate **5.2A** Exercises   Do the exercises, and check your work.

*Additional answers can be found in the Answers to Exercises appendix.

Mixed Exercises: Objectives 1 and 2

(24) 1) True or False: Raising a positive base to a negative exponent will give a negative result. (Example: 2^{-4})   false

2) True or False: $8^0 = 1$.   true

3) True or False: The reciprocal of 4 is $\dfrac{1}{4}$.   true

4) True or False: $3^{-2} - 2^{-2} = 1^{-2}$.   false

Evaluate.

5) 2^0 1

6) $(-4)^0$ 1

 (24) 7) -5^0 -1

8) -1^0 -1

9) 0^8 0

10) $-(-9)^0$ -1

11) $(5)^0 + (-5)^0$ 2

12) $\left(\dfrac{4}{7}\right)^0 - \left(\dfrac{7}{4}\right)^0$ 0

(24) 13) 6^{-2} $\dfrac{1}{36}$

14) 9^{-2} $\dfrac{1}{81}$

15) 2^{-4} $\dfrac{1}{16}$

16) 11^{-2} $\dfrac{1}{121}$

17) 5^{-3} $\dfrac{1}{125}$

18) 2^{-5} $\dfrac{1}{32}$

19) $\left(\dfrac{1}{8}\right)^{-2}$ 64

20) $\left(\dfrac{1}{10}\right)^{-3}$ 1000

(24) 21) $\left(\dfrac{1}{2}\right)^{-5}$ 32

22) $\left(\dfrac{1}{4}\right)^{-2}$ 16

23) $\left(\dfrac{4}{3}\right)^{-3}$ $\dfrac{27}{64}$

24) $\left(\dfrac{2}{5}\right)^{-3}$ $\dfrac{125}{8}$

25) $\left(\dfrac{9}{7}\right)^{-2}$ $\dfrac{49}{81}$

26) $\left(\dfrac{10}{3}\right)^{-2}$ $\dfrac{9}{100}$

27) $\left(-\dfrac{1}{4}\right)^{-3}$ -64

28) $\left(-\dfrac{1}{12}\right)^{-2}$ 144

29) $\left(-\dfrac{3}{8}\right)^{-2}$ $\dfrac{64}{9}$

30) $\left(-\dfrac{5}{2}\right)^{-3}$ $-\dfrac{8}{125}$

31) -2^{-6} $-\dfrac{1}{64}$

32) -4^{-3} $-\dfrac{1}{64}$

33) -1^{-5} -1

34) -9^{-2} $-\dfrac{1}{81}$

35) $2^{-3} - 4^{-2}$ $\dfrac{1}{16}$

36) $5^{-2} + 2^{-2}$ $\dfrac{29}{100}$

37) $2^{-2} + 3^{-2}$ $\dfrac{13}{36}$

38) $4^{-1} - 6^{-2}$ $\dfrac{2}{9}$

39) $-9^{-2} + 3^{-3} + (-7)^0$ $\dfrac{83}{81}$

40) $6^0 - 9^{-1} + 4^0 + 3^{-2}$ 2

R Rethink

R1) When evaluating an expression with negative exponents, when do you get a negative answer?

R2) Explain the importance of fully understanding and using the order of operations in this chapter.

5.2B Integer Exponents: Variable Bases

P Prepare

O Organize

What are your objectives for Section 5.2B?	How can you accomplish each objective?
1 Use 0 as an Exponent	• Use the same definition of *zero as an exponent* used in the previous section. • Complete the given example on your own. • Complete You Try 1.
2 Rewrite an Exponential Expression with Positive Exponents	• Follow the explanation to understand how to rewrite an expression with positive exponents. • Learn and apply the definition of $\dfrac{a^{-m}}{b^{-n}} = \dfrac{b^n}{a^m}$. • Complete the given examples on your own. • Complete You Trys 2 and 3.

W Work

Read the explanations, follow the examples, take notes, and complete the You Trys.

1   Use 0 as an Exponent

We can apply 0 as an exponent to bases containing variables.

EXAMPLE 1

In-Class Example 1

Evaluate. Assume that the variable does not equal zero.
a) k^0 b) $(-m)^0$ c) $-3q^0$

Answer: a) 1   b) 1   c) -3

Evaluate each expression. Assume that the variable does not equal zero.

a) t^0 b) $(-k)^0$ c) $-(11p)^0$

Solution

a) $t^0 = 1$ b) $(-k)^0 = 1$ c) $-(11p)^0 = -1 \cdot (11p)^0 = -1 \cdot 1 = -1$

YOU TRY 1

Evaluate. Assume that the variable does not equal zero.

a) p^0 b) $(10x)^0$ c) $-(7s)^0$

2 Rewrite an Exponential Expression with Positive Exponents

Next, let's apply the definition of a negative exponent to bases containing variables. As in Example 1, we will assume that the variable does not equal zero since having zero in the denominator of a fraction will make the fraction undefined.

Recall that $2^{-4} = \left(\dfrac{1}{2}\right)^4 = \dfrac{1}{16}$. That is, to rewrite the expression with a positive exponent, we take the reciprocal of the base.

What is the reciprocal of x? The reciprocal is $\dfrac{1}{x}$.

EXAMPLE 2

In-Class Example 2

Rewrite the expression with positive exponents. Assume that the variable does not equal zero.

a) y^{-4} b) $\left(\dfrac{3}{d}\right)^{-4}$ c) $7f^{-7}$

Answer: a) $\dfrac{1}{y^4}$ b) $\dfrac{d^4}{81}$
c) $\dfrac{7}{f^7}$

W Hint

Before working out these examples, be sure you correctly identify the base.

Rewrite the expression with positive exponents. Assume that the variable does not equal zero.

a) x^{-6} b) $\left(\dfrac{2}{n}\right)^{-6}$ c) $3a^{-2}$

Solution

a) $x^{-6} = \left(\dfrac{1}{x}\right)^6 = \dfrac{1^6}{x^6} = \dfrac{1}{x^6}$ b) $\left(\dfrac{2}{n}\right)^{-6} = \left(\dfrac{n}{2}\right)^6$ The reciprocal of $\dfrac{2}{n}$ is $\dfrac{n}{2}$.

$= \dfrac{n^6}{2^6} = \dfrac{n^6}{64}$

c) $3a^{-2} = 3 \cdot \left(\dfrac{1}{a}\right)^2$ Remember, the base is a, *not* $3a$, since there are no parentheses. Therefore, the exponent of -2 applies only to a.

$= 3 \cdot \dfrac{1}{a^2} = \dfrac{3}{a^2}$

YOU TRY 2

Rewrite the expression with positive exponents. Assume that the variable does not equal zero.

a) m^{-4} b) $\left(\dfrac{1}{z}\right)^{-7}$ c) $-2y^{-3}$

How could we rewrite $\dfrac{x^{-2}}{y^{-2}}$ with only positive exponents? One way would be to apply the power rule for exponents: $\dfrac{x^{-2}}{y^{-2}} = \left(\dfrac{x}{y}\right)^{-2} = \left(\dfrac{y}{x}\right)^2 = \dfrac{y^2}{x^2}$

Notice that to rewrite the original expression with only positive exponents, the terms with the negative exponents "switch" their positions in the fraction. We can generalize this way:

Definition

If m and n are any integers and a and b are real numbers not equal to zero, then

$$\frac{a^{-m}}{b^{-n}} = \frac{b^n}{a^m}$$

EXAMPLE 3

In-Class Example 3

Rewrite the expression with positive exponents. Assume that the variables do not equal zero.

a) $\dfrac{f^{-1}}{g^{-5}}$ b) $\dfrac{9s^{-10}}{t^4}$ c) $c^{-2}d^{-7}$

Answer: a) $\dfrac{g^5}{f}$ b) $\dfrac{9}{s^{10}t^4}$

c) $\dfrac{1}{c^2d^7}$

W Hint

If the exponent is already positive, do not change the position of the expression in the fraction.

Rewrite the expression with positive exponents. Assume that the variables do not equal zero.

a) $\dfrac{c^{-8}}{d^{-3}}$　　b) $\dfrac{5p^{-6}}{q^7}$　　c) $t^{-2}u^{-1}$　　d) $\dfrac{2xy^{-3}}{3z^{-2}}$　　e) $\left(\dfrac{ab}{4c}\right)^{-3}$

Solution

a) $\dfrac{c^{-8}}{d^{-3}} = \dfrac{d^3}{c^8}$　　　To make the exponents positive, "switch" the positions of the terms in the fraction.

b) $\dfrac{5p^{-6}}{q^7} = \dfrac{5}{p^6q^7}$　　　Since the exponent on q is positive, we do not change its position in the expression.

c) $t^{-2}u^{-1} = \dfrac{t^{-2}u^{-1}}{1}$

$= \dfrac{1}{t^2u^1} = \dfrac{1}{t^2u}$　　Move $t^{-2}u^{-1}$ to the denominator to write with positive exponents.

d) $\dfrac{2xy^{-3}}{3z^{-2}} = \dfrac{2xz^2}{3y^3}$　　To make the exponents positive, "switch" the positions of the factors with negative exponents in the fraction.

e) $\left(\dfrac{ab}{4c}\right)^{-3} = \left(\dfrac{4c}{ab}\right)^3$　　To make the exponent positive, use the reciprocal of the base.

$= \dfrac{4^3c^3}{a^3b^3}$　　Power rule

$= \dfrac{64c^3}{a^3b^3}$　　Simplify.

[YOU TRY 3]

Rewrite the expression with positive exponents. Assume that the variables do not equal zero.

a) $\dfrac{n^{-6}}{y^{-2}}$　　b) $\dfrac{z^{-9}}{3k^{-4}}$　　c) $8x^{-5}y$　　d) $\dfrac{8d^{-4}}{6m^2n^{-1}}$　　e) $\left(\dfrac{3x^2}{y}\right)^{-2}$

ANSWERS TO [YOU TRY] EXERCISES

1) a) 1 b) 1 c) −1   2) a) $\dfrac{1}{m^4}$ b) z^7 c) $-\dfrac{2}{y^3}$ 3) a) $\dfrac{y^2}{n^6}$ b) $\dfrac{k^4}{3z^9}$

c) $\dfrac{8y}{x^5}$ d) $\dfrac{4n}{3m^2d^4}$ e) $\dfrac{y^2}{9x^4}$

*Additional answers can be found in the Answers to Exercises appendix.

Objective 1: Use 0 as an Exponent

1) Identify the base in each expression.

a) w^0 w

b) $-3n^{-5}$ n

c) $(2p)^{-3}$ $2p$

d) $4c^0$ c

2) True or False: $6^0 - 4^0 = (6-4)^0$ false

Evaluate. Assume that the variables do not equal zero.

3) r^0 1

4) $(5m)^0$ 1

5) $-2k^0$ -2

6) $-z^0$ -1

7) $x^0 + (2x)^0$ 2

8) $\left(\dfrac{7}{8}\right)^0 - \left(\dfrac{3}{5}\right)^0$ 0

Objective 2: Rewrite an Exponential Expression with Positive Exponents

Rewrite each expression with only positive exponents. Assume that the variables do not equal zero.

9) d^{-3} $\dfrac{1}{d^3}$

10) y^{-7} $\dfrac{1}{y^7}$

11) p^{-1} $\dfrac{1}{p}$

12) a^{-5} $\dfrac{1}{a^5}$

13) $\dfrac{a^{-10}}{b^{-3}}$ $\dfrac{b^3}{a^{10}}$

14) $\dfrac{h^{-2}}{k^{-1}}$ $\dfrac{k}{h^2}$

15) $\dfrac{y^{-8}}{x^{-5}}$ $\dfrac{x^5}{y^8}$

16) $\dfrac{v^{-2}}{w^{-7}}$ $\dfrac{w^7}{v^2}$

17) $\dfrac{t^5}{8u^{-3}}$ $\dfrac{t^5 u^3}{8}$

18) $\dfrac{9x^{-4}}{y^5}$ $\dfrac{9}{x^4 y^5}$

19) $5m^6 n^{-2}$ $\dfrac{5m^6}{n^2}$

20) $\dfrac{1}{9}a^{-4}b^3$ $\dfrac{b^3}{9a^4}$

21) $\dfrac{2}{t^{-11}u^{-5}}$ $2t^{11}u^5$

22) $\dfrac{7r}{2t^{-9}u^2}$ $\dfrac{7rt^9}{2u^2}$

23) $\dfrac{8a^6 b^{-1}}{5c^{-10}d}$ $\dfrac{8a^6 c^{10}}{5bd}$

24) $\dfrac{17k^{-8}h^5}{20m^{-7}n^{-2}}$ $\dfrac{17h^5 m^7 n^2}{20k^8}$

25) $\dfrac{2z^4}{x^{-7}y^{-6}}$ $2x^7 y^6 z^4$

26) $\dfrac{1}{a^{-2}b^{-2}c^{-1}}$ $a^2 b^2 c$

27) $\left(\dfrac{a}{6}\right)^{-2}$ $\dfrac{36}{a^2}$

28) $\left(\dfrac{3}{y}\right)^{-4}$ $\dfrac{y^4}{81}$

29) $\left(\dfrac{2n}{q}\right)^{-5}$ $\dfrac{q^5}{32n^5}$

30) $\left(\dfrac{w}{5v}\right)^{-3}$ $\dfrac{125v^3}{w^3}$

31) $\left(\dfrac{12b}{cd}\right)^{-2}$ $\dfrac{c^2 d^2}{144b^2}$

32) $\left(\dfrac{2tu}{v}\right)^{-6}$ $\dfrac{v^6}{64t^6 u^6}$

33) $-9k^{-2}$ $-\dfrac{9}{k^2}$

34) $3g^{-5}$ $\dfrac{3}{g^5}$

35) $3t^{-3}$ $\dfrac{3}{t^3}$

36) $8h^{-4}$ $\dfrac{8}{h^4}$

37) $-m^{-9}$ $-\dfrac{1}{m^9}$

38) $-d^{-5}$ $-\dfrac{1}{d^5}$

39) $\left(\dfrac{1}{z}\right)^{-10}$ z^{10}

40) $\left(\dfrac{1}{k}\right)^{-6}$ k^6

41) $\left(\dfrac{1}{j}\right)^{-1}$ j

42) $\left(\dfrac{1}{c}\right)^{-7}$ c^7

43) $5\left(\dfrac{1}{n}\right)^{-2}$ $5n^2$

44) $7\left(\dfrac{1}{t}\right)^{-8}$ $7t^8$

45) $c\left(\dfrac{1}{d}\right)^{-3}$ cd^3

46) $x^2\left(\dfrac{1}{y}\right)^{-2}$ $x^2 y^2$

For Exercises 47–50, answer *always, sometimes,* or *never*.

47) If a is any integer value except zero, then the exponential expression $-a^{-2}$ will *always, sometimes,* or *never* give a negative result.   always

48) If b is any integer value except zero, then the exponential expression $(-b)^{-3}$ will *always, sometimes,* or *never* give a positive result.   sometimes

49) If a and b are any integer values except zero, then the exponential expression $a^0 - b^0$ will *always, sometimes,* or *never* equal zero.   always

50) If a and b are any integer values except zero, then the exponential expression $(a^0 b^0)^{-2}$ will *always, sometimes,* or *never* equal zero.   never

Determine whether the equation is *true* or *false*.

51) $2 \div 2t^5 = t^{-5}$ true

52) $\dfrac{6x^5}{7} \div \dfrac{6x^8}{7} = \dfrac{7}{6}x^{-3}$ false

53) $(p^{-1} \div 3q^{-1})^{-2} = \left(\dfrac{q}{3p}\right)^2$ false

54) $(h^{-2} \div 4k^{-2})^2 = \left(\dfrac{2h}{k}\right)^{-4}$ true

R1) If a variable is raised to the power zero, the answer will always be what number? Assume the variable does not equal zero.

R2) Why is it useful to write expressions without negative exponents?

R3) Were there any problems you were unable to do? If so, write them down or circle them and ask your instructor for help.

5.3 The Quotient Rule

P Prepare

O Organize

What is your objective for Section 5.3?	How can you accomplish the objective?
1 Use the Quotient Rule for Exponents	• Learn the property for the **Quotient Rule for Exponents,** and write an example in your notes. • Complete the given examples on your own. • Complete You Trys 1 and 2.

W Work

Read the explanations, follow the examples, take notes, and complete the You Trys.

1 Use the Quotient Rule for Exponents

In this section, we will discuss how to simplify the quotient of two exponential expressions with the same base. Let's begin by simplifying $\dfrac{8^6}{8^4}$. One way to simplify this expression is to write the numerator and denominator without exponents:

$$\frac{8^6}{8^4} = \frac{8 \cdot 8 \cdot 8 \cdot 8 \cdot 8 \cdot 8}{8 \cdot 8 \cdot 8 \cdot 8} \qquad \text{Divide out common factors.}$$
$$= 8 \cdot 8 = 8^2 = 64$$

Therefore, $\dfrac{8^6}{8^4} = 8^2 = 64$.

Do you notice a relationship between the exponents in the original expression and the exponent we get when we simplify?

$$\frac{8^6}{8^4} = 8^{6-4} = 8^2 = 64$$

That's right. We *subtracted* the exponents.

Property Quotient Rule for Exponents

If m and n are any integers and $a \neq 0$, then

$$\frac{a^m}{a^n} = a^{m-n}$$

To divide expressions with the same base, keep the base the same and subtract the denominator's exponent from the numerator's exponent.

EXAMPLE 1

Simplify. Assume that the variables do not equal zero.

a) $\dfrac{2^9}{2^3}$ b) $\dfrac{t^{10}}{t^4}$ c) $\dfrac{3}{3^{-2}}$ d) $\dfrac{n^5}{n^7}$ e) $\dfrac{3^2}{2^4}$

Solution

a) $\dfrac{2^9}{2^3} = 2^{9-3} = 2^6 = 64$ Since the bases are the same, subtract the exponents.

b) $\dfrac{t^{10}}{t^4} = t^{10-4} = t^6$ Because the bases are the same, subtract the exponents.

c) $\dfrac{3}{3^{-2}} = \dfrac{3^1}{3^{-2}} = 3^{1-(-2)}$ Since the bases are the same, subtract the exponents.

$\phantom{\dfrac{3}{3^{-2}}} = 3^3 = 27$ Be careful when subtracting the negative exponent!

d) $\dfrac{n^5}{n^7} = n^{5-7} = n^{-2}$ Same base; subtract the exponents.

$\phantom{\dfrac{n^5}{n^7}} = \left(\dfrac{1}{n}\right)^2 = \dfrac{1}{n^2}$ Write with a positive exponent.

e) $\dfrac{3^2}{2^4} = \dfrac{9}{16}$ Because the bases are not the same, we cannot apply the quotient rule. Evaluate the numerator and denominator separately.

W Hint

Be careful when you subtract exponents, especially when you are working with negative numbers!

[YOU TRY 1]

Simplify. Assume that the variables do not equal zero.

a) $\dfrac{5^7}{5^4}$ b) $\dfrac{c^4}{c^{-1}}$ c) $\dfrac{k^2}{k^{10}}$ d) $\dfrac{2^3}{2^7}$

We can apply the quotient rule to expressions containing more than one variable. Here are more examples.

EXAMPLE 2

Simplify. Assume that the variables do not equal zero.

a) $\dfrac{x^8 y^7}{x^3 y^4}$ b) $\dfrac{12a^{-5} b^{10}}{8a^{-3} b^2}$

Solution

a) $\dfrac{x^8 y^7}{x^3 y^4} = x^{8-3} y^{7-4}$ Subtract the exponents.

$\phantom{\dfrac{x^8 y^7}{x^3 y^4}} = x^5 y^3$

b) $\dfrac{12a^{-5}b^{10}}{8a^{-3}b^2}$ We will simplify $\dfrac{12}{8}$ in addition to applying the quotient rule.

$$\dfrac{\overset{3}{\cancel{12}}a^{-5}b^{10}}{\underset{2}{\cancel{8}}a^{-3}b^2} = \dfrac{3}{2}a^{-5-(-3)}b^{10-2} \quad \text{Subtract the exponents.}$$

$$= \dfrac{3}{2}a^{-5+3}b^8 = \dfrac{3}{2}a^{-2}b^8 = \dfrac{3b^8}{2a^2}$$

[YOU TRY 2] Simplify. Assume that the variables do not equal zero.

a) $\dfrac{r^4s^{10}}{rs^3}$ b) $\dfrac{30m^6n^{-8}}{42m^4n^{-3}}$

ANSWERS TO [YOU TRY] EXERCISES

1) a) 125 b) c^5 c) $\dfrac{1}{k^8}$ d) $\dfrac{1}{16}$ 2) a) r^3s^7 b) $\dfrac{5m^2}{7n^5}$

E Evaluate **5.3** Exercises   Do the exercises, and check your work.

*Additional answers can be found in the Answers to Exercises appendix.

Objective 1: Use the Quotient Rule for Exponents

State what is wrong with the following steps and then simplify correctly.

1) $\dfrac{a^5}{a^3} = a^{3-5} = a^{-2} = \dfrac{1}{a^2}$ You must subtract the denominator's exponent from the numerator's exponent; a^2.

2) $\dfrac{4^3}{2^6} = \left(\dfrac{4}{2}\right)^{3-6} = 2^{-3} = \dfrac{1}{2^3} = \dfrac{1}{8}$ You must have the same base in order to use the quotient rule; 1.

Determine whether the equation is *true* or *false*.

3) $3^{-3} \div 3^{-4} = \dfrac{3^3}{3^4}$ false   4) $2^{-4} \div 2^{-3} = \dfrac{2^3}{2^4}$ true

5) $t^7 \div t^{-5} = t^2$ false   6) $m^{10} \div m^{-3} = m^7$ false

7) $r^{-6} \div r^{-2} = r^{-4}$ true   8) $k^{-5} \div k^{-10} = k^{-5}$ false

Simplify using the quotient rule. Assume that the variables do not equal zero.

9) $\dfrac{d^{10}}{d^5}$ d^5 10) $\dfrac{z^{11}}{z^7}$ z^4

11) $\dfrac{m^9}{m^5}$ m^4 12) $\dfrac{a^6}{a}$ a^5

13) $\dfrac{8t^{15}}{t^8}$ $8t^7$ 14) $\dfrac{4k^4}{k^2}$ $4k^2$

15) $\dfrac{6^{12}}{6^{10}}$ 36   16) $\dfrac{4^4}{4}$ 64

17) $\dfrac{3^{12}}{3^8}$ 81   18) $\dfrac{2^7}{2^4}$ 8

19) $\dfrac{2^5}{2^9}$ $\dfrac{1}{16}$ 20) $\dfrac{9^5}{9^7}$ $\dfrac{1}{81}$

21) $\dfrac{5^6}{5^9}$ $\dfrac{1}{125}$ 22) $\dfrac{8^4}{8^6}$ $\dfrac{1}{64}$

23) $\dfrac{10d^4}{d^2}$ $10d^2$ 24) $\dfrac{3x^6}{x^2}$ $3x^4$

25) $\dfrac{20c^{11}}{30c^6}$ $\dfrac{2}{3}c^5$ 26) $\dfrac{35t^7}{56t^2}$ $\dfrac{5}{8}t^5$

27) $\dfrac{y^3}{y^8}$ $\dfrac{1}{y^5}$ 28) $\dfrac{m^4}{m^{10}}$ $\dfrac{1}{m^6}$

29) $\dfrac{x^{-3}}{x^6}$ $\dfrac{1}{x^9}$ 30) $\dfrac{u^{-20}}{u^{-9}}$ $\dfrac{1}{u^{11}}$

31) $\dfrac{t^{-6}}{t^{-3}}$ $\dfrac{1}{t^3}$ 32) $\dfrac{y^8}{y^{15}}$ $\dfrac{1}{y^7}$

33) $\dfrac{a^{-1}}{a^9}$ $\dfrac{1}{a^{10}}$ 34) $\dfrac{m^{-9}}{m^{-3}}$ $\dfrac{1}{m^6}$

35) $\dfrac{t^4}{t}$ t^3 36) $\dfrac{c^7}{c^{-1}}$ c^8

37) $\dfrac{15w^2}{w^{10}}$ $\dfrac{15}{w^8}$ 38) $\dfrac{-7p^3}{p^{12}}$ $-\dfrac{7}{p^9}$

39) $\dfrac{-6k}{k^4}$ $-\dfrac{6}{k^3}$

40) $\dfrac{21h^3}{h^7}$ $\dfrac{21}{h^4}$

47) $\dfrac{6v^{-1}w}{54v^2w^{-5}}$ $\dfrac{w^6}{9v^3}$

48) $\dfrac{3a^2b^{-11}}{18a^{-10}b^6}$ $\dfrac{a^{12}}{6b^{17}}$

41) $\dfrac{a^4b^9}{ab^2}$ a^3b^7

42) $\dfrac{p^5q^7}{p^2q^3}$ p^3q^4

49) $\dfrac{3c^5d^{-2}}{8cd^{-3}}$ $\dfrac{3}{8}c^4d$

50) $\dfrac{9x^{-5}y^2}{4x^{-2}y^6}$ $\dfrac{9}{4x^3y^4}$

43) $\dfrac{10k^{-2}l^{-6}}{15k^{-5}l^2}$ $\dfrac{2k^3}{3l^8}$

44) $\dfrac{28tu^{-2}}{14t^5u^{-9}}$ $\dfrac{2u^7}{t^4}$

51) $\dfrac{(x+y)^9}{(x+y)^2}$ $(x+y)^7$

52) $\dfrac{(a+b)^9}{(a+b)^4}$ $(a+b)^5$

45) $\dfrac{300x^7y^3}{30x^{12}y^8}$ $\dfrac{10}{x^5y^5}$

46) $\dfrac{63a^{-3}b^2}{7a^7b^8}$ $\dfrac{9}{a^{10}b^6}$

53) $\dfrac{(c+d)^{-5}}{(c+d)^{-11}}$ $(c+d)^6$

54) $\dfrac{(a+2b)^{-3}}{(a+2b)^{-4}}$ $a+2b$

R Rethink

R1) How could you change a positive exponent to a negative exponent?

R2) When do you add exponents?

R3) When do you subtract exponents?

R4) When do you multiply exponents?

R5) In what other courses have you seen exponents used?

Putting It All Together

P Prepare

O Organize

What is your objective?	How can you accomplish the objective?
1 Combine the Rules of Exponents	• Understand all of the **Rules of Exponents,** and summarize them in your notes. • Complete the given example on your own. • Complete You Try 1.

W Work

Read the explanations, follow the examples, take notes, and complete the You Try.

W Hint

Summarize all of the rules of exponents in your notes.

1 Combine the Rules of Exponents

Let's see how we can combine the rules of exponents to simplify expressions.

EXAMPLE 1

In-Class Example 1

Simplify using the rules of exponents. Assume that all variables represent nonzero real numbers.

a) $(4h^{-3})^2(2h)^3$ b) $\dfrac{z^9 \cdot z^{-6}}{z^4}$

c) $\left(\dfrac{21p^{-7}q^3}{28p^8q^{-4}}\right)^{-2}$

Answer: a) $\dfrac{128}{h^3}$ b) $\dfrac{1}{z}$

c) $\dfrac{16p^{30}}{9q^{14}}$

Simplify using the rules of exponents. Assume that all variables represent nonzero real numbers.

a) $(2t^{-6})^3(3t^2)^2$ b) $\dfrac{w^{-3} \cdot w^4}{w^6}$ c) $\left(\dfrac{12a^{-2}b^9}{30ab^{-2}}\right)^{-3}$

Solution

a) $(2t^{-6})^3(3t^2)^2$ We must follow the order of operations. Therefore, evaluate the exponents first.

$$(2t^{-6})^3 \cdot (3t^2)^2 = 2^3 t^{(-6)(3)} \cdot 3^2 t^{(2)(2)}$$ Apply the power rule.
$$= 8t^{-18} \cdot 9t^4$$ Simplify.
$$= 72t^{-18+4}$$ Multiply $8 \cdot 9$ and add the exponents.
$$= 72t^{-14}$$
$$= \frac{72}{t^{14}}$$ Write the answer using a positive exponent.

W Hint

Remember to use the order of operations when simplifying expressions like these.

b) $\dfrac{w^{-3} \cdot w^4}{w^6}$ Let's begin by simplifying the numerator:

$$\frac{w^{-3} \cdot w^4}{w^6} = \frac{w^{-3+4}}{w^6}$$ Add the exponents in the numerator.
$$= \frac{w^1}{w^6}$$

Now, we can apply the quotient rule:

$$= w^{1-6} = w^{-5}$$ Subtract the exponents.
$$= \frac{1}{w^5}$$ Write the answer using a positive exponent.

W Hint

Are you writing out the examples as you are reading them?

c) $\left(\dfrac{12a^{-2}b^9}{30ab^{-2}}\right)^{-3}$ Eliminate the negative exponent *outside* the parentheses by taking the reciprocal of the base. Notice that we have *not* eliminated the negatives on the exponents *inside* the parentheses.

$$\left(\frac{12a^{-2}b^9}{30ab^{-2}}\right)^{-3} = \left(\frac{30ab^{-2}}{12a^{-2}b^9}\right)^3$$

Next, we *could* apply the exponent of 3 to the quantity inside the parentheses. However, we could also simplify the expression inside the parentheses before cubing it.

$$\left(\frac{30ab^{-2}}{12a^{-2}b^9}\right)^3 = \left(\frac{5}{2}a^{1-(-2)}b^{-2-9}\right)^3$$ Simplify $\frac{30}{12}$ and subtract the exponents.
$$= \left(\frac{5}{2}a^3b^{-11}\right)^3$$
$$= \frac{125}{8}a^9b^{-33}$$ Apply the power rule.
$$= \frac{125a^9}{8b^{33}}$$ Write the answer using positive exponents.

Putting It All Together Exercises

Do the exercises, and check your work.

*Additional answers can be found in the Answers to Exercises appendix.

Objective 1: Combine the Rules of Exponents

Use the rules of exponents to evaluate.

1) $\left(\dfrac{2}{3}\right)^4$ $\dfrac{16}{81}$

2) $(2^2)^3$ 64

3) $\dfrac{3^9}{3^5 \cdot 3^4}$ 1

4) $\dfrac{(-5)^6 \cdot (-5)^2}{(-5)^5}$ -125

5) $\left(\dfrac{10}{3}\right)^{-2}$ $\dfrac{9}{100}$

6) $\left(\dfrac{3}{7}\right)^{-2}$ $\dfrac{49}{9}$

7) $(9-6)^2$ 9

8) $(3-8)^3$ -125

9) 10^{-2} $\dfrac{1}{100}$

10) 2^{-3} $\dfrac{1}{8}$

11) $\dfrac{2^7}{2^{12}}$ $\dfrac{1}{32}$

12) $\dfrac{3^{19}}{3^{15}}$ 81

13) $\left(-\dfrac{5}{3}\right)^{-7} \cdot \left(-\dfrac{5}{3}\right)^4$ $-\dfrac{27}{125}$

14) $\left(\dfrac{1}{8}\right)^{-2}$ 64

15) $3^{-2} - 12^{-1}$ $\dfrac{1}{36}$

16) $2^{-2} + 3^{-2}$ $\dfrac{13}{36}$

Simplify. Assume that all variables represent nonzero real numbers. The final answer should not contain negative exponents.

17) $-10(-3g^4)^3$ $270g^{12}$

18) $7(2d^3)^3$ $56d^9$

19) $\dfrac{33s}{s^{12}}$ $\dfrac{33}{s^{11}}$

20) $\dfrac{c^{-7}}{c^{-2}}$ $\dfrac{1}{c^5}$

21) $\left(\dfrac{2xy^4}{3x^{-9}y^{-2}}\right)^4$ $\dfrac{16}{81}x^{40}y^{24}$

22) $\left(\dfrac{a^6 b^5}{10a^3}\right)^3$ $\dfrac{a^9 b^{15}}{1000}$

23) $\left(\dfrac{9m^8}{n^3}\right)^{-2}$ $\dfrac{n^6}{81m^{16}}$

24) $\left(\dfrac{3s^{-6}}{r^2}\right)^{-4}$ $\dfrac{r^8 s^{24}}{81}$

25) $(-b^5)^3$ $-b^{15}$

26) $(h^{11})^8$ h^{88}

27) $(-3m^5 n^2)^3$ $-27m^{15}n^6$

28) $(13a^6 b)^2$ $169a^{12}b^2$

29) $\left(-\dfrac{9}{4}z^5\right)\left(\dfrac{8}{3}z^{-2}\right)$ $-6z^3$

30) $(15w^3)\left(-\dfrac{3}{5}w^6\right)$ $-9w^9$

31) $\left(\dfrac{s^7}{t^3}\right)^{-6}$ $\dfrac{t^{18}}{s^{42}}$

32) $\dfrac{m^{-3}}{n^{14}}$ $\dfrac{1}{m^3 n^{14}}$

33) $(-ab^3 c^5)^2\left(\dfrac{a^4}{bc}\right)^3$ $a^{14}b^3 c^7$

34) $\dfrac{(4v^3)^2}{(6v^8)^2}$ $\dfrac{4}{9v^{10}}$

35) $\left(\dfrac{48u^{-7}v^2}{36u^3 v^{-5}}\right)^{-3}$ $\dfrac{27u^{30}}{64v^{21}}$

36) $\left(\dfrac{xy^5}{9x^{-2}y}\right)^{-2}$ $\dfrac{81}{x^6 y^8}$

37) $\left(\dfrac{-3t^4 u}{t^2 u^{-4}}\right)^3$ $-27t^6 u^{15}$

38) $\left(\dfrac{k^7 m^7}{12k^{-1}m^6}\right)^2$ $\dfrac{1}{144}k^{16}m^2$

39) $(h^{-3})^6$ $\dfrac{1}{h^{18}}$

40) $(-d^4)^{-5}$ $-\dfrac{1}{d^{20}}$

41) $\left(\dfrac{h}{2}\right)^4$ $\dfrac{h^4}{16}$

42) $13f^{-2}$ $\dfrac{13}{f^2}$

43) $-7c^4(-2c^2)^3$ $56c^{10}$

44) $5p^3(4p^6)^2$ $80p^{15}$

45) $(12a^7)^{-1}(6a)^2$ $\dfrac{3}{a^5}$

46) $(9r^2 s^2)^{-1}$ $\dfrac{1}{9r^2 s^2}$

47) $\left(\dfrac{9}{20}r^4\right)(4r^{-3})\left(\dfrac{2}{33}r^9\right)$ $\dfrac{6}{55}r^{10}$

48) $\left(\dfrac{f^8 \cdot f^{-3}}{f^2 \cdot f^9}\right)^6$ $\dfrac{1}{f^{36}}$

49) $\dfrac{(a^2 b^{-5}c)^{-3}}{(a^4 b^{-3}c)^{-2}}$ $\dfrac{a^2 b^9}{c}$

50) $\dfrac{(x^{-1}y^7 z^4)^3}{(x^4 yz^{-5})^{-3}}$ $\dfrac{x^9 y^{24}}{z^3}$

51) $\dfrac{(2mn^{-2})^3(5m^2n^{-3})^{-1}}{(3m^{-3}n^3)^{-2}}$ $\quad \dfrac{72n^3}{5m^5}$

52) $\dfrac{(4s^3t^{-1})^2(5s^2t^{-3})^{-2}}{(4s^3t^{-1})^3}$ $\quad \dfrac{t^7}{100s^7}$

53) $\left(\dfrac{4n^{-3}m}{n^8m^2}\right)^0$ $\quad 1$

54) $\left(\dfrac{7qr^4}{37r^{-19}}\right)^0$ $\quad 1$

55) $\left(\dfrac{49c^4d^8}{21c^4d^5}\right)^{-2}$ $\quad \dfrac{9}{49d^6}$

56) $\dfrac{(2x^4y)^{-2}}{(5xy^3)^2}$ $\quad \dfrac{1}{100x^{10}y^8}$

Simplify. Assume that the variables represent nonzero integers. Write your final answer so that the exponents have positive coefficients.

57) $(p^{2c})^6$ $\quad p^{12c}$

58) $(5d^{4t})^2$ $\quad 25d^{8t}$

59) $y^m \cdot y^{3m}$ $\quad y^{4m}$

60) $x^{-5c} \cdot x^{9c}$ $\quad x^{4c}$

61) $t^{5b} \cdot t^{-8b}$ $\quad \dfrac{1}{t^{3b}}$

62) $a^{-4y} \cdot a^{-3y}$ $\quad \dfrac{1}{a^{7y}}$

63) $\dfrac{25c^{2x}}{40c^{9x}}$ $\quad \dfrac{5}{8c^{7x}}$

64) $-\dfrac{3y^{-10a}}{8y^{-2a}}$ $\quad -\dfrac{3}{8y^{8a}}$

R Rethink

R1) Did you remember to read all the directions before starting each section of the exercises? Why is it important to read all the directions?

R2) Were you able to complete the exercises without needing much prompting or help?

R3) Were there any problems you were unable to do? If so, write them down or circle them and ask your instructor for help.

5.4 Scientific Notation

P Prepare

O Organize

What are your objectives for Section 5.4?	How can you accomplish each objective?
1 Multiply a Number by a Power of 10	• Learn how to multiply a number by a positive power of 10. • Learn how to multiply a number by a negative power of 10. • Complete the given examples on your own. • Complete You Trys 1 and 2.
2 Understand Scientific Notation	• Write the definition of *scientific notation* in your own words, and write an example. • Complete the given example on your own. • Complete You Try 3.
3 Write a Number in Scientific Notation	• Write the procedure for **Writing a Number in Scientific Notation** in your own words. • Complete the given example on your own. • Complete You Try 4.
4 Perform Operations with Numbers in Scientific Notation	• Use the rules of exponents and the order of operations to complete the given example on your own. • Complete You Try 5.

Work | Read the explanations, follow the examples, take notes, and complete the You Trys.

The distance from Earth to the sun is approximately 150,000,000 km. A single rhinovirus (cause of the common cold) measures 0.00002 mm across. Performing operations on very large or very small numbers like these can be difficult. This is why scientists and economists, for example, often work with such numbers in a shorthand form called *scientific notation*. Writing numbers in scientific notation together with applying rules of exponents can simplify calculations with very large and very small numbers.

1 Multiply a Number by a Power of 10

Before discussing scientific notation further, we need to understand some principles behind the notation. Let's look at multiplying numbers by positive powers of 10.

EXAMPLE 1

In-Class Example 1

Multiply.
a) 9.5×10^1
b) $0.003271 \cdot 10^5$

Answer: a) 95 b) 327.1

Multiply.

a) 3.4×10^1 b) $0.0857 \cdot 10^3$

Solution

a) $3.4 \times 10^1 = 3.4 \times 10 = 34$

b) $0.0857 \cdot 10^3 = 0.0857 \cdot 1000 = 85.7$

Notice that when we multiply each of these numbers by a positive power of 10, the result is *larger* than the original number. In fact, the exponent determines how many places to the *right* the decimal point is moved.

$$3.40 \times 10^1 = 3.4 \times 10^1 = 34 \qquad 0.0857 \cdot 10^3 = 85.7$$
1 place to the right   3 places to the right

YOU TRY 1

Multiply by moving the decimal point the appropriate number of places.

a) 6.2×10^2 b) $5.31 \cdot 10^5$

What happens to a number when we multiply by a *negative* power of 10?

EXAMPLE 2

In-Class Example 2

Multiply.
a) $54 \cdot 10^{-3}$
b) 495×10^{-7}

Answer: a) 0.054
b) 0.0000495

Multiply.

a) $41 \cdot 10^{-2}$ b) 367×10^{-4}

Solution

a) $41 \cdot 10^{-2} = 41 \cdot \dfrac{1}{100} = \dfrac{41}{100} = 0.41$

b) $367 \times 10^{-4} = 367 \times \dfrac{1}{10,000} = \dfrac{367}{10,000} = 0.0367$

When we multiply each of these numbers by a negative power of 10, the result is *smaller* than the original number. The exponent determines how many places to the *left* the decimal point is moved:

$$41 \cdot 10^{-2} = 41. \cdot 10^{-2} = 0.41 \qquad 367 \times 10^{-4} = 0367. \times 10^{-4} = 0.0367$$
2 places to the left   4 places to the left

Multiply.

a) $83 \cdot 10^{-2}$ b) 45×10^{-3}

It is important to understand the previous concepts to understand how to use scientific notation.

2 Understand Scientific Notation

Definition

A number is in **scientific notation** if it is written in the form $a \times 10^n$, where $1 \le |a| < 10$ and n is an integer.

Note

Multiplying $|a|$ by a *positive* power of 10 will result in a number that is *larger* than $|a|$. Multiplying $|a|$ by a *negative* power of 10 will result in a number that is *smaller* than $|a|$. The double inequality $1 \le |a| < 10$ means that a is a number that has *one* nonzero digit to the left of the decimal point.

In other words, a number in scientific notation has one digit to the left of the decimal point and the number is multiplied by a power of 10.

Here are some examples of numbers written in scientific notation:

$$3.82 \times 10^{-5}, \quad 1.2 \cdot 10^3, \quad \text{and} \quad 7 \cdot 10^{-2}$$

The following numbers are *not* in scientific notation:

$$51.94 \times 10^4 \qquad 0.61 \cdot 10^{-3} \qquad 300 \cdot 10^6$$
$$\uparrow \qquad\qquad\qquad \uparrow \qquad\qquad\qquad \uparrow$$

| 2 digits to left of decimal point | Zero is to left of decimal point | 3 digits to left of decimal point |

Now let's convert a number written in scientific notation to a number without exponents.

EXAMPLE 3

In-Class Example 3

Rewrite without exponents.
a) $1.745 \cdot 10^3$ b) 1.07×10^{-2}
c) 3.904×10^4

Answer: a) 1745 b) 0.0107
c) 39,040

W Hint

How does this process compare to Examples 1 and 2?

Rewrite without exponents.

a) $5.923 \cdot 10^4$ b) 7.4×10^{-3} c) 1.8875×10^3

Solution

a) $5.923 \cdot 10^4 \rightarrow 5.9230 = 59{,}230$
   4 places to the right

   Remember, multiplying by a positive power of 10 will make the result *larger* than 5.923.

b) $7.4 \times 10^{-3} \rightarrow 007.4 = 0.0074$
   3 places to the left

   Multiplying by a negative power of 10 will make the result *smaller* than 7.4.

c) $1.8875 \times 10^3 \rightarrow 1.8875 = 1887.5$
   3 places to the right

---

3   Write a Number in Scientific Notation

Let's write the number 48,000 in scientific notation. First locate its decimal point: 48,000.

Next, determine where the decimal point will be when the number is in scientific notation:

48,000.

Decimal point
will be here.

Therefore, $48{,}000 = 4.8 \times 10^n$, where n is an integer. Will n be positive or negative? We can see that 4.8 must be multiplied by a *positive* power of 10 to make it larger, 48,000.

48000.

Decimal point   Decimal point
will be here.   starts here.

Now we count four places between the original and the final decimal place locations.

48000.
1 2 3 4

We use the number of spaces, 4, as the exponent of 10.

$$48{,}000 = 4.8 \times 10^4$$

EXAMPLE 4

Write each number in scientific notation.

Solution

a) The distance from Earth to the sun is approximately 150,000,000 km.

150,000,000.   150,000,000.   Move decimal point 8 places.

Decimal point   Decimal point
will be here.   is here.

$150{,}000{,}000 \text{ km} = 1.5 \times 10^8 \text{ km}$

b) A single rhinovirus measures 0.00002 mm across.

0.00002 mm $0.00002 \text{ mm} = 2 \times 10^{-5} \text{ mm}$

Decimal point
will be here.

Procedure How to Write a Number in Scientific Notation

1) Locate the decimal point in the original number.

2) Determine where the decimal point will be when converting to scientific notation. Remember, there will be *one* nonzero digit to the left of the decimal point.

3) Count how many places you must move the decimal point to take it from its original place to its position for scientific notation.

4) If the absolute value of the resulting number is *smaller* than the absolute value of the original number, you will multiply the result by a *positive* power of 10.

$$\text{Example: } 350.9 = 3.509 \times 10^2$$

If the absolute value of the resulting number is *larger* than the absolute value of the original number, you will multiply the result by a *negative* power of 10.

$$\text{Example: } 0.0000068 = 6.8 \times 10^{-6}$$

[YOU TRY 4] Write each number in scientific notation.

a) The gross domestic product of the United States in 2010 was approximately $14,582,400,000,000. (www.worldbank.org)

b) The diameter of a human hair is approximately 0.001 in.

---

4 Perform Operations with Numbers in Scientific Notation

We use the rules of exponents to perform operations with numbers in scientific notation.

EXAMPLE 5

In-Class Example 5

Simplify $\dfrac{5 \times 10^4}{2 \times 10^9}$.

Answer:
0.000025 or 2.5×10^{-5}

Simplify $\dfrac{3 \times 10^3}{4 \times 10^5}$.

Solution

$$\frac{3 \times 10^3}{4 \times 10^5} = \frac{3}{4} \times \frac{10^3}{10^5}$$

$$= 0.75 \times 10^{-2} \qquad \text{Write } \tfrac{3}{4} \text{ in decimal form.}$$

$$= 7.5 \times 10^{-3} \qquad \text{Use scientific notation.}$$

$$\text{or } 0.0075$$

[YOU TRY 5] Perform the operations, and simplify.

a) $(2.6 \cdot 10^2)(5 \cdot 10^4)$

b) $\dfrac{7.2 \times 10^{-9}}{6 \times 10^{-5}}$

We can use a graphing calculator to convert a very large or very small number to scientific notation, or to convert a number in scientific notation to a number written without an exponent. Suppose we are given a very large number such as 35,000,000,000. If you enter any number with more than 10 digits on the home screen on your calculator and press ENTER, the number will automatically be displayed in scientific notation as shown on the screen below. A small number with more than two zeros to the right of the decimal point (such as .000123) will automatically be displayed in scientific notation as shown.

The E shown in the screen refers to a power of 10, so 3.5E10 is the number 3.5×10^{10} in scientific notation. 1.23E-4 is the number 1.23×10^{-4} in scientific notation.

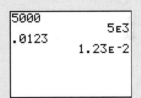

If a large number has 10 or fewer digits, or if a small number has fewer than three zeros to the right of the decimal point, then the number will not automatically be displayed in scientific notation. To display the number using scientific notation, press MODE, select SCI, and press ENTER. When you return to the home screen, all numbers will be displayed in scientific notation as shown below.

A number written in scientific notation can be entered directly into your calculator. For example, the number 2.38×10^7 can be entered directly on the home screen by typing 2.38 followed by 2nd , 7 ENTER as shown here. If you wish to display this number without an exponent, change the mode back to NORMAL and enter the number on the home screen as shown.

Write each number without an exponent, using a graphing calculator.

1. 3.4×10^5 2. 9.3×10^7 3. 1.38×10^{-3}

Write each number in scientific notation, using a graphing calculator.

4. 186,000   5. 5280   6. 0.0469

ANSWERS TO [YOU TRY] EXERCISES

1) a) 620 b) 531,000   2) a) 0.83 b) 0.045   3) a) 30,500 b) 0.000083 c) 6918.53
4) a) 1.45824×10^{13} dollars b) 1.0×10^{-3} in.   5) a) 13,000,000 b) 0.00012

ANSWERS TO TECHNOLOGY EXERCISES

1) 314,000   2) 93,000,000   3) .00138   4) 1.86×10^5 5) 5.28×10^3 6) 4.69×10^{-2}

*Additional answers can be found in the Answers to Exercises appendix.

Mixed Exercises: Objectives 1 and 2
Determine whether each number is in scientific notation.

1) 7.23×10^5 yes

2) 24.0×10^{-3} no

3) $0.16 \cdot 10^{-4}$ no

4) $-2.8 \cdot 10^4$ yes

5) -37×10^{-2} no

6) 0.9×10^{-1} no

7) $-5 \cdot 10^6$ yes

8) $7.5 \cdot 2^{-10}$ no

9) Explain, in your own words, how to determine whether a number is expressed in scientific notation.   Answers may vary.

10) Explain, in your own words, how to write 4.1×10^{-3} without an exponent.   Answers may vary.

Multiply.

11) $980.2 \cdot 10^4$ 9,802,000

12) $71.765 \cdot 10^2$ 7176.5

13) $0.1502 \cdot 10^8$ 15,020,000

14) $40.6 \cdot 10^{-3}$ 0.0406

15) 0.0674×10^{-1} 0.00674

16) $1,200,006 \times 10^{-7}$ 0.1200006

Objective 1: Multiply a Number by a Power of 10
Write each number without an exponent.

17) 1.92×10^6 1,920,000

18) -6.8×10^{-5} −0.000068

19) $2.03449 \cdot 10^3$ 2034.49

20) $-5.26 \cdot 10^4$ −52,600

21) $-7 \cdot 10^{-4}$ −0.0007

22) $8 \cdot 10^{-6}$ 0.000008

23) -9.5×10^{-3} −0.0095

24) 6.021967×10^5 602,196.7

25) 6×10^4 60,000

26) 3×10^6 3,000,000

27) $-9.815 \cdot 10^{-2}$ −0.09815

28) $-7.44 \cdot 10^{-4}$ −0.000744

Write the following quantities without an exponent.

29) About $2.4428 \cdot 10^7$ Americans played golf at least 2 times in a year. (Statistical Abstract of the U.S., www.census.gov)   24,428,000

30) In 2011, Facebook claimed that over 2.5×10^8 photos were uploaded each day. (www.facebook.com)   250,000,000 photos

31) The radius of one hydrogen atom is about 2.5×10^{-11} meter.   0.000000000025 meter

32) The length of a household ant is 2.54×10^{-3} meter.   0.00254 meter

Objective 3: Write a Number in Scientific Notation
Write each number in scientific notation.

33) 2110.5 2.1105×10^3

34) 38.25 3.825×10^1

35) 0.000096 9.6×10^{-5}

36) 0.00418 4.18×10^{-3}

37) −7,000,000 -7×10^6

38) 62,000 6.2×10^4

39) 3400 3.4×10^3

40) −145,000 -1.45×10^5

41) 0.0008 8×10^{-4}

42) −0.00000022 -2.2×10^{-7}

43) −0.076 -7.6×10^{-2}

44) 990 9.9×10^2

45) 6000 6×10^3

46) −500,000 -5×10^5

Write each number in scientific notation.

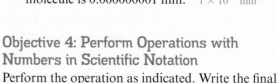

47) The total weight of the Golden Gate Bridge is 380,800,000 kg. (www.goldengatebridge.com) 3.808×10^8 kg

48) A typical hard drive may hold approximately 160,000,000,000 bytes of data. 1.6×10^{11} bytes

49) The diameter of an atom is about 0.00000001 cm. 1×10^{-8} cm

50) The oxygen-hydrogen bond length in a water molecule is 0.000000001 mm. 1×10^{-9} mm

Objective 4: Perform Operations with Numbers in Scientific Notation
Perform the operation as indicated. Write the final answer without an exponent.

51) $\dfrac{6 \cdot 10^9}{2 \cdot 10^5}$ 30,000

52) $(7 \cdot 10^2)(2 \cdot 10^4)$ 14,000,000

53) $(2.3 \times 10^3)(3 \times 10^2)$ 690,000

54) $\dfrac{8 \times 10^7}{4 \times 10^4}$ 2000

55) $\dfrac{8.4 \times 10^{12}}{-7 \times 10^9}$ −1200

56) $\dfrac{-4.5 \times 10^{-6}}{-1.5 \times 10^{-8}}$ 300

57) $(-1.5 \cdot 10^{-8})(4 \cdot 10^6)$ −0.06

58) $(-3 \cdot 10^{-2})(-2.6 \cdot 10^{-3})$ 0.000078

59) $\dfrac{-3 \cdot 10^5}{6 \cdot 10^8}$ −0.0005

60) $\dfrac{2 \cdot 10^1}{5 \cdot 10^4}$ 0.0004

61) $(9.75 \times 10^4) + (6.25 \times 10^4)$ 160,000

62) $(4.7 \times 10^{-3}) + (8.8 \times 10^{-3})$ 0.0135

63) $(3.19 \cdot 10^{-5}) + (9.2 \cdot 10^{-5})$ 0.0001239

64) $(2 \cdot 10^2) + (9.7 \cdot 10^2)$ 1170

For each problem, express each number in scientific notation, then solve the problem.

65) Humans shed about $1.44 \cdot 10^7$ particles of skin every day. How many particles would be shed in a year? (Assume 365 days in a year.)   5,256,000,000 particles

66) Scientists send a lunar probe to land on the moon and send back data. How long will it take for pictures to reach Earth if the distance between Earth and the moon is 360,000 km and if the speed of light is $3 \cdot 10^5$ km/sec?   1.2 sec

67) In Wisconsin in 2001, approximately 1,300,000 cows produced 2.21×10^{10} lb of milk. On average, how much milk did each cow produce? (www.nass.usda.gov)   17,000 lb/cow

68) The average snail can move 1.81×10^{-3} mi in 5 hours. What is its rate of speed in miles per hour?   0.000362 mph

69) A photo printer delivers approximately 1.1×10^6 droplets of ink per square inch. How many droplets of ink would a 4 in. × 6 in. photo contain?   26,400,000 droplets

70) In 2007, 3,500,000,000,000 prescription drug orders were filled in the United States. If the average price of each prescription was roughly $65.00, how much did the United States pay for prescription drugs last year? (National Conference of State Legislatures, www.ncsl.org) $227,500,000,000,000

71) In 2006, American households spent a total of about $7.3 \cdot 10^{11}$ dollars on food. If there were roughly 120,000,000 households in 2006, how much money did the average household spend on food? (Round to the closest dollar.) (www.census.gov) $6083

72) Find the population density of Australia if the estimated population in 2009 was about 22,000,000 people and the country encompasses about 2,900,000 sq mi. (Australian Bureau of Statistics, www.abs.gov.au)   7.6 people/sq mi

73) According to Nielsen Media Research, over 92,000,000 people watched Super Bowl XLIII in 2009 between the Pittsburgh Steelers and the Arizona Cardinals. The California Avocado Commission estimates that about 736,000,000 ounces of avocados were eaten during that Super Bowl, mostly in the form of guacamole. On average, how many ounces of avocados did each viewer eat during the Super Bowl?   8 ounces per person

74) In 2009, the United States produced about 5.5×10^9 metric tons of carbon emissions. The U.S. population that year was about 307 million. Find the amount of carbon emissions produced per person that year. Round to the nearest tenth. (http://epa.gov, U.S. Census Bureau)   17.9 metric tons

R Rethink

R1) When would you want to write a number in scientific notation?

R2) In which college courses, other than math, do you think scientific notation is used?

Group Activity — The Group Activity can be found online on Connect.

Before taking your next math test, complete the following test preparation checklist.

Test Preparation Checklist

❑ I checked whether it's a quiz, test, or exam.

❑ I began preparation at least three days before the test.

❑ I understand what material will be covered.

❑ I know how long I will have to take the test.

❑ I know what kinds of questions will be on the test.

❑ I formed or participated in a study group.

❑ I used my class notes and homework as study tools.

❑ I reviewed the relevant chapters of my textbook.

❑ I practiced by doing problems similar to those I will be tested on.

❑ I identified areas that were particularly challenging for me, and sought extra help.

❑ I got a good night's sleep before the test.

❑ I plan to make time to do some warm-up problems the day of the test.

❑ I have the materials I will need to complete the test: a pen or pencil, scratch paper, and a calculator (if allowed).

Chapter 5:   Summary

Definition/Procedure	Example

5.1A The Product Rule and Power Rules

Exponential Expression:

$$a^n = \underbrace{a \cdot a \cdot a \cdot \cdots \cdot a}_{n \text{ factors of } a}$$

a is the **base**, n is the exponent. **(p. 311)**

$5^4 = 5 \cdot 5 \cdot 5 \cdot 5$
5 is the **base**, 4 is the exponent.

Product Rule: $a^m \cdot a^n = a^{m+n}$ **(p. 312)**

$x^8 \cdot x^2 = x^{10}$

Basic Power Rule: $(a^m)^n = a^{mn}$ **(p. 313)**

$(t^3)^5 = t^{15}$

Power Rule for a Product: $(ab)^n = a^n b^n$ **(p. 313)**

$(2c)^4 = 2^4 c^4 = 16c^4$

Power Rule for a Quotient:

$\left(\dfrac{a}{b}\right)^n = \dfrac{a^n}{b^n}$, where $b \neq 0$. **(p. 314)**

$\left(\dfrac{w}{5}\right)^3 = \dfrac{w^3}{5^3} = \dfrac{w^3}{125}$

5.1B Combining the Rules

Remember to follow the order of operations. **(p. 317)**

Simplify $(3y^4)^2(2y^9)^3$.
$= 9y^8 \cdot 8y^{27}$ Exponents come before multiplication.
$= 72y^{35}$ Use the product rule, and multiply coefficients.

5.2A Real-Number Bases

Zero Exponent: If $a \neq 0$, then $a^0 = 1$. **(p. 320)**

$(-9)^0 = 1$

Negative Exponent:

For $a \neq 0$, $a^{-n} = \left(\dfrac{1}{a}\right)^n = \dfrac{1}{a^n}$. **(p. 320)**

Evaluate. $\left(\dfrac{5}{2}\right)^{-3} = \left(\dfrac{2}{5}\right)^3 = \dfrac{2^3}{5^3} = \dfrac{8}{125}$

5.2B Variable Bases

If $a \neq 0$ and $b \neq 0$, then $\left(\dfrac{a}{b}\right)^{-m} = \left(\dfrac{b}{a}\right)^m$. **(p. 323)**

Rewrite p^{-10} with a positive exponent (assume $p \neq 0$).

$p^{-10} = \left(\dfrac{1}{p}\right)^{10} = \dfrac{1}{p^{10}}$

If $a \neq 0$ and $b \neq 0$, then $\dfrac{a^{-m}}{b^{-n}} = \dfrac{b^n}{a^m}$. **(p. 324)**

Rewrite each expression with positive exponents. Assume that the variables represent nonzero real numbers.

a) $\dfrac{x^{-3}}{y^{-7}} = \dfrac{y^7}{x^3}$ b) $\dfrac{14m^{-6}}{n^{-1}} = \dfrac{14n}{m^6}$

5.3 The Quotient Rule

Quotient Rule: If $a \neq 0$, then $\dfrac{a^m}{a^n} = a^{m-n}$. **(p. 327)**

Simplify.

$\dfrac{4^9}{4^6} = 4^{9-6} = 4^3 = 64$

Putting It All Together

Combine the Rules of Exponents (p. 329)

Simplify.

$\left(\dfrac{a^4}{2a^7}\right)^{-5} = \left(\dfrac{2a^7}{a^4}\right)^5 = (2a^3)^5 = 32a^{15}$

5.4 Scientific Notation

Scientific Notation A number is in **scientific notation** if it is written in the form $a \times 10^n$, where $1 \le |a| < 10$ and n is an integer. That is, a is a number that has one nonzero digit to the left of the decimal point. **(p. 334)**	Write in scientific notation. a) $78{,}000 \rightarrow 78{,}000 \rightarrow 7.8 \times 10^4$ b) $0.00293 \rightarrow 0.00293 \rightarrow 2.93 \times 10^{-3}$
Converting from Scientific Notation (p. 334)	Write without exponents. a) $5 \times 10^{-4} \rightarrow 0005. \rightarrow 0.0005$ b) $1.7 \cdot 10^6 = 1.700000 \rightarrow 1{,}700{,}000$
Performing Operations (p. 336)	Multiply $(4 \times 10^2)(2 \times 10^4)$. $\quad = (4 \times 2)(10^2 \times 10^4)$ $\quad = 8 \times 10^6$ $\quad = 8{,}000{,}000$

Chapter 5:   Review Exercises

*Additional answers can be found in the Answers to Exercises appendix.

(5.1A)

1) Write in exponential form.

   a) $8 \cdot 8 \cdot 8 \cdot 8 \cdot 8 \cdot 8$ 8^6

   b) $(-7)(-7)(-7)(-7)$ $(-7)^4$

2) Identify the base and the exponent.

   a) -6^5 base: 6; exponent: 5   b) $(4t)^3$ base: $4t$; exponent: 3

   c) $4t^3$ base: t; exponent: 3   d) $-4t^3$ base: t; exponent: 3

3) Use the rules of exponents to simplify.

   a) $2^3 \cdot 2^2$ 32

   b) $\left(\dfrac{1}{3}\right)^2 \cdot \left(\dfrac{1}{3}\right)$ $\dfrac{1}{27}$

   c) $(7^3)^4$ 7^{12}

   d) $(k^5)^6$ k^{30}

4) Use the rules of exponents to simplify.

   a) $(3^2)^2$ 81

   b) $8^3 \cdot 8^7$ 8^{10}

   c) $(m^4)^9$ m^{36}

   d) $p^9 \cdot p^7$ p^{16}

5) Simplify using the rules of exponents.

   a) $(5y)^3$ $125y^3$

   b) $(-7m^4)(2m^{12})$ $-14m^{16}$

   c) $\left(\dfrac{a}{b}\right)^6$ $\dfrac{a^6}{b^6}$

   d) $6(xy)^2$ $6x^2y^2$

   e) $\left(\dfrac{10}{9}c^4\right)(2c)\left(\dfrac{15}{4}c^3\right)$ $\dfrac{25}{3}c^8$

6) Simplify using the rules of exponents.

   a) $\left(\dfrac{x}{y}\right)^{10}$ $\dfrac{x^{10}}{y^{10}}$

   b) $(-2z)^5$ $-32z^5$

   c) $(6t^7)\left(-\dfrac{5}{8}t^5\right)\left(\dfrac{2}{3}t^2\right)$ $-\dfrac{5}{2}t^{14}$ d) $-3(ab)^4$ $-3a^4b^4$

   e) $(10j^6)(4j)$ $40j^7$

(5.1B)

7) Simplify using the rules of exponents.

   a) $(z^5)^2(z^3)^4$ z^{22}

   b) $-2(3c^5d^8)^2$ $-18c^{10}d^{16}$

   c) $(9-4)^3$ 125

   d) $\dfrac{(10t^3)^2}{(2u^7)^3}$ $\dfrac{25t^6}{2u^{21}}$

8) Simplify using the rules of exponents.

   a) $\left(\dfrac{-20d^4c}{5b^3}\right)^3$ $-\dfrac{64d^{12}c^3}{b^9}$

   b) $(-2y^8z)^3(3yz^2)^2$ $-72y^{26}z^7$

   c) $\dfrac{x^7 \cdot (x^2)^5}{(2y^3)^4}$ $\dfrac{x^{17}}{16y^{12}}$

   d) $(6-8)^2$ 4

(5.2A)

9) Evaluate.

   a) 8^0 1

   b) -3^0 -1

   c) 9^{-1} $\dfrac{1}{9}$

   d) $3^{-2} - 2^{-2}$ $-\dfrac{5}{36}$

   e) $\left(\dfrac{4}{5}\right)^{-3}$ $\dfrac{125}{64}$

10) Evaluate.

   a) $(-12)^0$ 1

   b) $5^0 + 4^0$ 2

   c) -6^{-2} $-\dfrac{1}{36}$

   d) 2^{-4} $\dfrac{1}{16}$

   e) $\left(\dfrac{10}{3}\right)^{-2}$ $\dfrac{9}{100}$

(5.2B)

11) Rewrite the expression with positive exponents. Assume that the variables do not equal zero.

   a) v^{-9} $\dfrac{1}{v^9}$

   b) $\left(\dfrac{9}{c}\right)^{-2}$ $\dfrac{c^2}{81}$

   c) $\left(\dfrac{1}{y}\right)^{-8}$ y^8

   d) $-7k^{-9}$ $-\dfrac{7}{k^9}$

e) $\dfrac{19z^{-4}}{a^{-1}}$ $\dfrac{19a}{z^4}$

f) $20m^{-6}n^5$ $\dfrac{20n^5}{m^6}$

g) $\left(\dfrac{2j}{k}\right)^{-5}$ $\dfrac{k^5}{32j^5}$

12) Rewrite the expression with positive exponents. Assume that the variables do not equal zero.

a) $\left(\dfrac{1}{x}\right)^{-5}$ x^5

b) $3p^{-4}$ $\dfrac{3}{p^4}$

c) $a^{-8}b^{-3}$ $\dfrac{1}{a^8b^3}$

d) $\dfrac{12k^{-3}r^5}{16mn^{-6}}$ $\dfrac{3r^5n^6}{4k^3m}$

e) $\dfrac{c^{-1}d^{-1}}{15}$ $\dfrac{1}{15cd}$

f) $\left(-\dfrac{m}{4n}\right)^{-3}$ $-\dfrac{64n^3}{m^3}$

g) $\dfrac{10b^4}{a^{-9}}$ $10a^9b^4$

(5.3) In Exercises 13–16, assume that the variables represent nonzero real numbers. The answers should not contain negative exponents.

13) Simplify using the rules of exponents.

a) $\dfrac{3^8}{3^6}$ 9

b) $\dfrac{r^{11}}{r^3}$ r^8

c) $\dfrac{48t^{-2}}{32t^3}$ $\dfrac{3}{2t^5}$

d) $\dfrac{21xy^2}{35x^{-6}y^3}$ $\dfrac{3x^7}{5y}$

14) Simplify using the rules of exponents.

a) $\dfrac{2^9}{2^{15}}$ $\dfrac{1}{64}$

b) $\dfrac{d^4}{d^{-10}}$ d^{14}

c) $\dfrac{m^{-5}n^3}{mn^8}$ $\dfrac{1}{m^6n^5}$

d) $\dfrac{100a^8b^{-1}}{25a^7b^{-4}}$ $4ab^3$

15) Simplify by applying one or more of the rules of exponents.

a) $(-3s^4t^5)^4$ $81s^{16}t^{20}$

b) $\dfrac{(2a^6)^5}{(4a^7)^2}$ $2a^{16}$

c) $\left(\dfrac{z^4}{y^3}\right)^{-6}$ $\dfrac{y^{18}}{z^{24}}$

d) $(-x^3y)^5(6x^{-2}y^3)^2$ $-36x^{11}y^{11}$

e) $\left(\dfrac{cd^{-4}}{c^8d^{-9}}\right)^5$ $\dfrac{d^{25}}{c^{35}}$

f) $\left(\dfrac{14m^5n^5}{7m^4n}\right)^3$ $8m^3n^{12}$

g) $\left(\dfrac{3k^{-1}t}{5k^{-7}t^4}\right)^{-3}$ $\dfrac{125t^9}{27k^{18}}$

h) $\left(\dfrac{40}{21}x^{10}\right)(3x^{-12})\left(\dfrac{49}{20}x^2\right)$ 14

16) Simplify by applying one or more of the rules of exponents.

a) $\left(\dfrac{4}{3}\right)^8\left(\dfrac{4}{3}\right)^{-2}\left(\dfrac{4}{3}\right)^{-3}$ $\dfrac{64}{27}$

b) $\left(\dfrac{k^{10}}{k^4}\right)^3$ k^{18}

c) $\left(\dfrac{x^{-4}y^{11}}{xy^2}\right)^{-2}$ $\dfrac{x^{10}}{y^{18}}$

d) $(-9z^5)^{-2}$ $\dfrac{1}{81z^{10}}$

e) $\left(\dfrac{g^2\cdot g^{-1}}{g^{-7}}\right)^{-4}$ $\dfrac{1}{g^{32}}$

f) $(12p^{-3})\left(\dfrac{10}{3}p^5\right)\left(\dfrac{1}{4}p^2\right)^2$ $\dfrac{5}{2}p^6$

g) $\left(\dfrac{30u^2v^{-3}}{40u^7v^{-7}}\right)^{-2}$ $\dfrac{16u^{10}}{9v^8}$

h) $-5(3h^4k^9)^2$ $-45h^8k^{18}$

Simplify. Assume that the variables represent nonzero integers. Write your final answer so that the exponents have positive coefficients.

17) $y^{3k}\cdot y^{7k}$ y^{10k}

18) $\dfrac{z^{12c}}{z^{5c}}$ z^{7c}

(5.4) Write each number without an exponent.

19) 9.38×10^5 $938,000$

20) -4.185×10^2 -418.5

21) $9\cdot10^3$ 9000

22) $6.7\cdot10^{-4}$ 0.00067

23) 1.05×10^{-6} 0.00000105

24) 2×10^4 $20,000$

Write each number in scientific notation.

25) 0.0000575 5.75×10^{-5}

26) $36,940$ 3.694×10^4

27) $32,000,000$ 3.2×10^7

28) 0.0000004 4×10^{-7}

29) 0.0009315 9.315×10^{-4}

30) 66 6.6×10^1

Write the number without exponents.

31) Before 2010, golfer Tiger Woods earned over 7×10^7 dollars per year in product endorsements. (www.forbes.com)
$70,000,000$

Perform the operation as indicated. Write the final answer without an exponent.

32) $\dfrac{8\cdot10^6}{2\cdot10^{13}}$ 0.0000004

33) $\dfrac{-1\cdot10^9}{5\cdot10^{12}}$ -0.0002

34) $(9\times10^{-8})(4\times10^7)$ 3.6

35) $(5\cdot10^3)(3.8\cdot10^{-8})$ 0.00019

36) $\dfrac{-3\times10^{10}}{-4\times10^6}$ 7500

37) $(-4.2\times10^2)(3.1\times10^3)$ $-1,302,000$

For each problem, write each of the numbers in scientific notation, then solve the problem. Write the answer without exponents.

38) Eight porcupines have a total of about $2.4\cdot10^5$ quills on their bodies. How many quills would one porcupine have? $30,000$ quills

39) In 2010, North Dakota had approximately 4.0×10^7 acres of farmland and about 32,000 farms. What was the average size of a Nebraska farm in 2010? (www.census.gov) 1250 acres

40) One molecule of water has a mass of 2.99×10^{-23} g. Find the mass of 100,000,000 molecules. 0.00000000000000299 g

41) At the end of 2008, the number of SMS text messages sent in one month in the United States was 110.4 billion. If 270.3 million people used SMS text messaging, about how many did each person send that month? (Round to the nearest whole number.) (www.ctia.org/advocacy/research/index.cfm/AID/10323) 408 texts per person

42) When the polls closed on the west coast on November 4, 2008, and Barack Obama was declared the new president, there were about 143,000 visits per second to news websites. If the visits continued at that rate for 3 minutes, how many visits did the news websites receive during that time? (www.xconomy.com) $25,740,000$

Chapter 5:   Test

Additional answers can be found in the Answers to Exercises appendix.

Write in exponential form.

1) $(-3)(-3)(-3)$ $(-3)^3$

2) $x \cdot x \cdot x \cdot x \cdot x$ x^5

Use the rules of exponents to simplify.

3) $5^2 \cdot 5$ 125

4) $\left(\dfrac{1}{x}\right)^5 \cdot \left(\dfrac{1}{x}\right)^2$ $\dfrac{1}{x^7}$

5) $(8^3)^{12}$ 8^{36}

6) $p^7 \cdot p^{-2}$ p^5

Evaluate.

7) 3^4 81

8) 8^0 1

9) 2^{-5} $\dfrac{1}{32}$

10) $4^{-2} + 2^{-3}$ $\dfrac{3}{16}$

11) $\left(-\dfrac{3}{4}\right)^3$ $-\dfrac{27}{64}$

12) $\left(\dfrac{10}{7}\right)^{-2}$ $\dfrac{49}{100}$

Simplify using the rules of exponents. Assume that all variables represent nonzero real numbers. The final answer should not contain negative exponents.

13) $(5n^6)^3$ $125n^{18}$

14) $(-3p^4)(10p^8)$ $-30p^{12}$

15) $\dfrac{m^{10}}{m^4}$ m^6

16) $\dfrac{a^9 b}{a^5 b^7}$ $\dfrac{a^4}{b^6}$

17) $\left(\dfrac{-12t^{-6}u^8}{4t^5u^{-1}}\right)^{-3}$ $-\dfrac{t^{33}}{27u^{27}}$

18) $(2y^{-4})^6\left(\dfrac{1}{2}y^5\right)^3$ $\dfrac{8}{y^9}$

19) $\left(\dfrac{(9x^2y^{-2})^3}{4xy}\right)^0$ 1

20) $\dfrac{(2m+n)^3}{(2m+n)^2}$ $2m+n$

21) $\dfrac{12a^4b^{-3}}{20c^{-2}d^3}$ $\dfrac{3a^4c^2}{5b^3d^3}$

22) $\left(\dfrac{y^{-7} \cdot y^3}{y^5}\right)^{-2}$ y^{18}

23) Simplify $t^{10k} \cdot t^{3k}$. Assume that the variables represent nonzero integers. t^{13k}

24) Rewrite $7.283 \cdot 10^5$ without exponents. $728{,}300$

25) Write 0.000165 in scientific notation. 1.65×10^{-4}

26) Divide. Write the answer without exponents. $\dfrac{-7.5 \times 10^{12}}{1.5 \times 10^8}$ $-50{,}000$

27) Write the number without an exponent: In 2010, the population of Texas was about $2.51 \cdot 10^7$. (www.census.gov) $25{,}100{,}000$

28) An electron is a subatomic particle with a mass of 9.1×10^{-28} g. What is the mass of $2{,}000{,}000{,}000$ electrons? Write the answer without exponents. 0.0000000000000000182 g

Chapter 5:   Cumulative Review for Chapters 1–5

Additional answers can be found in the Answers to Exercises appendix.

Perform the indicated operations. Write the answer in lowest terms.

1) $\dfrac{2}{15} + \dfrac{1}{10} + \dfrac{7}{20}$ $\dfrac{7}{12}$

2) $\dfrac{4}{15} \div \dfrac{20}{21}$ $\dfrac{7}{25}$

3) $-26 + 5 - 7$ -28

4) $(5+1)^2 - 2[17 + 5(10-14)]$ 42

5) Glen Crest High School is building a new football field. The dimensions of a regulation-size field are $53\dfrac{1}{3}$ yd by 120 yd (including the 10 yd of end zone on each end). The sod for the field will cost \$1.80/yd^2.

   a) Find the perimeter of the field. $346\dfrac{2}{3}$ yd

   b) How much will it cost to sod the field?   \$11,520

6) Evaluate $2p^2 - 11q$ when $p = 3$ and $q = -4$. 62

7) Given this set of numbers $\left\{3, -4, -2.1\overline{3}, \sqrt{11}, 2\dfrac{2}{3}\right\}$, list the

   a) integers $-4, 3$

   b) irrational numbers $\sqrt{11}$

   c) natural numbers 3

   d) rational numbers $3, -4, -2.1\overline{3}, 2\dfrac{2}{3}$

   e) whole numbers 3

8) Combine like terms and simplify:
   $5(t^2 + 7t - 3) - 2(4t^2 - t + 5)$ $-3t^2 + 37t - 25$

9) Let x represent the unknown quantity, and write a mathematical expression for "thirteen less than half of a number." $\dfrac{1}{2}x - 13$

Solve.

10) $8t - 17 = 10t + 6$ $\left\{-\dfrac{23}{2}\right\}$

11) $\dfrac{x+3}{10} = \dfrac{2x-1}{4}$ $\left\{\dfrac{11}{8}\right\}$

Solve. Write the answer in interval notation.

12) $k - 2(3k - 7) \le 3(k + 4) - 6$ $[1, \infty)$

13) Write the equation of the line containing the points $(-8, -2)$ and $(6, 5)$. Write the answer in slope-intercept form. $y = \dfrac{1}{2}x + 2$

Determine whether the two lines are parallel, perpendicular, or neither.

14) $5y - 3x = 1$
    $3x - 5y = -8$ parallel

Solve each system of equations.

15) $9x - 2y = 8$
    $y = 2x + 1$ $(2, 5)$

16) $y = -6x + 5$
    $12x + 2y = 10$ $\{(x, y)\,|\, y = -6x + 5\}$

Simplify using the rules of exponents. The answer should not contain negative exponents. Assume that the variables represent nonzero real numbers.

17) $4^3 \cdot 4^7$ 4^{10}

18) $\left(\dfrac{x}{y}\right)^{-3}$ $\dfrac{y^3}{x^3}$

19) $\left(\dfrac{32x^3}{8x^{-2}}\right)^{-1}$ $\dfrac{1}{4x^5}$

20) $-(3rt^{-3})^4$ $-\dfrac{81r^4}{t^{12}}$

21) $(4z^3)(-7z^5)$ $-28z^8$

22) $\dfrac{n^2}{n^9}$ $\dfrac{1}{n^7}$

23) $(-2a^{-6}b)^5$ $-\dfrac{32b^5}{a^{30}}$

24) Write 0.000729 in scientific notation. 7.29×10^{-4}

25) Perform the indicated operation. Write the final answer without an exponent. $(6.2 \times 10^5)(9.4 \times 10^{-2})$ $58{,}280$

Polynomials

Math at Work:

Motorcycle Customization

Let's take a final look at how algebra is used to build motorcycles in a custom chopper shop. In this case, the support bracket for the fender of a custom motorcycle must be fabricated. To save money, Frank's boss tells him to use a piece of scrap metal and not a new piece. So, Frank has to figure out how big a piece of scrap metal he needs to be able to cut out the shape required to make the fender.

Frank draws the sketch on the right, which shows the shape and dimension of the piece of metal to be cut so that it could be bent into the correct shape and size for the fender. He knows that the height of the piece must be 2.84 in. To determine the width of the piece of metal that he needs, Jim analyzes the sketch and writes the equation

$$[(1.42)^2 - d^2] + (2.84 - d)^2 = (2.46)^2$$

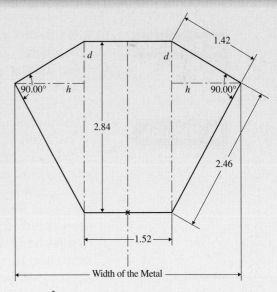

In order to solve this equation, Frank has to square the binomial $(2.84 - d)^2$, something we will learn how to do in this chapter. When he solves the equation, Frank determines that $d \approx 0.71$ in. He uses this value of d to find that the width of the piece of metal that he must use to cut the correct shape for the fender is 3.98 in.

Frank has a lot of responsibilities at work, and some days there simply isn't enough time to finish everything he needs to do. In those cases, he might bring some sketches or plans home with him and finish his work there. Frank says he doesn't mind doing a little bit of work at home. It helps keep his mind engaged for the next day of work, and he always makes sure he leaves enough time in the evening to relax with his family.

We will learn how to square binomials and perform other operations with polynomials in this chapter. We will also introduce strategies that will help you do well on the work you need to do at home—your math homework.

Learning happens in a variety of ways. You can learn by reading your textbook; you can learn by listening to your instructor in class. Yet when it comes to math, perhaps the deepest learning occurs when you work through problems on your own, taking the time to understand and apply the larger concepts you are studying. Homework, then, is an essential component to any math course. The strategies below will help you get the most out of the time you spend doing math homework.

- Choose a place to do your homework that is free of distractions, such as the library or a quiet room in your home.
- Before you start, make sure you have what you need to complete the homework: a pen, pencils, paper, your textbook, and class notes—and, of course, the assignment itself!

- Review the material in your textbook or class notes that cover the topics in your homework.
- Write down any formulas or other major pieces of information you will need to complete the homework so that you can reference them easily.

W Work

- Always show your work. If your instructor sees how you arrived at the solution to a problem, he or she can point out the errors in your thinking or calculations.
- Try to connect each problem you do to the larger concepts that it involves.
- If after a lot of trying, you just can't solve a problem, plan to get some help from a classmate, math tutor, or your instructor.

- Avoid careless errors by double-checking your work and reviewing all your calculations.
- Create a list of any topics you struggled with so you remember to ask about them in class.

R Rethink

- When your homework is returned to you, review it carefully. If it turns out you didn't understand a major concept, find time to seek out extra help.
- Never throw away your homework! Keep all your returned assignments for each class organized and in the same place. Homework represents an excellent study aid.

Chapter 6 **POWER** Plan

What are your goals for Chapter 6?	How can you accomplish each goal?
1 Be prepared before and during class.	• Don't stay out late the night before, and be sure to set your alarm clock! • Bring a pencil, notebook paper, and textbook to class. • Avoid distractions by turning off your cell phone during class. • Pay attention, take good notes, and ask questions. • Complete your homework on time, and ask questions on problems you do not understand.
2 Understand the homework to the point where you could do it without needing any help or hints.	• Read the directions, and show all your steps. • Go to the professor's office for help. • Rework homework and quiz problems, and find similar problems for practice.
3 Use the P.O.W.E.R. framework to help you do your math homework: *What's Different About College?*	• Read the Study Strategy that explains how to do your math homework. • Where can you begin using these techniques? • Complete the emPOWERme that appears before the Chapter Summary.
4 Write your own goal. _____ _____	• _____ _____

What are your objectives for Chapter 6?	How can you accomplish each objective?
1 Know how to add and subtract polynomials.	• Learn the basic terminology associated with polynomials. • Understand how to use like terms to add and subtract polynomials.

2 Know how to multiply any combination of polynomials.	• Use the distributive property to multiply polynomials.
	• Learn special techniques for multiplying binomials:
	  • FOIL
	  • Sum and difference of two terms
	  • Square of a binomial
	  • Higher powers

3 Know how to divide polynomials.	• Learn the procedure for dividing a polynomial by a monomial. • Review the process of long division, and apply it to the process for dividing a polynomial by a polynomial.
4 Write your own goal. _____ _____	• _____ _____

6.1 Addition and Subtraction of Polynomials

P Prepare   **O Organize**

What are your objectives for Section 6.1?	How can you accomplish each objective?
1 Learn the Vocabulary Associated with Polynomials	• Write the definition of *polynomial* in your own words. • Understand the defining features of a polynomial, such as descending powers of *x*, constant, degree of a term, and degree of a polynomial. • Be able to compare and contrast a *monomial,* a *binomial,* and a *trinomial.* • Complete the given example on your own. • Complete You Try 1.
2 Evaluate Polynomials	• Complete the given example on your own. • Complete You Try 2.
3 Add Polynomials	• Learn the procedure for **Adding Polynomials,** and write an example in your notes. • Complete the given examples on your own. • Complete You Try 3.
4 Subtract Polynomials	• Learn the procedure for **Subtracting Polynomials,** and write an example in your notes. • Complete the given examples on your own. • Complete You Try 4.
5 Add and Subtract Polynomials in More Than One Variable	• Know how to use the previous procedures, and line up the like terms. • Complete the given example on your own. • Complete You Try 5.

1 Learn the Vocabulary Associated with Polynomials

In Section 1.8, we defined an *algebraic expression* as a collection of numbers, variables, and grouping symbols connected by operation symbols such as $+$, $-$, $\times$, and $\div$.

An example of an algebraic expression is

$$5x^3 + \frac{7}{4}x^2 - x + 9$$

The *terms* of this algebraic expression are $5x^3$, $\frac{7}{4}x^2$, $-x$, and 9. A *term* is a number or a variable or a product or quotient of numbers and variables.

The expression $5x^3 + \frac{7}{4}x^2 - x + 9$ is also a *polynomial*.

Definition

A **polynomial in** x is the sum of a finite number of terms of the form ax^n, where n is a whole number and a is a real number. (The exponents must be whole numbers.)

Let's look more closely at the polynomial $5x^3 + \frac{7}{4}x^2 - x + 9$.

1) The polynomial is written **in descending powers of** x since the powers of x decrease from left to right. Generally, we write polynomials in descending powers of the variable.

2) Recall that the term without a variable is called a **constant**. The constant is 9. The **degree of a term** equals the exponent on its variable. (If a term has more than one variable, the degree equals the *sum* of the exponents on the variables.) We will list each term, its coefficient, and its degree.

Term	Coefficient	Degree	
$5x^3$	5	3	
$\frac{7}{4}x^2$	$\frac{7}{4}$	2	
$-x$	-1	1	
9	9	0	$(9 = 9x^0)$

3) The **degree of the polynomial** equals the highest degree of any nonzero term. The degree of $5x^3 + \frac{7}{4}x^2 - x + 9$ is 3. Or, we say that this is a **third-degree polynomial.**

EXAMPLE 1

Decide whether each expression is or is not a polynomial. If it is a polynomial, identify each term and the degree of each term. Then, find the degree of the polynomial.

a) $-8p^4 + 5.7p^3 - 9p^2 - 13$

b) $4c^2 - \dfrac{2}{5}c + 3 + \dfrac{6}{c^2}$

c) $a^3b^3 + 4a^3b^2 - ab + 1$

d) $7n^6$

Decide whether each expression is or is not a polynomial. If it is a polynomial, identify each term and the degree of each term. Then, find the degree of the polynomial.

a) $3a^2 - \dfrac{4}{5}a - 2 + \dfrac{8}{a^2}$

b) $-7k^4 + 1.3k^3 - 8k^2 - 11$

c) $x^3y^3 + 12x^3y^2 - xy + 5$

d) $9w^5$

Answer: a) not a polynomial
b) polynomial

Term	Degree
$-7k^4$	4
$1.3k^3$	3
$-8k^2$	2
-11	0

degree: 4

c) polynomial

Term	Degree
x^3y^3	6
$12x^3y^2$	5
$-xy$	2
5	0

degree: 6

d) polynomial

Term	Degree
$9w^5$	5

degree: 5

Solution

a) The expression $-8p^4 + 5.7p^3 - 9p^2 - 13$ *is* a polynomial in *p*. Its terms have whole-number exponents and real coefficients. The term with the highest degree is $-8p^4$, so the degree of the polynomial is 4.

Term	Degree
$-8p^4$	4
$5.7p^3$	3
$-9p^2$	2
-13	0

b) The expression $4c^2 - \dfrac{2}{5}c + 3 + \dfrac{6}{c^2}$ is *not* a polynomial because one of its terms has a variable in the denominator.

$$\left(\dfrac{6}{c^2} = 6c^{-2}; \text{ the exponent } -2 \text{ is not a whole number.}\right)$$

c) The expression $a^3b^3 + 4a^3b^2 - ab + 1$ *is* a polynomial because the variables have whole-number exponents and the coefficients are real numbers. Since this is a polynomial in two variables, we find the degree of each term by adding the exponents. The first term, a^3b^3, has the highest degree, 6, so the polynomial has degree 6.

Term	Degree
a^3b^3	6
$4a^3b^2$	5
$-ab$	2
1	0

Add the exponents to get the degree.

d) The expression $7n^6$ *is* a polynomial even though it has only one term. The degree of the term is 6, and that is the degree of the polynomial as well.

[YOU TRY 1]

Decide whether each expression *is* or *is not* a polynomial. If it is a polynomial, identify each term and the degree of each term. Then, find the degree of the polynomial.

a) $d^4 + 7d^3 + \dfrac{3}{d}$

b) $k^3 - k^2 - 3.8k + 10$

c) $5x^2y^2 + \dfrac{1}{2}xy - 6x - 1$

d) $2r + 3r^{1/2} - 7$

The polynomial in Example 1d) is $7n^6$ and has one term. We call $7n^6$ a *monomial*. A **monomial** is a polynomial that consists of one term (*mono* means one). Some other examples of monomials are

$$y^2, \qquad -4t^5, \qquad x, \qquad m^2n^2, \qquad \text{and} \qquad -3$$

A **binomial** is a polynomial that consists of exactly two terms (*bi* means two). Some examples are

$$w + 2, \qquad 4z^2 - 11, \qquad a^4 - b^4, \qquad \text{and} \qquad -8c^3d^2 + 3cd$$

A **trinomial** is a polynomial that consists of exactly three terms (*tri* means three). Here are some examples:

$$x^2 - 3x - 40, \qquad 2q^4 - 18q^2 + 10q, \qquad \text{and} \qquad 6a^4 + 29a^2b + 28b^2$$

Expressions have different values depending on the value of the variable(s). The same is true for polynomials.

2 Evaluate Polynomials

EXAMPLE 2

W Hint

Remember that *evaluate* means *find the value of the expression* for a given value of the variable. You are *not* solving for a variable.

Evaluate the trinomial $n^2 - 7n + 4$ when

a) $n = 3$ b) $n = -2$

Solution

a) Substitute 3 for n in $n^2 - 7n + 4$. Remember to put 3 in parentheses.

$$\begin{aligned} n^2 - 7n + 4 &= (3)^2 - 7(3) + 4 & \text{Substitute.} \\ &= 9 - 21 + 4 \\ &= -8 & \text{Add.} \end{aligned}$$

b) Substitute -2 for n in $n^2 - 7n + 4$. Put -2 in parentheses.

$$\begin{aligned} n^2 - 7n + 4 &= (-2)^2 - 7(-2) + 4 & \text{Substitute.} \\ &= 4 + 14 + 4 \\ &= 22 & \text{Add.} \end{aligned}$$

[YOU TRY 2]

Evaluate $t^2 - 9t - 6$ when

a) $t = 5$ b) $t = -4$

Recall in Section 1.8 we said that **like terms** contain the same variables with the same exponents. We add or subtract like terms by adding or subtracting the coefficients and leaving the variable(s) and exponent(s) the same. We use the same idea for adding and subtracting polynomials.

3 Add Polynomials

Procedure Adding Polynomials

To add polynomials, add like terms.

We can set up an addition problem vertically or horizontally.

EXAMPLE 3

In-Class Example 3

Add $(t^3 + 4t^2 - 7t - 5) + (6t^3 + t^2 - 3)$.

Answer: $7t^3 + 5t^2 - 7t - 8$

Add $(m^3 - 9m^2 + 5m - 4) + (2m^3 + 3m^2 - 1)$.

Solution

We will add these vertically. Line up like terms in columns and add.

$$
\begin{array}{r}
m^3 - 9m^2 + 5m - 4 \\
+ \ 2m^3 + 3m^2 \quad\ \ \ - 1 \\
\hline
3m^3 - 6m^2 + 5m - 5
\end{array}
$$

[YOU TRY 3] Add $(6b^3 - 11b^2 + 3b + 3) + (-2b^3 - 6b^2 + b - 8)$.

EXAMPLE 4

In-Class Example 4

Add $9a^2 - 11a + 3$ and $6a^2 + 9a + 5$.

Answer: $15a^2 - 2a + 8$

Add $10k^2 + 2k - 1$ and $5k^2 + 7k + 9$.

Solution

Let's add these horizontally. Put the polynomials in parentheses since each contains more than one term. Use the associative and commutative properties to rewrite like terms together.

$$(10k^2 + 2k - 1) + (5k^2 + 7k + 9) = (10k^2 + 5k^2) + (2k + 7k) + (-1 + 9)$$
$$= 15k^2 + 9k + 8 \qquad \text{Combine like terms.}$$

W Hint

Remember that the exponents remain unchanged when we add like terms.

4 Subtract Polynomials

To subtract two polynomials such as $(8x + 3) - (5x - 4)$, we will be using the distributive property to clear the parentheses in the second polynomial.

EXAMPLE 5

In-Class Example 5

Subtract $(6c + 1) - (4c - 9)$.

Answer: $2c + 10$

Subtract $(8x + 3) - (5x - 4)$.

Solution

$$
\begin{aligned}
(8x + 3) - (5x - 4) &= (8x + 3) - 1(5x - 4) \\
&= (8x + 3) + (-1)(5x - 4) && \text{Change } -1 \text{ to } + (-1). \\
&= (8x + 3) + (-5x + 4) && \text{Distribute.} \\
&= 3x + 7 && \text{Combine like terms.}
\end{aligned}
$$

In Example 5, notice that we changed the sign of each term in the second polynomial and then added it to the first.

Procedure Subtracting Polynomials

To subtract two polynomials, change the sign of each term in the second polynomial. Then, add the polynomials.

Let's see how we use this rule to subtract polynomials.

EXAMPLE 6

Subtract $(-6w^3 - w^2 + 10w + 1) - (2w^3 - 4w^2 + 9w - 5)$ vertically.

In-Class Example 6

Subtract
$(-5n^3 - n^2 + 7n + 7) -$
$(3n^3 - 8n^2 + 2n - 4)$.

Answer:
$-8n^3 + 7n^2 + 5n + 11$

Solution

To subtract vertically, line up like terms in columns.

$$
\begin{array}{l}
-6w^3 - w^2 + 10w + 1 \\
-\ (2w^3 - 4w^2 + 9w - 5)
\end{array}
$$

Change the signs in the second polynomial and add the polynomials.

$$
\begin{array}{l}
-6w^3 - w^2 + 10w + 1 \\
+\ -2w^3 + 4w^2 - 9w + 5 \\
\hline
-8w^3 + 3w^2 + w + 6
\end{array}
$$

[YOU TRY 4]

Subtract $(-7h^2 - 8h + 1) - (-3h^2 + h - 4)$.

5 Add and Subtract Polynomials in More Than One Variable

To add and subtract polynomials in more than one variable, remember that like terms contain the same variables with the same exponents.

EXAMPLE 7

Perform the indicated operation.

a) $(a^2b^2 + 2a^2b - 13ab - 4) + (9a^2b^2 - 5a^2b - ab + 17)$

b) $(6tu - t + 2u + 5) - (4tu + 8t - 2)$

In-Class Example 7

Perform the indicated operation.
a) $(6a^2b^2 + a^2b - 4ab - 13) +$ $(5a^2b^2 - 7a^2b - 2ab + 18)$
b) $(2xy - 3x + 4y - 1) -$ $(3xy + 6x - 4)$

Answer:
a) $11a^2b^2 - 6a^2b - 6ab + 5$
b) $-xy - 9x + 4y + 3$

Solution

a) $(a^2b^2 + 2a^2b - 13ab - 4) + (9a^2b^2 - 5a^2b - ab + 17)$
$= 10a^2b^2 - 3a^2b - 14ab + 13$ Combine like terms.

b) $(6tu - t + 2u + 5) - (4tu + 8t - 2) = (6tu - t + 2u + 5) - 4tu - 8t + 2$
$= 2tu - 9t + 2u + 7$ Combine like terms.

[YOU TRY 5]

Perform the indicated operation.

a) $(-12x^2y^2 + xy - 6y + 1) - (-4x^2y^2 - 10xy + 3y + 6)$

b) $(3.6m^3n^2 + 8.1mn - 10n) + (8.5m^3n^2 - 11.2mn + 4.3)$

ANSWERS TO [YOU TRY] EXERCISES

1) a) not a polynomial   b) polynomial of degree 3   c) polynomial of degree 4

Term	Degree
k^3	3
$-k^2$	2
$-3.8k$	1
10	0

Term	Degree
$5x^2y^2$	4
$\frac{1}{2}xy$	2
$-6x$	1
-1	0

d) not a polynomial   2) a) -26 b) 46   3) $4b^3 - 17b^2 + 4b - 5$ 4) $-4h^2 - 9h + 5$
5) a) $-8x^2y^2 + 11xy - 9y - 5$ b) $12.1m^3n^2 - 3.1mn - 10n + 4.3$

*Additional answers can be found in the Answers to Exercises appendix.

Objective 1: Learn the Vocabulary Associated with Polynomials

Is the given expression a polynomial? Why or why not?

 1) $-2p^2 - 5p + 6$ 2) $8r^3 + 7r^2 - t + \dfrac{4}{5}$

3) $c^3 + 5c^2 + 4c^{-1} - 8$ 4) $9a^5$

5) $f^{3/4} + 6f^{2/3} + 1$ 6) $7y - 1 + \dfrac{3}{y}$

Determine whether each is a monomial, a binomial, or a trinomial.

7) $4x - 1$ binomial   8) $-5q^2$ monomial

9) $m^2n^2 - mn + 13$ 10) $11c^2 + 3c$ binomial
  trinomial

11) 8 monomial   12) $k^5 + 2k^3 + 8k$ trinomial

For Exercises 13 and 14, answer *always, sometimes,* or *never.*

13) If a, b, and c are any integer values except zero, then the expression $ax^2 + bx + c$ can *always, sometimes,* or *never* be classified as a trinomial.   always

14) If a, b, and c are all different integer values, then the expression $ax^2 + bx + c$ can *always, sometimes,* or *never* be classified as a binomial.   sometimes

15) How do you determine the degree of a polynomial in one variable?   It is the same as the degree of the term in the polynomial with the highest degree.

16) Write a third-degree polynomial in one variable.
  Answers may vary.

17) How do you determine the degree of a term in a polynomial in more than one variable?
  Add the exponents on the variables.

18) Write a fourth-degree monomial in x and y.
  Answers may vary.

For each polynomial, identify each term in the polynomial, the coefficient and degree of each term, and the degree of the polynomial.

19) $3y^4 + 7y^3 - 2y + 8$ 20) $6a^2 + 2a - 11$

21) $-4x^2y^3 - x^2y^2 + \dfrac{2}{3}xy + 5y$

22) $3c^2d^2 + 0.7c^2d + cd - 1$

Objective 2: Evaluate Polynomials

Evaluate each polynomial when a) $r = 3$ and b) $r = -1$.

23) $2r^2 - 7r + 4$ a) 1   24) $2r^3 + 5r - 6$ a) 63
                        b) 13   b) −13

Evaluate each polynomial when $x = 5$ and $y = -2$.

25) $9x + 4y$ 37   26) $-2x + 3y + 16$ 0

27) $x^2y^2 - 5xy + 2y$ 146   28) $-2xy^2 + 7xy + 12y - 6$
                                      −140

29) $\dfrac{1}{2}xy - 4x - y$ −23   30) $x^2 - y^2$ 21

For Exercises 31 and 32, answer *always, sometimes,* or *never.*

31) If x is any real number, a polynomial of the form $x^4 + x^2 + 1$ will *always, sometimes,* or *never* have a negative value.   never

32) If x is any real number, a polynomial of the form $x^3 + x^2 + x + 1$ will *always, sometimes,* or *never* have a negative value.   sometimes

33) Bob will make a new gravel road from the highway to his house. The cost of building the road, y (in dollars), includes the cost of the gravel and is given by $y = 60x + 380$, where x is the number of hours he rents the equipment needed to complete the job.

a) Evaluate the binomial when $x = 5$, and explain what it means in the context of the problem. $y = 680$; if he rents the equipment for 5 hours, the cost of building the road will be $680.00.

b) If he keeps the equipment for 9 hours, how much will it cost to build the road? $920.00

c) If it cost $860.00 to build the road, for how long did Bob rent the equipment?   8 hours

34) An object is thrown upward so that its height, y (in feet), x seconds after being thrown is given by
$$y = -16x^2 + 48x + 64$$

a) Evaluate the polynomial when $x = 2$, and explain what it means in the context of the problem. $y = 96$; two seconds after the object is thrown, it will be 96 feet above the ground.

b) What is the height of the object 3 seconds after it is thrown?   64 ft

c) Evaluate the polynomial when $x = 4$, and explain what it means in the context of the problem.
$y = 0$; four seconds after the object is thrown, it will hit the ground.

Objective 3: Add Polynomials

Add like terms.

35) $-6z + 8z + 11z$ 13z

36) $m^2 + 7m^2 - 14m^2$ −6m²

37) $5c^2 + 9c - 16c^2 + c - 3c$ $\quad -11c^2 + 7c$

38) $-4y^3 + 3y^5 + 17y^5 + 6y^3 - 5y^5$ $\quad 15y^5 + 2y^3$

39) $6.7t^2 - 9.1t^6 - 2.5t^2 + 4.8t^6$ $\quad -4.3t^6 + 4.2t^2$

40) $\frac{5}{4}w^3 + \frac{3}{8}w^4 - \frac{2}{3}w^4 - \frac{5}{6}w^3$ $\quad -\frac{7}{24}w^4 + \frac{5}{12}w^3$

41) $7a^2b^2 + 4ab^2 - 16ab^2 - a^2b^2 + 5ab^2$ $\quad 6a^2b^2 - 7ab^2$

42) $x^5y^2 - 14xy + 6xy + 5x^5y^2 + 8xy$ $\quad 6x^5y^2$

Add the polynomials.

43) $\quad 5n - 8$ $\quad 9n - 5$
$\quad + \underline{4n + 3}$

44) $\quad 9d + 14$ $\quad 11d + 19$
$\quad + \underline{2d + 5}$

45) $\quad -7a^3 + 11a$ $\quad -5a^3 + 7a$
$\quad + \underline{2a^3 - 4a}$

46) $\quad -h^4 + 6h^2$ $\quad 4h^4 + 3h^2$
$\quad + \underline{5h^4 - 3h^2}$

47) $\quad 9r^2 + 16r + 2$
$\quad + \underline{3r^2 - 10r + 9}$
$\quad \quad 12r^2 + 6r + 11$

48) $\quad m^2 - 3m - 8$
$\quad + \underline{2m^2 + 7m + 1}$
$\quad \quad 3m^2 + 4m - 7$

49) $\quad b^2 - 8b - 14$
$\quad + \underline{3b^2 + 8b + 11}$
$\quad \quad 4b^2 - 3$

50) $\quad 8g^2 + g + 5$
$\quad + \underline{5g^2 - 6g - 5}$
$\quad \quad 13g^2 - 5g$

51) $\quad \frac{5}{6}w^4 - \frac{2}{3}w^2 \quad\quad + \frac{1}{2}$
$\quad + \underline{\frac{4}{9}w^4 + \frac{1}{6}w^2 - \frac{3}{8}w - 2}$
$\quad\quad\quad \frac{7}{18}w^4 - \frac{1}{2}w^2 - \frac{3}{8}w - \frac{3}{2}$

52) $\quad -1.7p^3 - 2p^2 + 3.8p - 6$ $\quad 4.5p^3 - 2p^2 + 2.6p + 8$
$\quad + \underline{6.2p^3 \quad\quad\quad - 1.2p + 14}$

53) $(6m^2 - 5m + 10) + (-4m^2 + 8m + 9)$ $\quad 2m^2 + 3m + 19$

54) $(3t^4 - 2t^2 + 11) + (t^4 + t^2 - 7)$ $\quad 4t^4 - t^2 + 4$

55) $\left(-2c^4 - \frac{7}{10}c^3 + \frac{3}{4}c - \frac{2}{9}\right)$ $\quad 10c^4 - \frac{1}{5}c^3 - \frac{1}{4}c + \frac{25}{9}$

$\quad + \left(12c^4 + \frac{1}{2}c^3 - c + 3\right)$

56) $\left(\frac{7}{4}y^3 - \frac{3}{8}\right) + \left(\frac{5}{6}y^3 + \frac{7}{6}y^2 - \frac{9}{16}\right)$ $\quad \frac{31}{12}y^3 + \frac{7}{6}y^2 - \frac{15}{16}$

57) $(2.7d^3 + 5.6d^2 - 7d + 3.1)$
$\quad + (-1.5d^3 + 2.1d^2 - 4.3d - 2.5)$
$\quad\quad 1.2d^3 + 7.7d^2 - 11.3d + 0.6$

58) $(0.2t^4 - 3.2t + 4.1)$
$\quad + (-2.7t^4 + 0.8t^3 - 6.4t + 3.9)$
$\quad\quad -2.5t^4 + 0.8t^3 - 9.6t + 8.0$

Objective 4: Subtract Polynomials

Subtract the polynomials.

59) $\quad 15w + 7$ $\quad 12w - 4$
$\quad - \underline{3w + 11}$

60) $\quad 12a - 8$ $\quad 10a - 17$
$\quad - \underline{2a + 9}$

61) $\quad y - 6$ $\quad -y + 2$
$\quad - \underline{2y - 8}$

62) $\quad 6p + 1$ $\quad -3p + 18$
$\quad - \underline{9p - 17}$

63) $\quad 3b^2 - 8b + 12$
$\quad - \underline{5b^2 + 2b - 7}$
$\quad\quad -2b^2 - 10b + 19$

64) $\quad -7d^2 + 15d + 6$
$\quad - \underline{8d^2 + 3d - 9}$
$\quad\quad -15d^2 + 12d + 15$

65) $\quad f^4 - 6f^3 + 5f^2 - 8f + 13$
$\quad - \underline{-3f^4 + 8f^3 - f^2 \quad\quad + 4}$
$\quad\quad 4f^4 - 14f^3 + 6f^2 - 8f + 9$

66) $\quad 11x^4 + x^3 - 9x^2 + 2x - 4$ $\quad 14x^4 - 9x^2 + 3x - 5$
$\quad - \underline{-3x^4 + x^3 \quad\quad - x + 1}$

67) $\quad 10.7r^2 + 1.2r + 9$ $\quad 5.8r^2 + 6.5r + 11.8$
$\quad - \underline{4.9r^2 - 5.3r - 2.8}$

68) $\quad -\frac{11}{10}m^3 + \frac{1}{2}m - \frac{5}{8}$ $\quad -\frac{3}{2}m^3 + \frac{5}{14}m + \frac{5}{24}$
$\quad - \underline{\frac{2}{5}m^3 + \frac{1}{7}m - \frac{5}{6}}$

69) $(j^2 + 16j) - (-6j^2 + 7j + 5)$ $\quad 7j^2 + 9j - 5$

70) $(-3p^2 + p + 4) - (4p^2 + p + 1)$ $\quad -7p^2 + 3$

71) $(17s^5 - 12s^2) - (9s^5 + 4s^4 - 8s^2 - 1)$ $\quad 8s^5 - 4s^4 - 4s^2 + 1$

72) $(-5d^4 - 8d^2 + d + 3) - (-3d^4 + 17d^3 - 6d^2 - 20)$
$\quad -2d^4 - 17d^3 - 2d^2 + d + 23$

73) $\left(-\frac{3}{8}r^2 + \frac{2}{9}r + \frac{1}{3}\right) - \left(-\frac{7}{16}r^2 - \frac{5}{9}r + \frac{7}{6}\right)$ $\quad \frac{1}{16}r^2 + \frac{7}{9}r - \frac{5}{6}$

74) $(3.8t^5 + 7.5t - 9.6) - (-1.5t^5 + 2.9t^2 - 1.1t + 3.4)$
$\quad 5.3t^5 - 2.9t^2 + 8.6t - 13.0$

75) Explain, in your own words, how to subtract two polynomials. $\quad$ Answers may vary.

76) Do you prefer adding and subtracting polynomials vertically or horizontally? Why? $\quad$ Answers may vary.

77) Will the sum of two trinomials always be a trinomial? Why or why not? Give an example.

78) Write a third-degree polynomial in x that does not contain a second-degree term. $\quad$ Answers may vary.

Mixed Exercises: Objectives 3 and 4

Perform the indicated operations.

79) $(8a^4 - 9a^2 + 17) - (15a^4 + 3a^2 + 3)$ $\quad -7a^4 - 12a^2 + 14$

80) $(-x + 15) + (-5x - 12)$ $\quad -6x + 3$

81) $(-11n^2 - 8n + 21) + (4n^2 + 15n - 3) + (7n^2 - 10)$
$\quad 7n + 8$

82) $(-15a^3 + 8) - (-7a^3 + 3a + 5) + (10a^3 - a + 17)$
$\quad 2a^3 - 4a + 20$

83) $(w^3 + 5w^2 + 3) - (6w^3 - 2w^2 + w + 12) + (9w^3 + 7)$
$\quad 4w^3 + 7w^2 - w - 2$

84) $(3r + 2) - (r^2 + 5r - 1) - (-9r^3 - r + 6)$
$\quad 9r^3 - r^2 - r - 3$

85) $\left(y^3 - \dfrac{3}{4}y^2 - 5y + \dfrac{3}{7}\right)$ $\dfrac{4}{3}y^3 - \dfrac{7}{4}y^2 + 3y - \dfrac{1}{14}$

    $+ \left(\dfrac{1}{3}y^3 - y^2 + 8y - \dfrac{1}{2}\right)$

86) $\left(\dfrac{3}{5}c^4 + c^3 - \dfrac{3}{2}c^2 + 1\right)$ $\dfrac{8}{5}c^4 - 5c^3 - \dfrac{7}{4}c^2 + 6c$

    $+ \left(c^4 - 6c^3 - \dfrac{1}{4}c^2 + 6c - 1\right)$

87) $(3m^3 - 5m^2 + m + 12) - [(7m^3 + 4m^2 - m + 11)$
    $+ (-5m^3 - 2m^2 + 6m + 8)]$ $m^3 - 7m^2 - 4m - 7$

88) $(j^2 - 13j - 9) - [(-7j^2 + 10j - 2) + (4j^2 - 11j - 6)]$
    $4j^2 - 12j - 1$

Perform the indicated operations.

89) Find the sum of $p^2 - 7$ and $8p^2 + 2p - 1$. $9p^2 + 2p - 8$

90) Add $12n - 15$ to $5n + 4$. $17n - 11$

91) Subtract $z^6 - 8z^2 + 13$ from $6z^6 + z^2 + 9$. $5z^6 + 9z^2 - 4$

92) Subtract $-7x^2 + 8x + 2$ from $2x^2 + x$. $9x^2 - 7x - 2$

93) Subtract the sum of $6p^2 + 1$ and $3p^2 - 8p + 4$ from $2p^2 + p + 5$. $-7p^2 + 9p$

94) Subtract $17g^3 + 2g - 10$ from the sum of $5g^3 + g^2 + g$ and $3g^3 - 2g - 7$. $-9g^3 + g^2 - 3g + 3$

For Exercises 95 and 96, answer *always, sometimes,* or *never*.

95) Finding the difference of two second-degree trinomials will *always, sometimes,* or *never* result in a second-degree polynomial.   sometimes

96) Finding the sum of two second-degree trinomials will *always, sometimes,* or *never* result in a polynomial having a degree greater than 2.   never

Objective 5: Add and Subtract Polynomials in More Than One Variable

Each of the polynomials is a polynomial in two variables. Perform the indicated operations.

97) $(5w + 17z) - (w + 3z)$ $4w + 14z$

98) $(-4g - 7h) + (9g + h)$ $5g - 6h$

99) $(ac + 8a + 6c) + (-6ac + 4a - c)$ $-5ac + 12a + 5c$

100) $(11rt - 6r + 2) - (10rt - 7r + 12t + 2)$ $rt + r - 12t$

101) $(-6u^2v^2 + 11uv + 14)$
    $- (-10u^2v^2 - 20uv + 18)$ $4u^2v^2 + 31uv - 4$

102) $(-7j^2k^2 + 9j^2k - 23jk^2 + 13)$
    $+ (10j^2k^2 + 5j^2k - 17)$ $3j^2k^2 + 14j^2k - 23jk^2 - 4$

103) $(12x^3y^2 - 5x^2y^2 + 9x^2y - 17) + (5x^3y^2 + x^2y - 1)$
    $- (6x^2y^2 + 10x^2y + 2)$ $17x^3y^2 - 11x^2y^2 - 20$

104) $(r^3s^2 + 2r^2s^2 + 10) - (7r^3s^2 + 18r^2s^2 - 9)$
    $+ (11r^3s^2 - 3r^2s^2 - 4)$ $5r^3s^2 - 19r^2s^2 + 15$

Find the polynomial that represents the perimeter of each rectangle.

105) $2x + 7$ $6x + 6$ units

    [rectangle] $x - 4$

106) $a^2 + 3a - 4$ $4a^2 - 4a - 6$ units

    [rectangle] $a^2 - 5a + 1$

107) $5p^2 - 2p + 3$ $10p^2 - 2p - 6$ units

    [rectangle] $p - 6$

108) $\dfrac{2}{3}m + 4$ $\dfrac{8}{3}m + 16$ units

    [rectangle] $\dfrac{2}{3}m + 4$

R Rethink

R1) What does it mean to evaluate a polynomial expression?

R2) What are the differences between a binomial, a trinomial, and a polynomial?

R3) Compare and contrast what you knew about adding like terms to adding and subtracting polynomials.

R4) Were there any problems you could not do? If so, write them down or circle them, then ask your instructor for help.

6.2 Multiplication of Polynomials

What are your objectives for Section 6.2?	How can you accomplish each objective?
1 Multiply a Monomial and a Polynomial	• Know which property to use to complete this objective. • Complete the given example on your own. • Complete You Try 1.
2 Multiply Two Polynomials	• Follow the explanation to fully understand the procedure for **Multiplying Polynomials,** and write an example in your notes. • Complete the given examples on your own. • Complete You Try 2.
3 Multiply Two Binomials Using FOIL	• Know what FOIL means, and be able to use it to multiply two binomials. • Complete the given example on your own. • Complete You Try 3.
4 Find the Product of Binomials of the Form $(a + b)(a - b)$	• Follow the explanation to fully understand the procedure for **The Product of the Sum and Difference of Two Terms,** and write an example in your notes. • Complete the given example on your own. • Complete You Try 4.
5 Square a Binomial	• Follow the explanation to fully understand the procedure for **The Square of a Binomial,** and write an example in your notes. • Complete the given example on your own. • Complete You Try 5.
6 Find Higher Powers of a Binomial	• Complete the given example on your own. • Complete You Try 6.

Read the explanations, follow the examples, take notes, and complete the You Trys.

We have already learned that when multiplying two monomials, we multiply the coefficients and add the exponents of the same bases:

$$4c^5 \cdot 3c^6 = 12c^{11} \qquad -3x^2y^4 \cdot 7xy^3 = -21x^3y^7$$

In this section, we will discuss how to multiply other types of polynomials.

1 Multiply a Monomial and a Polynomial

To multiply a monomial and a polynomial, we use the distributive property.

EXAMPLE 1

In-Class Example 1

Multiply $3y^2(8y^2 - 7y + 2)$.

Answer: $24y^4 - 21y^3 + 6y^2$

Multiply $2k^2(6k^2 + 5k - 3)$.

Solution

$$2k^2(6k^2 + 5k - 3) = (2k^2)(6k^2) + (2k^2)(5k) + (2k^2)(-3) \qquad \text{Distribute.}$$
$$= 12k^4 + 10k^3 - 6k^2 \qquad \text{Multiply.}$$

[**YOU TRY 1**] Multiply $5z^4(4z^3 - 7z^2 - z + 8)$.

2 Multiply Two Polynomials

To multiply two polynomials, we use the distributive property repeatedly. For example, to multiply $(2x - 3)(x^2 + 7x + 4)$, we multiply each term in the second polynomial by $(2x - 3)$.

$$(2x - 3)(x^2 + 7x + 4) = (2x - 3)(x^2) + (2x - 3)(7x) + (2x - 3)(4) \qquad \text{Distribute.}$$

Next, we distribute again.

W Hint

Count how many times the distributive property is used.

$$(2x - 3)(x^2) + (2x - 3)(7x) + (2x - 3)(4)$$
$$= (2x)(x^2) - (3)(x^2) + (2x)(7x) - (3)(7x) + (2x)(4) - (3)(4)$$
$$= 2x^3 - 3x^2 + 14x^2 - 21x + 8x - 12 \qquad \text{Multiply.}$$
$$= 2x^3 + 11x^2 - 13x - 12 \qquad \text{Combine like terms.}$$

This process of repeated distribution leads us to the following rule.

Procedure Multiplying Polynomials

To multiply two polynomials, multiply each term in the second polynomial by each term in the first polynomial. Then combine like terms. The answer should be written in descending powers.

Let's use this rule to multiply the polynomials in Example 2.

EXAMPLE 2

In-Class Example 2

Multiply $(c^2 + 4)(3c^3 + c - 8)$.

Answer:
$3c^5 + 13c^3 - 8c^2 + 4c - 32$

Multiply $(n^2 + 5)(2n^3 + n - 9)$.

Solution

Multiply each term in the second polynomial by each term in the first.

$$(n^2 + 5)(2n^3 + n - 9)$$
$$= (n^2)(2n^3) + (n^2)(n) + (n^2)(-9) + (5)(2n^3) + (5)(n) + (5)(-9) \qquad \text{Distribute.}$$
$$= 2n^5 + n^3 - 9n^2 + 10n^3 + 5n - 45 \qquad \text{Multiply.}$$
$$= 2n^5 + 11n^3 - 9n^2 + 5n - 45 \qquad \text{Combine like terms.}$$

Polynomials can be multiplied vertically as well.

Multiply $(a^3 - 4a^2 + 5a - 1)(3a + 7)$ vertically.

Solution

Set up the multiplication problem like you would for whole numbers:

$$
\begin{array}{r}
a^3 - 4a^2 + 5a - 1 \\
\times \qquad\qquad 3a + 7 \\
\hline
7a^3 - 28a^2 + 35a - 7 \\
3a^4 - 12a^3 + 15a^2 - 3a \qquad\quad \\
\hline
3a^4 - 5a^3 - 13a^2 + 32a - 7
\end{array}
$$

Multiply each term in $a^3 - 4a^2 + 5a - 1$ by 7.

Multiply each term in $a^3 - 4a^2 + 5a - 1$ by $3a$.

Line up like terms in the same column. Add.

[YOU TRY 2]   Multiply.

a) $(9x + 5)(2x^2 - x - 3)$ b) $\left(t^2 - \dfrac{2}{3}t - 4\right)(4t^2 + 6t - 5)$

---

3 Multiply Two Binomials Using FOIL

Multiplying two binomials is one of the most common types of polynomial multiplication used in algebra. A method called **FOIL** is one that is often used to multiply two binomials, and it comes from using the distributive property.

Let's use the distributive property to multiply $(x + 6)(x + 4)$.

$$
\begin{aligned}
(x + 6)(x + 4) &= (x + 6)(x) + (x + 6)(4) && \text{Distribute.} \\
&= x(x) + 6(x) + x(4) + 6(4) && \text{Distribute.} \\
&= x^2 + 6x + 4x + 24 && \text{Multiply.} \\
&= x^2 + 10x + 24 && \text{Combine like terms.}
\end{aligned}
$$

To be sure that each term in the first binomial has been multiplied by each term in the second binomial, we can use FOIL. **FOIL** stands for **First Outer Inner Last**. Let's see how we can apply FOIL to the example above:

$$
(x + 6)(x + 4) = (x + 6)(x + 4) = \underset{\text{F}}{x \cdot x} + \underset{\text{O}}{x \cdot 4} + \underset{\text{I}}{6 \cdot x} + \underset{\text{L}}{6 \cdot 4}
$$

$$
= x^2 + 4x + 6x + 24 \qquad \text{Multiply.}
$$
$$
= x^2 + 10x + 24 \qquad \text{Combine like terms.}
$$

You can see that we get the same result.

Use FOIL to multiply the binomials.

a) $(p + 5)(p - 2)$ b) $(4r - 3)(r - 1)$

c) $(a + 4b)(a - 3b)$ d) $(2x + 9)(3y + 5)$

Solution

a) $(p + 5)(p - 2) = (p + 5)(p - 2) = \underset{\text{F}}{p(p)} + \underset{\text{O}}{p(-2)} + \underset{\text{I}}{5(p)} + \underset{\text{L}}{5(-2)}$ Use FOIL.

$$
= p^2 - 2p + 5p - 10 \qquad \text{Multiply.}
$$
$$
= p^2 + 3p - 10 \qquad \text{Combine like terms.}
$$

Notice that the middle terms are like terms, so we can combine them.

b) $(4r - 3)(r - 1) = (4r - 3)(r - 1) = 4r(r) + 4r(-1) - 3(r) - 3(-1)$ Use FOIL.

$= 4r^2 - 4r - 3r + 3$ Multiply.

$= 4r^2 - 7r + 3$ Combine like terms.

Inner
Outer

The middle terms are like terms, so we can combine them.

         F   O   I   L

c) $(a + 4b)(a - 3b) = a(a) + a(-3b) + 4b(a) + 4b(-3b)$ Use FOIL.

$= a^2 - 3ab + 4ab - 12b^2$ Multiply.

$= a^2 + ab - 12b^2$ Combine like terms.

As in parts a) and b), we combined the middle terms.

         F   O   I   L

d) $(2x + 9)(3y + 5) = 2x(3y) + 2x(5) + 9(3y) + 9(5)$ Use FOIL.

$= 6xy + 10x + 27y + 45$ Multiply.

In this case, the middle terms were not like terms, so we could not combine them.

W Hint

Work out the example on your paper as you are reading it.

[YOU TRY 3]

Use FOIL to multiply the binomials.

a) $(n + 8)(n + 5)$ b) $(3k + 7)(k - 4)$

c) $(x - 2y)(x - 6y)$ d) $(5c - 8)(2d + 1)$

With practice, you should get to the point where you can find the product of two binomials "in your head." Remember that, as in the case of parts a)–c) in Example 4, it is often possible to combine the middle terms.

There are special types of binomial products that come up often in algebra. We will look at these next.

4   Find the Product of Binomials of the Form $(a + b)(a - b)$

W Hint

Do you see that when you multiply $(a + b)(a - b)$ using FOIL, the middle terms are always opposites of each other?

Let's find the product $(y + 6)(y - 6)$. Using FOIL, we get

$$(y + 6)(y - 6) = y^2 - 6y + 6y - 36$$
$$= y^2 - 36$$

Notice that the middle terms, the y-terms, drop out. In the result, $y^2 - 36$, the first term (y^2) is the square of y and the last term (36) is the square of 6. The resulting polynomial is a *difference of squares*. This pattern always holds when multiplying two binomials of the form $(a + b)(a - b)$.

Procedure   The Product of the Sum and Difference of Two Terms

$$(a + b)(a - b) = a^2 - b^2$$

EXAMPLE 5

Multiply.

a) $(z + 9)(z - 9)$ b) $(2 + c)(2 - c)$

c) $(5n^2 - 8)(5n^2 + 8)$ d) $\left(\dfrac{3}{4}t + u\right)\left(\dfrac{3}{4}t - u\right)$

Solution

a)   The product $(z + 9)(z - 9)$ is in the form $(a + b)(a - b)$, where $a = z$ and $b = 9$. Use the rule that says $(a + b)(a - b) = a^2 - b^2$.

$$(z + 9)(z - 9) = z^2 - 9^2$$
$$= z^2 - 81$$

b) $(2 + c)(2 - c) = 2^2 - c^2$ $a = 2$ and $b = c$
$$= 4 - c^2$$

Be very careful on a problem like this. The answer is $4 - c^2$, NOT $c^2 - 4$; subtraction is not commutative.

c) $(5n^2 - 8)(5n^2 + 8) = (5n^2 + 8)(5n^2 - 8)$ Commutative property
$$= (5n^2)^2 - 8^2$$ $a = 5n^2$ and $b = 8$; put $5n^2$ in parentheses.
$$= 25n^4 - 64$$

d) $\left(\dfrac{3}{4}t + u\right)\left(\dfrac{3}{4}t - u\right) = \left(\dfrac{3}{4}t\right)^2 - u^2$ $a = \dfrac{3}{4}t$ and $b = u$; put $\dfrac{3}{4}t$ in parentheses.
$$= \dfrac{9}{16}t^2 - u^2$$

YOU TRY 4

Multiply.

a) $(k + 7)(k - 7)$ b) $(3c^2 + 4)(3c^2 - 4)$

c) $(8 - p)(8 + p)$ d) $\left(\dfrac{5}{2}m + n\right)\left(\dfrac{5}{2}m - n\right)$

5 Square a Binomial

Another type of special binomial product is a **binomial square** such as $(x + 5)^2$. $(x + 5)^2$ means $(x + 5)(x + 5)$. Therefore, we can use FOIL to multiply.

$$(x + 5)^2 = (x + 5)(x + 5) = x^2 + 5x + 5x + 25$$ Use FOIL.
$$= x^2 + 10x + 25$$ Note that $10x = 2(x)(5)$.

Notice that the outer and inner products are the same. When we add those terms, we see that the middle term of the result is *twice* the product of the terms in each binomial.

In the expansion of $(x + 5)^2$, $10x$ is $2(x)(5)$.

The *first* term in the result is the square of the *first* term in the binomial, and the *last* term in the result is the square of the *last* term in the binomial. We can express these relationships with the following formulas:

Procedure The Square of a Binomial

$$(a + b)^2 = a^2 + 2ab + b^2$$
$$(a - b)^2 = a^2 - 2ab + b^2$$

We can think of the formulas in words as:

To square a binomial, you square the first term, square the second term, then multiply 2 times the first term times the second term and add.

Finding the products $(a + b)^2 = a^2 + 2ab + b^2$ and $(a - b)^2 = a^2 - 2ab + b^2$ is also called *expanding* the binomial squares $(a + b)^2$ and $(a - b)^2$.

 BE CAREFUL

$$(a + b)^2 \neq a^2 + b^2 \quad \text{and} \quad (a - b)^2 \neq a^2 - b^2$$

EXAMPLE 6

In-Class Example 6

Expand.
a) $(k + 8)^2$ b) $(r - 5)^2$

c) $(3a - 4b)^2$ d) $\left(\dfrac{1}{2}w + 5\right)^2$

Answer: a) $k^2 + 16k + 64$
b) $r^2 - 10r + 25$
c) $9a^2 - 24ab + 16b^2$
d) $\dfrac{1}{4}w^2 + 5w + 25$

Expand.

a) $(d + 7)^2$ b) $(m - 9)^2$ c) $(2x - 5y)^2$ d) $\left(\dfrac{1}{3}t + 4\right)^2$

Solution

a) $(d + 7)^2 = \underset{\uparrow}{d^2} \quad + \underset{\uparrow}{2(d)(7)} \quad + \underset{\uparrow}{(7)^2}$ $a = d, b = 7$

      Square the   Two times   Square the
      first term   first term   second term
          times second
           term

$$= d^2 + 14d + 49$$

Notice, $(d + 7)^2 \neq d^2 + 49$. Do not "distribute" the power of 2 to each term in the binomial!

b) $(m - 9)^2 = \underset{\uparrow}{m^2} \quad - \underset{\uparrow}{2(m)(9)} \quad + \underset{\uparrow}{(9)^2}$ $a = m, b = 9$

      Square the   Two times   Square the
      first term   first term   second term
          times second
           term

$$= m^2 - 18m + 81$$

c) $(2x - 5y)^2 = (2x)^2 - 2(2x)(5y) + (5y)^2$ $a = 2x, b = 5y$

$$= 4x^2 - 20xy + 25y^2$$

d) $\left(\dfrac{1}{3}t + 4\right)^2 = \left(\dfrac{1}{3}t\right)^2 + 2\left(\dfrac{1}{3}t\right)(4) + (4)^2$ $a = \dfrac{1}{3}t, b = 4$

$$= \dfrac{1}{9}t^2 + \dfrac{8}{3}t + 16$$

YOU TRY 5

Expand.

a) $(r + 10)^2$ b) $(h - 1)^2$ c) $(2p + 3q)^2$ d) $\left(\dfrac{3}{4}y - 5\right)^2$

---

6 Find Higher Powers of a Binomial

To find higher powers of binomials, we use techniques we have already discussed.

EXAMPLE 7

In-Class Example 7

Use Example 7.

Expand.

a) $(n + 2)^3$ b) $(3v - 2)^4$

Solution

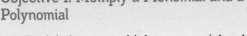

Hint

Could you make a general procedure for this objective?

a) Just as $x^2 \cdot x = x^3$, it is true that $(n + 2)^2 \cdot (n + 2) = (n + 2)^3$. So we can think of $(n + 2)^3$ as $(n + 2)^2(n + 2)$.

$$(n + 2)^3 = (n + 2)^2(n + 2)$$
$$= (n^2 + 4n + 4)(n + 2) \qquad \text{Square the binomial.}$$
$$= n^3 + 2n^2 + 4n^2 + 8n + 4n + 8 \qquad \text{Multiply.}$$
$$= n^3 + 6n^2 + 12n + 8 \qquad \text{Combine like terms.}$$

b) Since we can write $x^4 = x^2 \cdot x^2$, we can write $(3v - 2)^4 = (3v - 2)^2 \cdot (3v - 2)^2$.

$$(3v - 2)^4 = (3v - 2)^2 \cdot (3v - 2)^2$$
$$= (9v^2 - 12v + 4)(9v^2 - 12v + 4) \qquad \text{Square each binomial.}$$
$$= 81v^4 - 108v^3 + 36v^2 - 108v^3 + 144v^2 - 48v \qquad \text{Multiply.}$$
$$+ 36v^2 - 48v + 16$$
$$= 81v^4 - 216v^3 + 216v^2 - 96v + 16 \qquad \text{Combine like terms.}$$

[YOU TRY 6] Expand. a) $(k - 3)^3$ b) $(2h + 1)^4$

ANSWERS TO [YOU TRY] EXERCISES

1) $20z^7 - 35z^6 - 5z^5 + 40z^4$ 2) a) $18x^3 + x^2 - 32x - 15$ b) $4t^4 + \dfrac{10}{3}t^3 - 25t^2 - \dfrac{62}{3}t + 20$

3) a) $n^2 + 13n + 40$ b) $3k^2 - 5k - 28$ c) $x^2 - 8xy + 12y^2$ d) $10cd + 5c - 16d - 8$

4) a) $k^2 - 49$ b) $9c^4 - 16$ c) $64 - p^2$ d) $\dfrac{25}{4}m^2 - n^2$

5) a) $r^2 + 20r + 100$ b) $h^2 - 2h + 1$ c) $4p^2 + 12pq + 9q^2$ d) $\dfrac{9}{16}y^2 - \dfrac{15}{2}y + 25$

6) a) $k^3 - 9k^2 + 27k - 27$ b) $16h^4 + 32h^3 + 24h^2 + 8h + 1$

E Evaluate **6.2** Exercises   Do the exercises, and check your work.

*Additional answers can be found in the Answers to Exercises appendix.

Objective 1: Multiply a Monomial and a Polynomial

1) Explain how to multiply a monomial and a binomial.
   Answers may vary.

2) Explain how to multiply two binomials.
   Answers may vary.

Multiply.

3) $(3m^5)(8m^3)$ $24m^8$

4) $(2k^6)(7k^3)$ $14k^9$

5) $(-8c)(4c^5)$ $-32c^6$

6) $\left(-\dfrac{2}{9}z^3\right)\left(\dfrac{3}{4}z^9\right)$ $-\dfrac{1}{6}z^{12}$

Multiply.

7) $5a(2a - 7)$ $10a^2 - 35a$

8) $3y(10y - 1)$ $30y^2 - 3y$

9) $-6c(7c + 2)$ $-42c^2 - 12c$ 10) $-15d(11d - 2)$
   $-165d^2 + 30d$

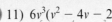 11) $6v^3(v^2 - 4v - 2)$ 12) $8f^5(f^2 - 3f - 6)$
   $6v^5 - 24v^4 - 12v^3$ $8f^7 - 24f^6 - 48f^5$

13) $-9b^2(4b^3 - 2b^2 - 6b - 9)$ $-36b^5 + 18b^4 + 54b^3 + 81b^2$

14) $-4h^7(5h^6 + 4h^3 + 11h - 3)$ $-20h^{13} - 16h^{10} - 44h^8 + 12h^7$

15) $3a^2b(ab^2 + 6ab - 13b + 7)$
    $3a^3b^3 + 18a^3b^2 - 39a^2b^2 + 21a^2b$

---

16) $4x^6y^2(-5x^2y + 11xy^2 - xy + 2y - 1)$
$-20x^8y^3 + 44x^7y^4 - 4x^7y^3 + 8x^6y^3 - 4x^6y^2$

17) $-\dfrac{3}{5}k^4(15k^2 + 20k - 3)$ $-9k^6 - 12k^5 + \dfrac{9}{5}k^4$

18) $\dfrac{3}{4}t^5(12t^3 - 20t^2 + 9)$ $-9t^8 - 15t^7 + \dfrac{27}{4}t^5$

Objective 2: Multiply Two Polynomials

For Exercises 19 and 20, answer *always, sometimes,* or *never*.

19) The product of two first-degree binomials will *always, sometimes,* or *never* be a second-degree polynomial. always

20) Multiplying a second-degree binomial by a third-degree trinomial will *always, sometimes,* or *never* result in a third-degree polynomial. never

Multiply.

21) $(c + 4)(6c + 7)$ $6c^2 + 31c + 28$

22) $(d + 8)(5d + 9)$ $5d^2 + 49d + 72$

23) $(f - 5)(3f^2 + 2f - 4)$ $3f^3 - 13f^2 - 14f + 20$

24) $(k - 2)(9k^2 - 4k - 12)$ $9k^3 - 22k^2 - 4k + 24$

25) $(4x^3 - x^2 + 6x + 2)(2x - 5)$
$8x^4 - 22x^3 + 17x^2 - 26x - 10$

26) $(3m^3 + 3m^2 - 4m - 9)(4m - 7)$
$12m^4 - 9m^3 - 37m^2 - 8m + 63$

27) $\left(\dfrac{1}{3}y^2 + 4\right)(12y^2 + 7y - 9)$
$4y^4 + \dfrac{7}{3}y^3 + 45y^2 + 28y - 36$

28) $\left(\dfrac{3}{5}q^2 - 1\right)(10q^2 - 7q + 20)$ $6q^4 - \dfrac{21}{5}q^3 + 2q^2 + 7q - 20$

29) $(s^2 - s + 2)(s^2 + 4s - 3)$ $s^4 + 3s^3 - 5s^2 + 11s - 6$

30) $(t^2 + 4t + 1)(2t^2 - t - 5)$ $2t^4 + 7t^3 - 7t^2 - 21t - 5$

31) $(4h^2 - h + 2)(-6h^3 + 5h^2 - 9h)$
$-24h^5 + 26h^4 - 53h^3 + 19h^2 - 18h$

32) $(n^4 + 8n^2 - 5)(n^2 - 3n - 4)$
$n^6 - 3n^5 + 4n^4 - 24n^3 - 37n^2 + 15n + 20$

Multiply both horizontally and vertically. Which method do you prefer and why?

33) $(3y - 2)(5y^2 - 4y + 3)$ $15y^3 - 22y^2 + 17y - 6$

34) $(2p^2 + p - 4)(5p + 3)$ $10p^3 + 11p^2 - 17p - 12$

Objective 3: Multiply Two Binomials Using FOIL

35) What do the letters in the word FOIL represent?
First, Outer, Inner, Last

36) Can FOIL be used to expand $(x + 8)^2$? Explain your answer. Yes. $(x + 8)^2 = (x + 8)(x + 8)$, so we can use FOIL to find the product.

Use FOIL to multiply.

37) $(w + 5)(w + 7)$ $w^2 + 12w + 35$

38) $(u + 5)(u + 3)$ $u^2 + 8u + 15$

39) $(r - 3)(r + 9)$ $r^2 + 6r - 27$

40) $(w - 12)(w - 4)$ $w^2 - 16w + 48$

41) $(y - 7)(y - 1)$ $y^2 - 8y + 7$

42) $(g + 4)(g - 8)$ $g^2 - 4g - 32$

43) $(3p + 7)(p - 2)$ $3p^2 + p - 14$

44) $(5u + 1)(u + 7)$ $5u^2 + 36u + 7$

45) $(7n^2 + 4)(3n + 1)$ $21n^3 + 7n^2 + 12n + 4$

46) $(4y^3 - 3)(7y + 6)$ $28y^4 + 24y^3 - 21y - 18$

47) $(5 - 4w)(3 - w)$ $4w^2 - 17w + 15$

48) $(2 - 3r)(4 - 5r)$ $15r^2 - 22r + 8$

49) $(4a - 5b)(3a + 4b)$ $12a^2 + ab - 20b^2$

50) $(3c + 2d)(c - 5d)$ $3c^2 - 13cd - 10d^2$

51) $(6x + 7y)(8x + 3y)$ $48x^2 + 74xy + 21y^2$

52) $(0.5p - 0.3q)(0.7p - 0.4q)$ $0.35p^2 - 0.41pq + 0.12q^2$

53) $\left(v + \dfrac{1}{3}\right)\left(v + \dfrac{3}{4}\right)$ $v^2 + \dfrac{13}{12}v + \dfrac{1}{4}$

54) $\left(t + \dfrac{5}{2}\right)\left(t + \dfrac{6}{5}\right)$ $t^2 + \dfrac{37}{10}t + 3$

55) $\left(\dfrac{1}{2}a + 5b^2\right)\left(\dfrac{2}{3}a - b^2\right)$ $\dfrac{1}{3}a^2 + \dfrac{17}{6}ab^2 - 5b^4$

56) $\left(\dfrac{3}{4}x - y^2\right)\left(\dfrac{1}{3}x + 4y^2\right)$ $\dfrac{1}{4}x^2 + \dfrac{8}{3}xy^2 - 4y^4$

Write an expression for a) the perimeter of each figure and b) the area of each figure.

57)
$y - 3$
$y + 5$
a) $4y + 4$ units
b) $y^2 + 2y - 15$ sq units

58)
$6w$
$5w + 4$
a) $22w + 8$ units
b) $30w^2 + 24w$ sq units

59)
$m^2 - 2m + 7$
$3m$
a) $2m^2 + 2m + 14$ units
b) $3m^3 - 6m^2 + 21m$ sq units

60)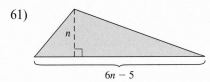

$3x^2 - 1$

$3x^2 - 1$

a) $12x^2 - 4$ units

b) $9x^4 - 6x^2 + 1$ sq units

Express the area of each triangle as a polynomial.

61)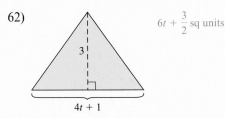

n

$6n - 5$

$3n^2 - \frac{5}{2}n$ sq units

62)

3

$4t + 1$

$6t + \frac{3}{2}$ sq units

63) Find the product $(2y + 3)(y - 5)(y - 4)$

  a) by first multiplying $(2y + 3)(y - 5)$ and then multiplying that result by $(y - 4)$.
    $2y^3 - 15y^2 + 13y + 60$

  b) by first multiplying $(y - 5)(y - 4)$ and then multiplying that result by $(2y + 3)$.
    $2y^3 - 15y^2 + 13y + 60$

  c) What do you notice about the results?
    They are the same.

64) To find the product $3(c + 5)(c - 1)$, Parth begins by multiplying $3(c + 5)$ and then he multiplies that result by $(c - 1)$. Yolanda begins by multiplying $(c + 5)(c - 1)$ and multiplies that result by 3. Who is right? Both are correct.

Multiply.

65) $2(n + 3)(4n - 5)$ $8n^2 + 14n - 30$

66) $-13(3p - 1)(p + 4)$ $-39p^2 - 143p + 52$

67) $-5z^2(z - 8)(z - 2)$ $-5z^4 + 50z^3 - 80z^2$

68) $11r^2(2r + 7)(-r + 1)$ $-22r^4 - 55r^3 + 77r^2$

69) $(c + 3)(c + 4)(c - 1)$ $c^3 + 6c^2 + 5c - 12$

70) $(2t + 3)(t + 1)(t + 4)$ $2t^3 + 13t^2 + 23t + 12$

Mixed Exercises: Objectives 4 and 5
For Exercises 71 and 72, answer *always, sometimes,* or *never.*

71) The product of three second-degree binomials will *always, sometimes,* or *never* result in an eighth-degree polynomial.   never

72) The product of three first-degree binomials will *always, sometimes,* or *never* result in a third-degree polynomial.   always

Find the following special products.

73) $(y + 5)(y - 5)$ $y^2 - 25$ 74) $(w + 2)(w - 2)$ $w^2 - 4$

75) $(a - 7)(a + 7)$ $a^2 - 49$ 76) $(f - 11)(f + 11)$ $f^2 - 121$

77) $(3 - p)(3 + p)$ $9 - p^2$ 78) $(12 + d)(12 - d)$
    $144 - d^2$

79) $\left(u^2 + \frac{1}{5}\right)\left(u^2 - \frac{1}{5}\right)$ 80) $\left(g^2 - \frac{1}{4}\right)\left(g^2 + \frac{1}{4}\right)$
    $u^4 - \frac{1}{25}$ $g^4 - \frac{1}{16}$

81) $\left(\frac{2}{3} - k\right)\left(\frac{2}{3} + k\right)$ $\frac{4}{9} - k^2$ 82) $\left(\frac{7}{4} + c\right)\left(\frac{7}{4} - c\right)$ $\frac{49}{16} - c^2$

83) $(2r + 7)(2r - 7)$ 84) $(4h - 3)(4h + 3)$
    $4r^2 - 49$ $16h^2 - 9$

85) $-(8j - k^2)(8j + k^2)$ $k^4 - 64j^2$

86) $-(3m + 5n^2)(3m - 5n^2)$ $25n^4 - 9m^2$

87) $(d + 4)^2$ $d^2 + 8d + 16$ 88) $(g + 2)^2$ $g^2 + 4g + 4$

89) $(n - 13)^2$ $n^2 - 26n + 169$ 90) $(b - 3)^2$ $b^2 - 6b + 9$

91) $(3u + 1)^2$ $9u^2 + 6u + 1$ 92) $(5n + 4)^2$ $25n^2 + 40n + 16$

93) $(2d^2 - 5)^2$ 94) $(4p^2 - 3)^2$
    $4d^4 - 20d^2 + 25$ $16p^4 - 24p^2 + 9$

95) $(3c + 2d)^2$ 96) $(2a - 5b)^2$
    $9c^2 + 12cd + 4d^2$ $4a^2 - 20ab + 25b^2$

97) $\left(\frac{3}{2}k + 8m\right)^2$ 98) $\left(\frac{4}{5}x - 3y\right)^2$
    $\frac{9}{4}k^2 + 24km + 64m^2$ $\frac{16}{25}x^2 - \frac{24}{5}xy + 9y^2$

99) $[(2a + b) + 3]^2$ 100) $[(3c - d) + 5]^2$
  $4a^2 + 4ab + b^2 + 12a + 6b + 9$ $9c^2 - 6cd + d^2 + 30c - 10d + 25$

For Exercises 101 and 102, answer *always, sometimes,* or *never.*

101) Multiplying two binomials together will *always, sometimes,* or *never* result in a trinomial.   sometimes

102) The square of a first-degree binomial is *always, sometimes,* or *never* a second-degree trinomial.   always

103) Does $3(r + 2)^2 = (3r + 6)^2$? Why or why not?

104) Explain how to find the product $4(a - 5)^2$, then find the product.   First find $(a - 5)^2$, then multiply by 4.
    $4(a - 5)^2 = 4a^2 - 40a + 100$

Find each product.

105) $7(y + 2)^2$ 106) $3(m + 4)^2$
    $7y^2 + 28y + 28$ $3m^2 + 24m + 48$

107) $4c(c + 3)^2$ 108) $-3a(a + 1)^2$
    $4c^3 + 24c^2 + 36c$ $-3a^3 - 6a^2 - 3a$

Objective 6: Find Higher Powers of a Binomial
Expand.

109) $(r + 5)^3$ 110) $(u + 3)^3$
    $r^3 + 15r^2 + 75r + 125$ $u^3 + 9u^2 + 27u + 27$

111) $(g - 4)^3$ 112) $(w - 5)^3$
    $g^3 - 12g^2 + 48g - 64$ $w^3 - 15w^2 + 75w - 125$

113) $(2a - 1)^3$ 114) $(3x + 4)^3$
    $8a^3 - 12a^2 + 6a - 1$ $27x^3 + 108x^2 + 144x + 64$

115) $(h + 3)^4$ 116) $(y + 5)^4$
    $h^4 + 12h^3 + 54h^2 + 108h + 81$ $y^4 + 20y^3 + 150y^2 + 500y + 625$

117) $(5t - 2)^4$
$625t^4 - 1000t^3 + 600t^2 - 160t + 16$

118) $(4c - 1)^4$
$256c^4 - 256c^3 + 96c^2 - 16c + 1$

119) Does $(x + 2)^2 = x^2 + 4$? Why or why not?
No; $(x + 2)^2 = x^2 + 4x + 4$

120) Does $(y - 1)^3 = y^3 - 1$? Why or why not?
No; $(y - 1)^3 = y^3 - 3y^2 + 3y - 1$

Mixed Exercises: Objectives 1–6

Find each product.

121) $(c - 12)(c + 7)$
$c^2 - 5c - 84$

122) $(11t^4)(-2t^6)$ $-22t^{10}$

123) $4(6 - 5a)(2a - 1)$
$-40a^2 + 68a - 24$

124) $(3y^2 - 8z)(3y^2 + 8z)$
$9y^4 - 64z^2$

125) $(2k - 9)(5k^2 + 4k - 1)$
$10k^3 - 37k^2 - 38k + 9$

126) $(m^3 + 12)^2$
$m^6 + 24m^3 + 144$

127) $\left(\dfrac{1}{6} - h\right)\left(\dfrac{1}{6} + h\right)$ $\dfrac{1}{36} - h^2$

128) $3(7p^3 + 4p^2 + 2) - (5p^3 - 18p - 11)$
$16p^3 + 12p^2 + 18p + 17$

129) $(3c + 1)^3$ $27c^3 + 27c^2 + 9c + 1$

130) $(4w - 5)(2w - 3)$ $8w^2 - 22w + 15$

131) $\left(\dfrac{3}{8}p^7\right)\left(\dfrac{3}{4}p^4\right)$ $\dfrac{9}{32}p^{11}$

132) $xy(2x - y)(x - 3y)(x - 2y)$
$2x^4y - 11x^3y^2 + 17x^2y^3 - 6xy^4$

133) $4(5y^2 - 8y - 3) - 3(2y^2 + y - 10)$ $14y^2 - 35y + 18$

134) $(r - 6)^3$ $r^3 - 18r^2 + 108r - 216$

135) $-5z(z - 3)^2$ $-5z^3 + 30z^2 - 45z$

136) $(2n^2 + 9n - 3)(4n^2 - n - 8)$
$8n^4 + 34n^3 - 37n^2 - 69n + 24$

137) Express the volume of the cube as a polynomial.

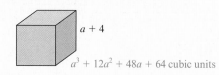

$a + 4$

$a^3 + 12a^2 + 48a + 64$ cubic units

Express the area of the shaded region as a polynomial.

138)

$5t + 3$

t t $3t - 2$ $14t^2 - t - 6$ sq units

R Rethink

R1) Write a paragraph that explains why you can use FOIL to multiply two binomials.

R2) How many special products did you encounter in this section? Have you mastered their formulas?

R3) What is the most common mistake you make when multiplying polynomials?

R4) Which exercises in this section do you find most challenging? Redo some of those problems to get more practice.

6.3 Dividing a Polynomial by a Monomial

P Prepare

O Organize

What is your objective for Section 6.3?	How can you accomplish the objective?
1 Divide a Polynomial by a Monomial	• Learn the procedure for **Dividing a Polynomial by a Monomial,** and write an example in your notes. • Complete the given examples on your own. • Complete You Trys 1 and 2.

W Work

Read the explanations, follow the examples, take notes, and complete the You Trys.

In this section we will learn how to divide a polynomial by a monomial.

1 Divide a Polynomial by a Monomial

The procedure for dividing a polynomial by a monomial is based on the procedure for adding or subtracting fractions.

To add $\frac{2}{9} + \frac{5}{9}$, we add the numerators and keep the denominator.

$$\frac{2}{9} + \frac{5}{9} = \frac{2+5}{9} = \frac{7}{9}$$

Reversing the process above, we can write $\frac{7}{9} = \frac{2+5}{9} = \frac{2}{9} + \frac{5}{9}$.

We can generalize this result and say that $\frac{a+b}{c} = \frac{a}{c} + \frac{b}{c}$ $(c \neq 0)$

This leads us to the following rule.

Procedure Dividing a Polynomial by a Monomial

To divide a polynomial by a monomial, divide *each term* in the polynomial by the monomial and simplify.

Example: $\dfrac{6k^2 + 10k - 18}{2} = \dfrac{6k^2}{2} + \dfrac{10k}{2} - \dfrac{18}{2} = 3k^2 + 5k - 9$

EXAMPLE 1

In-Class Example 1

Divide.
a) $\dfrac{16w^2 - 40w + 24}{8}$
b) $\dfrac{21r^3 + 56r^2 + 7r}{7r}$

Answer: a) $2w^2 - 5w + 3$
b) $3r^2 + 8r + 1$

W Hint

When dividing a trinomial by a monomial, what type of polynomial did we get as the quotient?

Divide.

a) $\dfrac{24x^2 - 8x + 20}{4}$ b) $\dfrac{12c^3 + 54c^2 + 6c}{6c}$

Solution

a) First, note that the polynomial is being divided by a *monomial*. Divide each term in the numerator by the monomial 4.

$$\frac{24x^2 - 8x + 20}{4} = \frac{24x^2}{4} - \frac{8x}{4} + \frac{20}{4} = 6x^2 - 2x + 5$$

Let's label the components of our division problem the same way as when we divide with integers.

$$\begin{array}{c} \text{Dividend} \to \\ \text{Divisor} \to \end{array} \frac{24x^2 - 8x + 20}{4} = 6x^2 - 2x + 5 \leftarrow \text{Quotient}$$

We can check our answer by multiplying the quotient by the divisor. The answer should be the dividend.

Check: $4(6x^2 - 2x + 5) = 24x^2 - 8x + 20$ ✓ The quotient is correct.

b) $\dfrac{12c^3 + 54c^2 + 6c}{6c} = \dfrac{12c^3}{6c} + \dfrac{54c^2}{6c} + \dfrac{6c}{6c} = 2c^2 + 9c + 1$

BE CAREFUL

Students will often incorrectly "cancel out" $\dfrac{6c}{6c}$ and get nothing. But $\dfrac{6c}{6c} = 1$ since a quantity divided by itself equals 1.

Check: $6c(2c^2 + 9c + 1) = 12c^3 + 54c^2 + 6c$ ✓ The quotient is correct.

Note

In Example 1b), c cannot equal zero because then the denominator of $\dfrac{12c^3 + 54c^2 + 6c}{6c}$ would equal zero. Remember, a fraction is undefined when its denominator equals zero!

[YOU TRY 1]

Divide $\dfrac{35n^5 + 20n^4 - 5n^2}{5n^2}$.

EXAMPLE 2

Divide $(6a - 7 - 36a^2 + 27a^3) \div (9a^2)$.

In-Class Example 2

Divide
$(36h^2 - 8 + 60h^3 - 7h) \div$
$(12h^2)$.

Answer:

$5h + 3 - \dfrac{7}{12h} - \dfrac{2}{3h^2}$

 Hint

The final answer should not have any negative exponents.

Solution

This is another example of a polynomial divided by a monomial. Notice, however, that the terms in the numerator are not written in descending powers. Rewrite them in descending powers before dividing.

$$\dfrac{6a - 7 - 36a^2 + 27a^3}{9a^2} = \dfrac{27a^3 - 36a^2 + 6a - 7}{9a^2}$$

$$= \dfrac{27a^3}{9a^2} - \dfrac{36a^2}{9a^2} + \dfrac{6a}{9a^2} - \dfrac{7}{9a^2} \qquad \text{Divide each term in the numerator by } 9a^2.$$

$$= 3a - 4 + \dfrac{2}{3a} - \dfrac{7}{9a^2} \qquad \text{Apply quotient rule and simplify.}$$

The quotient is *not* a polynomial since a and a^2 appear in denominators. The quotient of polynomials is not necessarily a polynomial.

[YOU TRY 2]

Divide $(30z^2 + 3 - 50z^3 + 18z) \div (10z^2)$.

ANSWERS TO [YOU TRY] EXERCISES

1) $7n^3 + 4n^2 - 1$ 2) $-5z + 3 + \dfrac{9}{5z} + \dfrac{3}{10z^2}$

*Additional answers can be found in the Answers to Exercises appendix.

Objective 1: Divide a Polynomial by a Monomial

1) Label the dividend, divisor, and quotient of

$$\frac{6c^3 + 15c^2 - 9c}{3c} = 2c^2 + 5c - 3.$$

dividend: $6c^3 + 15c^2 - 9c$; divisor: $3c$; quotient: $2c^2 + 5c - 3$

2) How would you check the answer to the division problem in Exercise 1?   Multiply the divisor by the quotient to get the dividend. $3c(2c^2 + 5c - 3) = 6c^3 + 15c^2 - 9c$

3) Explain, in your own words, how to divide a polynomial by a monomial.   Answers may vary.

4) Without dividing, determine what the degree of the quotient will be when performing the division

$$\frac{48y^5 - 16y^4 + 5y^3 - 32y^2}{16y^2}.$$ 3

Divide.

5) $\dfrac{49p^4 + 21p^3 + 28p^2}{7}$ $7p^4 + 3p^3 + 4p^2$

6) $\dfrac{10m^3 + 45m^2 + 30m}{5}$ $2m^3 + 9m^2 + 6m$

7) $\dfrac{12w^3 - 40w^2 - 36w}{4w}$ $3w^2 - 10w - 9$

8) $\dfrac{3a^5 - 27a^4 + 12a^3}{3a}$ $a^4 - 9a^3 + 4a^2$

9) $\dfrac{22z^6 + 14z^5 - 38z^3 + 2z}{2z}$ $11z^5 + 7z^4 - 19z^2 + 1$

10) $\dfrac{48u^7 - 18u^4 - 90u^2 + 6u}{6u}$ $8u^6 - 3u^3 - 15u + 1$

11) $\dfrac{9h^8 + 54h^6 - 108h^3}{9h^2}$ $h^6 + 6h^4 - 12h$

12) $\dfrac{72x^9 - 24x^7 - 56x^4}{8x^2}$ $9x^7 - 3x^5 - 7x^2$

13) $\dfrac{36r^7 - 12r^4 + 6}{12r}$ $3r^6 - r^3 + \dfrac{1}{2r}$

14) $\dfrac{20t^6 + 130t^2 + 2}{10t}$ $2t^5 + 13t + \dfrac{1}{5t}$

15) $\dfrac{8d^6 - 12d^5 + 18d^4}{2d^4}$ $4d^2 - 6d + 9$

16) $\dfrac{21p^4 - 15p^3 + 6p^2}{3p^2}$ $7p^2 - 5p + 2$

17) $\dfrac{28k^7 + 8k^5 - 44k^4 - 36k^2}{4k^2}$ $7k^5 + 2k^3 - 11k^2 - 9$

18) $\dfrac{42n^7 + 14n^6 - 49n^4 + 63n^3}{7n^3}$ $6n^4 + 2n^3 - 7n + 9$

19) $(35d^5 - 7d^2) \div (-7d^2)$ $-5d^3 + 1$

20) $(-30h^7 - 8h^5 + 2h^3) \div (-2h^3)$ $15h^4 + 4h^2 - 1$

21) $\dfrac{10w^5 + 12w^3 - 6w^2 + 2w}{6w^2}$ $\dfrac{5}{3}w^3 + 2w - 1 + \dfrac{1}{3w}$

22) $\dfrac{-48r^5 + 14r^3 - 4r^2 + 10}{4r}$ $-12r^4 + \dfrac{7}{2}r^2 - r + \dfrac{5}{2r}$

23) $(12k^8 - 4k^6 - 15k^5 - 3k^4 + 1) \div (2k^5)$ $6k^3 - 2k - \dfrac{15}{2} - \dfrac{3}{2k} + \dfrac{1}{2k^5}$

24) $(56m^6 + 4m^5 - 21m^2) \div (7m^3)$ $8m^3 + \dfrac{4}{7}m^2 - \dfrac{3}{m}$

25) Chandra divides $40p^3 - 10p^2 + 5p$ by $5p$ and gets a quotient of $8p^2 - 2p$. Is this correct? Why or why not?   The answer is incorrect. When you divide $5p$ by $5p$, you get 1. The quotient should be $8p^2 - 2p + 1$.

26) Ryan divides $\dfrac{20y^2 + 15y}{15y}$ and gets a quotient of $20y^2$. What was his mistake? What is the correct answer?   He canceled out the $15y$-terms so that he was left with $20y^2$. The correct answer is $\dfrac{4}{3}y + 1$.

For Exercises 27 and 28, answer *always, sometimes,* or *never.*

27) A fourth-degree polynomial divided by a second-degree monomial will *always, sometimes,* or *never* result in a second-degree polynomial.   sometimes

28) If you divide a trinomial by a monomial and the quotient is a polynomial, the dividend will *always, sometimes,* or *never* be of higher degree than the divisor.   always

Divide.

29) $\dfrac{48p^5q^3 + 60p^4q^2 - 54p^3q + 18p^2q}{6p^2q}$ $8p^3q^2 + 10p^2q - 9p + 3$

30) $(-45x^5y^4 - 27x^4y^5 + 9x^3y^5 + 63x^3y^3) \div (-9x^3y^2)$ $5x^2y^2 + 3xy^3 - y^3 - 7y$

31) $\dfrac{14s^6t^6 - 28s^5t^4 - s^3t^3 + 21st}{7s^2t}$ $2s^4t^5 - 4s^3t^3 - \dfrac{1}{7}st^2 + \dfrac{3}{s}$

32) $(4a^5b^4 - 32a^4b^4 - 48a^3b^4 + a^2b^3) \div (4ab^2)$ $a^4b^2 - 8a^3b^2 - 12a^2b^2 + \dfrac{1}{4}ab$

33) $\dfrac{50a^4b^4 + 30a^4b^3 - a^2b^2 + 2ab}{10a^2b^2}$ $5a^2b^2 + 3a^2b - \dfrac{1}{10} + \dfrac{1}{5ab}$

34) $(8p^2 + 4p - 32p^3 + 36p^4) \div (-4p)$ $-9p^3 + 8p^2 - 2p - 1$

35) $\dfrac{-27x^3y^3 + 9x^2y^3 + 36xy + 72}{9x^2y}$ $-3xy^2 + y^2 + \dfrac{4}{x} + \dfrac{8}{x^2y}$

36) $\dfrac{-20v^3 + 35v^2 - 15v}{-5v}$ $4v^2 - 7v + 3$

37) $\dfrac{45t^4 - 81t^2 - 27t^3 + 8t^6}{-9t^3}$ $-\dfrac{8}{9}t^3 - 5t + 3 + \dfrac{9}{t}$

38) $\dfrac{9q^2 + 42q^4 - 9 + 6q - 8q^3}{3q^2}$ $14q^2 - \dfrac{8}{3}q + 3 + \dfrac{2}{q} - \dfrac{3}{q^2}$

R Rethink

R1) Why does the quotient of a polynomial and a monomial always have the same number of terms as the polynomial?

R2) Were there any problems you could not do? If so, write them down or circle them and ask your instructor for help.

6.4 Dividing a Polynomial by a Polynomial

P Prepare

O Organize

What are your objectives for Section 6.4?	How can you accomplish each objective?
1 Review of Long Division	• Follow the detailed explanation of long division of whole numbers because it will help you understand the next objective. • Write a procedure for long division. • Complete the given example on your own. • Complete You Try 1.
2 Divide a Polynomial by a Polynomial	• Use the procedure developed for long division of whole numbers, and apply it to dividing a polynomial by a polynomial. • Complete the given examples on your own. • Complete You Trys 2–4. • Review the summary that outlines how to **Divide Polynomials** so that you know the difference between dividing by a monomial and dividing by a polynomial. Write examples in your notes.

W Work

Read the explanations, follow the examples, take notes, and complete the You Trys.

When dividing a polynomial by a polynomial containing two or more terms, we use long division of polynomials. This method is similar to long division of whole numbers, so let's look at a long division problem and compare it to the procedure with polynomial long division.

1 Review of Long Division

EXAMPLE 1

In-Class Example 1

Divide 947 by 4.

Answer: $236\frac{3}{4}$

W Hint

Write out the example as you are reading it.

Divide 854 by 3.

Solution

$$\begin{array}{r} 2 \\ 3\overline{)854} \\ -6\downarrow \\ \hline 25 \end{array}$$

1) How many times does 3 divide into 8 evenly?   2
2) Multiply $2 \times 3 = 6$.
3) Subtract $8 - 6 = 2$.
4) Bring down the 5.

Start the process again.

$$\begin{array}{r} 28 \\ 3\overline{)854} \\ -6 \\ \hline 25 \\ -24\downarrow \\ \hline 14 \end{array}$$

1) How many times does 3 divide into 25 evenly?   8
2) Multiply $8 \times 3 = 24$.
3) Subtract $25 - 24 = 1$.
4) Bring down the 4.

Do the procedure again.

$$\begin{array}{r} 284 \\ 3\overline{)854} \\ -6 \\ \hline 25 \\ -24 \\ \hline 14 \\ -12 \\ \hline 2 \end{array}$$

1) How many times does 3 divide into 14 evenly?   4
2) Multiply $4 \times 3 = 12$.
3) Subtract $14 - 12 = 2$.
4) There are no more numbers to bring down, so the remainder is 2.

Write the result.

$$854 \div 3 = 284\frac{2}{3} \quad \begin{array}{l} \leftarrow \text{Remainder} \\ \leftarrow \text{Divisor} \end{array}$$

Check: $(3 \times 284) + 2 = 852 + 2 = 854$ ✓

[**YOU TRY 1**] Divide 638 by 5.

2 Divide a Polynomial by a Polynomial

We can divide two polynomials using a long division process similar to that of Example 1.

EXAMPLE 2

In-Class Example 2

Divide $\dfrac{4a^2 + 17a + 15}{a + 3}$.

Answer: $4a + 5$

Divide $\dfrac{5x^2 + 13x + 6}{x + 2}$.

Solution

First, notice that we are dividing by a polynomial containing more than one term. That tells us to use long division of polynomials.

W Hint

Notice where to put the numerator and denominator of $\dfrac{5x^2 + 13x + 6}{x + 2}$ when you set up the long division problem.

We will work with the x in $x + 2$ like we worked with the 3 in Example 1.

$$
\begin{array}{r}
5x \\
x + 2 \overline{)\ 5x^2 + 13x + 6} \\
\underline{-(5x^2 + 10x)}\downarrow \\
3x + 6
\end{array}
$$

1) By what do we multiply x to get $5x^2$? $5x$
   Line up terms in the quotient according to exponents, so write $5x$ above $13x$.

2) Multiply $5x$ by $(x + 2)$: $5x(x + 2) = 5x^2 + 10x$

3) Subtract $(5x^2 + 13x) - (5x^2 + 10x) = 3x$.

4) Bring down the $+6$.

Start the process again. Remember, work with the x in $x + 2$ like we worked with the 3 in Example 1.

W Hint

Did you see that *like terms* are always lined up in the same columns?

$$
\begin{array}{r}
5x + 3 \\
x + 2 \overline{)\ 5x^2 + 13x + 6} \\
\underline{-(5x^2 + 10x)} \\
3x + 6 \\
\underline{-(3x + 6)} \\
0
\end{array}
$$

1) By what do we multiply x to get $3x$?   3
   Write $+3$ above $+6$.

2) Multiply 3 by $(x + 2)$: $3(x + 2) = 3x + 6$

3) Subtract $(3x + 6) - (3x + 6) = 0$.

4) There are no more terms. The remainder is 0.

Write the result.

$$
\frac{5x^2 + 13x + 6}{x + 2} = 5x + 3
$$

Check: $(x + 2)(5x + 3) = 5x^2 + 3x + 10x + 6 = 5x^2 + 13x + 6$ ✓

[YOU TRY 2]   Divide.

a) $\dfrac{r^2 + 11r + 28}{r + 4}$ b) $\dfrac{3k^2 + 17k + 10}{k + 5}$

Next, we will look at a division problem with a remainder.

EXAMPLE 3

Divide $\dfrac{-28n + 15n^3 + 41 - 17n^2}{3n - 4}$.

Solution

When we write our long division problem, the polynomial in the numerator must be rewritten so that the exponents are in descending order. Then, perform the long division.

$$
\begin{array}{r}
5n^2 \\
3n - 4 \overline{)\ 15n^3 - 17n^2 - 28n + 41} \\
\underline{-(15n^3 - 20n^2)}\downarrow \\
3n^2 - 28n
\end{array}
$$

1) By what do we multiply $3n$ to get $15n^3$? $5n^2$

2) Multiply $5n^2(3n - 4) = 15n^3 - 20n^2$

3) Subtract. $(15n^3 - 17n^2) - (15n^3 - 20n^2)$
   $= 15n^3 - 17n^2 - 15n^3 + 20n^2$
   $= 3n^2$

4) Bring down the $-28n$.

Repeat the process.

$$
\begin{array}{r}
5n^2 + n \\
3n - 4 \overline{)\; 15n^3 - 17n^2 - 28n + 41} \\
\underline{-(15n^3 - 20n^2)} \\
3n^2 - 28n \\
\underline{-(3n^2 - 4n)} \\
-24n + 41
\end{array}
$$

1) By what do we multiply $3n$ to get $3n^2$? $\quad n$

2) Multiply $n(3n - 4) = 3n^2 - 4n$.

3) Subtract. $(3n^2 - 28n) - (3n^2 - 4n)$
$= 3n^2 - 28n - 3n^2 + 4n$
$= -24n$

4) Bring down the $+ 41$.

Continue.

$$
\begin{array}{r}
5n^2 + n - 8 \\
3n - 4 \overline{)\; 15n^3 - 17n^2 - 28n + 41} \\
\underline{-(15n^3 - 20n^2)} \\
3n^2 - 28n \\
\underline{-(3n^2 - 4n)} \\
-24n + 41 \\
\underline{-(-24n + 32)} \\
9
\end{array}
$$

1) By what do we multiply $3n$ to get $-24n$? $\quad -8$

2) Multiply $-8(3n - 4) = -24n + 32$.

3) Subtract. $(-24n + 41) - (-24n + 32) = 9$

We are done with the long division process. How do we know that? Since the degree of 9 (degree zero) is less than the degree of $3n - 4$ (degree 1), we cannot divide anymore. *The remainder is 9.*

$$\frac{15n^3 - 17n^2 - 28n + 41}{3n - 4} = 5n^2 + n - 8 + \frac{9}{3n - 4}$$

Check: $(3n - 4)(5n^2 + n - 8) + 9 = 15n^3 + 3n^2 - 24n - 20n^2 - 4n + 32 + 9$
$= 15n^3 - 17n^2 - 28n + 41 \quad \checkmark$

[YOU TRY 3]   Divide $-34a^2 + 57 + 8a^3 - 21a$ by $2a - 9$.

In Example 3, we saw that we must write our polynomials in descending order. We have to watch out for something else as well—missing terms. If a polynomial is missing one or more terms, we put them into the polynomial with coefficients of zero.

EXAMPLE 4

Divide.

a) $x^3 + 64$ by $x + 4$ b) $t^4 + 3t^3 + 6t^2 + 11t + 5$ by $t^2 + 2$

Solution

a)   The polynomial $x^3 + 64$ is missing the x^2-term and the x-term. We will insert these terms into the polynomial by giving them coefficients of zero.

$$x^3 + 64 = x^3 + 0x^2 + 0x + 64$$

Divide.

$$
\begin{array}{r}
x^2 - 4x + 16 \\
x + 4 \overline{)\; x^3 + 0x^2 + 0x + 64} \\
\underline{-(x^3 + 4x^2)} \\
-4x^2 + 0x \\
\underline{-(-4x^2 - 16x)} \\
16x + 64 \\
\underline{-(16x + 64)} \\
0
\end{array}
$$

Therefore, $(x^3 + 64) \div (x + 4) = x^2 - 4x + 16$

Check: $(x + 4)(x^2 - 4x + 16) = x^3 - 4x^2 + 16x + 4x^2 - 16x + 64$
$= x^3 + 64 \quad \checkmark$

b) In this case, the divisor, $t^2 + 2$, is missing a t-term. Rewrite it as $t^2 + 0t + 2$ and divide.

$$
\begin{array}{r}
t^2 + 3t + 4 \\
t^2 + 0t + 2{\overline{\smash{\big)}\,t^4 + 3t^3 + 6t^2 + 11t + 5}} \\
\underline{-(t^4 + 0t^3 + 2t^2)} \\
3t^3 + 4t^2 + 11t \\
\underline{-(3t^3 + 0t^2 + 6t)} \\
4t^2 + 5t + 5 \\
\underline{-(4t^2 + 0t + 8)} \\
5t - 3 \leftarrow \text{Remainder}
\end{array}
$$

We are done with the long division process because the degree of $5t - 3$ (degree 1) is less than the degree of the divisor, $t^2 + 0t + 2$ (degree 2).

Write the answer as $t^2 + 3t + 4 + \dfrac{5t - 3}{t^2 + 2}$. The check is left to the student.

[YOU TRY 4] Divide.

a) $\dfrac{4m^3 + 17m^2 - 38}{m + 3}$

b) $\dfrac{p^4 + 6p^3 + 3p^2 + 10p + 1}{p^2 + 1}$

Summary Dividing Polynomials

Remember, when asked to divide two polynomials, first identify which type of division problem it is.

1) To divide a *polynomial* by a *monomial*, divide *each term* in the polynomial by the monomial and simplify.

$$
\text{Monomial} \rightarrow \quad \frac{56k^3 + 24k^2 - 8k + 2}{8k} = \frac{56k^3}{8k} + \frac{24k^2}{8k} - \frac{8k}{8k} + \frac{2}{8k}
$$

$$
= 7k^2 + 3k - 1 + \frac{1}{4k}
$$

2) To divide a *polynomial* by a *polynomial* containing two or more terms, use *long division*.

$$
\text{Binomial} \rightarrow \quad \frac{15x^3 + 34x^2 - 11x - 2}{5x - 2}
$$

$$
\begin{array}{r}
3x^2 + 8x + 1 \\
5x - 2{\overline{\smash{\big)}\,15x^3 + 34x^2 - 11x - 2}} \\
\underline{-(15x^3 - 6x^2)} \\
40x^2 - 11x \\
\underline{-(40x^2 - 16x)} \\
5x - 2 \\
\underline{-(5x - 2)} \\
0
\end{array}
$$

$$
\frac{15x^3 + 34x^2 - 11x - 2}{5x - 2} = 3x^2 + 8x + 1
$$

E Evaluate **6.4** Exercises

Do the exercises, and check your work.

*Additional answers can be found in the Answers to Exercises appendix.

Objective 1: Review of Long Division

Divide.

1) $3\overline{)747}$ 249

2) $5\overline{)1045}$ 209

3) $17\overline{)8976}$ 528

4) $14\overline{)5194}$ 371

5) $6\overline{)949}$ $158\frac{1}{6}$

6) $4\overline{)857}$ $214\frac{1}{4}$

7) $9\overline{)3937}$ $437\frac{4}{9}$

8) $8\overline{)4189}$ $523\frac{5}{8}$

Objective 2: Divide a Polynomial by a Polynomial

9) Label the dividend, divisor, and quotient of
$$3w + 1\overline{)\begin{array}{c}4w^2 - 2w - 7 \\ 12w^3 - 2w^2 - 23w - 7\end{array}}$$
dividend: $12w^3 - 2w^2 - 23w - 7$; divisor: $3w + 1$; quotient: $4w^2 - 2w - 7$

10) When do you use long division to divide polynomials?
Use long division when the divisor contains two or more terms.

11) If a polynomial of degree 3 is divided by a binomial of degree 1, determine the degree of the quotient.   2

12) How would you check the answer to the division problem in Exercise 9?   Multiply the divisor by the quotient to get the dividend. $(3w + 1)(4w^2 - 2w - 7) = 12w^3 - 2w^2 - 23w - 7$

Divide.

13) $\dfrac{g^2 + 9g + 20}{g + 5}$ $g + 4$

14) $\dfrac{m^2 + 8m + 15}{m + 3}$ $m + 5$

15) $\dfrac{a^2 + 13a + 42}{a + 7}$ $a + 6$

16) $\dfrac{w^2 + 5w + 4}{w + 1}$ $w + 4$

17) $\dfrac{k^2 - k - 30}{k + 5}$ $k - 6$

18) $\dfrac{v^2 - 6v - 16}{v + 2}$ $v - 8$

19) $\dfrac{x^2 + 3x - 40}{x - 5}$ $x + 8$

20) $\dfrac{c^2 - 13c + 36}{c - 4}$ $c - 9$

21) $\dfrac{6h^3 + 7h^2 - 17h - 4}{3h - 4}$ $2h^2 + 5h + 1$

22) $\dfrac{20f^3 - 23f^2 + 41f - 14}{5f - 2}$ $4f^2 - 3f + 7$

23) $(p + 23p^2 - 1 + 12p^3) \div (4p + 1)$ $3p^2 + 5p - 1$

24) $(16y^2 + 6 + 15y^3 + 13y) \div (3y + 2)$ $5y^2 + 2y + 3$

25) $(7m^2 - 16m - 41) \div (m - 4)$ $7m + 12 + \dfrac{7}{m - 4}$

26) $(2t^2 + 5t + 8) \div (t + 7)$ $2t - 9 + \dfrac{71}{t + 7}$

27) $\dfrac{24a + 20a^3 - 12 - 43a^2}{5a - 2}$ $4a^2 - 7a + 2 - \dfrac{8}{5a - 2}$

28) $\dfrac{17v^3 + 33v - 18 + 9v^4 - 56v^2}{9v - 1}$ $v^3 + 2v^2 - 6v + 3 - \dfrac{15}{9v - 1}$

29) $\dfrac{n^3 + 27}{n + 3}$ $n^2 - 3n + 9$

30) $\dfrac{d^3 - 8}{d - 2}$ $d^2 + 2d + 4$

31) $(8r^3 + 6r^2 - 25) \div (4r - 5)$ $2r^2 + 4r + 5$

32) $(12c^3 + 23c^2 + 61) \div (3c + 8)$ $4c^2 - 3c + 8 - \dfrac{3}{3c + 8}$

33) $\dfrac{12x^3 - 17x + 4}{2x + 3}$ $6x^2 - 9x + 5 - \dfrac{11}{2x + 3}$

34) $\dfrac{16h^3 - 106h + 15}{2h - 5}$ $8h^2 + 20h - 3$

35) $\dfrac{k^4 + k^3 + 9k^2 + 4k + 20}{k^2 + 4}$ $k^2 + k + 5$

36) $\dfrac{a^4 + 7a^3 + 6a^2 + 21a + 9}{a^2 + 3}$ $a^2 + 7a + 3$

37) $\dfrac{15t^4 - 40t^3 - 33t^2 + 10t + 2}{5t^2 - 1}$ $3t^2 - 8t - 6 + \dfrac{2t - 4}{5t^2 - 1}$

38) $\dfrac{18v^4 - 15v^3 - 18v^2 + 13v - 10}{3v^2 - 4}$ $6v^2 - 5v + 2 - \dfrac{7v + 2}{3v^2 - 4}$

39) Is the quotient of two polynomials always a polynomial? Explain.

40) Write a division problem that has a divisor of $2c - 1$ and a quotient of $8c - 5$. $\dfrac{16c^2 - 18c + 5}{2c - 1}$

Determine whether each statement is *true* or *false*.

41) If a first-degree binomial divides evenly into a second-degree trinomial, then the quotient is a first-degree binomial.   true

42) If a binomial does not divide evenly into a polynomial, then the result contains a remainder.   true

43) If a binomial divides evenly into a polynomial, then the quotient is a polynomial.   true

44) A fourth-degree polynomial divided by a second-degree polynomial will always result in a second-degree polynomial.   false

Mixed Exercises: Sections 6.3 and 6.4

Divide. This section contains division by polynomials *and* monomials.

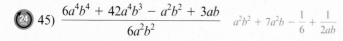

45) $\dfrac{6a^4b^4 + 42a^4b^3 - a^2b^2 + 3ab}{6a^2b^2}$ $a^2b^2 + 7a^2b - \dfrac{1}{6} + \dfrac{1}{2ab}$

46) $\dfrac{12n^2 - 37n + 16}{4n - 3}$ $3n - 7 - \dfrac{5}{4n - 3}$

47) $\dfrac{-15f^4 - 22f^2 + 5 + 49f + 36f^3}{5f - 2}$ $-3f^3 + 6f^2 - 2f + 9 + \dfrac{23}{5f - 2}$

48) $(12k^2 + 6k - 30k^3 + 48k^4) \div (-6k)$ $-8k^3 + 5k^2 - 2k - 1$

49) $\dfrac{8t^2 - 19t - 4}{t - 3}$ $8t + 5 + \dfrac{11}{t - 3}$

50) $\dfrac{-14x^3y^3 + 7x^2y^3 + 21xy + 56}{7x^2y}$ $-2xy^2 + y^2 + \dfrac{3}{x} + \dfrac{8}{x^2y}$

51) $(64p^3 - 27) \div (4p - 3)$ $16p^2 + 12p + 9$

52) $(6g^4 - g^3 - 43g^2 + 32g - 30) \div (2g - 5)$ $3g^3 + 7g^2 - 4g + 6$

53) $(11x^2 + x^3 - 21 + 6x^4 + 3x) \div (x^2 + 3)$ $6x^2 + x - 7$

54) $(12a^4 - 9a^3 - 10a^2 + 3a + 2) \div (3a^2 - 1)$ $4a^2 - 3a - 2$

55) $\dfrac{72c^2 + 32c + 8}{8}$ $9c^2 + 4c + 1$

56) $\dfrac{125t^3 + 8}{5t + 2}$ $25t^2 - 10t + 4$

57) $\dfrac{10h^4 - 6h^3 - 49h^2 + 27h + 19}{2h^2 - 9}$ $5h^2 - 3h - 2 + \dfrac{1}{2h^2 - 9}$

58) $\dfrac{24n^2 - 70 + 27n + 8n^4 + 6n^3}{2n^2 + 9}$ $4n^2 + 3n - 6 - \dfrac{16}{2n^2 + 9}$

59) $\dfrac{m^4 - 16}{m^2 + 4}$ $m^2 - 4$

60) $\dfrac{j^4 - 1}{j^2 - 1}$ $j^2 + 1$

61) $\dfrac{w^2 + 7w - 32}{w - 3}$ $w + 10 - \dfrac{2}{w - 3}$

62) $\dfrac{6t^4 + 20t^3 - 4t^2 - t}{4t^2}$ $\dfrac{3}{2}t^2 + 5t - 1 - \dfrac{1}{4t}$

63) $\left(x^2 + \dfrac{17}{2}x - 15\right) \div (2x - 3)$ $\dfrac{1}{2}x + 5$

64) $\left(3m^2 + \dfrac{44}{5}m - 11\right) \div (5m - 2)$ $\dfrac{3}{5}m + 2 - \dfrac{7}{5m - 2}$

65) $\dfrac{21p^4 - 29p^3 - 15p^2 + 28p + 16}{7p^2 + 2p - 4}$ $3p^2 - 5p + 1 + \dfrac{6p + 20}{7p^2 + 2p - 4}$

66) $\dfrac{8c^4 + 26c^3 + 29c^2 + 14c + 3}{2c^2 + 5c + 3}$ $4c^2 + 3c + 1$

For each rectangle, find a polynomial that represents the missing side.

67)    Find the width if the area is given by $6x^2 + 23x + 21$ sq units. $3x + 7$ units

2x + 3

68) 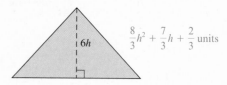   Find the width if the area is given by $20k^3 + 8k^5$ sq units. $k^4 + \dfrac{5}{2}k^2$ units

8k

69) Find the base of the triangle if the area is given by $15n^3 - 18n^2 + 6n$ sq units.

n $30n^2 - 36n + 12$ units

70) Find the base of the triangle if the area is given by $8h^3 + 7h^2 + 2h$ sq units.

6h $\dfrac{8}{3}h^2 + \dfrac{7}{3}h + \dfrac{2}{3}$ units

71) If Joelle travels $(3x^3 + 5x^2 - 26x + 8)$ miles in $(x + 4)$ hours, find an expression for her rate. $3x^2 - 7x + 2$ mph

72) If Lorenzo spent $(4a^3 - 11a^2 + 3a - 18)$ dollars on chocolates that cost $(a - 3)$ dollars per pound, find an expression for the number of pounds of chocolates he purchased. $4a^2 + a + 6$ pounds

R1) Explain how long division of whole numbers and long division of polynomials are similar.

R2) What is the most common mistake you make when performing polynomial long division?

R3) Write down or circle the problems you got wrong or could not do. Ask your instructor for help.

Group Activity — The Group Activity can be found online on Connect.

em POWER me What's Different About College?

Many new college students see homework as something that only high-school-aged kids and younger should have to do. By their way of thinking, adults don't do homework. This attitude is incorrect, as homework is an essential part of the education process, even at the highest levels of learning. Further, such mistaken ideas regarding what college is all about can make it more difficult to do well.

Read each statement and decide whether it would make most sense coming from a typical high school student or typical college student. Place a check in the appropriate column.

		High School	College
1.	My parents or teachers will remind me of my assignments and other responsibilities.		
2.	I have to manage my own time instead of "following the crowd" to classes and other places.		
3.	I have a lot of choices to make and the freedom to make mistakes.		
4.	At least they can't just throw me out if I do poorly.		
5.	I don't really need to be that careful about my spending— there's nothing to spend money on anyway.		
6.	I have to make sure that I take enough courses and fulfill all my requirements to graduate.		
7.	I have a lot of "down time" each day and free time during the week, and it's up to me to use it well.		
8.	Most of my academic work takes place in class, not outside class on homework, the computer, or at the library.		
9.	If I forget a homework assignment, the teacher will remind me that it's due.		

(Continued on next page)

	High School	College
10. My teachers are more like experts or scientists giving lectures, not teachers trying to reach me personally.		
11. Teachers help to identify the important material by writing key words or ideas on the board.		
12. If there's a textbook, the teacher may not follow it closely, but I'll still be expected to keep up with the reading on my own.		
13. The teaching is less about facts than concepts.		
14. The teacher will go over in class what you were assigned to read for homework.		
15. Teachers take account of school events and other teachers' assignments when they assign papers or schedule tests.		
16. On tests, students are expected to apply what they've learned to new situations or problems, not just repeat what they've learned.		

Note: The statements that are more accurate for high school are 1, 4, 5, 8, 9, 11, 14, and 15. The others (2, 3, 6, 7, 10, 12, 13, and 16) are more accurate for college. Carefully consider the items that you miscategorized to determine the kind of misconceptions that you might hold, and consider how that might affect your college success.

Adapted from www.smu.edu/~alec/whyhighschool.html.

Chapter 6: Summary

Definition/Procedure	Example

6.1 Addition and Subtraction of Polynomials

A **polynomial in x** is the sum of a finite number of terms of the form ax^n where n is a whole number and a is a real number.

The **degree of a term** equals the exponent on its variable. If a term has more than one variable, the degree equals the *sum* of the exponents on the variables.

The **degree of the polynomial** equals the highest degree of any nonzero term. **(p. 349)**

Identify each term in the polynomial, the coefficient and degree of each term, and the degree of the polynomial.
$3m^4n^2 - m^3n^2 + 2m^2n^3 + mn - 5n$

Term	Coeff.	Degree
$3m^4n^2$	3	6
$-m^3n^2$	-1	5
$2m^2n^3$	2	5
mn	1	2
$-5n$	-5	1

The degree of the polynomial is 6.

To **add polynomials,** add like terms. Polynomials may be added horizontally or vertically. **(p. 351)**

Add the polynomials.
$(4q^2 + 2q - 12) + (-5q^2 + 3q + 8)$
$= (4q^2 + (-5q^2)) + (2q + 3q) + (-12 + 8) = -q^2 + 5q - 4$

To **subtract two polynomials,** change the sign of each term in the second polynomial. Then add the polynomials. **(p. 352)**

Subtract the polynomials.
$(4t^3 - 7t^2 + 4t + 4) - (12t^3 - 8t^2 + 3t + 9)$
$= (4t^3 - 7t^2 + 4t + 4) + (-12t^3 + 8t^2 - 3t - 9)$
$= -8t^3 + t^2 + t - 5$

6.2 Multiplication of Polynomials

When multiplying a **monomial** and a **polynomial,** use the distributive property. **(p. 358)**

Multiply. $5y^3(-2y^2 + 8y - 3)$
$= (5y^3)(-2y^2) + (5y^3)(8y) + (5y^3)(-3)$
$= -10y^5 + 40y^4 - 15y^3$

To **multiply two polynomials,** multiply each term in the second polynomial by each term in the first polynomial. Then combine like terms. **(p. 358)**

Multiply.
$(5p + 2)(p^2 - 3p + 6)$
$= (5p)(p^2) + (5p)(-3p) + (5p)(6)$
$\quad\quad + (2)(p^2) + (2)(-3p) + (2)(6)$
$= 5p^3 - 15p^2 + 30p + 2p^2 - 6p + 12$
$= 5p^3 - 13p^2 + 24p + 12$

Multiplying Two Binomials
We can use FOIL to multiply two binomials. **FOIL** stands for First, Outer, Inner, Last. Then add like terms. **(p. 359)**

Use FOIL to multiply $(4a - 5)(a + 3)$.

$(4a - 5)(a + 3) = 4a^2 + 12a - 5a - 15$
$\quad\quad\quad\quad = 4a^2 + 7a - 15$

Special Products

1) $(a + b)(a - b) = a^2 - b^2$

2) $(a + b)^2 = a^2 + 2ab + b^2$

3) $(a - b)^2 = a^2 - 2ab + b^2$ **(p. 360)**

1) Multiply. $(c + 9)(c - 9) = c^2 - 9^2 = c^2 - 81$

2) Expand. $(x + 4)^2 = x^2 + 2(x)(4) + 4^2$
$\quad\quad\quad\quad = x^2 + 8x + 16$

3) Expand. $(6v - 7)^2 = (6v)^2 - 2(6v)(7) + 7^2$
$\quad\quad\quad\quad\quad = 36v^2 - 84v + 49$

Definition/Procedure	Example

6.3 Dividing a Polynomial by a Monomial

To **divide a polynomial by a monomial**, divide *each term* in the polynomial by the monomial and simplify. **(p. 367)**

Divide $\dfrac{22s^4 + 6s^3 - 7s^2 + 3s - 8}{4s^2}$.

$$= \frac{22s^4}{4s^2} + \frac{6s^3}{4s^2} - \frac{7s^2}{4s^2} + \frac{3s}{4s^2} - \frac{8}{4s^2}$$

$$= \frac{11s^2}{2} + \frac{3s}{2} - \frac{7}{4} + \frac{3}{4s} - \frac{2}{s^2}$$

6.4 Dividing a Polynomial by a Polynomial

To **divide a polynomial by another polynomial** containing two or more terms, use *long division*. **(p. 371)**

Divide $\dfrac{10w^3 + 2w^2 + 13w + 18}{5w + 6}$.

$$
\begin{array}{r}
2w^2 - 2w + 5 \\
5w + 6 \overline{) \; 10w^3 + 2w^2 + 13w + 18} \\
\underline{-(10w^3 + 12w^2)} \qquad\qquad \downarrow \\
-10w^2 + 13w \qquad\quad \\
\underline{-(-10w^2 - 12w)} \quad \downarrow \\
25w + 18 \\
\underline{-(25w + 30)} \\
-12 \;\; \rightarrow \text{Remainder}
\end{array}
$$

$$\frac{10w^3 + 2w^2 + 13w + 18}{5w + 6} = 2w^2 - 2w + 5 - \frac{12}{5w + 6}$$

Chapter 6:   Review Exercises

*Additional answers can be found in the Answers to Exercises appendix.

(6.1) For Exercises 1 and 2, identify each term in the polynomial, the coefficient and degree of each term, and the degree of the polynomial.

1) $7s^3 - 9s^2 + s + 6$

2) $a^2b^3 + 7ab^2 + 2ab + 9b$

3) Evaluate $2r^2 - 8r - 11$ for $r = -3$.   31

4) Evaluate $p^3q^2 + 4pq^2 - pq - 2q + 9$ for $p = -1$ and $q = 4$. -75

5) If $h(x) = 5x^2 - 3x - 6$, find

   a) $h(-2)$ 20

   b) $h(0)$ -6

6) $f(t) = \dfrac{2}{5}t + 4$. Find t so that $f(t) = \dfrac{4}{5}$. -8

Add or subtract as indicated.

7) $(6c^2 + 2c - 8) - (8c^2 + c - 13)$ $-2c^2 + c + 5$

8) $(-2m^2 - m + 11) + (6m^2 - 12m + 1)$ $4m^2 - 13m + 12$

9) $\begin{aligned} 6.7j^3 - 1.4j^2 + \quad\; j - 5.3 \\ + \underline{3.1j^3 + 5.7j^2 + 2.4j + 4.8} \\ 9.8j^3 + 4.3j^2 + 3.4j - 0.5 \end{aligned}$

10) $\begin{aligned} -4.2p^3 + 12.5p^2 - 7.2p + 6.1 \\ - \underline{\quad 1.3p^3 - \;\; 3.3p^2 + 2.5p + 4.3} \\ -5.5p^3 + 15.8p^2 - 9.7p + 1.8 \end{aligned}$

11) $\left(\dfrac{3}{5}k^2 + \dfrac{1}{2}k + 4\right) - \left(\dfrac{1}{10}k^2 + \dfrac{3}{2}k - 2\right)$ $\dfrac{1}{2}k^2 - k + 6$

12) $\left(\dfrac{2}{7}u^2 - \dfrac{5}{8}u + \dfrac{4}{3}\right) + \left(\dfrac{3}{7}u^2 + \dfrac{3}{8}u - \dfrac{11}{12}\right)$ $\dfrac{5}{7}u^2 - \dfrac{1}{4}u + \dfrac{5}{12}$

13) Subtract $4x^2y^2 - 7x^2y + xy + 5$ from $x^2y^2 + 2x^2y - 4xy + 11$. $-3x^2y^2 + 9x^2y - 5xy + 6$

14) Find the sum of $3c^3d^3 - 7c^2d^2 - c^2d + 8d + 1$ and $14c^3d^3 + 3c^2d^2 - 12cd - 2d - 6$. $17c^3d^3 - 4c^2d^2 - c^2d - 12cd + 6d - 5$

15) Find the sum of $6m + 2n - 17$ and $-3m + 2n + 14$. $3m + 4n - 3$

16) Subtract $-h^4 + 8j^4 - 2$ from $12h^4 - 3j^4 + 19$. $13h^4 - 11j^4 + 21$

17) Subtract $2x^2 + 3x + 18$ from the sum of $7x - 16$ and $8x^2 - 15x + 6$. $6x^2 - 11x - 28$

18) Find the sum of $7xy + 2x - 3y - 11$ and $-3xy + 5y + 1$ and subtract it from $-5xy - 9x + y + 4$. $-9xy - 11x - y + 14$

Find the polynomial that represents the perimeter of each rectangle.

19)

$d^2 + 6d + 2$

$d^2 - 3d + 1$ $4d^2 + 6d + 6$ units

20)

$7m - 1$

$3m + 5$ $20m + 8$ units

(6.2) Multiply.

21) $3r(8r - 13)$ $24r^2 - 39r$

22) $-5m^2(7m^2 - 4m + 8)$ $-35m^4 + 20m^3 - 40m^2$

23) $(4w + 3)(-8w^3 - 2w + 1)$ $-32w^4 - 24w^3 - 8w^2 - 2w + 3$

24) $\left(2t^2 - \frac{1}{3}\right)(-9t^2 + 7t - 12)$ $-18t^4 + 14t^3 - 21t^2 - \frac{7}{3}t + 4$

25) $(y - 3)(y - 9)$ $y^2 - 12y + 27$

26) $(f - 5)(f - 8)$ $f^2 - 13f + 40$

27) $(3n^2 - 4)(2n - 7)$ $6n^3 - 21n^2 - 8n + 28$

28) $(3p + 4)(3p + 1)$ $9p^2 + 15p + 4$

29) $-(a - 13)(a + 10)$ $-a^2 + 3a + 130$

30) $(5d^2 + 2)(6d + 5)$ $30d^3 + 25d^2 + 12d + 10$

31) $6pq^2(7p^3q^2 + 11p^2q^2 - pq + 4)$ $4p^4q^4 + 66p^3q^4 - 6p^2q^3 + 24pq^2$

32) $9x^3y^4(-6x^2y + 2xy^2 + 8x - 1)$
$-54x^5y^5 + 18x^4y^6 + 72x^4y^4 - 9x^3y^4$

33) $(2x - 9y)(2x + y)$ $4x^2 - 16xy - 9y^2$

34) $(7r + 3s)(r - s)$ $7r^2 - 4rs - 3s^2$

35) $(x^2 + 5x - 12)(10x^4 - 3x^2 + 6)$
$10x^6 + 50x^5 - 123x^4 - 15x^3 + 42x^2 + 30x - 72$

36) $(3m^2 - 4m + 2)(m^2 + m - 5)$ $3m^4 - m^3 - 17m^2 + 22m - 10$

37) $4f^2(2f - 7)(f - 6)$ $8f^4 - 76f^3 + 168f^2$

38) $-3(5u - 11)(u + 4)$ $-15u^2 - 27u + 132$

39) $(z + 3)(z + 1)(z + 4)$ $z^3 + 8z^2 + 19z + 12$

40) $(p + 2)(p + 5)(p + 4)$ $p^3 + 11p^2 + 38p + 40$

41) $\left(\frac{2}{7}d + 3\right)\left(\frac{1}{2}d - 8\right)$ $\frac{1}{7}d^2 - \frac{11}{14}d - 24$

42) $\left(\frac{3}{10}t - 6\right)\left(\frac{2}{9}t - 5\right)$ $\frac{1}{15}t^2 - \frac{17}{6}t + 30$

Expand.

43) $(c + 4)^2$ $c^2 + 8c + 16$

44) $(x - 12)^2$ $x^2 - 24x + 144$

45) $(4p - 3)^2$ $16p^2 - 24p + 9$

46) $(9 - 2y)^2$ $4y^2 - 36y + 81$

47) $(x - 3)^3$ $x^3 - 9x^2 + 27x - 27$

48) $(p + 4)^3$ $p^3 + 12p^2 + 48p + 64$

49) $[(m - 3) + n]^2$ $m^2 - 6m + 9 + 2mn - 6n + n^2$

Find the special products.

50) $(z + 7)(z - 7)$ $z^2 - 49$

51) $(p - 13)(p + 13)$ $p^2 - 169$

52) $\left(\frac{1}{4}n - 5\right)\left(\frac{1}{4}n + 5\right)$ $\frac{1}{16}n^2 - 25$

53) $\left(\frac{9}{2} + \frac{5}{6}x\right)\left(\frac{9}{2} - \frac{5}{6}x\right)$ $\frac{81}{4} - \frac{25}{36}x^2$

54) $\left(\frac{6}{11} - r^2\right)\left(\frac{6}{11} + r^2\right)$ $\frac{36}{121} - r^4$

55) $\left(3a^2 - \frac{1}{2}b\right)\left(3a^2 + \frac{1}{2}b\right)$ $9a^4 - \frac{1}{4}b^2$

56) $-4(2d - 7)^2$ $-16d^2 + 112d - 196$

57) $3u(u + 4)^2$ $3u^3 + 24u^2 + 48u$

58) $[(2p + 5) + q][(2p + 5) - q]$ $4p^2 + 20p + 25 - q^2$

59) Write an expression for the a) area and b) perimeter of the rectangle.

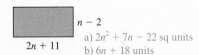

$n - 2$
$2n + 11$

a) $2n^2 + 7n - 22$ sq units
b) $6n + 18$ units

60) Express the volume of the cube as a polynomial.

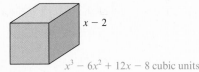

$x - 2$

$x^3 - 6x^2 + 12x - 8$ cubic units

61) When $3y^2 - 7y - 4$ is multiplied by a monomial, the result is $24y^5 - 56y^4 - 32y^3$. Find the monomial. $8y^3$

62) Is the product of two binomials always a binomial? Explain.   No. For example, $(x + 3)(x + 2) = x^2 + 5x + 6$, which is a trinomial.

(6.3 and 6.4) Divide.

63) $\dfrac{12t^6 - 30t^5 - 15t^4}{3t^4}$ $4t^2 - 10t - 5$

64) $\dfrac{42p^4 + 12p^3 - 18p^2 + 6p}{-6p}$ $-7p^3 - 2p^2 + 3p - 1$

65) $\dfrac{w^2 + 9w + 20}{w + 4}$ $w + 5$

66) $\dfrac{a^2 - 2a - 24}{a - 6}$ $a + 4$

67) $\dfrac{8r^3 + 22r^2 - r - 15}{2r + 5}$ $4r^2 + r - 3$

68) $\dfrac{-36h^3 + 99h^2 + 4h + 1}{12h - 1}$ $-3h^2 + 8h + 1$

69) $\dfrac{14t^4 + 28t^3 - 21t^2 + 20t}{14t^3}$ $t + 2 - \frac{3}{2t} + \frac{10}{7t^2}$

70) $\dfrac{48w^4 - 30w^3 + 24w^2 + 3w}{6w^2}$ $8w^2 - 5w + 4 + \frac{1}{2w}$

71) $(14v + 8v^2 - 3) \div (4v + 9)$ $2v - 1 + \dfrac{6}{4v + 9}$

72) $(-8 + 12r^2 - 19r) \div (3r - 1)$ $4r - 5 - \dfrac{13}{3r - 1}$

73) $\dfrac{6v^4 - 14v^3 + 25v^2 - 21v + 24}{2v^2 + 3}$ $3v^2 - 7v + 8$

74) $\dfrac{8t^4 + 20t^3 - 30t^2 - 65t + 13}{4t^2 - 13}$ $2t^2 + 5t - 1$

75) $\dfrac{c^3 - 8}{c - 2}$ $c^2 + 2c + 4$

76) $\dfrac{g^3 + 64}{g + 4}$ $g^2 - 4g + 16$

77) $\dfrac{-4 + 13k + 18k^3}{3k + 2}$ $6k^2 - 4k + 7 - \dfrac{18}{3k + 2}$

78) $\dfrac{10 + 12m^3 - 34m^2}{6m + 1}$ $2m^2 - 6m + 1 + \dfrac{9}{6m + 1}$

79) $(20x^4y^4 - 48x^2y^4 - 12xy^2 + 15x) \div (-12xy^2)$

80) $(3u^4 - 31u^3 - 13u^2 + 76u - 30) \div (u^2 - 11u + 5)$
   $3u^2 + 2u - 6$

Determine whether the statement is true or false.

81) The quotient of two polynomials is always a polynomial.
   false

82) A third-degree polynomial divided by a first-degree polynomial will always be a second-degree polynomial.
   false

83) Find the base of the triangle if the area is given by $12a^2 + 3a$ sq units.

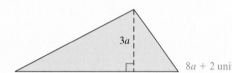

3a

8a + 2 units

84) Find the length of the rectangle if the area is given by $28x^3 - 51x^2 + 34x - 8$ sq units.

7x − 4 $4x^2 - 5x + 2$ units

Mixed Exercises

Perform the operations and simplify. Assume that all variables represent nonzero real numbers. The answer should not contain negative exponents.

85) $\begin{aligned} & 18c^3 + 7c^2 - 11c + 2 \\ + \; & \underline{2c^2 - 19c^2 \qquad\; - 1} \end{aligned}$ $20c^3 - 12c^2 - 11c + 1$

86) $\dfrac{15a - 11 + 14a^2}{7a - 3}$ $2a + 3 - \dfrac{2}{7a - 3}$

87) $(12 - 7w)(12 + 7w)$ $144 - 49w^2$

88) $(5p - 9)(2p^2 - 4p - 7)$ $10p^3 - 38p^2 + p + 63$

89) $5(-2r^7t^9)^3$ $-40r^{21}t^{27}$

90) $(7k^2 + k - 9) - (-4k^2 + 8k - 3)$ $11k^2 - 7k - 6$

91) $(39a^6b^6 + 21a^4b^5 - 5a^3b^4 + a^2b) \div (3a^3b^3)$
   $13a^3b^3 + 7ab^2 - \dfrac{5}{3}b + \dfrac{1}{3ab^2}$

92) $\dfrac{(6x^{-4}y^5)^{-2}}{(3xy^{-2})^{-4}}$ $\dfrac{9x^{12}}{4y^{18}}$

93) $(h - 5)^3$ $h^3 - 15h^2 + 75h - 125$

94) $\left(\dfrac{1}{8}m - \dfrac{2}{3}n\right)^2$ $\dfrac{1}{64}m^2 - \dfrac{1}{6}mn + \dfrac{4}{9}n^2$

95) $\dfrac{-23c + 41 + 2c^3}{c + 4}$ $2c^2 - 8c + 9 + \dfrac{5}{c + 4}$

96) $-7d^3(5d^2 + 12d - 8)$ $-35d^5 - 84d^4 + 56d^3$

97) $\left(\dfrac{5}{y^4}\right)^{-3}$ $\dfrac{y^{12}}{125}$

98) $(27q^3 + 8) \div (3q + 2)$ $9q^2 - 6q + 4$

99) $(6p^4 + 11p^3 - 20p^2 - 17p + 20) \div (3p^2 + p - 4)$
   $2p^2 + 3p - 5$

100) $\left(\dfrac{3b^{-2}c}{a^5}\right)^{-3}\left(\dfrac{4a^{-2}b}{c^4}\right)\left(\dfrac{2ab^3}{c^2}\right)$ $\dfrac{8a^{14}b^{10}}{27c^9}$

Chapter 6:   Test

*Additional answers can be found in the Answers to Exercises appendix.

1) Given the polynomial $6n^3 + 6n^2 - n - 7$,

   a) what is the coefficient of n? -1

   b) what is the degree of the polynomial?   3

2) What is the degree of the polynomial $6a^4b^5 + 11a^4b^3 - 2a^3b + 5ab^2 - 3$?   9

3) Evaluate $-2r^2 + 7s$ when $r = -4$ and $s = 5$.   3

Perform the indicated operation(s).

4) $4h^3(6h^2 - 3h + 1)$ $24h^5 - 12h^4 + 4h^3$

5) $(7a^3b^2 + 9a^2b^2 - 4ab + 8)$
   $+ (5a^3b^2 - 12a^2b^2 + ab + 1)$ $12a^3b^2 - 3a^2b^2 - 3ab + 9$

6) Subtract $6y^2 - 5y + 13$ from $15y^2 - 8y + 6$. $9y^2 - 3y - 7$

7) $3(-c^3 + 3c - 6) - 4(2c^3 + 3c^2 + 7c - 1)$
   $-11c^3 - 12c^2 - 19c - 14$

8) $(u - 5)(u - 9)$ $u^2 - 14u + 45$

9) $(4g + 3)(2g + 1)$ $8g^2 + 10g + 3$

10) $\left(v + \dfrac{2}{5}\right)\left(v - \dfrac{2}{5}\right)$ $v^2 - \dfrac{4}{25}$

11) $(3x - 7y)(2x + y)$ $6x^2 - 11xy - 7y^2$

12) $(4t^2 + 1)(9t - 7)$ $36t^3 - 28t^2 + 9t - 7$

13) $(5 - 6n)(2n^2 + 3n - 8)$ $-12n^3 - 8n^2 + 63n - 40$

14) $2y(y + 6)^2$ $2y^3 + 24y^2 + 72y$

Expand.

15) $(3m - 4)^2$ $9m^2 - 24m + 16$

16) $\left(\dfrac{4}{3}x + y\right)^2$ $\dfrac{16}{9}x^2 + \dfrac{8}{3}xy + y^2$

17) $(t - 2)^3$ $t^3 - 6t^2 + 12t - 8$

Divide.

18) $\dfrac{w^2 + 9w + 18}{w + 6}$ $w + 3$

19) $\dfrac{24m^6 - 40m^5 + 8m^4 - 6m^3}{8m^4}$ $3m^2 - 5m + 1 - \dfrac{3}{4m}$

20) $(22p - 50 + 18p^3 - 45p^2) \div (3p - 7)$ $6p^2 - p + 5 - \dfrac{15}{3p - 7}$

21) $\dfrac{y^3 - 27}{y - 3}$ $y^2 + 3y + 9$

22) $(2r^4 + 3r^3 + 6r^2 + 15r - 20) \div (r^2 + 5)$ $2r^2 + 3r - 4$

23) Write an expression for a) the area and b) the perimeter of the rectangle.

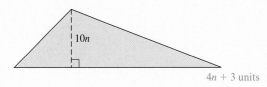
$d - 5$
$3d + 1$
a) $3d^2 - 14d - 5$ sq units
b) $8d - 8$ units

24) Write an expression for the base of the triangle if the area is given by $20n^2 + 15n$ sq units.

$10n$

$4n + 3$ units

Chapter 6:   Cumulative Review for Chapters 1–6

*Additional answers can be found in the Answers to Exercises appendix.

1) Given the set of numbers
$$\left\{\dfrac{3}{8}, -15, 2.1, \sqrt{17}, 41, 0.\overline{52}, 0, 9.32087326\ldots\right\}$$
list the

   a) whole numbers $\{41, 0\}$

   b) integers $\{-15, 41, 0\}$

   c) rational numbers $\left\{\dfrac{3}{8}, -15, 2.1, 41, 0.\overline{52}, 0\right\}$

2) Evaluate $-3^4 + 2 \cdot 9 \div (-3)$. -87

3) Divide $3\dfrac{1}{8} \div 1\dfrac{7}{24}$. $\dfrac{75}{31}$ or $2\dfrac{13}{31}$

Solve.

4) $-\dfrac{18}{7}m - 9 = 21$ $\left\{-\dfrac{35}{3}\right\}$

5) $5(u + 3) + 2u = 1 + 7(u - 2)$ $\varnothing$

6) $\dfrac{5}{6} - \dfrac{1}{2}(2p + 1) = \dfrac{1}{3}(p + 3)$ $\left\{-\dfrac{1}{2}\right\}$

7) Solve $5y - 16 \geq 8y - 1$. Write the answer in interval notation. $(-\infty, -5]$

8) *Write an equation in one variable, and solve.* In 2012, Anjani sold her house for \$175,000. This was 70% of the amount she paid for her house when she bought it in 2005. How much did she pay for the house in 2005?   \$250,000

9) Find the x- and y-intercepts of $3x - 8y = 24$, and sketch a graph of the equation.   x-int: $(8, 0)$; y-int: $(0, -3)$

10) Graph $y = -4$.

11) Write an equation of the line containing the points $(-4, 7)$ and $(2, -11)$. Express the answer in standard form.
    $3x + y = -5$

12) Write an equation of the line perpendicular to $4x - y = 1$ containing the point $(8, 1)$. Express the answer in slope-intercept form. $y = -\dfrac{1}{4}x + 3$

13) Solve this system using the elimination method.
$$\begin{array}{l} 3x - 4y = -17 \\ x + 2y = -4 \end{array}\qquad \left(-5, \dfrac{1}{2}\right)$$

14) *Write a system of two equations in two variables, and solve.* The length of a rectangle is 1 cm less than three times its width. The perimeter of the rectangle is 94 cm. What are the dimensions of the figure?   width: 12 cm; length: 35 cm

Simplify. The answers should not contain negative exponents.

15) $-8(2a^7)^2$ $-32a^{14}$

16) $c^{10} \cdot c^7$ c^{17}

17) $\left(\dfrac{4p^{-12}}{p^{-5}}\right)^3$ $\dfrac{64}{p^{21}}$

Perform the indicated operations.

18) $(6q^2 + 7q - 1) - 4(2q^2 - 5q + 8) + 3(-9q - 4)$ $-2q^2 - 45$

19) $(n - 7)(n + 8)$ $n^2 + n - 56$

20) $(3a - 11)(3a + 11)$ $9a^2 - 121$

21) $\dfrac{12a^4b^4 - 18a^3b + 60ab + 6b}{12a^3b^2}$ $ab^2 - \dfrac{3}{2b} + \dfrac{5}{a^2b} + \dfrac{1}{2a^3b}$

22) $(5p^3 - 14p^2 - 10p + 5) \div (p - 3)$ $5p^2 + p - 7 - \dfrac{16}{p - 3}$

23) $(4n^2 - 9)(3n^2 + n - 2)$ $12n^4 + 4n^3 - 35n^2 - 9n + 18$

24) $5c(c - 4)^2$ $5c^3 - 40c^2 + 80c$

25) $\dfrac{8z^3 + 1}{2z + 1}$ $4z^2 - 2z + 1$

Factoring Polynomials

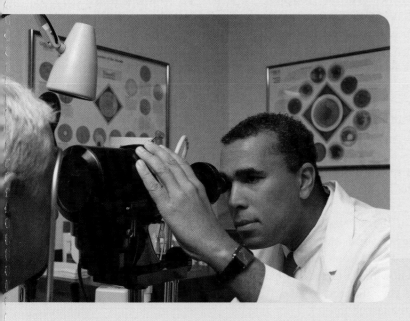

Math at Work:

Ophthalmologist

Mark Diamond is an ophthalmologist, a doctor specializing in the treatment of diseases of the eye. In medical school, he acquired a great deal of knowledge about human anatomy and about the anatomy of the eye in particular. Yet Mark also says he could not do his job without his background in mathematics. He explains that, in addition to the mathematical formulas he uses on a daily basis, the thinking he learned in math courses is invaluable to his work.

As a physician, Mark follows a very logical, analytical progression to form an accurate diagnosis and treatment plan. He examines a patient, performs tests, and then analyzes the results to settle on a diagnosis. Next, he thinks of different ways to solve the medical problem and decides on the course of treatment that is best for that patient. He says that this sort of careful, organized approach is precisely how he learned to deal with difficult math problems. Factoring, for example, requires the ability to think through and solve a problem in a thoughtful and logical manner—the same methods Mark applies in treating his patients.

In this chapter, we will learn different techniques for factoring polynomials. We also introduce strategies you can use to help you make important decisions in your academic career.

If you're like most students, considering the list of courses available to you brings a range of emotions: anticipation over what you'll learn; hope that the courses you choose will bring you closer to your dreams; fear that you won't be able to do well; and excitement that you're proceeding with your college career, taking another of the many small steps that will eventually add up to a complete journey through college.

Choosing the right set of courses can be intimidating. If you approach the problem thoughtfully, though, your final choices will make the best of the possibilities offered. Use the skills below to help you as you make your course decisions.

- Familiarize yourself with how to access your college's listing of courses. Usually the list can be found on the Web, but sometimes courses are included in an actual physical catalog.
- Make an appointment to see your college advisor. He or she will be able to give you good advice on the ins and outs of your college's requirements.

- Make a complete list of the courses you need to take in order to graduate. If you are not sure which courses you have taken and which requirements you have already fulfilled, get a copy of your transcript from the registrar.
- Think carefully about how many terms you have left before graduation and how many required courses you still need to take. It's never a good idea to leave all your requirements to the last term!

- Draft a personal schedule for the term. Start with the required courses, then fill in any electives (that is, courses that you can choose to take).
- Make sure your draft schedule does not conflict with work or family obligations.
- Complete the process by registering for classes.

- When you receive your course schedule, look it over to confirm that you are registered for the classes you signed up for.

- Take a step back and think about where you are in your course of study. Are you accomplishing the larger goals you have for your academic career? Are you getting the most out of your college experience?

Chapter 7 POWER Plan

What are your goals for Chapter 7?	How can you accomplish each goal? (Write in the steps you will take to succeed.)
1 Be prepared before and during class.	• _____ • _____ • _____ • _____
2 Understand the homework to the point where you could do it without needing any help or hints.	• _____ • _____ • _____
3 Use the P.O.W.E.R. framework to learn how to make good academic choices: *Identify Major Attractions*.	• _____ • _____ • _____
4 Write your own goal. _____ _____	• _____ _____

What are your objectives for Chapter 7?	How can you accomplish each objective?
1 Find the greatest common factor, and factor by grouping.	• Know how to find and factor out a GCF from a group of monomials or polynomials. • Understand that factoring a polynomial is the opposite of multiplying polynomials. • Learn the procedures for factoring out a GCF and factoring by grouping.
2 Learn how to factor trinomials.	• Practice the arithmetic skills needed to factor trinomials. • Be able to perform the procedure for factoring any polynomial of the form $x^2 + bx + c$. • Be able to perform the procedure for factoring any polynomial of the form $ax^2 + bx + c$ $(a \neq 1)$. • Understand how to factor $ax^2 + bx + c$ $(a \neq 1)$ by grouping. • Use the formulas for factoring a perfect square trinomial and the difference of two squares without needing hints.
3 Know how to solve quadratic equations by factoring.	• Understand the definition of a quadratic equation and the zero product rule. • Learn the procedure for solving a quadratic equation by factoring. • Update the **Five Steps for Solving Applied Problems** and use it to solve: • Geometry problems. • Consecutive integer problems. • Pythagorean theorem problems.
4 Write your own goal. _____ _____	• _____ _____

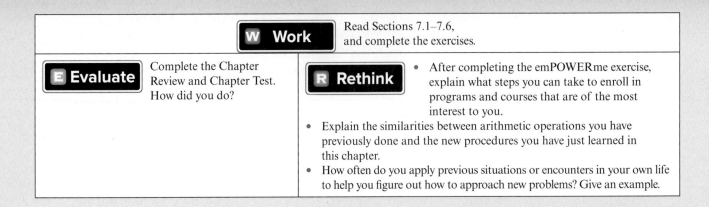

7.1 The Greatest Common Factor and Factoring by Grouping

P Prepare

O Organize

What are your objectives for Section 7.1?	How can you accomplish each objective?
1 Find the GCF of a Group of Monomials	• Write the definition of a GCF in your own words. • Notice the similarities between factoring a number and finding the GCF, and finding the GCF of monomials. • Complete the given examples on your own. • Complete You Trys 1 and 2.
2 Factor Out the Greatest Common Monomial Factor	• Understand that factoring a polynomial is the opposite of multiplying polynomials. • In your notes, write the steps for factoring out the GCF in your own words. • Work the given examples on your own by applying the procedure for factoring out the GCF. • Complete You Trys 3 and 4.
3 Factor Out the Greatest Common Binomial Factor	• Follow the same procedure you used for the previous examples. • Complete the examples on your own. • Complete You Try 5.
4 Factor by Grouping	• Complete the examples on your own, and notice the new steps added in each example. • Complete You Trys 6–8. • Write the **Procedure for Factoring by Grouping** in your own words.

Read the explanations, follow the examples, take notes, and complete the You Trys.

In Section 1.1, we discussed writing a number as the product of factors:

$$18 = 3 \cdot 6$$
$$\downarrow \qquad \downarrow \qquad \downarrow$$
$$\text{Product} \quad \text{Factor} \quad \text{Factor}$$

To **factor** an integer is to write it as the product of two or more integers. Therefore, 18 can also be factored in other ways:

$$18 = 1 \cdot 18 \qquad\qquad 18 = 2 \cdot 9 \qquad\qquad 18 = -1 \cdot (-18)$$
$$18 = -2 \cdot (-9) \qquad\quad 18 = -3 \cdot (-6) \qquad 18 = 2 \cdot 3 \cdot 3$$

The last **factorization**, $2 \cdot 3 \cdot 3$ or $2 \cdot 3^2$, is called the **prime factorization** of 18 since all of the factors are prime numbers. (See Section 1.1.) The factors of 18 are 1, 2, 3, 6, 9, 18, -1, -2, -3, -6, -9, and -18.

We can also write the factors as ± 1, ± 2, ± 3, ± 6, ± 9, and ± 18. (Read ± 1 as "plus or minus 1.")

In this chapter, we will learn how to factor polynomials, a skill that is used in many ways throughout algebra.

1 Find the GCF of a Group of Monomials

 Hint
Write the definition of GCF in your own words.

Definition

The **greatest common factor (GCF)** of a group of two or more integers is the *largest* common factor of the numbers in the group.

For example, if we want to find the GCF of 18 and 24, we can list their positive factors.

$$18: 1, 2, 3, 6, 9, 18$$
$$24: 1, 2, 3, 4, 6, 8, 12, 24$$

The greatest common factor of 18 and 24 is 6. We can also use prime factors.

We begin our study of factoring polynomials by discussing how to find the greatest common factor of a group of monomials.

EXAMPLE 1

In-Class Example 1

Find the greatest common factor of a^{10} and a^6.

Answer: a^6

Find the greatest common factor of x^4 and x^6.

Solution

We can write x^4 as $1 \cdot x^4$, and we can write x^6 as $x^4 \cdot x^2$. The largest power of x that is a factor of both x^4 and x^6 is x^4. Therefore, the GCF is x^4.

Notice that the power of 4 in the GCF is the smallest of the powers when comparing x^4 and x^6. This will always be true.

 Note

The exponent on the variable in the GCF will be the *smallest* exponent appearing on the variable in the group of terms.

Find the greatest common factor of y^5 and y^8.

EXAMPLE 2

In-Class Example 2

Find the greatest common factor for each group of terms.
a) $36k^4, 28k^9, 20k^8$
b) $-56a^8b, 21a^{10}b^3$
c) $20x^5y^4, 60y^2, 50x^3y$

Answer: a) $4k^4$ b) $7a^8b$
c) $10y$

Find the greatest common factor for each group of terms.

a) $24n^5, 8n^9, 16n^3$ b) $-15x^{10}y, 25x^6y^8$

c) $49a^4b^5, 21a^3, 35a^2b^4$

Solution

a) The GCF of the coefficients, 24, 8, and 16, is 8. The smallest exponent on n is 3, so n^3 is part of the GCF.

$$\text{The GCF is } 8n^3.$$

b) The GCF of the coefficients, -15 and 25, is 5. The smallest exponent on x is 6, so x^6 is part of the GCF. The smallest exponent on y is 1, so y is part of the GCF.

$$\text{The GCF is } 5x^6y.$$

c) The GCF of the coefficients is 7. The smallest exponent on a is 2, so a^2 is part of the GCF. There is no b in the term $21a^3$, so there will be no b in the GCF.

$$\text{The GCF is } 7a^2.$$

W Hint

Complete the example on your own and notice the two-step process.

YOU TRY 2
Find the greatest common factor for each group of terms.

a) $18w^6, 45w^{10}, 27w^5$ b) $-14hk^3, 18h^4k^2$ c) $54c^5d^5, 66c^8d^3, 24c^2$

---

2 Factor Out the Greatest Common Monomial Factor

Earlier we said that to **factor an integer** is to write it as the product of two or more integers. To **factor a polynomial** is to write it as a product of two or more polynomials. Throughout this chapter, we will study different factoring techniques. We will begin by discussing how to factor out the greatest common factor.

EXAMPLE 3

In-Class Example 3

Factor out the greatest common factor from $4c + 28$.

Answer: $4(c + 7)$

Factor out the greatest common factor from $3y + 15$.

Solution

We will use the distributive property to factor out the greatest common factor from $3y + 15$. First, identify the GCF of $3y$ and 15: The GCF is 3.

Then, rewrite each term as a product of two factors with one factor being 3.

$$3y + 15 = (3)(y) + (3)(5)$$
$$= 3(y + 5) \qquad \text{Distributive property}$$

When we factor $3y + 15$, we get $3(y + 5)$. We can check our result by multiplying.

$$3(y + 5) = 3 \cdot y + 3 \cdot 5 = 3y + 15$$

Procedure Steps for Factoring Out the Greatest Common Factor

1) Identify the GCF of all of the terms of the polynomial.

2) Rewrite each term as the product of the GCF and another factor.

3) Use the distributive property to factor out the GCF from the terms of the polynomial.

4) Check the answer by multiplying the factors. The result should be the original polynomial.

EXAMPLE 4

In-Class Example 4

Factor out the GCF.
a) $15v^5 + 10v^4 + 40v^3$
b) $d^6 - 11d^2$
c) $4x^3y^4 - 28x^3y^3 + 8x^2y^3 - 4xy^2$

Answer: a) $5v^3(3v^2 + 2v + 8)$
b) $d^2(d^4 - 11)$
c) $4xy^2(x^2y^2 - 7x^2y + 2xy - 1)$

Factor out the greatest common factor.

a) $28p^5 + 12p^4 + 4p^3$

b) $w^8 - 7w^6$

c) $6a^5b^3 + 30a^5b^2 - 54a^4b^2 - 6a^3b$

Solution

a) Identify the GCF of all of the terms: GCF $= 4p^3$.

$$28p^5 + 12p^4 + 4p^3 = (4p^3)(7p^2) + (4p^3)(3p) + (4p^3)(1)$$ Rewrite each term using the GCF as one of the factors.

$$= 4p^3(7p^2 + 3p + 1)$$ Distributive property

Check by multiplying: $4p^3(7p^2 + 3p + 1) = 28p^5 + 12p^4 + 4p^3$ ✓

b) The GCF of the two terms is w^6.

$$w^8 - 7w^6 = (w^6)(w^2) - (w^6)(7)$$ Rewrite each term using the GCF as one of the factors.
$$= w^6(w^2 - 7)$$ Distributive property

Check: $w^6(w^2 - 7) = w^8 - 7w^6$ ✓

c) The GCF of all of the terms is $6a^3b$.

$$6a^5b^3 + 30a^5b^2 - 54a^4b^2 - 6a^3b$$
$$= (6a^3b)(a^2b^2) + (6a^3b)(5a^2b) - (6a^3b)(9ab) - (6a^3b)(1)$$ Rewrite using the GCF.
$$= 6a^3b(a^2b^2 + 5a^2b - 9ab - 1)$$ Distributive property

Check: $6a^3b(a^2b^2 + 5a^2b - 9ab - 1) = 6a^5b^3 + 30a^5b^2 - 54a^4b^2 - 6a^3b$ ✓

[YOU TRY 3]   Factor out the greatest common factor.

a) $2w + 16$ b) $3u^6 + 36u^5 + 15u^4$ c) $z^5 - 9z^2$

d) $45r^4t^3 + 36r^4t^2 + 18r^3t^2 - 9r^2t$

Sometimes we need to take out a negative factor.

EXAMPLE 5

In-Class Example 5

Factor out $-3w$ from $-27w^3 + 12w^2 - 6w$.

Answer: $-3w(9w^2 - 4w + 2)$

Factor out $-2k$ from $-6k^4 + 10k^3 - 8k^2 + 2k$.

Solution

$-6k^4 + 10k^3 - 8k^2 + 2k$

$= (-2k)(3k^3) + (-2k)(-5k^2) + (-2k)(4k) + (-2k)(-1)$ Rewrite using $-2k$ as one of the factors.

$= -2k[3k^3 + (-5k^2) + 4k + (-1)]$ Distributive property

$= -2k(3k^3 - 5k^2 + 4k - 1)$ Rewrite $+(-5k^2)$ as $-5k^2$ and $+(-1)$ as -1.

Check: $-2k(3k^3 - 5k^2 + 4k - 1) = -6k^4 + 10k^3 - 8k^2 + 2k$ ✓

BE CAREFUL   When taking out a negative factor, be very careful with the signs!

[**YOU TRY 4**]   Factor out $-y^2$ from $-y^4 + 10y^3 - 8y^2$.

3 Factor Out the Greatest Common Binomial Factor

Until now, all of the GCFs have been monomials. Sometimes, however, the greatest common factor is a *binomial*.

EXAMPLE 6

In-Class Example 6

Factor out the GCF.
a) $p(q - 4) + 10(q - 4)$
b) $m(n + 8) - (n + 8)$

Answer: a) $(q - 4)(p + 10)$
b) $(n + 8)(m - 1)$

Factor out the greatest common factor.

a) $a(b + 5) + 8(b + 5)$ b) $x(y + 3) - (y + 3)$

Solution

a)   In the polynomial $\underbrace{a(b + 5)}_{\text{term}} + \underbrace{8(b + 5)}_{\text{term}}$, $a(b + 5)$ is a term and $8(b + 5)$ is a term.

What do these terms have in common? $b + 5$

The GCF of $a(b + 5)$ and $8(b + 5)$ is $b + 5$. Use the distributive property to factor out $b + 5$.

$a(b + 5) + 8(b + 5) = (b + 5)(a + 8)$ Distributive property

Check: $(b + 5)(a + 8) = (b + 5)a + (b + 5)8$ Distribute.

$= a(b + 5) + 8(b + 5)$ Commutative property

b)   Begin by writing $x(y + 3) - (y + 3)$ as $x(y + 3) - 1(y + 3)$. The GCF is $y + 3$.

$x(y + 3) - 1(y + 3) = (y + 3)(x - 1)$ Distributive property

The check is left to the student.

It is important to write -1 in front of $(y + 3)$. Otherwise, the following mistake is often made:

$$x(y + 3) - (y + 3) = (y + 3)x \quad \text{This is incorrect!}$$

The correct factor is $x - 1$, *not* x.

[YOU TRY 5]

Factor out the GCF.

a) $c(d - 8) + 2(d - 8)$ b) $k(k^2 + 15) - 7(k^2 + 15)$

c) $u(v + 2) - (v + 2)$

Taking out a binomial factor leads us to our next method of factoring—factoring by grouping.

4 Factor by Grouping

When we are asked to factor a polynomial containing four terms, we often try to **factor by grouping.**

EXAMPLE 7

Factor by grouping.

a) $rt + 7r + 2t + 14$ b) $3xz - 4yz + 18x - 24y$ c) $n^3 + 8n^2 - 5n - 40$

Solution

a) Begin by grouping terms together so that each group has a common factor.

$$\underbrace{rt + 7r}_{\downarrow} + \underbrace{2t + 14}_{\downarrow}$$

Factor out r to get $r(t + 7)$. $= r(t + 7) + 2(t + 7)$ Factor out 2 to get $2(t + 7)$.
$= (t + 7)(r + 2)$ Factor out $(t + 7)$.

Check by multiplying: $(t + 7)(r + 2) = rt + 7r + 2t + 14$ ✓

b) Group terms together so that each group has a common factor.

$$\underbrace{3xz - 4yz}_{\downarrow} + \underbrace{18x - 24y}_{\downarrow}$$

Factor out z to get $z(3x - 4y)$. $= z(3x - 4y) + 6(3x - 4y)$ Factor out 6 to get $6(3x - 4y)$.
$= (3x - 4y)(z + 6)$ Factor out $(3x - 4y)$.

Check by multiplying: $(3x - 4y)(z + 6) = 3xz - 4yz + 18x - 24$ ✓

c) Group terms together so that each group has a common factor.

$$\underbrace{n^3 + 8n^2}_{\downarrow} \underbrace{-5n - 40}_{\downarrow}$$

Factor out n^2 to get $n^2(n + 8)$. $= n^2(n + 8) - 5(n + 8)$ Factor out -5 to get $-5(n + 8)$.
$= (n + 8)(n^2 - 5)$ Factor out $(n + 8)$.

We must factor out -5, *not* 5, from the second group so that the binomial factors for both groups are the same! (If we had factored out 5, then the factorization of the second group would have been $5(-n - 8)$.)

Check: $(n + 8)(n^2 - 5) = n^3 + 8n^2 - 5n - 40$ ✓

[YOU TRY 6] Factor by grouping.

a) $xy + 4x + 10y + 40$ b) $5pr - 8qr + 10p - 16q$

c) $w^3 + 9w^2 - 6w - 54$

Sometimes we have to rearrange the terms before we can factor.

EXAMPLE 8

Factor $12c^2 - 2d + 3c - 8cd$ completely.

Solution

Group terms together so that each group has a common factor.

$$\underbrace{12c^2 - 2d}_{} + \underbrace{3c - 8cd}_{}$$

Factor out 2 to get $2(6c^2 - d)$. $= 2(6c^2 - d) + c(3 - 8d)$ Factor out c to get $c(3 - 8d)$.

These groups do not have common factors! Let's rearrange the terms in the original polynomial and group the terms differently.

$$\underbrace{12c^2 + 3c}_{} \; \underbrace{- 8cd - 2d}_{}$$

Factor out $3c$ to get $3c(4c + 1)$. $= 3c(4c + 1) - 2d(4c + 1)$ Factor out $-2d$ to get $-2d(4c + 1)$.

$= (4c + 1)(3c - 2d)$ Factor out $(4c + 1)$.

$12c^2 - 2d + 3c - 8cd$ factors to $(4c + 1)(3c - 2d)$. The check is left to the student.

Note

Often, there is more than one way to rearrange the terms so that the polynomial can be factored by grouping.

[YOU TRY 7] Factor $3k^2 - 48m + 8k - 18km$ completely.

Often, we have to combine the two factoring techniques we have learned here. That is, we begin by factoring out the GCF, and then we factor by grouping. Let's summarize how to factor a polynomial by grouping and then look at another example.

Procedure Steps for Factoring by Grouping

1) Before trying to factor by grouping, look at each term in the polynomial and ask yourself, *"Can I factor out a GCF first?"* If so, factor out the GCF from all of the terms.

2) Make two groups of two terms so that each group has a common factor.

3) Take out the common factor in each group of terms.

4) Factor out the common binomial factor using the distributive property.

5) Check the answer by multiplying the factors.

EXAMPLE 9

Factor $7h^4 + 7h^3 - 42h^2 - 42h$ completely.

Solution

Notice that this polynomial has four terms. This is a clue for us to try factoring by grouping. *However,* look at the polynomial carefully and ask yourself, *"Can I factor out a GCF?"* Yes! *Therefore, the first step in factoring this polynomial is to factor out 7h.*

$$7h^4 + 7h^3 - 42h^2 - 42h = 7h(h^3 + h^2 - 6h - 6) \qquad \text{Factor out the GCF, } 7h.$$

The polynomial in parentheses has four terms. Try to factor it by grouping.

$$7h(\underbrace{h^3 + h^2}\ \underbrace{-6h - 6})$$

$$= 7h[h^2(h + 1) - 6(h + 1)] \qquad \text{Take out the common factor in each group.}$$

$$= 7h(h + 1)(h^2 - 6) \qquad \text{Factor out } (h + 1) \text{ using the distributive property.}$$

Check by multiplying: $7h(h + 1)(h^2 - 6) = 7h(h^3 + h^2 - 6h - 6)$

$$= 7h^4 + 7h^3 - 42h^2 - 42h \ ✓$$

[YOU TRY 8] Factor $12t^3 + 12t^2 - 3t^2u - 3tu$ completely.

Remember, seeing a polynomial with four terms is a clue to try factoring by grouping. Not all polynomials will factor this way, however. We will learn other techniques later, and some polynomials must be factored using methods learned in later courses.

ANSWERS TO [YOU TRY] EXERCISES

1) y^5 2) a) $9w^5$ b) $2hk^2$ c) $6c^2$ 3) a) $2(w + 8)$ b) $3u^4(u^2 + 12u + 5)$ c) $z^2(z^3 - 9)$
d) $9r^2t(5r^2t^2 + 4r^2t + 2rt - 1)$ 4) $-y^2(y^2 - 10y + 8)$ 5) a) $(d - 8)(c + 2)$
b) $(k^2 + 15)(k - 7)$ c) $(v + 2)(u - 1)$ 6) a) $(y + 4)(x + 10)$ b) $(5p - 8q)(r + 2)$
c) $(w + 9)(w^2 - 6)$ 7) $(3k + 8)(k - 6m)$ 8) $3t(t + 1)(4t - u)$

*Additional answers can be found in the Answers to Exercises appendix.

Objective 1: Find the GCF of a Group of Monomials

Find the greatest common factor of each group of terms.

1) $28, 21c$ 7

2) $9t, 36$ 9

3) $18p^3, 12p^2$ $6p^2$

4) $32z^5, 56z^3$ $8z^3$

5) $12n^6, 28n^{10}, 36n^7$ $4n^6$

6) $63b^4, 45b^7, 27b$ $9b$

7) $35a^3b^2, 15a^2b$ $5a^2b$

8) $10x^5y^4, 2x^4y^4$ $2x^4y^4$

9) $21r^3s^6, 63r^3s^2, -42r^4s^5$ $21r^3s^2$

10) $-60p^2q^2, 36pq^5, 96p^3q^3$ $12pq^2$

11) $a^2b^2, 3ab^2, 6a^2b$ ab

12) $n^3m^4, -n^3m^4, -n^4$ n^3

13) $c(k-9), 5(k-9)$ $(k-9)$

14) $a^2(h+8), b^2(h+8)$ $(h+8)$

15) Explain how to find the GCF of a group of terms.
   Answers may vary.

16) What does it mean to factor a polynomial?
   Write it as a product of two or more polynomials.

Objective 2: Factor Out the Greatest Common Monomial Factor

Determine whether each expression is written in factored form.

17) $5p(p+9)$ yes

18) $8h^2 - 24h$ no

19) $18w^2 + 30w$ no

20) $-3z^2(2z+7)$ yes

21) $a^2b^2(-4ab)$ yes

22) $c^3d - (2c+d)$ no

Factor out the greatest common factor. Be sure to check your answer.

23) $2w + 10$ $2(w+5)$

24) $3y + 18$ $3(y+6)$

25) $18z^2 - 9$ $9(2z^2-1)$

26) $14h - 12h^2$ $2h(7-6h)$

27) $100m^3 - 30m$
   $10m(10m^2-3)$

28) $t^5 - t^4$ $t^4(t-1)$

29) $r^9 + r^2$ $r^2(r^7+1)$

30) $\frac{1}{2}a^2 + \frac{3}{2}a$ $\frac{1}{2}a(a+3)$

31) $\frac{1}{5}y^2 + \frac{4}{5}y$ $\frac{1}{5}y(y+4)$

32) $9a^3 + 2b^2$ does not factor

33) $s^7 - 4t^3$ does not factor

34) $14u^7 + 63u^6 - 42u^5$
   $7u^5(2u^2+9u-6)$

35) $10n^5 - 5n^4 + 40n^3$ $5n^3(2n^2-n+8)$

36) $3d^8 - 33d^7 - 24d^6 + 3d^5$ $3d^5(d^3-11d^2-8d+1)$

37) $40p^6 + 40p^5 - 8p^4 + 8p^3$ $8p^3(5p^3+5p^2-p+1)$

38) $44m^3n^3 - 24mn^4$ $4mn^3(11m^2-6n)$

39) $63a^3b^3 - 36a^3b^2 + 9a^2b$ $9a^2b(7ab^2-4ab+1)$

40) $8p^4q^3 + 8p^3q^3 - 72p^2q^2$ $8p^2q^2(p^2q+pq-9)$

41) Factor out -6 from $-30n - 42$. $-6(5n+7)$

42) Factor out $-c$ from $-9c^3 + 2c^2 - c$. $-c(9c^2-2c+1)$

43) Factor out $-4w^3$ from $-12w^5 - 16w^3$. $-4w^3(3w^2+4)$

44) Factor out $-m$ from $-6m^3 - 3m^2 + m$.
   $-m(6m^2+3m-1)$

45) Factor out -1 from $-k + 3$. $-1(k-3)$

46) Factor out -1 from $-p - 10$. $-1(p+10)$

Objective 3: Factor Out the Greatest Common Binomial Factor

Factor out the common binomial factor.

47) $u(t-5) + 6(t-5)$
   $(t-5)(u+6)$

48) $c(b+9) + 2(b+9)$
   $(b+9)(c+2)$

49) $y(6x+1) - z(6x+1)$
   $(6x+1)(y-z)$

50) $s(4r-3) - t(4r-3)$
   $(4r-3)(s-t)$

51) $p(q+12) + (q+12)$
   $(q+12)(p+1)$

52) $8x(y-2) + (y-2)$
   $(y-2)(8x+1)$

53) $5h^2(9k+8) - (9k+8)$ $(9k+8)(5h^2-1)$

54) $3a(4b+1) - (4b+1)$ $(4b+1)(3a-1)$

Objective 4: Factor by Grouping

Factor by grouping.

55) $ab + 2a + 7b + 14$ $(b+2)(a+7)$

56) $cd + 8c - 5d - 40$ $(d+8)(c-5)$

57) $3rt + 4r - 27t - 36$ $(3t+4)(r-9)$

58) $5pq + 15p - 6q - 18$ $(q+3)(5p-6)$

59) $8b^2 + 20bc + 2bc^2 + 5c^3$ $(2b+5c)(4b+c^2)$

60) $4a^3 - 12ab + a^2b - 3b^2$ $(a^2-3b)(4a+b)$

61) $fg - 7f + 4g - 28$ $(g-7)(f+4)$

62) $xy - 8y - 7x + 56$ $(x-8)(y-7)$

63) $st - 10s - 6t + 60$ $(t-10)(s-6)$

64) $cd + 3c - 11d - 33$ $(d+3)(c-11)$

65) $5tu + 6t - 5u - 6$ $(5u+6)(t-1)$

66) $qr + 3q - r - 3$ $(r+3)(q-1)$

67) $36g^4 + 3gh - 96g^3h - 8h^2$ $(12g^3+h)(3g-8h)$

68) $40j^3 + 72jk - 55j^2k - 99k^2$ $(5j^2+9k)(8j-11k)$

69) Explain, in your own words, how to factor by grouping.   Answers may vary.

70) What should be the first step in factoring $6ab + 24a + 18b + 54$?   Factor out 6 from each term.

Factor completely. You may need to begin by factoring out the GCF first or by rearranging terms.

Fill It In

Fill in the blanks with either the missing mathematical step or the reason for the given step.

71) $4xy + 12x + 20y + 60$

$4xy + 12x + 20y + 60$

$= \underline{4(xy + 3x + 5y + 15)}$ Factor out the GCF.

$= 4[x(y + 3) + 5(y + 3)]$ Group the terms and factor out the GCF from each group.

$= \underline{4(y + 3)(x + 5)}$ Take out the binomial factor.

72) $2m^2n - 4m^2 - 18mn + 36m$

$2m^2n - 4m^2 - 18mn + 36m$

$= \underline{2m(mn - 2m - 9n + 18)}$ Factor out the GCF.

$= 2m[m(n - 2) - 9(n - 2)]$ Group the terms and factor out the GCF from each group.

$= \underline{2m(n - 2)(m - 9)}$ Take out the binomial factor.

73) $3cd + 6c + 21d + 42$ $3(c + 7)(d + 2)$

74) $5xy + 15x - 5y - 15$ $5(x - 1)(y + 3)$

75) $2p^2q - 10p^2 - 8pq + 40p$ $2p(p - 4)(q - 5)$

76) $3uv^2 - 24uv + 3v^2 - 24v$ $3v(u + 1)(v - 8)$

77) $10st + 5s - 12t - 6$ $(5s - 6)(2t + 1)$

78) $8pq + 12p + 10q + 15$ $(4p + 5)(2q + 3)$

79) $3a^3 - 21a^2b - 2ab + 14b^2$ $(3a^2 - 2b)(a - 7b)$

80) $8c^3 + 32c^2d + cd + 4d^2$ $(8c^2 + d)(c + 4d)$

81) $8u^2v^2 + 16u^2v + 10uv^2 + 20uv$ $2uv(4u + 5)(v + 2)$

82) $10x^2y^2 - 5x^2y - 60xy^2 + 30xy$ $5xy(x - 6)(2y - 1)$

Mixed Exercises: Objectives 1–4

Factor completely.

83) $3mn + 21m + 10n + 70$ $(n + 7)(3m + 10)$

84) $4yz + 7z - 20y - 35$ $(4y + 7)(z - 5)$

85) $16b - 24$ $8(2b - 3)$

86) $2yz^3 + 14yz^2 + 3z^3 + 21z^2$ $z^2(2y + 3)(z + 7)$

87) $cd + 6c - 4d - 24$ $(d + 6)(c - 4)$

88) $5x^3 - 30x^2y^2 + xy - 6y^3$ $(x - 6y^2)(5x^2 + y)$

89) $6a^4b + 12a^4 - 8a^3b - 16a^3$ $2a^3(3a - 4)(b + 2)$

90) $6x^2 + 48x^3$ $6x^2(1 + 8x)$

91) $7cd + 12 + 28c + 3d$ $(d + 4)(7c + 3)$

92) $2uv + 12u - 7v - 42$ $(v + 6)(2u - 7)$

93) $dg - d + g - 1$ $(g - 1)(d + 1)$

94) $2ab - 10a - 12b + 60$ $2(a - 6)(b - 5)$

95) $x^4y^2 + 12x^3y^3$ $x^3y^2(x + 12y)$

96) $8u^2 - 16uv^2 + 3uv - 6v^3$ $(u - 2v^2)(8u + 3v)$

97) $4mn + 8m + 12n + 24$ $4(m + 3)(n + 2)$

98) $5c^2 - 20$ $5c(c - 4)$

99) Factor out -2 from $-6p^2 - 20p + 2$. $-2(3p^2 + 10p - 1)$

100) Factor out $-5g$ from $-5g^3 + 50g^2 - 25g$.

    $-5g(g^2 - 10g + 5)$

R Rethink

R1) Do you understand how to factor by grouping?

R2) Write down a problem you got wrong or that was difficult for you to do. Think about where you made a mistake or why it was hard, then redo the problem without looking at your previous work.

R3) How would you explain the process of factoring these expressions to a friend?

7.2 Factoring Trinomials of the Form $x^2 + bx + c$

What are your objectives for Section 7.2?	How can you accomplish each objective?
1 Practice Arithmetic Skills Needed for Factoring Trinomials	• Follow the approach in the examples to come up with a solution. • In your notes, make a chart that summarizes the approach used in the example. • Complete the given example on your own. • Complete You Try 1.
2 Factor a Trinomial of the Form $x^2 + bx + c$	• Notice that the process of factoring is the opposite of multiplying. • Write the procedure for **Factoring a Polynomial of the Form $x^2 + bx + c$** in your notes. How does it compare to the chart you made for Objective 1? • Complete the given example on your own. • Complete You Try 2.
3 More on Factoring a Trinomial of the Form $x^2 + bx + c$	• Add the step of *"Can I factor out a GCF?"* as the first step in the procedure for Objective 2. • Complete the given example on your own. • Complete You Try 3.
4 Factor a Trinomial Containing Two Variables	• Use the same procedure as before. • Complete the given example on your own. • Complete You Try 4.

W Work   **Read the explanations, follow the examples, take notes, and complete the You Trys.**

One of the factoring problems encountered most often in algebra is the factoring of trinomials. In this section, we will discuss how to factor a trinomial of the form $x^2 + bx + c$; notice that the coefficient of the squared term is 1.

Let's begin with arithmetic skills we need to be able to factor a trinomial of the form $x^2 + bx + c$.

1 Practice Arithmetic Skills Needed for Factoring Trinomials

EXAMPLE 1   Find two integers whose

a)   product is 15 and sum is 8.

b)   product is 24 and sum is -10.

c)   product is -28 and sum is 3.

In-Class Example 1

Find two integers whose
a) product is 16 and sum is 10.
b) product is 40 and sum is −13.
c) product is −36 and sum is 5.

Answer: a) 8 and 2
b) −8 and −5
c) −4 and 9

Solution

a) If the product of two numbers is *positive* 15 and the sum of the numbers is *positive* 8, *then the two numbers will be positive.* (The product of two positive numbers is positive, and their sum is positive as well.)

First, list the pairs of *positive* integers whose product is 15—the *factors* of 15. Then, find the *sum* of those factors.

Factors of 15	Sum of the Factors
1 · 15	1 + 15 = 16
3 · 5	3 + 5 = 8

The product of 3 and 5 is 15, and their sum is 8.

W Hint

Follow the approach used in the solution of each example.

b) If the product of two numbers is *positive* 24 and the sum of those numbers is *negative* 10, *then the two numbers will be negative.* (The product of two negative numbers is positive, while the sum of two negative numbers is negative.)

First, list the pairs of negative numbers that are factors of 24. Then, find the sum of those factors. You can stop making your list when you find the pair that works.

Factors of 24	Sum of the Factors
−1 · (−24)	−1 + (−24) = −25
−2 · (−12)	−2 + (−12) = −14
−3 · (−8)	−3 + (−8) = −11
−4 · (−6)	−4 + (−6) = −10

The product of −4 and −6 is 24, and their sum is −10.

c) If two numbers have a product of *negative* 28 and their sum is *positive* 3, *one number must be positive and one number must be negative.* (The product of a positive number and a negative number is negative, while the sum of the numbers can be either positive *or* negative.)

First, list pairs of factors of −28. Then, find the sum of those factors.

Factors of −28	Sum of the Factors
−1 · 28	−1 + 28 = 27
1 · (−28)	1 + (−28) = −27
−4 · 7	−4 + 7 = 3

The product of −4 and 7 is −28, and their sum is 3.

[**YOU TRY 1**] Find two integers whose

a) product is 21 and sum is 10.

b) product is −18 and sum is −3.

c) product is 20 and sum is −12.

Note

You should try to get to the point where you can come up with the correct numbers *in your head* without making a list.

2 Factor a Trinomial of the Form $x^2 + bx + c$

The process of factoring is the opposite of multiplying. Let's see how this will help us understand how to factor a trinomial of the form $x^2 + bx + c$.

Multiply $(x + 4)(x + 5)$ using FOIL.

$$(x + 4)(x + 5) = x^2 + 5x + 4x + 4 \cdot 5 \qquad \text{Multiply using FOIL.}$$

$$= x^2 + (5 + 4)x + 20 \qquad \text{Use the distributive property,}$$
$$\text{and multiply } 4 \cdot 5.$$

$$= x^2 + 9x + 20$$

$$(x + 4)(x + 5) = x^2 + 9x + 20 \quad \longleftarrow \quad \text{The } product \text{ of 4 and 5 is 20.}$$

$$\uparrow$$

The *sum* of 4 and 5 is 9.

W Hint

Does this look similar to the procedure in Example 1?

So, if we were asked to *factor* $x^2 + 9x + 20$, we need to think of two integers whose *product* is 20 and whose *sum* is 9. Those numbers are 4 and 5. The *factored form* of $x^2 + 9x + 20$ is $(x + 4)(x + 5)$.

W Hint

Compare the chart you made for Objective 1 to the procedure.

Procedure Factoring a Polynomial of the Form $x^2 + bx + c$

To factor a polynomial of the form $x^2 + bx + c$, find two integers m and n whose product is c and whose sum is b. Then, $x^2 + bx + c = (x + m)(x + n)$.

1) If b and c are positive, then both m and n must be positive.

2) If c is positive and b is negative, then both m and n must be negative.

3) If c is negative, then one integer, m, must be positive and the other integer, n, must be negative.

You can check the answer by multiplying the binomials. The result should be the original polynomial.

EXAMPLE 2

In-Class Example 2

Factor, if possible.
a) $n^2 + 9n + 20$
b) $t^2 - 14t + 33$
c) $x^2 + x - 72$
d) $k^2 - 11k - 12$
e) $y^2 - 10y + 25$
f) $d^2 + 6d + 28$

Answer: a) $(n + 4)(n + 5)$
b) $(t - 11)(t - 3)$
c) $(x + 9)(x - 8)$
d) $(k - 12)(k + 1)$
e) $(y - 5)(y - 5)$ or $(y - 5)^2$
f) prime

Factor, if possible.

a) $x^2 + 7x + 12$ b) $p^2 - 9p + 14$ c) $w^2 + w - 30$

d) $a^2 - 3a - 54$ e) $c^2 - 6c + 9$ f) $y^2 + 11y + 35$

Solution

a) To factor $x^2 + 7x + 12$, we must find two integers whose *product* is 12 and whose *sum* is 7. Both integers will be positive.

Factors of 12	Sum of the Factors
$1 \cdot 12$	$1 + 12 = 13$
$2 \cdot 6$	$2 + 6 = 8$
$3 \cdot 4$	$3 + 4 = 7$

The numbers are 3 and 4. Therefore, $x^2 + 7x + 12 = (x + 3)(x + 4)$.

Check: $(x + 3)(x + 4) = x^2 + 4x + 3x + 12 = x^2 + 7x + 12$ ✓

b) To factor $p^2 - 9p + 14$, find two integers whose *product* is 14 and whose *sum* is -9. Since 14 is positive and the coefficient of p is a negative number, -9, both integers will be negative.

Factors of 14	Sum of the Factors
$-1 \cdot (-14)$	$-1 + (-14) = -15$
$-2 \cdot (-7)$	$-2 + (-7) = -9$

The numbers are -2 and -7. So, $p^2 - 9p + 14 = (p - 2)(p - 7)$.

Check: $(p - 2)(p - 7) = p^2 - 7p - 2p + 14 = p^2 - 9p + 14$ ✓

c) $w^2 + w - 30$

The coefficient of w is 1, so we can think of this trinomial as $w^2 + 1w - 30$.

Find two integers whose *product* is -30 and whose *sum* is 1. Since the last term in the trinomial is negative, one of the integers must be positive and the other must be negative.

Try to find these integers mentally. Two numbers with a product of *positive* 30 are 5 and 6. We need a product of -30, so either the 5 is negative or the 6 is negative.

Factors of -30	Sum of the Factors
$-5 \cdot 6$	$-5 + 6 = 1$

The numbers are -5 and 6. Therefore, $w^2 + w - 30 = (w - 5)(w + 6)$.

Check: $(w - 5)(w + 6) = w^2 + 6w - 5w - 30 = w^2 + w - 30$ ✓

d) To factor $a^2 - 3a - 54$, find two integers whose *product* is -54 and whose *sum* is -3. Since the last term in the trinomial is negative, one of the integers must be positive and the other must be negative.

Find the integers mentally. First, think about two integers whose product is *positive* 54: 1 and 54, 2 and 27, 3 and 18, 6 and 9. One number must be positive and the other negative, however, to get our product of -54, and they must add up to -3.

Factors of -54	Sum of the Factors
$-6 \cdot 9$	$-6 + 9 = 3$
$6 \cdot (-9)$	$6 + (-9) = -3$

The numbers are 6 and -9: $a^2 - 3a - 54 = (a + 6)(a - 9)$.

The check is left to the student.

e) To factor $c^2 - 6c + 9$, notice that the *product*, 9, is positive and the *sum*, -6, is negative, so both integers must be negative. The numbers that multiply to 9 and add to -6 are the same number, -3 and -3: $(-3) \cdot (-3) = 9$ and $-3 + (-3) = -6$.

$$\text{So } c^2 - 6c + 9 = (c - 3)(c - 3) \text{ or } (c - 3)^2.$$

Either form of the factorization is correct.

f) To factor $y^2 + 11y + 35$, find two integers whose *product* is 35 and whose *sum* is 11. We are looking for two positive numbers.

Factors of 35	Sum of the Factors
$1 \cdot 35$	$1 + 35 = 36$
$5 \cdot 7$	$5 + 7 = 12$

There are no such factors! Therefore, $y^2 + 11y + 35$ does not factor using the methods we have learned here. We say that it is **prime.**

Note

We say that trinomials like $y^2 + 11y + 35$ are **prime** if they cannot be factored using the method presented here.

In later mathematics courses, you may learn how to factor such polynomials using other methods so that they are not considered prime.

[YOU TRY 2]

Factor, if possible.

a) $m^2 + 11m + 28$ b) $c^2 - 16c + 48$ c) $a^2 - 5a - 21$

d) $r^2 - 4r - 45$ e) $r^2 + 5r - 24$ f) $h^2 + 12h + 36$

3 More on Factoring a Trinomial of the Form $x^2 + bx + c$

Sometimes it is necessary to factor out the GCF before applying this method for factoring trinomials.

Note

From this point on, the *first* step in factoring *any* polynomial should be to ask yourself, "*Can I factor out a greatest common factor?*"

Since some polynomials can be factored more than once, after performing one factorization, ask yourself, "*Can I factor again?*" If so, factor again. If not, you know that the polynomial has been completely factored.

EXAMPLE 3

In-Class Example 3

Factor $2z^3 - 16z^2 - 18z$ completely.

Answer: $2z(z - 9)(z + 1)$

W Hint

Add *"Can I factor out a GCF?"* as the first step in this section's procedure.

Factor $4n^3 - 12n^2 - 40n$ completely.

Solution

Ask yourself, *"Can I factor out a GCF?"* Yes. The GCF is $4n$.

$$4n^3 - 12n^2 - 40n = 4n(n^2 - 3n - 10)$$

Look at the trinomial and ask yourself, *"Can I factor again?"* Yes. The integers whose product is -10 and whose sum is -3 are -5 and 2. Therefore,

$$4n^3 - 12n^2 - 40n = 4n(n^2 - 3n - 10)$$
$$= 4n(n - 5)(n + 2)$$

Ask yourself, *"Can I factor again?"* No. Therefore, $4n^3 - 12n^2 - 40n = 4n(n - 5)(n + 2)$.

$$\text{Check: } 4n(n - 5)(n + 2) = 4n(n^2 + 2n - 5n - 10)$$
$$= 4n(n^2 - 3n - 10)$$
$$= 4n^3 - 12n^2 - 40n \checkmark$$

[YOU TRY 3]

Factor completely.

a) $7p^4 + 42p^3 + 56p^2$ b) $3a^2b - 33ab + 90b$

4 Factor a Trinomial Containing Two Variables

If a trinomial contains two variables and we cannot take out a GCF, the trinomial may still be factored according to the method outlined in this section.

EXAMPLE 4

In-Class Example 4

Factor $a^2 + 7ab + 10b^2$ completely.

Answer: $(a + 2b)(a + 5b)$

Factor $x^2 + 12xy + 32y^2$ completely.

Solution

Ask yourself, *"Can I factor out a GCF?"* No. Notice that the first term is x^2. Let's rewrite the trinomial as

$$x^2 + 12yx + 32y^2$$

so that we can think of $12y$ as the coefficient of x. Find two expressions whose product is $32y^2$ and whose sum is $12y$. They are $4y$ and $8y$ since $4y \cdot 8y = 32y^2$ and $4y + 8y = 12y$.

$$x^2 + 12xy + 32y^2 = (x + 4y)(x + 8y)$$

We cannot factor $(x + 4y)(x + 8y)$ any more, so this is the complete factorization. The check is left to the student.

[YOU TRY 4]

Factor completely.

a) $m^2 + 10mn + 16n^2$ b) $5a^3 + 40a^2b - 45ab^2$

ANSWERS TO [YOU TRY] EXERCISES

1) a) $3, 7$ b) $-6, 3$ c) $-2, -10$ 2) a) $(m + 4)(m + 7)$ b) $(c - 12)(c - 4)$ c) prime
d) $(r - 9)(r + 5)$ e) $(r + 8)(r - 3)$ f) $(h + 6)(h + 6)$ or $(h + 6)^2$ 3) a) $7p^2(p + 4)(p + 2)$
b) $3b(a - 5)(a - 6)$ 4) a) $(m + 2n)(m + 8n)$ b) $5a(a - b)(a + 9b)$

*Additional answers can be found in the Answers to Exercises appendix.

Objective 1: Practice Arithmetic Skills Needed for Factoring Trinomials

1) Find two integers whose

	PRODUCT IS	and whose SUM IS	ANSWER
a)	10	7	
b)	−56	−1	
c)	−5	4	
d)	36	−13	

a) $5, 2$ b) $-8, 7$ c) $5, -1$ d) $-9, -4$

2) Find two integers whose

	PRODUCT IS	and whose SUM IS	ANSWER
a)	42	−13	
b)	−14	13	
c)	54	15	
d)	−21	−4	

a) $-7, -6$ b) $14, -1$ c) $6, 9$ d) $-7, 3$

Objective 2: Factor a Trinomial of the Form $x^2 + bx + c$

3) If $x^2 + bx + c$ factors to $(x + m)(x + n)$ and if c is positive and b is negative, what do you know about the signs of m and n?   They are negative.

4) If $x^2 + bx + c$ factors to $(x + m)(x + n)$ and if b and c are positive, what do you know about the signs of m and n?   They are positive.

5) When asked to factor a polynomial, what is the first question you should ask yourself?
Can I factor out a GCF?

6) What does it mean to say that a polynomial is prime?
It does not factor.

7) After factoring a polynomial, what should you ask yourself to be sure that the polynomial is completely factored?   Can I factor again?

8) How do you check the factorization of a polynomial?
Multiply the factors. The product should be the original polynomial.

Complete the factorization.

9) $n^2 + 7n + 10 = (n + 5)(n + 2)$

10) $p^2 + 11p + 28 = (p + 4)(p + 7)$

11) $c^2 - 16c + 60 = (c - 6)(c - 10)$

12) $t^2 - 12t + 27 = (t - 9)(t - 3)$

13) $x^2 + x - 12 = (x - 3)(x + 4)$

14) $r^2 - 8r - 9 = (r + 1)(r - 9)$

Factor completely, if possible. Check your answer.

15) $g^2 + 8g + 12$
$(g + 6)(g + 2)$

16) $p^2 + 9p + 14$
$(p + 7)(p + 2)$

17) $y^2 + 10y + 16$
$(y + 8)(y + 2)$

18) $a^2 + 11a + 30$
$(a + 6)(a + 5)$

19) $w^2 - 17w + 72$
$(w - 9)(w - 8)$

20) $d^2 - 14d + 33$
$(d - 3)(d - 11)$

21) $b^2 - 3b - 4$
$(b - 4)(b + 1)$

22) $t^2 + 2t - 48$
$(t - 6)(t + 8)$

23) $z^2 + 6z - 11$ prime

24) $x^2 - 7x - 15$ prime

25) $c^2 - 13c + 36$
$(c - 9)(c - 4)$

26) $h^2 - 13h + 12$
$(h - 1)(h - 12)$

27) $m^2 + 4m - 60$
$(m + 10)(m - 6)$

28) $v^2 - 4v - 45$
$(v - 9)(v + 5)$

29) $r^2 - 4r - 96$
$(r - 12)(r + 8)$

30) $a^2 - 21a + 110$
$(a - 11)(a - 10)$

31) $q^2 + 12q + 42$ prime

32) $d^2 - 15d + 32$ prime

33) $x^2 + 16x + 64$
$(x + 8)(x + 8)$ or $(x + 8)^2$

34) $c^2 - 10c + 25$
$(c - 5)(c - 5)$ or $(c - 5)^2$

35) $n^2 - 2n + 1$
$(n - 1)(n - 1)$ or $(n - 1)^2$

36) $w^2 + 20w + 100$
$(w + 10)(w + 10)$ or $(w + 10)^2$

37) $24 + 14d + d^2$
$(d + 12)(d + 2)$

38) $10 + 7k + k^2$
$(k + 5)(k + 2)$

39) $-56 + 12a + a^2$ prime

40) $63 + 21w + w^2$ prime

Objective 3: More on Factoring a Trinomial of the Form $x^2 + bx + c$

Factor completely, if possible. Check your answer.

41) $2k^2 - 22k + 48$
$2(k - 3)(k - 8)$

42) $6v^2 + 54v + 120$
$6(v + 4)(v + 5)$

43) $50h + 35h^2 + 5h^3$
$5h(h + 5)(h + 2)$

44) $3d^3 - 33d^2 - 36d$
$3d(d - 12)(d + 1)$

45) $r^4 + r^3 - 132r^2$
$r^2(r + 12)(r - 11)$

46) $2n^4 - 40n^3 + 200n^2$
$2n^2(n - 10)(n - 10)$ or $2n^2(n - 10)^2$

47) $7q^3 - 49q^2 - 42q$
$7q(q^2 - 7q - 6)$

48) $8b^4 + 24b^3 + 16b^2$
$8b^2(b + 2)(b + 1)$

49) $3z^4 + 24z^3 + 48z^2$
$3z^2(z + 4)(z + 4)$ or $3z^2(z + 4)^2$

50) $-36w + 6w^2 + 2w^3$
$2w(w - 3)(w + 6)$

51) $xy^3 - 2xy^2 - 63xy$
$xy(y - 9)(y + 7)$

52) $2c^3d - 14c^2d - 24cd$
$2cd(c^2 - 7c - 12)$

Factor completely by first taking out −1 and then factoring the trinomial, if possible. Check your answer.

53) $-m^2 - 12m - 35$
$-(m + 5)(m + 7)$

54) $-x^2 - 15x - 36$
$-(x + 12)(x + 3)$

55) $-c^2 - 3c + 28$
$-(c + 7)(c - 4)$

56) $-t^2 + 2t + 48$
$-(t - 8)(t + 6)$

57) $-z^2 + 13z - 30$
$-(z - 3)(z - 10)$

58) $-n^2 + 16n - 55$
$-(n - 11)(n - 5)$

59) $-p^2 + p + 56$
$-(p - 8)(p + 7)$

60) $-w^2 - 2w + 3$
$-(w - 1)(w + 3)$

Objective 4: Factor a Trinomial Containing Two Variables

Factor completely. Check your answer.

61) $x^2 + 7xy + 12y^2$
$(x + 4y)(x + 3y)$

62) $a^2 + 11ab + 18b^2$
$(a + 9b)(a + 2b)$

63) $c^2 - 7cd - 8d^2$
$(c - 8d)(c + d)$

64) $p^2 + 6pq - 72q^2$
$(p + 12q)(q - 6q)$

65) $u^2 - 14uv + 45v^2$
$(u - 5v)(u - 9v)$

66) $h^2 - 8hk + 7k^2$
$(h - 7k)(h - k)$

67) $m^2 + 4mn - 21n^2$
$(m - 3n)(m + 7n)$

68) $a^2 - 6ab - 40b^2$
$(a - 10b)(a + 4b)$

69) $a^2 + 24ab + 144b^2$
$(a + 12b)(a + 12b)$ or $(a + 12b)^2$

70) $g^2 + 6gh + 5h^2$
$(g + 5h)(g + h)$

79) $m^2 + 7mn - 44n^2$
$(m - 4n)(m + 11n)$

80) $a^2 + 10ab + 24b^2$
$(a + 6b)(a + 4b)$

81) $h^2 - 10h + 32$ prime

82) $z^2 + 9z + 36$ prime

83) $4q^3 - 28q^2 + 48q$
$4q(q - 3)(q - 4)$

84) $3w^3 - 9w^2 - 120w$
$3w(w - 8)(w + 5)$

85) $-k^2 - 18k - 81$
$-(k + 9)(k + 9)$ or $-(k + 9)^2$

86) $-y^2 + 8y - 16$
$-(y - 4)(y - 4)$ or $-(y - 4)^2$

87) $4h^5 + 32h^4 + 28h^3$
$4h^3(h + 7)(h + 1)$

88) $3r^4 - 6r^3 - 45r^2$
$3r^2(r - 5)(r + 3)$

89) $k^2 + 21k + 108$
$(k + 12)(k + 9)$

90) $j^2 - 14j - 15$
$(j - 15)(j + 1)$

91) $p^3q - 17p^2q^2 + 70pq^3$
$pq(p - 7q)(p - 10q)$

92) $u^3v^2 - 2u^2v^3 - 15uv^4$
$uv^2(u + 3v)(u - 5v)$

93) $a^2 + 9ab + 24b^2$ prime

94) $m^2 - 8mn - 35n^2$ prime

Determine whether each polynomial is factored completely. If it is not, explain why and factor it completely.

71) $3x^2 + 21x + 30 = (3x + 6)(x + 5)$ No; from $(3x + 6)$ you can factor out a 3. The correct answer is $3(x + 2)(x + 5)$.

72) $6a^2 + 24a - 72 = 6(a + 6)(a - 2)$ yes

73) $n^4 - 3n^3 - 108n^2 = n^2(n - 12)(n + 9)$ yes

74) $9y^3 - 45y^2 + 54y = (y - 2)(9y^2 - 27y)$
No; from $(9y^2 - 27y)$ you can factor out 9y. The correct answer is $9y(y - 3)(y - 2)$.

95) $x^2 - 13xy + 12y^2$
$(x - 12y)(x - y)$

96) $p^2 - 3pq - 40q^2$
$(p - 8q)(p + 5q)$

97) $5v^5 + 55v^4 - 45v^3$
$5v^3(v^2 + 11v - 9)$

98) $6t^4 + 42t^3 + 48t^2$
$6t^2(t^2 + 7t + 8)$

99) $6x^3y^2 - 48x^2y^2 - 54xy^2$ $6xy^2(x - 9)(x + 1)$

Mixed Exercises: Objectives 2–4

Factor completely. Begin by asking yourself, "Can I factor out a GCF?"

75) $2x^2 + 16x + 30$
$2(x + 5)(x + 3)$

76) $3c^2 + 21c + 18$
$3(c + 1)(c + 6)$

77) $n^2 - 6n + 8$
$(n - 4)(n - 2)$

78) $a^2 + a - 6$
$(a - 2)(a + 3)$

100) $2c^2d^4 - 18c^2d^3 + 28c^2d^2$ $2c^2d^2(d - 2)(d - 7)$

101) $36 - 13z + z^2$
$(z - 9)(z - 4)$

102) $121 + 22w + w^2$
$(w + 11)(w + 11)$ or $(w + 11)^2$

103) $a^2b^2 + 13ab + 42$
$(ab + 6)(ab + 7)$

104) $h^2k^2 + 8hk - 20$
$(hk + 10)(hk - 2)$

R Rethink

R1) Do you have quick recall of the multiplication facts from 1 to 12, or do you need more practice? If so, practice using flash cards.

R2) Were you able to do most of the factoring in your head?

R3) Could you complete similar exercises without looking at your notes?

7.3 Factoring Trinomials of the Form $ax^2 + bx + c$ $(a \neq 1)$

P Prepare

O Organize

What are your objectives for Section 7.3?	How can you accomplish each objective?
1 Factor $ax^2 + bx + c$ $(a \neq 1)$ by Grouping	• Write your own procedure for **Factoring** $ax^2 + bx + c$ $(a \neq 1)$ **by Grouping** in your notes by following Objective 1's introduction. • Complete the given examples on your own. • Complete You Trys 1 and 2.
2 Factor $ax^2 + bx + c$ $(a \neq 1)$ by Trial and Error	• Summarize in your notes how you would factor by trial and error. • Complete the given examples on your own. • Complete You Trys 3 and 4.

Read the explanations, follow the examples, take notes, and complete the You Trys.

In the previous section, we learned that we could factor $2x^2 + 10x + 8$ by first taking out the GCF of 2 and then factoring the trinomial.

$$2x^2 + 10x + 8 = 2(x^2 + 5x + 4) = 2(x + 4)(x + 1)$$

In this section, we will learn how to factor a trinomial like $2x^2 + 11x + 15$ where we *cannot* factor out the leading coefficient of 2.

1 Factor $ax^2 + bx + c$ ($a \neq 1$) by Grouping

Sum is 11.
$\downarrow$

To factor $2x^2 + 11x + 15$, first find the product of 2 and 15. Then, find two integers

Product: $2 \cdot 15 = 30$

whose *product* is 30 and whose *sum* is 11. The numbers are 6 and 5. Rewrite the middle term, $11x$, as $6x + 5x$, then factor by grouping.

$$2x^2 + 11x + 15 = \underbrace{2x^2 + 6x} + \underbrace{5x + 15}$$
$$= 2x(x + 3) + 5(x + 3) \qquad \text{Take out the common factor from each group.}$$
$$= (x + 3)(2x + 5) \qquad \text{Factor out } (x + 3).$$

Check: $(x + 3)(2x + 5) = 2x^2 + 5x + 6x + 15 = 2x^2 + 11x + 15$ ✓

EXAMPLE 1

Factor completely.

a) $8k^2 + 14k + 3$ b) $6c^2 - 17c + 12$ c) $7x^2 - 34xy - 5y^2$

Solution

a) Since we cannot factor out a GCF (the GCF = 1), we begin with a new method.

Sum is 14.
$\downarrow$

$8k^2 + 14k + 3$

Product: $8 \cdot 3 = 24$

Think of two integers whose *product* is 24 and whose *sum* is 14. The integers are 2 and 12. Rewrite the middle term, $14k$, as $2k + 12k$. Factor by grouping.

$$8k^2 + 14k + 3 = \underbrace{8k^2 + 2k} + \underbrace{12k + 3}$$
$$= 2k(4k + 1) + 3(4k + 1) \qquad \text{Take out the common factor from each group.}$$
$$= (4k + 1)(2k + 3) \qquad \text{Factor out } (4k + 1).$$

Check by multiplying: $(4k + 1)(2k + 3) = 8k^2 + 14k + 3$ ✓

b) *Sum* is -17.
$\downarrow$

$6c^2 - 17c + 12$

Product: $6 \cdot 12 = 72$

Think of two integers whose *product* is 72 and whose *sum* is -17. (Both numbers will be negative.) The integers are -9 and -8. Rewrite the middle term, $-17c$, as $-9c - 8c$. Factor by grouping.

$$6c^2 - 17c + 12 = \underbrace{6c^2 - 9c} - \underbrace{8c + 12}$$
$$= 3c(2c - 3) - 4(2c - 3) \qquad \text{Take out the common factor from each group.}$$
$$= (2c - 3)(3c - 4) \qquad \text{Factor out } (2c - 3).$$

Check: $(2c - 3)(3c - 4) = 6c^2 - 17c + 12$ ✓

c)

$$7x^2 - 34xy - 5y^2$$

Product: $7 \cdot (-5) = -35$

Sum is -34.

The integers whose *product* is -35 and whose *sum* is -34 are -35 and 1.
Rewrite the middle term, $-34xy$, as $-35xy + xy$.
Factor by grouping.

$$
\begin{aligned}
7x^2 - 34xy - 5y^2 &= 7x^2 - 35xy + xy - 5y^2 \\
&= 7x(x - 5y) + y(x - 5y) \quad \text{Take out the common factor from each group.} \\
&= (x - 5y)(7x + y) \quad \text{Factor out } (x - 5y).
\end{aligned}
$$

Check: $(x - 5y)(7x + y) = 7x^2 - 34xy - 5y^2$ ✓

> **[YOU TRY 1]** Factor completely.
>
> a) $4p^2 + 16p + 15$ 　　 b) $10y^2 - 13y + 4$ 　　 c) $5a^2 - 29ab - 6b^2$

Factor $6p^2 - 9p - 42$ completely.

Answer: $3(2p - 7)(p + 2)$

W Hint

Write down the steps on your own paper as you are reading the example.

Factor $12n^2 + 64n - 48$ completely.

Solution

It is tempting to jump right in and multiply $12 \cdot (-48) = -576$ and try to think of two integers with a product of -576 and a sum of 64. However, first ask yourself, *"Can I factor out a GCF?"* Yes! We can factor out 4.

$$12n^2 + 64n - 48 = 4(3n^2 + 16n - 12) \quad \text{Factor out 4.}$$

Product: $3 \cdot (-12) = -36$

Now factor $3n^2 + 16n - 12$ by finding two integers whose *product* is -36 and whose *sum* is 16. The numbers are 18 and -2.

$$
\begin{aligned}
&= 4(3n^2 + 18n - 2n - 12) \\
&= 4[3n(n + 6) - 2(n + 6)] \quad \text{Take out the common factor from each group.} \\
&= 4(n + 6)(3n - 2) \quad \text{Factor out } (n + 6).
\end{aligned}
$$

Check by multiplying: $4(n + 6)(3n - 2) = 4(3n^2 + 16n - 12)$
$$= 12n^2 + 64n - 48 \ ✓$$

> **[YOU TRY 2]** Factor completely.
>
> a) $24h^2 - 54h - 15$ 　　 b) $20d^3 + 38d^2 + 12d$

---

2 Factor $ax^2 + bx + c$ ($a \neq 1$) by Trial and Error

At the beginning of this section, we factored $2x^2 + 11x + 15$ by grouping. Now we will factor it by trial and error, which is just reversing the process of FOIL.

Factor $5t^2 + 31t + 6$ completely.

Answer: $(5t + 1)(t + 6)$

Factor $2x^2 + 11x + 15$ completely.

Solution

Can we factor out a GCF? No. So try to factor $2x^2 + 11x + 15$ as the product of two binomials. Notice that all terms are positive, so all factors will be positive.

www.mhhe.com/messersmith 　　　　　　 SECTION 7.3 　 **Factoring Trinomials of the Form $ax^2 + bx + c$ ($a \neq 1$)** 　 **407**

Begin with the squared term, $2x^2$. Which two expressions with integer coefficients can we multiply to get $2x^2$? $2x$ and x. Put these in the binomials.

$$2x^2 + 11x + 15 = (2x \quad)(x \quad) \qquad 2x \cdot x = 2x^2$$

Next, look at the last term, 15. What are the pairs of positive integers that multiply to 15? They are 15 and 1 as well as 5 and 3.

Try these numbers as the last terms of the binomials. The middle term, $11x$, comes from finding the sum of the products of the outer terms and inner terms.

First Try

$$2x^2 + 11x + 15 \overset{?}{=} (2x + 15)(x + 1) \quad \text{Incorrect!}$$

$15x$

These must both be $11x$. $\begin{array}{r} + \ 2x \\ \hline 17x \end{array}$

Try again. Switch the 15 and the 1.

$$2x^2 + 11x + 15 \overset{?}{=} (2x + 1)(x + 15) \quad \text{Incorrect!}$$

$1x$

These must both be $11x$. $\begin{array}{r} + \ 30x \\ \hline 31x \end{array}$

Try using 5 and 3. $2x^2 + 11x + 15 \overset{?}{=} (2x + 5)(x + 3)$ Correct!

$5x$

These must both be $11x$. $\begin{array}{r} + \ 6x \\ \hline 11x \end{array}$

Therefore, $2x^2 + 11x + 15 = (2x + 5)(x + 3)$. Check by multiplying.

W Hint

You will not always factor correctly on the first try. Don't worry, just try again!

EXAMPLE 4

Factor $3t^2 - 29t + 18$ completely.

In-Class Example 4

Factor $2m^2 - 13m + 20$ completely.

Answer: $(2m - 5)(m - 4)$

Solution

Can we factor out a GCF? No. To get a product of $3t^2$, we will use $3t$ and t.

$$3t^2 - 29t + 18 = (3t \quad)(t \quad) \qquad 3t \cdot t = 3t^2$$

Since the last term is positive and the middle term is negative, we want pairs of negative integers that multiply to 18. The pairs are -1 and -18, -2 and -9, and -3 and -6. Try these numbers as the last terms of the binomials. The middle term, $-29t$, comes from finding the sum of the products of the outer terms and inner terms.

$$3t^2 - 29t + 18 \overset{?}{=} (3t - 1)(t - 18) \quad \text{Incorrect!}$$

$-t$

These must both be $-29t$. $\begin{array}{r} + \ (-54t) \\ \hline -55t \end{array}$

Try again. Switch the -1 and the -18: $3t^2 - 29t + 18 \overset{?}{=} (3t - 18)(t - 1)$

Without multiplying, we know that this choice is incorrect. How? In the factor $(3t - 18)$, a 3 can be factored out to get $3(t - 6)$. But, we concluded that we could not factor out a GCF from the original polynomial, $3t^2 - 29t + 18$. Therefore, it will not be possible to take out a common factor from one of the binomial factors.

Note

If you cannot factor out a GCF from the original polynomial, then you cannot take out a factor from one of the binomial factors either.

Try using -2 and -9. $3t^2 - 29t + 18 = (3t - 2)(t - 9)$ Correct!

These must both be $-29t$. $-2t$

$\dfrac{+ (-27t)}{-29t}$

So, $3t^2 - 29t + 18 = (3t - 2)(t - 9)$. Check by multiplying.

[**YOU TRY 3**]   Factor completely.

a) $2k^2 + 17k + 8$ b) $6z^2 - 23z + 20$

EXAMPLE 5

In-Class Example 5

Factor completely.
a) $18k^2 + 21k - 30$
b) $-4h^2 + h + 18$

Answer: a) $3(6k - 5)(k + 2)$
b) $-(4h - 9)(h + 2)$

Factor completely.

a) $16a^2 + 62a - 8$ b) $-2c^2 + 3c + 20$

Solution

a) Ask yourself, *"Can I take out a common factor?"* Yes!

$$16a^2 + 62a - 8 = 2(8a^2 + 31a - 4)$$

Now, try to factor $8a^2 + 31a - 4$. To get a product of $8a^2$, we can try either $8a$ and a or $4a$ and $2a$. Let's start by trying $8a$ and a.

$$8a^2 + 31a - 4 = (8a \quad)(a \quad)$$

List pairs of integers that multiply to -4: 4 and -1, -4 and 1, 2 and -2.

Try 4 and -1. Do not put 4 in the same binomial as $8a$ since then it would be possible to factor out 2. But, 2 does not factor out of $8a^2 + 31a - 4$. Put the 4 in the same binomial as a.

$$8a^2 + 31a - 4 \overset{?}{=} (8a - 1)(a + 4)$$

$-a$

$\dfrac{+ 32a}{31a}$ Correct

Don't forget that the very first step was to factor out a 2. Therefore,

$$16a^2 + 62a - 8 = 2(8a^2 + 31a - 4) = 2(8a - 1)(a + 4)$$

Check by multiplying.

b) Since the coefficient of the squared term is negative, begin by factoring out -1. (There is no other common factor except 1.)

$$-2c^2 + 3c + 20 = -1(2c^2 - 3c - 20)$$

Try to factor $2c^2 - 3c - 20$. To get a product of $2c^2$, we will use $2c$ and c in the binomials.

$$2c^2 - 3c - 20 = (2c \quad)(c \quad)$$

We need pairs of integers whose product is -20. They are 1 and -20, -1 and 20, 2 and -10, -2 and 10, 4 and -5, -4 and 5.

Do *not* start with 1 and -20 or -1 and 20 because the middle term, $-3c$, is not very large. Using 1 and -20 or -1 and 20 would likely result in a larger middle term.

Think about 2 and -10 *and* -2 and 10. *These will not work because if we put any of these numbers in the factor containing $2c$, then it will be possible to factor out 2.*

W Hint

Be sure to read the reasoning in this example, and apply it to the You Try!

Try 4 and -5. Do not put 4 in the same binomial as $2c$ since then it would be possible to factor out 2.

$$2c^2 - 3c - 20 \overset{?}{=} (2c - 5)(c + 4)$$

$$\begin{array}{r} -5c \\ + \ 8c \\ \hline 3c \end{array} \quad \text{This must equal } -3c. \qquad \text{Incorrect}$$

Only the sign of the sum is incorrect. *Change the signs in the binomials to get the correct sum.*

$$2c^2 - 3c - 20 \overset{?}{=} (2c + 5)(c - 4)$$

$$\begin{array}{r} 5c \\ + \ (-8c) \\ \hline -3c \end{array} \qquad \text{Correct}$$

Remember that we factored out -1 to begin the problem.

$$-2c^2 + 3c + 20 = -1(2c^2 - 3c - 20) = -(2c + 5)(c - 4)$$

Check by multiplying.

[**YOU TRY 4**]   Factor completely.

a) $10y^2 - 58y + 40$ b) $-4n^2 - 5n + 6$

We have seen two methods for factoring $ax^2 + bx + c\ (a \neq 1)$:   factoring by grouping and factoring by trial and error. In either case, remember to begin by taking out a common factor from all terms whenever possible.

Using Technology

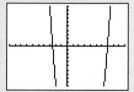

We found some ways to narrow down the possibilities when factoring $ax^2 + bx + c\ (a \neq 1)$ using the trial and error method.

We can also use a graphing calculator to help with the process. Consider the trinomial $2x^2 - 9x - 35$. Enter the trinomial into Y_1 and press ZOOM ; then enter 6 to display the graph in the standard viewing window.

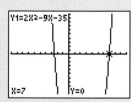

Look on the graph for the x-intercept (if any) that appears to be an integer. It appears that 7 is an x-intercept.

To check whether 7 is an x-intercept, press TRACE then 7 and press ENTER . As shown on the graph, when $x = 7$, $y = 0$, so 7 is an x-intercept.

When an x-intercept is an integer, then x minus that x-intercept is a factor of the trinomial. In this case, $x - 7$ is a factor of $2x^2 - 9x - 35$. We can then complete the factoring as $(2x + 5)(x - 7)$, since we must multiply -7 by 5 to obtain -35.

Find an x-intercept using a graphing calculator and factor the trinomial.

1) $3x^2 + 11x - 4$ 2) $2x^2 + x - 15$ 3) $5x^2 + 6x - 8$

4) $2x^2 - 5x + 3$ 5) $4x^2 - 3x - 10$ 6) $14x^2 - x - 4$

E Evaluate **7.3** Exercises

Do the exercises, and check your work.

*Additional answers can be found in the Answers to Exercises appendix.

Objective 1: Factor $ax^2 + bx + c$ $(a \neq 1)$ by Grouping

1) Find two integers whose

	PRODUCT IS	and whose SUM IS	ANSWER
a)	-50	5	
b)	27	-28	
c)	12	8	
d)	-72	-6	

a) $10, -5$ b) $-27, -1$ c) $6, 2$ d) $-12, 6$

2) Find two integers whose

	PRODUCT IS	and whose SUM IS	ANSWER
a)	18	19	
b)	-132	1	
c)	-30	-13	
d)	63	-16	

a) $18, 1$ b) $12, -11$ c) $-15, 2$ d) $-9, -7$

Factor by grouping.

 3) $3c^2 + 12c + 8c + 32$ $(3c + 8)(c + 4)$

4) $5y^2 + 15y + 2y + 6$ $(5y + 2)(y + 3)$

5) $6k^2 - 6k - 7k + 7$ $(6k - 7)(k - 1)$

6) $4r^2 - 4r + 9r - 9$ $(4r + 9)(r - 1)$

7) $6x^2 - 27xy + 8xy - 36y^2$ $(2x - 9y)(3x + 4y)$

8) $14p^2 - 8pq - 7pq + 4q^2$ $(7p - 4q)(2p - q)$

9) When asked to factor a polynomial, what is the first question you should ask yourself?
Can I factor out a GCF?

 10) After factoring a polynomial, what should you ask yourself to be sure that the polynomial is factored completely?   Can I factor again?

11) Find the polynomial that factors to $(4k + 9)(k + 2)$.
$4k^2 + 17k + 18$

12) Find the polynomial that factors to $(6m - 5)(2m - 3)$.
$12m^2 - 28m + 15$

Complete the factorization.

13) $5t^2 + 13t + 6 = (5t + 3)(\ t + 2\ )$

14) $4z^2 + 29z + 30 = (4z + 5)(\ z + 6\ )$

15) $6a^2 - 11a - 10 = (2a - 5)(\ 3a + 2\ )$

16) $15c^2 - 23c + 4 = (3c - 4)(\ 5c - 1\ )$

17) $12x^2 - 25xy + 7y^2 = (4x - 7y)(\ 3x - y\ )$

18) $12r^2 - 52rt - 9t^2 = (6r + t)(\ 2r - 9t\ )$

Factor by grouping. See Example 1.

19) $2h^2 + 13h + 15$ 20) $3z^2 + 13z + 14$
$(2h + 3)(h + 5)$ $(3z + 7)(z + 2)$
21) $7y^2 - 11y + 4$ 22) $5a^2 - 21a + 18$
$(7y - 4)(y - 1)$ $(5a - 6)(a - 3)$
23) $5b^2 + 9b - 18$ 24) $11m^2 - 18m - 8$
$(5b - 6)(b + 3)$ $(11m + 4)(m - 2)$
25) $6p^2 + p - 2$ 26) $8c^2 - 22c + 5$
$(3p + 2)(2p - 1)$ $(4c - 1)(2c - 5)$
27) $4t^2 + 16t + 15$ 28) $10k^2 + 23k + 12$
$(2t + 3)(2t + 5)$ $(5k + 4)(2k + 3)$
29) $9x^2 - 13xy + 4y^2$ 30) $6a^2 + ab - 5b^2$
$(9x - 4y)(x - y)$ $(6a - 5b)(a + b)$

Objective 2: Factor $ax^2 + bx + c$ $(a \neq 1)$ by Trial and Error

31) How do we know that $(2x - 4)$ cannot be a factor of $2x^2 + 13x - 24$?   because 2 can be factored out of $2x - 4$, but 2 cannot be factored out of $2x^2 + 13x - 24$

32) How do we know that $(3p + 2)$ cannot be a factor of $6p^2 - 25p + 14$?   Since the coefficient of the middle term in the trinomial is negative and the constant is positive, both factors will have a minus sign between the terms.

Factor by trial and error. See Examples 3 and 4.

33) $2r^2 + 9r + 10$
$(2r + 5)(r + 2)$

34) $3q^2 + 10q + 8$
$(3q + 4)(q + 2)$

35) $3u^2 - 23u + 30$
$(3u - 5)(u - 6)$

36) $7m^2 - 15m + 8$
$(7m - 8)(m - 1)$

37) $7a^2 + 31a - 20$
$(7a - 4)(a + 5)$

38) $5x^2 - 11x - 36$
$(5x + 9)(x - 4)$

39) $6y^2 + 23y + 10$
$(3y + 10)(2y + 1)$

40) $8u^2 + 18u + 7$
$(4u + 7)(2u + 1)$

41) $9w^2 + 20w - 21$
$(9w - 7)(w + 3)$

42) $10h^2 - 59h - 6$
$(10h + 1)(h - 6)$

43) $8c^2 - 42c + 27$
$(4c - 3)(2c - 9)$

44) $15v^2 - 16v + 4$
$(5v - 2)(3v - 2)$

45) $4k^2 + 40k + 99$
$(2k + 11)(2k + 9)$

46) $4n^2 - 41n + 10$
$(4n - 1)(n - 10)$

47) $20b^2 - 32b - 45$
$(10b + 9)(2b - 5)$

48) $14g^2 + 27g - 20$
$(7g - 4)(2g + 5)$

49) $2r^2 + 13rt - 24t^2$
$(2r - 3t)(r + 8t)$

50) $3c^2 - 17cd - 6d^2$
$(3c + d)(c - 6d)$

51) $6a^2 - 25ab + 4b^2$
$(6a - b)(a - 4b)$

52) $6x^2 + 31xy + 18y^2$
$(3x + 2y)(2x + 9y)$

Mixed Exercises: Objectives 1 and 2

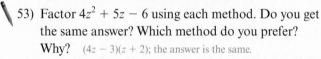

53) Factor $4z^2 + 5z - 6$ using each method. Do you get the same answer? Which method do you prefer? Why? $(4z - 3)(z + 2)$; the answer is the same.

54) Factor $10a^2 + 27a + 18$ using each method. Do you get the same answer? Which method do you prefer? Why? $(5a + 6)(2a + 3)$; the answer is the same.

Use the trinomial $ax^2 + bx + c$ ($a > 0$) to answer *always*, *sometimes*, or *never* to Exercises 55 and 56.

55) If the product of a and c is negative, both factors will have a minus sign between terms.   never

56) If the product of a and c is positive, both factors will have a minus sign between terms.   sometimes

Factor completely.

57) $3p^2 - 16p - 12$
$(3p + 2)(p - 6)$

58) $2t^2 - 19t + 24$
$(2t - 3)(t - 8)$

59) $4k^2 + 15k + 9$
$(4k + 3)(k + 3)$

60) $12x^3 + 15x^2 - 18x$
$3x(4x - 3)(x + 2)$

61) $30w^3 + 76w^2 + 14w$
$2w(5w + 1)(3w + 7)$

62) $12d^2 - 28d - 5$
$(2d - 5)(6d + 1)$

63) $21r^2 - 90r + 24$
$3(7r - 2)(r - 4)$

64) $45q^2 + 57q + 18$
$3(5q + 3)(3q + 2)$

65) $6y^2 - 10y + 3$
prime

66) $9z^2 + 14z + 8$
prime

67) $42b^2 + 11b - 3$
$(7b + 3)(6b - 1)$

68) $13u^2 + 17u - 18$
$(13u - 9)(u + 2)$

69) $7x^2 - 17xy + 6y^2$
$(7x - 3y)(x - 2y)$

70) $5a^2 + 23ab + 12b^2$
$(5a + 3b)(a + 4b)$

71) $2d^2 + 2d - 40$
$2(d + 5)(d - 4)$

72) $6c^2 + 42c + 72$
$6(c + 3)(c + 4)$

73) $30r^4t^2 + 23r^3t^2 + 3r^2t^2$
$r^2t^2(6r + 1)(5r + 3)$

74) $8m^2n^3 + 4m^2n^2 - 60m^2n$ $4m^2n(2n - 5)(n + 3)$

75) $9k^2 - 42k + 49$ $(3k - 7)^2$

76) $25p^2 + 20p + 4$ $(5p + 2)^2$

Factor completely by first taking out a negative common factor. See Example 5.

77) $-n^2 - 8n + 48$ $-(n + 12)(n - 4)$

78) $-c^2 - 16c - 63$ $-(c + 7)(c + 9)$

79) $-7a^2 + 4a + 3$ $-(7a + 3)(a - 1)$

80) $-3p^2 + 14p - 16$ $-(3p - 8)(p - 2)$

81) $-10z^2 + 19z - 6$ $-(5z - 2)(2z - 3)$

82) $-16k^3 + 48k^2 - 36k$ $-4k(2k - 3)^2$

83) $-20m^3 - 120m^2 - 135m$ $-5m(2m + 9)(2m + 3)$

84) $-3z^3 - 15z^2 + 198z$ $-3z(z + 11)(z - 6)$

85) $-6a^3b + 11a^2b^2 + 2ab^3$ $-ab(6a + b)(a - 2b)$

86) $-35u^4 - 203u^3v - 140u^2v^2$ $-7u^2(5u + 4v)(u + 5v)$

R Rethink

R1) How have your arithmetic skills helped you complete these exercises?

R2) Which method of factoring did you prefer while completing the exercises? Why?

R3) How much more practice will you need to master the objectives of this section?

7.4 Factoring Special Trinomials and Binomials

What are your objectives for Section 7.4?	How can you accomplish each objective?
1 Factor a Perfect Square Trinomial	• Notice, again, that factoring is the opposite of multiplication. • Learn the formula for **Factoring a Perfect Square Trinomial.** • Complete the given examples on your own. • Complete You Try 1.
2 Factor the Difference of Two Squares	• Notice, again, that factoring is the opposite of multiplication. • Learn the formula for **Factoring the Difference of Two Squares.** • Complete the given examples on your own. • Complete You Trys 2 and 3.

W Work

Read the explanations, follow the examples, take notes, and complete the You Trys.

1 Factor a Perfect Square Trinomial

Recall that we can square a binomial using the formulas

$$(a + b)^2 = a^2 + 2ab + b^2$$
$$(a - b)^2 = a^2 - 2ab + b^2$$

For example, $(x + 3)^2 = x^2 + 2x(3) + 3^2 = x^2 + 6x + 9$.

Since factoring a polynomial means writing the polynomial as a product of its factors, $x^2 + 6x + 9$ factors to $(x + 3)^2$.

The expression $x^2 + 6x + 9$ is a *perfect square trinomial*. A **perfect square trinomial** is a trinomial that results from squaring a binomial.

We can use the factoring method presented in Section 7.2 to factor a perfect square trinomial, or we can learn to recognize the special pattern that appears in these trinomials.

How are the terms of $x^2 + 6x + 9$ and $(x + 3)^2$ related?

x^2 is the square of x, the first term in the binomial.

9 is the square of 3, the last term in the binomial.

We get the term $6x$ by doing the following:

$$6x = \quad 2 \quad \cdot \quad x \quad \cdot \quad 3$$

Two times   First term in binomial   Last term in binomial

This follows directly from how we found $(x + 3)^2$ using the formula.

> ### Formula Factoring a Perfect Square Trinomial
>
> $$a^2 + 2ab + b^2 = (a + b)^2$$
> $$a^2 - 2ab + b^2 = (a - b)^2$$

> ### Note
> In order for a trinomial to be a perfect square, two of its terms must be perfect squares.

EXAMPLE 1

In-Class Example 1

Factor $t^2 + 12t + 36$ completely.

Answer: $(t + 6)^2$

Factor $k^2 + 18k + 81$ completely.

Solution

We cannot take out a common factor, so let's see whether this follows the pattern of a perfect square trinomial.

$$k^2 + 18k + 81$$

What do you square to get k^2? k $(k)^2$ $(9)^2$ What do you square to get 81? 9

Does the middle term equal $2 \cdot k \cdot 9$? Yes.

$$2 \cdot k \cdot 9 = 18k$$

> **Hint**
> Complete the examples by using the formula.

Therefore, $k^2 + 18k + 81 = (k + 9)^2$. Check by multiplying.

EXAMPLE 2

In-Class Example 2

Factor completely.
a) $r^2 - 14r + 49$
b) $8k^3 + 16k^2 + 8k$
c) $4u^2 - 28u + 49$
d) $4z^2 + 25z + 36$

Answer: a) $(r - 7)^2$
b) $8k(k + 1)^2$
c) $(2u - 7)^2$
d) $(4z + 9)(z + 4)$

Factor completely.

a) $c^2 - 16c + 64$ b) $4t^3 + 32t^2 + 64t$

c) $9w^2 + 12w + 4$ d) $4c^2 + 20c + 9$

Solution

a)   We cannot take out a common factor. However, since the middle term is negative and the first and last terms are positive, the sign in the binomial will be a minus $(-)$ sign. Does this fit the pattern of a perfect square trinomial?

$$c^2 - 16c + 64$$

What do you square to get c^2? c $(c)^2$ $(8)^2$ What do you square to get 64? 8

Does the middle term equal $2 \cdot c \cdot 8$? Yes. $2 \cdot c \cdot 8 = 16c$.

Notice that $c^2 - 16c + 64$ fits the pattern of $a^2 - 2ab + b^2 = (a - b)^2$ with $a = c$ and $b = 8$.

> **Hint**
> When you are trying to factor, ask yourself the same questions that you are being asked in this example.

Therefore, $c^2 - 16c + 64 = (c - 8)^2$. Check by multiplying.

b) From $4t^3 + 32t^2 + 64t$ we *can* begin by taking out the GCF of $4t$.

$$4t^3 + 32t^2 + 64t = 4t(t^2 + 8t + 16)$$

What do you square to get t^2? t $(t)^2$ $(4)^2$ What do you square to get 16? 4

Does the middle term of the trinomial in parentheses equal $2 \cdot t \cdot 4$? Yes. $2 \cdot t \cdot 4 = 8t$.

$$4t^3 + 32t^2 + 64t = 4t(t^2 + 8t + 16) = 4t(t + 4)^2$$

Check by multiplying.

c) We cannot take out a common factor. Since the first and last terms of $9w^2 + 12w + 4$ are perfect squares, let's see whether this is a perfect square trinomial.

$$9w^2 + 12w + 4$$

What do you square to get $9w^2$? $3w$ $(3w)^2$ $(2)^2$ What do you square to get 4? 2

Does the middle term equal $2 \cdot 3w \cdot 2$? Yes. $2 \cdot 3w \cdot 2 = 12w$.

Therefore, $9w^2 + 12w + 4 = (3w + 2)^2$. Check by multiplying.

d) We cannot take out a common factor. The first and last terms of $4c^2 + 20c + 9$ are perfect squares. Is this a perfect square trinomial?

$$4c^2 + 20c + 9$$

What do you square to get $4c^2$? $2c$ $(2c)^2$ $(3)^2$ What do you square to get 9? 3

Does the middle term equal $2 \cdot 2c \cdot 3$? No! $2 \cdot 2c \cdot 3 = 12c$.

This is *not* a perfect square trinomial. Applying a method from Section 7.3, we find that the trinomial does factor, however.

$$4c^2 + 20c + 9 = (2c + 9)(2c + 1). \text{ Check by multiplying.}$$

[YOU TRY 1]

Factor completely.

a) $g^2 + 14g + 49$ b) $6y^3 - 36y^2 + 54y$

c) $25v^2 - 10v + 1$ d) $9b^2 + 15b + 4$

---

2   Factor the Difference of Two Squares

Another common factoring problem is a **difference of two squares.** Some examples of these types of binomials are

$$c^2 - 36 \qquad 49x^2 - 25y^2 \qquad 64 - t^2 \qquad h^4 - 16$$

Notice that in each binomial, the terms are being *subtracted,* and each term is a perfect square.

In Section 6.2, Multiplication of Polynomials, we saw that

$$(a + b)(a - b) = a^2 - b^2$$

If we reverse the procedure, we get the factorization of the difference of two squares.

Formula Factoring the Difference of Two Squares

$$a^2 - b^2 = (a + b)(a - b)$$

Don't forget that we can check all factorizations by multiplying.

EXAMPLE 3

In-Class Example 3

Factor completely.
a) $h^2 - 64$ b) $25t^2 - 144u^2$

c) $m^2 - \dfrac{9}{49}$ d) $y^2 + 25$

Answer: a) $(h + 8)(h - 8)$
b) $(5t + 12u)(5t - 12u)$

c) $\left(m + \dfrac{3}{7}\right)\left(m - \dfrac{3}{7}\right)$

d) prime

Factor completely.

a) $c^2 - 36$ b) $49x^2 - 25y^2$ c) $t^2 - \dfrac{4}{9}$ d) $k^2 + 81$

Solution

a)   First, notice that $c^2 - 36$ is the difference of two terms *and* those terms are perfect squares. We can use the formula $a^2 - b^2 = (a + b)(a - b)$.

Identify a and b.

$$c^2 - 36$$

What do you square to get c^2? c $(c)^2$ $(6)^2$ What do you square to get 36? 6

Then, $a = c$ and $b = 6$. Therefore, $c^2 - 36 = (c + 6)(c - 6)$.

b)   Look carefully at $49x^2 - 25y^2$. Each term *is* a perfect square, and they are being subtracted.

Identify a and b.

$$49x^2 - 25y^2$$

What do you square to get $49x^2$? $7x$ $(7x)^2$ $(5y)^2$ What do you square to get $25y^2$? $5y$

Then, $a = 7x$ and $b = 5y$. So, $49x^2 - 25y^2 = (7x + 5y)(7x - 5y)$.

c)   Each term in $t^2 - \dfrac{4}{9}$ is a perfect square, and they are being subtracted.

$$t^2 - \dfrac{4}{9}$$

What do you square to get t^2? t $(t)^2$ $\left(\dfrac{2}{3}\right)^2$ What do you square to get $\dfrac{4}{9}$? $\dfrac{2}{3}$

So, $a = t$ and $b = \dfrac{2}{3}$. Therefore, $t^2 - \dfrac{4}{9} = \left(t + \dfrac{2}{3}\right)\left(t - \dfrac{2}{3}\right)$.

d)   Each term in $k^2 + 81$ is a perfect square, but the expression is the *sum* of two squares. This polynomial does not factor.

$k^2 + 81 \neq (k + 9)(k - 9)$ since $(k + 9)(k - 9) = k^2 - 81$.

$k^2 + 81 \neq (k + 9)(k + 9)$ since $(k + 9)(k + 9) = k^2 + 18k + 81$.

So, $k^2 + 81$ is prime.

W Hint

Are you writing out the steps as you are reading the example?

[YOU TRY 2] Factor completely.

a) $m^2 - 100$ b) $4c^2 - 81d^2$ c) $h^2 - \dfrac{64}{25}$ d) $p^2 + 49$

Remember that sometimes we can factor out a GCF first. And, after factoring once, ask yourself, *"Can I factor again?"*

EXAMPLE 4

In-Class Example 4

Factor completely.
a) $24 - 6x^2$
b) $3r^2 + 75$
c) $x^4 - 81$

Answer: a) $6(2 + x)(2 - x)$
b) $3(r^2 + 25)$
c) $(x^2 + 9)(x + 3)(x - 3)$

W Hint

What is the first question you should *always* ask yourself when you are trying to factor a polynomial?

Factor completely.

a) $300p - 3p^3$ b) $7w^2 + 28$ c) $x^4 - 81$

Solution

a) Ask yourself, *"Can I take out a common factor?"* Yes. Factor out $3p$.

$$300p - 3p^3 = 3p(100 - p^2)$$

Now ask yourself, *"Can I factor again?"* Yes. $100 - p^2$ is the difference of two squares. Identify a and b.

$$\begin{array}{cc} 100 & - \; p^2 \\ \downarrow & \downarrow \\ (10)^2 & (p)^2 \end{array}$$

So, $a = 10$ and $b = p$. $100 - p^2 = (10 + p)(10 - p)$.

Therefore, $300p - 3p^3 = 3p(10 + p)(10 - p)$.

BE CAREFUL

$(10 + p)(10 - p)$ is *not* the same as $(p + 10)(p - 10)$ because subtraction is not commutative. While $10 + p = p + 10$, $10 - p$ does *not* equal $p - 10$. You must write the terms in the correct order. Another way to see that they are not equivalent is to multiply $(p + 10)(p - 10)$. $(p + 10)(p - 10) = p^2 - 100$. This is not the same as $100 - p^2$.

b) Look at $7w^2 + 28$. Ask yourself, *"Can I take out a common factor?"* Yes. Factor out 7.

$$7(w^2 + 4)$$

"Can I factor again?" No, because $w^2 + 4$ is the *sum* of two squares.

Therefore, $7w^2 + 28 = 7(w^2 + 4)$.

c) The terms in $x^4 - 81$ have no common factors, but they are perfect squares. Identify a and b.

$$\begin{array}{cc} x^4 & - \; 81 \\ \downarrow & \downarrow \\ (x^2)^2 & (9)^2 \end{array}$$

What do you square to get x^4? x^2 What do you square to get 81? 9

So, $a = x^2$ and $b = 9$. $x^4 - 81 = (x^2 + 9)(x^2 - 9)$

Can we factor again?

$x^2 + 9$ is the *sum* of two squares. It will not factor.

$x^2 - 9$ is the difference of two squares, so it *will* factor.

$$x^2 - 9$$
$$\downarrow \quad \downarrow \qquad x^2 - 9 = (x + 3)(x - 3)$$
$$(x)^2 \ (3)^2$$
$$a = x \text{ and } b = 3$$

Therefore,

$$x^4 - 81 = (x^2 + 9)(x^2 - 9)$$
$$= (x^2 + 9)(x + 3)(x - 3)$$

$\left[\text{YOU TRY 3}\right]$ Factor completely.

a) $125d - 5d^3$ b) $3r^2 + 48$ c) $z^4 - 1$

Summary Special Factoring Rules

Perfect square trinomials: $a^2 + 2ab + b^2 = (a + b)^2$
$a^2 - 2ab + b^2 = (a - b)^2$

Difference of two squares: $a^2 - b^2 = (a + b)(a - b)$

ANSWERS TO $\left[\text{YOU TRY}\right]$ **EXERCISES**

1) a) $(g + 7)^2$ b) $6y(y - 3)^2$ c) $(5v - 1)^2$ d) $(3b + 4)(3b + 1)$

2) a) $(m + 10)(m - 10)$ b) $(2c + 9d)(2c - 9d)$ c) $\left(h + \dfrac{8}{5}\right)\left(h - \dfrac{8}{5}\right)$ d) prime

3) a) $5d(5 + d)(5 - d)$ b) $3(r^2 + 16)$ c) $(z^2 + 1)(z + 1)(z - 1)$

E Evaluate **7.4** Exercises   Do the exercises, and check your work.

Objective 1: Factor a Perfect Square Trinomial

1) Find the following.

a) 7^2 49   b) 9^2 81

c) 6^2 36   d) 10^2 100

e) 5^2 25   f) 4^2 16

g) 11^2 121   h) $\left(\dfrac{1}{3}\right)^2$ $\dfrac{1}{9}$

i) $\left(\dfrac{3}{8}\right)^2$ $\dfrac{9}{64}$

2) What is perfect square trinomial?
It is a trinomial that results from squaring a binomial.

3) Fill in the blank with a term that has a positive coefficient.

a) $(\underline{})^2 = c^4$ c^2 b) $(\underline{})^2 = 9r^2$ $3r$

c) $(\underline{})^2 = 81p^2$ $9p$ d) $(\underline{})^2 = 36m^4$ $6m^2$

e) $(\underline{})^2 = \dfrac{1}{4}$ $\dfrac{1}{2}$ f) $(\underline{})^2 = \dfrac{144}{25}$ $\dfrac{12}{5}$

4) If x^n is a perfect square, then n is divisible by what number? 2

5) What perfect square trinomial factors to $(y + 6)^2$?
$y^2 + 12y + 36$

6) What perfect square trinomial factors to $(3k - 8)^2$?
$9k^2 - 48k + 64$

7) Why isn't $4a^2 - 10a + 9$ a perfect square trinomial?

8) Why isn't $x^2 + 5x + 12$ a perfect square trinomial?

Factor completely.

9) $h^2 + 10h + 25$ $(h + 5)^2$ 10) $q^2 + 8q + 16$ $(q + 4)^2$

11) $b^2 - 14b + 49$ $(b - 7)^2$ 12) $t^2 - 24t + 144$ $(t - 12)^2$

13) $4w^2 + 4w + 1$ $(2w + 1)^2$ 14) $25m^2 + 20m + 4$
                                               $(5m + 2)^2$

15) $9k^2 - 24k + 16$ 16) $16a^2 - 56a + 49$
    $(3k - 4)^2$ $(4a - 7)^2$

17) $c^2 + c + \dfrac{1}{4}$ $\left(c + \dfrac{1}{2}\right)^2$ 18) $h^2 + \dfrac{1}{3}h + \dfrac{1}{36}$ $\left(h + \dfrac{1}{6}\right)^2$

19) $k^2 - \dfrac{14}{5}k + \dfrac{49}{25}$ $\left(k - \dfrac{7}{5}\right)^2$ 20) $p^2 - \dfrac{4}{3}h + \dfrac{4}{9}$ $\left(p - \dfrac{2}{3}\right)^2$

21) $a^2 + 8ab + 16b^2$ 22) $4x^2 - 12xy + 9y^2$
    $(a + 4b)^2$ $(2x - 3y)^2$

23) $25m^2 - 30mn + 9n^2$ 24) $49p^2 + 14pq + q^2$
    $(5m - 3n)^2$ $(7p + q)^2$

25) $4f^2 + 24f + 36$ 26) $8r^2 - 16r + 8$ $8(r - 1)^2$
    $4(f + 3)^2$

27) $5a^4 + 30a^3 + 45a^2$ 28) $3k^3 - 42k^2 + 147k$
    $5a^2(a + 3)^2$ $3k(k - 7)^2$

29) $-16y^2 - 80y - 100$ 30) $-81n^2 + 54n - 9$
    $-4(2y + 5)^2$ $-9(3n - 1)^2$

31) $75h^3 - 6h^2 + 12h$ 32) $98b^5 + 42b^4 + 18b^3$
    $3h(25h^2 - 2h + 4)$ $2b^3(49b^2 + 21b + 9)$

Objective 2: Factor the Difference of Two Squares

33) What binomial factors to

    a) $(x + 9)(x - 9)$?   b) $(9 + x)(9 - x)$?
      $x^2 - 81$ $81 - x^2$

34) What binomial factors to

    a) $(y - 10)(y + 10)$?   b) $(10 - y)(10 + y)$?
      $y^2 - 100$ $100 - y^2$

Complete the factorization.

35) $w^2 - 64 = (w + 8)($ $)$ $w - 8$

36) $t^2 - 1 = (t - 1)($ $)$ $t + 1$

37) $121 - p^2 = (11 + p)($ $)$ $11 - p$

38) $9h^2 - 49 = (3h + 7)($ $)$ $3h - 7$

39) $64c^2 - 25b^2 = (8c + 5b)($ $)$ $8c - 5b$

40) Does $n^2 + 9 = (n + 3)^2$? Explain.
    No; $(n + 3)^2 = n^2 + 6n + 9$. $n^2 + 9$ is prime.

Factor completely.

41) $k^2 - 4$ $(k + 2)(k - 2)$ 42) $z^2 - 100$ $(z + 10)(z - 10)$

43) $c^2 - 25$ $(c + 5)(c - 5)$ 44) $y^2 - 81$ $(y + 9)(y - 9)$

45) $w^2 + 49$ prime   46) $b^2 + 64$ prime

47) $x^2 - \dfrac{1}{9}$ $\left(x + \dfrac{1}{3}\right)\left(x - \dfrac{1}{3}\right)$ 48) $p^2 - \dfrac{1}{4}$ $\left(p + \dfrac{1}{2}\right)\left(p - \dfrac{1}{2}\right)$

49) $a^2 - \dfrac{4}{49}$ $\left(a + \dfrac{2}{7}\right)\left(a - \dfrac{2}{7}\right)$ 50) $t^2 - \dfrac{121}{64}$ $\left(t + \dfrac{11}{8}\right)\left(t - \dfrac{11}{8}\right)$

51) $144 - v^2$ $(12 + v)(12 - v)$ 52) $36 - r^2$ $(6 + r)(6 - r)$

53) $1 - h^2$ $(1 + h)(1 - h)$ 54) $169 - d^2$ $(13 + d)(13 - d)$

55) $\dfrac{36}{25} - b^2$ $\left(\dfrac{6}{5} + b\right)\left(\dfrac{6}{5} - b\right)$ 56) $\dfrac{9}{100} - q^2$ $\left(\dfrac{3}{10} + q\right)\left(\dfrac{3}{10} - q\right)$

57) $100m^2 - 49$ 58) $25a^2 - 121$
    $(10m + 7)(10m - 7)$ $(5a + 11)(5a - 11)$

59) $169k^2 - 1$ 60) $36p^2 - 1$ $(6p + 1)(6p - 1)$
    $(13k + 1)(13k - 1)$

61) $4y^2 + 49$ prime   62) $9d^2 + 25$ prime

63) $\dfrac{1}{9}t^2 - \dfrac{25}{4}$ 64) $\dfrac{16}{9}x^2 - \dfrac{1}{49}$

65) $u^4 - 100$ 66) $a^4 - 4$ $(a^2 + 2)(a^2 - 2)$
    $(u^2 + 10)(u^2 - 10)$

67) $36c^2 - d^4$ 68) $25y^2 - 144z^4$
    $(6c + d^2)(6c - d^2)$ $(5y + 12z^2)(5y - 12z^2)$

69) $r^4 - 1$ 70) $h^4 - 16$
    $(r^2 + 1)(r + 1)(r - 1)$ $(h^2 + 4)(h + 2)(h - 2)$

71) $r^4 - 81t^4$ 72) $y^4 - x^4$
    $(r^2 + 9t^2)(r + 3t)(r - 3t)$ $(x^2 + y^2)(x + y)(y - x)$

73) $5u^2 - 45$ 74) $3k^2 - 300$
    $5(u + 3)(u - 3)$ $3(k + 10)(k - 10)$

75) $2n^2 - 288$ 76) $11p^2 - 11$
    $2(n + 12)(n - 12)$ $11(p + 1)(p - 1)$

77) $12z^4 - 75z^2$ 78) $45b^5 - 245b^3$
    $3z^2(2z + 5)(2z - 5)$ $5b^3(3b + 7)(3b - 7)$

R Rethink

R1) How did your knowledge of common powers help you complete the exercises more quickly?

R2) Which concepts from the section do you need more help with?

R3) Write your own procedure for factoring the polynomials covered in this section.

Putting It All Together

What is your objective?	How can you accomplish the objective?
1 Learn Strategies for Factoring a Given Polynomial	• Be sure that you can apply the techniques you have learned in the previous sections. • Know which steps to take based on how many terms are in the polynomial. • Complete the given example on your own. • Complete You Try 1.

W Work Read the explanations, follow the examples, take notes, and complete the You Trys.

1 Learn Strategies for Factoring a Given Polynomial

W Hint

Look back in the chapter and make a list of the different types of factoring problems we have seen.

In this chapter, we have discussed several different types of factoring problems. We have practiced the factoring methods separately in each section, but how do we know which factoring method to use if we are given many different types of polynomials together? We will discuss some strategies in this section. First, recall the steps for factoring *any* polynomial:

> **Summary** To Factor a Polynomial
>
> 1) *Always* begin by asking yourself, *"Can I factor out a GCF?"* If so, factor it out.
> 2) Look at the expression to decide whether it will factor further. Apply the appropriate method to factor. If there are
>   a) *two terms,* see whether it is a difference of two squares, as in Section 7.4.
>   b) *three terms,* see whether it can be factored using the methods of Section 7.2 or Section 7.3, *or* determine whether it is a perfect square trinomial (Section 7.4).
>   c) *four terms,* see whether it can be factored by grouping as in Section 7.1.
> 3) After factoring, *always* look carefully at the result and ask yourself, *"Can I factor it again?"* If so, factor again.

Let's learn how to decide which factoring method should be used to factor a particular polynomial.

EXAMPLE 1

In-Class Example 1

Factor completely.
a) $45m^2 - 20n^2$
b) $w^2 - 10w - 24$
c) $xy^2 - x + 7y^2 - 7$
d) $b^2 - 8b + 16$
e) $8r^2 + 14r + 6$
f) $h^2 + 49$

Factor completely.

a) $8x^2 - 50y^2$ b) $t^2 - t - 56$ c) $a^2b - 9b + 4a^2 - 36$

d) $k^2 - 12k + 36$ e) $15p^2 + 51p + 18$ f) $c^2 + 4$

Solution

a) *"Can I factor out a GCF?"* is the first thing you should ask yourself. Yes. Factor out 2.

$$8x^2 - 50y^2 = 2(4x^2 - 25y^2)$$

Hint

After each step, what should you ask yourself?

Ask yourself, *"Can I factor again?"* Examine $4x^2 - 25y^2$. It has two terms that are being subtracted, and each term is a perfect square. $4x^2 - 25y^2$ is the difference of squares.

$$4x^2 - 25y^2 = (2x + 5y)(2x - 5y)$$
$$\qquad\qquad\quad (2x)^2 \quad (5y)^2$$

$$8x^2 - 50y^2 = 2(4x^2 - 25y^2) = 2(2x + 5y)(2x - 5y)$$

"Can I factor again?" No. It is completely factored.

b) Look at $t^2 - t - 56$. *"Can I factor out a GCF?"* No. Think of two numbers whose *product* is -56 and *sum* is -1. The numbers are -8 and 7.

$$t^2 - t - 56 = (t - 8)(t + 7)$$

"Can I factor again?" No. It is completely factored.

c) We have to factor $a^2b - 9b + 4a^2 - 36$. *"Can I factor out a GCF?"* No. Notice that this polynomial has *four terms*. When a polynomial has *four terms*, think about *factoring by grouping*.

$$\underbrace{a^2b - 9b}\ +\ \underbrace{4a^2 - 36}$$

$= b(a^2 - 9) + 4(a^2 - 9)$ Take out the common factor from each pair of terms.
$= (a^2 - 9)(b + 4)$ Factor out $(a^2 - 9)$ using the distributive property.

Examine $(a^2 - 9)(b + 4)$ and ask yourself, *"Can I factor again?"* Yes! $(a^2 - 9)$ is the difference of two squares. Factor again.

$$(a^2 - 9)(b + 4) = (a + 3)(a - 3)(b + 4)$$

"Can I factor again?" No. So, $a^2b - 9b + 4a^2 - 36 = (a + 3)(a - 3)(b + 4)$.

Note

Seeing four terms is a clue to try factoring by grouping.

d) We cannot take out a GCF from $k^2 - 12k + 36$. It is a trinomial, and notice that the first and last terms are perfect squares. *Is this a perfect square trinomial?*

$$k^2 - 12k + 36$$
$$(k)^2 \qquad\qquad (6)^2$$

Does the middle term equal $2 \cdot k \cdot 6$? Yes. $2 \cdot k \cdot 6 = 12k$

Use $a^2 - 2ab + b^2 = (a - b)^2$ with $a = k$ and $b = 6$.

Then, $k^2 - 12k + 36 = (k - 6)^2$.

"Can I factor again?" No. It is completely factored.

e) It is tempting to jump right in and try to factor $15p^2 + 51p + 18$ as the product of two binomials, but ask yourself, *"Can I take out a GCF?"* Yes! Factor out 3.

$$15p^2 + 51p + 18 = 3(5p^2 + 17p + 6)$$

"Can I factor again?" Yes.

$$3(5p^2 + 17p + 6) = 3(5p + 2)(p + 3)$$

"Can I factor again?" No. So, $15p^2 + 51p + 18 = 3(5p + 2)(p + 3)$.

f) Look at $c^2 + 4$ and ask yourself, *"Can I factor out a GCF?"* No.

The binomial $c^2 + 4$ is the sum of two squares, so it does not factor. This polynomial is prime.

[YOU TRY 1] Factor completely.

a) $6a^2 + 27a - 54$ b) $5h^3 - h^2 + 15h - 3$ c) $d^2 - 11d + 24$

d) $8 - 8t^4$ e) $4m^3 + 9$ f) $w^2 + 22w + 121$

ANSWERS TO [YOU TRY] EXERCISES

1) a) $3(2a - 3)(a + 6)$ b) $(h^2 + 3)(5h - 1)$ c) $(d - 3)(d - 8)$
d) $8(1 + t^2)(1 + t)(1 - t)$ e) prime   f) $(w + 11)^2$

Putting It All Together Exercises

 **Evaluate**   Do the exercises, and check your work.

*Additional answers can be found in the Answers to Exercises appendix.

Objective 1: Learn Strategies for Factoring a Given Polynomial

Factor completely.

1) $c^2 + 15c + 56$
$(c + 8)(c + 7)$

2) $r^2 - 100$
$(r + 10)(r - 10)$

3) $uv + 6u + 9v + 54$
$(u + 9)(v + 6)$

4) $5t^2 - 36t - 32$
$(5t + 4)(t - 8)$

5) $2p^2 - 13p + 21$
$(2p - 7)(p - 3)$

6) $h^2 - 22h + 121$
$(h - 11)^2$

7) $9v^5 + 90v^4 - 54v^3$
$9v^3(v^2 + 10v - 6)$

8) $m^2 + 6mn - 40n^2$
$(m + 10n)(m - 4n)$

9) $24q^3 + 52q^2 - 32q$
$4q(3q + 8)(2q - 1)$

10) $xy - x - 9y + 9$
$(x - 9)(y - 1)$

11) $144 - w^2$ $(12 + w)(12 - w)$

12) $z^2 - 11z + 42$ prime

13) $9r^2 + 12rt + 4t^2$ $(3r + 2t)^2$

14) $40b - 35$ $5(8b - 7)$

15) $7n^4 - 63n^3 - 70n^2$
$7n^2(n - 10)(n + 1)$

16) $4x^2 + 4x - 15$
$(2x - 3)(2x + 5)$

17) $9h^2 + 25$ prime

18) $4abc - 24ab + 12ac - 72a$ $4a(b + 3)(c - 6)$

19) $49c^2 + 56c + 16$ $(7c + 4)^2$

20) $m^2 - \dfrac{1}{100}$ $\left(m + \dfrac{1}{10}\right)\left(m - \dfrac{1}{10}\right)$

21) $p^2 + 10p + 14$ prime

22) $20x^2y + 6 - 24x^2 - 5y$
$(5y - 6)(2x + 1)(2x - 1)$

23) $100a^5b - 36ab^3$
$4ab(5a^2 + 3b)(5a^2 - 3b)$

24) $p^2 + 17pq + 30q^2$
$(p + 15q)(p + 2q)$

25) $t^2 - 2t - 16$ prime

26) $12g^4h^3 + 54g^3h + 30g^2h$ $6g^2h(2g^2h^2 + 9g + 5)$

27) $50n^2 - 40n + 8$ $2(5n - 2)^2$

28) $8a^2 - a - 9$ $(8a - 9)(a + 1)$

29) $36r^2 + 57rs + 21s^2$
$3(12r + 7s)(r + s)$

30) $t^2 - \dfrac{81}{169}$ $\left(t + \dfrac{9}{13}\right)\left(t - \dfrac{9}{13}\right)$

31) $81x^4 - y^4$
$(9x^2 + y^2)(3x + y)(3x - y)$

32) $v^2 - 23v + 132$
$(v - 12)(v - 11)$

33) $2a^2 - 10a - 72$
$2(a - 9)(a + 4)$

34) $p^2q - q - 6p^2 + 6$
$(p + 1)(p - 1)(q - 6)$

35) $h^2 - \dfrac{2}{5}h + \dfrac{1}{25}$ $\left(h - \dfrac{1}{5}\right)^2$

36) $16uv + 24u - 10v - 15$ $(8u - 5)(2v + 3)$

37) $-27r^3 + 144r^2 - 180r$ $-9r(3r - 10)(r - 2)$

38) $12b^2 + 36b + 27$
$3(2b + 3)^2$

39) $8b^2 - 14b - 15$
$(4b + 3)(2b - 5)$

40) $8y^4z^3 - 28y^3z^3 - 40y^3z^2 + 4y^2z^2$
$4y^2z^2(2y^2z - 7yz - 10y + 1)$

41) $49 - p^2$ $(7 + p)(7 - p)$

42) $2a^2 - 7a + 8$ prime

43) $6h^3k + 54h^2k^2 + 48hk^3$ $6hk(h + k)(h + 8k)$

44) $16u^2 + 40uv + 25v^2$
$(4u + 5v)^2$

45) $b^4 - 16$
$(b^2 + 4)(b + 2)(b - 2)$

46) $24k^2 + 31k - 15$
$(8k - 3)(3k + 5)$

47) $r^2 + 81$ prime

48) $36w^6 - 84w^4 + 12w^3$ $12w^3(3w^3 - 7w + 1)$

49) $ab - a - b + 1$
$(a - 1)(b - 1)$

50) $d^2 + 16d + 64$ $(d + 8)^2$

51) $7h^2 - 7$ $7(h + 1)(h - 1)$

52) $9p^2 - 18pq + 8p^2$
$(3p - 4q)(3p - 2q)$

53) $6m^2 - 60m + 150$
$6(m - 5)^2$

54) $100x^4 + 49y^2$ prime

55) $121z^2 - 169$
$(11z + 13)(11z - 13)$

56) $-12r^2 - 75r - 18$
$-3(4r + 1)(r + 6)$

57) $9c^2 + 54c + 72$
$9(c + 4)(c + 2)$

58) $16t^2 + 8t + 1$ $(4t + 1)^2$

59) $81u^4 - v^4$ $(9u^2 + v^2)(3u + v)(3u - v)$

60) $14v^3 + 12u^2 + 28uv^2 + 6uv$ $2(3u + 7v^2)(2u + v)$

61) $13h^2 + 15h + 2$
$(13h + 2)(h + 1)$

62) $2g^3 - 2g^2 - 112g$
$2g(g - 8)(g + 7)$

63) $5t^7 - 8t^4$
$t^4(5t^3 - 8)$

64) $m^2 - \dfrac{144}{25}$ $\left(m + \dfrac{12}{5}\right)\left(m - \dfrac{12}{5}\right)$

65) $d^2 - 7d - 30$
$(d - 10)(d + 3)$

66) $25k^2 - 60k + 36$
$(5k - 6)^2$

67) $z^2 + 144$ prime

68) $r^2 + 2r + 1$ $(r + 1)^2$

69) $b^2 - 19b + 84$
$(b - 12)(b - 7)$

70) $9y^4 - 81y^2$
$9y^2(y + 3)(y - 3)$

Extend the concepts of Sections 7.1–7.4 to factor completely.

71) $(2z + 1)y^2 + 6(2z + 1)y - 55(2z + 1)$
$(2z + 1)(y + 11)(y - 5)$

72) $(2k - 7)h^2 - 4(2k - 7)h - 45(2k - 7)$
$(2k - 7)(h + 5)(h - 9)$

73) $(t - 3)^2 + 3(t - 3) - 4$ $(t + 1)(t - 4)$

74) $(v + 8)^2 - 14(v + 8) + 48$ $v(v + 2)$

75) $(z + 7)^2 - 11(z + 7) + 28$ $z(z + 3)$

76) $(3n - 1)^2 - (3n - 1) - 72$ $(3n + 7)(3n - 10)$

77) $(a + b)^2 - (a - b)^2$ $4ab$

78) $(x + y)^2 - (x + 3y)^2$ $-4y(x + 2y)$

79) $(5p - 2q)^2 - (2p + q)^2$ $3(7p - q)(p - q)$

80) $(4s + t)^2 - (3s - 2t)^2$ $(7s - t)(s + 3t)$

81) $a^2 - 8a + 16 - b^2$ $(a + b - 4)(a - b - 4)$

82) $x^2 + 6x + 9 - y^2$ $(x + y + 3)(x - y + 3)$

83) $s^2 + 18s + 81 - t^2$ $(s + t + 9)(s - t + 9)$

84) $m^2 - 2m + 1 - n^2$ $(m + n - 1)(m - n - 1)$

R Rethink

R1) Some people think of factoring as solving a puzzle. Why do you think they feel that way?

R2) Have the procedures and tips you have learned help take most of the "guessing" out of factoring polynomials?

R3) Which critical thinking skills helped you factor polynomials?

7.5 Solving Quadratic Equations by Factoring

P Prepare

O Organize

What are your objectives for Section 7.5?	How can you accomplish each objective?
1 Solve a Quadratic Equation of the Form $ab = 0$	• Learn the definition of a *quadratic equation* and the *zero product rule*. • Complete the given example on your own. • Complete You Try 1.
2 Solve Quadratic Equations by Factoring	• Follow Example 2, and then write the procedure for **Solving a Quadratic Equation by Factoring** in your own words. • Notice that there is sometimes more than one way to solve a quadratic equation. • Complete the given examples on your own. • Complete You Trys 2 and 3.
3 Solve Higher-Degree Equations by Factoring	• A higher-degree equation may already be presented in $ab = 0$ form, but you may need to factor again before solving. • Complete the given example on your own. • Complete You Try 4.

Read the explanations, follow the examples, take notes, and complete the You Trys.

Earlier, we learned that a *linear equation in one variable* is an equation that can be written in the form $ax + b = 0$, where a and b are real numbers and $a \neq 0$. In this section, we will learn how to solve *quadratic equations*.

Definition

A **quadratic equation** can be written in the form $ax^2 + bx + c = 0$ where a, b, and c are real numbers and $a \neq 0$.

We say that a quadratic equation written in the form $ax^2 + bx + c = 0$ is in **standard form.** But quadratic equations can be written in other forms, too.

Some examples of quadratic equations are

$$x^2 + 13x + 36 = 0, \quad 5n(n - 3) = 0, \quad \text{and} \quad (z + 4)(z - 7) = -10.$$

Quadratic equations are also called *second-degree equations* because the highest power on the variable is 2.

There are many different ways to solve quadratic equations. In this section, we will learn how to solve them by factoring; other methods will be discussed later in this book.

Solving a quadratic equation by factoring is based on the *zero product rule:* If the product of two quantities is zero, then one or both of the quantities is zero.

For example, if $5y = 0$, then $y = 0$. If $p \cdot 4 = 0$, then $p = 0$. If $ab = 0$, then either $a = 0$, $b = 0$, or *both a and b* equal zero.

Definition

Zero product rule: If $ab = 0$, then $a = 0$ or $b = 0$.

We will use this idea to solve quadratic equations by factoring.

1 Solve a Quadratic Equation of the Form $ab = 0$

EXAMPLE 1

In-Class Example 1

Solve.
a) $y(y + 10) = 0$
b) $(4t + 1)(t - 5) = 0$

Answer: a) $\{0, 10\}$
b) $\left\{ -\dfrac{1}{4}, 5 \right\}$

W Hint

Where have you seen this last step before? Solving linear equations!

Solve.   a) $p(p + 4) = 0$ b) $(3x + 1)(x - 7) = 0$

Solution

a)   The zero product rule says that at least one of the factors on the left must equal zero in order for the *product* to equal zero.

$$p(p + 4) = 0$$

$$p = 0 \quad \text{or} \quad p + 4 = 0 \qquad \text{Set each factor equal to 0.}$$
$$p = -4 \qquad \text{Solve.}$$

Check the solutions in the original equation:

If $p = 0$:   If $p = -4$:
$$0(0 + 4) \overset{?}{=} 0 \qquad\qquad -4(-4 + 4) \overset{?}{=} 0$$
$$0(4) = 0 \ \checkmark \qquad\qquad\quad -4(0) = 0 \ \checkmark$$

The solution set is $\{-4, 0\}$.

b) At least one of the factors on the left must equal zero for the *product* to equal zero.

$$(3x + 1)(x - 7) = 0$$

$$3x + 1 = 0 \qquad x - 7 = 0 \qquad \text{Set each factor equal to 0.}$$
$$3x = -1 \qquad \text{or}$$
$$x = -\frac{1}{3} \qquad\qquad x = 7 \qquad \text{Solve each equation.}$$

Check in the original equation:

If $x = -\frac{1}{3}$:

$$\left[3\left(-\frac{1}{3}\right) + 1\right]\left(-\frac{1}{3} - 7\right) \overset{?}{=} 0$$

$$(-1 + 1)\left(-\frac{22}{3}\right) \overset{?}{=} 0$$

$$0\left(-\frac{22}{3}\right) = 0 \checkmark$$

If $x = 7$:

$$[3(7) + 1](7 - 7) \overset{?}{=} 0$$

$$22(0) = 0 \checkmark$$

The solution set is $\left\{-\frac{1}{3}, 7\right\}$.

YOU TRY 1

Solve

a) $k(k + 2) = 0$

b) $(2r - 3)(r + 6) = 0$

---

2 Solve Quadratic Equations by Factoring

If the equation is in standard form, $ax^2 + bx + c = 0$, begin by factoring the expression.

EXAMPLE 2

Solve $y^2 - 6y - 16 = 0$.

Solution

$$y^2 - 6y - 16 = 0$$
$$(y - 8)(y + 2) = 0 \qquad \text{Factor.}$$

$$y - 8 = 0 \quad \text{or} \quad y + 2 = 0 \qquad \text{Set each factor equal to zero.}$$
$$y = 8 \quad \text{or} \qquad y = -2 \qquad \text{Solve.}$$

Check in the original equation:

If $y = 8$:
$$(8)^2 - 6(8) - 16 \overset{?}{=} 0$$
$$64 - 48 - 16 = 0 \checkmark$$

If $y = -2$:
$$(-2)^2 - 6(-2) - 16 \overset{?}{=} 0$$
$$4 + 12 - 16 = 0 \checkmark$$

The solution set is $\{-2, 8\}$.

Here are the steps to use to solve a quadratic equation by factoring:

Procedure Solving a Quadratic Equation by Factoring

1) Write the equation in the form $ax^2 + bx + c = 0$ (standard form) so that all terms are on one side of the equal sign and zero is on the other side.
2) Factor the expression.
3) Set each factor equal to zero, and solve for the variable. (Use the zero product rule.)
4) Check the answer(s) in the original equation.

[YOU TRY 2] Solve $r^2 + 7r + 6 = 0$.

EXAMPLE 3

In-Class Example 3

Solve each equation by factoring.
a) $3d^2 - 13d = 10$
b) $4p^2 = -40p$
c) $r^2 = 8(r + 6)$
d) $(a - 5)(a - 2) = 18$

Answer: a) $\left\{ -\dfrac{2}{3}, 5 \right\}$
b) $\{-10, 0\}$ c) $\{-4, 12\}$
d) $\{-1, 8\}$

Solve each equation by factoring.

a) $2t^2 + 7t = 15$ b) $9v^2 = -54v$ c) $h^2 = -5(2h + 5)$

d) $(w - 4)(w - 5) = 2$

Solution

a) Begin by writing $2t^2 + 7t = 15$ in standard form, $at^2 + bt + c = 0$.

$$2t^2 + 7t - 15 = 0 \qquad \text{Standard form}$$
$$(2t - 3)(t + 5) = 0 \qquad \text{Factor.}$$

$$2t - 3 = 0 \quad \text{or} \quad t + 5 = 0 \qquad \text{Set each factor equal to zero.}$$
$$2t = 3$$
$$t = \frac{3}{2} \quad \text{or} \qquad t = -5 \qquad \text{Solve.}$$

Check in the original equation:

If $t = \dfrac{3}{2}$:

$$2\left(\frac{3}{2}\right)^2 + 7\left(\frac{3}{2}\right) \overset{?}{=} 15$$

$$2\left(\frac{9}{4}\right) + \frac{21}{2} \overset{?}{=} 15$$

$$\frac{9}{2} + \frac{21}{2} \overset{?}{=} 15$$

$$\frac{30}{2} = 15 \ \checkmark$$

If $t = -5$:
$$2(-5)^2 + 7(-5) \overset{?}{=} 15$$
$$2(25) - 35 \overset{?}{=} 15$$
$$50 - 35 = 15 \ \checkmark$$

The solution set is $\left\{ -5, \dfrac{3}{2} \right\}$.

b) Write $9v^2 = -54v$ in standard form.

$$9v^2 + 54v = 0 \qquad \text{Standard form}$$
$$9v(v + 6) = 0 \qquad \text{Factor.}$$

$$9v = 0 \quad \text{or} \quad v + 6 = 0 \qquad \text{Set each factor equal to zero.}$$
$$v = 0 \quad \text{or} \qquad v = -6 \qquad \text{Solve.}$$

Check. The solution set is $\{-6, 0\}$.

Since both terms in $9v^2 = -54v$ are divisible by 9, we could have started part b) by dividing by 9:

$$\frac{9v^2}{9} = \frac{-54v}{9} \qquad \text{Divide by 9.}$$
$$v^2 = -6v$$
$$v^2 + 6v = 0 \qquad \text{Write in standard form.}$$
$$v(v + 6) = 0 \qquad \text{Factor.}$$

$$v = 0 \quad or \quad v + 6 = 0 \qquad \text{Set each factor equal to zero.}$$
$$d = -6 \qquad \text{Solve.}$$

The solution set is $\{-6, 0\}$. We get the same result.

BE CAREFUL We cannot divide by v even though each term contains a factor of v. Doing so would eliminate the solution of zero. *In general, we can divide an equation by a nonzero real number but we cannot divide an equation by a variable because we may eliminate a solution, and we may be dividing by zero.*

c) To solve $h^2 = -5(2h + 5)$, begin by writing the equation in standard form.

$$h^2 = -10h - 25 \qquad \text{Distribute.}$$
$$h^2 + 10h + 25 = 0 \qquad \text{Write in standard form.}$$
$$(h + 5)^2 = 0 \qquad \text{Factor.}$$

Since $(h + 5)^2 = 0$ means $(h + 5)(h + 5) = 0$, setting each factor equal to zero will result in the same value for h.

$$h + 5 = 0 \qquad \text{Set } h + 5 = 0.$$
$$h = -5 \qquad \text{Solve.}$$

Check. The solution set is $\{-5\}$.

BE CAREFUL d) It is tempting to solve $(w - 4)(w - 5) = 2$ like this:

$$(w - 4)(w - 5) = 2$$

$$w - 4 = 2 \quad or \quad w - 5 = 2 \qquad \text{This is incorrect!}$$

One side of the equation must equal zero in order to use the zero product rule. Begin by multiplying on the left.

$$(w - 4)(w - 5) = 2$$
$$w^2 - 9w + 20 = 2 \qquad \text{Multiply using FOIL.}$$
$$w^2 - 9w + 18 = 0 \qquad \text{Standard form}$$
$$(w - 6)(w - 3) = 0 \qquad \text{Factor.}$$

$$w - 6 = 0 \quad or \quad w - 3 = 0 \qquad \text{Set each factor equal to zero.}$$
$$w = 6 \quad or \qquad w = 3 \qquad \text{Solve.}$$

The check is left to the student. The solution set is $\{3, 6\}$.

W Hint
Complete *all* the You Try problems. Ask for help if you get stuck.

YOU TRY 3   Solve.

a) $5c^2 = 6c + 8$ b) $3q^2 = -18q$ c) $(m - 5)(m - 10) = 6$

d) $z(z + 3) = 40$

3 Solve Higher-Degree Equations by Factoring

Sometimes, equations that are not quadratics can be solved by factoring as well.

EXAMPLE 4

In-Class Example 4

Solve.
a) $(7t + 2)(t^2 - 15t + 54) = 0$
b) $9y^3 - 900y = 0$

Answer: a) $\left\{-\dfrac{2}{7}, 6, 9\right\}$

b) $\{0, -10, 10\}$

Solve each equation.

a) $(4x - 1)(x^2 - 8x - 20) = 0$ b) $12n^3 - 108n = 0$

Solution

a) This is *not* a quadratic equation because if we multiplied the factors on the left we would get $4x^3 - 33x^2 - 72x + 20 = 0$. This is a *cubic* equation because the degree of the polynomial on the left is 3.

The original equation contains the product of two factors so we can use the zero product rule.

$$(4x - 1)(x^2 - 8x - 20) = 0$$
$$(4x - 1)(x - 10)(x + 2) = 0 \qquad \text{Factor.}$$

$4x - 1 = 0$ or $x - 10 = 0$ or $x + 2 = 0$ Set each factor equal to zero.
$\quad\; 4x = 1$

$\qquad x = \dfrac{1}{4}$ or $\qquad x = 10$ or $\qquad x = -2$ Solve.

The check is left to the student. The solution set is $\left\{-2, \dfrac{1}{4}, 10\right\}$.

b) The GCF of the terms in the equation $12n^3 - 108n = 0$ is $12n$. Remember, however, that *we can divide an equation by a constant but we cannot divide an equation by a variable.* Dividing by a variable may eliminate a solution and may mean we are dividing by zero. So let's begin by dividing each term by 12.

$$\frac{12n^3}{12} - \frac{108n}{12} = \frac{0}{12} \qquad \text{Divide by 12.}$$
$$n^3 - 9n = 0 \qquad \text{Simplify.}$$
$$n(n^2 - 9) = 0 \qquad \text{Factor out } n.$$
$$n(n + 3)(n - 3) = 0 \qquad \text{Factor } n^2 - 9.$$

$n = 0$ or $n + 3 = 0$ or $n - 3 = 0$ Set each factor equal to zero.
$\qquad\qquad\quad n = -3 \qquad\quad n = 3$ Solve.

W Hint

What do you think the "factored out variable" will always be equal to?

Check. The solution set is $\{0, -3, 3\}$.

YOU TRY 4   Solve.

a) $(5y + 3)(y^2 - 10y + 21) = 0$ b) $8k^3 - 32k = 0$

In this section, it was possible to solve all the equations by factoring. Below we show the relationship between solving a quadratic equation by factoring and solving it using a graphing calculator. In Chapter 10, we will learn other methods for solving quadratic equations.

Using Technology

Solve $x^2 - x - 6 = 0$ using a graphing calculator.

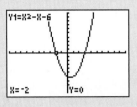

Recall from Chapter 4 that to find the x-intercepts of the graph of an equation, we let $y = 0$ and solve the equation for x. If we let $y = x^2 - x - 6$, then solving $x^2 - x - 6 = 0$ is the same as finding the x-intercepts of the graph of $y = x^2 - x - 6$. The x-intercepts are also called zeros of the equation since they are the values of x that make $y = 0$. Enter $y = x^2 - x - 6$ into the calculator, and display the graph using the standard viewing window. We obtain a graph called a *parabola,* and we can see that it has two x-intercepts.

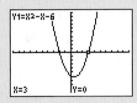

Since the scale for each tick mark on the graph is 1, it appears that the x-intercepts are -2 and 3. To verify this, press TRACE, type -2, and press ENTER as shown on the first screen. Since $x = -2$ and $y = 0$, $x = -2$ is an x-intercept. While still in "Trace" mode, type 3 and press ENTER as shown. Since $x = 3$ and $y = 0$, $x = 3$ is an x-intercept.

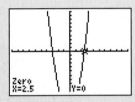

Sometimes an x-intercept is not an integer. Solve $2x^2 + x - 15 = 0$ using a graphing calculator.

Enter $2x^2 + x - 15$ into the calculator, and press GRAPH. The x-intercept on the right side of the graph is between two tick marks, so it is not an integer. To find this x-intercept press 2nd TRACE and select 2: zero. Now move the cursor to the left of one of the intercepts and press ENTER, then move the cursor again, so that it is to the right of the same intercept and press ENTER. Press ENTER one more time, and the calculator will reveal the intercept and, therefore, one solution to the equation as shown on the third screen.

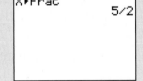

Press 2nd MODE to return to the home screen. Press ✕, T, θ, n MATH ENTER ENTER to display the x-intercept in fraction form: $x = \dfrac{5}{2}$, as shown on the final screen. Since the other x-intercept appears to be -3, press TRACE -3 ENTER to reveal that $x = -3$ and $y = 0$.

Solve using a graphing calculator.

1) $x^2 - 5x - 6 = 0$
2) $2x^2 - 9x - 5 = 0$
3) $x^2 + 4x - 21 = 0$
4) $5x^2 - 12x + 4 = 0$
5) $x^2 + 2x - 35 = 0$
6) $2x^2 - 11x + 12 = 0$

W Hint

Did you check all your answers? If necessary, get help before starting the exercise set!

ANSWERS TO [YOU TRY] EXERCISES

1) a) $\{-2, 0\}$ b) $\left\{-6, \frac{3}{2}\right\}$ 2) $\{-6, -1\}$ 3) a) $\left\{-\frac{4}{5}, 2\right\}$ b) $\{-6, 0\}$

c) $\{4, 11\}$ d) $\{-8, 5\}$ 4) a) $\left\{-\frac{3}{5}, 3, 7\right\}$ b) $\{0, -2, 2\}$

ANSWERS TO TECHNOLOGY EXERCISES

1) $\{-1, 6\}$ 2) $\left\{-\frac{1}{2}, 5\right\}$ 3) $\{-7, 3\}$ 4) $\left\{\frac{2}{5}, 2\right\}$ 5) $\{5, -7\}$ 6) $\left\{4, \frac{3}{2}\right\}$

E Evaluate 7.5 Exercises   Do the exercises, and check your work.

*Additional answers can be found in the Answers to Exercises appendix.

Objective 1: Solve a Quadratic Equation of the Form $ab = 0$

1) What is the standard form of an equation that is quadratic in x? $ax^2 + bx + c = 0$

2) A quadratic equation is also called a __second__-degree equation.

3) Identify the following equations as linear or quadratic.

   a) $5x^2 + 3x - 7 = 0$ quadratic

   b) $6(p + 1) = 0$ linear

   c) $(n + 4)(n - 9) = 8$ quadratic

   d) $2w + 3(w - 5) = 4w + 9$ linear

4) Which of the following equations are quadratic?

   a) $t^3 - 6t^2 - 4t + 24 = 0$ no

   b) $2(y^2 - 7) + 3y = 6y + 1$ yes

   c) $3a(a - 11) = 0$ yes

   d) $(c + 4)(2c^2 - 5c - 3) = 0$ no

5) Explain the zero product rule.

6) When Stephanie solves $m(m - 8) = 0$, she gets a solution set of $\{8\}$. Is this correct? Why or why not?

Solve each equation.

7) $(z + 11)(z - 4) = 0$ $\{-11, 4\}$

8) $(b + 1)(b + 7) = 0$ $\{-7, -1\}$

9) $(2r - 3)(r - 10) = 0$

10) $(5k - 4)(k + 9) = 0$

11) $d(d - 12) = 0$ $\{0, 12\}$

12) $6w(w + 2) = 0$ $\{-2, 0\}$

13) $(3x + 5)^2 = 0$ $\left\{-\frac{5}{3}\right\}$

14) $(c - 14)^2 = 0$ $\{14\}$

15) $(9h + 2)(2h + 1) = 0$ $\left\{-\frac{1}{2}, -\frac{2}{9}\right\}$

16) $(6q - 5)(2q - 3) = 0$ $\left\{\frac{5}{6}, \frac{3}{2}\right\}$

17) $\left(m + \frac{1}{4}\right)\left(m - \frac{2}{5}\right) = 0$ $\left\{-\frac{1}{4}, \frac{2}{5}\right\}$

18) $\left(v + \frac{7}{3}\right)\left(v + \frac{4}{3}\right) = 0$ $\left\{-\frac{7}{3}, -\frac{4}{3}\right\}$

19) $n(n - 4.6) = 0$ $\{0, 4.6\}$

20) $g(g + 0.7) = 0$ $\{-0.7, 0\}$

Objective 2: Solve Quadratic Equations by Factoring

21) Can we solve $(k - 4)(k - 8) = 5$ by setting each factor equal to 5 like this: $k - 4 = 5$ or $k - 8 = 5$? Why or why not?   No; the product of the factors must equal zero.

22) State two ways you could begin to solve $3x^2 + 18x + 24 = 0$.   i) Divide by 3 to get $x^2 + 6x + 8 = 0$; or ii) Factor out 3 to get $3(x^2 + 6x + 8) = 0$.

Solve each equation.

23) $p^2 + 8p + 12 = 0$ $\{-6, -2\}$

24) $c^2 + 3c - 28 = 0$ $\{-7, 4\}$

25) $t^2 - t - 110 = 0$ $\{-10, 11\}$

26) $w^2 - 17w + 72 = 0$ $\{8, 9\}$

27) $3a^2 - 10a + 8 = 0$

28) $2y^2 + 7y + 5 = 0$

29) $12z^2 + z - 6 = 0$

30) $8b^2 - 18b - 5 = 0$

31) $r^2 = 60 - 7r$ $\{-12, 5\}$

32) $h^2 + 20 = 12h$ $\{2, 10\}$

33) $d^2 - 15d = -54$ $\{6, 9\}$

34) $h^2 + 17h = -66$ $\{-11, -6\}$

35) $x^2 - 64 = 0$ $\{-8, 8\}$

36) $n^2 - 144 = 0$ $\{-12, 12\}$

37) $49 = 100u^2$

38) $81 = 4a^2$

39) $22k = -10k^2 - 12$

40) $4m - 48 = -24m^2$

41) $v^2 = 4v$ $\{0, 4\}$

42) $x^2 = x$ $\{0, 1\}$

43) $(z + 3)(z + 1) = 15$ $\{-6, 2\}$

44) $(c - 10)(c - 1) = -14$ $\{3, 8\}$

45) $t(19 - t) = 84$ $\{7, 12\}$

46) $48 = w(w - 2)$ $\{-6, 8\}$

47) $6k(k + 4) + 3 = 5(k^2 - 12) + 8k$ $\{-9, -7\}$

48) $7b(b + 1) + 15 = 6(b^2 + 2) + 11b$ $\{1, 3\}$

49) $3(n^2 - 15) + 4n = 4n(n - 3) + 19$ $\{8\}$

50) $8(p^2 - 6) + 9p = 3p(3p + 7) - 13$ $\{-7, -5\}$

Objective 3: Solve Higher-Degree Equations by Factoring

51) To solve $5t^3 - 20t = 0$, Julio begins by dividing the equation by $5t$ to get $t^2 - 4 = 0$. Is this correct? Why or why not?   No. You cannot divide an equation by a variable because you may eliminate a solution and may be dividing by zero.

52) What are two possible first steps for solving $5t^3 - 20t = 0$?   i) Divide by 5 to get $t^3 - 4t = 0$, then factor out t to get $t(t^2 - 4) = 0$; or ii) Factor out $5t$ to get $5t(t^2 - 4) = 0$.

Determine whether the statements in Exercises 55 and 56 are *always, sometimes,* or *never* true.

53) The equation $ax^2 + bx + c = 0$ has two different solutions.   sometimes

54) The solutions of $x^2 + bx + c = 0$ are opposites of the factors of c that have b as their sum.   always

The following equations are not quadratic but can be solved by factoring and applying the zero product rule. Solve each equation.

55) $7w(8w - 9)(w + 6) = 0$ $\left\{-6, 0, \dfrac{9}{8}\right\}$

56) $-5q(4q - 7)(q + 3) = 0$ $\left\{0, \dfrac{7}{4}, -3\right\}$

57) $(6m + 7)(m^2 - 5m + 6) = 0$ $\left\{-\dfrac{7}{6}, 2, 3\right\}$

58) $(9c - 2)(c^2 + 9c + 8) = 0$ $\left\{-8, -1, \dfrac{2}{9}\right\}$

59) $49h = h^3$ $\{0, -7, 7\}$ 60) $r^3 = 36r$ $\{0, -6, 6\}$

61) $5w^2 + 36w = w^3$ $\{-4, 0, 9\}$ 62) $10p^2 - 25p = p^3$ $\{0, 5\}$

63) $60a = 44a^2 - 8a^3$ $\left\{0, \dfrac{5}{2}, 3\right\}$ 64) $6z^3 + 16z = -50z^2$

65) $162b^3 - 8b = 0$ $\left\{0, -\dfrac{2}{9}, \dfrac{2}{9}\right\}$ 66) $75x = 27x^3$ $\left\{-8, -\dfrac{1}{3}, 0\right\}$ $\left\{0, -\dfrac{5}{3}, \dfrac{5}{3}\right\}$

Mixed Exercises: Objectives 1–3

Solve each equation.

67) $-63 = 4y(y - 8)$ $\left\{\dfrac{7}{2}, \dfrac{9}{2}\right\}$ 68) $-84 = g(g + 19)$ $\{-12, -7\}$

69) $6d - 1 = 9d^2$ $\left\{\dfrac{1}{3}\right\}$

70) $(9p - 2)(p^2 - 10p - 11) = 0$ $\left\{-1, \dfrac{2}{9}, 11\right\}$

71) $a^2 - a = 30$ $\{-5, 6\}$ 72) $45k + 27 = 18k^2$ $\left\{-\dfrac{1}{2}, 3\right\}$

73) $48t = 3t^3$ $\{0, -4, 4\}$

74) $(c - 8)(c - 6) = -1$ $\{7\}$

75) $104r + 36 = 12r^2$ $\left\{-\dfrac{1}{3}, 9\right\}$

76) $3t(t - 5) + 14 = 5 - t(t + 3)$ $\left\{\dfrac{3}{2}\right\}$

77) $w^2 - 121 = 0$ $\{-11, 11\}$

78) $h^2 + 15h + 54 = 0$ $\{-9, -6\}$

79) $(2n - 5)(n^2 - 6n + 9) = 0$ $\left\{\dfrac{5}{3}, 3\right\}$

80) $36b^2 + 60b = 0$ $\left\{-\dfrac{5}{3}, 0\right\}$

R Rethink

R1) Good factoring skills will help you solve quadratic equations quickly. Which techniques could you improve?

R2) Explain how the sequence of the objectives in this chapter has helped you develop the skills needed to solve quadratic equations.

R3) Do you think graphs of quadratic equations will look similar based on the number of solutions?

7.6 Applications of Quadratic Equations

P Prepare

O Organize

What are your objectives for Section 7.6?	How can you accomplish each objective?
1 Solve Problems Involving Geometry	• Review the geometry formulas you learned in Section 1.3. • Notice that you are using the same **Five Steps for Solving Applied Problems!** • Rewrite the procedure, and add more details to what you will do in Step 4 now that you are using quadratic equations. • Complete the given example on your own. • Complete You Try 1.
2 Solve Problems Involving Consecutive Integers	• Use the procedure box you developed in Objective 1 to solve this type of application. • Complete the given example on your own. • Complete You Try 2.
3 Solve Problems Using the Pythagorean Theorem	• Understand and memorize the Pythagorean theorem. • Note that you will substitute the information provided into the Pythagorean theorem and then solve for the variable. • Complete the given examples on your own. • Complete You Trys 3 and 4.
4 Solve Applied Problems Using Given Quadratic Equations	• Understand what is being asked in each step of the example. • Complete the given example on your own. • Complete You Try 5.

 **Work** **Read the explanations, follow the examples, take notes, and complete the You Trys.**

In Chapters 2 and 4, we learned how to solve applied problems involving linear equations. In this section, we will learn how to solve applications involving quadratic equations. Let's begin by reviewing the five steps for solving applied problems.

> **W Hint**
> Look at the previous section to get more details on what you are doing in Step 4.

Procedure Steps for Solving Applied Problems

Step 1: **Read** the problem carefully, more than once if necessary, until you understand it. Draw a picture, if applicable. Identify what you are being asked to find.

Step 2: **Choose a variable** to represent an unknown quantity. If there are any other unknowns, define them in terms of the variable.

Step 3: **Translate** the problem from English into an equation using the chosen variable.

Step 4: **Solve** the equation.

Step 5: **Check** the answer in the original problem, and **interpret** the solution as it relates to the problem. Be sure your answer makes sense in the context of the problem.

1 Solve Problems Involving Geometry

A builder must cut a piece of tile into a right triangle. The tile will have an area of 40 in^2, and the height will be 2 in. shorter than the base. Find the base and height.

Solution

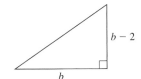

Step 1: **Read** the problem carefully. Draw a picture.

Step 2: **Choose a variable** to represent the unknown.

$$b = \text{the base}$$
$$b - 2 = \text{the height}$$

Step 3: **Translate** the information that appears in English into an algebraic equation. We are given the area of a triangular-shaped tile, so let's use the formula for the area of a triangle and substitute the expressions above for the base and the height and 40 for the area.

$$\text{Area} = \frac{1}{2}(\text{base})(\text{height})$$

$$40 = \frac{1}{2}(b)(b - 2) \qquad \text{Substitute Area} = 40, \text{ base} = b, \text{ height} = b - 2.$$

Step 4: **Solve** the equation. Eliminate the fraction first.

$$
\begin{aligned}
80 &= (b)(b - 2) & &\text{Multiply by 2.} \\
80 &= b^2 - 2b & &\text{Distribute.} \\
0 &= b^2 - 2b - 80 & &\text{Write the equation in standard form.} \\
0 &= (b - 10)(b + 8) & &\text{Factor.}
\end{aligned}
$$

$$b - 10 = 0 \quad \text{or} \quad b + 8 = 0 \qquad \text{Set each factor equal to zero.}$$
$$b = 10 \quad \text{or} \qquad b = -8 \qquad \text{Solve.}$$

Step 5: **Check** the answer, and **interpret** the solution as it relates to the problem. Since b represents the length of the base of the triangle, it cannot be a negative number. So, $b = -8$ cannot be a solution. Therefore, the length of the base is 10 in., which will make the height $10 - 2 = 8$ in. The area, then, is $\frac{1}{2}(10)(8) = 40$ in^2.

[YOU TRY 1] The height of a triangle is 3 cm more than its base. Find the height and base if its area is 35 cm^2.

---

2 Solve Problems Involving Consecutive Integers

In Chapter 2, we solved problems involving consecutive integers. Some applications involving consecutive integers lead to quadratic equations.

EXAMPLE 2

In-Class Example 2

Find three consecutive even integers such that the sum of the smaller two is one-fourth the product of the second and third integers.

Answer: 0, 2, 4 or 2, 4, 6

Twice the sum of three consecutive odd integers is five less than the product of the two larger integers. Find the numbers.

Solution

Step 1: Read the problem carefully, and identify what we are being asked to find.

We must find three consecutive odd integers.

Step 2: Choose a variable to represent an unknown, and define the other unknowns in terms of the variable.

$$x = \text{the first odd integer}$$
$$x + 2 = \text{the second odd integer}$$
$$x + 4 = \text{the third odd integer}$$

Step 3: Translate the information that appears in English into an algebraic equation. Read the problem slowly and carefully, breaking it into small parts.

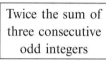

| *Statement:* | Twice the sum of three consecutive odd integers | is | 5 less than | the product of the two larger integers. |

Equation: $2[x + (x + 2) + (x + 4)] = (x + 2)(x + 4) - 5$

Step 4: Solve the equation.

$$2[x + (x + 2) + (x + 4)] = (x + 2)(x + 4) - 5$$

$2(3x + 6) = x^2 + 6x + 8 - 5$	Combine like terms; distribute.
$6x + 12 = x^2 + 6x + 3$	Combine like terms; distribute.
$0 = x^2 - 9$	Write in standard form.
$0 = (x + 3)(x - 3)$	Factor.

| $x + 3 = 0 \quad$ or $\quad x - 3 = 0$ | Set each factor equal to zero. |
| $x = -3 \qquad\qquad x = 3$ | Solve. |

Step 5: Check the answer, and **interpret** the solution as it relates to the problem.

Ⓦ Hint

Use your newly updated procedure box to solve.

We get two sets of solutions. If $x = -3$, then the other odd integers are -1 and 1. If $x = 3$, the other odd integers are 5 and 7.

Check these numbers in the original statement of the problem.

$$2[-3 + (-1) + 1] = (-1)(1) - 5 \qquad 2(3 + 5 + 7) = (5)(7) - 5$$
$$2(-3) = -1 - 5 \qquad\qquad\qquad 2(15) = 35 - 5$$
$$-6 = -6 \qquad\qquad\qquad\qquad\qquad 30 = 30$$

[YOU TRY 2] Find three consecutive odd integers such that the product of the smaller two is 15 more than four times the sum of the three integers.

3 Solve Problems Using the Pythagorean Theorem

Recall that a **right triangle** contains a 90° angle. We can label it as shown in the figure.

The side opposite the 90° angle is the longest side of the triangle and is called the **hypotenuse.** The other two

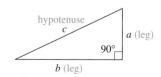

sides are called the **legs.** The Pythagorean theorem states a relationship between the lengths of the sides of a right triangle. This is a very important relationship in mathematics and is used in many different ways.

> **Definition** Pythagorean Theorem
>
> Given a right triangle with legs of length a and b and hypotenuse of length c, the **Pythagorean theorem** states that $a^2 + b^2 = c^2$ [or (leg)^2 + (leg)^2 = (hypotenuse)^2].
>
>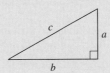
>
> The Pythagorean theorem is true *only* for right triangles.

In-Class Example 3

Find the length of the missing side.

Answer: 4

EXAMPLE 3   Find the length of the missing side.

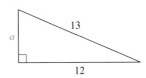

Solution

Since this is a right triangle, we can use the Pythagorean theorem to find the length of the side. Let a represent its length, and label the triangle.

The length of the hypotenuse is 13, so $c = 13$. a and 12 are legs. Let $b = 12$.

$$a^2 + b^2 = c^2 \qquad \text{Pythagorean theorem}$$
$$a^2 + (12)^2 = (13)^2 \qquad \text{Substitute values.}$$
$$a^2 + 144 = 169$$
$$a^2 - 25 = 0 \qquad \text{Write the equation in standard form.}$$
$$(a + 5)(a - 5) = 0 \qquad \text{Factor.}$$

$$a + 5 = 0 \quad \text{or} \quad a - 5 = 0 \qquad \text{Set each factor equal to 0.}$$
$$a = -5 \quad \text{or} \quad a = 5 \qquad \text{Solve.}$$

$a = -5$ does not make sense as an answer because the length of a side of a triangle cannot be negative. Therefore, $a = 5$.

Check: $5^2 + (12)^2 \stackrel{?}{=} (13)^2$
$$25 + 144 = 169 \ \checkmark$$

[**YOU TRY 3**]   Find the length of the missing side.

EXAMPLE 4

In-Class Example 4

A piece of wood is in the shape of a right triangle. One leg is 7 in. longer than the other leg, and the hypotenuse is 9 in. longer than the shortest side. Find the lengths of the sides of the piece of wood.

Answer: 8 in., 15 in., 17 in.

Ⓦ Hint

How did Example 4 compare with Example 3? What was different?

A community garden sits on a corner lot and is in the shape of a right triangle. One side is 10 ft longer than the shortest side, while the longest side is 20 ft longer than the shortest side. Find the lengths of the sides of the garden.

Solution

Step 1: **Read** the problem carefully, and identify what we are being asked to find. Draw a picture.

We must find the lengths of the sides of the garden.

Step 2: **Choose a variable** to represent an unknown, and define the other unknowns in terms of this variable. Draw and label the picture.

x = length of the shortest side (a leg)
$x + 10$ = length of the second side (a leg)
$x + 20$ = length of the third side (hypotenuse)

Step 3: **Translate** the information that appears in English into an algebraic equation. We will use the Pythagorean theorem.

$$a^2 + b^2 = c^2 \qquad \text{Pythagorean theorem}$$
$$x^2 + (x + 10)^2 = (x + 20)^2 \qquad \text{Substitute.}$$

Step 4: **Solve** the equation.

$$x^2 + (x + 10)^2 = (x + 20)^2$$
$$x^2 + x^2 + 20x + 100 = x^2 + 40x + 400 \qquad \text{Multiply using FOIL.}$$
$$2x^2 + 20x + 100 = x^2 + 40x + 400$$
$$x^2 - 20x - 300 = 0 \qquad \text{Write in standard form.}$$
$$(x - 30)(x + 10) = 0 \qquad \text{Factor.}$$

$$x - 30 = 0 \quad \text{or} \quad x + 10 = 0 \qquad \text{Set each factor equal to 0.}$$
$$x = 30 \quad \text{or} \qquad x = -10 \qquad \text{Solve.}$$

Step 5: **Check** the answer, and **interpret** the solution as it relates to the problem.

The length of the shortest side, x, cannot be a negative number, so x cannot equal -10. Therefore, the length of the shortest side must be 30 ft.

The length of the second side = $x + 10$, so $30 + 10 = 40$ ft.

The length of the longest side = $x + 20$, so $30 + 20 = 50$ ft.

Do these lengths satisfy the Pythagorean theorem? Yes.

$$a^2 + b^2 = c^2$$
$$(30)^2 + (40)^2 \stackrel{?}{=} (50)^2$$
$$900 + 1600 = 2500 \quad ✓$$

Therefore, the lengths of the sides are 30 ft, 40 ft, and 50 ft.

[YOU TRY 4] A wire is attached to the top of a pole. The pole is 2 ft shorter than the wire, and the distance from the wire on the ground to the bottom of the pole is 9 ft less than the length of the wire. Find the length of the wire and the height of the pole.

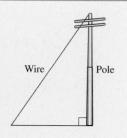

Wire　　Pole

4 Solve Applied Problems Using Given Quadratic Equations

Let's see how to use a quadratic equation to model a real-life situation.

EXAMPLE 5

In-Class Example 5

Use Example 5.

 Hint

Draw a picture to visualize what this situation might look like.

An object is launched from a platform with an initial velocity of 32 ft/sec. The height h (in feet) of the object t seconds after it is released is given by the quadratic equation

$$h = -16t^2 + 32t + 20$$

a) What is the initial height of the ball?

b) How long does it take the ball to reach a height of 32 feet?

c) How long does it take for the ball to hit the ground?

Solution

a) Since t represents the number of seconds after the ball is thrown, $t = 0$ at the time of release.

Let $t = 0$, and solve for h.

$$h = -16(0)^2 + 32(0) + 20 \qquad \text{Substitute 0 for } t.$$
$$h = 0 + 0 + 20$$
$$h = 20$$

The initial height of the ball is 20 ft.

b) We must find the *time* it takes for the ball to reach a height of 32 feet. Find t when $h = 32$.

$$h = -16t^2 + 32t + 20$$
$$32 = -16t^2 + 32t + 20 \qquad \text{Substitute 32 for } h.$$
$$0 = -16t^2 + 32t - 12 \qquad \text{Write in standard form.}$$
$$0 = 4t^2 - 8t + 3 \qquad \text{Divide by } -4.$$
$$0 = (2t - 1)(2t - 3) \qquad \text{Factor.}$$

$2t - 1 = 0 \quad \text{or} \quad 2t - 3 = 0 \qquad$ Set each factor equal to 0.

$\qquad 2t = 1 \qquad\qquad 2t = 3$

$\qquad t = \dfrac{1}{2} \quad \text{or} \qquad t = \dfrac{3}{2} \qquad$ Solve.

How can two answers be possible? After $\dfrac{1}{2}$ sec, the ball is 32 feet above the ground *on its way up,* and after $\dfrac{3}{2}$ sec, the ball is 32 feet above the ground *on its way down.*

The ball reaches a height of 32 feet after $\dfrac{1}{2}$ sec and after $\dfrac{3}{2}$ sec.

---

c) When the ball hits the ground, how high off the ground is it? *It is 0 feet high.* Find t when $h = 0$.

$$h = -16t^2 + 32t + 20$$
$$0 = -16t^2 + 32t + 20 \qquad \text{Substitute 0 for } h.$$
$$0 = 4t^2 - 8t - 5 \qquad \text{Divide by } -4.$$
$$0 = (2t + 1)(2t - 5) \qquad \text{Factor.}$$

$$2t + 1 = 0 \quad \text{or} \quad 2t - 5 = 0 \qquad \text{Set each factor equal to 0.}$$
$$2t = -1 \qquad\qquad 2t = 5$$
$$t = -\frac{1}{2} \quad \text{or} \qquad t = \frac{5}{2} \qquad \text{Solve.}$$

Since t represents time, t cannot equal $-\frac{1}{2}$. Therefore, $t = \frac{5}{2}$.

The ball will hit the ground after $\frac{5}{2}$ sec (or 2.5 sec).

[YOU TRY 5] An object is thrown upward from a building. The height h of the object (in feet) t sec after the object is released is given by the quadratic equation

$$h = -16t^2 + 36t + 36$$

a) What is the initial height of the object?

b) How long does it take the object to reach a height of 44 ft?

c) How long does it take for the object to hit the ground?

ANSWERS TO [YOU TRY] EXERCISES

1) base = 7 cm; height = 10 cm   2) 13, 15, 17 or −3, −1, 1   3) 3
4) length of wire = 17 ft; height of pole = 15 ft
5) a) 36 ft b) 0.25 sec and 2 sec c) 3 sec

E Evaluate **7.6** Exercises   Do the exercises, and check your work.

Additional answers can be found in the Answers to Exercises appendix.

Objective 1: Solve Problems Involving Geometry

Find the length and width of each rectangle.

 1) Area = 28 in²

length = 7 in.;
width = 4 in.

2) Area = 96 cm²

length = 12 cm;
width = 8 cm

Find the base and height of each triangle.

3) Area = 44 cm²

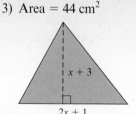

base = 11 cm;
height = 8 cm

4) Area = 12 ft²

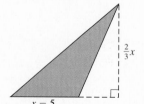

base = 4 ft;
height = 6 ft

Find the base and height of each parallelogram.

5) Area = 36 in^2

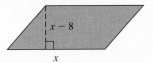

base = 9 in.;
height = 4 in.

6) Area = 240 mm^2

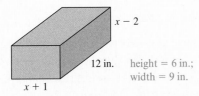

base = 20 mm;
height = 12 mm

7) The volume of the box is 648 in^3. Find its height and width.

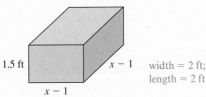

12 in.   height = 6 in.;
width = 9 in.

8) The volume of the box is 6 ft^3. Find its width and length.

1.5 ft $x - 1$ width = 2 ft;
length = 2 ft

$x - 1$

Write an equation and solve.

9) A rectangular sign is twice as long as it is wide. If its area is 8 ft^2, what are its length and width?
length = 4 ft; width = 2 ft

10) An ad in a magazine is in the shape of a rectangle and occupies 88 in^2. The length is three inches longer than the width. Find the dimensions of the ad.

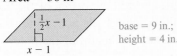

8 in. by 11 in.

11) The top of a kitchen island is a piece of granite that has an area of 15 ft^2. It is 3.5 ft longer than it is wide. Find the dimensions of the surface.   2.5 ft by 6 ft

12) To install an exhaust fan, a builder cuts a rectangular hole in the ceiling so that the width is 3 in. less than the length. The area of the hole is 180 in^2. Find the length and width of the hole.
length = 15 in.; width = 12 in.

13) A rectangular makeup case is 3 in. high and has a volume of 90 in^3. The width is 1 in. less than the length. Find the length and width of the case.
length = 6 in.; width = 5 in.

14) An artist's sketchbox is 4 in. high and shaped like a rectangular solid. The width is three-fourths as long as the length. Find the length and width of the box if its volume is 768 in^3.   length = 16 in.; width = 12 in.

15) The height of a triangle is 1 cm more than its base. Find the height and base if its area is 21 cm^2.
height = 7 cm; base = 6 cm

16) The area of a triangle is 24 in^2. Find the height and base if its height is one-third the length of the base.   height = 4 in.; base = 12 in.

Objective 2: Solve Problems Involving Consecutive Integers

Write an equation, and solve.

17) The product of two consecutive integers is 13 less than five times their sum. Find the integers.
8 and 9, or 1 and 2

18) The product of two consecutive integers is 10 less than four times their sum. Find the integers.
6 and 7, or 1 and 2

19) Find three consecutive even integers such that the product of the two smaller numbers is the same as twice the sum of all three integers.   6, 8, 10; or −2, 0, 2

20) Find three consecutive even integers such that five times the sum of the smallest and largest integers is the same as the square of the middle number.
8, 10, 12; or −2, 0, 2

21) Find three consecutive odd integers such that the product of the two larger numbers is 18 more than three times the sum of all three numbers.   7, 9, 11

22) Find three consecutive odd integers such that the square of the largest integer is 9 more than six times the sum of the two smaller numbers.   5, 7, 9; or −1, 1, 3

Objective 3: Solve Problems Using the Pythagorean Theorem

23) In your own words, explain the Pythagorean theorem.
Answers may vary.

24) Can the Pythagorean theorem be used to find a in this triangle? Why or why not?

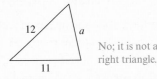

12   a

No; it is not a right triangle.

11

Use the Pythagorean theorem to find the length of the missing side.

25)

26)

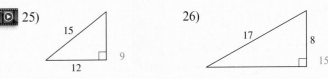

27)

12 ⟋13

5

28)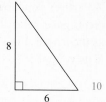

8

6 ⟍10

29)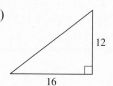

12

16
20

30)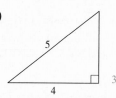

5

4
3

Find the lengths of the sides of each right triangle.

31)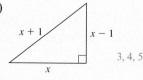

$x + 1$ $x - 1$

x

3, 4, 5

32)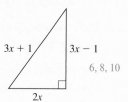

$3x + 1$ $3x - 1$

$2x$

6, 8, 10

33)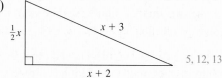

$\frac{1}{2}x$ $x + 3$

$x + 2$

5, 12, 13

34)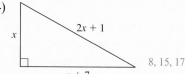

x $2x + 1$

$x + 7$

8, 15, 17

Write an equation, and solve.

35) The longer leg of a right triangle is 2 cm more than the shorter leg. The length of the hypotenuse is 4 cm more than the shorter leg. Find the length of the hypotenuse.   10 cm

36) The hypotenuse of a right triangle is 1 in. longer than the longer leg. The shorter leg measures 7 in. less than the longer leg. Find the measure of the longer leg of the triangle.   12 in.

37) A 13-ft ladder is leaning against a wall. The distance from the top of the ladder to the bottom of the wall is 7 ft more than the distance from the bottom of the ladder to the wall. Find the distance from the bottom of the ladder to the wall.   5 ft

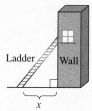

Ladder   Wall

x

38) A wire is attached to the top of a pole. The wire is 4 ft longer than the pole, and the distance from the wire on the ground to the bottom of the pole is 4 ft less than the height of the pole. Find the length of the wire and the height of the pole.

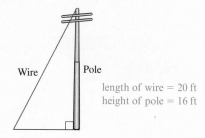

Wire   Pole

length of wire = 20 ft
height of pole = 16 ft

Write an equation, and solve.

39) Lance and Alberto leave the same location with Lance heading due north and Alberto riding due east. When Alberto has ridden 4 miles, the distance between him and Lance is 2 miles more than Lance's distance from the starting point. Find the distance between Lance and Alberto.   5 miles

40) A car heads east from an intersection while a motorcycle travels south. After 20 minutes, the car is 2 miles farther from the intersection than the motorcycle. The distance between the two vehicles is 4 miles more than the motorcycle's distance from the intersection. What is the distance between the car and the motorcycle?   10 miles

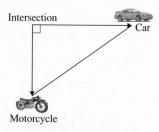

Intersection   Car

Motorcycle

Objective 4: Solve Applied Problems Using Given Quadratic Equations

Solve.

41) A rock is dropped from a cliff and into the ocean. The height h (in feet) of the rock after t sec is given by $h = -16t^2 + 144$.

a) What is the initial height of the rock?   144 ft

b) When is the rock 80 ft above the water?
after 2 sec

c) How long does it take the rock to hit the water?   3 sec

42) A Little League baseball player throws a ball upward. The height h of the ball (in feet) t seconds after the ball is released is given by $h = -16t^2 + 30t + 4$.

a) What is the initial height of the ball? 4 ft

b) When is the ball 18 feet above the ground? when $t = \frac{7}{8}$ sec and when $t = 1$ sec

c) How long does it take for the ball to hit the ground? 2 sec

Organizers of fireworks shows use quadratic and linear equations to help them design their programs. *Shells* contain the chemicals that produce the bursts we see in the sky. At a fireworks show, the shells are shot from *mortars* and when the chemicals inside the shells ignite, they explode, producing the brilliant bursts we see in the night sky.

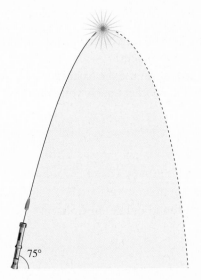

75°

43) At a fireworks show, a 3-in. shell is shot from a mortar at an angle of 75°. The height, y (in feet), of the shell t sec after being shot from the mortar is given by the quadratic equation

$$y = -16t^2 + 144t$$

and the horizontal distance of the shell from the mortar, x (in feet), is given by the linear equation

$$x = 39t$$

(http://library.thinkquest.org/15384/physics/physics.html)

a) How high is the shell after 3 sec? 288 ft

b) What is the shell's horizontal distance from the mortar after 3 sec? 117 ft

c) The maximum height is reached when the shell explodes. How high is the shell when it bursts after 4.5 sec? 324 ft

d) What is the shell's horizontal distance from its launching point when it explodes? (Round to the nearest foot.) 176 ft

44) When a 10-in. shell is shot from a mortar at an angle of 75°, the height, y (in feet), of the shell t sec after being shot from the mortar is given by

$$y = -16t^2 + 264t$$

and the horizontal distance of the shell from the mortar, x (in feet), is given by

$$x = 71t$$

a) How high is the shell after 3 sec? 648 ft

b) Find the shell's horizontal distance from the mortar after 3 sec. 213 ft

c) The shell explodes after 8.25 sec. What is its height when it bursts? 1089 ft

d) What is the shell's horizontal distance from its launching point when it explodes? (Round to the nearest foot.) 586 ft

e) Compare your answers to 43a) and 44a). What is the difference in their heights after 3 sec? 360 ft

f) Compare your answers to 43c) and 44c). What is the difference in the shells' heights when they burst? 765 ft

g) Use the information from Exercises 43 and 44. Assuming that the technicians timed the firings of the 3-in. shell and the 10-in. shell so that they exploded at the same time, how far apart would their respective mortars need to be so that the 10-in. shell would burst directly above the 3-in. shell?

45) An object is launched upward with an initial velocity of 96 ft/sec. The height h (in feet) of the object after t seconds is given by $h = -16t^2 + 96t$.

a) From what height is the object launched? 0 ft

b) When does the object reach a height of 128 ft? after 2 sec and after 4 sec

c) How high is the object after 3 sec? 144 ft

d) When does the object hit the ground? after 6 sec

46) An object is launched upward with an initial velocity of 128 ft/sec. The height h (in feet) of the object after t seconds is given by $h = -16t^2 + 128t$.

a) From what height is the object launched? 0 ft

b) Find the height of the object after 2 sec. 192 ft

c) When does the object hit the ground? after 8 sec

47) The equation $R = -9p^2 + 324p$ describes the relationship between the price of a ticket, p, in dollars, and the revenue, R, in dollars, from ticket sales at a music club.

a) Determine the club's revenue from ticket sales if the price of a ticket is $15. $2835

b) Determine the club's revenue from ticket sales if the price of a ticket is $20. $2880

c) If the club is expecting its revenue from ticket sales to be $2916, how much should it charge for each ticket? $18

48) The equation $R = -7p^2 + 700p$ describes the revenue, R, from ticket sales, in dollars, as a function of the price, p, in dollars, of a ticket to a fund-raising dinner.

a) Determine the revenue if the ticket price is $40.
$16,800

b) Determine the revenue if the ticket price is $70.
$14,700

c) If the goal of the organizers is to have ticket revenue of $17,500, how much should they charge for each ticket? $50

R **Rethink**

R1) Write a consecutive integer problem, and solve it.

R2) Where could a construction worker use geometry in his or her job? Take it a step further; where might a quadratic equation help to solve a problem?

R3) How could you use the techniques you've learned in another course you are taking?

Group Activity — The Group Activity can be found online on Connect.

Perhaps the biggest single decision you will need to make in your academic career is selecting a college major. It's not surprising, then, that many students struggle with this decision. Yet if you focus on what you value personally and in your academic life, you can pick a field of study that is right for you.

To complete the assessment below and on the next page, check off each of the characteristics that apply to you. Then use the pattern of results to determine how closely your interests and personality style match with the characteristics of others who are already in a particular field of study. Examining your responses may lead you toward some unexplored territory.

Characteristics	Does This Describe Me?	Possible Field of Study
High interest in creative expression. Appreciation of nonverbal communication. Understanding of aesthetics. Commitment to perfection. Ability to manipulate form and shape.	_____ _____ _____ _____ _____	Arts (e.g., dance, drama, music, art, creative writing)
Interest in organization and order. Ability to lead and manage people. Interest in practical problem solving. Ambition and interest in financial incentives. Can-do attitude. Ability to simplify complexity.	_____ _____ _____ _____ _____ _____	Business
Intense interest in solving real problems. "Tinkerer" mentality. Extreme ability to focus on minute details. Commitment to exactness and perfection. Strong logical ability. Ability to work alone for long stretches.	_____ _____ _____ _____ _____ _____	Engineering sciences (e.g., engineering, computer science)
Interest in people. Desire to solve real human problems. Commitment to people more than to money. Tolerance of "messy" situations with multiple, partial solutions. Insight and creativity. Ability to work with people.	_____ _____ _____ _____ _____ _____	Helping professions (e.g., nursing, counseling, teaching, many areas of medicine)
Interest in human emotions and motivations. Interest in cultural phenomena. Ability to integrate broad areas of study and inquiry. Good skills of human observation. Interest in the panorama of human life.	_____ _____ _____ _____ _____	Humanities (e.g., English literature, history, theater, film)
Interest in words, word origins, and speech. View of language as a science. View of literature as human expression. Appreciation of cultural differences as scientific phenomena.	_____ _____ _____ _____	Languages and linguistics

Characteristics	Does This Describe Me?	Possible Field of Study
Interest in physical performance.	_____	Physical education
Enjoyment of sports and athletics.	_____	
Commitment to helping others appreciate physicality.	_____	
Patience and perseverance.	_____	
Commitment to perfection through practice.	_____	
Enjoyment of research questions; high level of curiosity about natural phenomena.	_____	Physical, biological, and natural sciences (e.g., physics, astronomy, chemistry, biology, some areas of medicine)
Quantitative thinking a requirement; high comfort level with mathematics and statistics.	_____	
Minute problem-solving skills; attention at great level of detail.	_____	
Strong logical ability.	_____	
Ability to work with others.	_____	
Interest in people as individuals or groups.	_____	Social sciences (e.g., psychology, communication, sociology, education, political science, economics)
Ability to think quantitatively and qualitatively.	_____	
High comfort level with mathematics and statistics.	_____	
High level of creativity and curiosity.	_____	
Ability to work with others.	_____	
Interest in theory as much as problem solving.	_____	
Interest in the inner life.	_____	Spiritual and philosophical studies
Interest in highly theoretical questions.	_____	
Ability to think rigorously about abstract matters.	_____	
Appreciation of the human search for meaning.	_____	

After you complete the chart, consider how you can use the information. Did you learn anything new about yourself or about various courses of study? Do your responses direct you toward a particular major? Do they direct you away from any major?

Chapter 7: Summary

Definition/Procedure	Example

7.1 The Greatest Common Factor and Factoring by Grouping

To **factor a polynomial** is to write it as a product of two or more polynomials.

To factor out a greatest common factor (GCF):

1) Identify the GCF of all of the terms of the polynomial.
2) Rewrite each term as the product of the GCF and another factor.
3) Use the distributive property to factor out the GCF from the terms of the polynomial.
4) Check the answer by multiplying the factors. **(p. 391)**

Factor out the greatest common factor.

$$6k^6 - 27k^5 + 15k^4$$

The GCF is $3k^4$.

$$6k^6 - 27k^5 + 15k^4 = (3k^4)(2k^2) - (3k^4)(9k) + (3k^4)(5)$$
$$= 3k^4(2k^2 - 9k + 5)$$

Check: $3k^4(2k^2 - 9k + 5) = 6k^6 - 27k^5 + 15k^4$ ✓

The first step in factoring any polynomial is to ask yourself, *"Can I factor out a GCF?"*

The last step in factoring any polynomial is to ask yourself, *"Can I factor again?"*

Try to **factor by grouping** when you are asked to factor a polynomial containing four terms.

1) Make two groups of two terms so that each group has a common factor.
2) Take out the common factor from each group of terms.
3) Factor out the common factor using the distributive property.
4) Check the answer by multiplying the factors. **(p. 394)**

Factor $10xy + 5y - 8x - 4$ completely.

Since the four terms have a GCF of 1, we will not factor out a GCF. Begin by grouping two terms together so that each group has a common factor.

$$\underbrace{10xy + 5y}_{\downarrow} \; \underbrace{- 8x - 4}_{\downarrow}$$

$$= 5y(2x + 1) - 4(2x + 1) \quad \text{Take out the common factor.}$$
$$= (2x + 1)(5y - 4) \quad \text{Factor out } (2x + 1).$$

Check: $(2x + 1)(5y - 4) = 10xy + 5y - 8x - 4$ ✓

7.2 Factoring Trinomials of the Form $x^2 + bx + c$

Factoring $x^2 + bx + c$

If $x^2 + bx + c = (x + m)(x + n)$, then

1) if b and c are positive, then both m and n must be positive.
2) if c is positive and b is negative, then both m and n must be negative.
3) if c is negative, then one integer, m, must be positive and the other integer, n, must be negative. **(p. 400)**

Factor completely.

a) $t^2 + 9t + 20$

Think of two numbers whose *product* is 20 and whose *sum* is 9: **4 and 5.** Then,

$$t^2 + 9t + 20 = (t + 4)(t + 5)$$

b) $3s^3 - 33s^2 + 54s$

Begin by factoring out the GCF of $3s$.

$$3s^3 - 33s^2 + 54s = 3s(s^2 - 11s + 18) = 3s(s - 2)(s - 9)$$

7.3 Factoring Trinomials of the Form $ax^2 + bx + c$ ($a \neq 1$)

Factoring $ax^2 + bx + c$ by **grouping. (p. 406)**

Factor $5t^2 + 18t - 8$ completely.

$$\overset{\text{Sum is 18.}}{\underset{\downarrow}{}}$$
$$5t^2 + 18t - 8$$
$$\underset{\text{Product: } 5 \cdot (-8) = -40}{\underline{}}$$

Think of two integers whose *product* is -40 and whose *sum* is 18. **20 and -2**

Definition/Procedure	Example
	Factor by grouping. $$5t^2 + 18t - 8 = \underbrace{5t^2 + 20t}\ \underbrace{-\ 2t - 8}\quad\text{Write } 18t \text{ as } 20t - 2t.$$ $$= 5t(t + 4) - 2(t + 4)$$ $$= (t + 4)(5t - 2)$$
Factoring $ax^2 + bx + c$ by **trial and error.** When approaching a problem in this way, we must keep in mind that we are reversing the FOIL process. **(p. 407)**	Factor completely. $4x^2 - 16x + 15$ $$4x^2 - 16x + 15 = (2x - 3)(2x - 5)$$ $$-6x$$ $$+\ -10x$$ $$\overline{-16x}$$ $$4x^2 - 16x + 15 = (2x - 3)(2x - 5)$$

7.4 Factoring Special Trinomials and Binomials

A **perfect square trinomial** is a trinomial that results from squaring a binomial. **Factoring a Perfect Square Trinomial** $$a^2 + 2ab + b^2 = (a + b)^2$$ $$a^2 - 2ab + b^2 = (a - b)^2 \quad \text{(p. 413)}$$	Factor completely. a) $g^2 + 22g + 121 = (g + 11)^2$ $\quad a = g \qquad b = 11$ b) $16d^2 - 8d + 1 = (4d - 1)^2$ $\quad a = 4d \qquad b = 1$
Factoring the Difference of Two Squares $$a^2 - b^2 = (a + b)(a - b) \quad \text{(p. 415)}$$	Factor completely. $$w^2 - 64 = (w + 8)(w - 8)$$ $$(w)^2 \quad (8)^2 \quad a = w,\ b = 8$$

7.5 Solving Quadratic Equations by Factoring

A **quadratic equation** can be written in the form $ax^2 + bx + c = 0$, where a, b, and c are real numbers and $a \neq 0$. **(p. 424)**	Some examples of quadratic equations are $$5x^2 + 9 = 0, \quad y^2 = 4y + 21, \quad 4(p - 2)^2 = 8 - 7p$$
To solve a quadratic equation by factoring, use the **zero product rule:** If $ab = 0$, then $a = 0$ or $b = 0$. **(p. 424)**	Solve $(y + 7)(y - 4) = 0$ $y + 7 = 0 \quad \text{or} \quad y - 4 = 0 \qquad \text{Set each factor equal to zero.}$ $\quad y = -7 \quad \text{or} \qquad y = 4 \qquad \text{Solve.}$ The solution set is $\{-7, 4\}$.
Steps for **Solving a Quadratic Equation by Factoring** 1) Write the equation in the form $ax^2 + bx + c = 0$. 2) Factor the expression. 3) Set each factor equal to zero, and solve for the variable. 4) Check the answer(s) in the original equation. **(p. 426)**	Solve $2p^2 - 3p - 11 = -9$. $\quad 2p^2 - 3p - 2 = 0 \qquad \text{Standard form}$ $\quad (2p + 1)(p - 2) = 0 \qquad \text{Factor.}$ $2p + 1 = 0 \quad \text{or} \quad p - 2 = 0$ $\quad 2p = -1$ $\quad p = -\dfrac{1}{2} \quad \text{or} \qquad p = 2$ The solution set is $\left\{-\dfrac{1}{2}, 2\right\}$. Check the answers.

Definition/Procedure	Example

7.6 Applications of Quadratic Equations

Pythagorean Theorem

Given a right triangle with legs of length a and b and hypotenuse of length c,

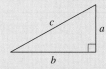

the Pythagorean theorem states that

$$a^2 + b^2 = c^2 \quad \textbf{(p. 435)}$$

Find the length of side a.

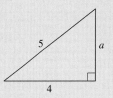

Let $b = 4$ and $c = 5$ in $\boldsymbol{a^2 + b^2 = c^2}$.

$$a^2 + (4)^2 = (5)^2$$
$$a^2 + 16 = 25$$
$$a^2 - 9 = 0$$
$$(a + 3)(a - 3) = 0$$
$$a + 3 = 0 \quad \text{or} \quad a - 3 = 0$$
$$a = 3 \quad \text{or} \quad a = -3$$

Reject -3 as a solution since the length of a side cannot be negative.

Therefore, $a = 3$.

Chapter 7:   Review Exercises

*Additional answers can be found in the Answers to Exercises appendix.

(7.1) Find the greatest common factor of each group of terms.

1) $40, 56$ 8

2) $36y, 12y^2, 54y^2$ $6y$

3) $15h^4, 45h^5, 20h^3$ $5h^3$

4) $4c^4d^3, 20c^4d^2, 28c^2d$ $4c^2d$

Factor out the greatest common factor.

5) $63t + 45$ $9(7t + 5)$

6) $21w^5 - 56w$ $7w(3w^4 - 8)$

7) $2p^6 - 20p^5 + 2p^4$
  $2p^4(p^2 - 10p + 1)$

8) $18a^3b^3 - 3a^2b^3 - 24ab^3$
  $3ab^2(6a^2b - ab - 8b)$

9) $n(m + 8) - 5(m + 8)$
  $(m + 8)(n - 5)$

10) $x(9y - 4) + w(9y - 4)$
  $(9y - 4)(x + w)$

11) Factor out $-5r$ from $-15r^3 - 40r^2 + 5r$. $-5r(3r^2 + 8r - 1)$

12) Factor out -1 from $-z^2 + 9z - 4$. $-(z^2 - 9z + 4)$

Factor by grouping.

13) $ab + 2a + 9b + 18$
  $(a + 9)(b + 2)$

14) $cd - 3c + 8d - 24$
  $(c + 8)(d - 3)$

15) $4xy - 28y - 3x + 21$
  $(x - 7)(4y - 3)$

16) $hk^2 + 6h - k^2 - 6$
  $(k^2 + 6)(h - 1)$

(7.2) Factor completely.

17) $q^2 + 10q + 24$
  $(q + 6)(q + 4)$

18) $t^2 - 12t + 27$
  $(t - 3)(t - 9)$

19) $z^2 - 6z - 72$
  $(z - 12)(z + 6)$

20) $h^2 + 6h - 7$
  $(h + 7)(h - 1)$

21) $m^2 - 13mn + 30n^2$
  $(m - 3n)(m - 10n)$

22) $a^2 + 11ab + 30b^2$
  $(a + 5b)(a + 6b)$

23) $4v^2 - 24v - 64$
  $4(v - 8)(v + 2)$

24) $7c^2 - 7c - 84$
  $7(c - 4)(c + 3)$

25) $9w^4 + 9w^3 - 18w^2$
  $9w^2(w - 1)(w + 2)$

26) $5x^3y - 25x^2y^2 + 20xy^3$
  $5xy(x - 4y)(x - y)$

(7.3) Factor completely.

27) $3r^2 - 23r + 14$
  $(3r - 2)(r - 7)$

28) $5k^2 + 11k + 6$ $(5k + 6)(k + 1)$

29) $4p^2 - 8p - 5$
  $(2p - 5)(2p + 1)$

30) $8d^2 + 29d - 12$ $(8d - 3)(d + 4)$

31) $12c^2 + 38c + 20$
  $2(3c + 2)(2c + 5)$

32) $21n^2 - 54n + 24$
  $3(7n - 4)(n - 2)$

33) $10x^2 + 39xy - 27y^2$ $(5x - 3y)(2x + 9y)$

34) $6a^2 - 19ab - 20b^2$ $(6a + 5b)(a - 4b)$

(7.4) Factor completely.

35) $w^2 - 49$ $(w + 7)(w - 7)$

36) $121 - p^2$ $(11 + p)(11 - p)$

37) $64t^2 - 25u^2$
  $(8t + 5u)(8t - 5u)$

38) $y^4 - 81$ $(y^2 + 9)(y + 3)(y - 3)$

39) $4b^2 + 9$ prime

40) $12c^2 - 48d^2$ $12(c + 2d)(c - 2d)$

41) $64x - 4x^3$
  $4x(4 + x)(4 - x)$

42) $\dfrac{25}{9} - h^2$ $\left(\dfrac{5}{3} + h\right)\left(\dfrac{5}{3} - h\right)$

43) $r^2 + 12r + 36$ $(r + 6)^2$

44) $9z^2 - 24z + 16$ $(3z - 4)^2$

45) $20k^2 - 60k + 45$
  $5(2k - 3)^2$

46) $25a^2 + 20ab + 4b^2$ $(5a + 2b)^2$

(7.1–7.4) Mixed Exercises

For Exercises 47–50, answer *always*, *sometimes*, or *never*.

47) A binomial of the form $ax^2 + bx$ will *always, sometimes,* or *never* factor into a difference of two squares.   never

48) A binomial that is the product of two binomials is *always, sometimes,* or *never* a difference of two squares.   always

49) If a, b, and c are any integer values except zero, then the terms in the trinomial $am^3 + bm^2 + cm$ will *always, sometimes,* or *never* have a greatest common factor.
always

50) If a, b, and c are any integer values except zero, then the expression $ap^2 + bp + c$ will *always, sometimes,* or *never* have a greatest common factor.   sometimes

Factor completely.

51) $10z^2 - 7z - 12$
$(5z + 4)(2z - 3)$

52) $4c^2 + 24c + 36$ $4(c + 3)^2$

53) $9k^4 - 16k^2$
$k^2(3k + 4)(3k - 4)$

54) $14m^5 + 63m^4 + 21m^3$
$7m^3(2m^2 + 9m + 3)$

55) $d^2 - 17d + 60$
$(d - 12)(d - 5)$

56) $\frac{4}{25}t^2 - \frac{1}{9}u^2$
$\left(\frac{2}{5}t + \frac{1}{3}u\right)\left(\frac{2}{5}t - \frac{1}{3}u\right)$

57) $3a^2b + a^2 - 12b - 4$
$(3b + 1)(a + 2)(a - 2)$

58) $8mn - 8m + 56n - 56$ $8(m + 7)(n - 1)$

59) $25c^2 - 20c + 4$ $(5c - 2)^2$

60) $12v^2 + 32v + 5$ $(6v + 1)(2v + 5)$

61) $p^2 + 16$ prime

62) $x^2 - 9x - 12$ prime

(7.5) Solve each equation.

63) $y(3y + 7) = 0$ $\left\{-\frac{7}{3}, 0\right\}$

64) $(2n - 3)^2 = 0$ $\left\{\frac{3}{2}\right\}$

65) $2k^2 + 18 = 13k$ $\left\{2, \frac{9}{2}\right\}$

66) $3t^2 - 75 = 0$ $\{-5, 5\}$

67) $h^2 + 17h + 72 = 0$
$\{-9, -8\}$

68) $21 = 8p^2 - 2p$ $\left\{-\frac{3}{2}, \frac{7}{4}\right\}$

69) $121 = 81r^2$ $\left\{-\frac{11}{9}, \frac{11}{9}\right\}$

70) $12c = -c^2$ $\{-12, 0\}$

71) $3m^2 - 120 = 18m$
$\{-4, 10\}$

72) $x(16 - x) = 63$ $\{7, 9\}$

73) $(w + 3)(w + 8) = -6$ $\{-6, -5\}$

74) $18 = 9b^2 + 9b$ $\{-2, 1\}$

75) $(5z + 4)(3z^2 - 7z + 4) = 0$ $\left\{-\frac{4}{5}, 1, \frac{4}{3}\right\}$

76) $6d^3 + 45d = 33d^2$ $\left\{0, \frac{5}{2}, 3\right\}$

77) $45p^3 - 20p = 0$ $\left\{0, -\frac{2}{3}, \frac{2}{3}\right\}$

78) $17v = 4(v^2 + 1)$ $\left\{\frac{1}{4}, 4\right\}$

(7.6)

79) Find the base and height if the area of the triangle is 18 cm^2.

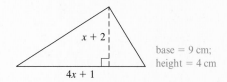
base = 9 cm;
height = 4 cm

80) Find the length and width of the rectangle if its area is 60 in^2.

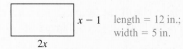
length = 12 in.;
width = 5 in.

81) Find the base and height of the parallelogram if its area is 12 ft^2.

base = 6 ft;
height = 2 ft

82) Find the height and length of the box if its volume is 480 in^3.

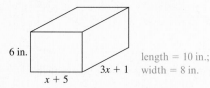
length = 10 in.;
width = 8 in.

83) Use the Pythagorean theorem to find the length of the missing side.   15

84) Find the length of the hypotenuse.   13

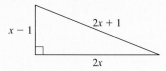

Write an equation, and solve.

85) A rectangular mirror has an area of 10 ft^2, and it is 1.5 ft longer than it is wide. Find the dimensions of the mirror.
length = 4 ft; width = 2.5 ft

86) The base of a triangular banner is 1 ft less than its height. If the area is 3 ft^2, find the base and height.
base = 2 ft; height = 3 ft

87) The sum of three consecutive integers is one less than the square of the smallest number. Find the integers.
-1, 0, 1; or 4, 5, 6

88) The product of two consecutive odd integers is 18 more than the square of the smaller number. Find the integers.
9 and 11

89) Desmond and Marcus leave an intersection with Desmond jogging north and Marcus jogging west. When Marcus is 1 mile farther from the intersection than Desmond, the distance between them is 2 miles more than Desmond's distance from the intersection. How far is Desmond from the intersection?   3 miles

90) An object is thrown upward with an initial velocity of 68 ft/sec. The height h (in feet) of the object t seconds after it is thrown is given by

$$h = -16t^2 + 68t + 60$$

a) How long does it take for the object to reach a height of 120 ft? $\frac{5}{4}$ sec and 3 sec

b) What is the initial height of the object?   60 ft

c) What is the height of the object after 2 seconds?   132 ft

d) How long does it take the object to hit the ground?
5 sec

Chapter 7:   Test

*Additional answers can be found in the Answers to Exercises appendix.

1) What is the first thing you should do when you are asked to factor a polynomial?   See whether you can factor out a GCF.

Factor completely.

2) $h^2 - 14h + 48$
   $(h - 8)(h - 6)$

3) $36 - v^2$ $(6 + v)(6 - v)$

4) $7p^2 + 6p - 16$
   $(7p - 8)(p + 2)$

5) $20a^3b^4 + 36a^2b^3 + 4ab^2$
   $4ab^2(5a^2b^2 + 9ab + 1)$

6) $k^2 + 81$ prime

7) $4z^3 + 28z^2 + 48z$
   $4z(z + 4)(z + 3)$

8) $36r^2 - 60r + 25$ $(6r - 5)^2$

9) $n^3 + 7n^2 - 4n - 28$
   $(n + 7)(n + 2)(n - 2)$

10) $x^2 - 3xy - 18y^2$
    $(x + 3y)(x - 6y)$

11) $81c^4 - d^4$
    $(9c^2 + d^2)(3c + d)(3c - d)$

12) $p^2 + 2p + 1$ $(p + 1)^2$

13) $32w^2 + 28w - 30$ $2(8w - 5)(2w + 3)$

Solve each equation.

14) $t^2 + 5t - 36 = 0$ $\{-9, 4\}$

15) $144r = r^3$ $\{0, -12, 12\}$

16) $49a^2 = 16$ $\left\{-\dfrac{4}{7}, \dfrac{4}{7}\right\}$

17) $(y - 7)(y - 5) = 3$ $\{4, 8\}$

18) $-9x = 2(3x^2 + 4) + 10x$ $\left\{-\dfrac{8}{3}, -\dfrac{1}{2}\right\}$

19) $20k^2 - 52k = 24$ $\left\{-\dfrac{2}{5}, 3\right\}$

Write an equation and solve.

20) Find the length and width of a rectangular ice cream sandwich if its volume is 9 in^3.   length = 6 in.; width = 2 in.

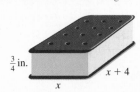

$\frac{3}{4}$ in.

$x + 4$

x

21) Find three consecutive odd integers such that the sum of the three numbers is 110 less than the product of the larger two integers.   9, 11, 13

22) Eric and Katrina leave home to go to work. Eric drives due east while his wife drives due south. At 8:30 A.M., Katrina is 3 miles farther from home than Eric, and the distance between them is 6 miles more than Eric's distance from home. Find Eric's distance from his house.   9 miles

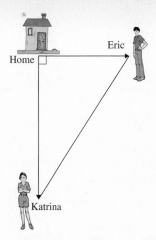

Home

Eric

Katrina

23) A rectangular cheerleading mat has an area of 252 ft^2. It is seven times longer than it is wide. Find the dimensions of the mat.   length = 42 ft; width = 6 ft

24) An object is launched upward from the ground with an initial velocity of 200 ft/sec. The height h (in feet) of the object after t seconds is given by

$$h = -16t^2 + 200t.$$

a) Find the height of the object after 1 second.   184 feet

b) Find the height of the object after 4 seconds.   544 feet

c) When is the object 400 feet above the ground? when $t = 2\dfrac{1}{2}$ sec and when $t = 10$ sec

d) How long does it take for the object to hit the ground? $12\dfrac{1}{2}$ sec

*Additional answers can be found in the Answers to Exercises appendix.

Perform the indicated operation(s) and simplify.

1) $\dfrac{2}{9} - \dfrac{5}{6} + \dfrac{1}{3}$ $-\dfrac{5}{18}$

2) $\dfrac{35}{48} \cdot \dfrac{32}{63}$ $\dfrac{10}{27}$

3) Solve $\dfrac{3}{8}t + \dfrac{1}{2} = \dfrac{1}{4}(4t + 1) - \dfrac{3}{4}t$ $\{-2\}$

4) Solve $A = \dfrac{1}{2}h(b_1 + b_2)$ for b_2. $b_2 = \dfrac{2A - hb_1}{h}$

5) Solve $5 - \dfrac{2}{3}k \le 13$. $[-12, \infty)$

6) Graph $y = -\dfrac{1}{4}x + 2$.

7) Write the equation of the line parallel to $5x - 3y = 7$ containing the point $(-5, -7)$. Express the answer in slope-intercept form. $y = \dfrac{5}{3}x + \dfrac{4}{3}$

8) Use any method to solve this system of equations.

$$4 + 2(1 - 3x) + 7y = 3x - 5y$$
$$10y + 9 = 3(2x + 5) + 2y \qquad \varnothing$$

9) Write a system of equations, and solve.
   Tickets to a high school football game cost $6.00 for an adult and $2.00 for a student. The revenue from the sale of 465 tickets was $1410. Find the number of adult tickets and number of student tickets sold.
   adults: 120; students: 345

Simplify. The answer should not contain any negative exponents.

10) $2(-3p^4q)^2$ $18p^8q^2$

11) $\dfrac{28a^8b^4}{40a^{-3}b^9}$ $\dfrac{7a^{11}}{10b^5}$

12) Write 0.0000839 in scientific notation. 8.39×10^{-5}

Multiply and simplify.

13) $(4w - 7)(2w + 3)$ $8w^2 - 2w - 21$

14) $(3n + 4)^2$ $9n^2 + 24n + 16$

15) $(6z - 5)(2z^2 + 7z - 3)$ $12z^3 + 32z^2 - 53z + 15$

16) Add $(11v^2 - 16v + 4) + (2v^2 + 3v - 9)$. $13v^2 - 13v - 5$

Divide.

17) $\dfrac{16x^3 - 57x + 14}{4x - 7}$ $4x^2 + 7x - 2$

18) $\dfrac{24m^2 - 8m + 4}{8m}$ $3m - 1 + \dfrac{1}{2m}$

Factor completely.

19) $6c^2 + 15c - 54$ $3(2c + 9)(c - 2)$

20) $r^2 + 25$ prime

21) $xy^2 + 4y^2 - x - 4$ $(x + 4)(y + 1)(y - 1)$

22) $\dfrac{1}{4} - b^2$ $\left(\dfrac{1}{2} + b\right)\left(\dfrac{1}{2} - b\right)$

23) $2h^4 + 28h^3 + 98h^2$ $2h^2(h + 7)^2$

Solve.

24) $3t^2 - 13t + 12 = 0$ $\left\{\dfrac{4}{3}, 3\right\}$

25) $24n^3 = 54n$ $\left\{0, -\dfrac{3}{2}, \dfrac{3}{2}\right\}$

Rational Expressions

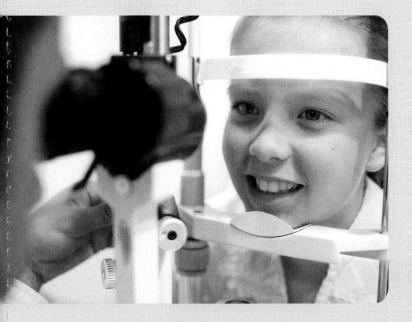

Math at Work:

Ophthalmologist

In the previous chapter, we saw how ophthalmologist Mark Diamond applies the same careful, analytical thinking he uses to solve difficult math problems to his work diagnosing and treating patients. Now let's take a look at how he uses math itself in his job. Indeed, those in the medical profession use mathematics on a daily basis to help care for their patients.

Some formulas in optics—an area of great attention for ophthalmologists—include rational expressions. If Mark determines that one of his patients needs glasses, for example, he would use the following formula to figure out the proper prescription:

$$P = \frac{1}{f}$$

where f is the focal length, in meters, and P is the power of the lens, in diopters. While computers now aid in these calculations, physicians like Mark believe that it is still important to double-check the calculations by hand.

Mark understands the value of maintaining physical fitness, and so he works hard to create a healthy lifestyle for himself and his family. Mathematics aids him in this, too. Mark regularly monitors his blood pressure and cholesterol, calculates appropriate serving sizes based on calories when he cooks, and keeps a log of the hours per week he works out at the gym.

In this chapter, we will learn how to perform operations with rational expressions and how to solve equations, like the one above, for a specific variable. We will also introduce some strategies you can use to address one of the biggest health issues in college—stress.

Coping with stress is one of the challenges that college students face. The many demands on your time can make you feel that you'll never finish what needs to get done. This pressure produces wear and tear on your body and mind, and it's easy to fall prey to ill health as a result. Further, many people find that when they graduate from college and enter the working world, they face a new set of pressures that can also be quite stressful.

However, stress and poor health are not inevitable. In fact, by following simple guidelines and deciding to make health a conscious priority, you can maintain good physical and mental health in any setting. The skills below will help you deal with stress.

- Keep fit. Being in good physical condition is the primary way to prepare for future stress. So jog, play sports, take walks, bike to campus—do whatever you can to get exercise.
- Reduce the caffeine in your diet. Coffee, soda, chocolate, and other caffeinated foods can make you feel jittery and anxious even without stress.

- Identify the causes of your stress. Once you understand why you are feeling stressed, you can take appropriate steps to reduce this feeling.

- Take charge of situations that make you feel stressed. Stress is most apt to arise when we are faced with situations in which we have little or no control. If you take control, you'll reduce the experience of stress.
- Reach out to friends. Turning to friends and family and simply talking about the stress in our lives can help us tolerate it more effectively.
- Keep perspective. We often overreact to events that ultimately have few consequences—for example, being cut off in traffic. Try to identify these events for what they are: worth forgetting.
- Relax! It may sound simple, but all of us know things we can do to unwind. Make time to do yoga, or play basketball, or go to the movies, or whatever it is that puts you in a better frame of mind.

- Consider carefully whether the steps you take to cope with stress are effective, and be open to new possibilities. If talking with friends only makes you more stressed, for example, consider learning meditation as an alternative stress reducer.

- Make peace with stress. A life without at least *some* stress would be one without challenges and without meaningful accomplishments. Think of stress as an inevitable but manageable consequence of working toward your goals.

Chapter 8 **POWER** Plan

P Prepare	**O** Organize
What are your goals for Chapter 8?	**How can you accomplish each goal? (Write in the steps you will take to succeed.)**
1 Be prepared before and during class.	• _____ • _____ • _____ • _____
2 Understand the homework to the point where you could do it without needing any help or hints.	• _____ • _____ • _____
3 Use the P.O.W.E.R. framework to learn how to cope with stress: *Determine Your Risk for Stress-Related Illness*	• _____ • _____ • _____
4 Write your own goal. _____ _____	• _____
What are your objectives for Chapter 8?	**How can you accomplish each objective?**
1 Learn how to simplify, multiply, and divide rational expressions.	• Understand the definition of a rational expression and find values that will make it undefined or equal to zero. • Use the Fundamental Property of Rational Expressions to write rational expressions in lowest terms. • Compare the process of multiplying fractions with that for multiplying rational expressions, and use the procedure for multiplying and dividing rational expressions.
2 Learn to find a least common denominator (LCD) and use it to add and subtract rational expressions.	• Compare the process of finding an LCD for a group of fractions with the procedure for finding an LCD for rational expressions. • Know how to rewrite a rational expression using the LCD. • Use the procedures for adding and subtracting rational expressions with common, different, and special denominators.
3 Learn to simplify complex fractions.	• Know how to identify a complex fraction by understanding the definition. • Use the procedure for simplifying a complex fraction with one term in the numerator and one term in the denominator. • Feel comfortable using both methods for simplifying a complex fraction with more than one term in the numerator and/or denominator.

| **4** Know how to solve rational equations and proportions. | • Understand the difference between an expression and an equation.
• Learn the procedure for solving a rational equation.
• Know how to identify and solve a proportion.
• Develop a procedure for solving an equation for a specific variable.
• Use the **Five Steps for Solving Applied Problems** to solve rational equations that involve
   • proportions.
   • distance, rate, and time.
   • work.
   • direct variation.
   • inverse variation. |
| **5** Write your own goal.
_____
_____ | • _____ |

 Work | Read Sections 8.1–8.7, and complete the exercises.

|  **Evaluate** Complete the Chapter Review and Chapter Test. How did you do? |  **Rethink**
• After reading the Study Strategy on coping with stress, what steps have you taken to help identify and deal with stress?
• This chapter covered many different types of procedures for performing operations with rational expressions and solving rational equations. How have these procedures compared with the ones you have learned for expressions and equations that contain fractions? Describe the process you are using to remember the different procedures. |

8.1 Simplifying Rational Expressions

P Prepare   **O Organize**

What are your objectives for Section 8.1?	How can you accomplish each objective?
1 Evaluate a Rational Expression	• Know the definition and be able to recognize a rational expression. • Complete the given example on your own. • Complete You Try 1.
2 Find the Values of the Variable That Make a Rational Expression Undefined or Equal to Zero	• Write down the steps you need to take to find the values of the variable that will make a rational expression equal to zero or undefined. • Complete the given example on your own. • Complete You Try 2.
3 Write a Rational Expression in Lowest Terms	• Learn the **Fundamental Property of Rational Expressions.** • Write the procedure for **Writing a Rational Expression in Lowest Terms** in your own words. • Complete the given example on your own. • Complete You Try 3.
4 Simplify a Rational Expression of the Form $\dfrac{a-b}{b-a}$	• Follow the explanation so you can understand the note that outlines the principle for this objective. • Complete the given example on your own. • Complete You Try 4.
5 Write Equivalent Forms of a Rational Expression	• Understand that there can be many different ways to write a rational expression. • Complete the given example on your own. • Complete You Try 5.

Read the explanations, follow the examples, take notes, and complete the You Trys.

1 Evaluate a Rational Expression

In Section 1.4, we defined a **rational number** as the quotient of two integers where the denominator does not equal zero. Some examples of rational numbers are

$$\frac{7}{8}, \qquad -\frac{2}{5}, \qquad 18 \left(\text{since } 18 = \frac{18}{1} \right)$$

We can define a rational expression in a similar way. A rational expression is a quotient of two polynomials provided that the denominator does not equal zero. We state the definition formally next.

W Hint

Why can't $Q = 0$?

Definition

A **rational expression** is an expression of the form $\dfrac{P}{Q}$, where P and Q are polynomials and where $Q \neq 0$.

Some examples of rational expressions are

$$\frac{4k^3}{7}, \quad \frac{2n-1}{n+6}, \quad \frac{5}{t^2-3t-28}, \quad -\frac{9x+2y}{x^2+y^2}$$

We can *evaluate* rational expressions for given values of the variables as long as the values do not make any denominators equal zero.

EXAMPLE 1

In-Class Example 1

Evaluate (if possible) for each value of x: $\frac{x^2-25}{x-1}$.

a) $x = 4$ b) $x = 5$ c) $x = 1$

Answer: a) -3 b) 0 c) undefined

Evaluate $\dfrac{x^2-9}{x+1}$ (if possible) for each value of x.

a) $x = 7$ b) $x = 3$ c) $x = -1$

Solution

a) $\dfrac{x^2-9}{x+1} = \dfrac{(7)^2-9}{(7)+1}$ Substitute 7 for x.

$\qquad = \dfrac{49-9}{7+1} = \dfrac{40}{8} = 5$

b) $\dfrac{x^2-9}{x+1} = \dfrac{(3)^2-9}{(3)+1}$ Substitute 3 for x.

$\qquad = \dfrac{9-9}{3+1} = \dfrac{0}{4} = 0$

c) $\dfrac{x^2-9}{x+1} = \dfrac{(-1)^2-9}{(-1)+1}$ Substitute -1 for x.

$\qquad = \dfrac{1-9}{0} = \dfrac{-8}{0}$

Undefined

Remember, a fraction is **undefined** when its denominator equals zero. Therefore, we say that $\dfrac{x^2-9}{x+1}$ is *undefined* when $x = -1$ since this value of x makes the denominator equal zero. So, x *cannot equal* -1 in this expression.

$\left[\text{YOU TRY 1}\right]$ Evaluate $\dfrac{k^2-1}{k+11}$ (if possible) for each value of k.

a) $k = 9$ b) $k = -2$ c) $k = -11$ d) $k = 1$

2 Find the Values of the Variable That Make a Rational Expression Undefined or Equal to Zero

Parts b) and c) in Example 1 remind us about two important aspects of fractions and rational expressions.

W Hint

Remember that zero divided by any nonzero number equals zero.

Note

1) A fraction (rational expression) *equals zero* when its numerator equals zero.

2) A fraction (rational expression) is *undefined* when its denominator equals zero.

EXAMPLE 2

In-Class Example 2

For each rational expression, for what values of the variable i) does the expression equal zero? ii) is the expression undefined?

a) $\dfrac{y-5}{y+9}$ b) $\dfrac{7r}{4}$

c) $\dfrac{2}{p^2+5p-24}$

Answer: a) i) 5   ii) -9
b) i) 0   ii) defined for all real numbers   c) i) never equals zero   ii) $-8, 3$

For each rational expression, for what values of the variable

  i)   does the expression equal zero?

  ii)   is the expression undefined?

a) $\dfrac{m+8}{m-3}$ b) $\dfrac{5c}{6}$ c) $\dfrac{4}{z^2-5z-36}$

Solution

a)   i) $\dfrac{m+8}{m-3} = 0$ when its *numerator* equals zero. Set the numerator equal to zero, and solve for m.

$$m+8=0$$
$$m=-8$$

Therefore, $\dfrac{m+8}{m-3} = 0$ when $m = -8$.

    ii) $\dfrac{m+8}{m-3}$ is *undefined* when its *denominator* equals zero. Set the denominator equal to zero, and solve for m.

$$m-3=0$$
$$m=3$$

$\dfrac{m+8}{m-3}$ is *undefined* when $m = 3$. This means that any real number *except* 3 can be substituted for m in this expression.

b)   i)   To determine the values of c that make $\dfrac{5c}{6} = 0$, set $5c = 0$ and solve.

$$5c=0$$
$$c=0 \qquad \text{Divide by 5.}$$

So, $\dfrac{5c}{6} = 0$ when $c = 0$.

    ii) $\dfrac{5c}{6}$ is *undefined* when the denominator equals zero. However, the denominator is 6, and $6 \neq 0$. Therefore, there *is no value of c* that makes $\dfrac{5c}{6}$ undefined. *We say that $\dfrac{5c}{6}$ is defined for all real numbers.*

c)   i) $\dfrac{4}{z^2-5z-36} = 0$ when the numerator equals zero. The numerator is 4, and $4 \neq 0$. Therefore, $\dfrac{4}{z^2-5z-36}$ will *never* equal zero.

    ii) $\dfrac{4}{z^2-5z-36}$ is *undefined* when $z^2-5z-36 = 0$. Solve for z.

$$z^2-5z-36=0$$
$$(z+4)(z-9)=0 \qquad \text{Factor.}$$
$$z+4=0 \quad \text{or} \quad z-9=0 \qquad \text{Set each factor equal to 0.}$$
$$z=-4 \quad \text{or} \qquad z=9 \qquad \text{Solve.}$$

$\dfrac{4}{z^2-5z-36}$ is undefined when $z = -4$ or $z = 9$. All real numbers *except* -4 and 9 can be substituted for z in this expression.

W Hint

Use your notes and follow the steps to answer You Try 2.

[YOU TRY 2] For each rational expression, for what values of the variable

    i) does the expression equal zero?

    ii) is the expression undefined?

a) $\dfrac{v-6}{v+11}$ b) $\dfrac{9w}{w^2-12w+20}$ c) $\dfrac{x^2-25}{8}$ d) $\dfrac{1}{5q+4}$

All of the operations that can be performed with fractions can also be done with rational expressions. We begin our study of these operations with rational expressions by learning how to write a rational expression in lowest terms.

3 Write a Rational Expression in Lowest Terms

One way to think about writing a fraction such as $\dfrac{8}{12}$ in lowest terms is

$$\frac{8}{12} = \frac{2 \cdot 4}{3 \cdot 4} = \frac{2}{3} \cdot \frac{4}{4} = \frac{2}{3} \cdot 1 = \frac{2}{3}$$

Since $\dfrac{4}{4} = 1$, we can also think of simplifying $\dfrac{8}{12}$ as $\dfrac{8}{12} = \dfrac{2 \cdot \cancel{4}}{3 \cdot \cancel{4}} = \dfrac{2}{3}$.

To write $\dfrac{8}{12}$ in lowest terms, we can *factor* the numerator and denominator, then *divide* the numerator and denominator by the common factor, 4. This is the approach we use to write a rational expression in lowest terms.

Definition Fundamental Property of Rational Expressions

If P, Q, and C are polynomials such that $Q \neq 0$ and $C \neq 0$, then

$$\frac{PC}{QC} = \frac{P}{Q}$$

This property mirrors the example above since

$$\frac{PC}{QC} = \frac{P}{Q} \cdot \frac{C}{C} = \frac{P}{Q} \cdot 1 = \frac{P}{Q}$$

Or, we can also think of the simplifying procedure as dividing the numerator and denominator by the common factor, C.

$$\frac{P\cancel{C}}{Q\cancel{C}} = \frac{P}{Q}$$

W Hint

Is this procedure any different from writing a fraction in lowest terms?

Procedure Writing a Rational Expression in Lowest Terms

1) Completely **factor** the numerator and denominator.
2) **Divide** the numerator and denominator by the greatest common factor.

EXAMPLE 3

Write each rational expression in lowest terms.

a) $\dfrac{21r^{10}}{3r^4}$ b) $\dfrac{8a + 40}{3a + 15}$ c) $\dfrac{5n^2 - 20}{n^2 + 5n + 6}$

Solution

a) We can simplify $\dfrac{21r^{10}}{3r^4}$ using the quotient rule presented in Chapter 5.

$$\dfrac{21r^{10}}{3r^4} = 7r^6$$
Divide 21 by 3, and use the quotient rule: $\dfrac{r^{10}}{r^4} = r^{10-4} = r^6$.

b) $\dfrac{8a + 40}{3a + 15} = \dfrac{8\cancel{(a + 5)}}{3\cancel{(a + 5)}}$ Factor.

$\phantom{\dfrac{8a + 40}{3a + 15}} = \dfrac{8}{3}$ Divide out the common factor, $a + 5$.

c) $\dfrac{5n^2 - 20}{n^2 + 5n + 6} = \dfrac{5(n^2 - 4)}{(n + 2)(n + 3)}$ Factor.

$\phantom{\dfrac{5n^2 - 20}{n^2 + 5n + 6}} = \dfrac{5\cancel{(n + 2)}(n - 2)}{\cancel{(n + 2)}(n + 3)}$ Factor completely.

$\phantom{\dfrac{5n^2 - 20}{n^2 + 5n + 6}} = \dfrac{5(n - 2)}{n + 3}$ Divide out the common factor, $n + 2$.

> **W Hint**
>
> Review the previous chapter if you forgot any steps for factoring.

BE CAREFUL

Notice that we divide by *factors* not *terms*.

$$\dfrac{\cancel{x + 5}}{2\cancel{(x + 5)}} = \dfrac{1}{2}$$

Divide by the *factor* $x + 5$.

$$\dfrac{x}{x + 5} \neq \dfrac{1}{5}$$

We cannot divide by x because the x in the denominator is a *term* in a sum.

[YOU TRY 3]

Write each rational expression in lowest terms.

a) $\dfrac{12b^8}{18b^3}$ b) $\dfrac{2h + 8}{7h + 28}$ c) $\dfrac{y^2 - 9y + 14}{9y^4 - 9y^3 - 18y^2}$

4 Simplify a Rational Expression of the Form $\dfrac{a - b}{b - a}$

Do you think that $\dfrac{x - 4}{4 - x}$ is in lowest terms? Let's look at it more closely to understand the answer.

$$\dfrac{x - 4}{4 - x} = \dfrac{x - 4}{-1(-4 + x)}$$
Factor -1 out of the denominator.

$$\phantom{\dfrac{x - 4}{4 - x}} = \dfrac{1\cancel{(x - 4)}}{-1\cancel{(x - 4)}}$$
Rewrite $-4 + x$ as $x - 4$.

$$\phantom{\dfrac{x - 4}{4 - x}} = -1$$
Divide out the common factor, $x - 4$.

Therefore, $\dfrac{x - 4}{4 - x} = -1$.

We can generalize this result as

EXAMPLE 4

Write each rational expression in lowest terms.

a) $\dfrac{7 - t}{t - 7}$ b) $\dfrac{4h^2 - 25}{15 - 6h}$ c) $\dfrac{x + 6}{x - 6}$

In-Class Example 4

Write each rational expression in lowest terms.

a) $\dfrac{p - 9}{9 - p}$ b) $\dfrac{16c^2 - 9}{3 - 4c}$

c) $\dfrac{a + 2}{a - 2}$

Answer: a) -1 b) $-4c - 3$

c) $\dfrac{a + 2}{a - 2}$

Solution

a) $\dfrac{7 - t}{t - 7} = -1$ since $\dfrac{7 - t}{t - 7} = \dfrac{-1(t - 7)}{(t - 7)} = -1$

b) $\dfrac{4h^2 - 25}{15 - 6h} = \dfrac{(2h + 5)(2h - 5)}{3(5 - 2h)}$ Factor.

$= \dfrac{(2h + 5)\overset{-1}{(2h - 5)}}{3(5 - 2h)}$

$= \dfrac{-1(2h + 5)}{3}$ $\dfrac{2h - 5}{5 - 2h} = -1$

$= \dfrac{-2h - 5}{3}$ or $-\dfrac{2h + 5}{3}$ Simplify.

c) The expression $\dfrac{x + 6}{x - 6}$ is already in lowest terms. Notice that it is *not* in the

form $\dfrac{a - b}{b - a}$.

[YOU TRY 4]

Write each rational expression in lowest terms.

a) $\dfrac{10 - m}{m - 10}$ b) $\dfrac{100 - 4k^2}{k^2 - 8k + 15}$ c) $\dfrac{c - 3}{c + 3}$

5 Write Equivalent Forms of a Rational Expression

The answer to Example 4b) can be written in several different ways. Recall that a fraction like $-\dfrac{1}{2}$ can also be written as $\dfrac{-1}{2}$ or $\dfrac{1}{-2}$. Likewise, when we are simplifying the expression $\dfrac{-1(2h + 5)}{3}$ we can distribute the -1 in the numerator to get $\dfrac{-2h - 5}{3}$.

Or, we can apply the -1 to the denominator to get $\dfrac{2h + 5}{-1 \cdot 3} = \dfrac{2h + 5}{-3}$. Or, we can write

$\dfrac{-1(2h + 5)}{3}$ as $-1 \cdot \dfrac{2h + 5}{3} = -\dfrac{2h + 5}{3}$. We can write the answer to Example 4b) in

any of these ways. You should be able to recognize equivalent forms of rational expressions because there isn't always just one way to write the correct answer.

Write $-\dfrac{5x-8}{3+x}$ in three different ways.

Solution

The negative sign in front of a fraction can be applied to the numerator or to the denominator. This can result in expressions that look quite different but that are, actually, equivalent.

i) Apply the negative sign to the denominator.

$$-\frac{5x-8}{3+x}=\frac{5x-8}{-1(3+x)}$$
$$=\frac{5x-8}{-3-x}\qquad\text{Distribute.}$$

ii) Apply the negative sign to the numerator.

$$-\frac{5x-8}{3+x}=\frac{-1(5x-8)}{3+x}=\frac{-5x+8}{3+x}$$

iii) Apply the negative sign to the numerator, distribute the -1, and rewrite addition as subtraction.

$$-\frac{5x-8}{3+x}=\frac{-1(5x-8)}{3+x}$$
$$=\frac{-5x+8}{3+x}\qquad\text{Distribute.}$$
$$=\frac{8-5x}{3+x}\qquad\text{Rewrite } -5x+8 \text{ as } 8-5x.$$

Therefore, $\dfrac{5x-8}{-3-x},\dfrac{-5x+8}{3+x}$, and $\dfrac{8-5x}{3+x}$ are *all* equivalent forms of $-\dfrac{5x-8}{3+x}$.

Keep this idea of equivalent forms of rational expressions in mind when checking your answers against the answers in the back of the book. Sometimes students believe their answer is wrong because it "looks different" when, in fact, it is an *equivalent form* of the given answer!

[YOU TRY 5]

Write $\dfrac{-(1-t)}{5t-8}$ in three different ways.

ANSWERS TO [YOU TRY] EXERCISES

1) a) 4 b) $\dfrac{1}{3}$ c) undefined d) 0   2) a) i) 6 ii) -11 b) i) 0 ii) 2, 10

c) i) $-5, 5$ ii) defined for all real numbers d) i) never equals zero ii) $-\dfrac{4}{5}$

3) a) $\dfrac{2b^5}{3}$ b) $\dfrac{2}{7}$ c) $\dfrac{y-7}{9y^2(y+1)}$ 4) a) -1 b) $-\dfrac{4(5+k)}{k-3}$ c) $\dfrac{c-3}{c+3}$

5) Some possibilities are $\dfrac{t-1}{5t-8},-\dfrac{1-t}{5t-8},\dfrac{1-t}{8-5t}$.

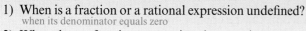
*Additional answers can be found in the Answers to Exercises appendix.

Objective 1: Evaluate a Rational Expression

1) When is a fraction or a rational expression undefined?
when its denominator equals zero

2) When does a fraction or a rational expression equal 0?   when its numerator equals zero

Evaluate (if possible) for a) $x = 3$ and b) $x = -2$.

3) $\dfrac{2x - 1}{5x + 2}$ a) $\dfrac{5}{17}$ b) $\dfrac{5}{8}$

4) $\dfrac{3(x^2 + 1)}{x^2 + 2x + 1}$ a) $\dfrac{15}{8}$ b) 15

Evaluate (if possible) for a) $z = 1$ and b) $z = -3$.

5) $\dfrac{(4z)^2}{z^2 - z - 12}$ a) $-\dfrac{4}{3}$ b) undefined

6) $\dfrac{3(z^2 - 9)}{z^2 + 8}$ a) $-\dfrac{8}{3}$ b) 0

7) $\dfrac{15 + 5z}{16 - z^2}$ a) $\dfrac{4}{3}$ b) 0

8) $\dfrac{4z - 3}{z^2 + 6z - 7}$ a) undefined   b) $\dfrac{15}{16}$

Objective 2: Find the Values of the Variable That Make a Rational Expression Undefined or Equal to Zero

9) How do you determine the values of the variable for which a rational expression is undefined?

10) If $x^2 + 9$ is the numerator of a rational expression, can that expression equal zero? Give a reason.

Determine the value(s) of the variable for which

a) the expression equals zero.

b) the expression is undefined.

11) $\dfrac{m + 4}{3m}$ a) -4 b) 0

12) $\dfrac{-y}{y + 3}$ a) 0   b) -3

13) $\dfrac{2w - 7}{4w + 1}$ a) $\dfrac{7}{2}$ b) $-\dfrac{1}{4}$

14) $\dfrac{3x + 13}{2x + 13}$ a) $-\dfrac{13}{3}$ b) $-\dfrac{13}{2}$

15) $\dfrac{11v - v^2}{5v - 9}$ a) 0, 11   b) $\dfrac{9}{5}$

16) $-\dfrac{r + 5}{r^2 - 100}$ a) -5 b) $-10, 10$

17) $\dfrac{8}{p}$ a) never equals 0   b) 0

18) $\dfrac{22}{m - 1}$ a) never equals 0   b) 1

19) $-\dfrac{7k}{k^2 + 9k + 20}$ a) 0   b) $-4, -5$

20) $\dfrac{a - 9}{a^2 + 8a - 9}$ a) 9   b) $-9, 1$

21) $\dfrac{c + 20}{2c^2 + 3c - 9}$ a) -20 b) $\dfrac{3}{2}, -3$

22) $\dfrac{4}{3f^2 - 13f + 10}$ a) never equals 0   b) 1, $\dfrac{10}{3}$

23) $\dfrac{g^2 + 9g + 18}{9g}$ a) $-6, -3$ b) 0

24) $\dfrac{6m - 11}{10}$

25) $\dfrac{4y}{y^2 + 9}$ a) 0   b) never undefined—any real number may be substituted for y

26) $\dfrac{q^2 + 49}{7}$ a) never equals 0   b) never undefined—any real number may be substituted for q

Objective 3: Write a Rational Expression in Lowest Terms

Write each rational expression in lowest terms.

27) $\dfrac{7x(x - 11)}{3(x - 11)}$ $\dfrac{7x}{3}$

28) $\dfrac{24(g + 3)}{6(g + 3)(g - 5)}$ $\dfrac{4}{g - 5}$

29) $\dfrac{24g^2}{56g^4}$ $\dfrac{3}{7g^2}$

30) $\dfrac{99d^7}{9d^3}$ $11d^4$

31) $\dfrac{4d - 20}{5d - 25}$ $\dfrac{4}{5}$

32) $\dfrac{12c - 3}{8c - 2}$ $\dfrac{3}{2}$

33) $\dfrac{-14h - 56}{6h + 24}$ $-\dfrac{7}{3}$

34) $\dfrac{-15v^2 + 12}{40v^2 - 32}$ $-\dfrac{3}{8}$

35) $\dfrac{39u^2 + 26}{30u^2 + 20}$ $\dfrac{13}{10}$

36) $\dfrac{3q + 15}{-7q - 35}$ $-\dfrac{3}{7}$

37) $\dfrac{g^2 - g - 56}{g + 7}$ $g - 8$

38) $\dfrac{b^2 + 9b + 20}{b^2 + b - 12}$ $\dfrac{b + 5}{b - 3}$

39) $\dfrac{t - 5}{t^2 - 25}$ $\dfrac{1}{t + 5}$

40) $\dfrac{r + 9}{r^2 + 7r - 18}$ $\dfrac{1}{r - 2}$

41) $\dfrac{3c^2 + 28c + 32}{c^2 + 10c + 16}$ $\dfrac{3c + 4}{c + 2}$

42) $\dfrac{3k^2 - 36k + 96}{k - 8}$ $3(k - 4)$

43) $\dfrac{q^2 - 25}{2q^2 - 7q - 15}$ $\dfrac{q + 5}{2q + 3}$

44) $\dfrac{6p^2 + 11p - 10}{9p^2 - 4}$ $\dfrac{2p + 5}{3p + 2}$

45) $\dfrac{4u^2 - 20u + 4uv - 20v}{13u + 13v}$ $\dfrac{4(u - 5)}{13}$

46) $\dfrac{ab + 3a - 6b - 18}{b^2 - 9}$ $\dfrac{a - 6}{b - 3}$

Objective 4: Simplify a Rational Expression of the Form $\dfrac{a - b}{b - a}$

47) Any rational expression of the form $\dfrac{a - b}{b - a}$ $(a \neq b)$ reduces to what? -1

48) Does $\dfrac{h + 4}{h - 4} = -1$?   no

Write each rational expression in lowest terms.

49) $\dfrac{8 - q}{q - 8}$ -1

50) $\dfrac{m - 15}{15 - m}$ -1

51) $\dfrac{m^2 - 121}{11 - m}$ $-m - 11$

52) $\dfrac{k - 9}{162 - 2k^2}$ $-\dfrac{1}{2(k + 9)}$

53) $\dfrac{36 - 42x}{7x^2 + 8x - 12}$ $-\dfrac{6}{x + 2}$

54) $\dfrac{a^2 - 6a - 27}{9 - a}$ $-a - 3$

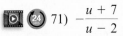 55) $\dfrac{16 - 4b^2}{b - 2}$ $-4(b + 2)$

56) $\dfrac{45 - 9v}{v^2 - 25}$ $-\dfrac{9}{v + 5}$

57) $\dfrac{y^3 - 3y^2 + 2y - 6}{21 - 7y}$ $-\dfrac{y^2 + 2}{7}$

58) $\dfrac{3t^3 - t^2 + 12t - 4}{1 - 9t^2}$ $-\dfrac{t^2 + 4}{3t + 1}$

Mixed Exercises: Objectives 1–4

Write each rational expression in lowest terms.

59) $\dfrac{18c + 45}{12c^2 + 18c - 30}$ $\dfrac{3}{2(c - 1)}$

60) $\dfrac{36n^3}{42n^9}$ $\dfrac{6}{7n^6}$

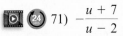 61) $\dfrac{4a^2 - 9}{4a^2 - 12a + 9}$ $\dfrac{2a + 3}{2a - 3}$

62) $\dfrac{k^2 - 16k + 64}{k^3 - 8k^2 + 9k - 72}$ $\dfrac{k - 8}{k^2 + 9}$

63) $\dfrac{b^2 + 6b - 72}{4b^2 + 52b + 48}$ $\dfrac{b - 6}{4(b + 1)}$

64) $\dfrac{5p^2 - 13p + 6}{32 - 8p^2}$ $-\dfrac{5p - 3}{8(p + 2)}$

65) $\dfrac{28h^4 - 56h^3 + 7h}{7h}$ $4h^3 - 8h^2 + 1$

66) $\dfrac{z^2 + 5z - 36}{-z^2 + 13z - 36}$ $\dfrac{z + 9}{9 - z}$

67) $\dfrac{14 - 6w}{12w^3 - 28w^2}$ $-\dfrac{1}{2w^2}$

68) $\dfrac{54d^6 + 6d^5 - 42d^3 - 18d^2}{6d^2}$ $9d^4 + d^3 - 7d - 3$

69) $\dfrac{-5v - 10}{v^3 - v^2 - 4v + 4}$ $-\dfrac{5}{(v - 2)(v - 1)}$

70) $\dfrac{38x^2 + 38}{-12x^2 - 12}$ $-\dfrac{19}{6}$

Objective 5: Write Equivalent Forms of a Rational Expression

Find three equivalent forms of each rational expression.

 71) $-\dfrac{u + 7}{u - 2}$

72) $-\dfrac{8y - 1}{2y + 5}$

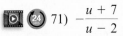 73) $-\dfrac{9 - 5t}{2t - 3}$

74) $\dfrac{w - 6}{-4w + 7}$

75) $\dfrac{-12m}{m^2 - 3}$

76) $\dfrac{-9x - 11}{18 - x}$

Reduce to lowest terms

    a) using long division.

    b) using the methods of this section.

77) $\dfrac{4y^2 - 11y + 6}{y - 2}$ $4y - 3$

78) $\dfrac{2x^2 + x - 28}{x + 4}$ $2x - 7$

Recall that the area of a rectangle is $A = lw$, where w = width and l = length. Solving for the width, we get $w = \dfrac{A}{l}$ and solving for the length gives us $l = \dfrac{A}{w}$.

Find the missing side in each rectangle.

79) Area = $3x^2 + 8x + 4$

 $x + 2$

Find the length. $3x + 2$

80) Area = $2y^2 - 3y - 20$

$2y + 5$

Find the width. $y - 4$

81) Area = $2c^3 + 4c^2 + 8c + 16$

$c^2 + 4$

Find the width. $2c + 4$

82) Area = $3n^3 - 12n^2 - n + 4$

 $n - 4$

Find the length. $3n^2 - 1$

Recall that the area of a triangle is $A = \dfrac{1}{2}bh$, where b = length of the base and h = height. Solving for the height, we get $h = \dfrac{2A}{b}$. Find the height of each triangle.

83) Area = $3k^2 + 13k + 4$ $3k + 1$

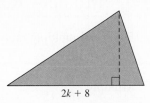

$2k + 8$

84) Area = $6p^2 + 52p + 32$ $p + 8$

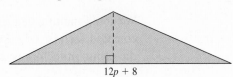

$12p + 8$

8.2 Multiplying and Dividing Rational Expressions

P Prepare   **O** Organize

What are your objectives for Section 8.2?	How can you accomplish each objective?
1 Multiply Rational Expressions	• Notice the steps you take to multiply fractions. • Understand how to **Multiply Rational Expressions** and compare to the steps taken in Example 1. • Complete the given examples on your own. • Complete You Trys 1 and 2.
2 Divide Rational Expressions	• Understand the principle behind the procedure for **Dividing Rational Expressions.** • Complete the given examples on your own. • Complete You Trys 3 and 4.

W Work   **Read the explanations, follow the examples, take notes, and complete the You Trys.**

1 Multiply Rational Expressions

Before we multiply rational expressions, let's review how to multiply fractions.

EXAMPLE 1

Multiply $\dfrac{9}{16} \cdot \dfrac{8}{15}$.

In-Class Example 1

Multiply $\dfrac{14}{25} \cdot \dfrac{5}{21}$.

Answer: $\dfrac{2}{15}$

Solution

We could multiply numerators, multiply denominators, then simplify by dividing out common factors. *Or* we can factor the numerators and denominators, then divide out the common factors before multiplying.

$$\frac{9}{16} \cdot \frac{8}{15} = \frac{\overset{1}{\cancel{3}} \cdot 3}{2 \cdot \cancel{8}} \cdot \frac{\overset{1}{\cancel{8}}}{\cancel{3} \cdot 5} \qquad \text{Factor, and divide out common factors.}$$

$$= \frac{3}{2 \cdot 5} \qquad \text{Multiply.}$$

$$= \frac{3}{10} \qquad \text{Simplify.}$$

Multiply $\dfrac{8}{35} \cdot \dfrac{5}{12}$.

Multiplying rational expressions works the same way.

Procedure Multiplying Rational Expressions

If $\dfrac{P}{Q}$ and $\dfrac{R}{T}$ are rational expressions, then $\dfrac{P}{Q} \cdot \dfrac{R}{T} = \dfrac{PR}{QT}$.

Note

We can also follow these steps to multiply rational expressions:

1) Factor the numerators and denominators.
2) Divide out common factors.
3) Multiply.

All products should be written in lowest terms.

W Hint

Is this the same procedure you used to multiply fractions?

EXAMPLE 2

Multiply.

a) $\dfrac{18}{y^7} \cdot \dfrac{y^4}{9}$ b) $\dfrac{9c + 45}{6c^{10}} \cdot \dfrac{c^6}{c^2 - 25}$

c) $\dfrac{2n^2 - 11n - 6}{n^2 - 2n - 24} \cdot \dfrac{n^2 + 8n + 16}{2n^2 + n}$

In-Class Example 2

Multiply.

a) $\dfrac{4}{t^2} \cdot \dfrac{t^5}{8}$

b) $\dfrac{k}{7k + 56} \cdot \dfrac{k^2 - 64}{8k^3}$

c) $\dfrac{3v^2 - 11v - 20}{v^2 - 4v - 5} \cdot \dfrac{v^2 + 2v + 1}{3v^2 + 4v}$

Answer: a) $\dfrac{t^3}{2}$ b) $\dfrac{k - 8}{56k^2}$

c) $\dfrac{v + 1}{v}$

Solution

a) $\dfrac{18}{y^7} \cdot \dfrac{y^4}{9} = \dfrac{\overset{2}{\cancel{18}}}{\cancel{y^4} \cdot y^3} \cdot \dfrac{\overset{1}{\cancel{y^4}}}{\cancel{9}}$ Factor; divide out common factors.

$= \dfrac{2}{y^3}$ Multiply.

b) $\dfrac{9c + 45}{6c^{10}} \cdot \dfrac{c^6}{c^2 - 25} = \dfrac{\overset{3}{\cancel{9}}(c + 5)}{\underset{2}{\cancel{6}} c^6 \cdot c^4} \cdot \dfrac{\cancel{c^6}}{\cancel{(c + 5)}(c - 5)}$ Factor; divide out common factors.

$= \dfrac{3}{2c^4(c - 5)}$ Multiply.

c) $\dfrac{2n^2 - 11n - 6}{n^2 - 2n - 24} \cdot \dfrac{n^2 + 8n + 16}{2n^2 + n} = \dfrac{(2n + 1)\cancel{(n - 6)}}{\cancel{(n + 4)}\cancel{(n - 6)}} \cdot \dfrac{\overset{(n+4)}{\cancel{(n + 4)}^2}}{n\cancel{(2n + 1)}}$ Factor; divide out common factors.

$= \dfrac{n + 4}{n}$ Multiply.

Multiply.

a) $\dfrac{n^7}{20} \cdot \dfrac{10}{n^2}$ b) $\dfrac{d^2}{d^2 - 4} \cdot \dfrac{4d - 8}{12d^5}$ c) $\dfrac{h^2 + 10h + 25}{3h^2 - 4h} \cdot \dfrac{3h^2 + 5h - 12}{h^2 + 8h + 15}$

2 Divide Rational Expressions

When we divide rational numbers, we multiply by a reciprocal. For example,

$\dfrac{7}{4} \div \dfrac{3}{8} = \dfrac{7}{\underset{1}{4}} \cdot \dfrac{\overset{2}{8}}{3} = \dfrac{14}{3}$. We divide rational expressions the same way.

To divide rational expressions, we multiply the first rational expression by the reciprocal of the second rational expression.

Procedure Dividing Rational Expressions

If $\dfrac{P}{Q}$ and $\dfrac{R}{T}$ are rational expressions with Q, R, and T not equal to zero, then

$$\frac{P}{Q} \div \frac{R}{T} = \frac{P}{Q} \cdot \frac{T}{R} = \frac{PT}{QR}$$

Multiply the first rational expression by the reciprocal of the second rational expression.

EXAMPLE 3

In-Class Example 3

Divide.

a) $\dfrac{42y^{10}}{z^4} \div \dfrac{7y^2}{z^6}$

b) $\dfrac{p^2 - 5p - 24}{p^3} \div (p - 8)^2$

c) $\dfrac{b^2 + 10b + 9}{b^2 - 49} \div \dfrac{b^2 - 3b - 4}{28 - 4b}$

Answer: a) $6y^8z^2$

b) $\dfrac{p + 3}{p^3(p - 8)}$

c) $-\dfrac{4(b + 9)}{(b + 7)(b - 4)}$

Divide.

a) $\dfrac{15a^7}{b^3} \div \dfrac{3a^4}{b^9}$ b) $\dfrac{t^2 - 16t + 63}{t^2} \div (t - 7)^2$

c) $\dfrac{x^2 - 9}{x^2 + 3x - 10} \div \dfrac{24 - 8x}{x^2 + 9x + 20}$

Solution

a) $\dfrac{15a^7}{b^3} \div \dfrac{3a^4}{b^9} = \dfrac{\overset{5a^3}{\cancel{15a^7}}}{b^3} \cdot \dfrac{\overset{b^6}{\cancel{b^9}}}{\cancel{3a^4}}$ Multiply by the reciprocal; divide out common factors.

$\qquad\qquad\qquad = 5a^3b^6$ Multiply.

Notice that we used the *quotient rule* for exponents to simplify:

$$\frac{a^7}{a^4} = a^3, \quad \frac{b^9}{b^3} = b^6$$

W Hint

Are you writing out all of the steps as you read the example?

b) $\dfrac{t^2 - 16t + 63}{t^2} \div (t - 7)^2 = \dfrac{(t - 9)\overset{1}{\cancel{(t - 7)}}}{t^2} \cdot \dfrac{1}{\underset{(t-7)}{\cancel{(t - 7)^2}}}$ Since $(t - 7)^2$ can be written as $\dfrac{(t - 7)^2}{1}$, its reciprocal is $\dfrac{1}{(t - 7)^2}$.

$\qquad\qquad\qquad = \dfrac{t - 9}{t^2(t - 7)}$ Divide out common factors, and multiply.

c) $\dfrac{x^2 - 9}{x^2 + 3x - 10} \div \dfrac{24 - 8x}{x^2 + 9x + 20} = \dfrac{x^2 - 9}{x^2 + 3x - 10} \cdot \dfrac{x^2 + 9x + 20}{24 - 8x}$ Multiply by the reciprocal.

$\qquad\qquad\qquad = \dfrac{(x + 3)\overset{-1}{\cancel{(x - 3)}}}{\underset{1}{\cancel{(x + 5)}}(x - 2)} \cdot \dfrac{(x + 4)\cancel{(x + 5)}}{8\cancel{(3 - x)}}$ Factor; $\dfrac{x - 3}{3 - x} = -1$.

$\qquad\qquad\qquad = -\dfrac{(x + 3)(x + 4)}{8(x - 2)}$ Divide out common factors, and multiply.

[**YOU TRY 3**] Divide.

a) $\dfrac{k^3}{12h^7} \div \dfrac{k^4}{16h^2}$ b) $\dfrac{w^2 + 4w - 45}{w} \div (w + 9)^2$

c) $\dfrac{2m^2 - m - 15}{1 - m^2} \div \dfrac{m^2 - 10m + 21}{12m - 12}$

Remember that a fraction, itself, represents division. That is, $\dfrac{30}{5} = 30 \div 5 = 6$. We can write division problems involving fractions and rational expressions in a similar way.

EXAMPLE 4

In-Class Example 4

Divide.

a) $\dfrac{\frac{10}{21}}{\frac{15}{28}}$ b) $\dfrac{\frac{4k + 1}{12}}{\frac{16k^2 - 1}{18}}$

Answer: a) $\dfrac{8}{9}$ b) $\dfrac{3}{2(4k - 1)}$

W Hint

Don't be intimidated! You've done problems similar to these before.

Divide.

a) $\dfrac{\frac{8}{35}}{\frac{16}{45}}$ b) $\dfrac{\frac{3w + 2}{5}}{\frac{9w^2 - 4}{10}}$

Solution

a) $\dfrac{\frac{8}{35}}{\frac{16}{45}}$ means $\dfrac{8}{35} \div \dfrac{16}{45}$. Then,

$$\dfrac{8}{35} \div \dfrac{16}{45} = \dfrac{8}{35} \cdot \dfrac{45}{16} \qquad \text{Multiply by the reciprocal.}$$

$$= \dfrac{\overset{1}{8}}{\underset{7}{35}} \cdot \dfrac{\overset{9}{45}}{\underset{2}{16}} \qquad \text{Divide 8 and 16 by 8. Divide 45 and 35 by 5.}$$

$$= \dfrac{9}{14} \qquad \text{Multiply.}$$

b) $\dfrac{\frac{3w + 2}{5}}{\frac{9w^2 - 4}{10}}$ means $\dfrac{3w + 2}{5} \div \dfrac{9w^2 - 4}{10}$. Then,

$$\dfrac{3w + 2}{5} \div \dfrac{9w^2 - 4}{10} = \dfrac{3w + 2}{5} \cdot \dfrac{10}{9w^2 - 4} \qquad \text{Multiply by the reciprocal.}$$

$$= \dfrac{\overset{1}{\cancel{3w + 2}}}{\underset{1}{\cancel{5}}} \cdot \dfrac{\overset{2}{\cancel{10}}}{\underset{1}{\cancel{(3w + 2)}}(3w - 2)} \qquad \text{Factor; divide out common factors.}$$

$$= \dfrac{2}{3w - 2} \qquad \text{Multiply.}$$

[**YOU TRY 4**] Divide.

a) $\dfrac{\frac{4}{45}}{\frac{20}{27}}$ b) $\dfrac{\frac{25u^2 - 9}{24}}{\frac{5u + 3}{16}}$

E Evaluate **8.2** Exercises

Do the exercises, and check your work.

*Additional answers can be found in the Answers to Exercises appendix.

Objective 1: Multiply Rational Expressions

Multiply.

1) $\dfrac{5}{6} \cdot \dfrac{7}{9}$ $\dfrac{35}{54}$

2) $\dfrac{4}{11} \cdot \dfrac{2}{3}$ $\dfrac{8}{33}$

3) $\dfrac{6}{15} \cdot \dfrac{25}{42}$ $\dfrac{5}{21}$

4) $\dfrac{12}{21} \cdot \dfrac{7}{4}$ 1

5) $\dfrac{16b^5}{3} \cdot \dfrac{4}{36b}$ $\dfrac{16b^4}{27}$

6) $\dfrac{26}{25r^3} \cdot \dfrac{15r^6}{2}$ $\dfrac{39r^3}{5}$

7) $\dfrac{21s^4}{15t^2} \cdot \dfrac{5t^4}{42s^{10}}$ $\dfrac{t^2}{6s^6}$

8) $\dfrac{15u^4}{14v^2} \cdot \dfrac{7v^7}{20u^8}$ $\dfrac{3v^5}{8u^4}$

9) $\dfrac{9c^4}{42c} \cdot \dfrac{35}{3c^3}$ $\dfrac{5}{2}$

10) $-\dfrac{10}{8x^7} \cdot \dfrac{24x^9}{9x}$ $-\dfrac{10x}{3}$

11) $\dfrac{5t^2}{(3t-2)^2} \cdot \dfrac{3t-2}{10t^3}$ $\dfrac{1}{2t(3t-2)}$

12) $\dfrac{11(z+5)^5}{6(z-4)} \cdot \dfrac{3}{22(z+5)}$ $\dfrac{(z+5)^4}{4(z-4)}$

13) $\dfrac{4u-5}{9u^3} \cdot \dfrac{6u^5}{(4u-5)^3}$ $\dfrac{2u^2}{3(4u-5)^2}$

14) $\dfrac{5k+6}{2k^3} \cdot \dfrac{12k^5}{(5k+6)^4}$

15) $\dfrac{6}{n+5} \cdot \dfrac{n^2+8n+15}{n+3}$ 6

16) $\dfrac{9p^2-1}{12} \cdot \dfrac{9}{9p+3}$ $\dfrac{3p-1}{4}$

17) $\dfrac{18y-12}{4y^2} \cdot \dfrac{y^2-4y-5}{3y^2+y-2}$ $\dfrac{3(y-5)}{2y^2}$

18) $\dfrac{12v-3}{8v+12} \cdot \dfrac{2v^2-5v-12}{3v-12}$ $\dfrac{4v-1}{4}$

19) $(c-6) \cdot \dfrac{5}{c^2-6c}$ $\dfrac{5}{c}$

20) $(r^2+r-2) \cdot \dfrac{18r^2}{3r^2+6r}$ $6r(r-1)$

21) $\dfrac{7x}{11-x} \cdot (x^2-121)$ $-7x(x+11)$

22) $\dfrac{4b}{2b^2-3b-5} \cdot (b+1)^2$ $\dfrac{4b(b+1)}{2b-5}$

Objective 2: Divide Rational Expressions

Divide.

23) $\dfrac{20}{9} \div \dfrac{10}{27}$ 6

24) $\dfrac{4}{5} \div \dfrac{12}{7}$ $\dfrac{7}{15}$

25) $42 \div \dfrac{9}{2}$ $\dfrac{28}{3}$

26) $\dfrac{18}{7} \div 6$ $\dfrac{3}{7}$

Divide.

27) $\dfrac{12}{5m^5} \div \dfrac{21}{8m^{12}}$ $\dfrac{32m^7}{35}$

28) $\dfrac{12k^3}{35} \div \dfrac{42k^6}{25}$ $\dfrac{10}{49k^3}$

29) $-\dfrac{50g}{7h^3} \div \dfrac{15g^4}{14h}$ $-\dfrac{20}{3g^3h^2}$

30) $-\dfrac{c^{12}}{b} \div \dfrac{c^2}{6b}$ $-6c^{10}$

31) $\dfrac{2(k-2)}{21k^6} \div \dfrac{(k-2)^2}{28}$ $\dfrac{8}{3k^6(k-2)}$

32) $\dfrac{18}{(x+4)^3} \div \dfrac{36(x-7)}{x+4}$

33) $\dfrac{16q^5}{p+7} \div \dfrac{2q^4}{(p+7)^2}$ $8q(p+7)$

34) $\dfrac{(2a-5)^2}{32a^5} \div \dfrac{2a-5}{8a^3}$

35) $\dfrac{q+8}{q} \div \dfrac{q^2+q-56}{5}$ $\dfrac{5}{q(q-7)}$

36) $\dfrac{4y^2-25}{10} \div \dfrac{18y-45}{18}$ $\dfrac{2y+5}{5}$

37) $\dfrac{z^2+18z+80}{2z+1} \div (z+8)^2$ $\dfrac{z+10}{(2z+1)(z+8)}$

38) $\dfrac{6w^2-30w}{7} \div (w-5)^2$ $\dfrac{6w}{7(w-5)}$

39) $\dfrac{36a-12}{16} \div (9a^2-1)$ $\dfrac{3}{4(3a+1)}$

40) $\dfrac{h^2-21h+108}{4h} \div (144-h^2)$ $\dfrac{9-h}{4h(h+12)}$

41) $\dfrac{7n^2-14n}{8n} \div \dfrac{n^2+4n-12}{4n+24}$ $\dfrac{7}{2}$

42) $\dfrac{4j+24}{9} \div \dfrac{j^2-36}{9j-54}$ 4

43) $\dfrac{4c-9}{2c^2-8c} \div \dfrac{12c-27}{c^2-3c-4}$ $\dfrac{c+1}{6c}$

44) $\dfrac{p+13}{p+3} \div \dfrac{p^3+13p^2}{p^2-5p-24}$ $\dfrac{p-8}{p^2}$

45) Explain how to multiply rational expressions.
Answers may vary.

46) Explain how to divide rational expressions.
Answers may vary.

47) Find the missing polynomial in the denominator of

$\dfrac{9h+45}{h^4} \cdot \dfrac{h^3}{} = \dfrac{9}{h(h-2)}.$ $h^2+3h-10$

48) Find the missing monomial in the numerator of

$$\dfrac{}{m^2 - 81} \cdot \dfrac{3m - 27}{2m^2} = \dfrac{15m^3}{m + 9}. \quad 10m^5$$

49) Find the missing binomial in the numerator of

$$\dfrac{4z^2 - 49}{z^2 - 3z - 40} \div \dfrac{}{z + 5} = \dfrac{2z + 7}{8 - z}. \quad 7 - 2z$$

50) Find the missing polynomial in the denominator of

$$\dfrac{12x^4}{50x^2 + 40x} \div \dfrac{x^3 + 2x^2 + x + 2}{} = \dfrac{6x^3}{5(x^2 + 1)}. \quad 5x^2 + 14x + 8$$

Divide.

51) $\dfrac{\frac{25}{42}}{\frac{8}{21}}$ $\quad \frac{25}{16}$

52) $\dfrac{\frac{9}{35}}{\frac{4}{15}}$ $\quad \frac{27}{28}$

53) $\dfrac{\frac{5}{24}}{\frac{15}{4}}$ $\quad \frac{1}{18}$

54) $\dfrac{\frac{4}{3}}{\frac{2}{9}}$ $\quad 6$

55) $\dfrac{\frac{3d + 7}{24}}{\frac{3d + 7}{6}}$ $\quad \frac{1}{4}$

56) $\dfrac{\frac{8s - 7}{4}}{\frac{8s - 7}{16}}$ $\quad 4$

57) $\dfrac{\frac{16r + 24}{r^3}}{\frac{12r + 18}{r}}$ $\quad \frac{4}{3r^2}$

58) $\dfrac{\frac{44m - 33}{3m^2}}{\frac{8m - 6}{m}}$ $\quad \frac{11}{6m}$

59) $\dfrac{\frac{a^2 - 25}{3a^{11}}}{\frac{4a + 20}{a^3}}$ $\quad \frac{a - 5}{12a^8}$

60) $\dfrac{\frac{4z - 8}{z^8}}{\frac{z^2 - 4}{z^6}}$ $\quad \frac{4}{z^2(z + 2)}$

61) $\dfrac{\frac{16x^2 - 25}{x^7}}{\frac{36x - 45}{6x^3}}$ $\quad \frac{2(4x + 5)}{3x^4}$

62) $-\dfrac{\frac{16a^2}{3a^2 + 2a}}{\frac{12}{9a^2 - 4}}$ $\quad -\frac{4a(3a - 2)}{3}$

Mixed Exercises: Objectives 1 and 2

63) $\dfrac{c^2 + c - 30}{9c + 9} \cdot \dfrac{c^2 + 2c + 1}{c^2 - 25}$ $\quad \frac{(c + 6)(c + 1)}{9(c + 5)}$

64) $\dfrac{d^2 + 3d - 54}{d - 12} \cdot \dfrac{d^2 - 10d - 24}{7d + 63}$ $\quad \frac{(d - 6)(d + 2)}{7}$

65) $\dfrac{3x + 2}{9x^2 - 4} \div \dfrac{4x}{15x^2 - 7x - 2}$ $\quad \frac{5x + 1}{4x}$

66) $\dfrac{b^2 - 10b + 25}{8b - 40} \div \dfrac{2b^2 - 5b - 25}{2b + 5}$ $\quad \frac{1}{8}$

67) $\dfrac{3k^2 - 12k}{12k^2 - 30k - 72} \cdot (2k + 3)^2$ $\quad \frac{k(2k + 3)}{2}$

68) $\dfrac{4a^3}{a^2 + a - 72} \cdot (a^2 - a - 56)$ $\quad \frac{4a^3(a + 7)}{a + 9}$

69) $\dfrac{7t^6}{t^2 - 4} \div \dfrac{14t^2}{3t^2 - 7t + 2}$ $\quad \frac{t^4(3t - 1)}{2(t + 2)}$

70) $\dfrac{4n^2 - 1}{10n^3} \div \dfrac{2n^2 - 7n - 4}{6n^5}$ $\quad \frac{3n^2(2n - 1)}{5(n - 4)}$

71) $\dfrac{4h^3}{h^2 - 64} \cdot \dfrac{8h - h^2}{12}$ $\quad -\frac{h^4}{3(h + 8)}$

72) $\dfrac{c^2 - 36}{c + 6} \div \dfrac{30 - 5c}{c - 9}$ $\quad -\frac{c - 9}{5}$

73) $\dfrac{54x^8}{22x^3y^2} \div \dfrac{36xy^5}{11x^2y}$ $\quad \frac{3x^6}{4y^6}$

74) $\dfrac{28cd^9}{2c^3d} \cdot \dfrac{5d^2}{84c^{10}d^2}$ $\quad \frac{5d^8}{6c^{12}}$

75) $\dfrac{a^2 - 4a}{6a + 54} \cdot \dfrac{a^2 + 13a + 36}{16 - a^2}$ $\quad -\frac{a}{6}$

76) $\dfrac{64 - u^2}{40 - 5u} \div \dfrac{u^2 + 10u + 16}{2u + 3}$ $\quad \frac{2u + 3}{5(u + 2)}$

77) $\dfrac{2a^2}{a^2 + a - 20} \cdot \dfrac{a^3 + 5a^2 + 4a + 20}{2a^2 + 8}$ $\quad \frac{a^2}{a - 4}$

78) $\dfrac{18x^4}{x^3 + 3x^2 - 9x - 27} \cdot \dfrac{6x^2 + 19x + 3}{18x^2 + 3x}$ $\quad \frac{6x^3}{x^2 - 9}$

79) $\dfrac{3m^2 + 8m + 4}{4} \div (12m + 8)$ $\quad \frac{m + 2}{16}$

80) $\dfrac{w^2 - 17w + 72}{3w} \div (w - 8)$ $\quad \frac{w - 9}{3w}$

Perform the operations and simplify.

81) $\dfrac{4j^2 - 21j + 5}{j^3} \div \left(\dfrac{3j + 2}{j^3 - j^2} \cdot \dfrac{j^2 - 6j + 5}{j} \right)$ $\quad \frac{4j - 1}{3j + 2}$

82) $\dfrac{2a}{a^2 + 18a + 81} \div \left(\dfrac{a^2 + 3a - 4}{a^2 + 9a + 20} \cdot \dfrac{a^2 + 5a}{a^2 + 8a - 9} \right)$ $\quad \frac{2}{a + 9}$

83) If the area of a rectangle is $\dfrac{3x}{2y^2}$ and the width is $\dfrac{y}{8x^4}$, what is the length of the rectangle? $\quad \frac{12x^5}{y^3}$

84) If the area of a triangle is $\dfrac{2n}{n^2 - 4n + 3}$ and the height is $\dfrac{n + 3}{n - 1}$, what is the length of the base of the triangle? $\quad \frac{4n}{n^2 - 9}$

R1) Why do you think you learned to multiply and divide rational expressions before learning to add and subtract rational expressions? (Hint: Compare this with adding and subtracting fractions.)

8.3 Finding the Least Common Denominator

P Prepare

O Organize

What are your objectives for Section 8.3?	How can you accomplish each objective?
1 Find the Least Common Denominator for a Group of Rational Expressions	• Notice that the skills you learned involving fractions apply again! • Write the procedure for **Finding the LCD** in your notes. • Complete the given examples on your own. • Complete You Trys 1–3.
2 Rewrite Rational Expressions Using Their LCD	• Compare the solution for Example 4 to the procedure for **Writing Rational Expressions as Equivalent Expressions with the LCD.** • Write the procedure in your own words. • Complete the given examples on your own. • Complete You Trys 4 and 5.

W Work

Read the explanations, follow the examples, take notes, and complete the You Trys.

1 Find the Least Common Denominator for a Group of Rational Expressions

Recall that to add or subtract fractions, they must have a common denominator. Similarly, rational expressions must have common denominators in order to be added or subtracted. In this section, we will discuss how to find the least common denominator (LCD) of rational expressions.

We begin by looking at the fractions $\frac{3}{8}$ and $\frac{5}{12}$. By inspection, we can see that the LCD = 24. But, *why* is that true? Let's write each of the denominators, 8 and 12, as the product of their prime factors:

$$8 = 2 \cdot 2 \cdot 2 = 2^3$$
$$12 = 2 \cdot 2 \cdot 3 = 2^2 \cdot 3$$

The LCD will contain each factor the *greatest* number of times it appears in any single factorization.

The LCD will contain 2^3 because 2 appears *three* times in the factorization of 8.

The LCD will contain 3 because 3 appears *one* time in the factorization of 12.

The LCD, then, is the product of the factors we have identified.

$$\text{LCD of } \frac{3}{8} \text{ and } \frac{5}{12} = 2^3 \cdot 3 = 8 \cdot 3 = 24$$

This is the same result that we obtained just by inspecting the two denominators.

We use the same procedure to find the least common denominator of rational expressions.

Procedure Finding the Least Common Denominator (LCD)

Step 1: Factor the denominators.

Step 2: The LCD will contain each unique factor the *greatest* number of times it appears in any single factorization.

Step 3: The LCD is the *product* of the factors identified in Step 2.

EXAMPLE 1

Find the LCD of each pair of rational expressions.

In-Class Example 1

Find the LCD of each pair of rational expressions.

a) $\dfrac{11}{18}, \dfrac{7}{24}$ b) $\dfrac{3}{16k}, \dfrac{9}{20k}$

c) $\dfrac{4}{9b^5}, \dfrac{5}{6b^2}$

Answer: a) 72 b) 80k
c) $18b^5$

a) $\dfrac{17}{24}, \dfrac{5}{36}$ b) $\dfrac{1}{12n}, \dfrac{10}{21n}$ c) $\dfrac{8}{49c^3}, \dfrac{13}{14c^2}$

Solution

a) Follow the steps for finding the least common denominator.

   Step 1: Factor the denominators.

   $$24 = 2 \cdot 2 \cdot 2 \cdot 3 = 2^3 \cdot 3$$
   $$36 = 2 \cdot 2 \cdot 3 \cdot 3 = 2^2 \cdot 3^2$$

   Step 2: The LCD will contain each unique factor the *greatest* number of times it appears in any factorization. *The LCD will contain* 2^3 *and* 3^2.

   Step 3: The LCD is the *product* of the factors in Step 2.

   $$\text{LCD} = 2^3 \cdot 3^2 = 8 \cdot 9 = 72$$

b) **Step 1:** Factoring the denominators of $\dfrac{1}{12n}$ and $\dfrac{10}{21n}$ gives us

   $$12n = 2 \cdot 2 \cdot 3 \cdot n = 2^2 \cdot 3 \cdot n$$
   $$21n = 3 \cdot 7 \cdot n$$

   Step 2: The LCD will contain each unique factor the *greatest* number of times it appears in any factorization. *It will contain* 2^2, 3, 7, *and* n.

   Step 3: The LCD is the *product* of the factors in Step 2.

   $$\text{LCD} = 2^2 \cdot 3 \cdot 7 \cdot n = 84n$$

c) **Step 1:** Factoring the denominators of $\dfrac{8}{49c^3}$ and $\dfrac{13}{14c^2}$ gives us

$$49c^3 = 7 \cdot 7 \cdot c^3 = 7^2 \cdot c^3$$
$$14c^2 = 2 \cdot 7 \cdot c^2$$

Step 2: The LCD will contain each unique factor the *greatest* number of times it appears in any factorization. *It will contain* 2, 7^2, *and* c^3.

Step 3: The LCD is the *product* of the factors in Step 2.

$$\text{LCD} = 2 \cdot 7^2 \cdot c^3 = 98c^3$$

[**YOU TRY 1**] Find the LCD of each pair of rational expressions.

    a) $\dfrac{14}{15}, \dfrac{11}{18}$ b) $\dfrac{3}{14}, \dfrac{7}{10}$ c) $\dfrac{20}{27h^2}, \dfrac{1}{6h^4}$

EXAMPLE 2

In-Class Example 2

Find the LCD of each group of rational expressions.

a) $\dfrac{19}{n}, \dfrac{2}{n-8}$

b) $\dfrac{3c}{c+12}, \dfrac{7}{c^2+7c-60}$

c) $\dfrac{9}{4d^2-12d}, \dfrac{4d}{d^2-6d+9}$

Answer: a) $n(n-8)$
b) $(c+12)(c-5)$
c) $4d(d-3)^2$

W Hint

Just because the problem looks complicated does *not* mean you can't do it. Break it down one step at a time, and it will be easier to manage.

Find the LCD of each group of rational expressions.

a) $\dfrac{4}{k}, \dfrac{6}{k+3}$ b) $\dfrac{7}{a-6}, \dfrac{2a}{a^2+3a-54}$ c) $\dfrac{3p}{p^2+4p+4}, \dfrac{1}{5p^2+10p}$

Solution

a) The denominators of $\dfrac{4}{k}$ and $\dfrac{6}{k+3}$ are already in simplest form. It is important to recognize that k *and* $k + 3$ *are different factors*.

The LCD will be the product of k and $k + 3$:   **LCD $= k(k + 3)$**

Usually, we leave the LCD in this form; we do not distribute.

b) **Step 1:** Factor the denominators of $\dfrac{7}{a-6}$ and $\dfrac{2a}{a^2+3a-54}$.

     $a - 6$ cannot be factored. $a^2 + 3a - 54 = (a - 6)(a + 9)$

Step 2: The LCD will contain each unique factor the *greatest* number of times it appears in any factorization. *It will contain* $a - 6$ *and* $a + 9$.

Step 3: The LCD is the *product* of the factors identified in Step 2.

$$\text{LCD} = (a - 6)(a + 9)$$

c) **Step 1:** Factor the denominators of $\dfrac{3p}{p^2+4p+4}$ and $\dfrac{1}{5p^2+10p}$.

     $p^2 + 4p + 4 = (p + 2)^2$ and $5p^2 + 10p = 5p(p + 2)$

Step 2: The unique factors are 5, p, and $p + 2$, with $p + 2$ *appearing at most twice. The factors we will use in the LCD are* 5, p, *and* $(p + 2)^2$.

Step 3: **LCD $= 5p(p + 2)^2$**

[**YOU TRY 2**] Find the LCD of each group of rational expressions.

    a) $\dfrac{6}{w}, \dfrac{9w}{w+1}$ b) $\dfrac{1}{r+8}, \dfrac{5r}{r^2+r-56}$ c) $\dfrac{4b}{b^2-18b+81}, \dfrac{3}{8b^2-72b}$

What is the least common denominator of $\dfrac{9}{x-7}$ and $\dfrac{5}{7-x}$? Is it $(x-7)(7-x)$? Read Example 3 to find out.

EXAMPLE 3

In-Class Example 3

Find the LCD of $\dfrac{6}{p-2}$ and $\dfrac{5}{2-p}$.

Answer: $p - 2$

Find the LCD of $\dfrac{9}{x-7}$ and $\dfrac{5}{7-x}$.

Solution

Recall from Section 8.1 that $a - b = -1(b - a)$. So, $7 - x = -(x - 7)$, and we can rewrite $\dfrac{5}{7-x}$ as $\dfrac{5}{-(x-7)} = -\dfrac{5}{x-7}$. Therefore, we can now think of our task as finding the LCD of $\dfrac{9}{x-7}$ and $-\dfrac{5}{x-7}$. The least common denominator of $\dfrac{9}{x-7}$ and $\dfrac{5}{7-x}$ is $x - 7$.

[YOU TRY 3]

Find the LCD of $\dfrac{2}{k-5}$ and $\dfrac{13}{5-k}$.

2 Rewrite Rational Expressions Using Their LCD

As we know from our previous work with fractions, after *determining* the least common denominator, we must *rewrite* those fractions as equivalent fractions with the LCD so that they can be added or subtracted.

EXAMPLE 4

In-Class Example 4

Identify the LCD of $\dfrac{5}{8}$ and $\dfrac{1}{12}$, and rewrite each as an equivalent fraction with the LCD as its denominator.

Answer: LCD = 24; $\dfrac{15}{24}, \dfrac{2}{24}$

Identify the LCD of $\dfrac{1}{6}$ and $\dfrac{8}{9}$, and rewrite each as an equivalent fraction with the LCD as its denominator.

Solution

The LCD of $\dfrac{1}{6}$ and $\dfrac{8}{9}$ is 18. We must rewrite each fraction with a denominator of 18.

$\dfrac{1}{6}$: By what number should we multiply 6 to get 18?   3

$$\dfrac{1}{6} \cdot \dfrac{3}{3} = \dfrac{3}{18} \qquad \text{Multiply the numerator \textit{and} denominator by 3 to obtain an equivalent fraction.}$$

$\dfrac{8}{9}$: By what number should we multiply 9 to get 18?   2

$$\dfrac{8}{9} \cdot \dfrac{2}{2} = \dfrac{16}{18} \qquad \text{Multiply the numerator \textit{and} denominator by 2 to obtain an equivalent fraction.}$$

Therefore, $\dfrac{1}{6} = \dfrac{3}{18}$ and $\dfrac{8}{9} = \dfrac{16}{18}$.

[YOU TRY 4]

Identify the LCD of $\dfrac{7}{12}$ and $\dfrac{2}{9}$, and rewrite each as an equivalent fraction with the LCD as its denominator.

The procedure for rewriting rational expressions as equivalent expressions with the least common denominator is very similar to the process used in Example 4.

Procedure Writing Rational Expressions as Equivalent Expressions with the Least Common Denominator

Step 1: Identify and write down the LCD.

Step 2: Look at each rational expression (with its denominator in factored form) and compare its denominator with the LCD. Ask yourself, *"What factors are missing?"*

Step 3: Multiply the numerator and denominator by the "missing" factors to obtain an equivalent rational expression with the desired LCD.

Note

Use the distributive property to multiply the terms in the numerator, but leave the denominator as the product of factors. (We will see why this is done in Section 8.4.)

EXAMPLE 5

 Hint

Be sure to complete the example on your own.

Identify the LCD of each pair of rational expressions, and rewrite each as an equivalent expression with the LCD as its denominator.

a) $\dfrac{5}{12z}, \dfrac{4}{9z^3}$ b) $\dfrac{m}{m-4}, \dfrac{3}{m+8}$ c) $\dfrac{2}{3x^2-6x}, \dfrac{9x}{x^2-7x+10}$

Solution

a) **Step 1:** Identify and write down the LCD of $\dfrac{5}{12z}$ and $\dfrac{4}{9z^3}$:   **LCD = $36z^3$**

     Step 2: Compare the denominators of $\dfrac{5}{12z}$ and $\dfrac{4}{9z^3}$ to the LCD and ask yourself, "What's missing?"

$\dfrac{5}{12z}$: $12z$ is "missing" the factors 3 and z^2. $\dfrac{4}{9z^3}$: $9z^3$ is "missing" the factor 4.

     Step 3: Multiply the numerator and denominator by $3z^2$.   Multiply the numerator and denominator by 4.

$\dfrac{5}{12z} \cdot \dfrac{3z^2}{3z^2} = \dfrac{15z^2}{36z^3}$ $\dfrac{4}{9z^3} \cdot \dfrac{4}{4} = \dfrac{16}{36z^3}$

$\dfrac{5}{12z} = \dfrac{15z^2}{36z^3}$ and $\dfrac{4}{9z^3} = \dfrac{16}{36z^3}$

b) **Step 1:** Identify and write down the LCD of $\dfrac{m}{m-4}$ and $\dfrac{3}{m+8}$:

     LCD = $(m-4)(m+8)$

Step 2: Compare the denominators of $\dfrac{m}{m-4}$ and $\dfrac{3}{m+8}$ to the LCD and ask yourself, "What's missing?"

$$\dfrac{m}{m-4}: \quad m-4 \text{ is "missing"}$$
$$\text{the factor } m+8.$$

$$\dfrac{3}{m+8}: \quad m+8 \text{ is "missing"}$$
$$\text{the factor } m-4.$$

Step 3: Multiply the numerator and denominator by $m+8$.

$$\dfrac{m}{m-4} \cdot \dfrac{m+8}{m+8} = \dfrac{m(m+8)}{(m-4)(m+8)}$$
$$= \dfrac{m^2+8m}{(m-4)(m+8)}$$

Multiply the numerator and denominator by $m-4$.

$$\dfrac{3}{m+8} \cdot \dfrac{m-4}{m-4} = \dfrac{3(m-4)}{(m+8)(m-4)}$$
$$= \dfrac{3m-12}{(m-4)(m+8)}$$

Notice that we multiplied the factors in the numerator but left the denominator in factored form.

$$\dfrac{m}{m-4} = \dfrac{m^2+8m}{(m-4)(m+8)} \quad \text{and} \quad \dfrac{3}{m+8} = \dfrac{3m-12}{(m-4)(m+8)}$$

c) **Step 1:** Identify and write down the LCD of $\dfrac{2}{3x^2-6x}$ and $\dfrac{9x}{x^2-7x+10}$.

First, we must factor the denominators.

$$\dfrac{2}{3x^2-6x} = \dfrac{2}{3x(x-2)}, \quad \dfrac{9x}{x^2-7x+10} = \dfrac{9x}{(x-2)(x-5)}$$

We will work with the factored forms of the expressions.

$$\textbf{LCD} = \mathbf{3x(x-2)(x-5)}$$

Step 2: Compare the denominators of $\dfrac{2}{3x(x-2)}$ and $\dfrac{9x}{(x-2)(x-5)}$ to the LCD and ask yourself, "What's missing?"

$$\dfrac{2}{3x(x-2)}: \quad 3x(x-2) \text{ is}$$
$$\text{"missing" the}$$
$$\text{factor } x-5.$$

$$\dfrac{9x}{(x-2)(x-5)}: \quad (x-2)(x-5) \text{ is}$$
$$\text{"missing" } 3x.$$

Step 3: Multiply the numerator and denominator by $x-5$.

$$\dfrac{2}{3x(x-2)} \cdot \dfrac{x-5}{x-5} = \dfrac{2(x-5)}{3x(x-2)(x-5)}$$
$$= \dfrac{2x-10}{3x(x-2)(x-5)}$$

Multiply the numerator and denominator by $3x$.

$$\dfrac{9x}{(x-2)(x-5)} \cdot \dfrac{3x}{3x} = \dfrac{27x^2}{3x(x-2)(x-5)}$$

$$\dfrac{2}{3x^2-6x} = \dfrac{2x-10}{3x(x-2)(x-5)} \quad \text{and} \quad \dfrac{9x}{x^2-7x+10} = \dfrac{27x^2}{3x(x-2)(x-5)}$$

[YOU TRY 5] Identify the least common denominator of each pair of rational expressions, and rewrite each as an equivalent expression with the LCD as its denominator.

a) $\dfrac{3}{10a^6}, \dfrac{7}{8a^5}$
b) $\dfrac{6}{n+10}, \dfrac{n}{2n-3}$
c) $\dfrac{v-9}{v^2+15v+44}, \dfrac{8}{5v^2+55v}$

E Evaluate 8.3 Exercises

Do the exercises, and check your work.

*Additional answers can be found in the Answers to Exercises appendix.

Objective 1: Find the Least Common Denominator for a Group of Rational Expressions

Find the LCD of each group of fractions.

1) $\dfrac{7}{12}, \dfrac{2}{15}$ 60

2) $\dfrac{3}{8}, \dfrac{9}{7}$ 56

3) $\dfrac{27}{40}, \dfrac{11}{10}, \dfrac{5}{12}$ 120

4) $\dfrac{19}{8}, \dfrac{1}{12}, \dfrac{3}{32}$ 96

5) $\dfrac{3}{n^7}, \dfrac{5}{n^{11}}$ n^{11}

6) $\dfrac{4}{c^2}, \dfrac{8}{c^3}$ c^3

7) $\dfrac{13}{14r^4}, \dfrac{3}{4r^7}$ $28r^7$

8) $\dfrac{11}{6p^4}, \dfrac{3}{10p^9}$ $30p^9$

9) $-\dfrac{5}{6z^5}, \dfrac{7}{36z^2}$ $36z^5$

10) $\dfrac{5}{24w^5}, -\dfrac{1}{4w^{10}}$ $24w^{10}$

11) $\dfrac{7}{10m}, \dfrac{9}{22m^4}$ $110m^4$

12) $-\dfrac{3}{2k^2}, \dfrac{5}{14k^5}$ $14k^5$

13) $\dfrac{4}{24x^3y^2}, \dfrac{11}{6x^3y}$ $24x^3y^2$

14) $\dfrac{3}{10a^4b^2}, \dfrac{8}{15ab^4}$ $30a^4b^4$

15) $\dfrac{4}{11}, \dfrac{8}{z - 3}$ $11(z - 3)$

16) $\dfrac{3}{n + 8}, \dfrac{1}{5}$ $5(n + 8)$

17) $\dfrac{10}{w}, \dfrac{6}{2w + 1}$ $w(2w + 1)$

18) $\dfrac{1}{y}, -\dfrac{6}{6y + 1}$ $y(6y + 1)$

19) What is the first step for finding the LCD of $\dfrac{9}{8t - 10}$ and $\dfrac{3t}{20t - 25}$?   Factor the denominators.

20) Is $(h - 9)(9 - h)$ the LCD of $\dfrac{2h}{h - 9}$ and $\dfrac{4}{9 - h}$?

Explain your answer.

No, because $\dfrac{4}{9 - h}$ can be written as $-\dfrac{4}{h - 9}$, so the LCD is $h - 9$.

Find the LCD of each group of fractions.

21) $\dfrac{8}{5c - 5}, \dfrac{9}{2c - 2}$ $10(c - 1)$

22) $\dfrac{5}{7k + 14}, \dfrac{11}{4k + 8}$ $28(k + 2)$

23) $\dfrac{2}{9p^4 - 6p^3}, \dfrac{3}{3p^6 - 2p^5}$ $3p^5(3p - 2)$

24) $\dfrac{21}{6a^2 - 8a}, \dfrac{13}{18a^3 - 24a^2}$ $6a^2(3a - 4)$

25) $\dfrac{4m}{m - 7}, \dfrac{2}{m - 3}$ $(m - 7)(m - 3)$

26) $\dfrac{5}{r + 9}, \dfrac{7}{r - 1}$ $(r + 9)(r - 1)$

27) $\dfrac{11}{z^2 + 11z + 24}, \dfrac{7z}{z^2 + 5z - 24}$ $(z + 3)(z + 8)(z - 3)$

28) $\dfrac{7x}{x^2 - 12x + 35}, \dfrac{x}{x^2 - x - 20}$ $(x - 5)(x - 7)(x + 4)$

29) $\dfrac{n - 1}{n + 4}, \dfrac{2n + 7}{n^2 + 8n + 16}$ $(n + 4)^2$

30) $\dfrac{k - 2}{k + 9}, \dfrac{5k + 1}{k^2 + 18k + 81}$ $(k + 9)^2$

31) $\dfrac{14t}{t^2 - 3t - 18}, -\dfrac{6}{t^2 - 36}, \dfrac{t}{t^2 + 9t + 18}$ $(t + 6)(t - 6)(t + 3)$

32) $\dfrac{6w}{w^2 - 10w + 16}, \dfrac{3}{w^2 - 7w - 8}, \dfrac{4w}{w^2 - w - 2}$ $(w - 8)(w - 2)(w + 1)$

33) $\dfrac{6}{a - 8}, \dfrac{7}{8 - a}$ $a - 8$ or $8 - a$

34) $\dfrac{6}{b - 3}, \dfrac{5}{3 - b}$ $b - 3$ or $3 - b$

35) $\dfrac{12}{y - x}, \dfrac{5y}{x - y}$ $x - y$ or $y - x$

36) $\dfrac{u}{v - u}, \dfrac{8}{u - v}$ $u - v$ or $v - u$

Objective 2: Rewrite Rational Expressions Using Their LCD

37) Explain, in your own words, how to rewrite $\dfrac{4}{x + 9}$ as an equivalent rational expression with a denominator of $(x + 9)(x - 3)$.   Answers may vary.

38) Explain, in your own words, how to rewrite $\dfrac{7}{5 - m}$ as an equivalent rational expression with a denominator of $m - 5$.   Answers may vary.

Rewrite each rational expression with the indicated denominator.

39) $\dfrac{7}{12} = \dfrac{}{48}$ $\quad \dfrac{28}{48}$

40) $\dfrac{5}{7} = \dfrac{}{42}$ $\quad \dfrac{30}{42}$

41) $\dfrac{8}{z} = \dfrac{}{9z}$ $\quad \dfrac{72}{9z}$

42) $\dfrac{-6}{b} = \dfrac{}{4b}$ $\quad \dfrac{-24}{4b}$

43) $\dfrac{3}{8k} = \dfrac{}{56k^4}$ $\quad \dfrac{21k^3}{56k^4}$

44) $\dfrac{5}{3p^4} = \dfrac{}{9p^6}$ $\quad \dfrac{15p^2}{9p^6}$

45) $\dfrac{6}{5t^5u^2} = \dfrac{}{10t^7u^5}$ $\quad \dfrac{12t^3u^3}{10t^7u^5}$

46) $\dfrac{13}{6cd^2} = \dfrac{}{24c^3d^3}$ $\quad \dfrac{52c^2d}{24c^3d^3}$

47) $\dfrac{7}{3r+4} = \dfrac{}{r(3r+4)}$ $\quad \dfrac{7r}{r(3r+4)}$

48) $\dfrac{8}{m-8} = \dfrac{}{m(m-8)}$ $\quad \dfrac{8m}{m(m-8)}$

49) $\dfrac{v}{4(v-3)} = \dfrac{}{16v^5(v-3)}$ $\quad \dfrac{4v^6}{16v^5(v-3)}$

50) $\dfrac{a}{5(2a+7)} = \dfrac{}{15a(2a+7)}$ $\quad \dfrac{3a^2}{15a(2a+7)}$

51) $\dfrac{9x}{x+6} = \dfrac{}{(x+6)(x-5)}$ $\quad \dfrac{9x^2-45x}{(x+6)(x-5)}$

52) $\dfrac{5b}{b+3} = \dfrac{}{(b+3)(b+7)}$ $\quad \dfrac{5b^2+35b}{(b+3)(b+7)}$

53) $\dfrac{z-3}{2z-5} = \dfrac{}{(2z-5)(z+8)}$ $\quad \dfrac{z^2+5z-24}{(2z-5)(z+8)}$

54) $\dfrac{w+2}{4w-1} = \dfrac{}{(4w-1)(w-4)}$ $\quad \dfrac{w^2-2w-8}{(4w-1)(w-4)}$

55) $\dfrac{5}{3-p} = \dfrac{}{p-3}$ $\quad -\dfrac{5}{p-3}$

56) $\dfrac{10}{10-n} = \dfrac{}{n-10}$ $\quad -\dfrac{10}{n-10}$

57) $-\dfrac{8c}{6c-7} = \dfrac{}{7-6c}$ $\quad \dfrac{8c}{7-6c}$

58) $-\dfrac{g}{3g-2} = \dfrac{}{2-3g}$ $\quad \dfrac{g}{2-3g}$

Identify the least common denominator of each pair of rational expressions, and rewrite each as an equivalent rational expression with the LCD as its denominator.

59) $\dfrac{8}{15}, \dfrac{1}{6}$ $\quad \dfrac{8}{15} = \dfrac{16}{30}; \dfrac{1}{6} = \dfrac{5}{30}$

60) $\dfrac{3}{8}, \dfrac{5}{12}$ $\quad \dfrac{3}{8} = \dfrac{9}{24}; \dfrac{5}{12} = \dfrac{10}{24}$

61) $\dfrac{4}{u}, \dfrac{8}{u^3}$ $\quad \dfrac{4}{u} = \dfrac{4u^2}{u^3}; \dfrac{8}{u^3} = \dfrac{8}{u^3}$

62) $\dfrac{9}{d^5}, \dfrac{7}{d^2}$ $\quad \dfrac{9}{d^5} = \dfrac{9}{d^5}; \dfrac{7}{d^2} = \dfrac{7d^3}{d^5}$

63) $\dfrac{9}{8n^6}, \dfrac{2}{3n^2}$

64) $\dfrac{5}{8a}, \dfrac{7}{10a^5}$

65) $\dfrac{6}{4a^3b^5}, \dfrac{6}{a^4b}$

66) $\dfrac{3}{x^3y}, \dfrac{6}{5xy^5}$

67) $\dfrac{r}{5}, \dfrac{2}{r-4}$

68) $\dfrac{t}{5t-1}, \dfrac{8}{7}$

69) $\dfrac{3}{d}, \dfrac{7}{d-9}$

70) $\dfrac{5}{c}, \dfrac{4}{c+2}$

71) $\dfrac{m}{m+7}, \dfrac{3}{m}$

72) $\dfrac{z}{z-4}, \dfrac{5}{z}$

73) $\dfrac{a}{30a-15}, \dfrac{1}{12a-6}$

74) $\dfrac{7}{24x-16}, \dfrac{x}{18x-12}$

75) $\dfrac{9}{k-9}, \dfrac{5k}{k+3}$

76) $\dfrac{6}{h+1}, \dfrac{11h}{h+7}$

77) $\dfrac{3}{a+2}, \dfrac{2a}{3a+4}$

78) $\dfrac{b}{6b-5}, \dfrac{8}{b-9}$

79) $\dfrac{9y}{y^2-y-42}, \dfrac{3}{2y^2+12y}$

80) $\dfrac{12q}{q^2-6q-16}, \dfrac{4}{2q^2-16q}$

81) $\dfrac{c}{c^2+9c+18}, \dfrac{11}{c^2+12c+36}$

82) $\dfrac{z}{z^2-8z+16}, \dfrac{9z}{z^2+4z-32}$

83) $\dfrac{11}{g-3}, \dfrac{4}{3-g}$

84) $\dfrac{6}{n-9}, \dfrac{1}{9-n}$

85) $\dfrac{4}{3x-4}, \dfrac{7x}{16-9x^2}$

86) $\dfrac{12}{5k-2}, \dfrac{4k}{4-25k^2}$

87) $\dfrac{2}{z^2+3z}, \dfrac{6}{3z^2+9z}, \dfrac{8}{z^2+6z+9}$

88) $\dfrac{4}{w^2-4w}, \dfrac{6}{7w^2-28w}, \dfrac{11}{w^2-8w+16}$

89) $\dfrac{t}{t^2-13t+30}, \dfrac{6}{t-10}, \dfrac{7}{t^2-9}$

90) $-\dfrac{2}{a+2}, \dfrac{a}{a^2-4}, \dfrac{15}{a^2-3a+2}$

R Rethink

R1) Where could you use some help to master the objectives presented in this section?

R2) When you got stuck on a "harder" problem, what did you do to help move forward on the problem?

8.4 Adding and Subtracting Rational Expressions

What are your objectives for Section 8.4?	How can you accomplish each objective?
1 Add and Subtract Rational Expressions with a Common Denominator	• Understand the procedure for **Adding and Subtracting Rational Expressions.** • Complete the given examples on your own. • Complete You Trys 1 and 2.
2 Add and Subtract Rational Expressions with Different Denominators	• Write the **Steps for Adding and Subtracting Rational Expressions with Different Denominators** in your own words. • Complete the given examples on your own. • Complete You Trys 3 and 4.
3 Add and Subtract Rational Expressions with Special Denominators	• Review the techniques you used to factor special trinomials and binomials in Section 7.4. • Complete the given example on your own. • Complete You Try 5.

W Work Read the explanations, follow the examples, take notes, and complete the You Trys.

We know that in order to add or subtract fractions, they must have a common denominator. The same is true for rational expressions.

1 Add and Subtract Rational Expressions with a Common Denominator

Let's first look at fractions and rational expressions with common denominators.

EXAMPLE 1

Add or subtract.

a) $\dfrac{8}{11} - \dfrac{5}{11}$ b) $\dfrac{2x}{4x-9} + \dfrac{5x+3}{4x-9}$

Solution

a) Since the fractions have the same denominator, subtract the terms in the numerator and keep the common denominator.

$$\frac{8}{11} - \frac{5}{11} = \frac{8-5}{11} = \frac{3}{11} \qquad \text{Subtract terms in the numerator.}$$

In-Class Example 1

Add or subtract.

a) $\dfrac{11}{15} - \dfrac{4}{15}$

b) $\dfrac{3d}{6d+1} + \dfrac{2d+9}{6d+1}$

Answer: a) $\dfrac{7}{15}$ b) $\dfrac{5d+9}{6d+1}$

b) Since the rational expressions have the same denominator, add the terms in the numerator and keep the common denominator.

$$\frac{2x}{4x-9} + \frac{5x+3}{4x-9} = \frac{2x+(5x+3)}{4x-9} \quad \text{Add terms in the numerator.}$$

$$= \frac{7x+3}{4x-9} \quad \text{Combine like terms.}$$

We can generalize the procedure for adding and subtracting rational expressions that have a common denominator as follows.

Procedure Adding and Subtracting Rational Expressions

If $\dfrac{P}{Q}$ and $\dfrac{R}{Q}$ are rational expressions with $Q \neq 0$, then

$$1) \quad \frac{P}{Q} + \frac{R}{Q} = \frac{P+R}{Q} \qquad \text{and} \qquad 2) \quad \frac{P}{Q} - \frac{R}{Q} = \frac{P-R}{Q}$$

[**YOU TRY 1**] Add or subtract.

a) $\dfrac{11}{12} - \dfrac{7}{12}$ b) $\dfrac{6h}{5h-2} + \dfrac{3h+8}{5h-2}$

All answers to a sum or difference of rational expressions should be in lowest terms. Sometimes it is necessary to simplify our result to lowest terms by factoring the numerator and dividing the numerator and denominator by the greatest common factor.

EXAMPLE 2

In-Class Example 2

Subtract $\dfrac{a^2-10}{a(a+8)} - \dfrac{14-5a}{a(a+8)}$.

Answer: $\dfrac{a-3}{a}$

Subtract $\dfrac{c^2-3}{c(c+4)} - \dfrac{5-2c}{c(c+4)}$.

Solution

$$\frac{c^2-3}{c(c+4)} - \frac{5-2c}{c(c+4)} = \frac{(c^2-3)-(5-2c)}{c(c+4)} \quad \text{Subtract terms in the numerator.}$$

$$= \frac{c^2-3-5+2c}{c(c+4)} \quad \text{Distribute.}$$

$$= \frac{c^2+2c-8}{c(c+4)} \quad \text{Combine like terms.}$$

$$= \frac{(c+4)(c-2)}{c(c+4)} \quad \text{Factor the numerator.}$$

$$= \frac{c-2}{c} \quad \text{Divide out the common factors.}$$

[**YOU TRY 2**] Add or subtract.

a) $\dfrac{19}{32w} + \dfrac{9}{32w}$ b) $\dfrac{m^2-5}{m(m+6)} - \dfrac{3m+49}{m(m+6)}$

SECTION 8.4 **Adding and Subtracting Rational Expressions**

After combining like terms in the numerator, ask yourself, "Can I factor the numerator?" If so, factor it. Sometimes, the expression can be simplified by dividing the numerator and denominator by the greatest common factor.

2 Add and Subtract Rational Expressions with Different Denominators

If we are asked to add or subtract rational expressions with different denominators, we must begin by rewriting each expression with the least common denominator. Then, add or subtract. Write the answer in lowest terms.

Using the procedure studied in Section 8.3, here are the steps to follow to add or subtract rational expressions with different denominators.

 Hint

Write this procedure in your own words.

Procedure Steps for Adding and Subtracting Rational Expressions with Different Denominators

1) Factor the denominators.

2) Write down the LCD.

3) Rewrite each rational expression as an equivalent rational expression with the LCD.

4) Add or subtract the numerators and keep the common denominator in factored form.

5) After combining like terms in the numerator ask yourself, *"Can I factor it?"* If so, factor.

6) Divide out common factors, if possible. The final answer should be written in lowest terms.

EXAMPLE 3

Add or subtract.

In-Class Example 3

Add or subtract.

a) $\dfrac{u - 17}{18} + \dfrac{u + 1}{9}$

b) $\dfrac{5}{4p} - \dfrac{1}{6p^2}$

c) $\dfrac{9r - 35}{r^2 - 49} + \dfrac{r}{r + 7}$

Answer: a) $\dfrac{u - 5}{6}$

b) $\dfrac{15p - 2}{12p^2}$ c) $\dfrac{r - 5}{r - 7}$

a) $\dfrac{t + 6}{4} + \dfrac{t - 8}{12}$ b) $\dfrac{3}{10a} - \dfrac{7}{8a^2}$ c) $\dfrac{7n - 30}{n^2 - 36} + \dfrac{n}{n + 6}$

Solution

a) The LCD is **12**. $\dfrac{t - 8}{12}$ already has the LCD.

Rewrite $\dfrac{t + 6}{4}$ with the LCD: $\dfrac{t + 6}{4} \cdot \dfrac{3}{3} = \dfrac{3(t + 6)}{12}$

$\dfrac{t + 6}{4} + \dfrac{t - 8}{12} = \dfrac{3(t + 6)}{12} + \dfrac{t - 8}{12}$ Write each expression with the LCD.

$= \dfrac{3(t + 6) + (t - 8)}{12}$ Add the numerators.

$= \dfrac{3t + 18 + t - 8}{12}$ Distribute.

$= \dfrac{4t + 10}{12}$ Combine like terms.

 Hint

Are you writing out the example as you read it?

Ask yourself, *"Can I factor the numerator?"* Yes.

$$= \frac{\overset{1}{\cancel{2}}(2t + 5)}{\underset{6}{\cancel{12}}} \qquad \text{Factor.}$$

$$= \frac{2t + 5}{6} \qquad \text{Divide out the common factor of 2.}$$

b) The LCD of $\dfrac{3}{10a}$ and $\dfrac{7}{8a^2}$ is $\mathbf{40a^2}$. Rewrite each expression with the LCD.

$$\frac{3}{10a} \cdot \frac{4a}{4a} = \frac{12a}{40a^2} \qquad \text{and} \qquad \frac{7}{8a^2} \cdot \frac{5}{5} = \frac{35}{40a^2}$$

$$\frac{3}{10a} - \frac{7}{8a^2} = \frac{12a}{40a^2} - \frac{35}{40a^2} \qquad \text{Write each expression with the LCD.}$$

$$= \frac{12a - 35}{40a^2} \qquad \text{Subtract the numerators.}$$

"Can I factor the numerator?" No. The expression is in simplest form since the numerator and denominator have no common factors.

c) First, factor the denominator of $\dfrac{7n - 30}{n^2 - 36}$ to get $\dfrac{7n - 30}{(n + 6)(n - 6)}$.

The LCD of $\dfrac{7n - 30}{(n + 6)(n - 6)}$ and $\dfrac{n}{n + 6}$ is $\mathbf{(n + 6)(n - 6)}$.

Rewrite $\dfrac{n}{n + 6}$ with the LCD: $\dfrac{n}{n + 6} \cdot \dfrac{n - 6}{n - 6} = \dfrac{n(n - 6)}{(n + 6)(n - 6)}$

$$\frac{7n - 30}{n^2 - 36} + \frac{n}{n + 6} = \frac{7n - 30}{(n + 6)(n - 6)} + \frac{n}{n + 6} \qquad \begin{array}{l}\text{Factor the}\\\text{denominator.}\end{array}$$

$$= \frac{7n - 30}{(n + 6)(n - 6)} + \frac{n(n - 6)}{(n + 6)(n - 6)} \qquad \begin{array}{l}\text{Write each expression}\\\text{with the LCD.}\end{array}$$

$$= \frac{7n - 30 + n(n - 6)}{(n + 6)(n - 6)} \qquad \text{Add the numerators.}$$

$$= \frac{7n - 30 + n^2 - 6n}{(n + 6)(n - 6)} \qquad \text{Distribute.}$$

$$= \frac{n^2 + n - 30}{(n + 6)(n - 6)} \qquad \text{Combine like terms.}$$

Ask yourself, *"Can I factor the numerator?"* Yes.

$$= \frac{\cancel{(n + 6)}(n - 5)}{\cancel{(n + 6)}(n - 6)} \qquad \text{Factor.}$$

$$= \frac{n - 5}{n - 6} \qquad \text{Write in lowest terms.}$$

[YOU TRY 3]   Add or subtract.

a) $\dfrac{t + 4}{5} + \dfrac{2t - 7}{15}$

b) $\dfrac{7}{12v} - \dfrac{9}{16v^2}$

c) $\dfrac{15h - 8}{h^2 - 64} + \dfrac{h}{h + 8}$

EXAMPLE 4

In-Class Example 4

Subtract
$$\frac{5k}{k^2 + 11k + 30} - \frac{2k + 1}{k^2 + 2k - 15}.$$

Answer:
$$\frac{3k^2 - 28k - 6}{(k + 6)(k + 5)(k - 3)}$$

Subtract $\dfrac{6r}{r^2 + 10r + 16} - \dfrac{3r + 4}{r^2 + 3r - 40}$.

Solution

Factor the denominators, then write down the LCD.

$$\frac{6r}{r^2 + 10r + 16} = \frac{6r}{(r + 8)(r + 2)}, \qquad \frac{3r + 4}{r^2 + 3r - 40} = \frac{3r + 4}{(r + 8)(r - 5)}$$

LCD = $(r + 8)(r + 2)(r - 5)$

Rewrite each expression with the LCD.

$$\frac{6r}{(r + 8)(r + 2)} \cdot \frac{r - 5}{r - 5} = \frac{6r(r - 5)}{(r + 8)(r + 2)(r - 5)}$$

$$\frac{3r + 4}{(r + 8)(r - 5)} \cdot \frac{r + 2}{r + 2} = \frac{(3r + 4)(r + 2)}{(r + 8)(r + 2)(r - 5)}$$

$$\frac{6r}{r^2 + 10r + 16} - \frac{3r + 4}{r^2 + 3r - 40}$$

$$= \frac{6r}{(r + 8)(r + 2)} - \frac{3r + 4}{(r + 8)(r - 5)} \qquad \text{Factor the denominators.}$$

$$= \frac{6r(r - 5)}{(r + 8)(r + 2)(r - 5)} - \frac{(3r + 4)(r + 2)}{(r + 8)(r + 2)(r - 5)} \qquad \text{Write each expression with the LCD.}$$

$$= \frac{6r(r - 5) - (3r + 4)(r + 2)}{(r + 8)(r + 2)(r - 5)} \qquad \text{Subtract the numerators.}$$

$$= \frac{6r^2 - 30r - (3r^2 + 10r + 8)}{(r + 8)(r + 2)(r - 5)} \qquad \text{Distribute. You } must \text{ use parentheses.}$$

$$= \frac{6r^2 - 30r - 3r^2 - 10r - 8}{(r + 8)(r + 2)(r - 5)} \qquad \text{Distribute.}$$

$$= \frac{3r^2 - 40r - 8}{(r + 8)(r + 2)(r - 5)} \qquad \text{Combine like terms.}$$

Ask yourself, *"Can I factor the numerator?"* No. The expression is in simplest form since the numerator and denominator have no common factors.

 BE CAREFUL In Example 4, when you move from

$$\frac{6r(r - 5) - (3r + 4)(r + 2)}{(r + 8)(r + 2)(r - 5)} \quad \text{to} \quad \frac{6r^2 - 30r - (3r^2 + 10r + 8)}{(r + 8)(r + 2)(r - 5)}$$

you *must* use parentheses since the entire quantity $3r^2 + 10r + 8$ is being subtracted from $6r^2 - 30r$.

[YOU TRY 4]

Subtract $\dfrac{4z}{z^2 + 10z + 21} - \dfrac{3z + 5}{z^2 - z - 12}$.

3 Add and Subtract Rational Expressions with Special Denominators

EXAMPLE 5

In-Class Example 5

Add or subtract.

a) $\dfrac{b}{b-5} - \dfrac{4}{5-b}$

b) $\dfrac{6}{1-x} + \dfrac{2}{x^2-1}$

Answer: a) $\dfrac{b+4}{b-5}$

b) $\dfrac{-2(3x+2)}{(x+1)(x-1)}$

W Hint

What did you learn in Sections 8.1 and 7.4?

Add or subtract.

a) $\dfrac{z}{z-9} - \dfrac{8}{9-z}$ b) $\dfrac{4}{7-w} + \dfrac{10}{w^2-49}$

Solution

a) Recall that $a - b = -(b - a)$. The least common denominator of $\dfrac{z}{z-9}$ and $\dfrac{8}{9-z}$ is $z - 9$ or $9 - z$. We will use **LCD = $z - 9$**.

Rewrite $\dfrac{8}{9-z}$ with the LCD:

$$\frac{8}{9-z} = \frac{8}{-(z-9)} = -\frac{8}{z-9}$$

$$\frac{z}{z-9} - \frac{8}{9-z} = \frac{z}{z-9} - \left(-\frac{8}{z-9}\right) \qquad \text{Write each expression with the LCD.}$$

$$= \frac{z}{z-9} + \frac{8}{z-9} \qquad \text{Distribute.}$$

$$= \frac{z+8}{z-9} \qquad \text{Add the numerators.}$$

b) Factor the denominator of $\dfrac{10}{w^2-49}$: $\dfrac{10}{w^2-49} = \dfrac{10}{(w+7)(w-7)}$

Rewrite $\dfrac{4}{7-w}$ with a denominator of $w - 7$:

$$\frac{4}{7-w} = \frac{4}{-(w-7)} = -\frac{4}{w-7}$$

Now we must find the LCD of $\dfrac{10}{(w+7)(w-7)}$ and $-\dfrac{4}{w-7}$.

$$\textbf{LCD} = (w+7)(w-7)$$

Rewrite $-\dfrac{4}{w-7}$ with the LCD.

$$-\frac{4}{w-7} \cdot \frac{w+7}{w+7} = -\frac{4(w+7)}{(w+7)(w-7)} = \frac{-4(w+7)}{(w+7)(w-7)}$$

$$\frac{4}{7-w} + \frac{10}{w^2-49} = -\frac{4}{w-7} + \frac{10}{(w+7)(w-7)}$$

$$= \frac{-4(w+7)}{(w+7)(w-7)} + \frac{10}{(w+7)(w-7)} \qquad \text{Write each expression with the LCD.}$$

$$= \frac{-4(w+7) + 10}{(w+7)(w-7)} \qquad \text{Add the numerators.}$$

$$= \frac{-4w - 28 + 10}{(w+7)(w-7)} \qquad \text{Distribute.}$$

$$= \frac{-4w - 18}{(w+7)(w-7)} \qquad \text{Combine like terms.}$$

Ask yourself, *"Can I factor the numerator?"* Yes.

$$= \frac{-2(2w + 9)}{(w + 7)(w - 7)} \quad \text{Factor.}$$

Although the numerator factors, the numerator and denominator do not contain any common factors. The result, $\frac{-2(2w + 9)}{(w + 7)(w - 7)}$, is in simplest form.

[YOU TRY 5] Add or subtract.

a) $\dfrac{n}{n - 12} - \dfrac{7}{12 - n}$ b) $\dfrac{15}{4 - t} + \dfrac{20}{t^2 - 16}$

W Hint

Did you check your answers? Were you able to figure out where you might have made any mistakes?

ANSWERS TO [YOU TRY] EXERCISES

1) a) $\dfrac{1}{3}$ b) $\dfrac{9h + 8}{5h - 2}$ 2) a) $\dfrac{7}{8w}$ b) $\dfrac{m - 9}{m}$ 3) a) $\dfrac{t + 1}{3}$ b) $\dfrac{28v - 27}{48v^2}$

c) $\dfrac{h - 1}{h - 8}$ 4) $\dfrac{z^2 - 42z - 35}{(z + 7)(z + 3)(z - 4)}$ 5) a) $\dfrac{n + 7}{n - 12}$ b) $\dfrac{-5(3t + 8)}{(t + 4)(t - 4)}$

E Evaluate **8.4** Exercises Do the exercises, and check your work.

*Additional answers can be found in the Answers to Exercises appendix.

Objective 1: Add and Subtract Rational Expressions with a Common Denominator

Add or subtract.

24 1) $\dfrac{5}{16} + \dfrac{9}{16}$ $\dfrac{7}{8}$

2) $\dfrac{5}{7} - \dfrac{3}{7}$ $\dfrac{2}{7}$

3) $\dfrac{11}{14} - \dfrac{3}{14}$ $\dfrac{4}{7}$

4) $\dfrac{1}{10} + \dfrac{9}{10}$ 1

5) $\dfrac{5}{p} - \dfrac{23}{p}$ $-\dfrac{18}{p}$

6) $\dfrac{6}{a} + \dfrac{3}{a}$ $\dfrac{9}{a}$

7) $\dfrac{7}{3c} + \dfrac{8}{3c}$ $\dfrac{5}{c}$

8) $\dfrac{10}{3k^2} - \dfrac{2}{3k^2}$ $\dfrac{8}{3k^2}$

9) $\dfrac{6}{z - 1} + \dfrac{z}{z - 1}$ $\dfrac{z + 6}{z - 1}$

10) $\dfrac{4n}{n + 9} - \dfrac{6}{n + 9}$ $\dfrac{4n - 6}{n + 9}$

24 11) $\dfrac{8}{x + 4} + \dfrac{2x}{x + 4}$ 2

12) $\dfrac{5m}{m + 7} + \dfrac{35}{m + 7}$ 5

13) $\dfrac{25t + 17}{t(4t + 3)} - \dfrac{5t + 2}{t(4t + 3)}$ $\dfrac{5}{t}$

14) $\dfrac{9w - 20}{w(2w - 5)} - \dfrac{20 - 7w}{w(2w - 5)}$ $\dfrac{8}{w}$

15) $\dfrac{d^2 + 15}{(d + 5)(d + 2)} + \dfrac{8d - 3}{(d + 5)(d + 2)}$ $\dfrac{d + 6}{d + 5}$

16) $\dfrac{2r + 15}{(r - 5)(r + 4)} + \dfrac{r^2 - 10r}{(r - 5)(r + 4)}$ $\dfrac{r - 3}{r + 4}$

Objective 2: Add and Subtract Rational Expressions with Different Denominators

17) For $\dfrac{4}{9b^2}$ and $\dfrac{5}{6b^4}$,

a) find the LCD. $18b^4$

b) explain, in your own words, how to rewrite each expression with the LCD.

c) rewrite each expression with the LCD.

$\dfrac{4}{9b^2} = \dfrac{8b^2}{18b^4}; \dfrac{5}{6b^4} = \dfrac{15}{18b^4}$

18) For $\dfrac{8}{x - 3}$ and $\dfrac{2}{x}$,

a) find the LCD. $x(x - 3)$

b) explain, in your own words, how to rewrite each expression with the LCD.

c) rewrite each expression with the LCD.

$\dfrac{8}{x - 3} = \dfrac{8x}{x(x - 3)}; \dfrac{2}{x} = \dfrac{2x - 6}{x(x - 3)}$

Add or subtract.

19) $\dfrac{3}{8} + \dfrac{2}{5}$ $\dfrac{31}{40}$

20) $\dfrac{5}{12} - \dfrac{1}{8}$ $\dfrac{7}{24}$

21) $\dfrac{4t}{3} + \dfrac{3}{2}$ $\dfrac{8t + 9}{6}$

22) $\dfrac{14x}{15} - \dfrac{5x}{6}$ $\dfrac{x}{10}$

23) $\dfrac{10}{3h^3} + \dfrac{2}{5h}$ $\dfrac{6h^2 + 50}{15h^3}$

24) $\dfrac{5}{8u} - \dfrac{2}{3u^2}$ $\dfrac{15u - 16}{24u^2}$

25) $\dfrac{3}{2f^2} - \dfrac{7}{f}$ $\dfrac{3 - 14f}{2f^2}$

26) $\dfrac{8}{5a} + \dfrac{2}{5a^2}$ $\dfrac{2(4a + 1)}{5a^2}$

27) $\dfrac{13}{y + 3} + \dfrac{3}{y}$ $\dfrac{16y + 9}{y(y + 3)}$

28) $\dfrac{3}{k} + \dfrac{11}{k + 9}$ $\dfrac{14k + 27}{k(k + 9)}$

29) $\dfrac{15}{d - 8} - \dfrac{4}{d}$ $\dfrac{11d + 32}{d(d - 8)}$

30) $\dfrac{14}{r - 5} - \dfrac{3}{r}$ $\dfrac{11r + 15}{r(r - 5)}$

31) $\dfrac{9}{c - 4} + \dfrac{6}{c + 8}$ $\dfrac{3(5c + 16)}{(c - 4)(c + 8)}$

32) $\dfrac{2}{z + 5} + \dfrac{1}{z + 2}$ $\dfrac{3(z + 3)}{(z + 5)(z + 2)}$

33) $\dfrac{m}{3m + 5} - \dfrac{2}{m - 10}$ $\dfrac{m^2 - 16m - 10}{(3m + 5)(m - 10)}$

34) $\dfrac{x}{x + 4} - \dfrac{3}{2x + 1}$ $\dfrac{2(x - 3)(x + 2)}{(x + 4)(2x + 1)}$

35) $\dfrac{8u + 2}{u^2 - 1} + \dfrac{3u}{u + 1}$ $\dfrac{3u + 2}{u - 1}$

36) $\dfrac{t}{t + 7} + \dfrac{9t - 35}{t^2 - 49}$ $\dfrac{t - 5}{t - 7}$

37) $\dfrac{7g}{g^2 + 10g + 16} + \dfrac{3}{g^2 - 64}$ $\dfrac{7g^2 - 53g + 6}{(g + 2)(g + 8)(g - 8)}$

38) $\dfrac{b}{b^2 - 25} + \dfrac{8}{b^2 - 3b - 40}$ $\dfrac{b^2 - 40}{(b + 5)(b - 5)(b - 8)}$

39) $\dfrac{5a}{a^2 - 6a - 27} - \dfrac{2a + 1}{a^2 + 2a - 3}$ $\dfrac{3(a + 1)}{(a - 9)(a - 1)}$

40) $\dfrac{3c}{c^2 + 4c - 12} - \dfrac{2c - 5}{c^2 + 2c - 24}$ $\dfrac{(c - 5)(c + 2)}{(c + 6)(c - 2)(c - 4)}$

41) $\dfrac{2x}{x^2 + x - 20} - \dfrac{4}{x^2 + 2x - 15}$ $\dfrac{2(x^2 - 5x + 8)}{(x - 4)(x + 5)(x - 3)}$

42) $\dfrac{3m}{m^2 + 10m + 24} - \dfrac{2}{m^2 + 3m - 4}$ $\dfrac{(3m + 4)(m - 3)}{(m + 4)(m - 1)(m + 6)}$

43) $\dfrac{4b + 1}{3b - 12} + \dfrac{5b}{b^2 - b - 12}$ $\dfrac{4b^2 + 28b + 3}{3(b - 4)(b + 3)}$

44) $\dfrac{k + 12}{2k - 18} + \dfrac{3k}{k^2 - 12k + 27}$ $\dfrac{k^2 + 15k - 36}{2(k - 9)(k - 3)}$

Objective 3: Add and Subtract Rational Expressions with Special Denominators

45) Is $(x - 6)(6 - x)$ the LCD for $\dfrac{9}{x - 6} + \dfrac{4}{6 - x}$?
Why or why not?

46) When Lamar adds $\dfrac{u}{7 - 2u} + \dfrac{5}{2u - 7}$, he gets $\dfrac{u - 5}{7 - 2u}$, but when he checks his answer in the back of the

book, it says that the answer is $\dfrac{5 - u}{2u - 7}$. Which is the correct answer?   Both are correct because $\dfrac{u - 5}{7 - 2u} = \dfrac{5 - u}{2u - 7}$.

Add or subtract.

47) $\dfrac{16}{q - 4} + \dfrac{10}{4 - q}$

48) $\dfrac{8}{z - 9} + \dfrac{4}{9 - z}$

49) $\dfrac{11}{f - 7} - \dfrac{15}{7 - f}$

50) $\dfrac{5}{a - b} - \dfrac{4}{b - a}$

51) $\dfrac{7}{x - 4} + \dfrac{x - 1}{4 - x}$

52) $\dfrac{10}{m - 5} + \dfrac{m + 21}{5 - m}$

53) $\dfrac{8}{3 - a} + \dfrac{a + 5}{a - 3}$ 1

54) $\dfrac{9}{6 - n} + \dfrac{n + 3}{n - 6}$ 1

55) $\dfrac{3}{2u - 3v} - \dfrac{6u}{3v - 2u}$

56) $\dfrac{3c}{11b - 5c} - \dfrac{9}{5c - 11b}$

57) $\dfrac{8}{x^2 - 9} + \dfrac{2}{3 - x}$

58) $\dfrac{4}{8 - y} + \dfrac{12}{y^2 - 64}$

59) $\dfrac{a}{4a^2 - 9} - \dfrac{4}{3 - 2a}$

60) $\dfrac{3b}{9b^2 - 25} - \dfrac{3}{5 - 3b}$

Mixed Exercises: Objectives 2 and 3
Perform the indicated operations.

61) $\dfrac{5}{a^2 - 2a} + \dfrac{8}{a} - \dfrac{10a}{a - 2}$ $\dfrac{-10a^2 + 8a - 11}{a(a - 2)}$

62) $\dfrac{3}{j^2 + 6j} + \dfrac{2j}{j + 6} - \dfrac{2}{3j}$ $\dfrac{6j^2 - 2j - 3}{3j(j + 6)}$

63) $\dfrac{3b - 1}{b^2 + 8b} + \dfrac{b}{3b^2 + 25b + 8} + \dfrac{2}{3b^2 + b}$ $\dfrac{10b^2 + 2b + 15}{b(b + 8)(3b + 1)}$

64) $\dfrac{2k + 7}{k^2 - 4k} + \dfrac{9k}{2k^2 - 15k + 28} + \dfrac{15}{2k^2 - 7k}$ $\dfrac{13k^2 + 15k - 109}{k(k - 4)(2k - 7)}$

65) $\dfrac{c}{c^2 - 8c + 16} - \dfrac{5}{c^2 - c - 12}$ $\dfrac{c^2 - 2c + 20}{(c - 4)^2(c + 3)}$

66) $\dfrac{n}{n^2 + 11n + 30} - \dfrac{6}{n^2 + 10n + 25}$ $\dfrac{n^2 - n - 36}{(n + 5)^2(n + 6)}$

67) $\dfrac{9}{4a + 4b} + \dfrac{8}{a - b} - \dfrac{6a}{a^2 - b^2}$ $\dfrac{17a + 23b}{4(a + b)(a - b)}$

68) $\dfrac{1}{x + y} + \dfrac{x}{x^2 - y^2} - \dfrac{10}{5x - 5y}$ $-\dfrac{3y}{(x + y)(x - y)}$

69) $\dfrac{2v + 1}{6v^2 - 29v - 5} - \dfrac{v - 2}{3v^2 - 13v - 10}$ $\dfrac{2(9v + 2)}{(6v + 1)(3v + 2)(v - 5)}$

70) $\dfrac{n + 2}{4n^2 + 11n - 3} - \dfrac{n - 3}{2n^2 + 7n + 3}$ $\dfrac{-2n^2 + 18n - 1}{(4n - 1)(2n + 1)(n + 3)}$

71) $\dfrac{g-5}{5g^2-30g}+\dfrac{g}{2g^2-17g+30}-\dfrac{6}{2g^2-5g}$ $\dfrac{7g^2-45g+205}{5g(g-6)(2g-5)}$

72) $\dfrac{y+6}{y^2-4y}+\dfrac{y}{2y^2-13y+20}-\dfrac{1}{2y^2-5y}$ $\dfrac{3y^2+6y-26}{y(y-4)(2y-5)}$

For each rectangle, find a rational expression in simplest form to represent its a) area and b) perimeter.

73)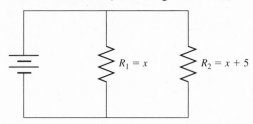

$\dfrac{k-4}{4}$ a) $\dfrac{2(k-4)}{k+1}$ b) $\dfrac{k^2-3k+28}{2(k+1)}$

$\dfrac{8}{k+1}$

74)

$\dfrac{4}{r-3}$ a) $\dfrac{2(r+1)}{3(r-3)}$ b) $\dfrac{r^2-2r+21}{3(r-3)}$

$\dfrac{r+1}{6}$

75)

$\dfrac{6}{h^2+9h+20}$ a) $\dfrac{6h}{(h+5)^2(h+4)}$

$\dfrac{h}{h+5}$ b) $\dfrac{2(h^2+4h+6)}{(h+5)(h+4)}$

76)

$\dfrac{1}{d^2-9}$ a) $\dfrac{d}{(d+3)^2(d-3)}$

$\dfrac{d}{d+3}$ b) $\dfrac{2(d^2-3d+1)}{(d+3)(d-3)}$

77) Find a rational expression in simplest form to represent the perimeter of the triangle. $\dfrac{49x+6}{4x^2}$

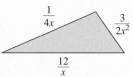

78) The total resistance of a set of resistors in parallel in an electrical circuit can be found using the formula $\dfrac{1}{R_T}=\dfrac{1}{R_1}+\dfrac{1}{R_2}$, where R_1 = the resistance in resistor 1, R_2 = the resistance in resistor 2, and R_T = the total resistance in the circuit. (Resistance is measured in ohms.) For the given circuit,

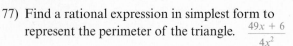

a) find the sum $\dfrac{1}{x}+\dfrac{1}{x+5}$. $\dfrac{2x+5}{x(x+5)}$

b) find an expression for the total resistance, R_T, in the circuit. $\dfrac{x(x+5)}{2x+5}$

c) if $R_1=10$ ohms, what is the total resistance in the circuit? 6 ohms

---

R Rethink

R1) For which problems could you use more practice?

R2) How often did you encounter a solution that could be simplified? What were some clues that alerted you that you may need to simplify?

Putting It All Together

P Prepare

O Organize

What is your objective?	How can you accomplish the objective?
1 Review the Concepts Presented in Sections 8.1–8.4	• Be sure that you can apply the techniques you have learned in the previous sections. • Understand when a common denominator is needed and when it is not. • Complete the given examples on your own. • Complete You Try 1.

In Section 8.1, we defined a rational expression, evaluated expressions, and discussed how to write a rational expression in lowest terms. We also learned that a rational expression *equals zero* when its *numerator equals zero*, and a rational expression is *undefined* when its *denominator equals zero*.

EXAMPLE 1

In-Class Example 1

Determine the values of h for which $\dfrac{h-2}{h^2-100}$
a) equals zero.
b) is undefined.

Answer: a) 2   b) $-10, 10$

Determine the values of c for which $\dfrac{c+8}{c^2-25}$

a) equals zero.   b) is undefined.

Solution

a) $\dfrac{c+8}{c^2-25}$ equals zero when its *numerator* equals zero.

Let $c+8=0$, and solve for c.

$$c + 8 = 0$$
$$c = -8$$

$\dfrac{c+8}{c^2-25}$ equals zero when $c = -8$.

b) $\dfrac{c+8}{c^2-25}$ is undefined when its *denominator* equals zero.

Let $c^2 - 25 = 0$, and solve for c.

$$c^2 - 25 = 0$$
$$(c+5)(c-5) = 0 \qquad \text{Factor.}$$
$$c + 5 = 0 \quad \text{or} \quad c - 5 = 0 \qquad \text{Set each factor equal to zero.}$$
$$c = -5 \quad \text{or} \qquad c = 5 \qquad \text{Solve.}$$

$\dfrac{c+8}{c^2-25}$ is undefined when $c = 5$ or $c = -5$. So, $c \neq 5$ and $c \neq -5$ in the expression.

W Hint

Be certain to follow all of the examples! The exercises will be a mixture of problems.

In Sections 8.2–8.4, we learned how to multiply, divide, add, and subtract rational expressions. Now we will practice these operations together so that we will learn to recognize the techniques needed to perform these operations.

EXAMPLE 2

In-Class Example 2

Divide $\dfrac{2m^2 - 11m + 12}{9m^2 - 1} \div \dfrac{m^3 - 4m^2}{5 - 15m}$.

Answer: $-\dfrac{5(2m-3)}{m^2(3m+1)}$

Divide $\dfrac{t^2 - 3t - 28}{16t^2 - 81} \div \dfrac{t^2 - 7t}{54 - 24t}$.

Solution

Do we need a common denominator to divide? *No.* A common denominator is needed to add or subtract but not to multiply or divide.

To divide, multiply the first rational expression by the reciprocal of the second expression, then factor, divide out common factors, and multiply.

$$\frac{t^2 - 3t - 28}{16t^2 - 81} \div \frac{t^2 - 7t}{54 - 24t} = \frac{t^2 - 3t - 28}{16t^2 - 81} \cdot \frac{54 - 24t}{t^2 - 7t} \qquad \text{Multiply by the reciprocal.}$$

$$= \frac{(t + 4)\cancel{(t - 7)}}{(4t + 9)\cancel{(4t - 9)}} \cdot \frac{6(\overset{-1}{\cancel{9 - 4t}})}{t\cancel{(t - 7)}} \qquad \text{Factor, and divide out common factors.}$$

$$= -\frac{6(t + 4)}{t(4t + 9)} \qquad \text{Multiply.}$$

Recall that $\dfrac{9 - 4t}{4t - 9} = -1$.

Add $\dfrac{x}{x + 2} + \dfrac{4}{3x - 1}$.

Solution

To add or subtract rational expressions, we need a common denominator.

$$\text{LCD} = (x + 2)(3x - 1)$$

Rewrite each expression with the LCD.

$$\frac{x}{x + 2} \cdot \frac{3x - 1}{3x - 1} = \frac{x(3x - 1)}{(x + 2)(3x - 1)}, \qquad \frac{4}{3x - 1} \cdot \frac{x + 2}{x + 2} = \frac{4(x + 2)}{(x + 2)(3x - 1)}$$

$$\frac{x}{x + 2} + \frac{4}{3x - 1} = \frac{x(3x - 1)}{(x + 2)(3x - 1)} + \frac{4(x + 2)}{(x + 2)(3x - 1)} \qquad \begin{array}{l}\text{Write each expression}\\\text{with the LCD.}\end{array}$$

$$= \frac{x(3x - 1) + 4(x + 2)}{(x + 2)(3x - 1)} \qquad \text{Add the numerators.}$$

$$= \frac{3x^2 - x + 4x + 8}{(x + 2)(3x - 1)} \qquad \text{Distribute.}$$

$$= \frac{3x^2 + 3x + 8}{(x + 2)(3x - 1)} \qquad \text{Combine like terms.}$$

Although this numerator will not factor, remember that sometimes it *is* possible to factor the numerator and simplify the result.

[YOU TRY 1]

a) Write in lowest terms: $\dfrac{3k^2 - 14k + 16}{4 - k^2}$ b) Subtract $\dfrac{b}{b + 10} - \dfrac{3}{b}$.

c) Multiply $\dfrac{r^3 + 9r^2 - r - 9}{12r + 12} \cdot \dfrac{5}{r^2 + 3r - 54}$.

d) Determine the values of w for which $\dfrac{6w - 1}{2w^2 + 16w}$ i) equals zero.   ii) is undefined.

ANSWERS TO [YOU TRY] EXERCISES

1) a) $\dfrac{8 - 3k}{k + 2}$ b) $\dfrac{b^2 - 3b - 30}{b(b + 10)}$ c) $\dfrac{5(r - 1)}{12(r - 6)}$ d) i) $\dfrac{1}{6}$ ii) $-8, 0$

Putting It All Together Exercises

*Additional answers can be found in the Answers to Exercises appendix.

Objective 1: Review the Concepts Presented in Sections 8.1–8.4

Evaluate, if possible, for a) $x = -3$ and b) $x = 2$.

1) $\dfrac{x + 3}{3x + 4}$ a) 0 b) $\dfrac{1}{2}$

2) $\dfrac{x}{x - 2}$ a) $\dfrac{3}{5}$ b) undefined

3) $\dfrac{5x - 3}{x^2 + 10x + 21}$ a) undefined b) $\dfrac{7}{45}$

4) $-\dfrac{x^2}{x^2 - 12}$ a) 3 b) $\dfrac{1}{2}$

Determine the values of the variable for which

 a) the expression is undefined.

 b) the expression equals zero.

5) $-\dfrac{5w}{w^2 - 36}$ a) $-6, 6$ b) 0

6) $\dfrac{m - 4}{2m^2 + 11m + 15}$ a) $-3, -\dfrac{5}{2}$ b) 4

7) $\dfrac{3 - 5b}{b^2 + 2b - 8}$ a) $-4, 2$ b) $\dfrac{3}{5}$

8) $\dfrac{5k - 8}{64 - k^2}$ a) $-8, 8$ b) $\dfrac{8}{5}$

9) $\dfrac{12}{5r}$ a) 0 b) never equals 0

10) $\dfrac{t - 15}{t^2 + 4}$ a) never undefined— any real number may be substituted for t b) 15

Write each rational expression in lowest terms.

11) $\dfrac{12w^{16}}{3w^5}$ $4w^{11}$

12) $\dfrac{42n^3}{18n^8}$ $\dfrac{7}{3n^5}$

13) $\dfrac{m^2 + 6m - 27}{2m^2 + 2m - 24}$ $\dfrac{m + 9}{2(m + 4)}$

14) $\dfrac{2j + 20}{2j^2 + 10j - 100}$ $\dfrac{1}{j - 5}$

15) $\dfrac{12 - 15n}{5n^2 + 6n - 8}$ $-\dfrac{3}{n + 2}$

16) $\dfrac{-x - y}{xy + y^2 + 5x + 5y}$ $-\dfrac{1}{y + 5}$

Perform the operations, and simplify.

17) $\dfrac{4c^2 + 4c - 24}{c + 3} \div \dfrac{3c - 6}{8}$ $\dfrac{32}{3}$

18) $\dfrac{6}{f + 11} - \dfrac{2}{f}$ $\dfrac{2(2f - 11)}{f(f + 11)}$

19) $\dfrac{4j}{j^2 - 81} + \dfrac{3}{j^2 - 3j - 54}$ $\dfrac{4j^2 + 27j + 2}{(j + 9)(j - 9)(j + 6)}$

20) $\dfrac{27a^4}{8b} \cdot \dfrac{40b^2}{81a^2}$ $\dfrac{5a^2b}{3}$

21) $\dfrac{12y^7}{4z^6} \cdot \dfrac{8z^4}{72y^6}$ $\dfrac{y}{3z^2}$

22) $\dfrac{3}{q^2 - q - 20} + \dfrac{8q}{q^2 + 11q + 28}$ $\dfrac{8q^2 - 37q + 21}{(q - 5)(q + 4)(q + 7)}$

23) $\dfrac{x}{2x^2 - 7x - 4} - \dfrac{x + 3}{4x^2 + 4x + 1}$ $\dfrac{x^2 + 2x + 12}{(2x + 1)^2(x - 4)}$

24) $\dfrac{n - 4}{4n - 44} \cdot \dfrac{121 - n^2}{n + 11}$ $-\dfrac{n - 4}{4}$ or $\dfrac{4 - n}{4}$

25) $\dfrac{16 - m^2}{m + 4} \div \dfrac{8m - 32}{m + 7}$ $-\dfrac{m + 7}{8}$

26) $\dfrac{16}{r - 7} + \dfrac{4}{7 - r}$ $\dfrac{12}{r - 7}$

27) $\dfrac{xy - 5x + 2y - 10}{y^2 - 25} \div \dfrac{x^3 + 2x^2}{19x}$ $\dfrac{19}{x(y + 5)}$

28) $\dfrac{\dfrac{10d}{d + 11}}{\dfrac{5d^7}{3d + 33}}$ $\dfrac{6}{d^6}$

29) $\dfrac{9}{d + 3} + \dfrac{8}{d^2}$ $\dfrac{9d^2 + 8d + 24}{d^2(d + 3)}$

30) $\dfrac{6a + 15}{5a - 10} \cdot \dfrac{a^2 - 4}{2a^2 + 9a + 10}$ $\dfrac{3}{5}$

31) $\dfrac{\dfrac{9k^2 - 1}{14k}}{\dfrac{3k - 1}{21k^4}}$ $\dfrac{3k^3(3k + 1)}{2}$

32) $\dfrac{13}{4z} - \dfrac{1}{3z}$ $\dfrac{35}{12z}$

33) $\dfrac{2w}{25 - w^2} + \dfrac{w - 3}{w^2 - 12w + 35}$ $-\dfrac{(w - 15)(w - 1)}{(w + 5)(w - 5)(w - 7)}$

34) $\dfrac{12a^4}{10a - 30} \div \dfrac{4a}{a^3 - 3a^2 + 5a - 15}$ $\dfrac{3a^3(a^2 + 5)}{10}$

35) $\dfrac{10}{x - 8} + \dfrac{4}{x + 3}$ $\dfrac{2(7x - 1)}{(x - 8)(x + 3)}$

36) $\dfrac{1}{4y} + \dfrac{8}{6y^4}$ $\dfrac{3y^3 + 16}{12y^4}$

37) $\dfrac{2h^2 + 11h + 5}{8} \div (2h + 1)^2$ $\dfrac{h + 5}{8(2h + 1)}$

38) $\dfrac{b^2 - 15b + 36}{b^2 - 8b - 48} \cdot (b + 4)^2$ $(b - 3)(b + 4)$

39) $\dfrac{3m}{7m - 4n} - \dfrac{20n}{4n - 7m}$ $\quad \dfrac{3m + 20n}{7m - 4n}$

40) $\dfrac{10d}{8c - 10d} + \dfrac{8c}{10d - 8c}$ $\quad -1$

41) $\dfrac{2p + 3}{p^2 + 7p} - \dfrac{4p}{p^2 - p - 56} + \dfrac{5}{p^2 - 8p}$ $\quad \dfrac{-2p^2 - 8p + 11}{p(p + 7)(p - 8)}$

42) $\dfrac{6u + 1}{3u^2 - 2u} - \dfrac{u}{3u^2 + u - 2} + \dfrac{10}{u^2 + u}$ $\quad \dfrac{5u^2 + 37u - 19}{u(3u - 2)(u + 1)}$

43) $\dfrac{6t + 6}{3t^2 - 24t} \cdot (t^2 - 7t - 8)$ $\quad \dfrac{2(t + 1)^2}{t}$

44) $\dfrac{3r^2 + r - 14}{5r^3 - 10r^2} \div (9r^2 - 49)$ $\quad \dfrac{1}{5r^2(3r - 7)}$

45) $\dfrac{\dfrac{3c^3}{8c + 40}}{\dfrac{9c}{c + 5}}$ $\quad \dfrac{c^2}{24}$

46) $\dfrac{\dfrac{6v - 30}{4}}{\dfrac{v - 5}{3}}$ $\quad \dfrac{9}{2}$

47) $\dfrac{f - 8}{f - 4} - \dfrac{4}{4 - f}$ $\quad 1$

48) $\dfrac{12p}{4p^2 + 11p + 6} - \dfrac{5}{p^2 - 4p - 12}$ $\quad \dfrac{12p^2 - 92p - 15}{(4p + 3)(p + 2)(p - 6)}$

49) $\left(\dfrac{3m}{3m - 1} - \dfrac{4}{m + 4} \right) \cdot \dfrac{9m^2 - 1}{21m^2 + 28}$ $\quad \dfrac{3m + 1}{7(m + 4)}$

50) $\left(\dfrac{2c}{c + 8} + \dfrac{4}{c - 2} \right) \div \dfrac{6}{4c + 32}$ $\quad \dfrac{4(c^2 + 16)}{3(c - 2)}$

51) $\dfrac{3}{k^2 + 3k} - \dfrac{4}{3k} + \dfrac{1}{k + 3}$ $\quad -\dfrac{1}{3k}$

52) $\dfrac{3}{w^2 - w} + \dfrac{4}{5w} - \dfrac{3}{w - 1}$ $\quad -\dfrac{11}{5w}$

53) Find a rational expression in simplest form to represent the a) area and b) perimeter of the rectangle.

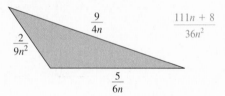

a) $\dfrac{6z}{(z + 5)(z + 2)}$

b) $\dfrac{2(z^2 + 8z + 30)}{(z + 5)(z + 2)}$

54) Find a rational expression in simplest form to represent the perimeter of the triangle.

R Rethink

R1) When you need to factor, does it feel comfortable?

R2) Discuss the importance of mastering each concept in 8.1–8.4, in order.

8.5 Simplifying Complex Fractions

P Prepare

O Organize

What are your objectives for Section 8.5?	How can you accomplish each objective?
1 Simplify a Complex Fraction with One Term in the Numerator and One Term in the Denominator	• Learn the definition of a *complex fraction,* and know how to identify a complex fraction. • Follow Example 1 to help develop a procedure you will use to simplify. • Write the procedure to **Simplify a Complex Fraction with One Term in the Numerator and One Term in the Denominator** in your own words. • Complete You Try 1.
2 Simplify a Complex Fraction with More Than One Term in the Numerator and/or Denominator by Rewriting It as a Division Problem	• Recognize that there are two ways to simplify these types of expressions. • Write the procedure to **Simplify a Complex Fraction Using Method 1** in your own words. • Complete the given example on your own. • Complete You Try 2.
3 Simplify a Complex Fraction with More Than One Term in the Numerator and/or Denominator by Multiplying by the LCD	• Write the procedure to **Simplify a Complex Fraction Using Method 2** in your own words. • Realize that when given a choice between using Method 1 and Method 2, you want to choose the method that will make the expression "easier" to simplify. • Complete the given examples on your own. • Complete You Trys 3 and 4.

W Work   Read the explanations, follow the examples, take notes, and complete the You Trys.

In algebra, we sometimes encounter fractions that contain fractions in their numerators, denominators, or both. These are called *complex fractions.* Some examples of complex fractions are

$$\frac{\dfrac{3}{7}}{\dfrac{9}{2}}, \qquad \frac{\dfrac{1}{8} + \dfrac{5}{6}}{2 - \dfrac{2}{3}}, \qquad \frac{\dfrac{4}{xy^2}}{\dfrac{1}{x} - \dfrac{1}{y}}, \qquad \frac{\dfrac{5a - 15}{4}}{\dfrac{a - 3}{a}}$$

Definition

A **complex fraction** is a rational expression that contains one or more fractions in its numerator, its denominator, or both.

www.mhhe.com/messersmith

SECTION 8.5   Simplifying Complex Fractions   **491**

A complex fraction is not considered to be an expression in simplest form. In this section, we will learn how to simplify two different types of complex fractions:

1) complex fractions with *one term* in the numerator and *one term* in the denominator

2) complex fractions that have *more than one term* in their numerators, their denominators, or both

1 Simplify a Complex Fraction with One Term in the Numerator and One Term in the Denominator

We studied these expressions in Section 8.2 when we learned how to divide fractions. We will look at another example.

EXAMPLE 1

In-Class Example 1

Simplify $\dfrac{\frac{4b-36}{7}}{\frac{b-9}{b}}$.

Answer: $\dfrac{4b}{7}$

Simplify. $\dfrac{\frac{5a-15}{4}}{\frac{a-3}{a}}$

← This is the numerator. ($\frac{5a-15}{4}$)
← This is the main fraction bar.
← This is the denominator. ($\frac{a-3}{a}$)

Solution

This complex fraction contains one term in the numerator and one term in the denominator. To simplify, rewrite as a division problem, multiply by the reciprocal, and simplify.

$$\frac{\frac{5a-15}{4}}{\frac{a-3}{a}} = \frac{5a-15}{4} \div \frac{a-3}{a} \qquad \text{Rewrite as a division problem.}$$

$$= \frac{5a-15}{4} \cdot \frac{a}{a-3} = \frac{5(a-3)}{4} \cdot \frac{a}{a-3} = \frac{5\overset{1}{\cancel{(a-3)}}}{4} \cdot \frac{a}{\underset{1}{\cancel{a-3}}} = \frac{5a}{4}$$

YOU TRY 1

Simplify $\dfrac{\frac{9}{z^2-64}}{\frac{3z+3}{z+8}}$.

Let's summarize how to simplify this first type of complex fraction.

2 Simplify a Complex Fraction with More Than One Term in the Numerator and/or Denominator by Rewriting It as a Division Problem

When a complex fraction has more than one term in the numerator and/or the denominator, we can use one of two methods to simplify.

EXAMPLE 2

Simplify.

a) $\dfrac{\dfrac{1}{4} + \dfrac{2}{3}}{2 - \dfrac{1}{2}}$ b) $\dfrac{\dfrac{5}{a^2 b}}{\dfrac{a}{b} + \dfrac{1}{a}}$

In-Class Example 2

Simplify.

a) $\dfrac{\dfrac{5}{6} - \dfrac{4}{9}}{3 + \dfrac{1}{2}}$ b) $\dfrac{\dfrac{6}{cd^2}}{\dfrac{c}{d} + \dfrac{1}{c}}$

Answer: a) $\dfrac{1}{9}$ b) $\dfrac{6}{d(c^2 + d)}$

Solution

a) The numerator, $\dfrac{1}{4} + \dfrac{2}{3}$, contains two terms; the denominator, $2 - \dfrac{1}{2}$, contains two terms. We will add the terms in the numerator and subtract the terms in the denominator so that the numerator and denominator will each contain one fraction.

$$\dfrac{\dfrac{1}{4} + \dfrac{2}{3}}{2 - \dfrac{1}{2}} = \dfrac{\dfrac{3}{12} + \dfrac{8}{12}}{\dfrac{4}{2} - \dfrac{1}{2}} = \dfrac{\dfrac{11}{12}}{\dfrac{3}{2}}$$

Add the fractions in the numerator.

Subtract the fractions in the denominator.

Rewrite as a division problem, multiply by the reciprocal, and simplify.

$$\dfrac{11}{12} \div \dfrac{3}{2} = \dfrac{11}{\overset{}{\underset{6}{12}}} \cdot \dfrac{\overset{1}{2}}{3} = \dfrac{11}{18}$$

W Hint

Compare the solutions to parts a) and b). Did it help you understand Method 1 better?

b)   The numerator, $\dfrac{5}{a^2b}$, contains one term; the denominator, $\dfrac{a}{b} + \dfrac{1}{a}$, contains two terms. We will add the terms in the denominator so that it, like the numerator, will contain only one term. The LCD of the expressions in the denominator is ab.

$$\dfrac{\dfrac{5}{a^2b}}{\dfrac{a}{b} + \dfrac{1}{a}} = \dfrac{\dfrac{5}{a^2b}}{\dfrac{a}{b} \cdot \dfrac{a}{a} + \dfrac{1}{a} \cdot \dfrac{b}{b}} = \dfrac{\dfrac{5}{a^2b}}{\dfrac{a^2}{ab} + \dfrac{b}{ab}} = \dfrac{\dfrac{5}{a^2b}}{\dfrac{a^2 + b}{ab}}$$

Rewrite as a division problem, multiply by the reciprocal, and simplify.

$$\dfrac{5}{a^2b} \div \dfrac{a^2 + b}{ab} = \dfrac{5}{a^2b} \cdot \dfrac{ab}{a^2 + b} = \dfrac{5}{a(a^2 + b)}$$

[YOU TRY 2]   Simplify.

a) $\dfrac{\dfrac{5}{8} - \dfrac{1}{2}}{1 + \dfrac{1}{4}}$ b) $\dfrac{\dfrac{6}{r^2 t^2}}{\dfrac{2}{r} - \dfrac{r}{t}}$

---

3 Simplify a Complex Fraction with More Than One Term in the Numerator and/or Denominator by Multiplying by the LCD

Another method we can use to simplify complex fractions involves multiplying the numerator and denominator of the complex fraction by the LCD of *all* of the fractions in the expression.

> **Procedure** Simplify a Complex Fraction Using Method 2
>
> 1) Identify and write down the LCD of *all* of the fractions in the complex fraction.
> 2) Multiply the numerator and denominator of the complex fraction by the LCD.
> 3) Simplify.

We will simplify the complex fractions we simplified in Example 2 using Method 2.

EXAMPLE 3

In-Class Example 3

Simplify using Method 2.

a) $\dfrac{\dfrac{5}{6} - \dfrac{4}{9}}{3 + \dfrac{1}{2}}$ b) $\dfrac{\dfrac{6}{cd^2}}{\dfrac{c}{d} + \dfrac{1}{c}}$

Answer: a) $\dfrac{1}{9}$ b) $\dfrac{6}{d(c^2 + d)}$

Simplify using Method 2.

a) $\dfrac{\dfrac{1}{4} + \dfrac{2}{3}}{2 - \dfrac{1}{2}}$ b) $\dfrac{\dfrac{5}{a^2b}}{\dfrac{a}{b} + \dfrac{1}{a}}$

Solution

a) List all of the fractions in the complex fraction: $\dfrac{1}{4}, \dfrac{2}{3}, \dfrac{1}{2}$. Write down their LCD:

LCD = 12.

Multiply the numerator and denominator of the complex fraction by the LCD, 12, then simplify.

$$\dfrac{12\left(\dfrac{1}{4} + \dfrac{2}{3}\right)}{12\left(2 - \dfrac{1}{2}\right)}$$ We are multiplying the expression by $\dfrac{12}{12}$, which equals 1.

$$= \dfrac{12 \cdot \dfrac{1}{4} + 12 \cdot \dfrac{2}{3}}{12 \cdot 2 - 12 \cdot \dfrac{1}{2}}$$ Distribute.

$$= \dfrac{3 + 8}{24 - 6} = \dfrac{11}{18}$$ Simplify.

This is the same result we obtained in Example 2 using Method 1.

Note

In the denominator, we multiplied the 2 by 12 even though 2 is not a fraction. Remember, *all* terms, not just the fractions, must be multiplied by the LCD.

b) List all of the fractions in the complex fraction: $\dfrac{5}{a^2b}, \dfrac{a}{b}, \dfrac{1}{a}$. Write down the LCD:

LCD = a^2b.

Multiply the numerator and denominator of the complex fraction by the LCD, a^2b, then simplify.

$$\dfrac{a^2b\left(\dfrac{5}{a^2b}\right)}{a^2b\left(\dfrac{a}{b} + \dfrac{1}{a}\right)}$$ We are multiplying the expression by $\dfrac{a^2b}{a^2b}$, which equals 1.

$$= \dfrac{a^2b \cdot \dfrac{5}{a^2b}}{a^2b \cdot \dfrac{a}{b} + a^2b \cdot \dfrac{1}{a}}$$ Distribute.

$$= \dfrac{5}{a^3 + ab} = \dfrac{5}{a(a^2 + b)}$$ Simplify.

If the numerator and denominator can be factored, factor them. Sometimes, you can divide by a common factor to simplify.

Notice that the result is the same as what we obtained in Example 2 using Method 1.

YOU TRY 3   Simplify using Method 2.

a) $\dfrac{\dfrac{5}{8} - \dfrac{1}{2}}{1 + \dfrac{1}{4}}$ b) $\dfrac{\dfrac{6}{r^2 t^2}}{\dfrac{2}{r} - \dfrac{r}{t}}$

W Hint

Be sure to do this You Try using Method 2 even though it is the same as You Try 2!

You should be familiar with both methods for simplifying complex fractions containing two terms in the numerator or denominator. After a lot of practice, you will be able to decide which method is better for a particular problem.

EXAMPLE 4

In-Class Example 4

Determine which method to use to simplify each complex fraction, then simplify.

a) $\dfrac{\dfrac{2}{v} + \dfrac{1}{v + 4}}{\dfrac{4}{v + 4} - \dfrac{1}{v}}$ b) $\dfrac{\dfrac{w^2 - 25}{8w + 72}}{\dfrac{4w - 20}{w^2 - 81}}$

Answer: a) $\dfrac{3v + 8}{3v - 4}$

b) $\dfrac{(w + 5)(w - 9)}{32}$

W Hint

Try doing this using Method 1. How "messy" and complicated does it become?

Determine which method to use to simplify each complex fraction, then simplify.

a) $\dfrac{\dfrac{4}{x} + \dfrac{1}{x + 3}}{\dfrac{2}{x + 3} - \dfrac{1}{x}}$ b) $\dfrac{\dfrac{n^2 - 1}{7n + 28}}{\dfrac{6n - 6}{n^2 - 16}}$

Solution

a) This complex fraction contains two terms in the numerator and two terms in the denominator. Let's use Method 2: Multiply the numerator and denominator by the LCD of all of the fractions.

LCD = $x(x + 3)$.

Multiply the numerator and denominator of the complex fraction by the LCD, $x(x + 3)$, then simplify.

$$\dfrac{x(x + 3)\left(\dfrac{4}{x} + \dfrac{1}{x + 3}\right)}{x(x + 3)\left(\dfrac{2}{x + 3} - \dfrac{1}{x}\right)} = \dfrac{x(x + 3) \cdot \dfrac{4}{x} + x(x + 3) \cdot \dfrac{1}{x + 3}}{x(x + 3) \cdot \dfrac{2}{x + 3} - x(x + 3) \cdot \dfrac{1}{x}}$$

Multiply the numerator and denominator by $x(x + 3)$ and distribute.

$$= \dfrac{4(x + 3) + x}{2x - (x + 3)}$$

Multiply.

$$= \dfrac{4x + 12 + x}{2x - x - 3} = \dfrac{5x + 12}{x - 3}$$

Distribute and combine like terms.

b) The complex fraction $\dfrac{\dfrac{n^2 - 1}{7n + 28}}{\dfrac{6n - 6}{n^2 - 16}}$ contains one term in the numerator, $\dfrac{n^2 - 1}{7n + 28}$, and

one term in the denominator, $\dfrac{6n - 6}{n^2 - 16}$. To simplify, rewrite it as a division problem, multiply by the reciprocal, and simplify.

$$\dfrac{\dfrac{n^2 - 1}{7n + 28}}{\dfrac{6n - 6}{n^2 - 16}} = \dfrac{n^2 - 1}{7n + 28} \div \dfrac{6n - 6}{n^2 - 16} \qquad \text{Rewrite as a division problem.}$$

$$= \dfrac{n^2 - 1}{7n + 28} \cdot \dfrac{n^2 - 16}{6n - 6} \qquad \text{Multiply by the reciprocal.}$$

$$= \dfrac{(n + 1)(n - 1)}{7(n + 4)} \cdot \dfrac{(n + 4)(n - 4)}{6(n - 1)} \qquad \text{Factor and divide out common factors.}$$

$$= \dfrac{(n + 1)(n - 4)}{42} \qquad \text{Multiply.}$$

[YOU TRY 4] Determine which method to use to simplify each complex fraction, then simplify.

a) $\dfrac{\dfrac{8}{k} - \dfrac{1}{k + 5}}{\dfrac{3}{k + 5} + \dfrac{5}{k}}$
b) $\dfrac{\dfrac{c^2 - 9}{8c - 56}}{\dfrac{2c + 6}{c^2 - 49}}$

ANSWERS TO [YOU TRY] EXERCISES

1) $\dfrac{3}{(z - 8)(z + 1)}$ 2) a) $\dfrac{1}{10}$ b) $\dfrac{6}{rt(2t - r^2)}$ 3) a) $\dfrac{1}{10}$ b) $\dfrac{6}{rt(2t - r^2)}$

4) a) $\dfrac{7k + 40}{8k + 25}$ b) $\dfrac{(c - 3)(c + 7)}{16}$

E Evaluate **8.5** Exercises

Do the exercises, and check your work.

*Additional answers can be found in the Answers to Exercises appendix.

Objective 1: Simplify a Complex Fraction with One Term in the Numerator and One Term in the Denominator

1) Explain, in your own words, two ways to simplify $\dfrac{\dfrac{2}{9}}{\dfrac{5}{18}}$.

Then simplify it both ways. Which method do you prefer and why?

2) Explain, in your own words, two ways to simplify $\dfrac{\dfrac{3}{2} - \dfrac{1}{5}}{\dfrac{1}{10} + \dfrac{3}{5}}$. Then simplify it both ways. Which method do you prefer and why?

Simplify completely.

3) $\dfrac{\frac{5}{9}}{\frac{7}{4}}$ $\dfrac{20}{63}$

4) $\dfrac{\frac{3}{10}}{\frac{5}{6}}$ $\dfrac{9}{25}$

(24) 5) $\dfrac{\frac{u^4}{v^2}}{\frac{u^3}{v}}$ $\dfrac{u}{v}$

6) $\dfrac{\frac{a^3}{b}}{\frac{a}{b^3}}$ a^2b^2

7) $\dfrac{\frac{x^4}{y}}{\frac{x^2}{y^2}}$ x^2y

8) $\dfrac{\frac{s^3}{t^3}}{\frac{s^4}{t}}$ $\dfrac{1}{st^2}$

9) $\dfrac{\frac{14m^5n^4}{9}}{\frac{35mn^6}{3}}$ $\dfrac{2m^4}{15n^2}$

10) $\dfrac{\frac{11b^4c^2}{4}}{\frac{55bc}{12}}$ $\dfrac{3b^3c}{5}$

11) $\dfrac{\frac{m-7}{m}}{\frac{m-7}{18}}$ $\dfrac{18}{m}$

12) $\dfrac{\frac{t-4}{9}}{\frac{t-4}{t^2}}$ $\dfrac{t^2}{9}$

(24) 13) $\dfrac{\frac{g^2-36}{20}}{\frac{g+6}{60}}$ $3(g-6)$

14) $\dfrac{\frac{6}{y^2-49}}{\frac{8}{y+7}}$ $\dfrac{3}{4(y-7)}$

15) $\dfrac{\frac{d^3}{16d-24}}{\frac{d}{40d-60}}$ $\dfrac{5d^2}{2}$

16) $\dfrac{\frac{45w-63}{w^5}}{\frac{30w-42}{w^2}}$ $\dfrac{3}{2w^3}$

17) $\dfrac{\frac{c^2-7c-8}{11c}}{\frac{c+1}{c}}$ $\dfrac{c-8}{11}$

18) $\dfrac{\frac{5x}{x-3}}{\frac{5}{x^2+4x-21}}$ $x(x+7)$

Objective 2: Simplify a Complex Fraction with More Than One Term in the Numerator and/or Denominator by Rewriting It as a Division Problem

Simplify using Method 1.

(24) 19) $\dfrac{\frac{7}{9}-\frac{2}{3}}{3+\frac{1}{9}}$ $\dfrac{1}{28}$

20) $\dfrac{\frac{1}{2}+\frac{3}{4}}{\frac{2}{3}+\frac{3}{2}}$ $\dfrac{15}{26}$

21) $\dfrac{\frac{r}{s}-4}{\frac{3}{s}+\frac{1}{r}}$ $\dfrac{r(r-4s)}{3r+s}$

22) $\dfrac{\frac{4}{c}-c^2}{1+\frac{8}{c}}$ $\dfrac{4-c^3}{c+8}$

23) $\dfrac{\frac{8}{r^2t}}{\frac{3}{r}-\frac{r}{t}}$ $\dfrac{8}{r(3t-r^2)}$

24) $\dfrac{\frac{9}{h^2k^3}}{\frac{6}{hk}-\frac{24}{k^2}}$ $\dfrac{3}{2hk(k-4h)}$

(24) 25) $\dfrac{\frac{5}{w-1}+\frac{3}{w+4}}{\frac{6}{w+4}+\frac{4}{w-1}}$ $\dfrac{8w+17}{10(w+1)}$

26) $\dfrac{\frac{5}{z-2}-\frac{2}{z+3}}{\frac{4}{z-2}+\frac{1}{z+3}}$ $\dfrac{3z+19}{5(z+2)}$

Objective 3: Simplify a Complex Fraction with More Than One Term in the Numerator and/or Denominator by Multiplying by the LCD

Simplify the complex fractions in Exercises 19–26 using Method 2. Think about which method you prefer. (You will discuss your preference in Exercises 35 and 36.)

27) Rework Exercise 19. $\dfrac{1}{28}$

28) Rework Exercise 20. $\dfrac{15}{26}$

29) Rework Exercise 21. $\dfrac{r(r-4s)}{3r+s}$

30) Rework Exercise 22. $\dfrac{4-c^3}{c+8}$

31) Rework Exercise 23. $\dfrac{8}{r(3t-r^2)}$

32) Rework Exercise 24. $\dfrac{3}{2hk(k-4h)}$

33) Rework Exercise 25. $\dfrac{8w+17}{10(w+1)}$

34) Rework Exercise 26. $\dfrac{3z+19}{5(z+2)}$

35) In Exercises 19–34, which types of complex fractions did you prefer to simplify using Method 1? Why?
Answers may vary.

36) In Exercises 19–34, which types of complex fractions did you prefer to simplify using Method 2? Why?
Answers may vary.

Mixed Exercises: Objectives 1–3

Simplify completely using any method.

37) $\dfrac{\frac{a-4}{12}}{\frac{a-4}{a}}$ $\dfrac{a}{12}$

38) $\dfrac{\frac{z^2+1}{5}}{z+\frac{1}{z}}$ $\dfrac{z}{5}$

39) $\dfrac{\frac{3}{n}-\frac{4}{n-2}}{\frac{1}{n-2}+\frac{5}{n}}$ $-\dfrac{n+6}{2(3n-5)}$

40) $\dfrac{\frac{1}{6}-\frac{5}{4}}{\frac{3}{5}+\frac{1}{3}}$ $-\dfrac{65}{56}$

41) $\dfrac{\dfrac{6}{w} - w}{1 + \dfrac{6}{w}}$ $\dfrac{6 - w^2}{w + 6}$ 42) $\dfrac{\dfrac{6t + 48}{t}}{\dfrac{9t + 72}{7}}$ $\dfrac{14}{3t}$

57) $\dfrac{1 + \dfrac{b}{a - b}}{\dfrac{b}{a^2 - b^2} + \dfrac{1}{a + b}}$ $a + b$ 58) $\dfrac{\dfrac{c}{c + 2} + \dfrac{1}{c^2 - 4}}{1 - \dfrac{3}{c + 2}}$ $\dfrac{c - 1}{c - 2}$

43) $\dfrac{\dfrac{6}{5}}{\dfrac{9}{15}}$ 2 44) $\dfrac{\dfrac{8}{k + 7} + \dfrac{1}{k}}{\dfrac{9}{k} + \dfrac{2}{k + 7}}$ $\dfrac{9k + 7}{11k + 63}$

59) $\dfrac{\dfrac{x^2 - x - 42}{2x - 14}}{\dfrac{x^2 - 36}{8x + 16}}$ $\dfrac{4(x + 2)}{x - 6}$ 60) $\dfrac{3b + \dfrac{1}{b}}{b - \dfrac{13}{b}}$ $\dfrac{3b^2 + 1}{b^2 - 13}$

45) $\dfrac{1 - \dfrac{4}{t + 5}}{\dfrac{4}{t^2 - 25} + \dfrac{t}{t - 5}}$ $\dfrac{t - 5}{t + 4}$ 46) $\dfrac{\dfrac{c^2}{d} + \dfrac{2}{c^2 d}}{\dfrac{d}{c} - \dfrac{c}{d}}$ $\dfrac{c^4 + 2}{c(d + c)(d - c)}$

61) $\dfrac{\dfrac{y^4}{z^3}}{\dfrac{y^6}{z^4}}$ $\dfrac{z}{y^2}$ 62) $\dfrac{\dfrac{k + 6}{k}}{\dfrac{k + 6}{5}}$ $\dfrac{5}{k}$

47) $\dfrac{\dfrac{9}{x} - \dfrac{9}{y}}{\dfrac{2}{x^2} - \dfrac{2}{y^2}}$ $\dfrac{9xy}{2(x + y)}$ 48) $\dfrac{\dfrac{m^2}{n^2}}{\dfrac{m^5}{n}}$ $\dfrac{1}{m^3 n}$

63) $\dfrac{7 - \dfrac{8}{m}}{\dfrac{7m - 8}{11}}$ $\dfrac{11}{m}$ 64) $\dfrac{\dfrac{7}{a} - \dfrac{7}{b}}{\dfrac{1}{a^2} - \dfrac{1}{b^2}}$ $\dfrac{7ab}{a + b}$

49) $\dfrac{\dfrac{24c - 60}{5}}{\dfrac{8c - 20}{c^2}}$ $\dfrac{3c^2}{5}$ 50) $\dfrac{\dfrac{3}{x^2 y^2}}{\dfrac{x}{y} + \dfrac{1}{x}}$ $\dfrac{3}{xy(x^2 + y)}$

65) $\dfrac{\dfrac{1}{h^2 - 4} + \dfrac{2}{h + 2}}{h - \dfrac{3}{2}}$ $\dfrac{2}{(h - 2)(h + 2)}$

51) $\dfrac{\dfrac{4}{9} + \dfrac{2}{5}}{\dfrac{1}{5} - \dfrac{2}{3}}$ $-\dfrac{38}{21}$ 52) $\dfrac{1 + \dfrac{4}{t - 3}}{\dfrac{t}{t - 3} + \dfrac{2}{t^2 - 9}}$ $\dfrac{t + 3}{t + 2}$

66) $\dfrac{\dfrac{w^2 + 10w + 25}{25 - w^2}}{\dfrac{8w + 40}{4w - 20}}$ $-\dfrac{1}{2}$

53) $\dfrac{\dfrac{1}{10}}{\dfrac{7}{8}}$ $\dfrac{4}{35}$ 54) $\dfrac{\dfrac{r^2 - 6}{40}}{r - \dfrac{6}{r}}$ $\dfrac{r}{40}$

67) $\dfrac{\dfrac{6}{v + 3} - \dfrac{4}{v - 1}}{\dfrac{2}{v - 1} + \dfrac{1}{v + 2}}$ $\dfrac{2(v - 9)(v + 2)}{3(v + 3)(v + 1)}$

55) $\dfrac{\dfrac{2}{uv^2}}{\dfrac{6}{v} - \dfrac{4v}{u}}$ $\dfrac{1}{v(3u - 2v^2)}$ 56) $\dfrac{\dfrac{y^2 - 9}{3y + 15}}{\dfrac{7y - 21}{y^2 - 25}}$ $\dfrac{(y - 5)(y + 3)}{21}$

68) $\dfrac{\dfrac{5}{r + 2} + \dfrac{7}{2r - 3}}{\dfrac{1}{r - 3} + \dfrac{3}{2r - 3}}$ $\dfrac{(17r - 1)(r - 3)}{(r + 2)(5r - 12)}$

R **Rethink**

R1) Discuss your thought process when deciding to use Method 1 or Method 2.

R2) Would you be able to complete exercises similar to these without any help?

8.6 Solving Rational Equations

What are your objectives for Section 8.6?	How can you accomplish each objective?
1 Differentiate Between Rational Expressions and Rational Equations	• Know the difference between an *expression* and an *equation*. • Complete the given example on your own. • Complete You Try 1.
2 Solve Rational Equations	• Write the procedure for **How to Solve a Rational Equation** in your own words. • Use your procedure to follow the examples, and complete them on your own. • Complete You Trys 2–5.
3 Solve a Proportion	• Be able to recognize a proportion and write the definition of a *proportion* (it should include the word *ratio*). • Complete the given example on your own. • Complete You Try 6.
4 Solve an Equation for a Specific Variable	• Follow the explanation to develop your own procedure to **Solve an Equation for a Specific Variable.** • Complete the given examples on your own. • Complete You Trys 7 and 8.

W **Work**

Read the explanations, follow the examples, take notes, and complete the You Trys.

A **rational equation** is an equation that contains a rational expression. Some examples of rational equations are

$$\frac{1}{2}a + \frac{7}{10} = \frac{3}{5}a - 4, \qquad \frac{8}{p + 7} - \frac{p}{p - 10} = 2, \qquad \frac{3n}{n^2 + 10n + 16} + \frac{5}{n + 8} = \frac{1}{n + 2}$$

1 Differentiate Between Rational Expressions and Rational Equations

In Chapter 2, we solved rational equations like the first one above, and we learned how to add and subtract rational expressions in Section 8.4. Let's summarize the difference between the two because this is often a point of confusion for students.

W **Hint**

Remember: An equation contains an = sign, but an expression does not.

Summary Expressions Versus Equations

1) *The sum or difference of rational expressions does* not *contain an = sign.* To add or subtract, rewrite each expression with the LCD, and *keep the denominator* while performing the operations.

2) *An equation contains an = sign.* To solve an equation containing rational expressions, *multiply* the equation by the LCD of all fractions to *eliminate* the denominators, then solve.

EXAMPLE 1

Determine whether each is an equation or is a sum or difference of expressions. Then, solve the equation or find the sum or difference.

a) $\dfrac{c - 5}{6} + \dfrac{c}{8} = \dfrac{3}{2}$ 　　　　 b) $\dfrac{c - 5}{6} + \dfrac{c}{8}$

Solution

a) This is an *equation* because it contains an $=$ sign. We will *solve* for c using the method we learned in Chapter 2: Eliminate the denominators by multiplying by the LCD of all of the expressions. **LCD = 24**

$$24\left(\dfrac{c - 5}{6} + \dfrac{c}{8}\right) = 24 \cdot \dfrac{3}{2} \qquad \text{Multiply by the LCD of 24 to eliminate the denominators.}$$

$$4(c - 5) + 3c = 36 \qquad \text{Distribute and eliminate denominators.}$$

$$4c - 20 + 3c = 36 \qquad \text{Distribute.}$$

$$7c - 20 = 36 \qquad \text{Combine like terms.}$$

$$7c = 56$$

$$c = 8$$

Check to verify that the solution set is $\{8\}$.

b) $\dfrac{c - 5}{6} + \dfrac{c}{8}$ is *not* an equation to be solved because it does *not* contain an $=$ sign.

It is a sum of rational expressions. Rewrite each expression with the LCD, then add, *keeping the denominators* while performing the operations.

LCD = 24

$$\dfrac{(c - 5)}{6} \cdot \dfrac{4}{4} = \dfrac{4(c - 5)}{24} \qquad\qquad \dfrac{c}{8} \cdot \dfrac{3}{3} = \dfrac{3c}{24}$$

$$\dfrac{c - 5}{6} + \dfrac{c}{8} = \dfrac{4(c - 5)}{24} + \dfrac{3c}{24} \qquad \text{Rewrite each expression with a denominator of 24.}$$

$$= \dfrac{4(c - 5) + 3c}{24} \qquad \text{Add the numerators.}$$

$$= \dfrac{4c - 20 + 3c}{24} \qquad \text{Distribute.}$$

$$= \dfrac{7c - 20}{24} \qquad \text{Combine like terms.}$$

[YOU TRY 1] Determine whether each is an equation or is a sum or difference of expressions. Then solve the equation or find the sum or difference.

a) $\dfrac{m + 1}{6} - \dfrac{m}{2}$ 　　　　 b) $\dfrac{m + 1}{6} - \dfrac{m}{2} = \dfrac{5}{6}$

2 Solve Rational Equations

Let's list the steps we use to solve a rational equation. Then we will look at more examples.

> **Procedure** How to Solve a Rational Equation
>
> 1) If possible, factor all denominators.
> 2) Write down the LCD of all of the expressions.
> 3) Multiply both sides of the equation by the LCD to *eliminate* the denominators.
> 4) Solve the equation.
> 5) Check the solution(s) in the original equation. If a proposed solution makes a denominator equal 0, then it is rejected as a solution.

EXAMPLE 2

Solve $\dfrac{t}{16} + \dfrac{2}{t} = \dfrac{3}{4}$.

In-Class Example 2

Solve $\dfrac{b}{18} - \dfrac{4}{b} = \dfrac{1}{3}$.

Answer: $\{-6, 12\}$

Solution

Since this is an *equation,* we will eliminate the denominators by multiplying the equation by the LCD of all of the expressions.   **LCD = 16t**

$$16t\left(\frac{t}{16} + \frac{2}{t}\right) = 16t\left(\frac{3}{4}\right) \qquad \text{Multiply both sides of the equation by the LCD, } 16t.$$

$$\cancel{16t}\left(\frac{t}{16}\right) + 16\cancel{t}\left(\frac{2}{\cancel{t}}\right) = \overset{4}{\cancel{16t}}\left(\frac{3}{\cancel{4}}\right) \qquad \text{Distribute and divide out common factors.}$$

$$t^2 + 32 = 12t$$

$$t^2 - 12t + 32 = 0 \qquad \text{Subtract } 12t.$$

$$(t - 8)(t - 4) = 0 \qquad \text{Factor.}$$

$$t - 8 = 0 \quad \text{or} \quad t - 4 = 0$$

$$t = 8 \quad \text{or} \qquad t = 4$$

Check:

$t = 8$	$t = 4$
$\dfrac{t}{16} + \dfrac{2}{t} \overset{?}{=} \dfrac{3}{4}$	$\dfrac{t}{16} + \dfrac{2}{t} \overset{?}{=} \dfrac{3}{4}$
$\dfrac{8}{16} + \dfrac{2}{8} \overset{?}{=} \dfrac{3}{4}$	$\dfrac{4}{16} + \dfrac{2}{4} \overset{?}{=} \dfrac{3}{4}$
$\dfrac{2}{4} + \dfrac{1}{4} = \dfrac{3}{4}$ ✓	$\dfrac{1}{4} + \dfrac{2}{4} = \dfrac{3}{4}$ ✓

W Hint
Remember when a rational expression is undefined.

The solution set is $\{4, 8\}$.

[**YOU TRY 2**]

Solve $\dfrac{d}{3} + \dfrac{4}{d} = \dfrac{13}{3}$.

It is *very* important to check the proposed solution. Sometimes, what appears to be a solution actually is not.

EXAMPLE 3

Solve $3 - \dfrac{7}{x+7} = \dfrac{x}{x+7}$.

Answer: $\varnothing$

Solve $2 - \dfrac{9}{k+9} = \dfrac{k}{k+9}$.

Solution

Since this is an *equation*, we will eliminate the denominators by multiplying the equation by the LCD of all of the expressions. **LCD = $k+9$**

$$(k+9)\left(2 - \dfrac{9}{k+9}\right) = (k+9)\left(\dfrac{k}{k+9}\right) \qquad \text{Multiply both sides of the equation by the LCD, } k+9.$$

$$(k+9)2 - (k+9)\cdot\dfrac{9}{k+9} = (k+9)\left(\dfrac{k}{k+9}\right) \qquad \text{Distribute and divide out common factors.}$$

$$2k + 18 - 9 = k \qquad \text{Multiply.}$$

$$2k + 9 = k$$

$$9 = -k \qquad \text{Subtract } 2k.$$

$$-9 = k \qquad \text{Divide by } -1.$$

Check: $2 - \dfrac{9}{(-9)+9} \stackrel{?}{=} \dfrac{-9}{(-9)+9}$ Substitute -9 for k in the original equation.

$$2 - \dfrac{9}{0} = \dfrac{-9}{0}$$

Since $k = -9$ makes the denominator equal zero, -9 cannot be a solution to the equation. Therefore, this equation has no solution. The solution set is $\varnothing$.

 BE CAREFUL *Always* check what *appears* to be the solution or solutions to an equation containing rational expressions. If one of these values makes a denominator zero, then it *cannot* be a solution to the equation.

[YOU TRY 3] Solve $\dfrac{3m}{m-4} - 1 = \dfrac{12}{m-4}$.

EXAMPLE 4

Solve $\dfrac{1}{2} - \dfrac{1}{n+3} = \dfrac{9-n}{2n^2-18}$.

Answer: $\{4\}$

Solve $\dfrac{1}{4} - \dfrac{1}{a+2} = \dfrac{a+18}{4a^2-16}$.

Solution

This is an *equation*. Eliminate the denominators by multiplying by the LCD. Begin by factoring the denominator of $\dfrac{a+18}{4a^2-16}$.

$$\dfrac{1}{4} - \dfrac{1}{a+2} = \dfrac{a+18}{4(a+2)(a-2)} \qquad \text{Factor the denominator.}$$

LCD = $4(a+2)(a-2)$

$$4(a+2)(a-2)\left(\dfrac{1}{4} - \dfrac{1}{a+2}\right) = 4(a+2)(a-2)\left(\dfrac{a+18}{4(a+2)(a-2)}\right) \qquad \text{Multiply both sides of the equation by the LCD.}$$

Distribute and divide out common factors.

$$4(a+2)(a-2)\left(\dfrac{1}{4}\right) - 4(a+2)(a-2)\left(\dfrac{1}{a+2}\right) = 4(a+2)(a-2)\left(\dfrac{a+18}{4(a+2)(a-2)}\right)$$

$$(a + 2)(a - 2) - 4(a - 2) = a + 18 \qquad \text{Multiply.}$$
$$a^2 - 4 - 4a + 8 = a + 18 \qquad \text{Distribute.}$$
$$a^2 - 4a + 4 = a + 18 \qquad \text{Combine like terms.}$$
$$a^2 - 5a - 14 = 0 \qquad \text{Subtract } a, \text{ and subtract 18.}$$
$$(a - 7)(a + 2) = 0 \qquad \text{Factor.}$$
$$a - 7 = 0 \quad \text{or} \quad a + 2 = 0 \qquad \text{Set each factor equal to zero.}$$
$$a = 7 \quad \text{or} \quad a = -2 \qquad \text{Solve.}$$

Look at the factored form of the equation. If $a = 7$, no denominator will equal zero. If $a = -2$, *however, two of the denominators will equal zero. Therefore, we must reject* $a = -2$ *as a solution.* Check only $a = 7$.

$$\text{Check:} \quad \frac{1}{4} - \frac{1}{7 + 2} \stackrel{?}{=} \frac{7 + 18}{4(7)^2 - 16} \qquad \text{Substitute } a = 7 \text{ into the original equation.}$$

$$\frac{1}{4} - \frac{1}{9} \stackrel{?}{=} \frac{25}{180} \qquad \text{Simplify.}$$

$$\frac{9}{36} - \frac{4}{36} \stackrel{?}{=} \frac{5}{36} \qquad \text{Get a common denominator, and simplify } \frac{25}{180}.$$

$$\frac{5}{36} = \frac{5}{36} \quad \checkmark \qquad \text{Subtract.}$$

The solution set is $\{7\}$.

This is a good example of why it is necessary to check all "solutions" to equations containing rational expressions.

[YOU TRY 4] Solve $\dfrac{1}{3} - \dfrac{1}{z + 2} = \dfrac{z + 14}{3z^2 - 12}$.

EXAMPLE 5

Solve $\dfrac{11}{6h^2 + 48h + 90} = \dfrac{h}{3h + 15} + \dfrac{1}{2h + 6}$.

In-Class Example 5

Solve $-\dfrac{3}{10a^2 + 20a - 80} =$

$\dfrac{a}{5a - 10} + \dfrac{1}{2a + 8}$.

Answer: $\left\{-7, \dfrac{1}{2}\right\}$

Solution

Since this is an *equation*, we will eliminate the denominators by multiplying by the LCD. Begin by factoring all denominators, then identify the LCD.

$$\frac{11}{6(h + 5)(h + 3)} = \frac{h}{3(h + 5)} + \frac{1}{2(h + 3)} \qquad \textbf{LCD} = \boldsymbol{6(h + 5)(h + 3)}$$

$$6(h + 5)(h + 3)\left(\frac{11}{6(h + 5)(h + 3)}\right) = 6(h + 5)(h + 3)\left(\frac{h}{3(h + 5)} + \frac{1}{2(h + 3)}\right) \qquad \text{Multiply by the LCD.}$$

$$\cancel{6(h + 5)(h + 3)}\left(\frac{11}{\cancel{6(h + 5)(h + 3)}}\right) = \overset{2}{\cancel{6}}(h + 5)(h + 3)\left(\frac{h}{3\cancel{(h + 5)}}\right) + \overset{3}{\cancel{6}}(h + 5)\cancel{(h + 3)}\left(\frac{1}{2\cancel{(h + 3)}}\right) \qquad \text{Distribute.}$$

$$11 = 2h(h + 3) + 3(h + 5) \qquad \text{Multiply.}$$
$$11 = 2h^2 + 6h + 3h + 15 \qquad \text{Distribute.}$$
$$11 = 2h^2 + 9h + 15 \qquad \text{Combine like terms.}$$
$$0 = 2h^2 + 9h + 4 \qquad \text{Subtract 11.}$$
$$0 = (2h + 1)(h + 4) \qquad \text{Factor.}$$

$$2h + 1 = 0 \quad \text{or} \quad h + 4 = 0$$

$$h = -\frac{1}{2} \quad \text{or} \quad h = -4 \qquad \text{Solve.}$$

Hint

Did you check all of your answers by hand for You Trys 2–5?

You can see from the factored form of the equation that neither $h = -\dfrac{1}{2}$ nor $h = -4$ will make a denominator zero. Check the values in the original equation to verify that the solution set is $\left\{-4, -\dfrac{1}{2}\right\}$.

YOU TRY 5

Solve $\dfrac{5}{6n^2 + 18n + 12} = \dfrac{n}{2n + 2} + \dfrac{1}{3n + 6}$.

3 Solve a Proportion

EXAMPLE 6

In-Class Example 6

Solve $\dfrac{6}{4c - 3} = \dfrac{8}{c + 9}$.

Answer: {3}

Solve $\dfrac{18}{r + 7} = \dfrac{6}{r - 1}$.

Solution

This rational equation is also a *proportion*. A **proportion** is a statement that two ratios are equal. We can solve this proportion as we have solved the other equations in this section, by multiplying both sides of the equation by the LCD. Or, recall from Section 2.6 that *we can solve a proportion by setting the cross products equal to each other.*

$$\dfrac{18}{r + 7} \diagdown\diagup \dfrac{6}{r - 1}$$

Multiply.   Multiply.

$18(r - 1) = 6(r + 7)$ Set the cross products equal to each other.

$18r - 18 = 6r + 42$ Distribute.

$12r = 60$

$r = 5$ Solve.

The proposed solution, $r = 5$, does *not* make a denominator equal zero. Check to verify that the solution set is {5}.

YOU TRY 6

Solve $\dfrac{7}{d + 3} = \dfrac{14}{3d + 5}$.

4 Solve an Equation for a Specific Variable

In Chapter 2, we learned how to solve an equation for a specific variable. For example, to solve $2l + 2w = P$ for w, we do the following:

$2l + 2\boxed{w} = P$ Put a box around w, the variable for which we are solving.

$2\boxed{w} = P - 2l$ Subtract $2l$.

$w = \dfrac{P - 2l}{2}$ Divide by 2.

Next we discuss how to solve for a specific variable in a rational expression.

www.mhhe.com/messersmith

SECTION 8.6 **Solving Rational Equations**   **505**

Solve $z = \dfrac{n}{d - D}$ for d.

Solution

Note that the equation contains a lowercase d and an uppercase D. These represent different quantities, so be sure to write them correctly. Put d in a box.

Since d is in the denominator of the rational expression, multiply both sides of the equation by $d - D$ to eliminate the denominator.

$$z = \frac{n}{\boxed{d} - D}$$
Put d in a box.

$$(\boxed{d} - D)z = (\boxed{d} - D)\left(\frac{n}{\boxed{d} - D}\right)$$
Multiply both sides by $d - D$ to eliminate the denominator.

$$\boxed{d}z - Dz = n$$
Distribute.

$$\boxed{d}z = n + Dz$$
Add Dz.

$$d = \frac{n + Dz}{z}$$
Divide by z.

[YOU TRY 7]

Solve $v = \dfrac{k}{m + M}$ for m.

EXAMPLE 8

Solve $\dfrac{1}{x} + \dfrac{1}{y} = \dfrac{1}{z}$ for y.

Solution

Put the y in a box. The LCD of all of the fractions is xyz. Multiply both sides of the equation by xyz.

$$\frac{1}{x} + \frac{1}{\boxed{y}} = \frac{1}{z}$$
Put y in a box.

$$x\,\boxed{y}\,z\left(\frac{1}{x} + \frac{1}{\boxed{y}}\right) = x\,\boxed{y}\,z\left(\frac{1}{z}\right)$$
Multiply both sides by xyz to eliminate the denominator.

$$x\,\boxed{y}\,z \cdot \frac{1}{x} + x\,\boxed{y}\,z \cdot \frac{1}{\boxed{y}} = x\,\boxed{y}\,z \cdot \left(\frac{1}{z}\right)$$
Distribute.

$$\boxed{y}\,z + xz = x\,\boxed{y}$$
Divide out common factors.

Since we are solving for y and there are terms containing y on each side of the equation, we must get yz and xy on one side of the equation and xz on the other side.

$$xz = x\,\boxed{y} - \boxed{y}\,z$$
Subtract yz from each side.

To isolate y, we will *factor* y out of each term on the right-hand side of the equation.

$$xz = \boxed{y}(x - z)$$
Factor out y.

$$\frac{xz}{x - z} = y$$
Divide by $x - z$.

[YOU TRY 8]

Solve $\dfrac{1}{x} + \dfrac{1}{y} = \dfrac{1}{z}$ for z.

We can use a graphing calculator to solve a rational equation in one variable. First, enter the left side of the equation in Y_1 and the right side of the equation in Y_2. Then enter $Y_1 - Y_2$ in Y_3. Then graph the equation in Y_3. The zeros or x-intercepts of the graph are the solutions to the equation.

We will solve $\dfrac{2}{x+5} - \dfrac{3}{x-2} = \dfrac{4x}{x^2+3x-10}$ using a graphing calculator.

1) Enter $\dfrac{2}{x+5} - \dfrac{3}{x-2}$ by entering $2/(x+5) - 3/(x-2)$ in Y_1.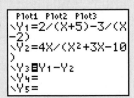

2) Enter $\dfrac{4x}{x^2+3x-10}$ by entering $4x/(x^2+3x-10)$ in Y_2.

3) Enter $Y_1 - Y_2$ in Y_3 as follows: press **VARS**, select Y-VARS using the right arrow key, and press ENTER ENTER to select Y_1. Then press −. Press **VARS**, select Y-VARS using the right arrow key, press ENTER 2 to select Y_2. Then press ENTER.

4) Move the cursor onto the = sign just right of $/Y_1$ and press ENTER to deselect Y_1. Repeat to deselect Y_2. Press GRAPH to graph $Y_1 - Y_2$.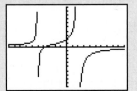

5) Press **2nd** TRACE 2:zero, move the cursor to the left of the zero, and press ENTER; move the cursor to the right of the zero, and press ENTER; and move the cursor close to the zero, and press ENTER to display the zero.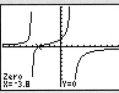

6) Press **X,T,Θ,n** MATH ENTER ENTER to display the zero $x = -\dfrac{19}{5}$.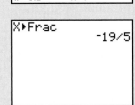

If there is more than one zero, repeat Steps 5 and 6 above for each zero.

Solve each equation using a graphing calculator.

1) $\dfrac{2x}{x-3} + \dfrac{1}{x+5} = \dfrac{4-2x}{x^2+2x-15}$

2) $\dfrac{4}{x-3} + \dfrac{5}{x+3} = \dfrac{15}{x^2-9}$

3) $\dfrac{2}{x+2} + \dfrac{4}{x-5} = \dfrac{16}{x^2-3x-10}$

4) $\dfrac{1}{x-7} + \dfrac{3}{x+4} = \dfrac{7}{x^2-3x-28}$

5) $\dfrac{6}{x+1} = \dfrac{5x+3}{x^2-x-2} - \dfrac{x}{x-2}$

6) $\dfrac{4}{x+3} - \dfrac{x}{x+2} = \dfrac{3x}{x^2+5x+6}$

E Evaluate **8.6** Exercises   Do the exercises, and check your work.

*Additional answers can be found in the Answers to Exercises appendix.

Objective 1: Differentiate Between Rational Expressions and Rational Equations

1) When solving an equation containing rational expressions, do you keep the LCD throughout the problem or do you eliminate the denominators?   Eliminate the denominators.

2) When adding or subtracting two rational expressions, do you keep the LCD throughout the problem or do you eliminate the denominators?   Keep the LCD.

Determine whether each is an equation or a sum or difference of expressions. Then solve the equation or find the sum or difference.

3) $\dfrac{3r + 5}{2} - \dfrac{r}{6}$ difference; $\dfrac{8r + 15}{6}$

4) $\dfrac{m}{12} + \dfrac{m - 8}{3}$ sum; $\dfrac{5m - 32}{12}$

5) $\dfrac{3h}{2} + \dfrac{4}{3} = \dfrac{2h + 3}{3}$ equation; $\left\{-\dfrac{2}{5}\right\}$

6) $\dfrac{7f - 24}{12} = f + \dfrac{1}{2}$ equation; $\{-6\}$

7) $\dfrac{3}{a^2} + \dfrac{1}{a + 11}$ sum; $\dfrac{a^2 + 3a + 33}{a^2(a + 11)}$

8) $\dfrac{z}{z - 5} - \dfrac{4}{z}$ difference; $\dfrac{z^2 - 4z + 20}{z(z - 5)}$

9) $\dfrac{8}{b - 11} - 5 = \dfrac{3}{b - 11}$ equation; $\{12\}$

10) $1 + \dfrac{2}{c + 5} = \dfrac{11}{c + 5}$ equation; $\{4\}$

Mixed Exercises: Objectives 2 and 3

Values that make the denominators equal zero cannot be solutions of an equation. Find *all* of the values that make the denominators zero and which, therefore, cannot be solutions of each equation. Do NOT solve the equation.

11) $\dfrac{k + 3}{k - 2} + 1 = \dfrac{7}{k}$ $0, 2$

12) $\dfrac{t}{t + 12} - \dfrac{5}{t} = 3$ $0, -12$

13) $\dfrac{8}{p + 3} - \dfrac{6}{p} = \dfrac{p}{p^2 - 9}$ $0, 3, -3$

14) $\dfrac{7}{d^2 - 64} + \dfrac{6}{d} = \dfrac{8}{d + 8}$ $0, 8, -8$

15) $\dfrac{9h}{h^2 - 5h - 36} + \dfrac{1}{h + 4} = \dfrac{h + 7}{3h - 27}$ $-4, 9$

16) $\dfrac{v + 8}{v^2 - 8v + 12} - \dfrac{5}{3v - 4} = \dfrac{2v}{v - 6}$ $\dfrac{4}{3}, 2, 6$

Solve each equation.

17) $\dfrac{a}{3} + \dfrac{7}{12} = \dfrac{1}{4}$ $\{-1\}$

18) $\dfrac{y}{2} - \dfrac{4}{3} = \dfrac{1}{6}$ $\{3\}$

19) $\dfrac{1}{4}j - j = -4$ $\left\{\dfrac{16}{3}\right\}$

20) $\dfrac{1}{3}h + h = -4$ $\{-3\}$

21) $\dfrac{8m - 5}{24} = \dfrac{m}{6} - \dfrac{7}{8}$ $\{-4\}$

22) $\dfrac{13u - 1}{20} = \dfrac{3u}{5} - 1$ $\{-19\}$

23) $\dfrac{8}{3x + 1} = \dfrac{2}{x + 3}$ $\{-11\}$

24) $\dfrac{4}{5t + 2} = \dfrac{2}{2t - 1}$ $\{-4\}$

25) $\dfrac{r+1}{2} = \dfrac{4r+1}{5}$ {1}

26) $\dfrac{w}{3} = \dfrac{6w-4}{9}$ $\left\{\dfrac{4}{3}\right\}$

53) $\dfrac{x^2}{2} = \dfrac{x^2-6x}{3}$ {0, −12}

27) $\dfrac{23}{z} + 8 = -\dfrac{25}{z}$ {−6}

28) $\dfrac{18}{a} - 2 = \dfrac{10}{a}$ {4}

54) $\dfrac{k^2}{3} = \dfrac{k^2+3k}{4}$ {0, 9}

29) $\dfrac{5q}{q+1} - 2 = \dfrac{5}{q+1}$ $\left\{\dfrac{7}{3}\right\}$

30) $\dfrac{n}{n+3} + 5 = \dfrac{12}{n+3}$ $\left\{-\dfrac{1}{2}\right\}$

55) $\dfrac{3}{t^2} = \dfrac{6}{t^2+8t}$ {8}

31) $\dfrac{2}{s+6} + 4 = \dfrac{2}{s+6}$ ∅

32) $\dfrac{u}{u-5} + 3 = \dfrac{5}{u-5}$ ∅

56) $\dfrac{5}{m^2-36} = \dfrac{4}{m^2+6m}$ {−24}

33) $\dfrac{3b}{b+7} - 6 = \dfrac{3}{b+7}$ {−15}

34) $\dfrac{c}{c-5} - 5 = \dfrac{20}{c-5}$ $\left\{\dfrac{5}{4}\right\}$

57) $\dfrac{b+3}{3b-18} - \dfrac{b+2}{b-6} = \dfrac{b}{3}$ {3, 1}

35) $\dfrac{8}{r} - 1 = \dfrac{6}{r}$ {2}

36) $\dfrac{11}{g} + 3 = -\dfrac{10}{g}$ {−7}

58) $\dfrac{3y-2}{y+2} = \dfrac{y}{4} + \dfrac{1}{4y+8}$ {1, 9}

37) $z + \dfrac{12}{z} = -8$ {−6, −2}

38) $y - \dfrac{28}{y} = 3$ {−4, 7}

59) $\dfrac{4}{n+1} = \dfrac{10}{n^2-1} - \dfrac{5}{n-1}$ ∅

39) $\dfrac{15}{b} = 8 - b$ {3, 5}

40) $n = 13 - \dfrac{12}{n}$ {1, 12}

60) $\dfrac{2}{c-6} - \dfrac{24}{c^2-36} = -\dfrac{3}{c+6}$ ∅

41) $\dfrac{8}{c+2} - \dfrac{12}{c-4} = \dfrac{2}{c+2}$ {−8}

61) $-\dfrac{a}{5} = \dfrac{3}{a+8}$ {−5, −3}

42) $\dfrac{2}{m-1} + \dfrac{1}{m+4} = \dfrac{4}{m+4}$ {11}

62) $\dfrac{u}{7} = \dfrac{2}{9-u}$ {2, 7}

43) $\dfrac{9}{c-8} - \dfrac{15}{c} = 1$ {−10, 12}

63) $\dfrac{8}{p+2} + \dfrac{p}{p+1} = \dfrac{5p+2}{p^2+3p+2}$ {−3}

44) $\dfrac{6}{r+5} - \dfrac{2}{r} = -1$ {−10, 1}

64) $\dfrac{6}{x-1} + \dfrac{x}{x+3} = \dfrac{2x+28}{x^2+2x-3}$ {−5, 2}

45) $\dfrac{3}{p-4} + \dfrac{8}{p+4} = \dfrac{13}{p^2-16}$ {3}

65) $\dfrac{-14}{3a^2+15a-18} = \dfrac{a}{a-1} + \dfrac{2}{3a+18}$ $\left\{-\dfrac{2}{3}\right\}$

46) $\dfrac{5}{w-7} - \dfrac{8}{w+7} = \dfrac{52}{w^2-49}$ {13}

66) $\dfrac{3}{2n^2+10n+8} = \dfrac{n}{2n+2} + \dfrac{1}{n+1}$ {−5}

47) $\dfrac{9}{k+5} - \dfrac{4}{k+1} = \dfrac{10}{k^2+6k+5}$ $\left\{\dfrac{21}{5}\right\}$

67) $\dfrac{3}{f+4} = \dfrac{f}{f+6} - \dfrac{2}{f^2+10f+24}$ {−5, 4}

48) $\dfrac{3}{a+2} + \dfrac{10}{a^2-6a-16} = \dfrac{5}{a-8}$ {−12}

68) $\dfrac{11}{c+9} = \dfrac{c}{c-4} - \dfrac{36-8c}{c^2+5c-36}$ {−4, −2}

49) $\dfrac{12}{g^2-9} + \dfrac{2}{g+3} = \dfrac{7}{g-3}$ ∅

69) $\dfrac{b}{b^2+b-6} + \dfrac{3}{b^2+9b+18} = \dfrac{8}{b^2+4b-12}$ {5}

50) $\dfrac{9}{t+4} + \dfrac{8}{t^2-16} = \dfrac{1}{t-4}$ ∅

70) $\dfrac{h}{h^2+2h-8} + \dfrac{4}{h^2+8h-20} = \dfrac{4}{h^2+14h+40}$ {−6}

51) $\dfrac{5}{p-3} - \dfrac{7}{p^2-7p+12} = \dfrac{8}{p-4}$ {−1}

71) $\dfrac{r}{r^2+8r+15} - \dfrac{2}{r^2+r-6} = \dfrac{2}{r^2+3r-10}$ {−2, 8}

52) $\dfrac{8}{x^2+2x-15} = \dfrac{6}{x-3} + \dfrac{4}{x+5}$ {−1}

72) $\dfrac{5}{t^2+5t-6} - \dfrac{t}{t^2+10t+24} = \dfrac{1}{t^2+3t-4}$ {−2, 7}

73) $\dfrac{k}{k^2 - 6k - 16} - \dfrac{12}{5k^2 - 65k + 200} = \dfrac{28}{5k^2 - 15k - 50}$ $\varnothing$

74) $\dfrac{q}{q^2 + 4q - 32} + \dfrac{2}{q^2 - 14q + 40} = \dfrac{6}{q^2 - 2q - 80}$ $\varnothing$

Objective 4: Solve an Equation for a Specific Variable

Solve for the indicated variable.

75) $W = \dfrac{CA}{m}$ for m $m = \dfrac{CA}{W}$

76) $V = \dfrac{nRT}{P}$ for P $P = \dfrac{nRT}{V}$

77) $a = \dfrac{rt}{2b}$ for b $b = \dfrac{rt}{2a}$

78) $y = \dfrac{kx}{z}$ for z $z = \dfrac{kx}{y}$

79) $B = \dfrac{t + u}{3x}$ for x $x = \dfrac{t + u}{3B}$

80) $Q = \dfrac{n - k}{5r}$ for r $r = \dfrac{n - k}{5Q}$

81) $d = \dfrac{t}{z - n}$ for n $n = \dfrac{dz - t}{d}$

82) $z = \dfrac{a}{b + c}$ for b $b = \dfrac{a - zc}{z}$

83) $h = \dfrac{3A}{r + s}$ for s $s = \dfrac{3A - hr}{h}$

84) $A = \dfrac{4r}{q - t}$ for t $t = \dfrac{Aq - 4r}{A}$

85) $r = \dfrac{kx}{y - az}$ for y $y = \dfrac{kx + raz}{r}$

86) $w = \dfrac{na}{kc + b}$ for c $c = \dfrac{na - wb}{wk}$

87) $\dfrac{1}{t} = \dfrac{1}{r} - \dfrac{1}{s}$ for r $r = \dfrac{st}{s + t}$

88) $\dfrac{1}{R_1} + \dfrac{1}{R_2} = \dfrac{1}{R_3}$ for R_2 $R_2 = \dfrac{R_1 R_3}{R_1 - R_3}$

89) $\dfrac{5}{x} = \dfrac{1}{y} - \dfrac{4}{z}$ for z $z = \dfrac{4xy}{x - 5y}$

90) $\dfrac{2}{A} + \dfrac{1}{C} = \dfrac{3}{B}$ for C $C = \dfrac{AB}{3A - 2B}$

R Rethink

R1) In your own words, explain the difference between a rational expression and a rational equation.

R2) Do you have a good understanding of how to solve rational equations?

R3) Select a problem that you thought was difficult, and write the reasoning for each step you took to solve it.

8.7 Applications of Rational Equations and Variation

What are your objectives for Section 8.7?	How can you accomplish each objective?
1 Solve Problems Involving Proportions	• Recall the **Five-Step Process for Solving Applied Problems,** and add additional steps for solving a proportion. • Complete the given example on your own. • Complete You Try 1.
2 Solve Problems Involving Distance, Rate, and Time	• Read the explanation regarding objects moving with or against another force. • Use the **Five-Step Process for Solving Applied Problems** again. • Complete the given example on your own. • Complete You Try 2.
3 Solve Problems Involving Work	• Follow the explanation of a problem involving work so that you can understand the procedure. • Use the **Five-Step Process for Solving Applied Problems** again. • Complete the given example on your own. • Complete You Try 3.
4 Solve Direct Variation Problems	• Understand what *direct variation* means, and write the definition in your own words. • Learn the **Steps for Solving a Variation Problem.** • Complete the given examples on your own. • Complete You Trys 4–6.
5 Solve Inverse Variation Problems	• Understand what *inverse variation* means, and write the definition in your own words. • Use the same **Steps for Solving a Variation Problem.** • Complete the given examples on your own. • Complete You Trys 7 and 8.

W Work   **Read the explanations, follow the examples, take notes, and complete the You Trys.**

We have studied applications of linear and quadratic equations. Now we turn our attention to applications involving equations with rational expressions. We will continue to use the five-step problem-solving method outlined in Section 2.3.

1 Solve Problems Involving Proportions

We first solved application problems involving proportions in Section 2.6. We begin this section with a problem involving a proportion.

W Hint

Write out the steps as you read the example.

Write an equation and solve.

At a small business, the ratio of employees who ride their bikes to work to those who drive a car is 4 to 3. The number of people who bike to work is three more than the number who drive. How many people bike to work, and how many drive their cars?

Solution

Step 1: **Read** the problem carefully, and identify what we are being asked to find.

We must find the number of people who bike to work and the number who drive.

Step 2: **Choose a variable** to represent the unknown, and define the other unknown in terms of this variable.

$$x = \text{the number of people who drive}$$
$$x + 3 = \text{the number of people who bike}$$

Step 3: **Translate** the information that appears in English into an algebraic equation.

Write a proportion. We will write our ratios in the form of $\dfrac{\text{number who bike}}{\text{number who drive}}$ so that the numerators contain the same quantities and the denominators contain the same quantities.

$$\text{Number who bike} \rightarrow \frac{4}{3} = \frac{x+3}{x} \leftarrow \text{Number who bike}$$
$$\text{Number who drive} \rightarrow \qquad\qquad \leftarrow \text{Number who drive}$$

The equation is $\dfrac{4}{3} = \dfrac{x+3}{x}$.

Step 4: **Solve** the equation.

$$\frac{4}{3} \diagup\!\!\!\!\diagdown \frac{x+3}{x}$$

Multiply.
Multiply.

$$4x = 3(x+3) \qquad \text{Set the cross products equal.}$$
$$4x = 3x + 9 \qquad \text{Distribute.}$$
$$x = 9 \qquad\qquad \text{Subtract } 3x.$$

Step 5: **Check** the answer, and **interpret** the solution as it relates to the problem.

Therefore, 9 people drive to work and $9 + 3 = 12$ people ride their bikes. The check is left to the student.

[YOU TRY 1]

Write an equation and solve.

In a classroom of college students, the ratio of students who have Internet access on their cell phones to those who do not is 3 to 5. The number of students who have Internet access is eight less than the number who do not. How many students have Internet access on their phones?

2 Solve Problems Involving Distance, Rate, and Time

In Section 4.4, we solved problems involving distance (d), rate (r), and time (t).

The basic formula is $d = rt$. We can solve this formula for r and then for t to obtain

$$r = \frac{d}{t} \quad \text{and} \quad t = \frac{d}{r}$$

The problems in this section involve boats going with and against a current, and planes going with and against the wind. Both situations use the same idea.

Suppose a boat's speed is 18 mph in still water. If that same boat had a 4-mph current pushing *against* it, how fast would it be traveling? (The current will cause the boat to slow down.)

Hint

How can you summarize the idea of an object moving with or against another force?

Speed *against* the current = 18 mph − 4 mph = 14 mph

$$\frac{\text{Speed } against}{\text{the current}} = \frac{\text{Speed in}}{\text{still water}} - \frac{\text{Speed of}}{\text{the current}}$$

If the speed of the boat in still water is 18 mph and a 4-mph current is *pushing* the boat, how fast would the boat be traveling *with* the current? (The current will cause the boat to travel faster.)

Speed *with* the current = 18 mph + 4 mph = 22 mph

$$\frac{\text{Speed } with}{\text{the current}} = \frac{\text{Speed in}}{\text{still water}} + \frac{\text{Speed of}}{\text{the current}}$$

A boat traveling *against* the current is said to be traveling *upstream*. A boat traveling *with* the current is said to be traveling *downstream*. We will use these ideas in Example 2.

EXAMPLE 2

Write an equation and solve.

A boat can travel 15 mi downstream in the same amount of time it can travel 9 mi upstream. If the speed of the current is 4 mph, what is the speed of the boat in still water?

In-Class Example 2

Write an equation and solve. A boat can travel 8 mi downstream in the same amount of time it can travel 6 mi upstream. If the speed of the current is 2 mph, what is the speed of the boat in still water?

Answer: 14 mph

Solution

Step 1: Read the problem carefully, and identify what we are being asked to find.

First, we must understand that "15 mi downstream" means *15 mi with the current,* and "9 mi upstream" means *9 miles against the current.*

We must find the speed of the boat in still water.

Hint

Be sure to define the unknowns—always write them on your paper!

Step 2: Choose a variable to represent the unknown, and define the other unknowns in terms of this variable.

x = the speed of the boat in still water
$x + 4$ = the speed of the boat *with* the current (downstream)
$x - 4$ = the speed of the boat *against* the current (upstream)

Step 3: **Translate** from English into an algebraic equation. Use a table to organize the information.

First, fill in the distances and the rates (or speeds).

	d	*r*	*t*
Downstream	15	$x + 4$	
Upstream	9	$x - 4$	

Next we must write expressions for the times it takes the boat to go downstream and upstream. We know that $d = rt$, so if we solve for t we get $t = \dfrac{d}{r}$.

Substitute the information from the table to get the expressions for the time.

$$\text{Downstream: } t = \frac{d}{r} = \frac{15}{x + 4} \qquad \text{Upstream: } t = \frac{d}{r} = \frac{9}{x - 4}$$

Put these values into the table.

	d	*r*	*t*
Downstream	15	$x + 4$	$\dfrac{15}{x + 4}$
Upstream	9	$x - 4$	$\dfrac{9}{x - 4}$

The problem states that it takes the boat the *same amount of time* to travel 15 mi downstream as it does to go 9 mi upstream. We can write an equation in English:

$$\frac{\text{Time for boat to go}}{\text{15 miles downstream}} = \frac{\text{Time for boat to go}}{\text{9 miles upstream}}$$

Looking at the table, we can write the algebraic equation using the expressions for time. The equation is $\dfrac{15}{x + 4} = \dfrac{9}{x - 4}$.

Step 4: **Solve** the equation.

$$\frac{15}{x + 4} \diagdown\diagup \frac{9}{x - 4} \qquad \text{Multiply.} \\ \text{Multiply.}$$

$$
\begin{aligned}
15(x - 4) &= 9(x + 4) &&\text{Set the cross products equal.} \\
15x - 60 &= 9x + 36 &&\text{Distribute.} \\
6x &= 96 && \\
x &= 16 &&\text{Solve.}
\end{aligned}
$$

Step 5: **Check** the answer, and **interpret** the solution as it relates to the problem.

The speed of the boat in still water is 16 mph.

Check: The speed of the boat going downstream is $16 + 4 = 20$ mph, so the time to travel downstream is

$$t = \frac{d}{r} = \frac{15}{20} = \frac{3}{4} \text{ hr}$$

The speed of the boat going upstream is $16 - 4 = 12$ mph, so the time to travel upstream is

$$t = \frac{d}{r} = \frac{9}{12} = \frac{3}{4} \text{ hr}$$

So, time upstream = time downstream. ✓

[YOU TRY 2] Write an equation and solve.

It takes a boat the same amount of time to travel 12 mi downstream as it does to travel 6 mi upstream. Find the speed of the boat in still water if the speed of the current is 3 mph.

---

3 Solve Problems Involving Work

Suppose it takes Tara 3 hr to paint a fence. What is the *rate* at which she does the job?

$$\text{Rate} = \frac{1 \text{ fence}}{3 \text{ hr}} = \frac{1}{3} \text{ fence/hr}$$

Tara works at a rate of $\frac{1}{3}$ of a fence per hour.

In general, we can say that if it takes t units of time to do a job, then the *rate* at which the job is done is $\frac{1}{t}$ job per unit of time.

This idea of *rate* is what we use to determine how long it can take for two or more people or things to do a job.

Let's assume, again, that Tara can paint the fence in 3 hr. At this rate, how much of the job can she do in 2 hr?

$$\begin{array}{c}
\text{Fractional part} \\
\text{of the job done}
\end{array} = \begin{array}{c}
\text{Rate of} \\
\text{work}
\end{array} \cdot \begin{array}{c}
\text{Amount of} \\
\text{time worked}
\end{array}$$

$$= \frac{1}{3} \cdot 2$$

$$= \frac{2}{3}$$

She can paint $\frac{2}{3}$ of the fence in 2 hr.

Procedure Solving Work Problems

The basic equation used to solve work problems is:

$$\begin{array}{c}
\text{Fractional part of a job} \\
\text{done by one person or thing}
\end{array} + \begin{array}{c}
\text{Fractional part of a job} \\
\text{done by another person or thing}
\end{array} = 1 \text{ (whole job)}$$

EXAMPLE 3

Write an equation and solve.

If Tara can paint the backyard fence in 3 hr but her sister, Grace, could paint the fence in 2 hr, how long would it take them to paint the fence together?

In-Class Example 3

Write an equation and solve. If Wei can paint the family's living room in 4 hr and his brother, Feng, can paint it in 5 hr, how long would it take for them to paint the room together?

Answer: $2\frac{2}{9}$ hr

Solution

Step 1: **Read** the problem carefully, and identify what we are being asked to find.

We must determine how long it would take Tara and Grace to paint the fence together.

Step 2: **Choose a variable** to represent the unknown.

t = the number of hours to paint the fence together

Step 3: **Translate** the information that appears in English into an algebraic equation.

Let's write down their rates:

$$\text{Tara's rate} = \frac{1}{3} \text{ fence/hr (since the job takes her 3 hours)}$$

$$\text{Grace's rate} = \frac{1}{2} \text{ fence/hr (since the job takes her 2 hours)}$$

It takes them t hours to paint the fence together. Recall that

$$\begin{array}{ccc} \text{Fractional part} & = & \text{Rate of} \cdot \text{Amount of} \\ \text{of job done} & & \text{work} \quad \text{time worked} \end{array}$$

$$\text{Tara's fractional part} = \frac{1}{3} \cdot t = \frac{1}{3}t$$

$$\text{Grace's fractional part} = \frac{1}{2} \cdot t = \frac{1}{2}t$$

The equation we can write comes from

$$\begin{array}{ccccc} \text{Fractional part of the} & + & \text{Fractional part of the} & = & \text{1 whole job} \\ \text{job done by Tara} & & \text{job done by Grace} & & \end{array}$$

$$\frac{1}{3}t \quad + \quad \frac{1}{2}t \quad = \quad 1$$

The equation is $\frac{1}{3}t + \frac{1}{2}t = 1$.

Step 4: **Solve** the equation.

$$6\left(\frac{1}{3}t + \frac{1}{2}t\right) = 6(1) \qquad \text{Multiply by the LCD, 6, to eliminate the fractions.}$$

$$6\left(\frac{1}{3}t\right) + 6\left(\frac{1}{2}t\right) = 6(1) \qquad \text{Distribute.}$$

$$2t + 3t = 6 \qquad \text{Multiply.}$$

$$5t = 6 \qquad \text{Combine like terms.}$$

$$t = \frac{6}{5} \qquad \text{Divide by 5.}$$

Step 5: **Check** the answer, and **interpret** the solution as it relates to the problem.

Tara and Grace could paint the fence together in $\frac{6}{5}$ hr or $1\frac{1}{5}$ hr.

Check:

Fractional part of the job done by Tara	+	Fractional part of the job done by Grace	=	1 whole job
$\frac{1}{3} \cdot \left(\frac{6}{5}\right)$	+	$\frac{1}{2} \cdot \left(\frac{6}{5}\right)$	$\overset{?}{=}$	1
$\frac{2}{5}$	+	$\frac{3}{5}$	=	1

$\left[\text{YOU TRY 3}\right]$ Write an equation and solve.

Javier can put up drywall in a house in 6 hr while it would take his coworker, Frank, 8 hr to drywall the same space. How long would it take them to install the drywall if they worked together?

4 Solve Direct Variation Problems

Suppose you are driving on a highway at a constant speed of 60 miles per hour. The distance you travel depends on the amount of time you drive.

Let $y =$ the distance traveled, in miles, and let $x =$ the number of hours you drive. An equation relating x and y is $y = 60x$.

A table relating x and y is shown here.

x	y
1	60
1.5	90
2	120
3	180

Notice that as the value of x increases, the value of y also increases. (The more hours you drive, the farther you will go.) Likewise, as the value of x decreases, the value of y also decreases.

We can say that the distance traveled, y, is *directly proportional to* the time spent traveling, x. Or y *varies directly as* x.

Definition

For $k > 0$, **y varies directly as x** (or **y is directly proportional to x**) means $y = kx$. k is called the **constant of variation.**

If two quantities vary directly, then as one quantity increases, the other increases as well. And, as one quantity decreases, the other decreases.

In our example of driving distance, $y = 60x$, 60 *is the constant of variation.*

Given information about how variables are related, we can write an equation and solve a variation problem.

EXAMPLE 4

Suppose y varies directly as x. If $y = 36$ when $x = 9$,

a) find the constant of variation, k.

b) write a variation equation relating x and y using the value of k found in part a).

c) find y when $x = 7$.

In-Class Example 4

Suppose y varies directly as x.
If $y = 42$ when $x = 7$,
a) find the constant of
variation, k.
b) write a variation equation
relating x and y using the
value of k found in part a).
c) find y when $x = 8$.

Answer: a) $k = 6$
b) $y = 6x$ c) $y = 48$

Solution

a) To find the constant of variation, write a *general* variation equation relating x and y. y varies directly as x means $y = kx$.

We are told that $y = 36$ when $x = 9$. Substitute these values into the equation, and solve for k.

$$y = kx$$
$$36 = k(9) \qquad \text{Substitute 9 for } x \text{ and 36 for } y.$$
$$4 = k \qquad \text{Divide by 9.}$$

b) The *specific* variation equation is the equation obtained when we substitute 4 for k in $y = kx$: $y = 4x$.

c) To find y when $x = 7$, substitute 7 for x in $y = 4x$ and evaluate.

$$y = 4x$$
$$= 4(7) \qquad \text{Substitute 7 for } x.$$
$$= 28 \qquad \text{Multiply.}$$

W Hint

Break down each variation problem you solve into these four steps.

Procedure Steps for Solving a Variation Problem

Step 1: Write the *general* variation equation.

Step 2: Find k by substituting the known values into the equation and solving for k.

Step 3: Write the *specific* variation equation by substituting the value of k into the *general* variation equation.

Step 4: Use the specific variation equation to solve the problem.

[YOU TRY 4]

Suppose y varies directly as x. If $y = 24$ when $x = 8$,

a) find the constant of variation, k.

b) write the specific variation equation relating x and y.

c) find y when $x = 5$.

EXAMPLE 5

Suppose h varies directly as the square of m. If $h = 80$ when $m = 4$, find h when $m = 3$.

In-Class Example 5

Suppose b varies directly as the square of n. If $b = 50$ when $n = 5$, find b when $n = 3$.

Answer: $b = 18$

Solution

Step 1: Write the *general* variation equation.

h varies directly as the *square* of m means $h = km^2$.

Step 2: Find k using the known values: $h = 80$ when $m = 4$.

$$h = km^2 \qquad \text{General variation equation}$$
$$80 = k(4)^2 \qquad \text{Substitute 4 for } m \text{ and 80 for } h.$$
$$80 = k(16)$$
$$5 = k \qquad \text{Divide by 16.}$$

Step 3: Substitute $k = 5$ into $h = km^2$ to get the *specific* variation equation, $h = 5m^2$.

Step 4: We are asked to find h when $m = 3$. Substitute $m = 3$ into $h = 5m^2$ to get h.

$$h = 5m^2 \qquad \text{Specific variation equation}$$
$$= 5(3)^2 \qquad \text{Substitute 3 for } m.$$
$$= 5(9)$$
$$= 45$$

$\begin{bmatrix} \textbf{YOU TRY 5} \end{bmatrix}$ Suppose v varies directly as the cube of z. If $v = 54$ when $z = 3$, find v when $z = 5$.

EXAMPLE 6

In-Class Example 6

Milton's earnings selling peanuts at a baseball game vary directly as the number of bags of peanuts he sells. If he sells 80 bags, he earns $200. How much would he earn if he sold 100 bags of peanuts?

Answer: $250

W Hint

Don't try this example until you have mastered the example and You Try before it.

Elzbieta's weekly earnings as a personal trainer vary directly as the number of hours she spends training clients each week. If she trains clients 25 hr in a week, she earns $875. How much would she earn if she trained clients for 30 hr?

Solution

Let h = the number of hours Elzbieta trains clients
  E = earnings

We will follow the four steps for solving a variation problem.

Step 1: Write the *general* variation equation, $E = kh$.

Step 2: Find k using the known values $E = 875$ when $h = 25$.

$$E = kh \qquad \text{General variation equation}$$
$$875 = k(25) \qquad \text{Substitute 25 for } h \text{ and 875 for } E.$$
$$35 = k \qquad \text{Divide by 25.}$$

Step 3: Substitute $k = 35$ into $E = kh$ to get the *specific* variation equation, $E = 35h$.

Step 4: We must find Elzbieta's earnings when she trains clients 30 hr in a week. Substitute $h = 30$ into $E = 35h$ to find E.

$$E = 35h \qquad \text{Specific variation equation}$$
$$E = 35(30) \qquad \text{Substitute 30 for } h.$$
$$E = 1050$$

Elzbieta would earn $1050 if she trained clients for 30 hr.

$\begin{bmatrix} \textbf{YOU TRY 6} \end{bmatrix}$ The amount of money raised by a booster club for its high school sports teams is directly proportional to the number of participants in its golf tournament. If they raised $2240 with 56 golfers, how much would they have raised if 80 people had participated?

5 Solve Inverse Variation Problems

If two quantities vary *inversely* (are *inversely* proportional), then as one value increases, the other decreases. Likewise, as one value decreases, the other increases.

> ## Definition
>
> For $k > 0$, **y varies inversely as x** (or **y is inversely proportional to x**) means $y = \dfrac{k}{x}$.
>
> k is the **constant of variation.**

A good example of inverse variation is the relationship between the time, t, it takes to travel a given distance, d, when driving at a certain rate (or speed), r. We can define this relationship as $t = \dfrac{d}{r}$. As the rate, r, increases, the time, t, that it takes to travel d mi decreases. Likewise, as r decreases, the time, t, that it takes to travel d mi increases. Therefore, t varies *inversely* as r.

EXAMPLE 7

In-Class Example 7

Suppose v varies inversely as w. If $v = 15$ when $w = 6$, find v when $w = 9$.

Answer: $w = 10$

W Hint

Are you writing out the steps as you read the example?

Suppose y varies inversely as x. If $y = 8$ when $x = 7$, find y when $x = 4$.

Solution

Step 1: Write the *general* variation equation, $y = \dfrac{k}{x}$.

Step 2: Find k using the known values: $y = 8$ when $x = 7$.

$$y = \frac{k}{x}$$

$$8 = \frac{k}{7} \qquad \text{Substitute 7 for } x \text{ and 8 for } y.$$

$$56 = k \qquad \text{Multiply by 7.}$$

Step 3: Substitute $k = 56$ into $y = \dfrac{k}{x}$ to get the *specific* variation equation, $y = \dfrac{56}{x}$.

Step 4: Substitute 4 for x in $y = \dfrac{56}{x}$ to find y.

$$y = \frac{56}{4}$$

$$y = 14$$

[YOU TRY 7] Suppose p varies inversely as the square of n. If $p = 7.5$ when $n = 4$, find p when $n = 2$.

EXAMPLE 8

If the voltage in an electrical circuit is held constant (stays the same), then the current in the circuit varies inversely as the resistance. If the current is 40 amps when the resistance is 3 ohms, find the current when the resistance is 8 ohms.

Solution

Let C = the current (in amps)

R = the resistance (in ohms)

We will follow the four steps for solving a variation problem.

Step 1: Write the *general* variation equation, $C = \dfrac{k}{R}$.

Step 2: Find k using the known values $C = 40$ when $R = 3$.

$$C = \frac{k}{R} \qquad \text{General variation equation}$$

$$40 = \frac{k}{3} \qquad \text{Substitute 40 for } C \text{ and 3 for } R.$$

$$120 = k \qquad \text{Multiply by 3.}$$

Step 3: Substitute $k = 120$ into $C = \dfrac{k}{R}$ to get the *specific* variation equation, $C = \dfrac{120}{R}$.

Step 4: We must find the current when the resistance is 8 ohms. Substitute $R = 8$ into $C = \dfrac{120}{R}$ to find C.

$$C = \frac{120}{R} \qquad \text{Specific variation equation}$$

$$C = \frac{120}{8} \qquad \text{Substitute 8 for } R.$$

$$C = 15$$

The current is 15 amps.

[YOU TRY 8]

The time it takes to drive between two cities taking the same route varies inversely as the speed of the car. If it takes Elisha 4 hr to drive between the cities when he goes 60 mph, how long would it take him to drive between the cities if he drives 50 mph?

W Hint

Did you check your
answers by hand before
looking at the provided
answers?

ANSWERS TO [YOU TRY] EXERCISES

1) 12   2) 9 mph   3) $3\frac{3}{7}$ hr   4) a) 3 b) $y = 3x$ c) 15

5) 250   6) \$3200   7) 30   8) 4.8 hr

*Additional answers can be found in the Answers to Exercises appendix.

Objective 1: Solve Problems Involving Proportions

Solve the following proportions.

1) $\dfrac{8}{15} = \dfrac{32}{x}$ {60}

2) $\dfrac{9}{12} = \dfrac{6}{a}$ {8}

3) $\dfrac{4}{7} = \dfrac{n}{n+9}$ {12}

4) $\dfrac{5}{3} = \dfrac{c}{c-10}$ {25}

Write an equation for each, and solve. See Example 1.

5) The scale on a blueprint is 2.5 in. to 10 ft in actual room length. Find the length of a room that is 3 in. long on the blueprint.   12 ft

6) A survey conducted by the U.S. Centers for Disease Control and Prevention revealed that, in Michigan, approximately 2 out of 5 adults between the ages of 18 and 24 smoke cigarettes. In a group of 400 Michigan citizens in this age group, how many would be expected to be smokers? (www.cdc.gov)   160

7) The ratio of employees at a small company who have their paychecks directly deposited into their bank accounts to those who do not is 9 to 2. If the number of people who have direct deposit is 14 more than the number who do not, how many employees do not have direct deposit?   4

8) In a gluten-free flour mixture, the ratio of potato-starch flour to tapioca flour is 2 to 1. If a mixture contains 3 more cups of potato-starch flour than tapioca flour, how much of each type of flour is in the mixture?
   3 cups of tapioca flour and 6 cups of potato-starch flour

9) At a state university, the ratio of the number of freshmen who graduated in four years to those who took longer was about 2 to 5. If the number of students who graduated in four years was 1200 less than the number who graduated in more than four years, how many students graduated in four years?   800

10) Francesca makes her own ricotta cheese for her restaurant. The ratio of buttermilk to whole milk in her recipe is 1 to 4. How much of each type of milk will she need if she uses 18 more cups of whole milk than buttermilk?
   whole milk: 24 cups; buttermilk: 6 cups

11) The ancient Greeks believed that the rectangle most pleasing to the eye, the golden rectangle, had sides in which the ratio of its length to its width was approximately 8 to 5. They erected many buildings, including the Parthenon, using this golden ratio.

The marble floor of a museum foyer is to be designed as a golden rectangle. If its width is to be 18 feet less than its length, find the length and width of the foyer.   length: 48 ft; width: 30 ft

12) The ratio of seniors at Central High School who drive to school to those who take the school bus is 7 to 2. If the number of students who drive is 320 more than the number who take the bus, how many students drive and how many take the bus?
   drive: 448; bus: 128

13) A math professor surveyed her class and found that the ratio of students who used the school's tutoring service to those who did not was 3 to 8. The number of students who did not use the tutoring lab was 15 more than the number who did. How many students used the tutoring service and how many did not?   used tutoring: 9; did not use tutoring: 24

14) An industrial cleaning solution calls for 5 parts water to 2 parts concentrated cleaner. If a worker uses 15 more quarts of water than concentrated cleaner to make a solution,

a) how much concentrated cleaner did she use?
   10 quarts

b) how much water did she use?   25 quarts

c) how much solution did she make?   35 quarts

Objective 2: Solve Problems Involving Distance, Rate, and Time

Answer the following questions about rates.

15) If the speed of a boat in still water is 8 mph,

a) what is its speed going *against* a 2-mph current?
   6 mph

b) what is its speed *with* a 2-mph current?   10 mph

16) If an airplane travels at a constant rate of 350 mph,

a) what is its speed going *into* a 50-mph wind?
   300 mph

b) what is its speed going *with* a 25-mph wind?
   375 mph

17) If an airplane travels at a constant rate of *x* mph,

a) what is its speed going *with* a 40-mph wind?
   $x + 40$ mph

b) what is its speed going *against* a 30-mph wind?
   $x - 30$ mph

18) If the speed of a boat in still water is 11 mph,

    a) what is its speed going *against* a current with a rate of x mph? $11 - x$ mph

    b) what is its speed going *with* a current with a rate of x mph? $11 + x$ mph

Write an equation for each, and solve. See Example 2.

19) A boat can travel 4 mi upstream in the same amount of time it can travel 6 mi downstream. If the speed of the current is 2 mph, what is the speed of the boat in still water?   10 mph

20) Flying at a constant speed, a plane can travel 800 miles with the wind in the same amount of time it can fly 650 miles against the wind. If the wind blows at 30 mph, what is the speed of the plane?   290 mph

21) When the wind is blowing at 25 mph, a plane flying at a constant speed can travel 500 miles with the wind in the same amount of time it can fly 400 miles against the wind. Find the speed of the plane.   225 mph

22) With a current flowing at 3 mph, a boat can travel 9 mi downstream in the same amount of time it can travel 6 mi upstream. What is the speed of the boat in still water?   15 mph

23) The speed of a boat in still water is 28 mph. The boat can travel 32 mi with the current in the same amount of time it can travel 24 mi against the current. Find the speed of the current.   4 mph

24) A boat can travel 20 mi downstream in the same amount of time it can travel 12 mi upstream. The speed of the boat in still water is 20 mph. Find the speed of the current.   5 mph

25) The speed of a plane in still air is 280 mph. Flying against the wind, it can fly 600 mi in the same amount of time it takes to go 800 mi with the wind. What is the speed of the wind?   40 mph

26) The speed of a boat in still water is 10 mph. If the boat can travel 9 mi downstream in the same amount of time it can travel 6 mi upstream, find the speed of the current.   2 mph

27) Bill drives 120 miles from his house in San Diego to Los Angeles for a business meeting. Afterward, he drives from LA to Las Vegas, a distance of 240 miles. If he averages the same speed on both legs of the trip and it takes him 2 hours less to go from San Diego to Los Angeles, what is his average driving speed?   60 mph

28) Rashard drives 80 miles from Detroit to Lansing, and later drives 60 miles more from Lansing to Grand Rapids. The trip from Lansing to Grand Rapids takes him a half hour less than the drive from Detroit to Lansing. Find his average driving speed if it is the same on both parts of the trip.   40 mph

Objective 3: Solve Problems Involving Work

Answer the following questions about work rate.

29) It takes Midori 3 hr to do her homework. What is her rate? $\frac{1}{3}$ homework/hour

30) It takes Signe 20 hr to complete her self-portrait for art class. How much of the job does she do in 12 hr? $\frac{3}{5}$ job

31) Tomasz can set up his new computer in t hours. What is the rate at which he does this job? $\frac{1}{t}$ job/hour

32) It takes Jesse twice as long to edit a chapter in a book as it takes Curtis. If it takes Curtis t hours to edit the chapter, at what rate does Jesse do the job? $\frac{1}{2t}$ job/hour

Write an equation for each, and solve. See Example 3.

33) It takes Rupinderjeet 4 hr to paint a room while the same job takes Sana 5 hr. How long would it take for them to paint the room together? $2\frac{2}{9}$ hr

34) A hot-water faucet can fill a sink in 9 min while it takes the cold-water faucet only 7 min. How long would it take to fill the sink if both faucets were on? $3\frac{15}{16}$ min

35) Wayne can clean the carpets in his house in 4 hr but it would take his son, Garth, 6 hr to clean them on his own. How long would it take them to clean the carpets together? $2\frac{2}{5}$ hr

36) Janice and Blanca have to type a report on a project they did together. Janice could type it in 40 min, and Blanca could type it in 1 hr. How long would it take them if they worked together?   24 min

37) A faucet can fill a tub in 12 minutes. The leaky drain can empty the tub in 30 minutes. If the faucet is on and the drain is leaking, how long would it take to fill the tub?   20 min

38) A pipe can fill a pool in 8 hr, and another pipe can empty a pool in 12 hr. If both pipes are accidentally left open, how long would it take to fill the pool?   24 hr

39) A new machine in a factory can do a job in 5 hr. When it is working together with an older machine, the job can be done in 3 hr. How long would it take the old machine to do the job by itself?   7.5 hr

40) It takes Lily 75 minutes to mow the lawn. When she works with her brother, Preston, it takes only 30 minutes. How long would it take Preston to mow the lawn himself?   50 min

41) It would take Mei twice as long as Ting to make decorations for a party. If they worked together, they could make the decorations in 40 min. How long would it take Mei to make the decorations by herself?   2 hr

42) It takes Lemar three times as long as his boss, Emilio, to build a custom shelving unit. Together they can build the unit in 4.5 hr. How long would it take Lemar to build the shelves by himself?   18 hr

Objectives 4 and 5: Solve Direct and Inverse Variation Problems

43) If u varies directly as v, then as v increases, the value of u _____.   increases

44) If m varies inversely as n, then as n increases, the value of m _____.   decreases

Decide whether each equation represents direct or inverse variation.

45) $y = 10x$ direct

46) $r = \dfrac{7}{s}$ inverse

47) $z = \dfrac{5}{a^3}$ inverse

48) $p = 8q^2$ direct

Write a general variation equation using k as the constant of variation.

49) A varies directly as w. $A = kw$

50) h varies directly as c. $h = kc$

51) x varies inversely as g. $x = \dfrac{k}{g}$

52) Q varies inversely as T. $Q = \dfrac{k}{T}$

53) C varies directly as the cube of d. $C = kd^3$

54) m varies directly as the square of p. $m = kp^2$

55) b varies inversely as the square of z. $b = \dfrac{k}{z^2}$

56) R varies inversely as the cube of x. $R = \dfrac{k}{x^3}$

Solve each step-by-step variation problem.

57) Suppose z varies directly as x. If $z = 63$ when $x = 7$,

   a) find the constant of variation.   9

   b) write the specific variation equation relating z and x. $z = 9x$

   c) find z when $x = 6$.   54

58) Suppose w varies directly as q. If $w = 30$ when $q = 5$,

   a) find the constant of variation.   6

   b) write the specific variation equation relating w and q. $w = 6q$

   c) find w when $q = 7$.   42

59) Suppose T varies inversely as n. If $T = 10$ when $n = 6$,

   a) find the constant of variation.   60

   b) write the specific variation equation relating T and n. $T = \dfrac{60}{n}$

   c) find T when $n = 5$.   12

60) Suppose y varies inversely as x. If $y = 15$ when $x = 5$,

   a) find the constant of variation.   75

   b) write the specific variation equation relating x and y. $y = \dfrac{75}{x}$

   c) find y when $x = 3$.   25

61) Suppose u varies inversely as the square of v. If $u = 36$ when $v = 4$,

   a) find the constant of variation.   576

   b) write the specific variation equation relating u and v. $u = \dfrac{576}{v^2}$

   c) find u when $v = 3$.   64

62) Suppose H varies directly as the cube of J. If $H = 500$ when $J = 5$,

   a) find the constant of variation.   4

   b) write the specific variation equation relating H and J. $H = 4J^3$

   c) find H when $J = 2$.   32

Solve.

63) If N varies directly as d, and $N = 28$ when $d = 4$, find N when $d = 11$.   77

64) If p varies directly as r, and $p = 30$ when $r = 6$, find p when $r = 2$.   10

65) If b varies inversely as a, and $b = 18$ when $a = 5$, find b when $a = 10$.   9

66) If W varies inversely as z, and $W = 6$ when $z = 9$, find W when $z = 27$.   2

67) If Q varies inversely as the cube of T, and $Q = 216$ when $T = 2$, find Q when $T = 3$.   64

68) If y varies directly as the cube of x, and $y = 192$ when $x = 4$, find y when $x = 5$.   375

69) If h varies directly as the square of v, and $h = 175$ when $v = 5$, find h when $v = 2$.   28

70) If C varies inversely as the square of D, and $C = 16$ when $D = 3$, find C when $D = 2$.   36

Write a variation equation and solve.

71) Hassan is paid hourly, and his weekly earnings vary directly as the number of hours he works. If Hassan earned $576.00 when he worked 32 hr, how much would he earn if he worked 40 hr? $720.00

72) If distance is held constant, the time it takes to travel that distance varies inversely as the speed at which one travels. If it takes 6 hr to travel the given distance at 70 mph, how long would it take to travel the same distance at 60 mph?   7 hr

 73) If the area is held constant, the width of a rectangle varies inversely as its length. If the width of a rectangle is 12 cm when the length is 24 cm, find the width of a rectangle with the same area when the length is 36 cm.   8 cm

74) The circumference of a circle is directly proportional to its radius. A circle of radius 5 in. has a circumference of approximately 31.4 in. What is the circumference of a circle of radius 7 in.?   43.96 in.

75) The weight of a ball varies directly as the cube of its radius. If a ball with radius 3 in. weighs 3.24 lb, how much would a ball made out of the same material weigh if it had a radius of 2 in.?   0.96 lb

76) If the force (weight) is held constant, then the work done by a bodybuilder in lifting a barbell varies directly as the distance the barbell is lifted. If Jay does 300 ft-lb of work lifting a barbell 2 ft off of the ground, how much work would it take for him to lift the same barbell 3 ft off the ground?   450 ft-lb

77) If the area is held constant, the height of a triangle varies inversely as the length of its base. If the height of a triangle is 10 in. when the base is 18 in., find the height of a triangle with the same area when the base is 12 in.   15 in.

78) The loudness of sound is inversely proportional to the square of the distance between the source of the sound and the listener. If the sound level measures 18 dB (decibels) 10 ft from a speaker, how loud is the sound 4 ft from the speaker?   112.5 dB

79) Hooke's law states that the force required to stretch a spring is proportional to the distance that the spring is stretched from its original length. A force of 120 lb is required to stretch a spring 4 in. from its natural length. How much force is needed to stretch the spring 6 in. beyond its natural length?   180 lb

80) The amount of garbage produced is proportional to the population. If a city of 60,000 people produces 9600 tons of garbage in a week, how much garbage would a city of 100,000 people produce in a week? 16,000 tons

81) The intensity of light, in lumens, varies inversely as the square of the distance from the source. If the intensity of the light is 40 lumens 5 ft from the source, what is the intensity of the light 4 ft from the source?   62.5 lumens

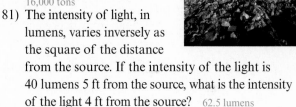 82) The weight of an object on Earth varies inversely as the square of its distance from the center of the Earth. If an object weighs 210 lb on the surface of the Earth (4000 mi from the center), what is the weight of the object if it is 200 mi above the Earth? (Round to the nearest pound.)   190 lb

R Rethink

R1) Where have you encountered a problem involving work in the last 2 weeks? Involving variation?

R2) Write a problem similar to the examples using the information from your experience.

Group Activity — The Group Activity can be found online on Connect.

Are you at risk for a stress-related illness? The more stress in your life, the more likely it is that you will experience a major illness. To determine the stress in your life, take the stressor value given beside each event you have experienced and multiply it by the number of occurrences over the past year (up to a maximum of four), and then add up these scores.

87	Experienced the death of a spouse
77	Getting married
77	Experienced the death of a close family member
76	Were divorced
74	Experienced a marital separation
68	Experienced the death of a close friend
68	Experienced pregnancy or fathered a pregnancy
65	Had a major personal injury or illness
62	Were fired from work
60	Ended a marital engagement or a steady relationship
58	Had sexual difficulties
58	Experienced a marital reconciliation
57	Had a major change in self-concept or self-awareness
56	Experienced a major change in the health or behavior of a family member
54	Became engaged to be married
53	Had a major change in financial status
52	Took on a mortgage or loan of more than $10,000
52	Had a major change in use of drugs
50	Had a major conflict or change in values
50	Had a major change in the number of arguments with your spouse
50	Gained a new family member
50	Entered college
50	Changed to a new school
50	Changed to a different line of work
49	Had a major change in amount of independence and responsibility
47	Had a major change in responsibilities at work
46	Experienced a major change in use of alcohol
45	Revised personal habits
44	Had trouble with school administration
43	Held a job while attending school
43	Had a major change in social activities
42	Had trouble with in-laws
42	Had a major change in working hours or conditions
42	Changed residence or living conditions

41	Had your spouse begin or cease work outside the home
41	Changed your choice of major field of study
41	Changed dating habits
40	Had an outstanding personal achievement
38	Had trouble with your boss
38	Had a major change in amount of participation in school activities
37	Had a major change in type and/or amount of recreation
36	Had a major change in religious activities
34	Had a major change of sleeping habits
33	Took a trip or vacation
30	Had a major change in eating habits
26	Had a major change in the number of family get-togethers
22	Were found guilty of minor violations of the law

Scoring: If your total score is above 1435, you are in a high-stress category and therefore more at risk for experiencing a stress-related illness. A high score does *not* mean that you are sure to get sick. Many other factors determine ill health, and high stress is only one cause. Other positive factors in your life, such as getting enough sleep and exercise, may prevent illness. Still, having an unusually high amount of stress in your life is a cause for concern, and you may want to take steps to reduce it.

Source of table: Marx, M. B., Garrity, T. F., & Bowers, F. R. (1975). The influence of recent life experience on the health of college freshmen. *Journal of Psychosomatic Research, 19*, 87–98.

Chapter 8: Summary

Definition/Procedure	Example

8.1 Simplifying Rational Expressions

A **rational expression** is an expression of the form $\dfrac{P}{Q}$, where P and Q are polynomials and where $Q \neq 0$.

We can *evaluate* rational expressions. **(p. 455)**

Evaluate $\dfrac{5a - 8}{a + 3}$ for $a = 2$.

$$\frac{5(2) - 8}{2 + 3} = \frac{10 - 8}{5} = \frac{2}{5}$$

How to Determine When a Rational Expression Equals Zero and When It Is Undefined

1) To determine what values of the variable make the expression equal zero, set the numerator equal to zero and solve for the variable.

2) To determine what values of the variable make the expression undefined, set the denominator equal to zero and solve for the variable. **(p. 456)**

For what value(s) of x is $\dfrac{x - 7}{x + 9}$

a) equal to zero?   b) undefined?

a) $\dfrac{x - 7}{x + 9} = 0$ when $x - 7 = 0$.

$$x - 7 = 0$$
$$x = 7$$

When $x = 7$, the expression equals zero.

b) $\dfrac{x - 7}{x + 9}$ is undefined when its denominator equals zero.

Solve $x + 9 = 0$.

$$x + 9 = 0$$
$$x = -9$$

When $x = -9$, the expression is undefined.

Writing a Rational Expression in Lowest Terms

1) Completely *factor* the numerator and denominator.

2) *Divide* the numerator and denominator by the greatest common factor. **(p. 458)**

Simplify $\dfrac{3r^2 - 10r + 8}{2r^2 - 8}$.

$$\frac{3r^2 - 10r + 8}{2r^2 - 8} = \frac{(3r - 4)(r - 2)}{2(r + 2)(r - 2)} = \frac{3r - 4}{2(r + 2)}$$

Simplifying $\dfrac{a - b}{b - a}$.

A rational expression of the form $\dfrac{a - b}{b - a}$ will simplify to -1. **(p. 459)**

Simplify $\dfrac{5 - w}{w^2 - 25}$.

$$\frac{5 - w}{w^2 - 25} = \frac{\overset{-1}{5 - w}}{(w + 5)(w - 5)} = -\frac{1}{w + 5}$$

8.2 Multiplying and Dividing Rational Expressions

Multiplying Rational Expressions

1) Factor numerators and denominators.

2) Divide out common factors.

3) Multiply.

All answers should be written in lowest terms. **(p. 465)**

Multiply $\dfrac{16v^4}{v^2 + 10v + 21} \cdot \dfrac{3v + 21}{4v}$.

$$\frac{16v^4}{v^2 + 10v + 21} \cdot \frac{3v + 21}{4v} = \frac{\overset{4}{16}v^3 \cdot v}{(v + 3)(v + 7)} \cdot \frac{3(v + 7)}{4v}$$

$$= \frac{12v^3}{v + 3}$$

Dividing Rational Expressions

To **divide** rational expressions, multiply the first expression by the reciprocal of the second. **(p. 466)**

Divide $\dfrac{2x^2 + 5x}{x + 4} \div \dfrac{4x^2 - 25}{12x - 30}$.

$$\frac{2x^2 + 5x}{x + 4} \div \frac{4x^2 - 25}{12x - 30} = \frac{2x^2 + 5x}{x + 4} \cdot \frac{12x - 30}{4x^2 - 25}$$

$$= \frac{x(2x + 5)}{x + 4} \cdot \frac{6(2x - 5)}{(2x + 5)(2x - 5)} = \frac{6x}{x + 4}$$

Definition/Procedure	Example

8.3 Finding the Least Common Denominator

How to Find the Least Common Denominator (LCD)

1) Factor the denominators.

2) The LCD will contain each unique factor the greatest number of times it appears in any single factorization.

3) The LCD is the *product* of the factors identified in Step 2. **(p. 471)**

Find the LCD of $\dfrac{9b}{b^2 + 8b}$ and $\dfrac{6}{b^2 + 16a + 64}$.

1) $b^2 + 8b = b(b + 8)$, $b^2 + 16a + 64 = (b + 8)^2$

2) The factors we will use in the LCD are b and $(b + 8)^2$.

3) LCD $= b(b + 8)^2$

8.4 Adding and Subtracting Rational Expressions

Adding and Subtracting Rational Expressions

1) Factor the denominators.

2) Write down the LCD.

3) Rewrite each rational expression as an equivalent rational expression with the LCD.

4) Add or subtract the numerators, and keep the common denominator in factored form.

5) After combining like terms in the numerator, ask yourself, *"Can I factor it?"* If so, factor.

6) Divide out common factors, if possible. The final answer should be written in lowest terms. **(p. 480)**

Add $\dfrac{y}{y + 7} + \dfrac{10y - 28}{y^2 - 49}$.

1) Factor the denominator of $\dfrac{10y - 28}{y^2 - 49}$.

$$\frac{10y - 28}{y^2 - 49} = \frac{10y - 28}{(y + 7)(y - 7)}$$

2) The LCD is $(y + 7)(y - 7)$.

3) Rewrite $\dfrac{y}{y + 7}$ with the LCD.

$$\frac{y}{y + 7} \cdot \frac{y - 7}{y - 7} = \frac{y(y - 7)}{(y + 7)(y - 7)}$$

4) $\dfrac{y}{y + 7} + \dfrac{10y - 28}{y^2 - 49} = \dfrac{y(y - 7)}{(y + 7)(y - 7)} + \dfrac{10y - 28}{(y + 7)(y - 7)}$

$$= \frac{y(y - 7) + 10y - 28}{(y + 7)(y - 7)}$$

$$= \frac{y^2 - 7y + 10y - 28}{(y + 7)(y - 7)}$$

$$= \frac{y^2 + 3y - 28}{(y + 7)(y - 7)}$$

5) $= \dfrac{\cancel{(y + 7)}(y - 4)}{\cancel{(y + 7)}(y - 7)}$ Factor.

6) $= \dfrac{y - 4}{y - 7}$ Divide out common factors.

Definition/Procedure	Example

8.5 Simplifying Complex Fractions

A **complex fraction** is a rational expression that contains one or more fractions in its numerator, its denominator, or both. **(p. 491)**

Some examples of complex fractions are

$$\frac{\dfrac{9}{16}}{\dfrac{3}{4}}, \qquad \frac{\dfrac{b+3}{2}}{\dfrac{6b+18}{7}}, \qquad \frac{\dfrac{1}{x}-\dfrac{1}{y}}{1-\dfrac{x}{y}}$$

To simplify a complex fraction containing one term in the numerator and one term in the denominator,

1) Rewrite the complex fraction as a division problem.

2) Perform the division by multiplying the first fraction by the reciprocal of the second. **(p. 492)**

Simplify $\dfrac{\dfrac{b+3}{2}}{\dfrac{6b+18}{7}}$.

$$\frac{\dfrac{b+3}{2}}{\dfrac{6b+18}{7}} = \frac{b+3}{2} \div \frac{6b+18}{7}$$

$$= \frac{b+3}{2} \cdot \frac{7}{6(b+3)} = \frac{\cancel{b+3}}{2} \cdot \frac{7}{6\cancel{(b+3)}} = \frac{7}{12}$$

To simplify complex fractions containing more than one term in the numerator and/or the denominator,

Method 1

1) Combine the terms in the numerator and combine the terms in the denominator so that each contains only one fraction.

2) Rewrite as a division problem.

3) Perform the division. **(p. 493)**

Method 1

Simplify $\dfrac{\dfrac{1}{x}-\dfrac{1}{y}}{1-\dfrac{x}{y}}$.

$$\frac{\dfrac{1}{x}-\dfrac{1}{y}}{1-\dfrac{x}{y}} = \frac{\dfrac{y}{xy}-\dfrac{x}{xy}}{\dfrac{y}{y}-\dfrac{x}{y}} = \frac{\dfrac{y-x}{xy}}{\dfrac{y-x}{y}} = \frac{y-x}{xy} \div \frac{y-x}{y}$$

$$= \frac{\cancel{y-x}}{xy} \cdot \frac{\cancel{y}}{\cancel{y-x}} = \frac{1}{x}$$

Method 2

1) Write down the LCD of *all* of the fractions in the complex fraction.

2) Multiply the numerator and denominator of the complex fraction by the LCD.

3) Simplify. **(p. 494)**

Method 2

Simplify $\dfrac{\dfrac{1}{x}-\dfrac{1}{y}}{1-\dfrac{x}{y}}$.

1) LCD $= xy$

2) Multiply the numerator and denominator by the LCD:

$$\frac{xy\left(\dfrac{1}{x}-\dfrac{1}{y}\right)}{xy\left(1-\dfrac{x}{y}\right)}$$

$$\frac{xy\left(\dfrac{1}{x}-\dfrac{1}{y}\right)}{xy\left(1-\dfrac{x}{y}\right)} = \frac{xy \cdot \dfrac{1}{x} - xy \cdot \dfrac{1}{y}}{xy \cdot 1 - xy \cdot \dfrac{x}{y}} \qquad \text{Distribute.}$$

3)

$$= \frac{y-x}{xy-x^2} \qquad \text{Simplify.}$$

$$= \frac{\cancel{y-x}}{x\cancel{(y-x)}} = \frac{1}{x}$$

Definition/Procedure	Example

8.6 Solving Rational Equations

An **equation** contains an $=$ sign, an **expression** does not.

How to Solve a Rational Equation

1) If possible, factor all denominators.

2) Write down the LCD of all of the expressions.

3) Multiply both sides of the equation by the LCD to *eliminate* the denominators.

4) Solve the equation.

5) Check the solution(s) in the original equation. If a proposed solution makes a denominator equal 0, then it is rejected as a solution. **(p. 502)**

Solve $\dfrac{n}{n + 6} + 1 = \dfrac{18}{n + 6}$.

This is an *equation* because it contains an $=$ sign. We must eliminate the denominators. Identify the LCD of all of the expressions in the equation.

LCD $= (n + 6)$

Multiply both sides of the equation by $(n + 6)$.

$$(n + 6)\left(\frac{n}{n + 6} + 1\right) = (n + 6)\left(\frac{18}{n + 6}\right)$$

$$(n + 6) \cdot \left(\frac{n}{n + 6}\right) + (n + 6) \cdot 1 = (n + 6) \cdot \frac{18}{n + 6}$$

$$n + n + 6 = 18$$
$$2n + 6 = 18$$
$$2n = 12$$
$$n = 6$$

The solution set is $\{6\}$.

The check is left to the student.

Solve an Equation for a Specific Variable (p. 505)

Solve $x = \dfrac{3b}{n + m}$ for n.

Since we are solving for n, put it in a box.

$$x = \frac{3b}{\boxed{n} + m}$$

$$(\boxed{n} + m)x = (\boxed{n} + m) \cdot \frac{3b}{\boxed{n} + m}$$

$$(\boxed{n} + m)x = 3b$$
$$\boxed{n}x + mx = 3b$$
$$\boxed{n}x = 3b - mx$$
$$n = \frac{3b - mx}{x}$$

8.7 Applications of Rational Equations and Variation

Use the **Five Steps for Solving Applied Problems** outlined in Section 2.3. **(p. 512)**

Write an equation and solve.

Jeff can wash and wax his car in 3 hours, but it takes his dad only 2 hours to wash and wax the car. How long would it take the two of them to wash and wax together?

Step 1: **Read** the problem carefully.

Step 2: **Choose a variable** to represent the unknown.
$t =$ number of hours to wash and wax the car together.

Definition/Procedure	Example

Step 3: **Translate** from English into an algebraic equation.

$$\text{Jeff's rate} = \frac{1}{3} \text{ wash/hr} \quad \text{Dad's rate} = \frac{1}{2} \text{ wash/hr}$$

$$\text{Fractional part} = \text{Rate} \cdot \text{Time}$$

$$\text{Jeff's part} = \frac{1}{3} \cdot t = \frac{1}{3}t$$

$$\text{Dad's part} = \frac{1}{2} \cdot t = \frac{1}{2}t$$

$$\begin{array}{ccc} \text{Fractional} & \text{Fractional} & \text{1 whole} \\ \text{job by Jeff} & + \text{ job by his dad} & = \text{ job} \\ \frac{1}{3}t & + \quad \frac{1}{2}t & = \quad 1 \end{array}$$

Equation: $\frac{1}{3}t + \frac{1}{2}t = 1$

Step 4: **Solve** the equation.

$$6\left(\frac{1}{3}t + \frac{1}{2}t\right) = 6(1) \qquad \text{Multiply by 6, the LCD.}$$

$$6 \cdot \frac{1}{3}t + 6 \cdot \frac{1}{2}t = 6(1) \qquad \text{Distribute.}$$

$$2t + 3t = 6 \qquad \text{Multiply.}$$

$$5t = 6$$

$$t = \frac{6}{5}$$

Step 5: **Interpret** the solution as it relates to the problem.

Jeff and his dad could wash and wax the car together in $\frac{6}{5}$ hours or $1\frac{1}{5}$ hours.

The **check** is left to the student.

Direct Variation

For $k > 0$, **y varies directly as x** (or **y is directly proportional to x**) means

$$y = kx$$

k is called the **constant of variation. (p. 517)**

The circumference, C, of a circle is given by $C = 2\pi r$. C varies directly as r where $k = 2\pi$.

Inverse Variation

For $k > 0$, **y varies inversely as x** (or **y is inversely proportional to x**) means

$$y = \frac{k}{x}$$

k is called the **constant of variation. (p. 520)**

The time, t (in hours), it takes to drive 600 miles is inversely proportional to the rate, r, at which you drive.

$$t = \frac{600}{r}$$

where $k = 600$.

Chapter 8: Review Exercises

*Additional answers can be found in the Answers to Exercises appendix.

(8.1) Evaluate, if possible, for a) $n = 5$ and b) $n = -2$.

1) $\dfrac{n^2 - 3n - 10}{3n + 2}$ a) 0 b) 0

2) $\dfrac{3n - 2}{n^2 - 4}$ a) $\dfrac{13}{21}$ b) undefined

Determine the value(s) of the variable for which
a) the expression equals zero.
b) the expression is undefined.

3) $\dfrac{2s}{4s + 11}$ a) 0 b) $-\dfrac{11}{4}$

4) $\dfrac{k + 3}{k - 4}$ a) -3 b) 4

5) $\dfrac{15}{4t^2 - 9}$ a) never equals 0 b) $-\dfrac{3}{2}, \dfrac{3}{2}$

6) $\dfrac{2c^2 - 3c - 9}{c^2 - 7c}$ a) $-\dfrac{3}{2}, 3$ b) 0, 7

7) $\dfrac{3m^2 - m - 10}{m^2 + 49}$

8) $\dfrac{15 - 5d}{d^2 + 25}$

Write each rational expression in lowest terms.

9) $\dfrac{77k^9}{7k^3}$ $11k^6$

10) $\dfrac{54a^3}{9a^{11}}$ $\dfrac{6}{a^8}$

11) $\dfrac{r^2 - 14r + 48}{4r^2 - 24r}$ $\dfrac{r - 8}{4r}$

12) $\dfrac{18c - 66}{39c - 143}$ $\dfrac{6}{13}$

13) $\dfrac{3z - 5}{6z^2 - 7z - 5}$ $\dfrac{1}{2z + 1}$

14) $\dfrac{y^2 + 8y - yz - 8z}{yz - 3y - z^2 + 3z}$ $\dfrac{y + 8}{z - 3}$

15) $\dfrac{11 - x}{x^2 - 121}$ $-\dfrac{1}{x + 11}$

16) $\dfrac{4t^3 - 16t}{3t^2 + 10t + 8}$ $\dfrac{4t(t - 2)}{3t + 4}$

Find three equivalent forms of each rational expression.

17) $-\dfrac{4n + 1}{5 - 3n}$

18) $-\dfrac{u - 8}{u + 2}$

Find the missing side in each rectangle.

19) Area = $2b^2 + 13b + 21$

2b + 7

Find the width. $b + 3$

20) Area = $3x^2 - 8x - 3$

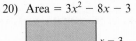

$x - 3$

Find the length. $3x + 1$

(8.2) Perform the operations and simplify.

21) $\dfrac{64}{45} \cdot \dfrac{27}{56}$ $\dfrac{24}{35}$

22) $\dfrac{6}{25} \div \dfrac{9}{10}$ $\dfrac{4}{15}$

23) $\dfrac{t + 6}{4} \cdot \dfrac{2(t + 2)}{(t + 6)^2}$ $\dfrac{t + 2}{2(t + 6)}$

24) $\dfrac{4m^3}{30n} \div \dfrac{20m^6}{3n^5}$ $\dfrac{n^4}{50m^3}$

25) $\dfrac{3x^2 + 11x + 8}{15x + 40} \div \dfrac{9x + 9}{x - 3}$ $\dfrac{x - 3}{45}$

26) $\dfrac{6w - 1}{6w^2 + 5w - 1} \cdot \dfrac{3w + 3}{12w}$ $\dfrac{1}{4w}$

27) $\dfrac{r^2 - 16r + 63}{2r^3 - 18r^2} \div (r - 7)^2$ $\dfrac{1}{2r^2(r - 7)}$

28) $(h^2 + 10h + 24) \cdot \dfrac{h}{h^2 + h - 12}$ $\dfrac{h(h + 6)}{h - 3}$

29) $\dfrac{3p^5}{20q^2} \cdot \dfrac{4q^3}{21p^7} \div \dfrac{q}{35p^2}$ $\dfrac{q}{35p^2}$

30) $\dfrac{25 - a^2}{9a^2 - 6a + 1} \div \dfrac{4a - 20}{3a - 1}$ $-\dfrac{a + 5}{4(3a - 1)}$

Divide.

31) $\dfrac{\dfrac{3s + 8}{12}}{\dfrac{3s + 8}{4}}$ $\dfrac{1}{3}$

32) $\dfrac{\dfrac{16m - 8}{m^2}}{\dfrac{12m - 6}{m^4}}$ $\dfrac{4m^2}{3}$

33) $\dfrac{\dfrac{9}{8}}{\dfrac{15}{4}}$ $\dfrac{3}{10}$

34) $\dfrac{\dfrac{2r + 10}{r^2}}{\dfrac{r^2 - 25}{4r}}$ $\dfrac{8}{r(r - 5)}$

(8.3) Find the LCD of each group of fractions.

35) $\dfrac{9}{10}, \dfrac{7}{15}, \dfrac{6}{5}$ 30

36) $\dfrac{3}{9x^2y}, \dfrac{13}{4xy^4}$ $36x^2y^4$

37) $\dfrac{3}{k^5}, \dfrac{11}{k^2}$ k^5

38) $\dfrac{3}{2m}, \dfrac{4}{m + 4}$ $2m(m + 4)$

39) $\dfrac{1}{4x + 9}, \dfrac{3x}{x - 7}$ $(4x + 9)(x - 7)$

40) $\dfrac{8}{3d^2 - d}, \dfrac{11}{9d - 3}$ $3d(3d - 1)$

41) $\dfrac{w}{w - 5}, \dfrac{11}{5 - w}$ $w - 5$ or $5 - w$

42) $\dfrac{6m}{m^2 - n^2}, \dfrac{n}{n - m}$ $(m + n)(m - n)$ or $(m + n)(n - m)$

43) $\dfrac{3c - 11}{c^2 + 9c + 20}, \dfrac{8c}{c^2 - 2c - 35}$ $(c + 4)(c + 5)(c - 7)$

44) $\dfrac{6}{x^2 + 7x}, \dfrac{1}{2x^2 + 14x}, \dfrac{13}{x^2 + 14x + 49}$ $2x(x + 7)^2$

Rewrite each rational expression with the indicated denominator.

45) $\dfrac{3}{5y} = \dfrac{}{20y^3}$ $\dfrac{12y^2}{20y^3}$

46) $\dfrac{4k}{k - 9} = \dfrac{}{(k - 6)(k - 9)}$

47) $\dfrac{6}{2z + 5} = \dfrac{}{z(2z + 5)}$

48) $\dfrac{n}{9 - n} = \dfrac{}{n - 9}$ $-\dfrac{n}{n - 9}$

49) $\dfrac{t - 3}{3t + 1} = \dfrac{}{(3t + 1)(t + 4)}$ $\dfrac{t^2 + t - 12}{(3t + 1)(t + 4)}$

Identify the LCD of each group of fractions, and rewrite each as an equivalent fraction with the LCD as its denominator.

50) $\dfrac{4}{5a^3b}, \dfrac{3}{8ab^5}$ $\dfrac{4}{5a^3b} = \dfrac{32b^4}{40a^3b^5}, \dfrac{3}{8ab^5} = \dfrac{15a^2}{40a^3b^5}$

51) $\dfrac{8c}{c^2 + 5c - 24}, \dfrac{5}{c^2 - 6c + 9}$

52) $\dfrac{6}{p + 9}, \dfrac{3}{p}$ $\dfrac{6}{p + 9} = \dfrac{6p}{p(p + 9)}; \dfrac{3}{p} = \dfrac{3p + 27}{p(p + 9)}$

53) $\dfrac{7}{2q^2 - 12q}, \dfrac{3q}{36 - q^2}, \dfrac{q - 5}{2q^2 + 12q}$

54) $\dfrac{1}{g - 12}, \dfrac{6}{12 - g}$ $\dfrac{1}{g - 12} = \dfrac{1}{g - 12}; \dfrac{6}{12 - g} = -\dfrac{6}{g - 12}$

(8.4) Add or subtract.

55) $\dfrac{5}{9c} + \dfrac{7}{9c}$ $\dfrac{4}{3c}$

56) $\dfrac{5}{6z^2} + \dfrac{9}{12z}$ $\dfrac{10 + 9z}{12z^2}$

57) $\dfrac{9}{10u^2v^2} - \dfrac{1}{8u^3v}$ $\dfrac{36u - 5v}{40u^3v^2}$

58) $\dfrac{3m}{m - 4} - \dfrac{1}{m - 4}$ $\dfrac{3m - 1}{m - 4}$

59) $\dfrac{n}{3n - 5} - \dfrac{4}{n}$ $\dfrac{n^2 - 12n + 20}{n(3n - 5)}$

60) $\dfrac{8}{t + 2} + \dfrac{8}{t}$ $\dfrac{16t + 16}{t(t + 2)}$

61) $\dfrac{9}{y + 2} - \dfrac{5}{y - 3}$ $\dfrac{4y - 37}{(y - 3)(y + 2)}$

62) $\dfrac{7d - 3}{d^2 + 3d - 28} + \dfrac{3d}{5d + 35}$ $\dfrac{3d^2 + 23d - 15}{5(d + 7)(d - 4)}$

63) $\dfrac{k - 3}{k^2 + 14k + 49} - \dfrac{2}{k^2 + 7k}$ $\dfrac{(k - 7)(k + 2)}{k(k + 7)^2}$

64) $\dfrac{8p + 3}{2p + 2} - \dfrac{6}{p^2 - 3p - 4}$ $\dfrac{8p^2 - 29p - 24}{2(p + 1)(p - 4)}$

65) $\dfrac{t + 9}{t - 18} - \dfrac{11}{18 - t}$ $\dfrac{t + 20}{t - 18}$

66) $\dfrac{1}{12 - r} + \dfrac{24}{r^2 - 144}$ $-\dfrac{1}{r + 12}$

67) $\dfrac{4w}{w^2 + 11w + 24} - \dfrac{3w - 1}{2w^2 - w - 21}$ $\dfrac{5w^2 - 51w + 8}{(2w - 7)(w + 8)(w + 3)}$

68) $\dfrac{2a + 7}{a^2 - 6a + 9} + \dfrac{6}{a^2 + 2a - 15}$ $\dfrac{2a^2 + 23a + 17}{(a + 5)(a - 3)^2}$

69) $\dfrac{b}{9b^2 - 4} + \dfrac{b + 1}{6b^2 - 4b} - \dfrac{1}{6b + 4}$ $\dfrac{2b^2 + 7b + 2}{2b(3b + 2)(3b - 2)}$

70) $\dfrac{d + 4}{d^2 + 3d} + \dfrac{d}{5d^2 + 12d - 9} - \dfrac{8}{5d^2 - 3d}$ $\dfrac{6d^2 + 9d - 36}{d(d + 3)(5d - 3)}$

71) Find a rational expression in simplest form to represent the a) area and b) perimeter of the rectangle.

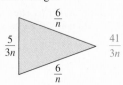

$\dfrac{2}{x^2}$ $\dfrac{x}{x + 2}$

a) $\dfrac{2}{x(x + 2)}$

b) $\dfrac{2x^3 + 4x + 8}{x^2(x + 2)}$

72) Find a rational expression in simplest form to represent the perimeter of the triangle.

$\dfrac{6}{n}$ $\dfrac{5}{3n}$ $\dfrac{41}{3n}$ $\dfrac{6}{n}$

(8.5) Simplify completely.

73) $\dfrac{\frac{x}{y}}{\frac{x^3}{y^2}}$ $\dfrac{y}{x^2}$

74) $\dfrac{\frac{a}{b} - \frac{2a}{b^2}}{\frac{4}{ab} - \frac{a}{b}}$ $\dfrac{a^2(b - 2)}{b(2 + a)(2 - a)}$

75) $\dfrac{p + \frac{4}{p}}{\frac{9}{p} + p}$ $\dfrac{p^2 + 4}{p^2 + 9}$

76) $\dfrac{\frac{n}{6n + 48}}{\frac{n^2}{8n + 64}}$ $\dfrac{4}{3n}$

77) $\dfrac{\frac{4}{5} - \frac{2}{3}}{\frac{1}{2} + \frac{1}{6}}$ $\dfrac{1}{5}$

78) $\dfrac{\frac{4q}{7q + 70}}{\frac{q^3}{8q + 80}}$ $\dfrac{32}{7q^2}$

79) $\dfrac{1 - \frac{1}{y - 8}}{\frac{2}{y + 4} + 1}$ $\dfrac{(y + 4)(y - 9)}{(y - 8)(y + 6)}$

80) $\dfrac{\frac{10}{21}}{\frac{16}{9}}$ $\dfrac{15}{56}$

81) $\dfrac{1 + \frac{1}{r - t}}{\frac{1}{r^2 - t^2} + \frac{1}{r + t}}$ $r + t$

82) $\dfrac{\frac{z}{z + 2} + \frac{1}{z^2 - 4}}{1 - \frac{3}{z + 2}}$ $\dfrac{z - 1}{z - 2}$

(8.6) Solve each equation.

83) $\dfrac{5a + 4}{15} = \dfrac{a}{5} + \dfrac{4}{5}$ $\{4\}$

84) $\dfrac{16}{9c - 27} + \dfrac{2c - 4}{c - 3} = \dfrac{c}{9}$ $\{1, 20\}$

85) $\dfrac{m}{7} = \dfrac{5}{m + 2}$ $\{-7, 5\}$

86) $\dfrac{2}{y - 7} = \dfrac{8}{y + 5}$ $\{11\}$

87) $\dfrac{r}{r + 5} + 4 = \dfrac{5}{r + 5}$ $\{-3\}$

88) $\dfrac{3}{j + 9} + \dfrac{j}{j - 3} = \dfrac{2j^2 + 2}{j^2 + 6j - 27}$ $\{1, 11\}$

89) $\dfrac{5}{t^2 + 10t + 24} + \dfrac{5}{t^2 + 3t - 18} = \dfrac{t}{t^2 + t - 12}$ $\{-1, 5\}$

90) $p - \dfrac{20}{p} = 8$ $\{-2, 10\}$

91) $\dfrac{3}{x + 1} = \dfrac{6x}{x^2 - 1} - \dfrac{4}{x - 1}$ $\varnothing$

92) $\dfrac{9}{4k^2 + 28k + 48} = \dfrac{k}{4k + 16} + \dfrac{9}{8k + 24}$ $\left\{-\dfrac{3}{2}, -6\right\}$

Solve for the indicated variable.

93) $R = \dfrac{s + T}{D}$ for D $D = \dfrac{s + T}{R}$

94) $A = \dfrac{2p}{c}$ for c $c = \dfrac{2p}{A}$

95) $w = \dfrac{N}{c - ak}$ for k $k = \dfrac{cw - N}{aw}$

96) $n = \dfrac{t}{a + b}$ for a $a = \dfrac{t - nb}{n}$

97) $\dfrac{1}{R_1} + \dfrac{1}{R_2} = \dfrac{1}{R_3}$ for R_1 $R_1 = \dfrac{R_2 R_3}{R_2 - R_3}$

98) $\dfrac{1}{r} = \dfrac{1}{s} + \dfrac{1}{t}$ for s $s = \dfrac{rt}{t - r}$

(8.7) Write an equation and solve.

99) A boat can travel 8 miles downstream in the same amount of time it can travel 6 miles upstream. If the speed of the boat in still water is 14 mph, what is the speed of the current?   2 mph

100) The ratio of saturated fat to total fat in a Starbucks tall Caramel Frappuccino is 2 to 3. If there are 4 more grams of total fat in the drink than there are grams of saturated fat, how much total fat is in 2 Caramel Frappuccinos? (Starbucks brochure)   24 g

101) Crayton and Flow must put together notebooks for each person attending a conference. Working alone, it would take Crayton 5 hours while it would take Flow 8 hours. How long would it take for them to assemble the notebooks together? $3\frac{1}{13}$ hr

102) An airplane flying at constant speed can fly 350 miles with the wind in the same amount of time it can fly 300 miles against the wind. What is the speed of the plane if the wind blows at 20 mph?   260 mph

103) Suppose c varies directly as m. If $c = 96$ when $m = 12$, find c when $m = 3$.   24

104) Suppose p varies inversely as the square of d. If $p = 40$ when $d = 2$, find p when $d = 4$.   10

Solve each problem by writing a variation equation.

105) The surface area of a cube varies directly as the square of the length of one of its sides. A cube has a surface area of 54 cm^2 when the length of each side is 3 cm. What is the surface area of a cube with a side of length 6 cm?   216 cm^2

106) The frequency of a vibrating piano string varies inversely as its length. If a 4-ft-long string vibrates at 125 cycles/second, what is the frequency of a piano string that is 2 feet long?   250 cycles/sec

Mixed Exercises: Sections 8.2–8.6

Perform the operations, and simplify.

107) $\dfrac{5n}{2n - 1} - \dfrac{2n + 3}{n + 2}$ $\dfrac{n^2 + 6n + 3}{(2n - 1)(n + 2)}$

108) $\dfrac{27w^3}{3w^2 + w - 4} \cdot \dfrac{2 - 2w}{15w}$ $-\dfrac{18w^2}{5(3w + 4)}$

109) $\dfrac{2a^2 + 9a + 10}{4a - 7} \div (2a + 5)^2$ $\dfrac{a + 2}{(4a - 7)(2a + 5)}$

110) $\dfrac{5}{8b} + \dfrac{2}{9b^4}$ $\dfrac{45b^3 + 16}{72b^4}$

111) $\dfrac{c^2}{c^2 - d^2} + \dfrac{c}{d - c} - \dfrac{cd}{(c + d)(c - d)}$

112) $\dfrac{\dfrac{7}{x} + \dfrac{8}{y}}{1 - \dfrac{6}{y}}$ $\dfrac{8x + 7y}{x(y - 6)}$

Solve.

113) $\dfrac{h}{5} = \dfrac{h - 3}{h + 1} + \dfrac{12}{5h + 5}$ $\{1, 3\}$

114) $\dfrac{5w}{6} - \dfrac{2}{3} = -\dfrac{1}{6}$ $\left\{\dfrac{3}{5}\right\}$

115) $\dfrac{8}{3g^2 - 7g - 6} - \dfrac{8}{3g + 2} = -\dfrac{4}{g - 3}$ $\{-10\}$

116) $\dfrac{4k}{k + 16} = \dfrac{4}{k + 1}$ $\{-4, 4\}$

Chapter 8:   Test

*Additional answers can be found in the Answers to Exercises appendix.

1) Evaluate, if possible, for $k = -4$.

$\dfrac{5k + 8}{k^2 + 16}$ $-\dfrac{3}{8}$

Determine the values of the variable for which
a) the expression is undefined.
b) the expression equals zero.

2) $\dfrac{2c - 9}{c + 10}$ a) -10 b) $\dfrac{9}{2}$

3) $\dfrac{n^2 + 1}{n^2 - 5n - 36}$
a) $-4, 9$ b) never equals zero

Write each rational expression in lowest terms.

4) $\dfrac{21t^8u^2}{63t^{12}u^5}$ $\dfrac{1}{3t^4u^3}$

5) $\dfrac{3h^2 - 25h + 8}{9h^2 - 1}$ $\dfrac{h - 8}{3h + 1}$

6) Write three equivalent forms of $\dfrac{7 - m}{4m - 5}$.   possible answers:
$\dfrac{m - 7}{4m - 5}, \dfrac{m - 7}{5 - 4m}, \dfrac{m - 7}{-4m + 5}$

7) Identify the LCD of $\dfrac{2z}{z + 6}$ and $\dfrac{9}{z}$. $z(z + 6)$

Perform the operations, and simplify.

8) $\dfrac{8}{15r} + \dfrac{2}{15r}$ $\dfrac{2}{3r}$

9) $\dfrac{28a^9}{b^2} \div \dfrac{20a^{15}}{b^3}$ $\dfrac{7b}{5a^6}$

10) $\dfrac{5h}{12} - \dfrac{7h}{9}$ $-\dfrac{13h}{36}$

11) $\dfrac{6}{c + 2} + \dfrac{c}{3c + 5}$ $\dfrac{c^2 + 20c + 30}{(3c + 5)(c + 2)}$

12) $\dfrac{k^3 - 9k^2 + 2k - 18}{4k - 24} \cdot \dfrac{k^2 + 3k - 54}{81 - k^2}$ $-\dfrac{k^2 + 2}{4}$

13) $\dfrac{8d^2 + 24d}{20} \div (d + 3)^2$ $\dfrac{2d}{5(d + 3)}$

14) $\dfrac{2t - 5}{t - 7} + \dfrac{t + 9}{7 - t}$ $\dfrac{t - 14}{t - 7}$ or $\dfrac{14 - t}{7 - t}$

15) $\dfrac{3}{2v^2 - 7v + 6} - \dfrac{v + 4}{v^2 + 7v - 18}$ $\dfrac{-2v^2 - 2v + 39}{(2v - 3)(v - 2)(v + 9)}$

Simplify completely.

16) $\dfrac{1 - \dfrac{1}{m+2}}{\dfrac{m}{m+2} - \dfrac{1}{m}}$ $\dfrac{m}{m-2}$

17) $\dfrac{\dfrac{5x+5y}{x^2y^2}}{\dfrac{20}{xy}}$ $\dfrac{x+y}{4xy}$

Solve each equation.

18) $\dfrac{3r+1}{2} + \dfrac{1}{10} = \dfrac{6r}{5}$ $\{-2\}$

19) $\dfrac{28}{w^2-4} = \dfrac{7}{w-2} - \dfrac{5}{w+2}$ $\varnothing$

20) $\dfrac{3}{x+8} + \dfrac{x}{x-4} = \dfrac{7x+9}{x^2+4x-32}$ $\{-7, 3\}$

21) Solve for b.

$\dfrac{1}{a} + \dfrac{1}{b} = \dfrac{1}{c}$ $b = \dfrac{ac}{a-c}$

Write an equation for each, and solve.

22) Every Sunday night, the equipment at a restaurant must be taken apart and cleaned. Ricardo can do this job twice as fast as Michael. When they work together, they can do the cleaning in 2 hr. How long would it take each man to do the job on his own?   Ricardo: 3 hr; Michael: 6 hr

23) A current flows at 4 mph. If a boat can travel 12 mi downstream in the same amount of time it can go 6 mi upstream, find the speed of the boat in still water.   12 mph

24) Suppose m varies directly as the square of n. If $m = 48$ when $n = 4$, find m when $n = 5$.   75

25) If the temperature remains the same, the volume of a gas is inversely proportional to the pressure. If the volume of a gas is 6.25 L (liters) at a pressure of 2 atm (atmospheres), what is the volume of the gas at 1.25 atm?   10 L

Chapter 8:   Cumulative Review for Chapters 1–8

*Additional answers can be found in the Answers to Exercises appendix.

1) Find the area of the triangle.

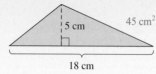

45 cm²

5 cm

18 cm

2) Evaluate $72 - 30 \div 6 + 4(3^2 - 10)$.   63

3) Write an equation and solve.

The length of a rectangular garden is 4 ft longer than the width. Find the dimensions of the garden if its perimeter is 28 ft.   5 ft by 9 ft

Solve each inequality. Write the answer in interval notation.

4) $19 - 8w > 5$ $\left(-\infty, \dfrac{7}{4}\right)$

5) $4 \le \dfrac{3}{5}t + 4 \le 13$ $[0, 15]$

6) Find the x- and y-intercepts of $4x - 3y = 6$, and graph the equation. x-int: $\left(\dfrac{3}{2}, 0\right)$; y-int: $(0, -2)$

7) Find the slope of the line containing the points $(4, 1)$ and $(-2, 9)$. $-\dfrac{4}{3}$

8) Solve the system.

$5x + 4y = 5$
$7x - 6y = 36$ $\left(3, -\dfrac{5}{2}\right)$

Simplify. The answer should not contain negative exponents.

9) $(2p^3)^5$ $32p^{15}$

10) $(5y^2)^{-3}$ $\dfrac{1}{125y^6}$

Multiply and simplify.

11) $(2n - 3)^2$ $4n^2 - 12n + 9$ 12) $(8a + b)(8a - b)$ $64a^2 - b^2$

Divide.

13) $\dfrac{45h^4 - 25h^3 + 15h^2 - 10}{15h^2}$ $3h^2 - \dfrac{5}{3}h + 1 - \dfrac{2}{3h^2}$

14) $\dfrac{5k^3 + 18k^2 - 11k - 8}{k+4}$ $5k^2 - 2k - 3 + \dfrac{4}{k+4}$

Factor completely.

15) $4d^2 + 4d - 15$ $(2d + 5)(2d - 3)$ 16) $3z^4 - 48$
   $3(z^2 + 4)(z + 2)(z - 2)$

17) $rt + 8t - r - 8$ $(r + 8)(t - 1)$

18) Solve $x(x + 16) = x - 36$. $\{-12, -3\}$

19) For what values of a is $\dfrac{7a+2}{a^2-6a}$

   a) undefined?   a) 0, 6
   b) equal to zero?   b) $-\dfrac{2}{7}$

20) Write $\dfrac{3c^2 + 21c - 54}{c^2 + 3c - 54}$ in lowest terms. $\dfrac{3(c-2)}{c-6}$

Perform the operations, and simplify.

21) $\dfrac{10n^2}{n^2 - 8n + 16} \cdot \dfrac{3n^2 - 14n + 8}{10n - 15n^2}$ $\dfrac{2n}{4-n}$

22) $\dfrac{6}{y+5} - \dfrac{3}{y}$ $\dfrac{3(y-5)}{y(y+5)}$

23) Simplify $\dfrac{\dfrac{2}{r-8} + 1}{1 - \dfrac{3}{r-8}} \cdot \dfrac{r-6}{r-11}$

24) Solve $\dfrac{1}{v-1} + \dfrac{2}{5v-3} = \dfrac{37}{5v^2 - 8v + 3}$. $\{6\}$

25) Suppose h varies inversely as the square of p. If $h = 12$ when $p = 2$, find h when $p = 4$.   3

Roots and Radicals

Math at Work:

Forensic Scientist

Forensic scientists like Eli Thomas use mathematics in many ways to help them analyze evidence and solve crimes. To help him reconstruct an accident scene, Eli can use this formula containing a radical to estimate the minimum speed of a vehicle when the accident occurred:

$$S = \sqrt{30\,fd}$$

where f = drag factor, based on the type of road surface

d = length of the skid, in feet

S = speed of the vehicle, in miles per hour

For instance, Eli is investigating an accident in a residential neighborhood where the speed limit is 25 mph. The car involved in the accident left skid marks 60 ft long. Tests conducted by members of Eli's team showed that the drag factor of the asphalt road was 0.80. Eli wants to figure out whether the driver was speeding at the time of the accident.

To find the answer, substitute the values into the equation and evaluate it to determine the minimum speed of the vehicle at the time of the accident:

$$S = \sqrt{30\,fd}$$
$$S = \sqrt{30(0.80)(60)}$$
$$S = \sqrt{1440} \approx 38 \text{ mph}$$

The driver was going at least 38 mph when the accident occurred. This is well over the speed limit of 25 mph.

Eli understands that by working with his colleagues, he can accomplish far more than he could alone. Collaboration allows him to conduct investigations quickly and effectively. Similarly, collaboration can help you succeed in your academic work, particularly within the context of a study group.

We will learn how to simplify radicals in this chapter as well as how to work with equations like the one given here. We will also introduce skills you can use when you participate in a study group.

There is a lot of truth to the old saying, "Two heads are better than one." Working with your fellow students enables you to pool your knowledge and skills as you study for a test, prepare a lab report, or work toward a similar academic goal. However, simply gathering with a few of your classmates is not enough. To get the most out of your study group, apply the skills outlined below:

P Prepare

- Decide on the membership of your study group. Three to five people is typically a good size.
- Identify the specific goal the study group has. Do you want to prepare for a test? Work together on a problem set?
- Decide whether you want your study group to meet only once or twice in order to collaborate on one specific task (e.g., studying for a midterm exam) or to meet regularly throughout the term.

O Organize

- Determine a schedule for when your study group will meet.
- If you are studying for a test, consider meeting at least twice—once to plan a strategy for studying, and then again to execute this strategy.
- Think about assigning each member of the study group a specific topic to teach the rest of the group.

W Work

- Meet somewhere quiet with minimal distractions.
- Compare ideas about the task you are trying to accomplish. One of the benefits of a study group is hearing diverse approaches to a problem.
- If you are studying for a test, brainstorm potential questions and quiz one another about the material.
- Stay motivated and stay focused. Don't let the opportunity to socialize detract from the work you are doing.

E Evaluate

- After you've gotten the results of your group project, meet again. Discuss whether working together was effective.
- Identify ways the study group might have operated differently and what strategies were most beneficial.

R Rethink

- Complete the survey on page 595 with regard to your study group to help assess its effectiveness.
- Ask yourself whether a study group is right for you. Some people prefer studying on their own.

Chapter 9 POWER Plan

P Prepare

What are your goals for Chapter 9?
1 Be prepared before and during class.
2 Understand the homework to the point where you could do it without needing any help or hints.
3 Use the P.O.W.E.R. framework to learn how to work with a study group: *Form a Study Group.*
4 Write your own goal.

O Organize

How can you accomplish each goal? (Write in the steps you will take to succeed.)
• _____
• _____
• _____
• _____
• _____
• _____
• _____
• _____
• _____
• _____
• _____

What are your objectives for Chapter 9?	How can you accomplish each objective?
1 Learn how to find roots of a rational number.	• Know how to find three different types of square roots. • Be able to approximate the square root of a whole number. • Use the Pythagorean theorem. • Find higher roots of rational numbers.
2 Learn how to simplify radicals; learn the product and quotient rules, and how to add and subtract radicals.	• Be able to use the product and quotient rules to multiply and divide square roots. • Use the rules and definitions from this chapter to simplify square roots and higher roots containing variables. • Understand like radicals, and use them in the procedure for adding and subtracting radicals.
3 Perform more operations on radicals.	• Learn to multiply radical expressions. • Learn to use FOIL to multiply radical expressions. • Be able to square a binomial containing radical expressions. • Know how to multiply two binomials of the form $(a + b)(a - b)$ containing radicals.
4 Learn the procedure for dividing radicals.	• Learn to rationalize a denominator containing one square root term and containing a higher root. • Learn to rationalize a denominator containing two terms.

5 Know how to solve radical equations.	• Understand the steps for solving a radical equation, and know to look for extraneous solutions.
	• Be able to solve radical equations:
	  • Containing one square root.
	  • Containing two square roots.
	  • By squaring a binomial.

6 Write your own goal. _____ _____	• _____ _____

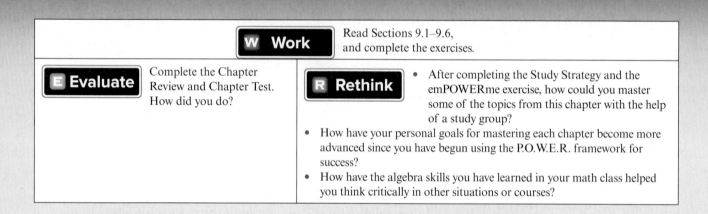

	W Work Read Sections 9.1–9.6, and complete the exercises.
E Evaluate Complete the Chapter Review and Chapter Test. How did you do?	**R Rethink** • After completing the Study Strategy and the emPOWERme exercise, how could you master some of the topics from this chapter with the help of a study group? • How have your personal goals for mastering each chapter become more advanced since you have begun using the P.O.W.E.R. framework for success? • How have the algebra skills you have learned in your math class helped you think critically in other situations or courses?

9.1 Finding Roots

What are your objectives for Section 9.1?	How can you accomplish each objective?
1 Find the Square Root of a Rational Number	• There are many definitions of new terms presented in this objective. Be sure to study them, and write them in your notes. • Understand the three different types of square roots in the **Summary,** and learn how to recognize them. • Complete the given examples on your own. • Complete You Trys 1–3.
2 Approximate the Square Root of a Whole Number	• Write down a procedure you would use to approximate the square root of a whole number. • Complete the given example on your own. • Complete You Try 4.
3 Use the Pythagorean Theorem	• Remember to use the **Five Steps for Solving Applied Problems.** • Complete the given examples on your own. • Complete You Trys 5 and 6.
4 Find Higher Roots of Rational Numbers	• Review the perfect cubes and perfect fourth powers. • Learn the definition and the property of roots of negative numbers. • Complete the examples on your own. • Complete You Trys 7 and 8.

W Work   **Read the explanations, follow the examples, take notes, and complete the You Trys.**

In Section 1.2, we introduced the idea of exponents as representing repeated multiplication. For example,

$$3^2 \text{ means } 3 \cdot 3, \text{ so } 3^2 = 9$$
$$(-3)^2 \text{ means } -3 \cdot (-3), \text{ so } (-3)^2 = 9$$
$$2^4 \text{ means } 2 \cdot 2 \cdot 2 \cdot 2, \text{ so } 2^4 = 16$$

In this chapter we will study the opposite procedure, finding *roots* of numbers.

1   Find the Square Root of a Rational Number

The **square root** of a number, like 9, is a number that, when squared, results in the given number. So, 3 and -3 are square roots of 9 since $3^2 = 9$ and $(-3)^2 = 9$.

EXAMPLE 1   Find all square roots of 36.

In-Class Example 1

Find all square roots of 49.

Answer: $-7, 7$

Solution

To find a *square* root of 36, ask yourself, "What number do I *square* to get 36?" Or, "What number multiplied by itself equals 36?" One number is 6 since $6^2 = 36$. Another number is -6 since $(-6)^2 = 36$. So, the square roots of 36 are 6 and -6.

[YOU TRY 1] Find all square roots of 144.

The **positive** or **principal square root** of a number is represented with the $\sqrt{}$ symbol. Therefore, $\sqrt{36} = 6$.

 BE CAREFUL $\sqrt{36} = 6$ but $\sqrt{36} \neq -6$. The $\sqrt{}$ symbol represents *only* the positive square root.

To find the negative square root of a number, we must put a $-$ in front of the $\sqrt{}$. For example, $-\sqrt{36} = -6$.

We call $\sqrt{}$ the **square root symbol** or the **radical sign.** The number under the radical sign is the **radicand.**

$$\text{Radical sign} \longrightarrow \underbrace{\sqrt{25}}_{\uparrow} \longleftarrow \text{Radicand}$$
$$\text{Radical}$$

The entire expression, $\sqrt{36}$, is called a **radical.**

EXAMPLE 2

In-Class Example 2

Find each square root.

a) $\sqrt{49}$ b) $-\sqrt{25}$

c) $\sqrt{\dfrac{4}{81}}$ d) $-\sqrt{\dfrac{144}{121}}$

e) $\sqrt{0.16}$

Answer: a) 7 b) -5

c) $\dfrac{2}{9}$ d) $-\dfrac{12}{11}$ e) 0.4

Find each square root.

a) $\sqrt{81}$ b) $-\sqrt{49}$ c) $\sqrt{\dfrac{9}{16}}$ d) $-\sqrt{\dfrac{121}{4}}$ e) $\sqrt{0.64}$

Solution

a) $\sqrt{81} = 9$ since $(9)^2 = 81$.

b) $-\sqrt{49}$ means $-1 \cdot \sqrt{49}$. Therefore,
$$-\sqrt{49} = -1 \cdot \sqrt{49} = -1 \cdot (7) = -7$$

c) Since $\sqrt{9} = 3$ and $\sqrt{16} = 4$, $\sqrt{\dfrac{9}{16}} = \dfrac{3}{4}$.

d) $-\sqrt{\dfrac{121}{4}} = -1 \cdot \sqrt{\dfrac{121}{4}} = -1 \cdot \left(\dfrac{11}{2}\right) = -\dfrac{11}{2}$

e) $\sqrt{0.64} = 0.8$

[YOU TRY 2] Find each square root.

a) $-\sqrt{64}$ b) $\sqrt{\dfrac{81}{25}}$ c) $-\sqrt{\dfrac{1}{144}}$ d) $\sqrt{0.36}$

EXAMPLE 3

In-Class Example 3

Find $\sqrt{-100}$.

Answer: There is no such real number.

Find $\sqrt{-4}$.

Solution

Recall that to find $\sqrt{-4}$, you can ask yourself, "What number do I *square* to get -4?" or "What number multiplied by itself equals -4?" *There is no such real number* since $2^2 = 4$ and $(-2)^2 = 4$. Therefore, $\sqrt{-4}$ *is not a real number.*

Let's review what we know about a square root and the radicand and add a third fact.

W Hint

Be familiar with these different types. It will help you do your homework more easily!

Summary Types of Square Roots

1) If the radicand is a perfect square, the *square* root is a *rational* number.

    Example: $\sqrt{25} = 5$ 25 is a perfect square.

    $\sqrt{\dfrac{4}{81}} = \dfrac{2}{9}$ $\dfrac{4}{81}$ is a perfect square.

2) If the radicand is a negative number, the square root is *not* a real number.

    Example: $\sqrt{-9}$ is *not* a real number.

3) If the radicand is positive and *not* a perfect square, then the square root is an *irrational* number.

    Example: $\sqrt{13}$ is irrational.   13 is not a perfect square.

The square root of such a number is a real number that is a nonrepeating, nonterminating decimal. It is important to be able to approximate such square roots because sometimes it is necessary to estimate their places on a number line or on a Cartesian coordinate system when graphing.

For the purposes of graphing, approximating a radical to the nearest tenth is sufficient. A calculator with a $\sqrt{}$ key will give a better approximation of the radical.

2 Approximate the Square Root of a Whole Number

EXAMPLE 4

In-Class Example 4

Approximate $\sqrt{17}$ to the nearest tenth, and plot it on a number line.

Answer: 4.1

W Hint

Summarize the steps you would take to approximate the square root of a number that is not a perfect square.

Approximate $\sqrt{13}$ to the nearest tenth, and plot it on a number line.

Solution

What is the largest perfect square that is *less than* 13? **9**

What is the smallest perfect square that is *greater than* 13? **16**

    Since 13 is between 9 and 16 ($9 < 13 < 16$), it is true that $\sqrt{13}$ is between $\sqrt{9}$ and $\sqrt{16}$.

$$(\sqrt{9} < \sqrt{13} < \sqrt{16})$$
$$\sqrt{9} = 3$$
$$\sqrt{13} = ?$$
$$\sqrt{16} = 4$$

$\sqrt{13}$ must be between 3 and 4. Numerically, 13 is closer to 16 than it is to 9. So, $\sqrt{13}$ will be closer to $\sqrt{16}$ than to $\sqrt{9}$. Check to see if 3.6 is a good approximation of $\sqrt{13}$. ($\approx$ means approximately equal to.)

$$\text{If } \sqrt{13} \approx 3.6, \text{ then } (3.6)^2 \approx 13.$$
$$(3.6)^2 = (3.6) \cdot (3.6) = 12.96$$

Is 3.7 a better approximation of $\sqrt{13}$?

$$\text{If } \sqrt{13} \approx 3.7, \text{ then } (3.7)^2 \approx 13.$$
$$(3.7)^2 = (3.7) \cdot (3.7) = 13.69$$

3.6 is a better approximation of $\sqrt{13}$.

$$\sqrt{13} \approx 3.6$$

√13

← | | | | •| | | →
0 1 2 3 4 5 6

A calculator evaluates $\sqrt{13}$ as 3.6055513. Remember that this is only an approximation. We will discuss how to approximate radicals using a calculator later in this chapter.

[**YOU TRY 4**]
  Approximate $\sqrt{29}$ to the nearest tenth, and plot it on a number line.

---

3 Use the Pythagorean Theorem

Recall from Section 7.6 that we can apply the Pythagorean theorem to right triangles. If a and b are the lengths of the legs of a right triangle, and if c is the length of the hypotenuse, then

$$a^2 + b^2 = c^2$$

EXAMPLE 5

In-Class Example 5

Use
a) $a = 5$, $b = 12$, find c.
b) $a = 2$, $c = 3$, find b.

Answer: a) 13 b) $\sqrt{5}$

Let a and b represent the lengths of the legs of a right triangle, and let c represent the length of the hypotenuse. Use the Pythagorean theorem to find the length of the missing side.

a) $a = 4$, $b = 3$, find c.   b) $a = 7$, $c = 10$, find b.

Solution

a) Substitute the values into the Pythagorean theorem and solve.

$a^2 + b^2 = c^2$	Pythagorean theorem
$(4)^2 + (3)^2 = c^2$	Let $a = 4$ and $b = 3$.
$16 + 9 = c^2$	Square.
$25 = c^2$	Add.

To solve $25 = c^2$, or $c^2 = 25$, we must find c so that the square of c is 25. Additionally, c must be a positive number since it represents the length of the hypotenuse. In other words, one solution of $c^2 = 25$ is the *positive* square root of 25. Therefore, $c = \sqrt{25} = 5$.

Note

If, $k > 0$, then the *positive* solution of $x^2 = k$ is $x = \sqrt{k}$.

b) Substitute the values into the Pythagorean theorem and solve.

$a^2 + b^2 = c^2$	Pythagorean theorem
$(7)^2 + b^2 = (10)^2$	Let $a = 7$ and $c = 10$.
$49 + b^2 = 100$	Square.
$b^2 = 51$	Subtract 49 from each side.
$b = \sqrt{51}$	Solve for b.

Let a and b represent the lengths of the legs of a right triangle, and let c represent the length of the hypotenuse. Use the Pythagorean theorem to find the length of the missing side.

a) $a = 6$, $b = 8$, find c.   b) $b = 4$, $c = 9$, find a.

EXAMPLE 6

In-Class Example 6

Use Example 6.

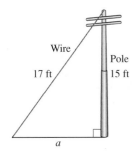

A 17-ft wire is attached to the top of a 15-ft pole so that the wire, the pole, and the ground form a right triangle as shown in the figure. Find the distance from the base of the pole to the point where the wire is attached to the ground.

Solution

Step 1: **Read** the problem carefully, and identify what we are being asked to find.

We must find the distance from the base of the pole to the wire.

Step 2: **Choose a variable** to represent the unknown. Label the picture.

a = distance from the base of the pole to the wire

Also notice that the wire is the hypotenuse of the triangle, so $c = 17$. Therefore, $b = 15$.

Step 3: **Translate** the information into an algebraic equation. Use the Pythagorean theorem.

$$a^2 + b^2 = c^2 \qquad \text{Pythagorean theorem}$$
$$a^2 + (15)^2 = (17)^2 \qquad \text{Let } b = 15 \text{ and } c = 17.$$

Step 4: **Solve** the equation.

$$a^2 + 225 = 289 \qquad \text{Square.}$$
$$a^2 = 64 \qquad \text{Subtract 225 from each side.}$$
$$a = \sqrt{64} \qquad \text{Solve.}$$
$$a = 8 \qquad \text{Simplify.}$$

When we solve $a^2 = 64$, only the *positive* square root of 64 makes sense because a represents a length in this problem.

Step 5: **Check** the answer, and **interpret** the solution as it relates to the problem.

The distance from the base of the pole to the wire is 8 ft.

Do these lengths satisfy the Pythagorean theorem? Yes.

$$a^2 + b^2 = c^2$$
$$8^2 + 15^2 \overset{?}{=} 17^2$$
$$64 + 225 = 289 \quad \checkmark$$

[YOU TRY 6]

A 10-ft ladder is leaning against a house. The base of the ladder is 6 ft from the house. Find the distance from the ground to the top of the ladder.

---

4 Find Higher Roots of Rational Numbers

We saw in Example 2a) that $\sqrt{81} = 9$ since $9^2 = 81$. Finding a square root is the *opposite* of squaring a number. Similarly, we can find higher roots of numbers like $\sqrt[3]{a}$ (read as "the **cube root** of a"), $\sqrt[4]{a}$ (read as "the **fourth root** of a"), $\sqrt[5]{a}$ (the **fifth root** of a), etc.

Find each root.

a) $\sqrt[3]{125}$ b) $\sqrt[4]{81}$ c) $\sqrt[5]{32}$

Solution

a) To find $\sqrt[3]{125}$ (read as "the cube root of 125"), ask yourself, "What number do I *cube* to get 125?" That number is 5 since $5^3 = 125$. Therefore, $\sqrt[3]{125} = 5$.

Finding the cube root of a number is the *opposite* of cubing a number.

Hint

Be sure to know your powers! Make a chart from memory to see where you need help.

b) To find $\sqrt[4]{81}$ (read as "the fourth root of 81") ask yourself, "What number do I raise to the *fourth power* to get 81?" That number is 3 since $3^4 = 81$. Therefore, $\sqrt[4]{81} = 3$.

Finding the fourth root of a number is the *opposite* of raising a number to the fourth power.

c) To find $\sqrt[5]{32}$ (read as "the fifth root of 32") ask yourself, "What number do I raise to the *fifth power* to get 32?" That number is 2 since $2^5 = 32$. Therefore, $\sqrt[5]{32} = 2$.

Finding the fifth root of a number is the *opposite* of raising a number to the fifth power.

[YOU TRY 7]

Find each root.

a) $\sqrt[4]{16}$ b) $\sqrt[3]{27}$

We can use a general notation for writing roots of numbers.

Definition

The expression $\sqrt[n]{a}$ is read as "the *nth* root of *a*." If $\sqrt[n]{a} = b$, then $a = b^n$.

n is the **index** or **order** of the radical. (The plural of *index* is **indices.**)

Note

When finding square roots, we do not write $\sqrt[2]{a}$. The square root of *a* is written as $\sqrt{a}$, and the index is understood to be 2.

In Section 1.2, we first presented the powers of numbers that students are expected to know ($2^2 = 4$, $2^3 = 8$, etc.). Use of these powers was first necessary in the study of the rules of exponents in Chapter 5. Knowing these powers is necessary for finding roots as well, so you can refer to p. 19 to review this list of powers.

While it is true that the square root of a negative number is not a real number, sometimes it *is* possible to find a *higher* root of a negative number.

EXAMPLE 8

Find each root, if possible.

a) $\sqrt[3]{-64}$　　b) $\sqrt[5]{-32}$　　c) $-\sqrt[4]{16}$　　d) $\sqrt[4]{-16}$

Solution

a) To find $\sqrt[3]{-64}$, ask yourself, "What number do I *cube* to get -64?" That number is -4.

$$\sqrt[3]{-64} = -4 \text{ since } (-4)^3 = -64$$

b) To find $\sqrt[5]{-32}$, ask yourself, "What number do I raise to the *fifth power* to get -32?" That number is -2.

$$\sqrt[5]{-32} = -2 \text{ since } (-2)^5 = -32$$

c) $-\sqrt[4]{16}$ means $-1 \cdot \sqrt[4]{16}$. Therefore, $-\sqrt[4]{16} = -1 \cdot \sqrt[4]{16} = -1 \cdot 2 = -2$.

d) To find $\sqrt[4]{-16}$, ask yourself, "What number do I raise to the *fourth power* to get -16?" *There is no such real number* since $2^4 = 16$ and $(-2)^4 = 16$.

$$\sqrt[4]{-16} \text{ is not a real number}$$

We can summarize what we have seen in Example 8 as follows:

W Hint

Learn this property! Why does this make sense?

Property Roots of Negative Numbers

1) The *odd root* of a negative number is a negative number.

2) The *even root* of a negative number is not a real number.

[YOU TRY 8]

Find each root, if possible.

a) $\sqrt[6]{-64}$　　b) $-\sqrt[3]{125}$　　c) $\sqrt[4]{81}$

Using Technology

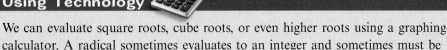

We can evaluate square roots, cube roots, or even higher roots using a graphing calculator. A radical sometimes evaluates to an integer and sometimes must be approximated using a decimal.

To evaluate a square root:

For example, to evaluate $\sqrt{9}$, press $\boxed{2^{nd}}$ $\boxed{x^2}$, enter the radicand $\boxed{9}$, and then press $\boxed{)}$ $\boxed{ENTER}$. The result is 3 as shown on the screen on the left below. When the radicand is a perfect square such as 9, 16, or 25, then the square root evaluates to a whole number. For example, $\sqrt{16}$ evaluates to 4 and $\sqrt{25}$ evaluates to 5 as shown.

If the radicand of a square root is not a perfect square, then the result is a decimal approximation. For example, to evaluate $\sqrt{19}$, press $\boxed{2^{nd}}$ $\boxed{x^2}$, enter the radicand $\boxed{1}$ $\boxed{9}$, and then press $\boxed{)}$ $\boxed{ENTER}$. The result is approximately 4.3589, rounded to four decimal places.

```
√(9)
              3
√(16)
              4
√(25)
              5
```

```
√(19)
    4.358898944
```

To evaluate a cube root:

For example, to evaluate $\sqrt[3]{27}$, press **MATH** [4], enter the radicand [2][7], and then press **)** [ENTER]. The result is 3 as shown.

If the radicand is a perfect cube such as 27, then the cube root evaluates to an integer. Since 28 is not a perfect cube, the cube root evaluates to approximately 3.0366.

To evaluate radicals with an index greater than 3:

For example, to evaluate $\sqrt[4]{16}$, enter the index [4], press **MATH** [5], enter the radicand [1][6], and press [ENTER]. The result is 2. Since the fifth root of 18 evaluates to a decimal, the result is an approximation of 1.7826 rounded to four decimal places as shown.

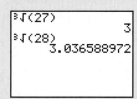

Evaluate each root using a graphing calculator. If necessary, approximate to the nearest tenth.

1) $\sqrt{25}$ 2) $\sqrt[3]{216}$ 3) $\sqrt{29}$ 4) $\sqrt{324}$ 5) $\sqrt[5]{1024}$ 6) $\sqrt[3]{343}$

ANSWERS TO [YOU TRY] EXERCISES

1) -12 and 12 2) a) -8 b) $\dfrac{9}{5}$ c) $-\dfrac{1}{12}$ d) 0.6 3) not a real number

4) 5.4

$\xleftarrow{\qquad}\underset{\;0\;\;\;1\;\;\;2\;\;\;3\;\;\;4\;\;\;5\;\;\;6}{|\;+\;+\;+\;+\;+\overset{\sqrt{29}}{\bullet}|}\xrightarrow{\qquad}$ 5) a) 10 b) $\sqrt{65}$ 6) 8 ft

7) a) 2 b) 3 8) a) not a real number   b) -5 c) 3

ANSWERS TO TECHNOLOGY EXERCISES

1) 5 2) 6 3) 5.4 4) 18 5) 4 6) 7

E Evaluate **9.1** Exercises   Do the exercises, and check your work.

*Additional answers can be found in the Answers to Exercises appendix.

Objective 1: Find the Square Root of a Rational Number

Decide whether each statement is *true* or *false*. If it is false, explain why.

1) $\sqrt{100} = 10$ and -10 2) $\sqrt{49} = 7$ true

3) The square root of a negative number is a negative number.
   False; the square root of a negative number is not a real number.

4) The square root of a nonnegative number is always positive.   False; $\sqrt{0} = 0$

Find all square roots of each number.

 5) 81   9 and -9 6) 121   11 and -11

7) 4   2 and -2 8) 64   8 and -8

9) 900   30 and -30 10) 400   20 and -20

11) $\dfrac{1}{36}$ $\dfrac{1}{6}$ and $-\dfrac{1}{6}$ 12) $\dfrac{25}{9}$ $\dfrac{5}{3}$ and $-\dfrac{5}{3}$

13) 0.25   0.5 and -0.5 14) 0.01   0.1 and -0.1

Find each square root, if possible.

15) $\sqrt{144}$ 12   16) $\sqrt{16}$ 4

17) $\sqrt{9}$ 3   18) $\sqrt{25}$ 5

19) $\sqrt{-64}$ not real   20) $\sqrt{-1}$ not real

21) $-\sqrt{36}$ -6 22) $-\sqrt{169}$ -13

23) $\sqrt{\dfrac{64}{81}}$ $\dfrac{8}{9}$

24) $\sqrt{\dfrac{1}{100}}$ $\dfrac{1}{10}$

25) $\sqrt{\dfrac{4}{9}}$ $\dfrac{2}{3}$

26) $\sqrt{\dfrac{49}{25}}$ $\dfrac{7}{5}$

 27) $-\sqrt{\dfrac{1}{16}}$ $-\dfrac{1}{4}$

28) $-\sqrt{\dfrac{9}{121}}$ $-\dfrac{3}{11}$

29) $\sqrt{0.49}$ 0.7

30) $\sqrt{0.04}$ 0.2

31) $-\sqrt{0.0144}$ -0.12

32) $-\sqrt{0.81}$ -0.9

Objective 2: Approximate the Square Root of a Whole Number

Approximate each square root to the nearest tenth, and plot it on a number line.

 33) $\sqrt{11}$ 3.3

34) $\sqrt{3}$ 1.7

35) $\sqrt{2}$ 1.4

36) $\sqrt{5}$ 2.2

37) $\sqrt{33}$ 5.7

38) $\sqrt{39}$ 6.2

39) $\sqrt{55}$ 7.4

40) $\sqrt{72}$ 8.5

Objective 3: Use the Pythagorean Theorem

Use the Pythagorean theorem to find the length of the missing side.

41)

42)
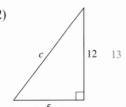

43)
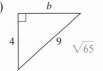

44)
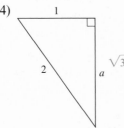

Let a and b represent the lengths of the legs of a right triangle, and let c represent the length of the hypotenuse. Use the Pythagorean theorem to find the length of the missing side.

45) If $b = 5$ and $c = 13$, find a. 12

46) If $a = 3$ and $c = 5$, find b. 4

47) If $a = 1$ and $b = 1$, find c. $\sqrt{2}$

48) If $a = 7$ and $b = 3$, find c. $\sqrt{58}$

Write an equation and solve. (Hint: Draw a picture.)

49) The width of a rectangle is 3 in., and its diagonal is 8 in. long. What is the length of the rectangle? $\sqrt{55}$ in.

50) The length of a rectangle is 7 cm, and its width is 5 cm. Find the length of the diagonal. $\sqrt{74}$ cm

Write an equation and solve.

51) A 13-ft ladder is leaning against a wall so that the base of the ladder is 5 ft away from the wall. How high on the wall does the ladder reach? 12 ft

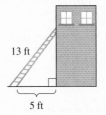

52) Salma is flying a kite. It is 30 ft from her horizontally, and it is 40 ft above her hand. How long is the kite string?

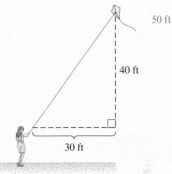

53) Martha digs a garden in the corner of her yard so that it is in the shape of a right triangle. One side of the garden is next to the garage, and the other side is against the back fence. How much fencing will she need to enclose the remaining side of the garden? 17 ft

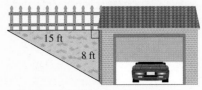

54) The Ramirez family's property is on a corner in the shape of a right triangle. How long is the property on Cooper Ave.? 80 ft

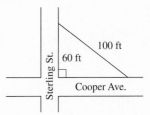

55) A laptop screen has a 13-in. diagonal and a length of 11.2 in. Find the width of the screen to the nearest tenth.   6.6 in.

56) The screen of an LCD TV has a 40-in. diagonal and a length of 34.8 in. Find the height of the screen to the nearest tenth.   19.7 in.

Objective 4: Find Higher Roots of Rational Numbers

Decide whether each statement is *true* or *false*. If it is false, explain why.

57) The cube root of a negative number is a negative number.   true

58) The even root of a negative number is a negative number.
   False; the even root of a negative number is not a real number.

59) The odd root of a negative number is not a real number.
   False; the odd root of a negative number is a negative number.

60) Every nonnegative real number has two real, even roots.   False; the only even root of zero is zero.

61) Explain how to find $\sqrt[3]{64}$. $\sqrt[3]{64}$ is the number you cube to get 64. $\sqrt[3]{64} = 4$

62) Explain how to find $\sqrt[4]{16}$.
   $\sqrt[4]{16}$ is the number you raise to the fourth power to get 16. $\sqrt[4]{16} = 2$

63) Does $\sqrt[4]{81} = -3$? Why or why not?
   No; the even root of a negative number is not a real number.

64) Does $\sqrt[3]{-8} = -2$? Why or why not?
   Yes; $\sqrt[3]{-8} = -2$ because $(-2)^3 = -8$.

Find each root, if possible.

65) $\sqrt[3]{8}$ 2

66) $\sqrt[3]{1}$ 1

67) $\sqrt[3]{125}$ 5

68) $\sqrt[3]{27}$ 3

69) $\sqrt[3]{-1}$ -1

70) $\sqrt[3]{-8}$ -2

71) $\sqrt[4]{81}$ 3

72) $\sqrt[4]{16}$ 2

73) $\sqrt[4]{-1}$ not real

74) $\sqrt[4]{-81}$ not real

75) $-\sqrt[4]{16}$ -2

76) $-\sqrt[4]{1}$ -1

77) $\sqrt[5]{-32}$ -2

78) $-\sqrt[6]{64}$ -2

79) $-\sqrt[3]{-27}$ 3

80) $-\sqrt[3]{-1000}$ 10

81) $\sqrt[6]{-64}$ not real

82) $\sqrt[4]{-16}$ not real

83) $\sqrt[3]{\dfrac{8}{125}}$ $\dfrac{2}{5}$

84) $\sqrt[4]{\dfrac{81}{16}}$ $\dfrac{3}{2}$

85) $\sqrt{60 - 11}$ 7

86) $\sqrt{100 + 21}$ 11

87) $\sqrt[3]{100 + 25}$ 5

88) $\sqrt[3]{9 - 36}$ -3

89) $\sqrt{1 - 9}$ not real

90) $\sqrt{25 - 36}$ not real

R Rethink

R1) Explain how finding roots is the opposite of evaluating powers.

R2) Explain why the square root of a negative number is not a real number.

R3) Where could you use the Pythagorean theorem to help you with a project at home or elsewhere?

9.2 Simplifying Radicals: The Product and Quotient Rules

What are your objectives for Section 9.2?	How can you accomplish each objective?
1 Multiply Square Roots	• Understand and learn the **Product Rule for Square Roots.** • Complete the given example on your own. • Complete You Try 1.
2 Simplify the Square Root of a Whole Number	• Learn the property **When Is a Square Root Simplified?** • Notice that simplifying requires that you use the **Product Rule for Square Roots.** • Complete the given examples on your own. • Complete You Trys 2 and 3.
3 Use the Quotient Rule for Square Roots	• Understand and learn the **Quotient Rule for Square Roots.** • Complete the given examples on your own, and notice that you will need to know when to apply the various definitions learned throughout the section. • Complete You Trys 4 and 5.
4 Simplify Square Root Expressions Containing Variables	• Be sure to use all of the properties and definitions from the section to master this concept. • Complete the given example on your own. • Complete You Trys 6 and 7.
5 Simplify Higher Roots	• Understand and learn the **Product and Quotient Rule for Higher Roots.** • Complete the given examples on your own. • Complete You Trys 8–10.

 Work **Read the explanations, follow the examples, take notes, and complete the You Trys.**

In this section, we will introduce rules for finding the product and quotient of square roots as well as for simplifying expressions containing square roots.

1 Multiply Square Roots

Let's begin with the product $\sqrt{4} \cdot \sqrt{25}$. $\sqrt{4} \cdot \sqrt{25} = 2 \cdot 5 = 10$. Also notice that $\sqrt{4} \cdot \sqrt{25} = \sqrt{4 \cdot 25} = \sqrt{100} = 10$.

We obtain the same result. This leads us to the product rule for multiplying expressions containing square roots.

Definition Product Rule for Square Roots

Let a and b be nonnegative real numbers. Then,

$$\sqrt{a} \cdot \sqrt{b} = \sqrt{a \cdot b}$$

In other words, the product of two square roots equals the square root of the product.

EXAMPLE 1

In-Class Example 1

Multiply. Assume that the variable represents a nonnegative real number.
a) $\sqrt{5} \cdot \sqrt{3}$ b) $\sqrt{10} \cdot \sqrt{w}$

Answer: a) $\sqrt{15}$ b) $\sqrt{10w}$

Multiply. Assume that the variable represents a nonnegative real number.

a) $\sqrt{7} \cdot \sqrt{2}$ b) $\sqrt{5} \cdot \sqrt{x}$

Solution

a) $\sqrt{7} \cdot \sqrt{2} = \sqrt{7 \cdot 2} = \sqrt{14}$ b) $\sqrt{5} \cdot \sqrt{x} = \sqrt{5 \cdot x} = \sqrt{5x}$

 BE CAREFUL We can multiply radicals this way *only if* the indices are the same. In future math courses, we will learn how to multiply radicals with different indices such as $\sqrt{5} \cdot \sqrt[3]{t}$.

[YOU TRY 1]

Multiply. Assume that the variable represents a nonnegative real number.

a) $\sqrt{2} \cdot \sqrt{3}$ b) $\sqrt{6} \cdot \sqrt{z}$

---

2 Simplify the Square Root of a Whole Number

Knowing how to simplify radicals is very important in the study of algebra. We begin by discussing how to simplify expressions containing square roots.

How do we know when a square root is simplified?

W Hint

Why isn't a square root expression simplified if it contains exponents greater than or equal to 2?

Property When Is a Square Root Simplified?

An expression containing a square root is simplified when all of the following conditions are met:

1) The radicand does not contain any factors (other than 1) that are perfect squares.

2) The radicand does not contain any fractions.

3) There are no radicals in the denominator of a fraction.

Note: Condition 1) implies that the radical cannot contain variables with exponents greater than or equal to 2, the index of the square root.

We will discuss higher roots later in this section.

To simplify expressions containing square roots, we reverse the process of multiplying. That is, we use the product rule that says

$$\sqrt{a \cdot b} = \sqrt{a} \cdot \sqrt{b}$$

where a or b is a perfect square.

Simplify completely.

a) $\sqrt{12}$ b) $\sqrt{21}$ c) $\sqrt{72}$

Solution

a)   The radical $\sqrt{12}$ is not in simplest form since 12 contains a factor (other than 1) that is a perfect square. Think of two numbers that multiply to 12 so that at least one of the numbers is a perfect square: $12 = 4 \cdot 3$.

(While it is true that $12 = 6 \cdot 2$, neither 6 nor 2 is a perfect square.)

Rewrite $\sqrt{12}$:

$$\begin{aligned} \sqrt{12} &= \sqrt{4 \cdot 3} && \text{4 is a perfect square.} \\ &= \sqrt{4} \cdot \sqrt{3} && \text{Product rule} \\ &= 2\sqrt{3} && \sqrt{4} = 2 \end{aligned}$$

$2\sqrt{3}$ is completely simplified because 3 does not have any factors that are perfect squares.

b)   Try to write 21 as the product of two numbers so that one number is a perfect square other than 1.

$21 = 3 \cdot 7$ Neither 3 nor 7 is a perfect square.
$21 = 1 \cdot 21$ While 1 is a perfect square, it will not help us simplify $\sqrt{21}$.

$\sqrt{21}$ is in simplest form because 21 does not contain any factors that are perfect squares.

c)   There are different ways to simplify $\sqrt{72}$. We will look at two of them.

i)   Two numbers that multiply to 72 are 36 and 2 with 36 being a perfect square. We can write

$$\begin{aligned} \sqrt{72} &= \sqrt{36 \cdot 2} && \text{36 is a perfect square.} \\ &= \sqrt{36} \cdot \sqrt{2} && \text{Product rule} \\ &= 6\sqrt{2} && \sqrt{36} = 6 \end{aligned}$$

ii)   We can also think of 72 as $9 \cdot 8$ since 9 is a perfect square. We can write

$$\begin{aligned} \sqrt{72} &= \sqrt{9 \cdot 8} && \text{9 is a perfect square.} \\ &= \sqrt{9} \cdot \sqrt{8} && \text{Product rule} \\ &= 3\sqrt{8} && \sqrt{9} = 3 \end{aligned}$$

Therefore, $\sqrt{72} = 3\sqrt{8}$. Is $\sqrt{8}$ in simplest form? *No, because $8 = 4 \cdot 2$ and 4 is a perfect square.* We must continue to simplify.

$$\begin{aligned} \sqrt{72} &= 3\sqrt{8} \\ &= 3\sqrt{4 \cdot 2} && \text{4 is a perfect square.} \\ &= 3\sqrt{4} \cdot \sqrt{2} && \text{Product rule} \\ &= 3 \cdot 2 \cdot \sqrt{2} && \sqrt{4} = 2 \\ &= 6\sqrt{2} && \text{Multiply } 3 \cdot 2. \end{aligned}$$

$6\sqrt{2}$ is completely simplified because 2 does not have any factors that are perfect squares.

Example 2c) shows that using either $\sqrt{72} = \sqrt{36 \cdot 2}$ or $\sqrt{72} = \sqrt{9 \cdot 8}$ leads us to the same result. Furthermore, this example illustrates that a radical is not always *completely* simplified after just one iteration of the simplification process. It is necessary to always examine the radical to determine whether or not it can be simplified more.

Note

After simplifying a radical, look at the result and ask yourself, "*Is the radical in simplest form?*" If it is not, simplify again. Asking yourself this question will help you to be sure that the radical *is* completely simplified.

[YOU TRY 2] Simplify completely.

a) $\sqrt{20}$ b) $\sqrt{150}$ c) $\sqrt{33}$ d) $\sqrt{24}$

EXAMPLE 3

Multiply and simplify.

a) $\sqrt{5} \cdot \sqrt{10}$ b) $\sqrt{18} \cdot \sqrt{3}$

Solution

a) $\sqrt{5} \cdot \sqrt{10} = \sqrt{50}$ Product rule
$\phantom{\sqrt{5} \cdot \sqrt{10}} = \sqrt{25 \cdot 2}$ 25 is a perfect square.
$\phantom{\sqrt{5} \cdot \sqrt{10}} = \sqrt{25} \cdot \sqrt{2} = 5\sqrt{2}$

b) We can find the product $\sqrt{18} \cdot \sqrt{3}$ in two different ways.

   i)   Begin by multiplying the radicands to obtain one radical.

$$\sqrt{18} \cdot \sqrt{3} = \sqrt{18 \cdot 3} \qquad \text{Product rule}$$
$$\phantom{\sqrt{18} \cdot \sqrt{3}} = \sqrt{54} \qquad \text{Multiply.}$$
$$\phantom{\sqrt{18} \cdot \sqrt{3}} = \sqrt{9 \cdot 6} \qquad \text{9 is a perfect square.}$$
$$\phantom{\sqrt{18} \cdot \sqrt{3}} = \sqrt{9} \cdot \sqrt{6} = 3\sqrt{6}$$

   ii) Simplify $\sqrt{18}$ *before* multiplying the radicals.

$$\sqrt{18} = \sqrt{9 \cdot 2} = \sqrt{9} \cdot \sqrt{2} = 3\sqrt{2}$$

Then, substitute $3\sqrt{2}$ for $\sqrt{18}$.

$$\sqrt{18} \cdot \sqrt{3} = 3\sqrt{2} \cdot \sqrt{3} \qquad \text{Substitute } 3\sqrt{2} \text{ for } \sqrt{18}.$$
$$\phantom{\sqrt{18} \cdot \sqrt{3}} = 3\sqrt{2 \cdot 3} \qquad \text{Product rule}$$
$$\phantom{\sqrt{18} \cdot \sqrt{3}} = 3\sqrt{6} \qquad \text{Multiply.}$$

Either way, we get the same result.

[YOU TRY 3] Multiply and simplify.

a) $\sqrt{3} \cdot \sqrt{15}$ b) $\sqrt{20} \cdot \sqrt{2}$

---

3 Use the Quotient Rule for Square Roots

Let's simplify $\dfrac{\sqrt{36}}{\sqrt{9}}$. We can say $\dfrac{\sqrt{36}}{\sqrt{9}} = \dfrac{6}{3} = 2$. It is also true that

$\dfrac{\sqrt{36}}{\sqrt{9}} = \sqrt{\dfrac{36}{9}} = \sqrt{4} = 2$.

This leads us to the quotient rule for dividing expressions containing square roots.

> ### Definition Quotient Rule for Square Roots
>
> Let a and b be nonnegative real numbers such that $b \neq 0$. Then
>
> $$\sqrt{\dfrac{a}{b}} = \dfrac{\sqrt{a}}{\sqrt{b}}$$
>
> The square root of a quotient equals the quotient of the square roots.

EXAMPLE 4

Simplify completely.

a) $\sqrt{\dfrac{9}{25}}$ b) $\dfrac{\sqrt{60}}{\sqrt{5}}$ c) $\sqrt{\dfrac{2}{49}}$ d) $\dfrac{15\sqrt{12}}{3\sqrt{2}}$

Solution

a) Since 9 and 25 are each perfect squares, simplify the expression by finding the square root of each separately.

$$\sqrt{\dfrac{9}{25}} = \dfrac{\sqrt{9}}{\sqrt{25}} \qquad \text{Quotient rule}$$

$$= \dfrac{3}{5} \qquad \sqrt{9} = 3 \text{ and } \sqrt{25} = 5$$

b) We can simplify $\dfrac{\sqrt{60}}{\sqrt{5}}$ using two different methods.

i) Begin by applying the quotient rule to obtain a fraction under *one* radical, simplify the fraction, then simplify the radical.

$$\dfrac{\sqrt{60}}{\sqrt{5}} = \sqrt{\dfrac{60}{5}} \qquad \text{Quotient rule}$$

$$= \sqrt{12} \qquad \text{Simplify } \dfrac{60}{5}.$$

$$= \sqrt{4 \cdot 3} \qquad \text{4 is a perfect square.}$$

$$= \sqrt{4} \cdot \sqrt{3} \qquad \text{Product rule}$$

$$= 2\sqrt{3} \qquad \sqrt{4} = 2$$

ii) We can apply the product rule to rewrite $\sqrt{60}$ and then simplify the fraction.

$$\dfrac{\sqrt{60}}{\sqrt{5}} = \dfrac{\sqrt{5} \cdot \sqrt{12}}{\sqrt{5}} \qquad \text{Product rule}$$

$$= \dfrac{\overset{1}{\cancel{\sqrt{5}}} \cdot \sqrt{12}}{\underset{1}{\cancel{\sqrt{5}}}} \qquad \text{Divide out the common factor.}$$

$$= \sqrt{12} \qquad \text{Simplify.}$$

$$= \sqrt{4 \cdot 3} = \sqrt{4} \cdot \sqrt{3} = 2\sqrt{3}$$

Either method will produce the same result.

c) We cannot simplify the fraction $\dfrac{2}{49}$, but 49 *is* a perfect square.

$$\sqrt{\dfrac{2}{49}} = \dfrac{\sqrt{2}}{\sqrt{49}} \qquad \text{Use the quotient rule.}$$

$$= \dfrac{\sqrt{2}}{7} \qquad \sqrt{49} = 7$$

d) Think of $\dfrac{15\sqrt{12}}{3\sqrt{2}}$ as $\dfrac{15}{3} \cdot \dfrac{\sqrt{12}}{\sqrt{2}}$. Simplify $\dfrac{15}{3}$ and use the quotient rule.

$$\dfrac{15}{3} \cdot \dfrac{\sqrt{12}}{\sqrt{2}} = 5 \cdot \sqrt{\dfrac{12}{2}} = 5\sqrt{6}$$

[YOU TRY 4] Simplify completely.

a) $\sqrt{\dfrac{49}{144}}$ b) $\sqrt{\dfrac{360}{10}}$ c) $\dfrac{\sqrt{120}}{\sqrt{3}}$ d) $\sqrt{\dfrac{6}{25}}$ e) $\dfrac{28\sqrt{30}}{4\sqrt{6}}$

Let's look at another example that requires us to use both the product and quotient rules.

EXAMPLE 5

Simplify $\sqrt{\dfrac{7}{12}} \cdot \sqrt{\dfrac{1}{3}}$.

In-Class Example 5

Simplify $\sqrt{\dfrac{1}{5}} \cdot \sqrt{\dfrac{11}{20}}$.

Answer: $\dfrac{\sqrt{11}}{10}$

Ⓦ Hint

Try multiplying these radicals by simplifying first.

Solution

Begin by using the product rule to multiply the radicands. Then simplify.

$$\sqrt{\dfrac{7}{12}} \cdot \sqrt{\dfrac{1}{3}} = \sqrt{\dfrac{7}{12} \cdot \dfrac{1}{3}} \qquad \text{Product rule}$$

$$= \sqrt{\dfrac{7}{36}} \qquad \text{Multiply.}$$

$$= \dfrac{\sqrt{7}}{\sqrt{36}} \qquad \text{Quotient rule}$$

$$= \dfrac{\sqrt{7}}{6} \qquad \text{Simplify.}$$

[YOU TRY 5] Simplify $\sqrt{\dfrac{1}{8}} \cdot \sqrt{\dfrac{3}{2}}$.

---

4 Simplify Square Root Expressions Containing Variables

Recall that a square root expression in simplified form cannot contain any factors (other than 1) that are perfect squares. Let's see what this means when the radicand contains variables by first examining some expressions containing only numbers.

It is true that

$$\sqrt{4^2} = \sqrt{16} = 4 \qquad \text{and} \qquad \sqrt{(-4)^2} = \sqrt{16} = 4$$

Using a variable, we can say that if a is any real number, then $\sqrt{a^2} = |a|$.

Using this property on the numerical examples above, we can say

$$\sqrt{4^2} = |4| = 4 \qquad \text{and} \qquad \sqrt{(-4)^2} = |-4| = 4$$

Notice that if a is a nonnegative number as in the example on the left, we do not need to use absolute values.

> **Property** Simplifying $\sqrt{a^2}$ for $a \geq 0$
>
> If $a \geq 0$, then $\sqrt{a^2} = a$ and $\sqrt{a} \cdot \sqrt{a} = \sqrt{a^2} = a$.

In the rest of this book, we will assume that all variables represent positive real numbers so that absolute values are not needed when simplifying square root expressions.

Let's simplify some square root expressions containing variables.

EXAMPLE 6

In-Class Example 6

Simplify completely.
a) $\sqrt{y^4}$ b) $\sqrt{144k^6}$
c) $\sqrt{50a^{12}}$ d) $\sqrt{\dfrac{45}{n^8}}$
e) $\sqrt{d^7}$

Answer: a) y^2 b) $12k^3$
c) $5a^6\sqrt{2}$ d) $\dfrac{3\sqrt{5}}{n^4}$
e) $d^3\sqrt{d}$

Simplify completely.

a) $\sqrt{t^4}$ b) $\sqrt{81p^6}$ c) $\sqrt{45c^{18}}$ d) $\sqrt{\dfrac{8}{w^{10}}}$ e) $\sqrt{x^7}$

Solution

a) Ask yourself, *"What do I square to get t^4?"* t^2. Therefore, $\sqrt{t^4} = t^2$.

Or, we can think of simplifying in this way:

$$\begin{aligned} \sqrt{t^4} &= \sqrt{(t^2)^2} & \text{Write } t^4 \text{ as a perfect square.} \\ &= t^2 & \text{Simplify.} \end{aligned}$$

b) Ask yourself, *"What do I square to get $81p^6$?"* $9p^3$. Therefore, $\sqrt{81p^6} = 9p^3$.

Or, we can begin by using the product rule:

$$\begin{aligned} \sqrt{81p^6} &= \sqrt{81} \cdot \sqrt{p^6} & \text{Product rule} \\ &= 9p^3 & \text{Simplify; } \sqrt{p^6} = \sqrt{(p^3)^2} = p^3 \end{aligned}$$

c) $$\begin{aligned} \sqrt{45c^{18}} &= \sqrt{45} \cdot \sqrt{c^{18}} & \text{Product rule} \\ &= \sqrt{9} \cdot \sqrt{5} \cdot \sqrt{(c^9)^2} & \text{Product rule; write } \sqrt{c^{18}} \text{ as a perfect square.} \\ &= 3\sqrt{5} \cdot c^9 & \text{Simplify.} \\ &= 3c^9\sqrt{5} & \text{Use the commutative property to write the radical last.} \end{aligned}$$

d) $$\begin{aligned} \sqrt{\frac{8}{w^{10}}} &= \frac{\sqrt{8}}{\sqrt{w^{10}}} & \text{Quotient rule} \\[2mm] &= \frac{\sqrt{4} \cdot \sqrt{2}}{\sqrt{(w^5)^2}} & 4 \text{ is a perfect square; } w^{10} = (w^5)^2 \\[2mm] &= \frac{2\sqrt{2}}{w^5} & \text{Simplify.} \end{aligned}$$

e) The exponent in the radicand of $\sqrt{x^7}$ is odd, so we must take a different approach to simplifying this radical. Since this is a *square* root expression, we want to write x^7 as a product so that one of the factors is a perfect *square*. We will write x^7 as $x^6 \cdot x$ since x^6 is the largest perfect square that is a factor of x^7.

$$\begin{aligned} \sqrt{x^7} &= \sqrt{x^6 \cdot x} \\ &= \sqrt{x^6} \cdot \sqrt{x} & \text{Product rule} \\ &= x^3\sqrt{x} & \sqrt{x^6} = x^3 \end{aligned}$$

> **W Hint**
>
> Have you been completing all the You Trys?

Simplify completely.

a) $\sqrt{n^8}$ b) $\sqrt{64z^{14}}$ c) $\sqrt{63b^{10}}$ d) $\sqrt{\dfrac{12}{r^{12}}}$ e) $\sqrt{h^5}$

Notice in Example 6a)–d) that to find the square root of a variable with an *even* exponent, we can just divide the exponent by 2. For example,

$$\sqrt{t^4} = t^{4/2} = t^2 \qquad\qquad \sqrt{p^6} = p^{6/2} = p^3$$

We can use the property $\sqrt{a} \cdot \sqrt{a} = a$ to find a product more easily.

EXAMPLE 7

In-Class Example 7

Simplify completely.
a) $\sqrt{9} \cdot \sqrt{9}$
b) $\sqrt{23k} \cdot \sqrt{23k}$

Answer: a) 9 b) 23k

Simplify completely.

a) $\sqrt{7} \cdot \sqrt{7}$ b) $\sqrt{19p} \cdot \sqrt{19p}$

Solution

a) $\sqrt{7} \cdot \sqrt{7} = 7$ Use the property $\sqrt{a} \cdot \sqrt{a} = a$.

  We can verify the result by multiplying the radicands: $\sqrt{7} \cdot \sqrt{7} = \sqrt{49} = 7$

b) $\sqrt{19p} \cdot \sqrt{19p} = 19p$ Use the property $\sqrt{a} \cdot \sqrt{a} = a$.

YOU TRY 7

Simplify completely.

a) $\sqrt{10} \cdot \sqrt{10}$ b) $\sqrt{34n} \cdot \sqrt{34n}$

---

5 Simplify Higher Roots

In Section 9.1, we found higher roots like $\sqrt[4]{16} = 2$ and $\sqrt[3]{-27} = -3$. Now we will extend what we learned about multiplying, dividing, and simplifying *square* roots to doing the same with higher roots.

> **Definition** Product and Quotient Rules for Higher Roots
>
> If a and b are real numbers such that the roots exist, then
>
> $$\sqrt[n]{a} \cdot \sqrt[n]{b} = \sqrt[n]{a \cdot b} \qquad \text{and} \qquad \sqrt[n]{\dfrac{a}{b}} = \dfrac{\sqrt[n]{a}}{\sqrt[n]{b}} \quad (b \neq 0)$$

These rules enable us to multiply, divide, and simplify radicals with any index in a way that is similar to multiplying, dividing, and simplifying *square* roots.

EXAMPLE 8

In-Class Example 8

Multiply.
a) $\sqrt[3]{3} \cdot \sqrt[3]{2}$ b) $\sqrt[4]{5} \cdot \sqrt[4]{z}$

Answer: a) $\sqrt[3]{6}$ b) $\sqrt[4]{5z}$

Multiply.

a) $\sqrt[3]{2} \cdot \sqrt[3]{5}$ b) $\sqrt[4]{7} \cdot \sqrt[4]{y}$

Solution

a) $\sqrt[3]{2} \cdot \sqrt[3]{5} = \sqrt[3]{2 \cdot 5} = \sqrt[3]{10}$ b) $\sqrt[4]{7} \cdot \sqrt[4]{y} = \sqrt[4]{7 \cdot y} = \sqrt[4]{7y}$

BE CAREFUL

Remember that we can apply the product rule *only* if the indices of the radicals are the same.

Let's see how we can use the product rule to simplify higher roots.

EXAMPLE 9

In-Class Example 9

Simplify completely.
a) $\sqrt[3]{56}$ b) $\sqrt[4]{32}$

c) $\dfrac{\sqrt[3]{120}}{\sqrt[3]{3}}$

Answer: a) $2\sqrt[3]{7}$ b) $2\sqrt[4]{2}$
c) $2\sqrt[3]{5}$

Ⓦ Hint

Write a few sentences to explain how simplifying higher roots is similar to simplifying square roots.

Simplify completely.

a) $\sqrt[3]{40}$ b) $\sqrt[4]{48}$ c) $\dfrac{\sqrt[3]{72}}{\sqrt[3]{3}}$

Solution

a) Because we are simplifying a *cube* root, $\sqrt[3]{40}$, think of two numbers whose product is 40 with one of those factors being a perfect *cube*. Those numbers are 8 and 5.

$$\begin{aligned} \sqrt[3]{40} &= \sqrt[3]{8 \cdot 5} \qquad &\text{Rewrite 40 as } 8 \cdot 5. \\ &= \sqrt[3]{8} \cdot \sqrt[3]{5} \qquad &\text{Product rule} \\ &= 2\sqrt[3]{5} \qquad &\text{Simplify.} \end{aligned}$$

b) To simplify a *fourth* root, $\sqrt[4]{48}$, think of two numbers whose product is 48 with one of those factors being a perfect *fourth power*. Those numbers are 16 and 3.

$$\begin{aligned} \sqrt[4]{48} &= \sqrt[4]{16 \cdot 3} \qquad &\text{Rewrite 48 as } 16 \cdot 3. \\ &= \sqrt[4]{16} \cdot \sqrt[4]{3} \qquad &\text{Product rule} \\ &= 2\sqrt[4]{3} \qquad &\text{Simplify.} \end{aligned}$$

c) Begin by applying the quotient rule.

$$\begin{aligned} \frac{\sqrt[3]{72}}{\sqrt[3]{3}} &= \sqrt[3]{\frac{72}{3}} \qquad &\text{Quotient rule} \\ &= \sqrt[3]{24} \qquad &\text{Divide.} \end{aligned}$$

Is $\sqrt[3]{24}$ completely simplified? If it is in simplest form, then 24 will not contain any factors that are perfect cubes. But $24 = 8 \cdot 3$, and 8 is a perfect cube. Therefore, $\sqrt[3]{24}$ is not in simplest form. We must continue to simplify.

$$\begin{aligned} &= \sqrt[3]{8} \cdot \sqrt[3]{3} \qquad &\text{Product rule} \\ &= 2\sqrt[3]{3} \qquad &\text{Simplify.} \end{aligned}$$

YOU TRY 9

Simplify completely.

a) $\sqrt[3]{80}$ b) $\sqrt[4]{32}$ c) $\dfrac{\sqrt[3]{54}}{\sqrt[3]{2}}$

Note

The expression $\sqrt[n]{a}$ is simplified when a does not contain any factors (other than 1) that are perfect *n*th powers and when the radicand does not contain any fractions. This implies that the radicand cannot contain variables with exponents greater than or equal to *n*.

Next let's simplify higher roots containing variables. Recall that we are assuming that $a > 0$ so that $\sqrt[n]{a^n} = a$.

EXAMPLE 10

Simplify completely.

a) $\sqrt[3]{k^6}$ b) $\sqrt[3]{125h^{12}}$ c) $\sqrt[3]{\dfrac{z^3}{64}}$

d) $\sqrt[3]{m^5}$ e) $\sqrt[3]{54x^{13}}$ f) $\sqrt[4]{d^8}$

In-Class Example 10

Simplify completely.

a) $\sqrt[3]{x^6}$ b) $\sqrt[3]{64t^{15}}$

c) $\sqrt[3]{\dfrac{c^3}{125}}$ d) $\sqrt[3]{r^4}$

e) $\sqrt[3]{81a^{17}}$ f) $\sqrt[4]{n^8}$

Answer: a) x^2 b) $4t^5$

c) $\dfrac{c}{5}$ d) $r\sqrt[3]{r}$

e) $3a^5\sqrt[3]{3a^2}$ f) n^2

Ⓦ Hint

Remember that the best way to read a math book is to write out the examples as you are reading them!

Solution

a) Since we are trying to simplify a *cube* root, ask yourself, *"What do I cube to get k^6?"* k^2. Therefore, $\sqrt[3]{k^6} = k^2$.

Or, we can think of simplifying in this way:

$$\sqrt[3]{k^6} = \sqrt[3]{(k^2)^3} \qquad \text{Write } k^6 \text{ as a perfect cube.}$$
$$= k^2 \qquad \text{Simplify.}$$

b) $\sqrt[3]{125h^{12}} = \sqrt[3]{125} \cdot \sqrt[3]{h^{12}}$ Product rule

$\qquad\qquad = 5 \cdot \sqrt[3]{(h^4)^3}$ Simplify; write h^{12} as a perfect cube.

$\qquad\qquad = 5h^4$ Simplify.

c) $\sqrt[3]{\dfrac{z^3}{64}} = \dfrac{\sqrt[3]{z^3}}{\sqrt[3]{64}}$ Quotient rule

$\qquad\quad = \dfrac{z}{4}$ Simplify.

d) We are trying to simplify a *cube root* expression, $\sqrt[3]{m^5}$, but m^5 is not a perfect *cube*. Therefore, write m^5 as a product so that one of the factors is a perfect *cube*. We can write m^5 as $m^3 \cdot m^2$.

$$\sqrt[3]{m^5} = \sqrt[3]{m^3 \cdot m^2}$$
$$= \sqrt[3]{m^3} \cdot \sqrt[3]{m^2} \qquad \text{Product rule}$$
$$= m\sqrt[3]{m^2} \qquad \text{Simplify.}$$

The radicand in the expression $m\sqrt[3]{m^2}$ is completely simplified because the exponent is less than the index of the radical.

e) In the radicand of $\sqrt[3]{54x^{13}}$, x^{13} is not a perfect *cube*. We will need to write x^{13} as a product so that one of the factors is a perfect *cube*. $x^{13} = x^{12} \cdot x$

$$\sqrt[3]{54x^{13}} = \sqrt[3]{54} \cdot \sqrt[3]{x^{13}} \qquad \text{Product rule}$$
$$= (\sqrt[3]{27} \cdot \sqrt[3]{2}) \cdot \sqrt[3]{x^{12} \cdot x} \qquad \text{Product rule}$$
$$= 3\sqrt[3]{2} \cdot \sqrt[3]{x^{12}} \cdot \sqrt[3]{x} \qquad \text{Simplify; product rule}$$
$$= 3\sqrt[3]{2} \cdot \sqrt[3]{(x^4)^3} \cdot \sqrt[3]{x} \qquad \text{Write } x^{12} \text{ as a perfect cube.}$$
$$= 3\sqrt[3]{2} \cdot x^4 \cdot \sqrt[3]{x} \qquad \text{Simplify.}$$
$$= 3x^4\sqrt[3]{2x} \qquad \text{Product rule}$$

f) $\sqrt[4]{d^8} = \sqrt[4]{(d^2)^4}$ Write d^8 as a perfect fourth power.

$\qquad\quad = d^2$ Simplify.

[YOU TRY 10] Simplify completely.

a) $\sqrt[3]{y^9}$ b) $\sqrt[3]{1000t^{18}}$ c) $\sqrt[3]{\dfrac{a^6}{8}}$ d) $\sqrt[3]{w^7}$

e) $\sqrt[3]{72b^{17}}$ f) $\sqrt[4]{q^{12}}$

ANSWERS TO [YOU TRY] EXERCISES

1) a) $\sqrt6$ b) $\sqrt{6z}$ 2) a) $2\sqrt5$ b) $5\sqrt6$ c) $\sqrt{33}$ d) $2\sqrt6$ 3) a) $3\sqrt5$ b) $2\sqrt{10}$

4) a) $\dfrac{7}{12}$ b) 6 c) $2\sqrt{10}$ d) $\dfrac{\sqrt6}{5}$ e) $7\sqrt5$ 5) $\dfrac{\sqrt3}{4}$

6) a) n^4 b) $8z^7$ c) $3b^5\sqrt7$ d) $\dfrac{2\sqrt3}{r^6}$ e) $h^2\sqrt h$ 7) a) 10 b) $34n$

8) a) $\sqrt[3]{12}$ b) $\sqrt[4]{9d}$ 9) a) $2\sqrt[3]{10}$ b) $2\sqrt[4]{2}$ c) 3

10) a) y^3 b) $10t^6$ c) $\dfrac{a^2}{2}$ d) $w^2\sqrt[3]{w}$ e) $2b^5\sqrt[3]{9b^2}$ f) q^3

E Evaluate **9.2** Exercises   Do the exercises, and check your work.

Additional answers can be found in the Answers to Exercises appendix.

Objective 1: Multiply Square Roots

Multiply and simplify.

1) $\sqrt2\cdot\sqrt3$ $\sqrt6$ 2) $\sqrt5\cdot\sqrt7$ $\sqrt{35}$

3) $\sqrt{13}\cdot\sqrt{10}$ $\sqrt{130}$ 4) $\sqrt2\cdot\sqrt k$ $\sqrt{2k}$

5) $\sqrt{15}\cdot\sqrt n$ $\sqrt{15n}$ 6) $\sqrt7\cdot\sqrt r$ $\sqrt{7r}$

11) $\sqrt{20}$ $2\sqrt5$ 12) $\sqrt8$ $2\sqrt2$

13) $\sqrt{90}$ $3\sqrt{10}$ 14) $\sqrt{24}$ $2\sqrt6$

15) $\sqrt{21}$ simplified   16) $\sqrt{42}$ simplified

17) $-\sqrt{75}$ $-5\sqrt3$ 18) $-\sqrt{54}$ $-3\sqrt6$

19) $3\sqrt{80}$ $12\sqrt5$ 20) $-5\sqrt{96}$ $-20\sqrt6$

21) $-\sqrt{1600}$ -40 22) $\sqrt{400}$ 20

Objective 2: Simplify the Square Root of a Whole Number

Label each statement as *true* or *false*. Give a reason for your answer.

7) $\sqrt{24}$ is in simplest form.
False; 24 contains a factor of 4, which is a perfect square.

8) $\sqrt{30}$ is in simplest form.
True; 30 does not have any factors (other than 1) that are perfect squares.

Simplify completely. If the radical is already simplified, then say so.

Fill It In

Fill in the blanks with either the missing mathematical step or reason for the given step.

9) $\sqrt{60} = \sqrt{4\cdot15}$ Factor.

   $= \sqrt4\cdot\sqrt{15}$ Product rule

   $= 2\sqrt{15}$ Simplify.

10) $\sqrt{200} = \sqrt{100\cdot2}$ Factor.

   $= \sqrt{100}\cdot\sqrt2$ Product rule

   $= 10\sqrt2$ Simplify.

Choose from *always, sometimes,* or *never.*

23) If a is a whole number and a perfect square, and if $a > 1$, then a is *always, sometimes,* or *never* a prime number.   never

24) A number that is a perfect square is *always, sometimes,* or *never* an odd number.   sometimes

Multiply and simplify.

25) $\sqrt6\cdot\sqrt2$ $2\sqrt3$ 26) $\sqrt5\cdot\sqrt{15}$ $5\sqrt3$

27) $\sqrt{12}\cdot\sqrt3$ 6 28) $\sqrt8\cdot\sqrt2$ 4

29) $\sqrt{20}\cdot\sqrt3$ $2\sqrt{15}$ 30) $\sqrt{21}\cdot\sqrt3$ $3\sqrt7$

31) $7\sqrt6\cdot\sqrt{12}$ $42\sqrt2$ 32) $5\sqrt6\cdot\sqrt{20}$ $10\sqrt{30}$

33) $\sqrt{20}\cdot\sqrt{18}$ $6\sqrt{10}$ 34) $\sqrt{50}\cdot\sqrt{12}$ $10\sqrt6$

Objective 3: Use the Quotient Rule for Square Roots

Simplify completely.

35) $\sqrt{\dfrac{9}{16}}$ $\dfrac{3}{4}$

36) $\sqrt{\dfrac{2}{72}}$ $\dfrac{1}{6}$

37) $\dfrac{\sqrt{32}}{\sqrt{2}}$ 4

38) $\dfrac{\sqrt{54}}{\sqrt{6}}$ 3

39) $\sqrt{\dfrac{60}{5}}$ $2\sqrt{3}$

40) $\dfrac{\sqrt{140}}{\sqrt{7}}$ $2\sqrt{5}$

41) $-\sqrt{\dfrac{3}{64}}$ $-\dfrac{\sqrt{3}}{8}$

42) $-\sqrt{\dfrac{6}{49}}$ $-\dfrac{\sqrt{6}}{7}$

43) $\sqrt{\dfrac{52}{49}}$ $\dfrac{2\sqrt{13}}{7}$

44) $\sqrt{\dfrac{45}{121}}$ $\dfrac{3\sqrt{5}}{11}$

45) $\dfrac{16\sqrt{63}}{2\sqrt{3}}$ $8\sqrt{21}$

46) $\dfrac{48\sqrt{50}}{12\sqrt{5}}$ $4\sqrt{10}$

47) $\dfrac{10\sqrt{80}}{15\sqrt{2}}$ $\dfrac{4\sqrt{10}}{3}$

48) $\dfrac{28\sqrt{54}}{35\sqrt{3}}$ $\dfrac{12\sqrt{2}}{5}$

49) $\sqrt{\dfrac{1}{2}} \cdot \sqrt{\dfrac{5}{8}}$ $\dfrac{\sqrt{5}}{4}$

50) $\sqrt{\dfrac{2}{3}} \cdot \sqrt{\dfrac{1}{27}}$ $\dfrac{\sqrt{2}}{9}$

51) $\sqrt{\dfrac{2}{7}} \cdot \sqrt{\dfrac{50}{7}}$ $\dfrac{10}{7}$

52) $\sqrt{\dfrac{12}{11}} \cdot \sqrt{\dfrac{3}{11}}$ $\dfrac{6}{11}$

Objective 4: Simplify Square Root Expressions Containing Variables

Simplify completely.

53) $\sqrt{c^2}$ c

54) $\sqrt{a^4}$ a^2

55) $\sqrt{t^6}$ t^3

56) $\sqrt{w^8}$ w^4

57) $\sqrt{121b^{10}}$ $11b^5$

58) $\sqrt{9r^{12}}$ $3r^6$

59) $\sqrt{36q^{14}}$ $6q^7$

60) $\sqrt{144n^{16}}$ $12n^8$

61) $\sqrt{28r^4}$ $2r^2\sqrt{7}$

62) $\sqrt{75n^4}$ $5n^2\sqrt{3}$

63) $\sqrt{18z^{12}}$ $3z^6\sqrt{2}$

64) $\sqrt{500q^{22}}$ $10q^{11}\sqrt{5}$

65) $\sqrt{\dfrac{y^8}{169}}$ $\dfrac{y^4}{13}$

66) $\sqrt{\dfrac{k^6}{49}}$ $\dfrac{k^3}{7}$

67) $\sqrt{\dfrac{99}{w^2}}$ $\dfrac{3\sqrt{11}}{w}$

68) $\dfrac{\sqrt{45}}{\sqrt{m^{18}}}$ $\dfrac{3\sqrt{5}}{m^9}$

69) $\sqrt{r^4s^{12}}$ r^2s^6

70) $\sqrt{c^8d^2}$ c^4d

71) $\sqrt{36x^{10}y^2}$ $6x^5y$

72) $\sqrt{81a^6b^{16}}$ $9a^3b^8$

73) $\sqrt{\dfrac{m^{18}n^8}{49}}$ $\dfrac{m^9n^4}{7}$

74) $\sqrt{\dfrac{u^{14}v^4}{900}}$ $\dfrac{u^7v^2}{30}$

Simplify completely.

75) $\sqrt{11} \cdot \sqrt{11}$ 11

76) $\sqrt{5} \cdot \sqrt{5}$ 5

77) $\sqrt{93} \cdot \sqrt{93}$ 93

78) $\sqrt{87} \cdot \sqrt{87}$ 87

79) $\sqrt{42c} \cdot \sqrt{42c}$ $42c$

80) $\sqrt{61y} \cdot \sqrt{61y}$ $61y$

81) $\sqrt{185h} \cdot \sqrt{185h}$ $185h$

82) $\sqrt{194w} \cdot \sqrt{194w}$ $194w$

Simplify completely.

Fill It In

Fill in the blanks with either the missing mathematical step or reason for the given step.

83) $\sqrt{p^9} = \sqrt{p^8 \cdot p^1}$ Factor.

    $= \sqrt{p^8} \cdot \sqrt{p^1}$ Product rule

    $= p^4\sqrt{p}$ Simplify.

84) $\sqrt{x^{19}} = \sqrt{x^{18} \cdot x^1}$ Factor.

    $= \sqrt{x^{18}} \cdot \sqrt{x^1}$ Product rule

    $= x^9\sqrt{x}$ Simplify.

85) $\sqrt{h^3}$ $h\sqrt{h}$

86) $\sqrt{a^5}$ $a^2\sqrt{a}$

87) $\sqrt{g^{13}}$ $g^6\sqrt{g}$

88) $\sqrt{k^{11}}$ $k^5\sqrt{k}$

89) $\sqrt{100w^5}$ $10w^2\sqrt{w}$

90) $\sqrt{144z^3}$ $12z\sqrt{z}$

91) $\sqrt{75t^{11}}$ $5t^5\sqrt{3t}$

92) $\sqrt{20c^9}$ $2c^4\sqrt{5c}$

93) $\sqrt{x^2y^9}$ $xy^4\sqrt{y}$

94) $\sqrt{a^4b^3}$ $a^2b\sqrt{b}$

95) $\sqrt{4t^9u^5}$ $2t^4u^2\sqrt{tu}$

96) $\sqrt{36m^7n^{11}}$ $6m^3n^5\sqrt{mn}$

Objective 5: Simplify Higher Roots

97) How do you know that a radical expression containing a cube root is completely simplified?

98) How do you know that a radical expression containing a fourth root is completely simplified?

Multiply.

99) $\sqrt[3]{2} \cdot \sqrt[3]{6}$ $\sqrt[3]{12}$

100) $\sqrt[3]{3} \cdot \sqrt[3]{7}$ $\sqrt[3]{21}$

101) $\sqrt[4]{9} \cdot \sqrt[4]{n}$ $\sqrt[4]{9n}$

102) $\sqrt[4]{13} \cdot \sqrt[4]{p^3}$ $\sqrt[4]{13p^3}$

Simplify completely.

103) $\sqrt[3]{24}$ $2\sqrt[3]{3}$

104) $\sqrt[3]{48}$ $2\sqrt[3]{6}$

105) $\sqrt[3]{72}$ $2\sqrt[3]{9}$

106) $\sqrt[3]{81}$ $3\sqrt[3]{3}$

107) $\sqrt[4]{48}$ $2\sqrt[4]{3}$

108) $\sqrt[4]{64}$ $2\sqrt[4]{4}$

109) $\sqrt[4]{162}$ $3\sqrt[4]{2}$

110) $\sqrt[4]{243}$ $3\sqrt[4]{3}$

111) $\sqrt[3]{\dfrac{1}{64}}$ $\dfrac{1}{4}$

112) $\sqrt[3]{-\dfrac{54}{2}}$ -3

113) $\sqrt[4]{\dfrac{1}{16}}$ $\dfrac{1}{2}$

114) $\sqrt[4]{\dfrac{64}{4}}$ 2

115) $\dfrac{\sqrt[3]{500}}{\sqrt[3]{2}}$ $5\sqrt[3]{2}$

116) $\dfrac{\sqrt[3]{120}}{\sqrt[3]{5}}$ $2\sqrt[3]{3}$

Simplify completely.

117) $\sqrt[3]{r^3}$ r

118) $\sqrt[3]{h^9}$ h^3

119) $\sqrt[3]{w^{12}}$ w^4

120) $\sqrt[3]{x^6}$ x^2

121) $\sqrt[3]{27d^{15}}$ $3d^5$

122) $\sqrt[3]{1000p^9}$ $10p^3$

123) $\sqrt[3]{125a^{18}b^6}$ $5a^6b^2$

124) $\sqrt[3]{8m^3n^{12}}$ $2mn^4$

125) $\sqrt[3]{\dfrac{t^9}{8}}$ $\dfrac{t^3}{2}$

126) $\sqrt[3]{\dfrac{k^6}{64}}$ $\dfrac{k^2}{4}$

127) $\sqrt[3]{t^4}$ $t\sqrt[3]{t}$

128) $\sqrt[3]{a^8}$ $a^2\sqrt[3]{a^2}$

129) $\sqrt[3]{x^{11}}$ $x^3\sqrt[3]{x^2}$

130) $\sqrt[3]{p^{13}}$ $p^4\sqrt[3]{p}$

131) $\sqrt[3]{b^{17}}$ $b^5\sqrt[3]{b^2}$

132) $\sqrt[3]{w^{10}}$ $w^3\sqrt[3]{w}$

133) $\sqrt[3]{27x^3y^{12}}$ $3xy^4$

134) $\sqrt[3]{1000p^{15}q^9}$ $10p^5q^3$

135) $\sqrt[3]{24k^7}$ $2k^2\sqrt[3]{3k}$

136) $\sqrt[3]{81r^{11}}$ $3r^3\sqrt[3]{3r^2}$

137) $\sqrt[4]{n^4}$ n

138) $\sqrt[4]{z^{12}}$ z^3

139) $\sqrt[4]{m^5}$ $m\sqrt[4]{m}$

140) $\sqrt[4]{y^9}$ $y^2\sqrt[4]{y}$

141) If A is the area of an equilateral triangle, then the length of a side, s, is given by $s = \sqrt{\dfrac{4\sqrt{3}A}{3}}$. If an equilateral triangle has an area of $2\sqrt{3}$ in^2, how long is each side of the triangle? $2\sqrt{2}$ in.

142) If V is the volume of a sphere, then the radius of the sphere, r, is given by $r = \sqrt[3]{\dfrac{3V}{4\pi}}$. If a ball has a volume of 36π in^3, find its radius.   3 in.

R Rethink

R1) Do you prefer to simplify before multiplying/dividing or to multiply/divide first? Why?

R2) Which objectives have you mastered? Which objective(s) might you still need help with?

9.3 Adding and Subtracting Radicals

P Prepare

O Organize

What are your objectives for Section 9.3?	How can you accomplish each objective?
1 Add and Subtract Radical Expressions	• Write down the definition of *like radicals.* • Write the procedure for **Adding and Subtracting Radicals** in your own words. • Complete the given examples on your own. • Complete You Trys 1–3.
2 Add and Subtract Radical Expressions Containing Variables	• Use the same procedure for **Adding and Subtracting Radicals,** and apply it to radical expressions containing variables. • Complete the given example on your own. • Complete You Try 4.

W Work   **Read the explanations, follow the examples, take notes, and complete the You Trys.**

Just as we can add and subtract like terms such as $2x + 5x = 7x$, we can add and subtract *like radicals* such as $2\sqrt{3} + 5\sqrt{3}$.

Note

Like radicals have the same index and the same radicand.

Some examples of like radicals are

$$2\sqrt{3} \text{ and } 5\sqrt{3}, \qquad -\sqrt[3]{2} \text{ and } 9\sqrt[3]{2}, \qquad \sqrt{x} \text{ and } 4\sqrt{x}, \qquad 6\sqrt[3]{a^2} \text{ and } 8\sqrt[3]{a^2}$$

1 Add and Subtract Radical Expressions

We add and subtract like radicals in the same way we add and subtract like terms—add or subtract the "coefficients" of the radicals, and multiply that result by the radical. Recall that we are using the distributive property when we are combining like terms in this way.

EXAMPLE 1

Add or subtract.

a) $2x + 5x$ b) $2\sqrt{3} + 5\sqrt{3}$ c) $\sqrt{6} - 4\sqrt{6}$

d) $\sqrt[3]{7} + \sqrt[3]{7}$ e) $\sqrt{5} - \sqrt{2}$

Solution

a) First notice that $2x$ and $5x$ are like terms. Therefore, they can be added.

$$2x + 5x = (2 + 5)x \qquad \text{Distributive property}$$
$$= 7x \qquad \text{Simplify.}$$

Or, we can say that by just adding the coefficients, $2x + 5x = 7x$.

b) We can add $2\sqrt{3}$ and $5\sqrt{3}$ because they are like radicals.

$$2\sqrt{3} + 5\sqrt{3} = (2 + 5)\sqrt{3} \qquad \text{Distributive property}$$
$$= 7\sqrt{3} \qquad \text{Simplify.}$$

c) $\sqrt{6} - 4\sqrt{6}$

$= 1\sqrt{6} - 4\sqrt{6}$
$= (1 - 4)\sqrt{6}$
$= -3\sqrt{6}$

d) $\sqrt[3]{7} + \sqrt[3]{7}$

$= 1\sqrt[3]{7} + 1\sqrt[3]{7} = (1 + 1)\sqrt[3]{7} = 2\sqrt[3]{7}$

e) We cannot find the difference $\sqrt{5} - \sqrt{2}$ using the distributive property because $\sqrt{5}$ and $\sqrt{2}$ are not like radicals.

YOU TRY 1

Add or subtract.

a) $4t + 6t$ b) $4\sqrt{11} + 6\sqrt{11}$ c) $\sqrt{5} - 9\sqrt{5}$

d) $\sqrt{7} - \sqrt{3}$ e) $\sqrt[3]{4} + \sqrt[3]{4}$

EXAMPLE 2

Perform the operations and simplify: $3 + 9\sqrt{2} - 11 + \sqrt{2}$

Solution

Begin by writing like terms together.

$$\begin{aligned}
3 + 9\sqrt{2} - 11 + \sqrt{2} &= 3 - 11 + 9\sqrt{2} + \sqrt{2} && \text{Commutative property}\\
&= -8 + (9 + 1)\sqrt{2} && \text{Subtract; distributive property}\\
&= -8 + 10\sqrt{2} && \text{Add.}
\end{aligned}$$

Is $-8 + 10\sqrt{2}$ in simplest form? *Yes.* The terms are not like so they cannot be combined further, *and* $\sqrt{2}$ is in simplest form.

[**YOU TRY 2**] Perform the operations and simplify: $-5 + 10\sqrt{7} - 1 - \sqrt{7}$

Sometimes it looks like two radicals cannot be added or subtracted. But if the radicals can be *simplified* and they turn out to be *like* radicals, then we can add or subtract them.

Procedure Steps for Adding and Subtracting Radicals

1) Write each radical expression in simplest form.
2) Combine like radicals.

EXAMPLE 3

Perform the operations and simplify.

a) $\sqrt{12} + 5\sqrt{3}$ b) $6\sqrt{18} + 3\sqrt{50} - \sqrt{45}$ c) $-7\sqrt[3]{40} + \sqrt[3]{5}$

In-Class Example 3

Perform the operations and simplify.
a) $\sqrt{27} - \sqrt{3}$
b) $3\sqrt{20} + 2\sqrt{12} + 6\sqrt{5}$
c) $\sqrt[4]{6} + 4\sqrt[4]{96}$

Answer: a) $2\sqrt{3}$
b) $12\sqrt{5} + 4\sqrt{3}$ c) $9\sqrt[4]{6}$

Solution

a) $\sqrt{12}$ and $5\sqrt{3}$ are not like radicals. Can either radical be simplified? *Yes.* We can simplify $\sqrt{12}$.

$$\begin{aligned}
\sqrt{12} + 5\sqrt{3} &= \sqrt{4 \cdot 3} + 5\sqrt{3} && \text{4 is a perfect square.}\\
&= \sqrt{4} \cdot \sqrt{3} + 5\sqrt{3} && \text{Product rule}\\
&= 2\sqrt{3} + 5\sqrt{3} && \sqrt{4} = 2\\
&= 7\sqrt{3} && \text{Add like radicals.}
\end{aligned}$$

b) $6\sqrt{18}, 3\sqrt{50}$, and $\sqrt{45}$ are not like radicals. In this case, *each* radical should be simplified to determine whether they can be combined.

$$\begin{aligned}
6\sqrt{18} + 3\sqrt{50} - \sqrt{45} &= 6\sqrt{9 \cdot 2} + 3\sqrt{25 \cdot 2} - \sqrt{9 \cdot 5} && \text{Factor.}\\
&= 6\sqrt{9} \cdot \sqrt{2} + 3\sqrt{25} \cdot \sqrt{2} - \sqrt{9} \cdot \sqrt{5} && \text{Product rule}\\
&= 6 \cdot 3 \cdot \sqrt{2} + 3 \cdot 5 \cdot \sqrt{2} - 3\sqrt{5} && \text{Simplify radicals.}\\
&= 18\sqrt{2} + 15\sqrt{2} - 3\sqrt{5} && \text{Multiply.}\\
&= 33\sqrt{2} - 3\sqrt{5} && \text{Add like radicals.}
\end{aligned}$$

$33\sqrt{2} - 3\sqrt{5}$ is in simplest form since they are not like expressions.

c) $$\begin{aligned}
-7\sqrt[3]{40} + \sqrt[3]{5} &= -7\sqrt[3]{8 \cdot 5} + \sqrt[3]{5} && \text{8 is a perfect cube.}\\
&= -7\sqrt[3]{8} \cdot \sqrt[3]{5} + \sqrt[3]{5} && \text{Product rule}\\
&= -7 \cdot 2 \cdot \sqrt[3]{5} + \sqrt[3]{5} && \sqrt[3]{8} = 2\\
&= -14\sqrt[3]{5} + \sqrt[3]{5} && \text{Multiply.}\\
&= -13\sqrt[3]{5} && \text{Add like radicals.}
\end{aligned}$$

W Hint

Write out each example on your paper, and pay close attention to every step along the way.

[YOU TRY 3] Perform the operations and simplify.

a) $7\sqrt{3} - \sqrt{12}$ b) $2\sqrt{63} - 11\sqrt{28} + 2\sqrt{21}$ c) $\sqrt[3]{54} + 5\sqrt[3]{16}$

2 Add and Subtract Radical Expressions Containing Variables

Next we will add and subtract radicals containing variables.

EXAMPLE 4

In-Class Example 4

Perform the operations and simplify.
a) $10\sqrt{t} - \sqrt{t}$
b) $\sqrt{12n} + \sqrt{3n}$
c) $\sqrt{5z^2} - 3z\sqrt{20}$
d) $5x\sqrt[3]{16x^3} + 4\sqrt[3]{2x^6}$

Answer: a) $9\sqrt{t}$ b) $3\sqrt{3n}$
c) $-5z\sqrt{5}$ d) $14x^2\sqrt[3]{2}$

 Hint

Use the same procedure as before.

Perform the operations and simplify.

a) $6\sqrt{x} - \sqrt{x}$ b) $\sqrt{8k} + 3\sqrt{2k}$ c) $\sqrt{3c^2} - 5c\sqrt{12}$

d) $9r\sqrt[3]{24r^3} + 2\sqrt[3]{3r^6}$

Solution

a) These are like radicals, so use the distributive property.

$$6\sqrt{x} - \sqrt{x} = (6-1)\sqrt{x} = 5\sqrt{x}$$

b) Simplify the first radical to see whether the two radicals can then be added.

$$
\begin{aligned}
\sqrt{8k} + 3\sqrt{2k} &= \sqrt{4}\cdot\sqrt{2k} + 3\sqrt{2k} &&\text{Product rule; 4 is a perfect square.} \\
&= 2\sqrt{2k} + 3\sqrt{2k} &&\sqrt{4} = 2 \\
&= 5\sqrt{2k} &&\text{Add like radicals.}
\end{aligned}
$$

c) Simplify each radical to determine whether they can be combined.

$$
\begin{aligned}
\sqrt{3c^2} - 5c\sqrt{12} &= \sqrt{3}\cdot\sqrt{c^2} - 5c\cdot\sqrt{4}\cdot\sqrt{3} &&\text{Product rule} \\
&= \sqrt{3}\cdot c - 5c\cdot 2\cdot\sqrt{3} &&\text{Simplify.} \\
&= c\sqrt{3} - 10c\sqrt{3} &&\text{Multiply.} \\
&= -9c\sqrt{3} &&\text{Subtract like radicals.}
\end{aligned}
$$

d) Begin by simplifying each radical.

$$
\begin{aligned}
9r\sqrt[3]{24r^3} &= 9r\cdot\sqrt[3]{24}\cdot\sqrt[3]{r^3} &&\text{Product rule} \\
&= 9r\cdot\sqrt[3]{8}\cdot\sqrt[3]{3}\cdot r &&\text{Product rule; simplify.} \\
&= 9r\cdot 2\sqrt[3]{3}\cdot r &&\sqrt[3]{8} = 2 \\
&= 18r^2\sqrt[3]{3} &&\text{Multiply.}
\end{aligned}
$$

$$
\begin{aligned}
2\sqrt[3]{3r^6} &= 2\sqrt[3]{3}\cdot\sqrt[3]{r^6} &&\text{Product rule} \\
&= 2\sqrt[3]{3}\cdot r^2 &&\text{Simplify.} \\
&= 2r^2\sqrt[3]{3} &&\text{Multiply.}
\end{aligned}
$$

Substitute the simplified radicals into the original expression.

$$
\begin{aligned}
9r\sqrt[3]{24r^3} + 2\sqrt[3]{3r^6} &= 18r^2\sqrt[3]{3} + 2r^2\sqrt[3]{3} &&\text{Substitute.} \\
&= 20r^2\sqrt[3]{3} &&\text{Add like radicals.}
\end{aligned}
$$

[YOU TRY 4] Perform the operations and simplify.

a) $3\sqrt{p} - 5\sqrt{p}$ b) $\sqrt{45a} + 7\sqrt{5a}$

c) $y\sqrt{24} - \sqrt{6y^2}$ d) $10\sqrt[3]{2n^3} + 4n\sqrt[3]{16}$

ANSWERS TO [YOU TRY] EXERCISES

1) a) $10t$ b) $10\sqrt{11}$ c) $-8\sqrt{5}$ d) $\sqrt{7} - \sqrt{3}$ e) $2\sqrt[4]{4}$
2) $-6 + 9\sqrt{7}$ 3) a) $5\sqrt{3}$ b) $-16\sqrt{7} + 2\sqrt{21}$ c) $13\sqrt[3]{2}$
4) a) $-2\sqrt{p}$ b) $10\sqrt{5a}$ c) $y\sqrt{6}$ d) $18n\sqrt[3]{2}$

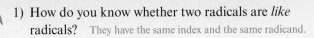

*Additional answers can be found in the Answers to Exercises appendix.

Objective 1: Add and Subtract Radical Expressions

1) How do you know whether two radicals are *like* radicals?   They have the same index and the same radicand.

2) What are the steps for adding or subtracting radicals?
   1) Write each radical expression in simplest form.
   2) Combine like radicals.

Perform the operations and simplify.

3) $4\sqrt{11} + 5\sqrt{11}$ $9\sqrt{11}$ 4) $14\sqrt{3} + 7\sqrt{3}$ $21\sqrt{3}$

5) $4\sqrt{2} - 9\sqrt{2}$ $-5\sqrt{2}$ 6) $14\sqrt{7} - 23\sqrt{7}$ $-9\sqrt{7}$

7) $9\sqrt[3]{5} - 2\sqrt[3]{5}$ $7\sqrt[3]{5}$ 8) $4\sqrt[3]{4} + 6\sqrt[3]{4}$ $10\sqrt[3]{4}$

9) $11\sqrt[3]{2} + 7\sqrt[3]{2}$ $18\sqrt[3]{2}$

10) $\sqrt[3]{10} - 9\sqrt[3]{10}$ $-8\sqrt[3]{10}$

11) $4 - \sqrt{13} + 8 - 6\sqrt{13}$ $12 - 7\sqrt{13}$

12) $-3 + 5\sqrt{6} - 4\sqrt{6} + 9$ $6 + \sqrt{6}$

Fill It In

Fill in the blanks with either the missing mathematical step or reason for the given step.

13) $\sqrt{24} + \sqrt{6} = \sqrt{4 \cdot 6} + \sqrt{6}$ Factor.

   $= \underline{\sqrt{4} \cdot \sqrt{6} + \sqrt{6}}$ Product rule

   $= 2\sqrt{6} + \sqrt{6}$ Simplify.

   $= \underline{3\sqrt{6}}$ Add like radicals.

14) $\sqrt{50} - 9\sqrt{2} = \sqrt{25 \cdot 2} - 9\sqrt{2}$ Factor.

   $= \sqrt{25} \cdot \sqrt{2} - 9\sqrt{2}$ Product rule

   $= \underline{5\sqrt{2} - 9\sqrt{2}}$ Simplify.

   $= \underline{-4\sqrt{2}}$ Subtract like radicals.

15) $6\sqrt{3} - \sqrt{12}$ $4\sqrt{3}$ 16) $\sqrt{45} + 6\sqrt{5}$ $9\sqrt{5}$

17) $\sqrt{75} + \sqrt{3}$ $6\sqrt{3}$ 18) $\sqrt{44} - 8\sqrt{11}$ $-6\sqrt{11}$

19) $\sqrt{28} - 3\sqrt{63}$ $-7\sqrt{7}$ 20) $3\sqrt{45} + \sqrt{20}$ $11\sqrt{5}$

21) $3\sqrt{72} - 4\sqrt{8}$ $10\sqrt{2}$ 22) $3\sqrt{98} + 4\sqrt{50}$ $41\sqrt{2}$

23) $\sqrt{32} - 3\sqrt{18}$ $-5\sqrt{2}$ 24) $\sqrt{96} + 4\sqrt{24}$ $12\sqrt{6}$

25) $\frac{5}{2}\sqrt{40} + \frac{1}{3}\sqrt{90}$ $6\sqrt{10}$ 26) $\frac{2}{3}\sqrt{180} - \frac{6}{7}\sqrt{245}$ $-2\sqrt{5}$

27) $\frac{4}{3}\sqrt{18} - \frac{5}{8}\sqrt{128}$ $-\sqrt{2}$ 28) $\frac{4}{5}\sqrt{150} - \frac{4}{3}\sqrt{54}$ 0

29) $\sqrt{50} - \sqrt{2} + \sqrt{98}$ $11\sqrt{2}$ 30) $\sqrt{3} - \sqrt{12} + \sqrt{75}$ $4\sqrt{3}$

31) $\sqrt{96} + 3\sqrt{24} - \sqrt{54}$ $7\sqrt{6}$

32) $\sqrt{20} - \sqrt{45} - 2\sqrt{80}$ $-9\sqrt{5}$

33) $2\sqrt[3]{11} + 5\sqrt[3]{88}$ $12\sqrt[3]{11}$ 34) $6\sqrt[3]{9} + \sqrt[3]{72}$ $8\sqrt[3]{9}$

35) $8\sqrt[3]{3} - 3\sqrt[3]{81}$ $-\sqrt[3]{3}$ 36) $2\sqrt[3]{81} - 12\sqrt[3]{3}$ $-6\sqrt[3]{3}$

37) $11\sqrt[3]{16} + 10\sqrt[3]{2}$ $32\sqrt[3]{2}$ 38) $\sqrt[3]{6} - \sqrt[3]{48}$ $-\sqrt[3]{6}$

Objective 2: Add and Subtract Radical Expressions Containing Variables

Perform the operations and simplify.

39) $12\sqrt{c} + 3\sqrt{c}$ $15\sqrt{c}$ 40) $-7\sqrt{y} - 2\sqrt{y}$ $-9\sqrt{y}$

41) $5\sqrt{3a} - 9\sqrt{3a}$ $-4\sqrt{3a}$ 42) $6\sqrt{5t} + \sqrt{5t}$ $7\sqrt{5t}$

43) $\sqrt{5b} + \sqrt{45b}$ $4\sqrt{5b}$ 44) $9\sqrt{2w} + \sqrt{32w}$ $13\sqrt{2w}$

45) $\sqrt{50n} - \sqrt{18n}$ $2\sqrt{2n}$ 46) $\sqrt{12k} + \sqrt{48k}$ $6\sqrt{3k}$

47) $11\sqrt{3z} + 2\sqrt{12z}$ $15\sqrt{3z}$ 48) $8\sqrt{5r} - 5\sqrt{20r}$ $-2\sqrt{5r}$

49) $5\sqrt{63v} + 6\sqrt{7v}$ $21\sqrt{7v}$ 50) $2\sqrt{8p} - 6\sqrt{2p}$ $-2\sqrt{2p}$

51) $\sqrt{4h} - 8\sqrt{8p} + \sqrt{4h} + 6\sqrt{8p}$ $4\sqrt{h} - 4\sqrt{2p}$

52) $6\sqrt{3m} + 10\sqrt{2m} + 8\sqrt{3m} - \sqrt{2m}$ $9\sqrt{2m} + 14\sqrt{3m}$

53) $9z\sqrt{12} - \sqrt{3z^2}$ $17z\sqrt{3}$ 54) $2m\sqrt{45} - \sqrt{5m^2}$ $5m\sqrt{5}$

55) $8\sqrt{7r^2} + 3r\sqrt{28}$ $14r\sqrt{7}$ 56) $\sqrt{2x^2} - 7x\sqrt{50}$ $-34x\sqrt{2}$

57) $6q\sqrt{q} + 7\sqrt{q^3}$ $13q\sqrt{q}$ 58) $9\sqrt{r^3} + 3r\sqrt{r}$ $12r\sqrt{r}$

59) $16m^3\sqrt{m} - 13\sqrt{m^7}$ $3m^3\sqrt{m}$ 60) $9d^2\sqrt{d} - 25\sqrt{d^5}$ $-16d^2\sqrt{d}$

61) $8w\sqrt{w^5} + 4\sqrt{w^7}$ $12w^3\sqrt{w}$ 62) $18\sqrt{q^5} - 4q\sqrt{q^3}$ $14q^2\sqrt{q^3}$

63) $\sqrt{xy^3} + 6y\sqrt{xy}$ $7y\sqrt{xy}$ 64) $7a\sqrt{ab} + 3\sqrt{a^3b}$ $10a\sqrt{ab}$

65) $9v\sqrt{6u^3} - 2u\sqrt{54uv^2}$ $3uv\sqrt{5u}$

66) $6c^2\sqrt{8d^3} - 9d\sqrt{2c^4d}$ $3c^2d\sqrt{2d}$

67) $3\sqrt{75m^3n} + m\sqrt{12mn}$ $17m\sqrt{3mn}$

68) $y\sqrt{54xy} - 6\sqrt{24xy^3}$ $-9y\sqrt{6xy}$

69) $15\sqrt[3]{t^2} - 25\sqrt[3]{t^2}$ $-10\sqrt[3]{t^2}$

70) $10\sqrt[3]{m} - 6\sqrt[3]{m}$ $4\sqrt[3]{m}$

71) $5\sqrt[3]{27x^2} + \sqrt[3]{8x^2}$ $17\sqrt[3]{x^2}$

72) $3\sqrt[3]{64k^2} - 9\sqrt[3]{125k^2}$ $-33\sqrt[3]{k^2}$

73) $7r^5\sqrt[3]{r} - 14\sqrt[3]{r^{16}}$ $-7r^5\sqrt[3]{r}$

74) $6t^3\sqrt[3]{t} - 5\sqrt[3]{t^{10}}$ $t^3\sqrt[3]{t}$

75) $7\sqrt[3]{81a^5} + 4a\sqrt[3]{3a^2}$ $25a\sqrt[3]{3a^2}$

76) $3\sqrt[3]{40x} - 12\sqrt[3]{5x}$ $-6\sqrt[3]{5x}$

77) $2c^2\sqrt[3]{108c} - 12\sqrt[3]{32c^7}$ $-18c^2\sqrt[3]{4c}$

78) $9\sqrt[3]{128h^2} + 4\sqrt[3]{16h^2}$ $44\sqrt[3]{2h^2}$

79) $3\sqrt[4]{x^4} + \sqrt[4]{16x^4}$ $5x$

80) $\sqrt[4]{81q^4} - 2\sqrt[4]{q^4}$ q

81) $-3\sqrt[4]{c^{15}} + 8c^3\sqrt[4]{c^3}$ $5c^3\sqrt[4]{c^3}$

82) $5a\sqrt[4]{a^7} + \sqrt[4]{a^{11}}$ $6a^2\sqrt[4]{a^3}$

Find the perimeter of each figure.

83)

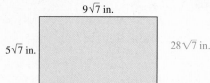

9√7 in.

5√7 in.   28√7 in.

84)

5√3 cm   3√2 cm   (12√3 + 3√2) cm

7√3 cm

R Rethink

R1) How are adding and subtracting radical expressions similar to adding and subtracting polynomials?

R2) Write two expressions that show this similarity.

9.4 Combining Operations on Radicals

P Prepare

O Organize

What are your objectives for Section 9.4?	How can you accomplish each objective?
1 Multiply Radical Expressions	• If possible, simplify radical expressions first. • Complete the given example on your own. • Complete You Try 1.
2 Multiply Radical Expressions Using FOIL	• Review how to use FOIL. • Complete the given example on your own. • Complete You Try 2.
3 Square a Binomial Containing Radical Expressions	• Use the formulas developed in Chapter 6 to square a binomial containing radicals. • Complete the given example on your own. • Complete You Try 3.
4 Multiply Two Binomials of the Form $(a + b)(a - b)$ Containing Radicals	• Use the formulas developed in Chapter 6 to multiply. • Complete the given example on your own. • Complete You Try 4.

Read the explanations, follow the examples, take notes, and complete the You Trys.

In Section 9.2, we learned to multiply radicals like $\sqrt{6} \cdot \sqrt{2}$. In this section, we will learn how to simplify expressions that combine multiplication, addition, and subtraction of radicals. Remember, we will assume that all variables represent positive real numbers.

1 Multiply Radical Expressions

EXAMPLE 1

In-Class Example 1

Multiply and simplify.
a) $8(\sqrt{2} - \sqrt{18})$
b) $\sqrt{3}(\sqrt{15} + \sqrt{11})$
c) $\sqrt{p}(\sqrt{p} + \sqrt{20q})$

Answer: a) $-16\sqrt{2}$
b) $3\sqrt{5} + \sqrt{33}$
c) $p + 2\sqrt{5pq}$

Multiply and simplify.

a) $6(\sqrt{3} - \sqrt{12})$ b) $\sqrt{2}(\sqrt{14} + \sqrt{5})$ c) $\sqrt{a}(\sqrt{a} + \sqrt{50b})$

Solution

a) Since $\sqrt{12}$ can be simplified, we will do that first.
$$\sqrt{12} = \sqrt{4 \cdot 3} = \sqrt{4} \cdot \sqrt{3} = 2\sqrt{3}$$
Substitute $2\sqrt{3}$ for $\sqrt{12}$ in the original expression.

$$
\begin{aligned}
6(\sqrt{3} - \sqrt{12}) &= 6(\sqrt{3} - 2\sqrt{3}) && \text{Substitute } 2\sqrt{3} \text{ for } \sqrt{12}. \\
&= 6(-\sqrt{3}) && \text{Subtract.} \\
&= -6\sqrt{3} && \text{Multiply.}
\end{aligned}
$$

 Hint

In this section, you will apply skills and techniques you have already used!

b) Neither $\sqrt{14}$ nor $\sqrt{5}$ can be simplified. Apply the distributive property.
$$
\begin{aligned}
\sqrt{2}(\sqrt{14} + \sqrt{5}) &= \sqrt{2} \cdot \sqrt{14} + \sqrt{2} \cdot \sqrt{5} && \text{Distribute.} \\
&= \sqrt{28} + \sqrt{10} && \text{Product rule}
\end{aligned}
$$

Is $\sqrt{28} + \sqrt{10}$ in simplest form? *No.* $\sqrt{28}$ can be simplified.

$$
\begin{aligned}
&= \sqrt{4 \cdot 7} + \sqrt{10} && \text{4 is a perfect square.} \\
&= \sqrt{4} \cdot \sqrt{7} + \sqrt{10} && \text{Product rule} \\
&= 2\sqrt{7} + \sqrt{10} && \sqrt{4} = 2
\end{aligned}
$$

The sum is in simplest form. The radicals are not like, so they cannot be combined.

c) Since $\sqrt{50b}$ can be simplified, we will do that first.

$$
\begin{aligned}
\sqrt{50b} &= \sqrt{50} \cdot \sqrt{b} && \text{Product rule} \\
&= \sqrt{25} \cdot \sqrt{2} \cdot \sqrt{b} && \text{Product rule; 25 is a perfect square.} \\
&= 5\sqrt{2b} && \text{Simplify; multiply } \sqrt{2} \cdot \sqrt{b}.
\end{aligned}
$$

Substitute $5\sqrt{2b}$ for $\sqrt{50b}$ in the original expression.

$$
\begin{aligned}
\sqrt{a}(\sqrt{a} + \sqrt{50b}) &= \sqrt{a}(\sqrt{a} + 5\sqrt{2b}) && \text{Substitute } 5\sqrt{2b} \text{ for } \sqrt{50b}. \\
&= \sqrt{a} \cdot \sqrt{a} + \sqrt{a} \cdot 5\sqrt{2b} && \text{Distribute.} \\
&= a + 5\sqrt{2ab} && \text{Multiply.}
\end{aligned}
$$

[YOU TRY 1]

Multiply and simplify.

a) $7(\sqrt{3} - \sqrt{75})$ b) $\sqrt{5}(\sqrt{10} + \sqrt{6})$ c) $\sqrt{r}(\sqrt{r} + \sqrt{72t})$

2 Multiply Radical Expressions Using FOIL

In Chapter 6, we first multiplied binomials using **FOIL** (First Outer Inner Last).

$$(3x + 4)(x + 2) = 3x \cdot x + 3x \cdot 2 + 4 \cdot x + 4 \cdot 2$$

$$\begin{array}{cccc} \text{F} & \text{O} & \text{I} & \text{L} \end{array}$$

$$= 3x^2 + 6x + 4x + 8$$
$$= 3x^2 + 10x + 8$$

We can multiply binomials containing radicals the same way.

EXAMPLE 2

 Hint

Are you writing down the steps as you are reading the examples?

Multiply and simplify.

a) $(2\sqrt{5} + \sqrt{3})(\sqrt{5} - 8\sqrt{3})$ b) $(\sqrt{x} + \sqrt{2})(\sqrt{x} + 6\sqrt{2})$

Solution

a) Since we must multiply two binomials, we will use FOIL.

$$(2\sqrt{5} + \sqrt{3})(\sqrt{5} - 8\sqrt{3})$$

$$\begin{array}{cccc} \text{F} & \text{O} & \text{I} & \text{L} \end{array}$$

$$= 2\sqrt{5} \cdot (\sqrt{5}) + 2\sqrt{5} \cdot (-8\sqrt{3}) + \sqrt{3} \cdot (\sqrt{5}) + \sqrt{3} \cdot (-8\sqrt{3}) \quad \text{Use FOIL.}$$
$$= 2 \cdot 5 + (-16\sqrt{15}) + \sqrt{15} + (-8 \cdot 3) \quad \text{Multiply.}$$
$$= 10 - 16\sqrt{15} + \sqrt{15} - 24 \quad \text{Multiply.}$$
$$= -14 - 15\sqrt{15} \quad \text{Combine like terms.}$$

b) $(\sqrt{x} + \sqrt{2})(\sqrt{x} + 6\sqrt{2})$

$$\begin{array}{cccc} \text{F} & \text{O} & \text{I} & \text{L} \end{array}$$

$$= \sqrt{x} \cdot (\sqrt{x}) + \sqrt{x} \cdot (6\sqrt{2}) + \sqrt{2} \cdot (\sqrt{x}) + \sqrt{2} \cdot (6\sqrt{2}) \quad \text{Use FOIL.}$$
$$= x + 6\sqrt{2x} + \sqrt{2x} + 6 \cdot 2 \quad \text{Multiply.}$$
$$= x + 6\sqrt{2x} + 1\sqrt{2x} + 12 \quad \text{Multiply.}$$
$$= x + 7\sqrt{2x} + 12 \quad \text{Combine like terms.}$$

[YOU TRY 2] Multiply and simplify.

a) $(6\sqrt{7} + \sqrt{5})(\sqrt{7} - 9\sqrt{5})$ b) $(\sqrt{a} + \sqrt{10})(\sqrt{a} + 4\sqrt{10})$

3 Square a Binomial Containing Radical Expressions

Recall again, from Chapter 6, that we can use FOIL to square a binomial or we can use these special formulas:

$$(a + b)^2 = a^2 + 2ab + b^2$$
$$(a - b)^2 = a^2 - 2ab + b^2$$

For example,

$$(n + 6)^2 = (n)^2 + 2(n)(6) + (6)^2$$
$$= n^2 + 12n + 36$$

and

$$(2y - 3)^2 = (2y)^2 - 2(2y)(3) + (3)^2$$
$$= 4y^2 - 12y + 9$$

To square a binomial containing radicals, we can either use FOIL or we can use the formulas above. The formulas will help us solve radical equations in Section 9.6.

EXAMPLE 3

Multiply and simplify.

a) $(\sqrt{6} + 4)^2$ b) $(3\sqrt{2} - 7)^2$ c) $(\sqrt{x} + \sqrt{5})^2$

Solution

a) Use $(a + b)^2 = a^2 + 2ab + b^2$.

$$(\sqrt{6} + 4)^2 = (\sqrt{6})^2 + 2(\sqrt{6})(4) + (4)^2 \qquad \text{Substitute } \sqrt{6} \text{ for } a \text{ and 4 for } b.$$
$$= 6 + 8\sqrt{6} + 16 \qquad \text{Multiply.}$$
$$= 22 + 8\sqrt{6} \qquad \text{Combine like terms.}$$

b) Use $(a - b)^2 = a^2 - 2ab + b^2$.

$$(3\sqrt{2} - 7)^2 = (3\sqrt{2})^2 - 2(3\sqrt{2})(7) + (7)^2 \qquad \text{Substitute } 3\sqrt{2} \text{ for } a \text{ and 7 for } b.$$
$$= (9 \cdot 2) - (6\sqrt{2})(7) + 49 \qquad \text{Multiply.}$$
$$= 18 - 42\sqrt{2} + 49 \qquad \text{Multiply.}$$
$$= 67 - 42\sqrt{2} \qquad \text{Combine like terms.}$$

c) $(\sqrt{x} + \sqrt{5})^2 = (\sqrt{x})^2 + 2(\sqrt{x})(\sqrt{5}) + (\sqrt{5})^2 \qquad \text{Use } (a + b)^2 = a^2 + 2ab + b^2.$
$$= x + 2\sqrt{5x} + 5 \qquad \text{Square; product rule}$$

[YOU TRY 3]   Multiply and simplify.

a) $(\sqrt{5} + 8)^2$ b) $(2\sqrt{3} - 9)^2$ c) $(\sqrt{h} + \sqrt{10})^2$

4   Multiply Two Binomials of the Form $(a + b)(a - b)$ Containing Radicals

We will review one last rule from Chapter 6 on multiplying binomials. We will use this in Section 9.5 when we divide radicals.

$$(a + b)(a - b) = a^2 - b^2$$

For example, $(c + 9)(c - 9) = (c)^2 - (9)^2 = c^2 - 81$. The same rule applies when we multiply binomials containing radicals.

EXAMPLE 4

Multiply and simplify.

a) $(4 + \sqrt{3})(4 - \sqrt{3})$ b) $(\sqrt{x} + \sqrt{y})(\sqrt{x} - \sqrt{y})$

Solution

a) Use $(a + b)(a - b) = a^2 - b^2$.

$$(4 + \sqrt{3})(4 - \sqrt{3}) = (4)^2 - (\sqrt{3})^2 \qquad \text{Substitute 4 for } a \text{ and } \sqrt{3} \text{ for } b.$$
$$= 16 - 3 \qquad \text{Square each term.}$$
$$= 13 \qquad \text{Subtract.}$$

b) $(\sqrt{x} + \sqrt{y})(\sqrt{x} - \sqrt{y}) = (\sqrt{x})^2 - (\sqrt{y})^2 \qquad \text{Use } (a + b)(a - b) = a^2 - b^2.$
$$= x - y \qquad \text{Square each term.}$$

Note

Notice in Example 4 that when we multiply expressions containing square roots of the form $(a + b)(a - b)$, the radicals are eliminated. *This will always be true.*

[YOU TRY 4] Multiply and simplify.

a) $(5 - \sqrt{7})(5 + \sqrt{7})$ b) $(\sqrt{p} + \sqrt{q})(\sqrt{p} - \sqrt{q})$

ANSWERS TO [YOU TRY] EXERCISES

1) a) $-28\sqrt{3}$ b) $5\sqrt{2} + \sqrt{30}$ c) $r + 6\sqrt{2rt}$
2) a) $-3 - 53\sqrt{35}$ b) $a + 5\sqrt{10a} + 40$
3) a) $69 + 16\sqrt{5}$ b) $93 - 36\sqrt{3}$ c) $h + 2\sqrt{10h} + 10$
4) a) 18 b) $p - q$

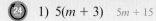 **9.4** Exercises   Do the exercises, and check your work.

Additional answers can be found in the Answers to Exercises appendix.

Objective 1: Multiply Radical Expressions

Multiply and simplify.

1) $5(m + 3)$ $5m + 15$

2) $7(p + 4)$ $7p + 28$

3) $7(\sqrt{2} + 6)$ $7\sqrt{2} + 42$

4) $4(5 - \sqrt{6})$ $20 - 4\sqrt{6}$

5) $\sqrt{2}(\sqrt{5} - 8)$ $\sqrt{10} - 8\sqrt{2}$

6) $\sqrt{10}(4 + \sqrt{3})$ $4\sqrt{10} + \sqrt{30}$

7) $-6(\sqrt{32} + \sqrt{2})$ $-30\sqrt{2}$

8) $4(\sqrt{27} - \sqrt{3})$ $8\sqrt{3}$

9) $6(\sqrt{45} - \sqrt{20})$ $6\sqrt{5}$

10) $-5(\sqrt{18} + \sqrt{50})$ $-40\sqrt{2}$

11) $\sqrt{5}(\sqrt{54} - \sqrt{96})$ $-\sqrt{30}$

12) $\sqrt{3}(\sqrt{45} + \sqrt{20})$ $5\sqrt{15}$

13) $\sqrt{2}(7 + \sqrt{6})$ $7\sqrt{2} + 2\sqrt{3}$

14) $\sqrt{18}(\sqrt{8} - 8)$ $12 - 24\sqrt{2}$

15) $\sqrt{a}(\sqrt{a} - \sqrt{64b})$ $a - 8\sqrt{ab}$

16) $\sqrt{d}(\sqrt{12c} + \sqrt{7d})$ $2\sqrt{3cd} + d\sqrt{7}$

Objective 2: Multiply Radical Expressions Using FOIL

17) How are the problems *Multiply* $(x + 2)(x + 7)$ and *Multiply* $(\sqrt{10} + 2)(\sqrt{10} + 7)$ similar? What method can be used to multiply each of them? Both are examples of multiplication of two binomials. They can be multiplied using FOIL.

18) a) Multiply $(x + 2)(x + 7)$. $x^2 + 9x + 14$

   b) Multiply $(\sqrt{10} + 2)(\sqrt{10} + 7)$. $24 + 9\sqrt{10}$

Multiply and simplify.

Fill It In

Fill in the blanks with either the missing mathematical step or reason for the given step.

19) $(4 + \sqrt{5})(3 + \sqrt{5})$

   $= \underline{4 \cdot 3 + 4\sqrt{5} + 3\sqrt{5} + \sqrt{5} \cdot \sqrt{5}}$ Use FOIL.

   $= 12 + 4\sqrt{5} + 3\sqrt{5} + 5$ Multiply.

   $= \underline{17 + 7\sqrt{5}}$ Combine like terms.

20) $(1 + \sqrt{6})(8 + \sqrt{6})$

   $= 1 \cdot 8 + 1\sqrt{6} + 8\sqrt{6} + \sqrt{6} \cdot \sqrt{6}$ Use FOIL.

   $= \underline{8 + 1\sqrt{6} + 8\sqrt{6} + 6}$ Multiply.

   $= \underline{14 + 9\sqrt{6}}$ Combine like terms.

Multiply and simplify.

21) $(k + 3)(k + 6)$ $k^2 + 9k + 18$

22) $(b - 8)(b + 2)$ $b^2 - 6b - 16$

23) $(6 + \sqrt{7})(2 + \sqrt{7})$ $19 + 8\sqrt{7}$

24) $(5 + \sqrt{3})(1 + \sqrt{3})$ $8 + 6\sqrt{3}$

25) $(\sqrt{2} + 7)(\sqrt{2} - 4)$ $-26 + 3\sqrt{2}$

26) $(\sqrt{5} - 7)(\sqrt{5} + 2)$ $-9 - 5\sqrt{5}$

27) $(\sqrt{3} - 4\sqrt{5})(2\sqrt{3} - \sqrt{5})$ $26 - 9\sqrt{15}$

28) $(5\sqrt{2} - \sqrt{3})(2\sqrt{3} - \sqrt{2})$ $-16 + 11\sqrt{6}$

29) $(3\sqrt{6} - 2\sqrt{2})(\sqrt{2} + 5\sqrt{6})$ $86 - 14\sqrt{3}$

30) $(2\sqrt{10} + 3\sqrt{2})(\sqrt{10} - 2\sqrt{2})$ $8 - 2\sqrt{5}$

---

31) $(3 + 2\sqrt{5})(\sqrt{7} + \sqrt{2})$
$3\sqrt{7} + 3\sqrt{2} + 2\sqrt{35} + 2\sqrt{10}$

32) $(\sqrt{2} + 4)(\sqrt{3} - 6\sqrt{5})$ $\sqrt{6} - 6\sqrt{10} + 4\sqrt{3} - 24\sqrt{5}$

33) $(\sqrt{m} + \sqrt{7})(\sqrt{m} + 5\sqrt{7})$ $m + 6\sqrt{7m} + 35$

34) $(\sqrt{x} + \sqrt{3})(\sqrt{x} + 4\sqrt{3})$ $x + 5\sqrt{3x} + 12$

35) $(\sqrt{6p} - 2\sqrt{q})(8\sqrt{q} + 5\sqrt{6p})$ $-2\sqrt{6pq} + 30p - 16q$

36) $(4\sqrt{3r} + \sqrt{s})(3\sqrt{s} - 2\sqrt{3r})$ $10\sqrt{3rs} + 3s - 24r$

Objective 3: Square a Binomial Containing Radical Expressions

37) How are the problems *Multiply* $(p - 4)^2$ and *Multiply* $(\sqrt{7} - 4)^2$ similar? What method can be used to multiply each of them?   Both are examples of the square of a binomial. We can multiply them using the formula $(a - b)^2 = a^2 - 2ab + b^2$.

38) a) Multiply $(p - 4)^2$. $p^2 - 8p + 16$

   b) Multiply $(\sqrt{7} - 4)^2$. $23 - 8\sqrt{7}$

Multiply and simplify.

39) $(2b - 11)^2$ $4b^2 - 44b + 121$

40) $(3d + 5)^2$ $9d^2 + 30d + 25$

41) $(\sqrt{3} + 1)^2$ $4 + 2\sqrt{3}$

42) $(\sqrt{2} + 9)^2$ $83 + 18\sqrt{2}$

43) $(\sqrt{13} - \sqrt{5})^2$ $18 - 2\sqrt{65}$

44) $(\sqrt{3} + \sqrt{11})^2$
$14 + 2\sqrt{33}$

45) $(2\sqrt{7} + \sqrt{10})^2$ $38 + 4\sqrt{70}$

46) $(2\sqrt{3} - \sqrt{2})^2$
$14 - 4\sqrt{6}$

47) $(\sqrt{6} - 4\sqrt{2})^2$ $38 - 16\sqrt{3}$

48) $(\sqrt{3} + 2\sqrt{15})^2$
$63 + 12\sqrt{5}$

49) $(\sqrt{k} + \sqrt{11})^2$
$k + 2\sqrt{11k} + 11$

50) $(\sqrt{m} + \sqrt{7})^2$
$m + 2\sqrt{7m} + 7$

51) $(\sqrt{x} - \sqrt{y})^2$
$x - 2\sqrt{xy} + y$

52) $(\sqrt{b} - \sqrt{a})^2$
$a - 2\sqrt{ab} + b$

Objective 4: Multiply Two Binomials of the Form $(a + b)(a - b)$ Containing Radicals

53) What formula can be used to multiply $(5 + \sqrt{6})(5 - \sqrt{6})$? $(a + b)(a - b) = a^2 - b^2$

54) What happens to the radical terms whenever we multiply $(a + b)(a - b)$ where the binomials contain square roots?   The radicals are eliminated.

55) $(a + 9)(a - 9)$ $a^2 - 81$

56) $(g - 8)(g + 8)$ $g^2 - 64$

57) $(\sqrt{5} + 3)(\sqrt{5} - 3)$ -4

58) $(\sqrt{3} + 2)(\sqrt{3} - 2)$ -1

59) $(6 - \sqrt{2})(6 + \sqrt{2})$ 34

60) $(4 - \sqrt{11})(4 + \sqrt{11})$ 5

61) $(4\sqrt{3} + \sqrt{2})(4\sqrt{3} - \sqrt{2})$ 46

62) $(2\sqrt{2} - 2\sqrt{7})(2\sqrt{2} + 2\sqrt{7})$ -20

63) $(\sqrt{11} + 5\sqrt{2})(\sqrt{11} - 5\sqrt{2})$ -39

64) $(\sqrt{15} + 5\sqrt{3})(\sqrt{15} - 5\sqrt{3})$ -60

65) $(\sqrt{c} + \sqrt{d})(\sqrt{c} - \sqrt{d})$ $c - d$

66) $(\sqrt{2y} + \sqrt{z})(\sqrt{2y} - \sqrt{z})$ $2y - z$

67) $(6 - \sqrt{t})(6 + \sqrt{t})$ $36 - t$

68) $(4 - \sqrt{q})(4 + \sqrt{q})$ $16 - q$

69) $(8\sqrt{f} - \sqrt{g})(8\sqrt{f} + \sqrt{g})$ $64f - g$

70) $(\sqrt{a} + 6\sqrt{b})(\sqrt{a} - 6\sqrt{b})$ $a - 36b$

Mixed Exercises: Objectives 1–4
Multiply and simplify.

71) $(7 + \sqrt{2})(11 + \sqrt{2})$ $79 + 18\sqrt{2}$

72) $\sqrt{7}(\sqrt{14} + \sqrt{2})$ $7\sqrt{2} + \sqrt{14}$

73) $\sqrt{n}(\sqrt{98m} + \sqrt{5n})$ $7\sqrt{2mn} + n\sqrt{5}$

74) $(\sqrt{x} + \sqrt{15})^2$ $x + 2\sqrt{15x} + 15$

75) $(3\sqrt{5} + 4)(3\sqrt{5} - 4)$ 29

76) $(1 + 4\sqrt{7})(2\sqrt{3} + \sqrt{5})$ $2\sqrt{3} + \sqrt{5} + 8\sqrt{21} + 4\sqrt{35}$

77) $(2 - \sqrt{7})^2$ $11 - 4\sqrt{7}$

78) $-3(\sqrt{24} - \sqrt{5})$ $-6\sqrt{6} + 3\sqrt{5}$

79) $(\sqrt{5} - \sqrt{b})(3\sqrt{5} - \sqrt{b})$ $15 - 4\sqrt{5b} + b$

80) $(8 - \sqrt{11})(8 + \sqrt{11})$ 53

81) $(\sqrt{5} + \sqrt{10})(\sqrt{2} - 3\sqrt{3})$ $\sqrt{10} - 3\sqrt{15} + 2\sqrt{5} - 3\sqrt{30}$

82) $(\sqrt{3} + 2)(\sqrt{3} - 10)$ $-17 - 8\sqrt{3}$

R Rethink

R1) Discuss how these exercises compared to those in Chapter 6.

R2) Were there certain types of problems you consistently got wrong? Look closely at them to figure out (and correct) the mistakes.

9.5 Dividing Radicals

What are your objectives for Section 9.5?	How can you accomplish each objective?
1 Rationalize a Denominator Containing One Square Root Term	• Recognize that a radical expression is not in simplest form if the denominator contains a radical. • Write a procedure for simplifying expressions with denominators containing a square root. • Complete the given examples on your own. • Complete You Trys 1–3.
2 Rationalize a Denominator Containing a Higher Root	• Understand the procedure used to rationalize a denominator containing a higher root. • Complete the given examples on your own. • Complete You Trys 4 and 5.
3 Rationalize a Denominator Containing Two Terms	• Learn the definition of *conjugate*. • Learn the procedure for **Rationalizing a Denominator Containing Two Terms.** • Complete the given examples on your own. • Complete You Trys 6 and 7.
4 Divide Out Common Factors in Radical Quotients	• Read the explanation, and note that you need to factor first. • Complete the given example on your own. • Complete You Try 8.

W Work

Read the explanations, follow the examples, take notes, and complete the You Trys.

It is generally agreed that a radical expression is *not* in simplest form if its denominator contains a radical. For example, $\dfrac{1}{\sqrt{2}}$ is not simplified, but an equivalent form, $\dfrac{\sqrt{2}}{2}$, is simplified.

Later we will show that $\dfrac{1}{\sqrt{2}} = \dfrac{\sqrt{2}}{2}$. The process of eliminating radicals from the denominator of an expression is called **rationalizing the denominator.** We will look at two types of rationalizing problems:

1) Rationalizing a denominator containing one term

2) Rationalizing a denominator containing two terms

To rationalize a denominator, we will use the fact that multiplying the numerator and denominator of a fraction by the same quantity results in an equivalent fraction:

$$\frac{2}{3} \cdot \frac{4}{4} = \frac{8}{12} \qquad \frac{2}{3} \text{ and } \frac{8}{12} \text{ are equivalent because } \frac{4}{4} = 1.$$

We use the same idea to rationalize the denominator of a radical expression.

1 Rationalize a Denominator Containing One Square Root Term

The goal of rationalizing is to eliminate the radical from the denominator. With regard to square roots, recall that $\sqrt{a} \cdot \sqrt{a} = \sqrt{a^2} = a$ for $a \geq 0$. For example,

$$\sqrt{2} \cdot \sqrt{2} = \sqrt{2^2} = 2, \quad \sqrt{14} \cdot \sqrt{14} = \sqrt{(14)^2} = 14, \quad \sqrt{x} \cdot \sqrt{x} = \sqrt{x^2} = x \, (x \geq 0)$$

We will use this property to rationalize the denominators of the following expressions.

EXAMPLE 1

Rationalize the denominator of each expression.

a) $\dfrac{1}{\sqrt{2}}$ b) $\dfrac{30}{\sqrt{12}}$ c) $\dfrac{7\sqrt{3}}{\sqrt{5}}$

In-Class Example 1

Rationalize the denominator of each expression.

a) $\dfrac{1}{\sqrt{5}}$ b) $\dfrac{24}{\sqrt{18}}$ c) $\dfrac{3\sqrt{2}}{\sqrt{7}}$

Answer: a) $\dfrac{\sqrt{5}}{5}$ b) $4\sqrt{2}$

c) $\dfrac{3\sqrt{14}}{7}$

Solution

a) To eliminate the square root from the denominator of $\dfrac{1}{\sqrt{2}}$, ask yourself, "By what do I multiply $\sqrt{2}$ to get a *perfect square* under the square root?" The answer is $\sqrt{2}$ since $\sqrt{2} \cdot \sqrt{2} = \sqrt{2^2} = \sqrt{4} = 2$. Multiply by $\sqrt{2}$ in the numerator *and* denominator.

W Hint

$\dfrac{\sqrt{2}}{\sqrt{2}} = 1$, so in a) we are multiplying $\dfrac{1}{\sqrt{2}}$ by 1.

$$\dfrac{1}{\sqrt{2}} = \dfrac{1}{\sqrt{2}} \cdot \dfrac{\sqrt{2}}{\sqrt{2}} \qquad \text{Rationalize the denominator.}$$

$$= \dfrac{\sqrt{2}}{\sqrt{2^2}} \qquad \text{Multiply.}$$

$$= \dfrac{\sqrt{2}}{2} \qquad \sqrt{2^2} = \sqrt{4} = 2$$

BE CAREFUL

$\dfrac{\sqrt{2}}{2}$ is in simplest form. We cannot divide out terms inside and outside of the radical.

Wrong: $\dfrac{\sqrt{2}}{2} = \dfrac{\sqrt{2}^{\,1}}{2_1} = \sqrt{1} = 1$

W Hint

Simplify the denominator first!

b) First, simplify the denominator of $\dfrac{30}{\sqrt{12}}$: $\sqrt{12} = \sqrt{4} \cdot \sqrt{3} = 2\sqrt{3}$.

$$\dfrac{30}{\sqrt{12}} = \dfrac{30}{2\sqrt{3}} \qquad \text{Substitute } 2\sqrt{3} \text{ for } \sqrt{12}.$$

$$= \dfrac{15}{\sqrt{3}} \qquad \text{Simplify } \dfrac{30}{2}.$$

Rationalize the denominator. Ask yourself, "By what do I multiply $\sqrt{3}$ to get a *perfect square* under the square root?" The answer is $\sqrt{3}$.

$$= \frac{15}{\sqrt{3}} \cdot \frac{\sqrt{3}}{\sqrt{3}} \qquad \text{Rationalize the denominator;} \ \frac{\sqrt{3}}{\sqrt{3}} = 1.$$

$$= \frac{15\sqrt{3}}{\sqrt{3^2}}$$

$$= \frac{15\sqrt{3}}{3} \qquad \sqrt{3^2} = 3$$

$$= 5\sqrt{3} \qquad \text{Simplify} \ \frac{15}{3}.$$

c) To rationalize $\dfrac{7\sqrt{3}}{\sqrt{5}}$, multiply the numerator and denominator by $\sqrt{5}$.

$$\frac{7\sqrt{3}}{\sqrt{5}} = \frac{7\sqrt{3}}{\sqrt{5}} \cdot \frac{\sqrt{5}}{\sqrt{5}} \qquad \text{Rationalize the denominator;} \ \frac{\sqrt{5}}{\sqrt{5}} = 1.$$

$$= \frac{7\sqrt{15}}{5}$$

[**YOU TRY 1**] Rationalize the denominator of each expression.

a) $\dfrac{1}{\sqrt{3}}$ b) $\dfrac{60}{\sqrt{20}}$ c) $\dfrac{11\sqrt{6}}{\sqrt{5}}$

Sometimes we will apply the quotient or product rule before rationalizing.

EXAMPLE 2

Ⓦ Hint

Remember . . . write out the steps as you read the example!

Simplify completely.

a) $\sqrt{\dfrac{2}{16}}$ b) $\sqrt{\dfrac{7}{8}} \cdot \sqrt{\dfrac{4}{3}}$

Solution

a) First, simplify $\dfrac{2}{16}$ under the radical.

$$\sqrt{\frac{2}{16}} = \sqrt{\frac{1}{8}} \qquad \text{Simplify.}$$

$$= \frac{\sqrt{1}}{\sqrt{8}} \qquad \text{Quotient rule}$$

$$= \frac{1}{\sqrt{4} \cdot \sqrt{2}} \qquad \text{Product rule}$$

$$= \frac{1}{2\sqrt{2}} \qquad \sqrt{4} = 2$$

$$= \frac{1}{2\sqrt{2}} \cdot \frac{\sqrt{2}}{\sqrt{2}} \qquad \text{Rationalize the denominator.}$$

$$= \frac{\sqrt{2}}{2 \cdot 2} \qquad \sqrt{2} \cdot \sqrt{2} = 2$$

$$= \frac{\sqrt{2}}{4} \qquad \text{Multiply.}$$

b) First, use the product rule to multiply the radicands.

$$\sqrt{\frac{7}{8}} \cdot \sqrt{\frac{4}{3}} = \sqrt{\frac{7}{8} \cdot \frac{4}{3}} \qquad \text{Product rule}$$

$$= \sqrt{\frac{7}{8} \cdot \frac{\overset{1}{4}}{3}} \qquad \text{Multiply the fractions under the radical.}$$

$$= \sqrt{\frac{7}{6}} \qquad \text{Multiply.}$$

$$= \frac{\sqrt{7}}{\sqrt{6}} \qquad \text{Quotient rule}$$

$$= \frac{\sqrt{7}}{\sqrt{6}} \cdot \frac{\sqrt{6}}{\sqrt{6}} \qquad \text{Rationalize the denominator.}$$

$$= \frac{\sqrt{42}}{6} \qquad \text{Multiply.}$$

YOU TRY 2

Simplify completely.

a) $\sqrt{\dfrac{2}{54}}$ b) $\sqrt{\dfrac{11}{15}} \cdot \sqrt{\dfrac{3}{2}}$

We work with radical expressions containing variables the same way. **Remember, we are assuming that all variables represent positive real numbers.**

EXAMPLE 3

In-Class Example 3

Simplify completely.

a) $\dfrac{12}{\sqrt{t}}$ b) $\sqrt{\dfrac{48x^3}{5y}}$

Answer: a) $\dfrac{12\sqrt{t}}{t}$

b) $\dfrac{4x\sqrt{15xy}}{5y}$ c) $\dfrac{\sqrt{7b}}{b}$

Simplify completely.

a) $\dfrac{11}{\sqrt{x}}$ b) $\sqrt{\dfrac{75a^3}{7b}}$

Solution

a) Ask yourself, "By what do I multiply $\sqrt{x}$ to get a *perfect square* under the square root?" The perfect square we want to get is $\sqrt{x^2}$.

$$\sqrt{x} \cdot \sqrt{?} = \sqrt{x^2} = x$$
$$\sqrt{x} \cdot \sqrt{x} = \sqrt{x^2} = x$$

$$\dfrac{11}{\sqrt{x}} = \dfrac{11}{\sqrt{x}} \cdot \dfrac{\sqrt{x}}{\sqrt{x}} \qquad \text{Rationalize the denominator; } \dfrac{\sqrt{x}}{\sqrt{x}} = 1.$$

$$= \dfrac{11\sqrt{x}}{\sqrt{x^2}} \qquad \text{Multiply.}$$

$$= \dfrac{11\sqrt{x}}{x} \qquad \sqrt{x^2} = x$$

W Hint

What questions do you ask yourself while simplifying?

b) Before rationalizing, apply the quotient rule and simplify the numerator.

$$\sqrt{\dfrac{75a^3}{7b}} = \dfrac{\sqrt{75a^3}}{\sqrt{7b}} = \dfrac{\sqrt{25} \cdot \sqrt{3} \cdot \sqrt{a^2} \cdot \sqrt{a}}{\sqrt{7b}} = \dfrac{5a\sqrt{3a}}{\sqrt{7b}}$$

Next, rationalize the denominator. "By what do I multiply $\sqrt{7b}$ to get a *perfect square* under the square root?" The perfect square we want to get is $\sqrt{7^2b^2}$ or $\sqrt{49b^2}$.

$$\sqrt{7b} \cdot \sqrt{?} = \sqrt{7^2b^2} = 7b$$
$$\sqrt{7b} \cdot \sqrt{7b} = \sqrt{7^2b^2} = 7b$$

$$\dfrac{5a\sqrt{3a}}{\sqrt{7b}} = \dfrac{5a\sqrt{3a}}{\sqrt{7b}} \cdot \dfrac{\sqrt{7b}}{\sqrt{7b}} \qquad \text{Rationalize the denominator.}$$

$$= \dfrac{5a\sqrt{21ab}}{7b} \qquad \text{Multiply.}$$

YOU TRY 3

Simplify completely.

a) $\dfrac{8}{\sqrt{w}}$ b) $\sqrt{\dfrac{40h^3}{3k}}$

---

2 Rationalize a Denominator Containing a Higher Root

Many students assume that to rationalize denominators like we have up until this point, we always multiply the numerator and denominator of the expression by the denominator as in

$$\frac{5}{\sqrt{6}} = \frac{5}{\sqrt{6}} \cdot \frac{\sqrt{6}}{\sqrt{6}} = \frac{5\sqrt{6}}{6}$$

We will see, however, why this reasoning is incorrect.

To rationalize an expression like $\dfrac{5}{\sqrt{6}}$ we asked ourselves, "By what do I multiply $\sqrt{6}$ to get a *perfect square* under the *square root*?"

To rationalize an expression like $\dfrac{3}{\sqrt[3]{2}}$ we must ask ourselves, "By what do I multiply $\sqrt[3]{2}$ to get a *perfect cube* under the *cube root*?" The perfect cube we want is 2^3 (since we began with 2) so that $\sqrt[3]{2} \cdot \sqrt[3]{2^2} = \sqrt[3]{2^3} = 2$.

We will practice some fill-in-the-blank problems to eliminate radicals before we move on to rationalizing.

EXAMPLE 4

In-Class Example 4

Fill in the blank.
a) $\sqrt[3]{6} \cdot \sqrt[3]{?} = \sqrt[3]{6^3} = 6$
b) $\sqrt[3]{49} \cdot \sqrt[3]{?} = \sqrt[3]{7^3} = 7$
c) $\sqrt[3]{?} \cdot \sqrt[3]{m} = \sqrt[3]{m^3} = m$

Answer: a) 6^2 b) 7 c) m^2

Fill in the blank.

a) $\sqrt[3]{5} \cdot \sqrt[3]{?} = \sqrt[3]{5^3} = 5$ b) $\sqrt[3]{3^2} \cdot \sqrt[3]{?} = \sqrt[3]{3^3} = 3$

c) $\sqrt[3]{x^2} \cdot \sqrt[3]{?} = \sqrt[3]{x^3} = x$

Solution

a) Ask yourself, "By what do I multiply $\sqrt[3]{5}$ to get $\sqrt[3]{5^3}$?" The answer is $\sqrt[3]{5^2}$.

$$\sqrt[3]{5} \cdot \sqrt[3]{?} = \sqrt[3]{5^3} = 5$$
$$\sqrt[3]{5} \cdot \sqrt[3]{5^2} = \sqrt[3]{5^3} = 5$$

b) "By what do I multiply $\sqrt[3]{3^2}$ to get $\sqrt[3]{3^3}$?" $\sqrt[3]{3}$

$$\sqrt[3]{3^2} \cdot \sqrt[3]{?} = \sqrt[3]{3^3} = 3$$
$$\sqrt[3]{3^2} \cdot \sqrt[3]{3} = \sqrt[3]{3^3} = 3$$

c) "By what do I multiply $\sqrt[3]{x^2}$ to get $\sqrt[3]{x^3}$?" $\sqrt[3]{x}$

$$\sqrt[3]{x^2} \cdot \sqrt[3]{?} = \sqrt[3]{x^3} = x$$
$$\sqrt[3]{x^2} \cdot \sqrt[3]{x} = \sqrt[3]{x^3} = x$$

[YOU TRY 4]

Fill in the blank.

a) $\sqrt[3]{2} \cdot \sqrt[3]{?} = \sqrt[3]{2^3} = 2$ b) $\sqrt[3]{25} \cdot \sqrt[3]{?} = \sqrt[3]{5^2} = 5$

c) $\sqrt[3]{n} \cdot \sqrt[3]{?} = \sqrt[3]{n^3} = n$

We will use the technique presented in Example 4 to rationalize denominators with indices higher than 2.

EXAMPLE 5

Rationalize the denominator.

a) $\dfrac{3}{\sqrt[3]{2}}$ b) $\sqrt[3]{\dfrac{5}{9}}$ c) $\dfrac{6}{\sqrt[3]{4k}}$

Solution

a) To rationalize the denominator of $\dfrac{3}{\sqrt[3]{2}}$, *first* identify what we want to get as the denominator *after* multiplying. **We want to obtain** $\sqrt[3]{2^3}$ **since** $\sqrt[3]{2^3} = 2.$

$$\frac{3}{\sqrt[3]{2}} \cdot \frac{\quad}{\quad} = \frac{}{\sqrt[3]{2^3}} \quad \leftarrow \text{This is what we want to get.}$$
$$\uparrow$$
$$\text{What is needed here?}$$

Ask yourself, "By what do I multiply $\sqrt[3]{2}$ to get $\sqrt[3]{2^3}$?" $\sqrt[3]{2^2}$

$$\frac{3}{\sqrt[3]{2}} \cdot \frac{\sqrt[3]{2^2}}{\sqrt[3]{2^2}} = \frac{3\sqrt[3]{2^2}}{\sqrt[3]{2^3}} \qquad \text{Multiply.}$$

$$= \frac{3\sqrt[3]{4}}{2} \qquad \text{Simplify.}$$

b) $\sqrt[3]{\dfrac{5}{9}} = \dfrac{\sqrt[3]{5}}{\sqrt[3]{9}}$ \qquad Quotient rule

$\qquad = \dfrac{\sqrt[3]{5}}{\sqrt[3]{3^2}}$ \qquad Rewrite 9 as 3^2.

To rationalize the denominator, we want to obtain $\sqrt[3]{3^3}$ since $\sqrt[3]{3^3} = 3$. "By what do I multiply $\sqrt[3]{3^2}$ to get $\sqrt[3]{3^3}$?" $\sqrt[3]{3}$

$$\frac{\sqrt[3]{5}}{\sqrt[3]{3^2}} \cdot \frac{\sqrt[3]{3}}{\sqrt[3]{3}} = \frac{\sqrt[3]{5} \cdot \sqrt[3]{3}}{\sqrt[3]{3^3}} \qquad \text{Multiply.}$$

$$= \frac{\sqrt[3]{15}}{3} \qquad \text{Multiply and simplify.}$$

c) To rationalize the denominator of $\dfrac{6}{\sqrt[3]{4k}}$, first identify what we want to get as the denominator *after* multiplying. Remember, we need a radicand that is a perfect cube. We want to obtain $\sqrt[3]{8k^3}$ since $\sqrt[3]{8k^3} = \sqrt[3]{(2k)^3} = 2k$.

Ask yourself, "By what do I multiply $\sqrt[3]{4k}$ to get $\sqrt[3]{8k^3}$?" $\sqrt[3]{2k^2}$ Now, rationalize the denominator.

$$\frac{6}{\sqrt[3]{4k}} \cdot \frac{\sqrt[3]{2k^2}}{\sqrt[3]{2k^2}} = \frac{6\sqrt[3]{2k^2}}{\sqrt[3]{8k^3}} \qquad \text{Multiply.}$$

$$= \frac{6\sqrt[3]{2k^2}}{2k} \qquad \text{Simplify } \sqrt[3]{8k^3}.$$

$$= \frac{3\sqrt[3]{2k^2}}{k} \qquad \text{Simplify } \frac{6}{2}.$$

[YOU TRY 5] Rationalize the denominator.

a) $\dfrac{6}{\sqrt[3]{5}}$ b) $\sqrt[3]{\dfrac{11}{4}}$ c) $\dfrac{30}{\sqrt[3]{3z^2}}$

3 Rationalize a Denominator Containing Two Terms

To rationalize the denominator of an expression like $\dfrac{9}{4 + \sqrt{6}}$, we multiply the numerator and the denominator of the expression by the *conjugate* of $4 + \sqrt{6}$.

Definition

The **conjugate** of a binomial is the binomial obtained by changing the sign between the two terms.

Expression	Conjugate
$4 + \sqrt{6}$	$4 - \sqrt{6}$
$\sqrt{10} - 2\sqrt{3}$	$\sqrt{10} + 2\sqrt{3}$
$\sqrt{a} + \sqrt{b}$	$\sqrt{a} - \sqrt{b}$

In Section 9.4, we applied the formula $(a + b)(a - b) = a^2 - b^2$ to multiply binomials containing square roots. Recall that the terms containing the square roots were eliminated.

EXAMPLE 6

Multiply $7 - \sqrt{2}$ by its conjugate.

Solution

The conjugate of $7 - \sqrt{2}$ is $7 + \sqrt{2}$. We will first multiply using FOIL to show *why* the radical drops out, then we will multiply using the formula $(a + b)(a - b) = a^2 - b^2$.

i) Use FOIL to multiply.

$$\overset{\overset{\text{F}\qquad\quad\text{O}\qquad\quad\text{I}\qquad\qquad\text{L}}{}}{(7 - \sqrt{2})(7 + \sqrt{2})} = 7 \cdot 7 + 7 \cdot \sqrt{2} - 7 \cdot \sqrt{2} - \sqrt{2} \cdot \sqrt{2}$$
$$= 49 - 2$$
$$= 47$$

W Hint

Which method do you prefer? Why?

ii) Use $(a + b)(a - b) = a^2 - b^2$.

$$(7 - \sqrt{2})(7 + \sqrt{2}) = (7)^2 - (\sqrt{2})^2 \qquad \text{Substitute 7 for } a \text{ and } \sqrt{2} \text{ for } b.$$
$$= 49 - 2$$
$$= 47$$

Each method gives the same result.

[YOU TRY 6] Multiply $5 + \sqrt{6}$ by its conjugate.

Procedure Rationalizing a Denominator Containing Two Terms

To rationalize the denominator of an expression in which the denominator contains two terms, multiply the numerator and denominator of the expression by the conjugate of the denominator.

EXAMPLE 7

Rationalize the denominator, and simplify completely.

a) $\dfrac{9}{4 + \sqrt{6}}$ b) $\dfrac{8 + \sqrt{3}}{\sqrt{3} - 5}$ c) $\dfrac{5}{\sqrt{x} + 2}$

Solution

a) The denominator of $\dfrac{9}{4 + \sqrt{6}}$ has two terms, so we must multiply the numerator and denominator by the conjugate, $4 - \sqrt{6}$.

$$\dfrac{9}{4 + \sqrt{6}} \cdot \dfrac{4 - \sqrt{6}}{4 - \sqrt{6}} \qquad \text{Multiply by the conjugate.}$$

$$= \dfrac{9(4 - \sqrt{6})}{(4)^2 - (\sqrt{6})^2} \qquad (a + b)(a - b) = a^2 - b^2$$

$$= \dfrac{36 - 9\sqrt{6}}{16 - 6} \qquad \text{Simplify.}$$

$$= \dfrac{36 - 9\sqrt{6}}{10} \qquad \text{Subtract.}$$

b) $\dfrac{8 + \sqrt{3}}{\sqrt{3} - 5} = \dfrac{8 + \sqrt{3}}{\sqrt{3} - 5} \cdot \dfrac{\sqrt{3} + 5}{\sqrt{3} + 5}$ Multiply by the conjugate of the denominator.

$$= \dfrac{8\sqrt{3} + 40 + 3 + 5\sqrt{3}}{(\sqrt{3})^2 - (5)^2} \qquad \text{Use FOIL to multiply the numerators.} \\ (a + b)(a - b) = a^2 - b^2$$

$$= \dfrac{13\sqrt{3} + 43}{3 - 25} \qquad \text{Combine like terms; simplify.}$$

$$= \dfrac{13\sqrt{3} + 43}{-22} = -\dfrac{13\sqrt{3} + 43}{22} \qquad \text{Simplify.}$$

c) $\dfrac{5}{\sqrt{x} + 2} = \dfrac{5}{\sqrt{x} + 2} \cdot \dfrac{\sqrt{x} - 2}{\sqrt{x} - 2}$ Multiply by the conjugate.

$$= \dfrac{5(\sqrt{x} - 2)}{(\sqrt{x})^2 - (2)^2} \qquad \text{Multiply.} \\ (a + b)(a - b) = a^2 - b^2$$

$$= \dfrac{5\sqrt{x} - 10}{x - 4} \qquad \text{Multiply; simplify.}$$

YOU TRY 7

Rationalize the denominator, and simplify completely.

a) $\dfrac{4}{7 + \sqrt{2}}$ b) $\dfrac{6 + \sqrt{5}}{\sqrt{5} - 9}$ c) $\dfrac{11}{\sqrt{n} + 8}$

4 Divide Out Common Factors in Radical Quotients

Sometimes it is necessary to simplify a radical expression by dividing out common factors from the numerator and denominator. This is a skill we will need in Chapter 10 when we are solving quadratic equations, so we will look at an example here.

EXAMPLE 8

Simplify $\dfrac{5\sqrt{2} + 45}{5}$ **completely.**

Solution

 It is tempting to do one of the following:

$$\frac{5\sqrt{2} + 45}{5} = \sqrt{2} + 45 \qquad \textbf{Incorrect!}$$

or

$$\frac{5\sqrt{2} + \overset{9}{\cancel{45}}}{\cancel{5}} = 5\sqrt{2} + 9 \qquad \textbf{Incorrect!}$$

Each is incorrect because $5\sqrt{2}$ is a *term* in a sum, and 45 is a *term* in a sum.

The correct way to simplify $\dfrac{5\sqrt{2} + 45}{5}$ is to begin by factoring out a 5 in the numerator and *then* divide the numerator and denominator by any common factors.

$$\frac{5\sqrt{2} + 45}{5} = \frac{5(\sqrt{2} + 9)}{5} \qquad \text{Factor out 5 from the numerator.}$$

$$= \frac{\overset{1}{\cancel{5}}(\sqrt{2} + 9)}{\underset{1}{\cancel{5}}} \qquad \text{Divide by 5, a \textit{factor} of the numerator.}$$

$$= \sqrt{2} + 9 \qquad \text{Simplify.}$$

 Hint

You must factor the numerator first!

Note

In Example 8, we can divide numerator and denominator by 5 in $\dfrac{5(\sqrt{2} + 9)}{5}$ because the 5 in the numerator is part of a *product*, not a sum or difference.

[**YOU TRY 8**]

Simplify $\dfrac{4\sqrt{6} - 20}{4}$ **completely.**

ANSWERS TO [**YOU TRY**] **EXERCISES**

1) a) $\dfrac{\sqrt{3}}{3}$ b) $6\sqrt{5}$ c) $\dfrac{11\sqrt{30}}{5}$ 2) a) $\dfrac{\sqrt{3}}{9}$ b) $\dfrac{\sqrt{110}}{10}$

3) a) $\dfrac{8\sqrt{w}}{w}$ b) $\dfrac{2h\sqrt{30hk}}{3k}$ 4) a) 2^2 or 4 b) 5 c) n^2

5) a) $\dfrac{6\sqrt[3]{25}}{5}$ b) $\dfrac{\sqrt[3]{22}}{2}$ c) $\dfrac{10\sqrt[3]{9z}}{z}$ 6) 19

7) a) $\dfrac{28 - 4\sqrt{2}}{47}$ b) $-\dfrac{15\sqrt{5} + 59}{76}$ c) $\dfrac{11\sqrt{n} - 88}{n - 64}$ 8) $\sqrt{6} - 5$

*Additional answers can be found in the Answers to Exercises appendix.

Objective 1: Rationalize a Denominator Containing One Square Root Term

1) What does it mean to rationalize the denominator of a radical expression?
   Eliminate the radical from the denominator.

2) In your own words, explain how to rationalize the denominator of an expression containing one term in the denominator. Answers may vary.

Rationalize the denominator of each expression.

3) $\dfrac{1}{\sqrt{6}}$ $\dfrac{\sqrt{6}}{6}$

4) $\dfrac{1}{\sqrt{10}}$ $\dfrac{\sqrt{10}}{10}$

5) $\dfrac{7}{\sqrt{3}}$ $\dfrac{7\sqrt{3}}{3}$

6) $\dfrac{9}{\sqrt{2}}$ $\dfrac{9\sqrt{2}}{2}$

7) $\dfrac{8}{\sqrt{2}}$ $4\sqrt{2}$

8) $\dfrac{15}{\sqrt{5}}$ $3\sqrt{5}$

9) $\dfrac{45}{\sqrt{10}}$ $\dfrac{9\sqrt{10}}{2}$

10) $\dfrac{32}{\sqrt{6}}$ $\dfrac{16\sqrt{6}}{3}$

11) $\dfrac{4}{\sqrt{72}}$ $\dfrac{\sqrt{2}}{3}$

12) $\dfrac{15}{\sqrt{75}}$ $\sqrt{3}$

13) $-\dfrac{20}{\sqrt{8}}$ $-5\sqrt{2}$

14) $-\dfrac{6}{\sqrt{24}}$ $-\dfrac{\sqrt{6}}{2}$

15) $\dfrac{\sqrt{11}}{\sqrt{7}}$ $\dfrac{\sqrt{77}}{7}$

16) $\dfrac{\sqrt{5}}{\sqrt{3}}$ $\dfrac{\sqrt{15}}{5}$

17) $\dfrac{2\sqrt{10}}{\sqrt{3}}$ $\dfrac{2\sqrt{30}}{3}$

18) $\dfrac{9\sqrt{7}}{\sqrt{2}}$ $\dfrac{9\sqrt{14}}{2}$

19) $\dfrac{\sqrt{5}}{\sqrt{18}}$ $\dfrac{\sqrt{10}}{6}$

20) $-\dfrac{21\sqrt{5}}{\sqrt{98}}$ $-\dfrac{3\sqrt{10}}{2}$

21) $\dfrac{6\sqrt{8}}{\sqrt{27}}$ $\dfrac{4\sqrt{6}}{3}$

22) $\dfrac{10\sqrt{12}}{\sqrt{32}}$ $\dfrac{5\sqrt{6}}{2}$

23) $\dfrac{35\sqrt{6}}{14\sqrt{5}}$ $\dfrac{\sqrt{30}}{2}$

24) $\dfrac{27\sqrt{7}}{45\sqrt{3}}$ $\dfrac{\sqrt{21}}{5}$

25) $\sqrt{\dfrac{7}{14}}$ $\dfrac{\sqrt{2}}{2}$

26) $\sqrt{\dfrac{10}{30}}$ $\dfrac{\sqrt{3}}{3}$

27) $\sqrt{\dfrac{12}{80}}$ $\dfrac{\sqrt{15}}{10}$

28) $\sqrt{\dfrac{14}{600}}$ $\dfrac{\sqrt{21}}{30}$

29) $\sqrt{\dfrac{50}{2000}}$ $\dfrac{\sqrt{10}}{20}$

30) $\sqrt{\dfrac{88}{180}}$ $\dfrac{\sqrt{110}}{15}$

Multiply and simplify.

31) $\sqrt{\dfrac{10}{7}} \cdot \sqrt{\dfrac{7}{3}}$ $\dfrac{\sqrt{30}}{3}$

32) $\sqrt{\dfrac{11}{6}} \cdot \sqrt{\dfrac{2}{11}}$ $\dfrac{\sqrt{3}}{3}$

33) $\sqrt{\dfrac{8}{15}} \cdot \sqrt{\dfrac{3}{2}}$ $\dfrac{2\sqrt{5}}{5}$

34) $\sqrt{\dfrac{20}{21}} \cdot \sqrt{\dfrac{14}{15}}$ $\dfrac{2\sqrt{2}}{3}$

35) $\sqrt{\dfrac{1}{13}} \cdot \sqrt{\dfrac{8}{3}}$ $\dfrac{2\sqrt{78}}{39}$

36) $\sqrt{\dfrac{7}{5}} \cdot \sqrt{\dfrac{1}{10}}$ $\dfrac{\sqrt{14}}{10}$

37) $\sqrt{\dfrac{3}{8}} \cdot \sqrt{\dfrac{27}{2}}$ $\dfrac{9}{4}$

38) $\sqrt{\dfrac{5}{12}} \cdot \sqrt{\dfrac{20}{3}}$ $\dfrac{5}{3}$

Simplify completely.

39) $\dfrac{5}{\sqrt{k}}$ $\dfrac{5\sqrt{k}}{k}$

40) $\dfrac{7}{\sqrt{r}}$ $\dfrac{7\sqrt{r}}{r}$

41) $\dfrac{\sqrt{6}}{\sqrt{d}}$ $\dfrac{\sqrt{6d}}{d}$

42) $\dfrac{\sqrt{3}}{\sqrt{u}}$ $\dfrac{\sqrt{3u}}{u}$

43) $\sqrt{\dfrac{16}{a}}$ $\dfrac{4\sqrt{a}}{a}$

44) $\sqrt{\dfrac{81}{z}}$ $\dfrac{9\sqrt{z}}{z}$

45) $\sqrt{\dfrac{x}{y}}$ $\dfrac{\sqrt{xy}}{y}$

46) $\sqrt{\dfrac{m}{n}}$ $\dfrac{\sqrt{mn}}{n}$

47) $\sqrt{\dfrac{40p^3}{q}}$ $\dfrac{2p\sqrt{10pq}}{q}$

48) $\sqrt{\dfrac{63r^3}{t}}$ $\dfrac{3r\sqrt{7rt}}{t}$

49) $\sqrt{\dfrac{64v^7}{5w}}$ $\dfrac{8v^3\sqrt{5vw}}{5w}$

50) $\sqrt{\dfrac{49t^5}{2u}}$ $\dfrac{7t^2\sqrt{2tu}}{2u}$

51) $\dfrac{\sqrt{18h^5}}{\sqrt{2h}}$ $3h^2$

52) $\dfrac{\sqrt{84k^7}}{\sqrt{21k}}$ $2k^3$

53) $\dfrac{\sqrt{15x^6}}{\sqrt{9y}}$ $\dfrac{x^3\sqrt{15y}}{3y}$

54) $\dfrac{\sqrt{11a^4}}{\sqrt{100b}}$ $\dfrac{a^2\sqrt{11b}}{10b}$

Objective 2: Rationalize a Denominator Containing a Higher Root

Fill in the blank.

55) $\sqrt[3]{2} \cdot \sqrt[3]{?} = \sqrt[3]{2^3} = 2$ 2^2 or 4

56) $\sqrt[3]{5} \cdot \sqrt[3]{?} = \sqrt[3]{5^3} = 5$ 5^2 or 25

57) $\sqrt[3]{9} \cdot \sqrt[3]{?} = \sqrt[3]{3^3} = 3$ 3

58) $\sqrt[3]{4} \cdot \sqrt[3]{?} = \sqrt[3]{2^3} = 2$ 2

59) $\sqrt[3]{c} \cdot \sqrt[3]{?} = \sqrt[3]{c^3} = c$ c^2

60) $\sqrt[3]{m^2} \cdot \sqrt[3]{?} = \sqrt[3]{m^3} = m$ m

61) To rationalize the denominator of $\dfrac{9}{\sqrt[3]{2}}$, Jason did the following: $\dfrac{9}{\sqrt[3]{2}} \cdot \dfrac{\sqrt[3]{2}}{\sqrt[3]{2}} = \dfrac{9\sqrt[3]{2}}{2}$.

What did he do wrong? What should he have done?

62) To rationalize the denominator of $\dfrac{9}{\sqrt[3]{25}}$, we should multiply the numerator and denominator by $\sqrt[3]{5}$:

$$\frac{9}{\sqrt[3]{25}} \cdot \frac{\sqrt[3]{5}}{\sqrt[3]{5}} = \frac{9\sqrt[3]{5}}{\sqrt[3]{125}} = \frac{9\sqrt[3]{5}}{5}$$

When we multiply the numerator and denominator by $\sqrt[3]{5}$, we are actually multiplying the fraction by what number?   1

Rationalize the denominator of each expression.

63) $\dfrac{4}{\sqrt[3]{3}}$ $\dfrac{4\sqrt[3]{9}}{3}$

64) $\dfrac{11}{\sqrt[3]{5}}$ $\dfrac{11\sqrt[3]{25}}{5}$

65) $\dfrac{1}{\sqrt[3]{4}}$ $\dfrac{\sqrt[3]{2}}{2}$

66) $\dfrac{1}{\sqrt[3]{9}}$ $\dfrac{\sqrt[3]{3}}{3}$

67) $\dfrac{35}{\sqrt[3]{5}}$ $7\sqrt[3]{25}$

68) $\dfrac{26}{\sqrt[3]{2}}$ $13\sqrt[3]{4}$

69) $\sqrt[3]{\dfrac{8}{9}}$ $\dfrac{2\sqrt[3]{3}}{3}$

70) $\sqrt[3]{\dfrac{27}{4}}$ $\dfrac{3\sqrt[3]{2}}{2}$

71) $\sqrt[3]{\dfrac{11}{25}}$ $\dfrac{\sqrt[3]{55}}{5}$

72) $\sqrt[3]{\dfrac{1}{81}}$ $\dfrac{\sqrt[3]{9}}{9}$

73) $\dfrac{6}{\sqrt[3]{x}}$ $\dfrac{6\sqrt[3]{x^2}}{x}$

74) $\dfrac{10}{\sqrt[3]{p}}$ $\dfrac{10\sqrt[3]{p^2}}{p}$

75) $\dfrac{7}{\sqrt[3]{m^2}}$ $\dfrac{7\sqrt[3]{m}}{m}$

76) $\dfrac{3}{\sqrt[3]{h^2}}$ $\dfrac{3\sqrt[3]{h}}{h}$

77) $\sqrt[3]{\dfrac{1000}{c}}$ $\dfrac{10\sqrt[3]{c^2}}{c}$

78) $\sqrt[3]{\dfrac{64}{z}}$ $\dfrac{4\sqrt[3]{z^2}}{z}$

79) $\sqrt[3]{\dfrac{4}{w^2}}$ $\dfrac{\sqrt[3]{4w}}{w}$

80) $\sqrt[3]{\dfrac{9}{r^2}}$ $\dfrac{\sqrt[3]{9r}}{r}$

81) $\dfrac{16}{\sqrt[3]{2t}}$ $\dfrac{8\sqrt[3]{4t^2}}{t}$

82) $\dfrac{12}{\sqrt[3]{2v^2}}$ $\dfrac{6\sqrt[3]{4v}}{v}$

83) $\dfrac{\sqrt[3]{5a}}{\sqrt[3]{36b}}$ $\dfrac{\sqrt[3]{30ab^2}}{6b}$

84) $\dfrac{\sqrt[3]{3v}}{\sqrt[3]{5u^2}}$ $\dfrac{\sqrt[3]{75uv}}{5u}$

85) $\dfrac{7}{\sqrt[4]{2}}$ $\dfrac{7\sqrt[4]{8}}{2}$

86) $\dfrac{5}{\sqrt[4]{3}}$ $\dfrac{5\sqrt[4]{27}}{3}$

87) $\dfrac{2}{\sqrt[4]{x^3}}$ $\dfrac{2\sqrt[4]{x}}{x}$

88) $\dfrac{13}{\sqrt[4]{t}}$ $\dfrac{13\sqrt[4]{t^3}}{t}$

Objective 3: Rationalize a Denominator Containing Two Terms

89) How do you find the conjugate of a binomial?
Change the sign between the two terms.

90) When you multiply a binomial containing a square root by its conjugate, what happens to the radical?
The radical is eliminated.

Find the conjugate of each binomial. Then multiply the binomial by its conjugate.

91) $(4 + \sqrt{3})$
$(4 - \sqrt{3});\ 13$

92) $(\sqrt{7} - 9)$
$(\sqrt{7} + 9);\ -74$

93) $(\sqrt{5} - \sqrt{11})$
$(\sqrt{5} + \sqrt{11});\ -6$

94) $(\sqrt{13} + \sqrt{2})$
$(\sqrt{13} - \sqrt{2});\ 11$

95) $(\sqrt{p} - 6)$
$(\sqrt{p} + 6);\ p - 36$

96) $(\sqrt{r} + 7)$
$(\sqrt{r} - 7);\ r - 49$

Rationalize the denominator, and simplify completely.

Fill It In

Fill in the blanks with either the missing mathematical step or reason for the given step.

97) $\dfrac{5}{4 - \sqrt{3}} = \dfrac{5}{4 - \sqrt{3}} \cdot \dfrac{4 + \sqrt{3}}{4 + \sqrt{3}}$ Multiply by the conjugate.

$\qquad = \dfrac{5(4 + \sqrt{3})}{(4)^2 - (\sqrt{3})^2}$ $(a + b)(a - b) = a^2 - b^2$

$\qquad = \dfrac{20 + 5\sqrt{3}}{16 - 3}$ Multiply terms in numerator; square terms in denominator.

$\qquad = \dfrac{20 + 5\sqrt{3}}{13}$ Simplify.

98) $\dfrac{\sqrt{2}}{\sqrt{11} + \sqrt{6}}$

$\qquad = \dfrac{\sqrt{2}}{\sqrt{11} + \sqrt{6}} \cdot \dfrac{\sqrt{11} - \sqrt{6}}{\sqrt{11} - \sqrt{6}}$ Multiply by the conjugate.

$\qquad = \dfrac{\sqrt{2}(\sqrt{11} - \sqrt{6})}{(\sqrt{11})^2 - (\sqrt{6})^2}$ $(a + b)(a - b) = a^2 - b^2$

$\qquad = \dfrac{\sqrt{22} - \sqrt{12}}{11 - 6}$ Multiply terms in numerator; square terms in denominator.

$\qquad = \dfrac{\sqrt{22} - 2\sqrt{3}}{5}$ Simplify.

99) $\dfrac{6}{3 + \sqrt{2}}$ $\dfrac{18 - 6\sqrt{2}}{7}$

100) $\dfrac{7}{4 - \sqrt{5}}$ $\dfrac{28 + 7\sqrt{5}}{11}$

101) $\dfrac{5}{7 - \sqrt{3}}$ $\dfrac{35 + 5\sqrt{3}}{46}$

102) $\dfrac{1}{6 + \sqrt{7}}$ $\dfrac{6 - \sqrt{7}}{29}$

103) $\dfrac{10}{\sqrt{5} - 6}$ $-\dfrac{10\sqrt{5} + 60}{31}$

104) $\dfrac{\sqrt{6}}{\sqrt{3} + 1}$ $\dfrac{3\sqrt{2} - \sqrt{6}}{2}$

105) $\dfrac{\sqrt{2}}{\sqrt{10} - 3}$ $2\sqrt{5} + 3\sqrt{2}$

106) $\dfrac{9 + \sqrt{7}}{5 - \sqrt{7}}$ $\dfrac{26 + 7\sqrt{7}}{9}$

107) $\dfrac{5 + \sqrt{3}}{4 + \sqrt{3}}$ $\dfrac{17 - \sqrt{3}}{13}$

108) $\dfrac{6 + \sqrt{2}}{\sqrt{2} + 1}$ $5\sqrt{2} - 4$

109) $\dfrac{1 - 2\sqrt{3}}{\sqrt{2} - \sqrt{3}}$ $6 + 2\sqrt{6} - \sqrt{2} - \sqrt{3}$

110) $\dfrac{3\sqrt{6} - \sqrt{2}}{\sqrt{7} - \sqrt{5}}$

111) $\dfrac{12}{\sqrt{w} + 3}$ $\dfrac{12\sqrt{w} - 36}{w - 9}$

112) $\dfrac{7}{\sqrt{h} - 1}$ $\dfrac{7\sqrt{h} + 7}{h - 1}$

113) $\dfrac{9}{5 - \sqrt{n}}$ $\dfrac{45 + 9\sqrt{n}}{25 - n}$

114) $\dfrac{4}{2 + \sqrt{x}}$ $\dfrac{8 - 4\sqrt{x}}{4 - x}$

115) $\dfrac{\sqrt{m}}{\sqrt{m} + \sqrt{n}}$ $\dfrac{m - \sqrt{mn}}{m - n}$

116) $\dfrac{\sqrt{x} + \sqrt{y}}{\sqrt{x} - \sqrt{y}}$ $\dfrac{x + 2\sqrt{xy} + y}{x - y}$

Objective 4: Divide Out Common Factors in Radical Quotients

Simplify completely.

117) $\dfrac{9 + 36\sqrt{5}}{9}$ $1 + 4\sqrt{5}$

118) $\dfrac{21 - 14\sqrt{11}}{7}$ $3 - 2\sqrt{11}$

119) $\dfrac{20 - 44\sqrt{3}}{12}$ $\dfrac{5 - 11\sqrt{3}}{3}$

120) $\dfrac{12 + 21\sqrt{2}}{15}$ $\dfrac{4 + 7\sqrt{2}}{5}$

121) $\dfrac{\sqrt{45} + 6}{9}$ $\dfrac{\sqrt{5} + 2}{3}$

122) $\dfrac{\sqrt{60} - 18}{4}$ $\dfrac{\sqrt{15} - 9}{2}$

123) $\dfrac{-24 - \sqrt{800}}{4}$ $-6 - 5\sqrt{2}$

124) $\dfrac{-18 + \sqrt{96}}{8}$ $\dfrac{-9 + 2\sqrt{6}}{4}$

Mixed Exercises: Objectives 1–4

Rationalize the denominator, and simplify completely.

125) $\dfrac{\sqrt{11}}{\sqrt{50}}$ $\dfrac{\sqrt{22}}{10}$

126) $\dfrac{5\sqrt{2}}{\sqrt{t}}$ $\dfrac{5\sqrt{2t}}{t}$

127) $\sqrt[3]{\dfrac{4}{3}}$ $\dfrac{\sqrt[3]{36}}{3}$

128) $\dfrac{9}{\sqrt{2} + 4}$ $\dfrac{36 - 9\sqrt{2}}{14}$

129) $\sqrt{\dfrac{125a^3}{6b}}$ $\dfrac{5a\sqrt{30ab}}{6b}$

130) $\sqrt{\dfrac{54}{8}}$ $\dfrac{3\sqrt{3}}{2}$

131) $\dfrac{3 + \sqrt{10}}{\sqrt{10} - 9}$ $-\dfrac{37 + 12\sqrt{10}}{71}$

132) $-\dfrac{12\sqrt{3}}{\sqrt{20}}$ $-\dfrac{6\sqrt{15}}{5}$

133) $\sqrt{\dfrac{108}{63}}$ $\dfrac{2\sqrt{21}}{7}$

134) $\sqrt{\dfrac{20c^3}{11d}}$ $\dfrac{2c\sqrt{55cd}}{11d}$

135) $\dfrac{\sqrt{256}}{\sqrt{2y}}$ $\dfrac{8\sqrt{2y}}{y}$

136) $\sqrt[3]{\dfrac{1}{81}}$ $\dfrac{\sqrt[3]{9}}{9}$

137) The equation $r = \sqrt{\dfrac{A}{\pi}}$ describes the radius of a circle, r, in terms of its area, A. If the area of a circle is measured in square inches, find r when $A = 12\pi$. $r = 2\sqrt{3}$ in.

138) The equation $r = \sqrt{\dfrac{V}{\pi h}}$ describes the radius of a right circular cylinder, r, in terms of its volume, V, and height, h. If the volume of a cylinder is measured in cubic inches, find r when the volume is 20π in^3 and the cylinder is 5 in. tall. $r = 2$ in.

R Rethink

R1) Why do we rationalize the denominator? $\left(\text{Hint: Think of } \dfrac{1}{\sqrt{2}} \text{ versus } \dfrac{\sqrt{2}}{2}.\right)$

R2) Are there any problems you could not do? If so, write them down on your paper or circle them so that you can ask your instructor for help.

9.6 Solving Radical Equations

P Prepare

O Organize

What are your objectives for Section 9.6?	How can you accomplish each objective?
1 Understand the Steps for Solving a Radical Equation	• Learn the **Steps for Solving Radical Equations Containing Square Roots.** • Understand that extraneous solutions can occur and that you *must* check all potential answers.
2 Solve a Radical Equation Containing One Square Root	• Follow Example 1, and notice that both parts give one solution. • Follow Example 2, and notice that there is no solution. • Complete You Trys 1 and 2.
3 Solve a Radical Equation Containing Two Square Roots	• Use the procedure to solve radical equations, and always start by getting a radical on one side by itself. • Complete the given example on your own, and notice that you are squaring monomials. • Complete You Try 3.
4 Solve a Radical Equation by Squaring a Binomial	• Use the procedure to solve radical equations, and always start by getting a radical on one side by itself. • Complete the given examples on your own, and notice that these examples produce a binomial on one side. • Complete You Trys 4–6.

W Work

Read the explanations, follow the examples, take notes, and complete the You Trys.

In this section, we will learn how to solve *radical equations*.

An equation containing a variable in the radicand is a **radical equation.** Some examples of radical equations are

$$\sqrt{x} = 5, \qquad \sqrt{4w - 3} + 2 = w, \qquad \sqrt{3y + 1} - \sqrt{y - 1} = 2$$

1 Understand the Steps for Solving a Radical Equation

Let's review what happens when we square a square root expression: If $x \geq 0$, then $(\sqrt{x})^2 = x$. That is, to eliminate the radical from $\sqrt{x}$, we *square* the expression.

Therefore, to solve equations like those above containing *square roots,* we *square* both sides of the equation to obtain new equations. The solutions of the new equations contain all of the solutions of the original equation and may also contain *extraneous solutions.*

An **extraneous solution** is a value that satisfies one of the new equations but does not satisfy the original equation. Extraneous solutions occur frequently when solving radical equations, so we *must* check all possible solutions in the original equation and discard any that are extraneous.

 Hint

Always check your answer in the original equation!

Procedure Steps for Solving Radical Equations Containing Square Roots

Step 1: Get a radical on a side by itself.

Step 2: Square both sides of the equation to eliminate a radical.

Step 3: Combine like terms on each side of the equation.

Step 4: If the equation still contains a radical, repeat Steps 1–3.

Step 5: Solve the equation.

Step 6: Check the proposed solutions *in the original equation,* and discard extraneous solutions.

2 Solve a Radical Equation Containing One Square Root

EXAMPLE 1

Solve.

a) $\sqrt{x + 5} = 4$ b) $\sqrt{p - 9} + 6 = 8$

In-Class Example 1

Solve.

a) $\sqrt{y + 7} = 5$

b) $\sqrt{a - 1} + 9 = 12$

Answer: a) {18} b) {10}

Solution

a) *Step 1:* The radical *is* on a side by itself: $\sqrt{x + 5} = 4$

Step 2: *Square* both sides to eliminate the *square root.*

$$(\sqrt{x + 5})^2 = 4^2 \quad \text{Square both sides.}$$
$$x + 5 = 16$$

Steps 3 and 4 do not apply because there are no like terms to combine and no radicals remain.

Step 5: Solve the equation.

$$x = 11 \quad \text{Subtract 5 from each side.}$$

Step 6: Check $x = 11$ in the *original* equation.

$$\sqrt{x + 5} = 4$$
$$\sqrt{11 + 5} \overset{?}{=} 4$$
$$\sqrt{16} = 4 \; \checkmark$$

The solution set is {11}.

b) The first step is to get the radical on a side by itself.

$$\sqrt{p - 9} + 6 = 8$$

$$\sqrt{p - 9} = 2 \qquad \text{Subtract 6 from each side.}$$

$$(\sqrt{p - 9})^2 = 2^2 \qquad \text{Square both sides to eliminate the radical.}$$

$$p - 9 = 4 \qquad \text{The square root has been eliminated.}$$

$$p = 13 \qquad \text{Add 9 to each side.}$$

Check $p = 13$ in the *original* equation.

$$\sqrt{p - 9} + 6 = 8$$

$$\sqrt{13 - 9} + 6 \overset{?}{=} 8$$

$$2 + 6 = 8 \ \checkmark$$

The solution set is $\{13\}$.

[YOU TRY 1]   Solve.

a) $\sqrt{t + 10} = 3$ b) $\sqrt{w - 3} + 7 = 11$

EXAMPLE 2   Solve $t = \sqrt{t^2 + 2t + 3}$.

In-Class Example 2

Solve

$x = \sqrt{x^2 + 6x + 2}.$

Answer: $\varnothing$

Solution

The radical is already on a side by itself, so we will square both sides.

$$t = \sqrt{t^2 + 2t + 3}$$

$$t^2 = \left(\sqrt{t^2 + 2t + 3}\right)^2 \qquad \text{Square both sides to eliminate the radical.}$$

$$t^2 = t^2 + 2t + 3 \qquad \text{The square root has been eliminated.}$$

Solve the equation.

$$0 = 2t + 3 \qquad \text{Subtract } t^2 \text{ from each side.}$$

$$-2t = 3 \qquad \text{Subtract } 2t \text{ from each side.}$$

$$t = -\frac{3}{2} \qquad \text{Divide by } -2.$$

 Hint

This example shows why you **must** check possible solutions in the original equation!

Check $t = -\dfrac{3}{2}$ in the *original* equation.

$$t = \sqrt{t^2 + 2t + 3}$$

$$-\frac{3}{2} \overset{?}{=} \sqrt{\left(-\frac{3}{2}\right)^2 + 2\left(-\frac{3}{2}\right) + 3}$$

$$-\frac{3}{2} \overset{?}{=} \sqrt{\frac{9}{4} - 3 + 3}$$

$$-\frac{3}{2} \overset{?}{=} \sqrt{\frac{9}{4}}$$

$$-\frac{3}{2} = \frac{3}{2} \qquad \text{FALSE}$$

Because $t = -\dfrac{3}{2}$ does not satisfy the *original* equation, it is an extraneous solution.

There is no real solution to this equation. The solution is $\varnothing$.

[YOU TRY 2] Solve $c = \sqrt{c^2 - 8c - 4}$.

3 Solve a Radical Equation Containing Two Square Roots

Next, we will take our first look at solving an equation containing two square roots.

EXAMPLE 3

In-Class Example 3

Solve

$\sqrt{3k + 1} - 2\sqrt{k - 1} = 0$.

Answer: {5}

Solve $\sqrt{7n + 1} - 4\sqrt{n - 5} = 0$.

Solution

Begin by getting a radical on a side by itself.

$$\sqrt{7n + 1} = 4\sqrt{n - 5} \qquad \text{Add } 4\sqrt{n - 5} \text{ to each side.}$$
$$\left(\sqrt{7n + 1}\right)^2 = \left(4\sqrt{n - 5}\right)^2 \qquad \text{Square both sides to eliminate the radicals.}$$
$$7n + 1 = 16(n - 5) \qquad 4^2 = 16$$
$$7n + 1 = 16n - 80 \qquad \text{Distribute.}$$
$$-9n = -81$$
$$n = 9 \qquad \text{Solve.}$$

Check $n = 9$ in the *original* equation.

$$\sqrt{7n + 1} - 4\sqrt{n - 5} = 0$$
$$\sqrt{7(9) + 1} - 4\sqrt{9 - 5} = 0$$
$$\sqrt{63 + 1} - 4\sqrt{4} \stackrel{?}{=} 0$$
$$\sqrt{64} - 4(2) \stackrel{?}{=} 0$$
$$8 - 8 = 0 \ \checkmark \qquad \text{The solution set is } \{9\}.$$

[YOU TRY 3] Solve $\sqrt{3z + 5} - 2\sqrt{z - 2} = 0$.

4 Solve a Radical Equation by Squaring a Binomial

Sometimes, we have to square a binomial in order to solve a radical equation.

EXAMPLE 4

In-Class Example 4

Solve

$\sqrt{2n - 7} + 5 = n$.

Answer: {8}

W Hint

Squaring a binomial will sometimes yield two possible answers.

Solve $\sqrt{4w - 3} + 2 = w$.

Solution

As usual, start by getting the radical on a side by itself.

$$\sqrt{4w - 3} = w - 2 \qquad \text{Subtract 2 from each side.}$$
$$(\sqrt{4w - 3})^2 = (w - 2)^2 \qquad \text{Square both sides to eliminate the radicals.}$$
$$4w - 3 = w^2 - 4w + 4 \qquad \text{Simplify; square the binomial.}$$
$$0 = w^2 - 8w + 7 \qquad \text{Subtract } 4w; \text{ add 3.}$$
$$0 = (w - 1)(w - 7) \qquad \text{Factor.}$$

$$w - 1 = 0 \quad \text{or} \quad w - 7 = 0 \qquad \text{Set each factor equal to zero.}$$
$$w = 1 \quad \text{or} \qquad w = 7 \qquad \text{Solve.}$$

Check $w = 1$ and $w = 7$ in the *original* equation.

$$w = 1: \quad \sqrt{4w - 3} + 2 = w$$
$$\sqrt{4(1) - 3} + 2 \overset{?}{=} 1$$
$$\sqrt{1} + 2 \overset{?}{=} 1$$
$$3 \overset{?}{=} 1 \quad \text{FALSE}$$

$$w = 7: \quad \sqrt{4w - 3} + 2 = w$$
$$\sqrt{4(7) - 3} + 2 \overset{?}{=} 7$$
$$\sqrt{25} + 2 \overset{?}{=} 7$$
$$5 + 2 = 7 \quad \text{TRUE}$$

$w = 7$ *is* a solution but $w = 1$ is *not* because $w = 1$ does not satisfy the original equation. The solution set is $\{7\}$.

[YOU TRY 4] Solve $\sqrt{3r - 5} + 3 = r$.

Recall from Section 9.4 that we can square binomials containing radical expressions just like we squared the binomial $(w - 2)^2$ in Example 4. We can use the formulas

$$(a + b)^2 = a^2 + 2ab + b^2 \quad \text{and} \quad (a - b)^2 = a^2 - 2ab + b^2$$

EXAMPLE 5

In-Class Example 5

Square the binomial, and simplify $(5 - \sqrt{k + 4})^2$.

Answer: $k + 29 - 10\sqrt{k + 4}$

Square the binomial, and simplify $(3 - \sqrt{h + 6})^2$.

Solution

Use the formula $(a - b)^2 = a^2 - 2ab + b^2$.

$$(3 - \sqrt{h + 6})^2 = (3)^2 - 2(3)(\sqrt{h + 6}) + (\sqrt{h + 6})^2 \qquad \text{Substitute 3 for } a \text{ and } \sqrt{h + 6} \text{ for } b.$$
$$= 9 - 6\sqrt{h + 6} + (h + 6)$$
$$= h + 15 - 6\sqrt{h + 6} \qquad \text{Combine like terms.}$$

[YOU TRY 5] Square each binomial, and simplify.

a) $(\sqrt{t} - 5)^2$ b) $(2 + \sqrt{n - 8})^2$

To solve the next two equations, we will have to square both sides of the equation twice to eliminate the radicals. Be very careful when you are squaring the binomials that contain a radical.

EXAMPLE 6

In-Class Example 6

Solve each equation.
a) $\sqrt{x + 12} + \sqrt{x} = 6$
b) $\sqrt{3c - 5} - \sqrt{c - 3} = 2$

Answer: a) $\{4\}$ b) $\{3, 7\}$

Solve each equation.

a) $\sqrt{a + 7} + \sqrt{a} = 7$ b) $\sqrt{3y + 1} - \sqrt{y - 1} = 2$

Solution

a) This equation contains two radicals *and* a constant. Get one of the radicals on a side by itself, then square both sides.

$$\sqrt{a + 7} = 7 - \sqrt{a} \qquad \text{Subtract } \sqrt{a} \text{ from each side.}$$
$$(\sqrt{a + 7})^2 = (7 - \sqrt{a})^2 \qquad \text{Square both sides.}$$
$$a + 7 = (7)^2 - 2(7)(\sqrt{a}) + (\sqrt{a})^2 \qquad \text{Use the formula } (a - b)^2 = a^2 - 2ab + b^2.$$
$$a + 7 = 49 - 14\sqrt{a} + a \qquad \text{Simplify.}$$

W Hint

Be sure you are writing out the example as you read it!

The equation still contains a radical. Therefore, repeat Steps 1–3. Begin by getting the radical on a side by itself.

$$7 = 49 - 14\sqrt{a} \qquad \text{Subtract } a \text{ from each side.}$$
$$-42 = -14\sqrt{a} \qquad \text{Subtract 49 from each side.}$$
$$3 = \sqrt{a} \qquad \text{Divide by } -14.$$
$$3^2 = (\sqrt{a})^2 \qquad \text{Square both sides.}$$
$$9 = a \qquad \text{Solve.}$$

The check is left to the student. The solution set is {9}.

b) *Step 1:* Get a radical on a side by itself.

$$\sqrt{3y + 1} - \sqrt{y - 1} = 2$$
$$\sqrt{3y + 1} = 2 + \sqrt{y - 1} \qquad \text{Add } \sqrt{y - 1} \text{ to each side.}$$

Step 2: Square both sides of the equation to eliminate a radical.

$$(\sqrt{3y + 1})^2 = (2 + \sqrt{y - 1})^2 \qquad \text{Square both sides.}$$
$$3y + 1 = (2)^2 + 2(2)(\sqrt{y - 1}) + (\sqrt{y - 1})^2 \qquad \text{Use the formula}$$
$$3y + 1 = 4 + 4\sqrt{y - 1} + y - 1 \qquad (a + b)^2 = a^2 + 2ab + b^2.$$

Step 3: Combine like terms on the right side.

$$3y + 1 = y + 3 + 4\sqrt{y - 1} \qquad \text{Combine like terms.}$$

Step 4: The equation still contains a radical, so repeat Steps 1–3.

Step 1: Get the radical on a side by itself.

$$3y + 1 = y + 3 + 4\sqrt{y - 1}$$
$$2y - 2 = 4\sqrt{y - 1} \qquad \text{Subtract } y \text{ and subtract 3.}$$
$$y - 1 = 2\sqrt{y - 1} \qquad \text{Divide by 2.}$$

(We do not *need* to eliminate the 2 from in front of the radical before squaring both sides. The radical must not be a part of a *sum* or *difference* when we square.)

Step 2: Square both sides of the equation to eliminate the radical.

$$(y - 1)^2 = (2\sqrt{y - 1})^2 \qquad \text{Square both sides.}$$
$$y^2 - 2y + 1 = 4(y - 1) \qquad \text{Square the binomial; } 2^2 = 4.$$

Steps 3 and 4 no longer apply.

Step 5: Solve the equation.

$$y^2 - 2y + 1 = 4y - 4 \qquad \text{Distribute.}$$
$$y^2 - 6y + 5 = 0 \qquad \text{Subtract } 4y; \text{ add 4.}$$
$$(y - 1)(y - 5) = 0 \qquad \text{Factor.}$$

$$y - 1 = 0 \quad \text{or} \quad y - 5 = 0 \qquad \text{Set each factor equal to zero.}$$
$$y = 1 \quad \text{or} \qquad y = 5 \qquad \text{Solve.}$$

Step 6: The check is left to the student. Verify that $y = 1$ and $y = 5$ each satisfy the original equation. The solution set is {1, 5}.

[**YOU TRY 6**] Solve each equation.

a) $\sqrt{r + 9} + \sqrt{r} = 9$ b) $\sqrt{5z + 6} - \sqrt{4z + 1} = 1$

BE CAREFUL Watch out for two common mistakes that students make when solving an equation like the one in Example 6b.

1) Do not square both sides before getting a radical on a side by itself.

This is incorrect:

$$(\sqrt{3y+1} - \sqrt{y-1})^2 = 2^2$$
$$3y + 1 - (y - 1) = 4$$

On the left we must multiply using FOIL or use the formula $(a - b)^2 = a^2 - 2ab + b^2$.

2) The *second* time we perform Step 2, watch out for this common error:

This is incorrect:

$$(y - 1)^2 = (2\sqrt{y-1})^2$$
$$y^2 - 1 = 2(y - 1)$$

On the left we must multiply using FOIL or use the formula $(a - b)^2 = a^2 - 2ab + b^2$, and on the right we must remember to square the 2.

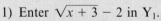

Using Technology

We can use a graphing calculator to solve a radical equation in one variable. First subtract every term on the right side of the equation from both sides of the equation and enter the result in Y_1. Graph the equation in Y_1. The zeros or x-intercepts of the graph are the solutions to the equation.

We will solve $\sqrt{x+3} = 2$ using a graphing calculator.

1) Enter $\sqrt{x+3} - 2$ in Y_1.

2) Press $\boxed{\text{ZOOM}}\,\boxed{6}$ to graph the function in Y_1 as shown.

3) Press $\boxed{\text{2nd}}\,\boxed{\text{TRACE}}$ 2:zero, move the cursor to the left of the zero and press $\boxed{\text{ENTER}}$, move the cursor to the right of the zero and press $\boxed{\text{ENTER}}$, and move the cursor close to the zero and press $\boxed{\text{ENTER}}$ to display the zero. The solution to the equation is $x = 1$.

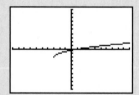

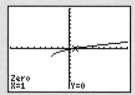

Solve each equation using a graphing calculator.

1) $\sqrt{x-2} = 1$ 2) $\sqrt{3x-2} = 5$

3) $\sqrt{3x-2} = \sqrt{x} + 2$ 4) $\sqrt{4x-5} = \sqrt{x} + 4$

5) $\sqrt{2x-7} = \sqrt{x} - 1$ 6) $\sqrt{\sqrt{x}-1} = 1$

ANSWERS TO $\left[\text{YOU TRY}\right]$ EXERCISES

1) a) $\{-1\}$ b) $\{19\}$ 2) $\varnothing$ 3) $\{13\}$ 4) $\{7\}$
5) a) $t - 10\sqrt{t} + 25$ b) $n + 4\sqrt{n - 8} - 4$
6) a) $\{16\}$ b) $\{2, 6\}$

ANSWERS TO TECHNOLOGY EXERCISES

1) $\{3\}$ 2) $\{9\}$ 3) $\{9\}$ 4) $\{3\}$ 5) $\{4\}$ 6) $\{4\}$

E Evaluate   9.6   Exercises

Do the exercises, and check your work.

Additional answers can be found in the Answers to Exercises appendix.

Objective 1: Understand the Steps for Solving a Radical Equation

1) Why is it necessary to check the proposed solutions to a radical equation in the original equation?
   Sometimes there are extraneous solutions.

2) How do you know, without actually solving and checking the solution, that $\sqrt{x} = -2$ has no solution?
   The principal square root of a number cannot equal a negative number.

Objective 2: Solve a Radical Equation Containing One Square Root

Solve.

3) $\sqrt{d} = 12$ $\{144\}$

4) $\sqrt{w} = 5$ $\{25\}$

5) $\sqrt{x} = \dfrac{1}{3}$ $\left\{\dfrac{1}{9}\right\}$

6) $\sqrt{t} = \dfrac{5}{8}$ $\left\{\dfrac{25}{64}\right\}$

7) $\sqrt{b - 11} - 3 = 0$ $\{20\}$

8) $\sqrt{z + 9} - 1 = 0$ $\{-8\}$

9) $\sqrt{c - 6} + 3 = 4$ $\{7\}$

10) $\sqrt{y + 7} - 2 = 3$ $\{18\}$

11) $-9 = \sqrt{3k + 1} - 11$ $\{1\}$

12) $8 = \sqrt{2n + 1} - 1$ $\{40\}$

13) $\sqrt{5p + 8} + 7 = 5$ $\varnothing$

14) $\sqrt{2b - 1} + 3 = 0$ $\varnothing$

15) $\sqrt{5 - 2r} - 13 = -7$ $\left\{-\dfrac{31}{2}\right\}$

16) $8 = 11 - \sqrt{4 - 3v}$ $\left\{-\dfrac{5}{3}\right\}$

17) $g = \sqrt{g^2 + 8g - 24}$ $\{3\}$

18) $x = \sqrt{x^2 + 9x - 36}$ $\{4\}$

19) $\sqrt{v^2 - 7v - 42} = v$ $\varnothing$

20) $c = \sqrt{c^2 + c + 9}$ $\varnothing$

21) $2p = \sqrt{4p^2 - 3p + 6}$ $\{2\}$

22) $\sqrt{9r^2 - 2r + 10} = 3r$ $\{5\}$

Objective 3: Solve a Radical Equation Containing Two Square Roots

Solve.

23) $\sqrt{5y + 4} = \sqrt{y + 8}$ $\{1\}$

24) $\sqrt{3a + 1} = \sqrt{5a - 9}$ $\{5\}$

25) $\sqrt{z} = 3\sqrt{7}$ $\{63\}$

26) $4\sqrt{3} = \sqrt{a}$ $\{48\}$

27) $\sqrt{3 - 11h} - 3\sqrt{h + 7} = 0$ $\{-3\}$

28) $3\sqrt{t + 32} - 5\sqrt{t + 16} = 0$ $\{-7\}$

29) $\sqrt{2p - 1} + 2\sqrt{p + 4} = 0$ $\varnothing$

30) $2\sqrt{3c + 4} + \sqrt{c - 6} = 0$ $\varnothing$

Objective 4: Solve a Radical Equation by Squaring a Binomial

Solve.

31) $\sqrt{x} = x - 6$ $\{9\}$

32) $\sqrt{h} = h - 2$ $\{4\}$

33) $m + 4 = 5\sqrt{m}$ $\{1, 16\}$

34) $b + 5 = 6\sqrt{b}$ $\{1, 25\}$

35) $c = \sqrt{5c - 9} + 3$ $\{9\}$

36) $\sqrt{2k + 11} + 2 = k$ $\{7\}$

37) $6 + \sqrt{w^2 + 3w - 9} = w$ $\varnothing$

38) $\sqrt{z^2 + 5z - 8} = z + 4$ $\varnothing$

39) $\sqrt{7u - 5} - u = 1$ $\{2, 3\}$

40) $3 = \sqrt{10z + 6} - z$ $\{1, 3\}$

41) $a - 2\sqrt{a + 1} = -1$ $\{-1, 3\}$

42) $r - 3\sqrt{r + 2} = 2$ $\{14\}$

Square each binomial, and simplify.

43) $(y + 7)^2$ $y^2 + 14y + 49$

44) $(t - 5)^2$ $t^2 - 10t + 25$

45) $(\sqrt{n} + 7)^2$ $n + 14\sqrt{n} + 49$

46) $(\sqrt{w} - 5)^2$ $w - 10\sqrt{w} + 25$

47) $(4 - \sqrt{c + 3})^2$ $19 - 8\sqrt{c + 3} + c$

48) $(3 + \sqrt{x - 2})^2$ $7 + 6\sqrt{x - 2} + x$

49) $(2\sqrt{3p - 4} + 1)^2$ $12p + 4\sqrt{3p - 4} - 15$

50) $(3\sqrt{2k + 1} - 4)^2$ $18k - 24\sqrt{2k + 1} + 25$

Solve.

51) $\sqrt{w + 5} = 5 - \sqrt{w}$ {4}

52) $\sqrt{p + 8} = 4 - \sqrt{p}$ {1}

53) $\sqrt{y} - \sqrt{y - 13} = 1$ {49}

54) $\sqrt{d} - \sqrt{d - 16} = 2$ {25}

55) $\sqrt{3t + 10} - \sqrt{2t} = 2$ {2, 18}

56) $\sqrt{4z - 3} - \sqrt{5z + 1} = -1$ {3, 7}

57) $\sqrt{4c + 5} + \sqrt{2c - 1} = 4$ {1}

58) $4 = \sqrt{3a} + \sqrt{a - 2}$ {3}

59) $3 - \sqrt{3h + 1} = \sqrt{3h - 14}$ $\varnothing$

60) $\sqrt{7x - 6} - 1 = \sqrt{7x - 3}$ $\varnothing$

61) $\sqrt{5a + 19} - \sqrt{a + 12} = 1$ $\left\{\dfrac{1}{4}\right\}$

62) $\sqrt{v + 7} - \sqrt{2v + 7} = 1$ {−3}

63) $3 = \sqrt{5 - 2y} + \sqrt{y + 2}$ {−2, 2}

64) $4 + \sqrt{r - 4} = \sqrt{3r + 4}$ {4, 20}

65) $\sqrt{2n - 5} = \sqrt{2n + 3} - 2$ {3}

66) $-3 = \sqrt{5k - 9} - \sqrt{5k + 6}$ {2}

67) $\sqrt{3b + 4} - \sqrt{2b + 9} = -1$ {0}

68) $\sqrt{z - 4} + \sqrt{5z} = 6$ {5}

Write an equation and solve.

69) The sum of 7 and the square root of a number is 12. Find the number.   25

70) The sum of 6 and the square root of a number equals the number. Find the number.   9

71) Three times the square root of a number is the sum of 4 and the square root of the number. Find the number.   4

72) Twice the square root of a number is 8 less than the number. Find the number.   16

Solve.

73) If the area of a square is A and each side has length l, then the length of a side is given by $l = \sqrt{A}$. A square rug has an area of 36 ft². Find the dimensions of the rug.   6 ft × 6 ft

74) If the area of a circle is A and the radius is r, then the radius of the circle is given by $r = \sqrt{\dfrac{A}{\pi}}$. The center circle on a basketball court has an area of 113.1 ft². What is the radius of the center circle to the nearest tenth?   6.0 ft

Forensic scientists can use the formula $S = \sqrt{30\,fd}$ to determine the speed, S, in mph, of a vehicle in an accident based on the drag factor, f, of the road and the length of the skid marks, d, in feet. Use this formula for Exercises 75 and 76.

75) Accident investigators respond to a car accident on a highway where the speed limit is 65 mph. They determine that the concrete highway has a drag factor of 0.90. The skid marks left by the car are 142 ft long. Find the speed of the car, to the nearest unit, at the time of the accident to determine whether the car was exceeding the speed limit.
62 mph; the car was not speeding.

76) A car involved in an accident on a busy street leaves skid marks 80 ft long. Accident investigators determine that the asphalt street has a drag factor of 0.78. If the speed limit is 35 mph, was the driver speeding at the time of the accident? Support your answer by finding the speed of the car to the nearest unit.   The speed of the car was about 43 mph, so it was speeding at the time of the accident.

Use the following information for Exercises 77 and 78.

The distance a person can see to the horizon is approximated by $d = 1.2\sqrt{h}$, where d is the number of miles a person can see to the horizon from a height of h feet.

77) Andrew is in a hot air balloon and can see 48 miles to the horizon. Find his height above the ground.   1600 ft

78) On a clear day, visitors to Chicago's Willis Tower (formerly the Sears Tower) can see about 44.1 miles to the horizon. Find the height of the observation deck at Willis Tower to the nearest foot. (www.the-skydeck.com)   1351 ft

R1) Do you think it is important to follow a procedure to solve problems? Have you noticed that they are similar?

R2) Were you able to notice any similarities among the equations that yielded no solution? Describe them.

Group Activity — The Group Activity can be found online on Connect.

em POWER me Form a Study Group

Form an informal study group with two or three of your classmates and, using the skills introduced on page 538, review the content of this chapter collaboratively. Ask one another sample questions, help teach each other major concepts, and push one another to work hard. Once you've finished, complete the following survey to see how effective your study group was. (Note: You can use this survey to help assess any study group, not just the one you formed for this exercise.)

Statement	Strongly Agree	Agree	Disagree	Strongly Disagree
1. I know the material better than I did before.				
2. I helped the members of my group to succeed.				
3. Other group members offered me new insights that helped me to succeed.				
4. My group stayed focused throughout our time working together.				
5. Everyone in the group tried to contribute as much as he or she could.				
6. By working as a team, we ensured that we did not overlook any major topics.				
7. Studying with my classmates was more fun than studying alone.				
8. We had the right number of people in our study group.				
9. I enjoyed the experience of helping members of my group to learn.				
10. I was motivated to help the others in my group to do well.				

Based on the results of this survey, try to identify what worked in your study group and what did not. Consider whether the issues apply to this particular study group or to your group work in general. Apply your conclusions to your future group work, both in academic contexts and beyond.

Chapter 9:   Summary

Definition/Procedure	Example

9.1 Finding Roots

If the radicand is a perfect square, then the square root is a *rational* number. **(p. 543)**	$\sqrt{25} = 5$ since $5^2 = 25$.
If the radicand is a negative number, then the square root is *not* a real number. **(p. 543)**	$\sqrt{-49}$ is not a real number.
If the radicand is positive and not a perfect square, then the square root is an *irrational* number. **(p. 543)**	$\sqrt{3}$ is irrational because 3 is not a perfect square.
The $\sqrt[n]{a}$ is read as "the *n*th root of *a*." If $\sqrt[n]{a} = b$, then $b^n = a$. *n* is the **index** of the radical. **(p. 546)**	$\sqrt[4]{81} = 3$ since $3^4 = 81$.

The Pythagorean Theorem If the lengths of the legs of a right triangle are *a* and *b* and the hypotenuse has length *c*, then $a^2 + b^2 = c^2$. **(p. 544)** 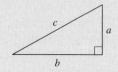	Find the length of the missing side. 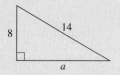 Use the Pythagorean theorem, $a^2 + b^2 = c^2$. $\qquad b = 8, c = 14 \qquad$ Find *a*. $a^2 + (8)^2 = (14)^2 \qquad$ Substitute values. $a^2 + 64 = 196$ $\qquad a^2 = 132 \qquad$ Subtract 64 from each side. $\qquad a = \sqrt{132} \qquad$ Square root property $\qquad a = 2\sqrt{33} \qquad \sqrt{132} = \sqrt{4} \cdot \sqrt{33} = 2\sqrt{33}$ When we solve $a^2 = 132$, only the *positive* square root makes sense because *a* represents a length.
The *odd root* of a negative number is a negative number. **(p. 547)**	$\sqrt[3]{-8} = -2$ since $(-2)^3 = -8$.
The *even root* of a negative number is not a real number. **(p. 547)**	$\sqrt[4]{-16}$ is not a real number.

9.2 Simplifying Radicals: The Product and Quotient Rules

Product Rule for Square Roots Let *a* and *b* be nonnegative real numbers. Then $\sqrt{a} \cdot \sqrt{b} = \sqrt{ab}$. **(p. 552)**	$\sqrt{5} \cdot \sqrt{6} = \sqrt{5 \cdot 6} = \sqrt{30}$
An expression containing a square root is simplified when all of the following conditions are met: 1) The radicand does not contain any factors (other than 1) which are perfect squares. 2) The radicand does not contain any fractions. 3) There are no radicals in the denominator of a fraction. To *simplify square roots,* reverse the process of multiplying radicals, where *a* or *b* is a perfect square: $\sqrt{ab} = \sqrt{a} \cdot \sqrt{b}$. After simplifying a radical, look at the result and ask yourself, *"Is the radical in simplest form?"* If it is not, simplify again. **(p. 552)**	Simplify $\sqrt{24}$. $\quad \sqrt{24} = \sqrt{4 \cdot 6} \qquad$ 4 is a perfect square. $\qquad\quad = \sqrt{4} \cdot \sqrt{6} \qquad$ Product rule $\qquad\quad = 2\sqrt{6} \qquad \sqrt{4} = 2$ Simplify $\sqrt{12} \cdot \sqrt{2}$. $\sqrt{12} \cdot \sqrt{2} = \sqrt{24} \qquad\qquad \sqrt{12} \cdot \sqrt{2} = \sqrt{4 \cdot 3} \cdot \sqrt{2}$ $\qquad\qquad = \sqrt{4 \cdot 6} \qquad$ **or** $\qquad\qquad = 2\sqrt{3} \cdot \sqrt{2}$ $\qquad\qquad = \sqrt{4} \cdot \sqrt{6} \qquad\qquad\qquad\qquad = 2\sqrt{3 \cdot 2}$ $\qquad\qquad = 2\sqrt{6} \qquad\qquad\qquad\qquad\quad = 2\sqrt{6}$

Definition/Procedure	Example

Quotient Rule for Square Roots

Let a and b be nonnegative real numbers such that $b \neq 0$. Then $\sqrt{\dfrac{a}{b}} = \dfrac{\sqrt{a}}{\sqrt{b}}$. **(p. 555)**

Simplify $\sqrt{\dfrac{72}{25}} = \dfrac{\sqrt{72}}{\sqrt{25}}$ Quotient rule

$\qquad\qquad = \dfrac{\sqrt{36} \cdot \sqrt{2}}{5}$ Product rule; $\sqrt{25} = 5$

$\qquad\qquad = \dfrac{6\sqrt{2}}{5}$ $\sqrt{36} = 6$

Simplifying Square Root Expressions Containing Variables

If a is a nonnegative real number, then $\sqrt{a^2} = a$. In the rest of the book, we will assume that all variables represent positive real numbers.

To simplify a square root expression containing variables, use the product rule and factor out all perfect squares. **(p. 556)**

Simplify.

a) $\sqrt{w^2} = w$

b) $\sqrt{49t^8} = \sqrt{(7t^4)^2} = 7t^4$

c) $\sqrt{n^5} = \sqrt{n^4 \cdot n}$

$\qquad = \sqrt{n^4} \cdot \sqrt{n}$ Product rule

$\qquad = n^2 \sqrt{n}$ Simplify.

Product Rule for Higher Roots

If a and b are real numbers such that the roots exist, then $\sqrt[n]{a} \cdot \sqrt[n]{b} = \sqrt[n]{a \cdot b}$. **(p. 558)**

Multiply $\sqrt[3]{3} \cdot \sqrt[3]{7}$.

$\qquad \sqrt[3]{3} \cdot \sqrt[3]{7} = \sqrt[3]{3 \cdot 7} = \sqrt[3]{21}$ Product rule

Simplify $\sqrt[3]{40}$.

$\qquad \sqrt[3]{40} = \sqrt[3]{8 \cdot 5}$

$\qquad\qquad = \sqrt[3]{8} \cdot \sqrt[3]{5}$ Product rule

$\qquad\qquad = 2\sqrt[3]{5}$ $\sqrt[3]{8} = 2$

Quotient Rule for Higher Roots

If a and b are real numbers such that the roots exist and $b \neq 0$, then

$$\sqrt[n]{\dfrac{a}{b}} = \dfrac{\sqrt[n]{a}}{\sqrt[n]{b}} \quad \textbf{(p. 558)}$$

Simplify $\sqrt[4]{\dfrac{64}{81}}$.

$$\sqrt[4]{\dfrac{64}{81}} = \dfrac{\sqrt[4]{64}}{\sqrt[4]{81}} = \dfrac{\sqrt[4]{16} \cdot \sqrt[4]{4}}{3} = \dfrac{2\sqrt[4]{4}}{3}$$

Simplifying Higher Roots with Variables in the Radicand

To simplify $\sqrt[n]{a^m}$, use the product rule and factor out all perfect nth powers. **(p. 559)**

Simplify each expression.

a) $\sqrt[3]{k^{12}} = k^4$

b) $\sqrt[3]{d^{16}} = \sqrt[3]{d^{15} \cdot d}$

$\qquad = \sqrt[3]{d^{15}} \cdot \sqrt[3]{d}$ Product rule

$\qquad = d^5 \sqrt[3]{d}$ Simplify.

9.3 Adding and Subtracting Radicals

Like radicals have the same index and the same radicand. In order to add or subtract radicals, they must be like radicals.

Steps for Adding and Subtracting Radicals

1) Write each radical expression in simplest form.

2) Combine like radicals. **(p. 564)**

Perform the operations and simplify.

a) $6\sqrt{5} + 2\sqrt{7} - 3\sqrt{5} + 9\sqrt{7} = 3\sqrt{5} + 11\sqrt{7}$

b) $\sqrt{72} - \sqrt{45} + \sqrt{18}$

$\qquad = \sqrt{36} \cdot \sqrt{2} - \sqrt{9} \cdot \sqrt{5} + \sqrt{9} \cdot \sqrt{2}$

$\qquad = \quad 6\sqrt{2} \quad - \quad 3\sqrt{5} \quad + \quad 3\sqrt{2}$

$\qquad = 9\sqrt{2} - 3\sqrt{5}$

9.4 Combining Operations on Radicals

Multiply expressions containing radicals using the same techniques that are used for multiplying polynomials. **(p. 569)**

Multiply and simplify.

a) $\sqrt{2}(\sqrt{5} + 9) = \sqrt{2} \cdot \sqrt{5} + \sqrt{2} \cdot 9 = \sqrt{10} + 9\sqrt{2}$

b) $(\sqrt{g} + \sqrt{6})(\sqrt{g} - \sqrt{3})$

Use FOIL to multiply two binomials.

$\qquad (\sqrt{g} + \sqrt{6})(\sqrt{g} - \sqrt{3})$

$\qquad = \sqrt{g} \cdot \sqrt{g} - \sqrt{3}\sqrt{g} + \sqrt{6} \cdot \sqrt{g} - \sqrt{6} \cdot \sqrt{3}$

$\qquad\qquad\quad\text{F}\qquad\quad\text{O}\qquad\quad\text{I}\qquad\quad\text{L}$

$\qquad = \quad g \quad - \quad \sqrt{3g} \quad + \quad \sqrt{6g} \quad - \quad \sqrt{18}$

$\qquad = g - \sqrt{3g} + \sqrt{6g} - 3\sqrt{2}$ $\sqrt{18} = 3\sqrt{2}$

Definition/Procedure	Example
To *square a binomial*, we can use either FOIL or one of the special formulas from Chapter 6: $$(a + b)^2 = a^2 + 2ab + b^2$$ $$(a - b)^2 = a^2 - 2ab + b^2 \ \textbf{(p. 570)}$$	$$(\sqrt{3} + 4)^2 = (\sqrt{3})^2 + 2(\sqrt{3})(4) + (4)^2$$ $$= 3 + 8\sqrt{3} + 16$$ $$= 19 + 8\sqrt{3}$$
To multiply binomials of the form $(a + b)(a - b)$, use the formula $(a + b)(a - b) = a^2 - b^2$. **(p. 571)**	$$(2 + \sqrt{11})(2 - \sqrt{11}) = (2)^2 - (\sqrt{11})^2$$ $$= 4 - 11$$ $$= -7$$

9.5 Dividing Radicals

The process of eliminating radicals from the denominator of an expression is called **rationalizing the denominator.** First, we give examples of rationalizing denominators containing one term. **(p. 575)**	Rationalize the denominator of each expression. a) $\dfrac{7}{\sqrt{5}} = \dfrac{7}{\sqrt{5}} \cdot \dfrac{\sqrt{5}}{\sqrt{5}} = \dfrac{7\sqrt{5}}{5}$ b) $\dfrac{9}{\sqrt[3]{4}} = \dfrac{9}{\sqrt[3]{2^2}} \cdot \dfrac{\sqrt[3]{2}}{\sqrt[3]{2}} = \dfrac{9\sqrt[3]{2}}{\sqrt[3]{2^3}} = \dfrac{9\sqrt[3]{2}}{2}$
The **conjugate** of a binomial is the binomial obtained by changing the sign between the two terms. **(p. 580)**	The conjugate of $\sqrt{21} - 1$ is $\sqrt{21} + 1$. The conjugate of $-14 + \sqrt{5}$ is $-14 - \sqrt{5}$.
Rationalizing a Denominator with Two Terms To rationalize the denominator of an expression containing two terms, multiply the numerator and denominator of the expression by the conjugate of the denominator. **(p. 580)**	Rationalize the denominator of $\dfrac{7}{\sqrt{5} - 3}$. $\dfrac{7}{\sqrt{5} - 3} = \dfrac{7}{\sqrt{5} - 3} \cdot \dfrac{\sqrt{5} + 3}{\sqrt{5} + 3}$ Multiply by the conjugate. $\quad = \dfrac{7(\sqrt{5} + 3)}{(\sqrt{5})^2 - (3)^2}$ $(a + b)(a - b) = a^2 - b^2$ $\quad = \dfrac{7(\sqrt{5} + 3)}{5 - 9}$ Square the terms. $\quad = \dfrac{7(\sqrt{5} + 3)}{-4}$ Subtract. $\quad = -\dfrac{7\sqrt{5} + 21}{4}$ Distribute.

9.6 Solving Radical Equations

Steps for Solving Radical Equations Containing Square Roots *Step 1:* Get a radical on a side by itself. *Step 2:* Square both sides of the equation to eliminate a radical. *Step 3:* Combine like terms on each side of the equation. *Step 4:* If the equation still contains a radical, repeat Steps 1–3. *Step 5:* Solve the equation. *Step 6:* Check the proposed solutions *in the original equation,* and discard extraneous solutions. **(p. 587)**	Solve $n = 1 + \sqrt{2n + 1}$. $\quad n - 1 = \sqrt{2n + 1}$ Get the radical by itself. $\quad (n - 1)^2 = (\sqrt{2n + 1})^2$ Square both sides. $n^2 - 2n + 1 = 2n + 1$ $\quad n^2 - 4n = 0$ Get all terms on the same side. $\quad n(n - 4) = 0$ Factor. $n = 0 \quad \text{or} \quad n - 4 = 0$ $\qquad \text{or} \qquad n = 4$ Check $n = 0$ and $n = 4$ in the *original* equation. $n = 0$: $\qquad\qquad\qquad$ $n = 4$: $n = 1 + \sqrt{2n + 1}$ $\qquad$ $n = 1 + \sqrt{2n + 1}$ $0 \overset{?}{=} 1 + \sqrt{2(0) + 1}$ $\qquad$ $4 \overset{?}{=} 1 + \sqrt{2(4) + 1}$ $0 \overset{?}{=} 1 + \sqrt{1}$ $\qquad\qquad$ $4 \overset{?}{=} 1 + \sqrt{9}$ $0 = 2$ $\qquad\qquad\qquad\quad$ $4 = 1 + 3$ $\quad$ FALSE $\qquad\qquad\qquad$ TRUE $n = 4$ *is* a solution, but $n = 0$ is *not* because $n = 0$ does not satisfy the original equation. The solution set is $\{4\}$.

Chapter 9: Review Exercises

*Additional answers can be found in the Answers to Exercises appendix.

(9.1) Find each root, if possible.

1) $\sqrt{49}$ 7

2) $\sqrt{-25}$ not real

3) $-\sqrt{16}$ -4

4) $\sqrt{\dfrac{169}{9}}$ $\dfrac{13}{3}$

5) $\sqrt[3]{-125}$ -5

6) $\sqrt[5]{32}$ 2

7) $\sqrt[4]{81}$ 3

8) Approximate $\sqrt{34}$ to the nearest tenth and plot it on a number line. 5.8

9) Use the Pythagorean theorem to find the length of the missing side. 6

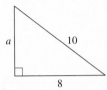

10) The length of a rectangle is 9 cm, and its width is 4 cm. Find the length of its diagonal. $\sqrt{97}$ cm

Choose from *always, sometimes,* or *never*.

11) The square root of a prime number is *always, sometimes,* or *never* equal to an integer. never

12) The square root of a positive, proper fraction is *always, sometimes,* or *never* less than 1. always

(9.2) Simplify completely.

13) $\sqrt{45}$ $3\sqrt{5}$

14) $\sqrt{108}$ $6\sqrt{3}$

15) $\sqrt[3]{81}$ $3\sqrt[3]{3}$

16) $\sqrt[3]{250}$ $5\sqrt[3]{2}$

17) $\sqrt[4]{80}$ $2\sqrt[4]{5}$

18) $-\sqrt{\dfrac{125}{5}}$ -5

19) $\sqrt{\dfrac{48}{6}}$ $2\sqrt{2}$

20) $\dfrac{\sqrt{48}}{\sqrt{121}}$ $\dfrac{4\sqrt{3}}{11}$

21) $\sqrt[3]{-\dfrac{16}{2}}$ -2

22) $\sqrt[4]{\dfrac{162}{2}}$ 3

23) $\sqrt{h^8}$ h^4

24) $-\sqrt{\dfrac{128}{p^{12}}}$ $-\dfrac{8\sqrt{2}}{p^6}$

25) $\sqrt[3]{b^{24}}$ b^8

26) $\sqrt[3]{64u^{15}}$ $4u^5$

27) $\sqrt{w^{15}}$ $w^7\sqrt{w}$

28) $\sqrt[3]{r^{20}}$ $r^6\sqrt[3]{r^2}$

29) $\sqrt{63d^4}$ $3d^2\sqrt{7}$

30) $\sqrt{72h^3}$ $6h\sqrt{2h}$

31) $\sqrt{44x^{12}y^5}$ $2x^6y^2\sqrt{11y}$

32) $\sqrt[3]{-27p^6q^{21}}$ $-3p^2q^7$

33) $\sqrt[3]{48m^{17}}$ $2m^5\sqrt[3]{6m^2}$

34) $\dfrac{\sqrt{54c^{17}}}{\sqrt{6c^9}}$ $3c^4$

35) $\dfrac{40\sqrt{150}}{56\sqrt{2}}$ $\dfrac{25\sqrt{3}}{7}$

36) $\sqrt{\dfrac{10}{27}}\cdot\sqrt{\dfrac{2}{3}}$ $\dfrac{2\sqrt{5}}{9}$

Perform the indicated operation and simplify.

37) $\sqrt{6}\cdot\sqrt{5}$ $\sqrt{30}$

38) $\sqrt{8}\cdot\sqrt{10}$ $4\sqrt{5}$

39) $\sqrt{5}\cdot\sqrt{10}$ $5\sqrt{2}$

40) $\sqrt{40}\cdot\sqrt{30}$ $20\sqrt{3}$

41) $\sqrt{18}\cdot\sqrt{18}$ 18

42) $\sqrt{51}\cdot\sqrt{51}$ 51

43) $\sqrt{67r}\cdot\sqrt{67r}$ $67r$

44) $\sqrt{33k}\cdot\sqrt{33k}$ $33k$

45) $\sqrt{d^3}\cdot\sqrt{d^{11}}$ d^7

46) $\sqrt{n^9}\cdot\sqrt{n^7}$ n^8

47) $\sqrt[3]{3}\cdot\sqrt[3]{10}$ $\sqrt[3]{30}$

48) $\sqrt[3]{4}\cdot\sqrt[3]{7}$ $\sqrt[3]{28}$

49) $\sqrt[3]{6}\cdot\sqrt[3]{t}$ $\sqrt[3]{6t}$

50) $\sqrt[3]{9}\cdot\sqrt[3]{x}$ $\sqrt[3]{9x}$

(9.3) Perform the operations and simplify.

51) $8\sqrt{6}+4\sqrt{6}$ $12\sqrt{6}$

52) $\sqrt{80}+\sqrt{150}$ $4\sqrt{5}+5\sqrt{6}$

53) $\sqrt{80}-\sqrt{20}+\sqrt{48}$ $2\sqrt{5}+4\sqrt{3}$

54) $\sqrt{28}+2\sqrt{45}-5\sqrt{63}$ $6\sqrt{5}-13\sqrt{7}$

55) $4\sqrt[3]{9}-9\sqrt[3]{72}$ $-14\sqrt[3]{9}$

56) $8\sqrt{p^3}-3p\sqrt{p}$ $5p\sqrt{p}$

57) $10n^2\sqrt{8n}-45n\sqrt{2n^3}$ $-25n^2\sqrt{2n}$

58) $4y^3\sqrt[3]{y}+6\sqrt[3]{y^{10}}$ $10y^3\sqrt[3]{y}$

(9.4) Multiply and simplify.

59) $\sqrt{7}(\sqrt{7}-\sqrt{5})$ $7-\sqrt{35}$

60) $4\sqrt{t}(\sqrt{20t}+\sqrt{2})$ $8t\sqrt{5}+4\sqrt{2t}$

61) $(5-\sqrt{2})(3+\sqrt{2})$ $13+2\sqrt{2}$

62) $(\sqrt{2}+5\sqrt{s})(3\sqrt{s}+4\sqrt{2})$ $23\sqrt{2s}+8+15s$

63) $(2\sqrt{6}-5)^2$ $49-20\sqrt{6}$

64) $(1+\sqrt{k})^2$ $1+2\sqrt{k}+k$

65) $(\sqrt{7}-\sqrt{6})(\sqrt{7}+\sqrt{6})$ 1

66) $(3+\sqrt{n})(3-\sqrt{n})$ $9-n$

(9.5) Rationalize the denominator of each expression.

67) $\dfrac{18}{\sqrt{3}}$ $6\sqrt{3}$

68) $\dfrac{20\sqrt{3}}{\sqrt{6}}$ $10\sqrt{2}$

69) $\sqrt{\dfrac{98}{h}}$ $\dfrac{7\sqrt{2h}}{h}$

70) $\dfrac{\sqrt{63}}{\sqrt{2k}}$ $\dfrac{3\sqrt{14k}}{2k}$

71) $\dfrac{15}{\sqrt[3]{3}}$ $5\sqrt[3]{9}$

72) $-\dfrac{9}{\sqrt[3]{4}}$ $-\dfrac{9\sqrt[3]{2}}{2}$

73) $\dfrac{7}{\sqrt[3]{2c}}$ $\dfrac{7\sqrt[3]{4c^2}}{2c}$

74) $\dfrac{\sqrt[3]{x^2}}{\sqrt[3]{y}}$ $\dfrac{\sqrt[3]{x^2y^2}}{y}$

75) $\dfrac{2}{3+\sqrt{2}}$ $\dfrac{6-2\sqrt{2}}{11}$

76) $\dfrac{-1+\sqrt{5}}{8-\sqrt{5}}$ $\dfrac{7\sqrt{5}-3}{59}$

77) $\dfrac{6}{9-\sqrt{x}}$ $\dfrac{54+6\sqrt{x}}{81-x}$

78) $\dfrac{z-4}{\sqrt{z}+2}$ $\sqrt{z}-2$

Simplify completely.

79) $\dfrac{16-24\sqrt{3}}{8}$ $2-3\sqrt{3}$

80) $\dfrac{-\sqrt{48}-6}{12}$ $\dfrac{-2\sqrt{3}-3}{6}$

(9.6) Solve.

81) $\sqrt{r}-9=0$ $\{81\}$

82) $\sqrt{3w-4}=6$ $\left\{\dfrac{40}{3}\right\}$

83) $3\sqrt{2v+7}-15=0$ $\{9\}$

84) $m=\sqrt{m^2+9m+18}$ $\varnothing$

85) $\sqrt{a + 11} - 2\sqrt{a + 8} = 0$ {−7}

86) $k + \sqrt{k + 7} = 5$ {2}

87) $1 = \sqrt{5x - 1} - x$ {1, 2}

88) $\sqrt{t + 8} + \sqrt{5 - t} = 5$ {−4, 1}

89) $\sqrt{4c + 21} - \sqrt{c} = -6$ ∅

90) $\sqrt{3b + 1} + \sqrt{b} = 3$ {1}

Mixed Exercises
Simplify each expression.

91) $9d\sqrt{d} - 4\sqrt{d^3}$ $5d\sqrt{d}$

92) $\sqrt{\dfrac{3}{24}}$ $\dfrac{\sqrt{2}}{4}$

93) $3(5\sqrt{8} + \sqrt{2})$ $33\sqrt{2}$

94) $\sqrt[3]{-1000y^{12}}$ $-10y^4$

95) $\dfrac{32\sqrt{13} + 8}{8}$ $4\sqrt{13} + 1$

96) $\dfrac{\sqrt{5}}{6 + \sqrt{10}}$ $\dfrac{6\sqrt{5} - 5\sqrt{2}}{26}$

97) $\dfrac{4}{\sqrt[3]{25m}}$ $\dfrac{4\sqrt[3]{5m^2}}{5m}$

98) $\sqrt{\dfrac{2}{15}} \cdot \sqrt{\dfrac{6}{5}}$ $\dfrac{2}{5}$

99) $(4 - \sqrt{3})^2$ $19 - 8\sqrt{3}$

100) $\dfrac{\sqrt{20}}{\sqrt{6x}}$ $\dfrac{\sqrt{30x}}{3x}$

101) $\sqrt{12} \cdot \sqrt{8}$ $4\sqrt{6}$

102) $\sqrt{82w} \cdot \sqrt{82w}$ $82w$

103) $\sqrt{128p^{15}}$ $8p^7\sqrt{2p}$

104) If the area of a circle is A and the radius is r, then the radius of the circle is given by $r = \sqrt{\dfrac{A}{\pi}}$. A circle has an area of 28π cm^2. What is the radius of the circle? $2\sqrt{7}$ cm

Chapter 9:   Test
*Additional answers can be found in the Answers to Exercises appendix.

Find each root, if possible.

1) $\sqrt{121}$ 11

2) $\sqrt{-81}$ not real

3) $\sqrt[3]{-1000}$ −10

4) $\sqrt[4]{81}$ 3

5) Find the length of the missing side. $\sqrt{39}$ cm

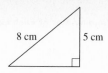
8 cm   5 cm

Simplify completely.

6) $\sqrt{125}$ $5\sqrt{5}$

7) $\sqrt[3]{48}$ $2\sqrt[3]{6}$

8) $\sqrt{\dfrac{192}{6}}$ $4\sqrt{2}$

9) $\sqrt{m^6}$ m^3

10) $\sqrt{h^7}$ $h^3\sqrt{h}$

11) $\sqrt[3]{k^{12}}$ k^4

12) $\sqrt[3]{b^{14}}$ $b^4\sqrt[3]{b^2}$

13) $\sqrt{63m^5n^8}$ $3m^2n^4\sqrt{7m}$

14) $\sqrt[3]{\dfrac{x^{15}y^7}{8}}$ $\dfrac{x^5y^2\sqrt[3]{y}}{2}$

15) $\dfrac{\sqrt{120}}{\sqrt{2}}$ $2\sqrt{15}$

Perform the operations and simplify.

16) $\sqrt{3y} \cdot \sqrt{12y}$ $6y$

17) $\sqrt[3]{z^4} \cdot \sqrt[3]{z^6}$ $z^3\sqrt[3]{z}$

18) $2\sqrt{3} - \sqrt{18} + \sqrt{108}$ $8\sqrt{3} - 3\sqrt{2}$

19) $\sqrt{18a^5} - 4a\sqrt{8a^3}$ $-5a^2\sqrt{2a}$

20) $\sqrt{\dfrac{8}{3}} \cdot \sqrt{\dfrac{10}{27}}$ $\dfrac{4\sqrt{5}}{9}$

Multiply and simplify.

21) $\sqrt{5}(\sqrt{10} - 9)$ $5\sqrt{2} - 9\sqrt{5}$

22) $(4 - 3\sqrt{7})(\sqrt{3} + 9)$ $4\sqrt{3} + 36 - 3\sqrt{21} - 27\sqrt{7}$

23) $(\sqrt{5} + 4)(\sqrt{5} - 4)$ −11

24) $(7 - 2\sqrt{t})^2$ $49 - 28\sqrt{t} + 4t$

Rationalize the denominator of each expression.

25) $\dfrac{2}{\sqrt{6}}$ $\dfrac{\sqrt{6}}{3}$

26) $\dfrac{11}{\sqrt{2} + 3}$ $\dfrac{33 - 11\sqrt{2}}{7}$

27) $\sqrt{\dfrac{7}{5k}}$ $\dfrac{\sqrt{35k}}{5k}$

28) $\dfrac{12}{\sqrt[3]{4}}$ $6\sqrt[3]{2}$

29) Simplify $\dfrac{8 - \sqrt{48}}{4}$ completely. $2 - \sqrt{3}$

Choose from always, sometimes, or never.

30) The square root of a variable term with an even exponent can *always, sometimes,* or *never* be simplified. always

31) A radical equation *always, sometimes,* or *never* has a negative number as a solution. sometimes

32) How do you know, without actually solving and checking the solution, that $\sqrt{x} = -4$ has no real number solution? The principal square root of a number cannot equal a negative number.

Solve.

33) $\sqrt{4n - 1} = 3$ $\left\{\dfrac{5}{2}\right\}$

34) $y = 2 + \sqrt{y - 2}$ {2, 3}

35) $\sqrt{3p} + \sqrt{p - 2} = 4$ {3}

36) In the formula $r = \sqrt{\dfrac{V}{\pi h}}$, V represents the volume of a right circular cylinder, h represents the height of the cylinder, and r represents the radius. A cylindrical container has a volume of 64π cubic inches. It is 4 inches high. What is the radius of the container? 4 in.

*Additional answers can be found in the Answers to Exercises appendix.

1) Divide $\dfrac{5}{8} \div \dfrac{7}{12}$. $\dfrac{15}{14}$

2) Write an expression for "twice the sum of -9 and 4" and simplify. $2(-9 + 4); -10$

3) Combine like terms. $5a - 7b + 3 - 3a + \dfrac{5}{2}b - 2$ $2a - \dfrac{9}{2}b + 1$

4) Write in scientific notation. 0.000941 9.41×10^{-4}

5) Solve $4(5k - 6) + 9 = 3k + 2(k + 5)$. $\left\{\dfrac{5}{3}\right\}$

6) Graph $-x + 4y = 8$.

7) Write the slope-intercept form of the line containing the points $(-4, 2)$ and $(1, -4)$. $y = -\dfrac{6}{5}x - \dfrac{14}{5}$

8) Solve the system. $\begin{aligned} 3x + 2y &= 7 \\ 6x - y &= -6 \end{aligned}$ $\left(-\dfrac{1}{3}, 4\right)$

9) Simplify $2^{-3} - 4^{-2}$. $\dfrac{1}{16}$

Perform the operations and simplify.

10) $(4t - 3)(2t^2 - 9t - 5)$ $8t^3 - 42t^2 + 7t + 15$

11) $(7n^2 + 10n - 1) - (8n^2 + n - 5)$ $-n^2 + 9n + 4$

12) $\dfrac{12a^3 - 17a + 7}{2a - 1}$ $6a^2 + 3a - 7$

Factor completely.

13) $p^2 - 10p + 25$ $(p - 5)^2$

14) $6y^2 - 21y + 18$ $3(2y - 3)(y - 2)$

15) $h^3 - 9h^2 + 4h - 36$ $(h - 9)(h^2 + 4)$

16) Solve $5(r^2 - 5) - 9r = 4r^2 - 6r + 3$. $\{-4, 7\}$

17) *Write an equation and solve.* The width of a rectangle is 3 in. less than its length. The area is 40 in². Find the dimensions of the rectangle.   length = 8 in., width = 5 in.

Perform the operations and simplify.

18) $\dfrac{m - 1}{m + 3} - \dfrac{5}{4m}$ $\dfrac{4m^2 - 9m - 15}{4m(m + 3)}$

19) $\dfrac{k^2 - 2k - 63}{64 - k^2} \cdot \dfrac{3k^2 - 24k}{k + 7}$ $-\dfrac{3k(k - 9)}{k + 8}$

20) Solve $\dfrac{4}{x - 5} = \dfrac{x}{x^2 - 25} - \dfrac{2}{3}$. $\left\{-\dfrac{5}{2}, -2\right\}$

Simplify.

21) $4\sqrt{27} - 8\sqrt{75}$ $-28\sqrt{3}$

22) $\dfrac{10\sqrt{80}}{15\sqrt{12}}$ $\dfrac{4\sqrt{15}}{9}$

23) $\sqrt[3]{72x^6}$ $2x^2\sqrt[3]{9}$

24) Rationalize the denominator of $\dfrac{\sqrt{2}}{\sqrt{6} - 4}$. $-\dfrac{\sqrt{3} + 2\sqrt{2}}{5}$

25) *Write an equation and solve.* Twice the square root of a number is 3 less than the number. Find the number.   9

10 Quadratic Equations

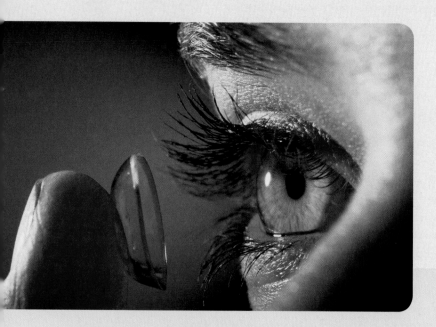

Math at Work:

Ophthalmologist

We have already seen two uses of mathematics in Mark Diamond's work as an ophthalmologist, and here we have a third. Mark can use a quadratic equation to convert between a prescription for glasses and a prescription for contact lenses. Specifically, after having reexamined a patient for contact lens use, Mark can use the following quadratic equation to double-check the prescription for the contact lenses based on the prescription his patient currently has for her glasses:

$$D_c = s(D_g)^2 + D_g$$

where D_g = power of the glasses, in diopters
s = distance of the glasses to the eye, in meters
D_c = power of the contact lenses, in diopters

If the power of a patient's eyeglasses is 9.00 diopters and the glasses rest 1 cm or 0.01 m from the eye, the power the patient would need in her contact lenses would be

$$D_c = 0.01(9)^2 + 9$$
$$D_c = 0.01(81) + 9$$
$$D_c = 0.81 + 9$$
$$D_c = 9.81 \text{ diopters}$$

An eyeglass power of 9.00 diopters would convert to a contact lens power of 9.81 diopters.

Mark realizes that his work is complex. Nonetheless, he wants his patients to understand what is happening with their eyes and how he is treating them. To help explain, he often refers to diagrams of the eye and other visuals that he keeps in his office. These visuals can make ideas that might otherwise be confusing simple and straightforward.

In this chapter, we will learn different ways to solve quadratic equations and introduce strategies for creating visuals.

As Mark's use of visuals in his work as an ophthalmologist suggests, visuals have enormous power when it comes to presenting complex information in a clear, organized way. Creating a sketch or a graph is a great approach when you are stuck on a math problem: Your visual may allow you to "see" a solution that has eluded you before. In addition, visuals are an essential tool when you are presenting information to others, in a classroom or a conference room. Charts, graphs, slides, and the like can help you make your points vividly and powerfully. The skills below will help you in creating effective visuals, either for your own personal use or to share with an audience.

- Determine the overall goal of your visual. Do you want to find an approach to a math problem? Do you want to bring numbers to life as part of a presentation?

- Decide the specific type of visual you will use. Some numbers, such as changes over time, are best communicated in the form of a line graph. Other sets of data are best seen in a bar chart, in a sketch, and so on.
- Gather the tools you will need to make your visual: a pen, ruler, protractor, compass, and so on, if you are making your sketch by hand, or the software you will use if you are creating your visual electronically.

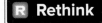

- Take your time creating the visual, ensuring that it is neat and accurate.
- Be sure to label all the major parts of your visual, paying special attention to units (inches, miles, tons, etc.).
- If you are presenting your visual to others, make sure it is large enough to be visible to the whole audience and that it has a title.
- If you are creating your visual from a math word problem, ensure that your visual includes all the key elements you will need to solve the problem.

- Consider whether your visual accurately represents the material on which it is based.
- Imagine that your visual was created by a friend or classmate. What suggestions might you make to improve it?

R Rethink

- Consider whether a different form of visual would be more effective. For instance, did you use a bar graph when the material might have been better illustrated as a pie chart?
- Ask yourself what new insights your visual offers you. What do you know about the material that you perhaps did not recognize before?

Chapter 10 **POWER** Plan

P Prepare

What are your goals for Chapter 10?	**O** Organize How can you accomplish each goal? (Write in the steps you will take to succeed.)
1 Be prepared before and during class.	• _____ • _____ • _____ • _____
2 Understand the homework to the point where you could do it without needing any help or hints.	• _____ • _____ • _____
3 Use the P.O.W.E.R. framework to learn how to create effective visuals: *What's Your Learning Style?*	• _____ • _____ • _____
4 Write your own goal. _____ _____	• _____

What are your objectives for Chapter 10?	How can you accomplish each objective?
1 Learn how to solve quadratic equations.	• Understand the square root property, and use it to solve quadratic equations of the form $x^2 = k$ and $(ax + b)^2 = k$. • Understand how to complete the square for quadratic equations of the form $ax^2 + bx + c = 0$. • Know how the quadratic formula is derived, and use it to solve a quadratic equation.
2 Learn to graph a quadratic equation.	• Understand how to use the graph of a quadratic equation of the form $y = ax^2 + bx + c$. • Be able to quickly recall and use the procedures for finding the vertex and graphing a parabola.
3 Learn the basic components of a function.	• Understand, identify, and use the definitions of *relation, function, domain,* and *range*. • Learn the procedure for using the vertical line test. • Be able to find the domain and range of a function by using its equation. • Understand function notation. • Solve problems and applied problems that involve functions.
4 Write your own goal. _____ _____	• _____

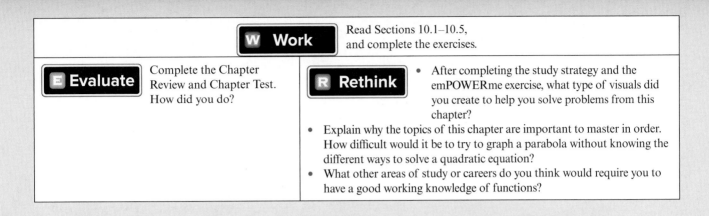

E Evaluate Complete the Chapter Review and Chapter Test. How did you do?	**R Rethink**	• After completing the study strategy and the emPOWERme exercise, what type of visuals did you create to help you solve problems from this chapter?
		• Explain why the topics of this chapter are important to master in order. How difficult would it be to try to graph a parabola without knowing the different ways to solve a quadratic equation?
		• What other areas of study or careers do you think would require you to have a good working knowledge of functions?

10.1 Solving Quadratic Equations Using the Square Root Property

P Prepare
O Organize

What are your objectives for Section 10.1?	How can you accomplish each objective?
1 Solve an Equation of the Form $x^2 = k$	• Learn the definition of *quadratic equation*. • Learn the **Square Root Property.** • Complete the given examples on your own. • Complete You Try 1.
2 Solve an Equation of the Form $(ax + b)^2 = k$	• Follow Example 2, and create a procedure for solving an equation of the form $(ax + b)^2 = k$. • Complete the given example on your own. • Complete You Trys 2 and 3.
3 Solve a Formula for a Specific Variable	• Complete the given example on your own. • Complete You Try 4.

 **Work**   **Read the explanations, follow the examples, take notes, and complete the You Trys.**

We defined a quadratic equation in Chapter 7. Let's restate the definition:

> ## Definition
>
> A **quadratic equation** can be written in the form $ax^2 + bx + c = 0$, where a, b, and c are real numbers and $a \neq 0$.

In Section 7.5, we learned how to solve quadratic equations by factoring. For example, we can use the zero product rule to solve $x^2 - 3x - 28 = 0$.

$$x^2 - 3x - 28 = 0$$
$$(x - 7)(x + 4) = 0 \qquad \text{Factor.}$$
$$x - 7 = 0 \quad \text{or} \quad x + 4 = 0 \qquad \text{Set each factor equal to zero.}$$
$$x = 7 \quad \text{or} \qquad x = -4 \qquad \text{Solve.}$$

The solution set is $\{-4, 7\}$.

It is not easy to solve all quadratic equations by factoring, however. Therefore, we need to learn other methods for solving quadratic equations. In this chapter, we will discuss three more methods for solving quadratic equations. Let's begin with the square root property.

1   Solve an Equation of the Form $x^2 = k$

Look at the equation $x^2 = 4$, for example. We can solve by factoring like this:

$$x^2 = 4$$
$$x^2 - 4 = 0 \qquad \text{Get all terms on the same side.}$$
$$(x + 2)(x - 2) = 0 \qquad \text{Factor.}$$
$$x + 2 = 0 \quad \text{or} \quad x - 2 = 0 \qquad \text{Set each factor equal to zero.}$$
$$x = -2 \quad \text{or} \qquad x = 2 \qquad \text{Solve.}$$

giving us a solution set of $\{-2, 2\}$.

Or, we can solve an equation like $x^2 = 4$ using the **square root property,** as we will see in Example 1a).

Definition The Square Root Property

Let k be a constant. If $x^2 = k$, then $x = \sqrt{k}$ or $x = -\sqrt{k}$.

The solution is often written as $x = \pm\sqrt{k}$, read as "x equals plus or minus the square root of k."

Note

We can use the square root property to solve an equation containing a squared quantity and a constant. To do so, we will get the squared quantity containing the variable on one side of the equal sign and the constant on the other side.

EXAMPLE 1

In-Class Example 1

Solve using the square root property.
a) $g^2 = 36$
b) $t^2 - 60 = 0$
c) $4r^2 - 25 = 1$
d) $h^2 + 33 = 9$

Answer: a) $\{-6, 6\}$
b) $\{-2\sqrt{15}, 2\sqrt{15}\}$
c) $\left\{ -\dfrac{\sqrt{26}}{2}, \dfrac{\sqrt{26}}{2} \right\}$ d) $\varnothing$

 Hint

Remember to write out the steps in the example as you are reading it.

Solve using the square root property.

a) $x^2 = 4$ b) $c^2 - 45 = 0$ c) $2n^2 + 7 = 19$ d) $y^2 + 15 = 6$

Solution

a)
$$x^2 = 4$$
$$x = \sqrt{4} \quad \text{or} \quad x = -\sqrt{4} \qquad \text{Square root property}$$
$$x = 2 \quad \text{or} \quad x = -2$$

Check:

$$x = 2: \quad x^2 = 4 \qquad\qquad x = -2: \quad x^2 = 4$$
$$(2)^2 \stackrel{?}{=} 4 \qquad\qquad\qquad (-2)^2 \stackrel{?}{=} 4$$
$$4 = 4 \ \checkmark \qquad\qquad\qquad 4 = 4 \ \checkmark$$

The solution set is $\{-2, 2\}$. We can also write it as $\{\pm 2\}$.

An equivalent way to solve $x^2 = 4$ is to write it as

$$x^2 = 4$$
$$x = \pm\sqrt{4} \qquad \text{Square root property}$$
$$x = \pm 2$$

We will use this approach when solving equations using the square root property.

b) To solve $c^2 - 45 = 0$, begin by getting c^2 on a side by itself.

$$c^2 - 45 = 0$$
$$c^2 = 45 \qquad\qquad \text{Add 45 to each side.}$$
$$c = \pm\sqrt{45} \qquad\quad \text{Square root property}$$
$$c = \pm\sqrt{9} \cdot \sqrt{5} \qquad \text{Product rule for radicals}$$
$$c = \pm 3\sqrt{5} \qquad\quad \sqrt{9} = 3$$

The check is left to the student. The solution set is $\{-3\sqrt{5}, 3\sqrt{5}\}$ or $\{\pm 3\sqrt{5}\}$.

c) To solve $2n^2 + 7 = 19$, begin by getting $2n^2$ on a side by itself.

$$2n^2 + 7 = 19$$
$$2n^2 = 12 \qquad \text{Subtract 7 from each side.}$$
$$n^2 = 6 \qquad\quad \text{Divide by 2.}$$
$$n = \pm\sqrt{6} \qquad \text{Square root property}$$

The check is left to the student. The solution set is $\{-\sqrt{6}, \sqrt{6}\}$ or $\{\pm\sqrt{6}\}$.

d)
$$y^2 + 15 = 6$$
$$y^2 = -9 \qquad\quad \text{Subtract 15 from each side.}$$
$$y = \pm\sqrt{-9} \qquad \text{Square root property}$$

Since $\sqrt{-9}$ is not a real number, there is no real number solution to $y^2 + 15 = 6$. The solution set is $\varnothing$.

[YOU TRY 1]

Solve using the square root property.

a) $m^2 = 25$ b) $h^2 - 28 = 0$ c) $4a^2 + 9 = 49$ d) $p^2 + 30 = 14$

Can we solve $(a - 5)^2 = 9$ using the square root property? Yes. The equation has a *squared quantity* and a *constant*.

2 Solve an Equation of the Form $(ax + b)^2 = k$

EXAMPLE 2

Solve $x^2 = 9$ and $(a - 5)^2 = 9$ using the square root property.

In-Class Example 2

Solve $n^2 = 49$ and $(p - 11)^2 = 49$ using the square root property.

Answer: $\{-7, 7\}$ and $\{4, 18\}$

Solution

While the equation $(a - 5)^2 = 9$ has a *binomial* that is being squared, the two equations are actually in the same form.

$$x^2 = 9 \qquad\qquad (a - 5)^2 = 9$$
$$\uparrow \quad \uparrow \qquad\qquad\qquad \uparrow \qquad \uparrow$$

x squared $=$ constant $\qquad (a - 5)$ squared $=$ constant

W Hint

Can you write a procedure for solving equations of the form $(ax + b)^2 = k$?

Solve $x^2 = 9$:

$$x^2 = 9$$
$$x = \pm\sqrt{9} \qquad \text{Square root property}$$
$$x = \pm 3$$

The solution set is $\{-3, 3\}$ or $\{\pm 3\}$.

We solve $(a - 5)^2 = 9$ in the same way with some additional steps.

$$(a - 5)^2 = 9$$
$$a - 5 = \pm\sqrt{9} \qquad \text{Square root property}$$
$$a - 5 = \pm 3$$

This means $a - 5 = 3$ or $a - 5 = -3$. Solve both equations.

$$a - 5 = 3 \quad \text{or} \quad a - 5 = -3$$
$$a = 8 \quad \text{or} \qquad a = 2 \qquad \text{Add 5 to each side.}$$

Check:

$$a = 8: \quad (a - 5)^2 = 9 \qquad\qquad a = 2: \quad (a - 5)^2 = 9$$
$$(8 - 5)^2 \stackrel{?}{=} 9 \qquad\qquad\qquad (2 - 5)^2 \stackrel{?}{=} 9$$
$$3^2 \stackrel{?}{=} 9 \qquad\qquad\qquad\qquad (-3)^2 \stackrel{?}{=} 9$$
$$9 = 9 \ \checkmark \qquad\qquad\qquad\qquad 9 = 9 \ \checkmark$$

The solution set is $\{2, 8\}$.

EXAMPLE 3

In-Class Example 3

Solve.
a) $(k - 6)^2 = 5$
b) $(2f + 5)^2 = 16$
c) $(4d - 7)^2 = 80$

Answer:
a) $\{6 - \sqrt{5}, 6 + \sqrt{5}\}$
b) $\left\{-\dfrac{9}{2}, -\dfrac{1}{2}\right\}$
c) $\left\{\dfrac{7 - 4\sqrt{5}}{4}, \dfrac{7 + 4\sqrt{5}}{4}\right\}$

W Hint

Be sure that you understand what is being done in the problem as it goes from step to step.

Solve.

a) $(w - 4)^2 = 3$ b) $(3r + 2)^2 = 25$ c) $(2x - 7)^2 = 18$

Solution

a) $(w - 4)^2 = 3$
$$w - 4 = \pm\sqrt{3} \qquad \text{Square root property}$$
$$w = 4 \pm \sqrt{3} \qquad \text{Add 4 to each side.}$$

Check: $w = 4 + \sqrt{3}$: $\quad (w - 4)^2 = 3 \qquad\qquad w = 4 - \sqrt{3}$: $\quad (w - 4)^2 = 3$
$(4 + \sqrt{3} - 4)^2 \overset{?}{=} 3 \qquad\qquad\qquad (4 - \sqrt{3} - 4)^2 \overset{?}{=} 3$
$(\sqrt{3})^2 \overset{?}{=} 3 \qquad\qquad\qquad\qquad (-\sqrt{3})^2 \overset{?}{=} 3$
$3 = 3 \ \checkmark \qquad\qquad\qquad\qquad\qquad 3 = 3 \ \checkmark$

The solution set is $\{4 - \sqrt{3}, 4 + \sqrt{3}\}$ or $\{4 \pm \sqrt{3}\}$.

b) $(3r + 2)^2 = 25$
$$3r + 2 = \pm\sqrt{25} \qquad \text{Square root property}$$
$$3r + 2 = \pm 5$$

This means $3r + 2 = 5$ or $3r + 2 = -5$. Solve both equations.

$3r + 2 = 5 \quad$ or $\quad 3r + 2 = -5$
$3r = 3 \qquad\qquad\quad 3r = -7 \qquad \text{Subtract 2 from each side.}$
$r = 1 \quad$ or $\quad r = -\dfrac{7}{3} \qquad \text{Divide by 3.}$

The check is left to the student. The solution set is $\left\{-\dfrac{7}{3}, 1\right\}$.

c) $(2x - 7)^2 = 18$
$$2x - 7 = \pm\sqrt{18} \qquad \text{Square root property}$$
$$2x - 7 = \pm 3\sqrt{2} \qquad \text{Simplify } \sqrt{18}.$$
$$2x = 7 \pm 3\sqrt{2} \qquad \text{Add 7 to each side.}$$
$$x = \dfrac{7 \pm 3\sqrt{2}}{2} \qquad \text{Divide by 2.}$$

One solution is $\dfrac{7 + 3\sqrt{2}}{2}$, and the other is $\dfrac{7 - 3\sqrt{2}}{2}$.

The solution set is $\left\{\dfrac{7 - 3\sqrt{2}}{2}, \dfrac{7 + 3\sqrt{2}}{2}\right\}$. This can also be written as $\left\{\dfrac{7 \pm 3\sqrt{2}}{2}\right\}$. The check is left to the student.

Solve.

a) $(z - 10)^2 = 7$ b) $(4d + 1)^2 = 16$ c) $(5p - 2)^2 = 27$

3 Solve a Formula for a Specific Variable

Sometimes, we need to use the square root property to solve a formula for a specific variable.

EXAMPLE 4

In-Class Example 4

Use the formula in Example 4 to find the radius of the base of a cone with a height of 12 in. and a volume of 36π in^3.

Answer: 3 in.

The formula for the volume, V, of a right circular cone is $V = \dfrac{1}{3}\pi r^2 h$, where r is the radius of the base and h is the height. Find the radius of the base of a right circular cone if it is 9 in. high and its volume is 12π in^3.

Solution

We will substitute the given values into the formula and solve for r.

$$V = \frac{1}{3}\pi r^2 h$$

$$12\pi = \frac{1}{3}\pi r^2 (9) \qquad \text{Substitute } 12\pi \text{ for } V \text{ and } 9 \text{ for } h.$$

$$12\pi = 3\pi r^2 \qquad \text{Multiply.}$$

$$\frac{12\pi}{3\pi} = \frac{3\pi r^2}{3\pi} \qquad \text{Divide both sides by } 3\pi.$$

$$4 = r^2 \qquad \text{Simplify.}$$

$$\pm\sqrt{4} = r \qquad \text{Square root property}$$

$$\pm 2 = r \qquad \text{Simplify.}$$

The radius of the cone cannot be negative, so we discard $r = -2$. The radius is 2 in.

YOU TRY 4

The formula for the volume, V, of a right circular cylinder is $V = \pi r^2 h$, where r is the radius and h is the height. Find the radius of a cylinder if it is 15 cm high and its volume is 240π cm^3.

ANSWERS TO YOU TRY EXERCISES

1) a) $\{-5, 5\}$ b) $\{-2\sqrt{7}, 2\sqrt{7}\}$ c) $\{-\sqrt{10}, \sqrt{10}\}$ d) $\varnothing$ 2) $\{-5, -1\}$

3) a) $\{10 - \sqrt{7}, 10 + \sqrt{7}\}$ b) $\left\{-\dfrac{5}{4}, \dfrac{3}{4}\right\}$ c) $\left\{\dfrac{2 - 3\sqrt{3}}{5}, \dfrac{2 + 3\sqrt{3}}{5}\right\}$ 4) 4 cm

*Additional answers can be found in the Answers to Exercises appendix.

Objective 1: Solve an Equation of the Form $x^2 = k$

1) What are two methods that can be used to solve $x^2 - 81 = 0$? Solve the equation using both methods. factoring and the square root property; $\{-9, 9\}$

2) If k is a negative number and $x^2 = k$, what can you conclude about the solution to the equation? There is no real number solution. The solution set is $\varnothing$.

Solve using the square root property.

3) $b^2 = 16$ $\{-4, 4\}$

4) $h^2 = 100$ $\{-10, 10\}$

5) $w^2 = 11$ $\{-\sqrt{11}, \sqrt{11}\}$

6) $x^2 = 23$ $\{-\sqrt{23}, \sqrt{23}\}$

7) $p^2 = -49$ $\varnothing$

8) $s^2 = -81$ $\varnothing$

9) $x^2 = \dfrac{25}{9}$ $\left\{-\dfrac{5}{3}, \dfrac{5}{3}\right\}$

10) $m^2 = \dfrac{16}{121}$ $\left\{-\dfrac{4}{11}, \dfrac{4}{11}\right\}$

11) $y^2 = 0.04$ $\{-0.2, 0.2\}$

12) $d^2 = 0.25$ $\{-0.5, 0.5\}$

13) $r^2 - 144 = 0$ $\{-12, 12\}$

14) $a^2 - 1 = 0$ $\{-1, 1\}$

15) $c^2 - 19 = 0$ $\{-\sqrt{19}, \sqrt{19}\}$

16) $a^2 - 6 = 0$ $\{-\sqrt{6}, \sqrt{6}\}$

17) $v^2 - 54 = 0$ $\{-3\sqrt{6}, 3\sqrt{6}\}$

18) $g^2 - 75 = 0$ $\{-5\sqrt{3}, 5\sqrt{3}\}$

19) $t^2 - \dfrac{5}{64} = 0$ $\left\{-\dfrac{\sqrt{5}}{8}, \dfrac{\sqrt{5}}{8}\right\}$

20) $c^2 - \dfrac{14}{81} = 0$ $\left\{-\dfrac{\sqrt{14}}{9}, \dfrac{\sqrt{14}}{9}\right\}$

21) $z^2 + 5 = 19$ $\{-\sqrt{14}, \sqrt{14}\}$

22) $x^2 + 9 = 17$ $\{-2\sqrt{2}, 2\sqrt{2}\}$

23) $n^2 + 10 = 6$ $\varnothing$

24) $y^2 + 11 = 9$ $\varnothing$

25) $3d^2 + 14 = 41$ $\{-3, 3\}$

26) $2m^2 - 5 = 67$ $\{-6, 6\}$

27) $4p^2 - 9 = 39$ $\{-2\sqrt{3}, 2\sqrt{3}\}$

28) $3j^2 + 7 = 31$ $\{-2\sqrt{2}, 2\sqrt{2}\}$

29) $3 = 35 - 8h^2$ $\{-2, 2\}$

30) $145 = 2w^2 - 55$ $\{-10, 10\}$

31) $10 = 14 + 2x^2$ $\varnothing$

32) $6 = 24 + 3k^2$ $\varnothing$

33) $4y^2 + 15 = 24$ $\left\{-\dfrac{3}{2}, \dfrac{3}{2}\right\}$

34) $9n^2 + 17 = 18$ $\left\{-\dfrac{1}{3}, \dfrac{1}{3}\right\}$

35) $9w^2 - 5 = 5$

36) $16a^2 + 2 = 13$

37) $-7 = 4 - 5b^2$

38) $-1 = 13 - 6t^2$

Objective 2: Solve an Equation of the Form $(ax + b)^2 = k$

Choose from *always, sometimes,* or *never.*

39) The solutions of $(ax + b)^2 = k$ (where $k > 0$) are *always, sometimes,* or *never* positive.   sometimes

40) The equation $(ax + b)^2 = 0$ will *always, sometimes,* or *never* have exactly one solution.   always

Solve using the square root property.

41) $(r + 6)^2 = 25$ $\{-11, -1\}$

42) $(x - 1)^2 = 16$ $\{-3, 5\}$

43) $(q - 8)^2 = 1$ $\{7, 9\}$

44) $(c + 11)^2 = 49$ $\{-18, -4\}$

45) $(a + 2)^2 = 13$ $\{-2 - \sqrt{13}, -2 + \sqrt{13}\}$

46) $(t - 5)^2 = 7$ $\{5 - \sqrt{7}, 5 + \sqrt{7}\}$

47) $(k - 10)^2 = 45$ $\{10 - 3\sqrt{5}, 10 + 3\sqrt{5}\}$

48) $(b + 4)^2 = 20$ $\{-4 - 2\sqrt{5}, -4 + 2\sqrt{5}\}$

49) $(m + 7)^2 = -18$ $\varnothing$

50) $(y - 3)^2 = -100$ $\varnothing$

51) $0 = (p + 3)^2 - 68$ $\{-3 - 2\sqrt{17}, -3 + 2\sqrt{17}\}$

52) $0 = (d + 5)^2 - 72$ $\{-5 - 6\sqrt{2}, -5 + 6\sqrt{2}\}$

53) $(2z - 1)^2 = 9$ $\{-1, 2\}$

54) $(5h + 9)^2 = 36$ $\left\{-3, -\dfrac{3}{5}\right\}$

55) $121 = (4q + 5)^2$ $\left\{-4, \dfrac{3}{2}\right\}$

56) $64 = (3c - 4)^2$ $\left\{-\dfrac{4}{3}, 4\right\}$

57) $(3g - 10)^2 = 24$

58) $(2w - 7)^2 = 63$

59) $125 = (5u + 8)^2$

60) $44 = (4a + 5)^2$

61) $(2x + 3)^2 - 54 = 0$

62) $(6t - 1)^2 - 90 = 0$

63) $(7h - 8)^2 + 32 = 0$ $\varnothing$

64) $(2b + 9)^2 + 18 = 0$ $\varnothing$

65) $(5y - 2)^2 + 6 = 22$

66) $29 = 4 + (3m + 1)^2$

67) $1 = (6r + 7)^2 - 8$

68) $(3 - 4k)^2 - 18 = -2$

69) $(2z - 11)^2 + 3 = 17$

70) $(5x + 8)^2 - 2 = 6$

71) $\left(1 - \dfrac{1}{2}c\right)^2 - 6 = -5$ $\{0, 4\}$

72) $\left(\dfrac{2}{3}p + 5\right)^2 + 7 = 56$ $\{-18, 3\}$

Objective 3: Solve a Formula for a Specific Variable

Solve each problem.

73) The area of a circle is 81π cm^2. Find the radius of the circle.   9 cm

74) The volume of a right circular cylinder is 28π in^3. Find the radius of the cylinder if it is 7 in. tall.   2 in.

75) The surface area, S, of a sphere is given by $S = 4\pi r^2$, where r is the radius. Find the radius of the sphere with a surface area of π m^2. $\dfrac{1}{2}$ m

76) The surface area, S, of a cube is given by $S = 6L^2$, where L is the length of one of its sides. Find the length of a side of a cube with a surface area of 150 in^2.   5 in.

77) The illuminance E (the measure of light emitted, in lux) of a light source is given by $E = \dfrac{I}{d^2}$, where I is the luminous intensity (measured in candela) and d is the distance, in meters, from the light source. Find the distance, d, from the light source when $E = 300$ lux and $I = 2700$ candela.   3 m

78) The power, P (in watts), in an electrical system is given by $P = \dfrac{V^2}{R}$, where V is the voltage, in volts, and R is the resistance, in ohms. Find the voltage if the power is 192 watts and the resistance is 3 ohms.   24 volts

The kinetic energy K, in joules, of an object is given by $K = \dfrac{1}{2}mv^2$, where m is the mass of the object, in kg, and v is the velocity of the object in m/sec. Use this formula for Exercises 79 and 80.

79) Find the velocity of a roller coaster car with a mass of 1200 kg and kinetic energy of 153,600 joules.
16 m/sec

80) Find the velocity of a boat with a mass of 25,000 kg and kinetic energy of 28,125 joules.   1.5 m/sec

R Rethink

R1) Can equations that are solved using the square root property have two solutions that are exactly the same? Explain your thought process.

R2) Think of other formulas you have used that could be solved for a specific variable using the square root property. Where could that be helpful?

10.2 Solving Quadratic Equations by Completing the Square

P Prepare

O Organize

What are your objectives for Section 10.2?	How can you accomplish each objective?
1 Complete the Square for an Expression of the Form $x^2 + bx$	<ul><li>Recall the definition of a *perfect square trinomial,* and refer back to Section 7.4 if needed.</li><li>Follow the explanation and write the steps for **Completing the Square for $x^2 + bx$** in your own words.</li><li>Complete the given examples on your own.</li><li>Complete You Try 1.</li></ul>
2 Solve an Equation of the Form $ax^2 + bx + c = 0$ by Completing the Square	<ul><li>Write the **Steps for Solving a Quadratic Equation ($ax^2 + bx + c = 0$) by Completing the Square** in your own words.</li><li>Complete the given examples on your own.</li><li>Complete You Trys 2 and 3.</li></ul>

Read the explanations, follow the examples, take notes, and complete the You Trys.

The next method we will learn for solving a quadratic equation is **completing the square.** But first we need to review an idea presented in Section 7.4.

A **perfect square trinomial** is a trinomial whose factored form is the square of a binomial. Some examples of perfect square trinomials are

Perfect Square Trinomials	Factored Form
$x^2 + 6x + 9$	$(x + 3)^2$
$d^2 - 14d + 49$	$(d - 7)^2$

In the trinomial $x^2 + 6x + 9$, x^2 is called the *quadratic term,* $6x$ is called the *linear term,* and 9 is called the *constant.*

1 Complete the Square for an Expression of the Form $x^2 + bx$

In a perfect square trinomial where the coefficient of the quadratic term is 1, the constant term is related to the coefficient of the linear term in the following way: *If you find half of the linear coefficient and square the result, you will get the constant term.*

W Hint

Be sure you understand the two explanations to the right before moving on.

$x^2 + 6x + 9$: The constant, 9, is obtained by

1) finding half of the coefficient of x

$$\frac{1}{2}(6) = 3$$

2) then squaring the result.

$$3^2 = 9 \text{ (the constant)}$$

$d^2 - 14d + 49$: The constant, 49, is obtained by

1) finding half of the coefficient of d

$$\frac{1}{2}(-14) = -7$$

2) then squaring the result.

$$(-7)^2 = 49 \text{ (the constant)}$$

We can generalize this procedure so that we can find the constant needed to obtain the perfect square trinomial for any quadratic expression of the form $x^2 + bx$. Finding this perfect square trinomial is called **completing the square** because the trinomial will factor to the square of a binomial.

Procedure Completing the Square for $x^2 + bx$

To find the constant needed to complete the square for $x^2 + bx$:

Step 1: Find half of the coefficient of x: $\frac{1}{2}b$

Step 2: Square the result: $\left(\frac{1}{2}b\right)^2$

Step 3: Then add it to $x^2 + bx$ to get $x^2 + bx + \left(\frac{1}{2}b\right)^2$

The coefficient of the squared term *must* be 1 before you complete the square!

EXAMPLE 1

In-Class Example 1

Complete the square for each expression to obtain a perfect square trinomial. Then, factor.
a) $n^2 + 12n$ b) $t^2 - 16t$

Answer:
a) $n^2 + 12n + 36$; $(n + 6)^2$
b) $t^2 - 16t + 64$; $(t - 8)^2$

Complete the square for each expression to obtain a perfect square trinomial. Then, factor.

a) $k^2 + 10k$ b) $p^2 - 8p$

Solution

a) Find the constant needed to complete the square for $k^2 + 10k$.

> **Step 1:** Find half of the coefficient of k:
> $$\frac{1}{2}(10) = 5$$
>
> **Step 2:** Square the result:
> $$5^2 = 25$$
>
> **Step 3:** Add 25 to $k^2 + 10k$:
> $$k^2 + 10k + 25$$

The perfect square trinomial is $k^2 + 10k + 25$. The factored form is $(k + 5)^2$.

b) Find the constant needed to complete the square for $p^2 - 8p$.

> **Step 1:** Find half of the coefficient of p:
> $$\frac{1}{2}(-8) = -4$$
>
> **Step 2:** Square the result:
> $$(-4)^2 = 16$$
>
> **Step 3:** Add 16 to $p^2 - 8p$:
> $$p^2 - 8p + 16$$

The perfect square trinomial is $p^2 - 8p + 16$. The factored form is $(p - 4)^2$.

[YOU TRY 1] Complete the square for each expression to obtain a perfect square trinomial. Then, factor.

a) $w^2 + 2w$ b) $r^2 - 18r$

At the beginning of the section and in Example 1, we saw the following perfect square trinomials and their factored forms. We will look at the relationship between the constant in the factored form and the coefficient of the linear term.

Perfect Square Trinomial		Factored Form
$x^2 + 6x + 9$	3 is $\frac{1}{2}(6)$	$(x + 3)^2$
$d^2 - 14d + 49$	-7 is $\frac{1}{2}(-14)$	$(d - 7)^2$
$k^2 + 10k + 25$	5 is $\frac{1}{2}(10)$	$(k + 5)^2$
$p^2 - 8p + 16$	-4 is $\frac{1}{2}(-8)$	$(p - 4)^2$
$t^2 + 3t + \dfrac{9}{4}$	$\frac{3}{2}$ is $\frac{1}{2}(3)$	$\left(t + \dfrac{3}{2}\right)^2$

This pattern will always hold true and can be helpful in factoring some perfect square trinomials.

2 Solve an Equation of the Form $ax^2 + bx + c = 0$ by Completing the Square

Any quadratic equation of the form $ax^2 + bx + c = 0$ $(a \neq 0)$ can be written in the form $(x - h)^2 = k$ by completing the square. Once an equation is in this form, we can use the square root property to solve for the variable.

Procedure Steps for Solving a Quadratic Equation $(ax^2 + bx + c = 0)$ by Completing the Square

Step 1: **The coefficient of the squared term must be 1.** If it is not 1, divide both sides of the equation by a to obtain a leading coefficient of 1.

Step 2: **Get the variables on one side of the equal sign and the constant on the other side.**

Step 3: **Complete the square.** Find half of the linear coefficient, then square the result. Add that quantity to *both* sides of the equation.

Step 4: **Factor.**

Step 5: **Solve using the square root property.**

EXAMPLE 2

In-Class Example 2

Solve by completing the square.
a) $v^2 + 10v + 16 = 0$
b) $z^2 - 4z + 13 = 0$

Answer: a) $\{-8, -2\}$ b) $\varnothing$

Solve by completing the square.

a) $x^2 + 12x + 27 = 0$ b) $k^2 - 2k + 5 = 0$

Solution

a) $x^2 + 12x + 27 = 0$

     Step 1: The coefficient of x^2 is already 1.

     Step 2: Get the variables on one side of the equal sign and the constant on the other side: $x^2 + 12x = -27$

     Step 3: Complete the square: $\dfrac{1}{2}(12) = 6$

$$6^2 = 36$$

           Add 36 to both sides of the equation: $x^2 + 12x + 36 = -27 + 36$
$$x^2 + 12x + 36 = 9$$

     Step 4: Factor: $(x + 6)^2 = 9$

     Step 5: Solve using the square root property.

$$(x + 6)^2 = 9$$
$$x + 6 = \pm\sqrt{9}$$
$$x + 6 = \pm 3$$

$$x + 6 = 3 \quad \text{or} \quad x + 6 = -3$$
$$x = -3 \quad \text{or} \quad x = -9$$

The check is left to the student. The solution set is $\{-9, -3\}$.

Notice that we would have obtained the same result in Example 2a) if we had solved the equation by factoring.

$$x^2 + 12x + 27 = 0$$
$$(x + 9)(x + 3) = 0$$

$$x + 9 = 0 \quad \text{or} \quad x + 3 = 0$$
$$x = -9 \quad \text{or} \quad x = -3$$

b) $k^2 - 2k + 5 = 0$

Step 1: The coefficient of k^2 is already 1.

Step 2: Get the variables on one side of the equal sign and the constant on the other side: $k^2 - 2k = -5$

Step 3: Complete the square: $\dfrac{1}{2}(-2) = -1$

$$(-1)^2 = 1$$

Add 1 to *both* sides of the equation: $k^2 - 2k + 1 = -5 + 1$
$$k^2 - 2k + 1 = -4$$

Step 4: Factor: $(k - 1)^2 = -4$

Step 5: Solve using the square root property.

$$(k - 1)^2 = -4$$
$$k - 1 = \pm\sqrt{-4}$$

Since $\sqrt{-4}$ is not a real number, there is no real number solution to $k^2 - 2k + 5 = 0$. The solution set is $\varnothing$.

W Hint

Compare the signs of the constant on the right side of the equal sign in Step 4 of a) and b). What do you notice?

[YOU TRY 2]

Solve by completing the square.

a) $w^2 + 4w - 21 = 0$ b) $a^2 - 6a + 34 = 0$

Remember that in order to complete the square, the coefficient of the quadratic term must be 1.

EXAMPLE 3

Solve by completing the square.

a) $4n^2 - 16n + 15 = 0$ b) $10y + 2y^2 = 3$

In-Class Example 3

Solve by completing the square.
a) $4x^2 - 24x + 35 = 0$
b) $6h + 2h^2 = 7$

Answer: a) $\left\{\dfrac{5}{2}, \dfrac{7}{2}\right\}$

b) $\left\{-\dfrac{3}{2} - \dfrac{\sqrt{23}}{2}, -\dfrac{3}{2} + \dfrac{\sqrt{23}}{2}\right\}$

Solution

a) $4n^2 - 16n + 15 = 0$

Step 1: Since the coefficient of n^2 is *not* 1, divide the whole equation by 4.

$$\frac{4n^2}{4} - \frac{16n}{4} + \frac{15}{4} = \frac{0}{4} \qquad \text{Divide by 4.}$$

$$n^2 - 4n + \frac{15}{4} = 0 \qquad \text{Simplify.}$$

Step 2: Get the constant on a side by itself: $n^2 - 4n = -\dfrac{15}{4}$

 Hint

Are you writing out the steps as you read the example?

Step 3: Complete the square: $\dfrac{1}{2}(-4) = -2$

$$(-2)^2 = 4$$

Add 4 to both sides of the equation.

$$n^2 - 4n + 4 = -\dfrac{15}{4} + 4$$

$$n^2 - 4n + 4 = -\dfrac{15}{4} + \dfrac{16}{4} \qquad \text{Get a common denominator.}$$

$$n^2 - 4n + 4 = \dfrac{1}{4} \qquad \text{Add.}$$

Step 4: Factor: $(n - 2)^2 = \dfrac{1}{4}$

Step 5: Solve using the square root property.

$$(n - 2)^2 = \dfrac{1}{4}$$

$$n - 2 = \pm\sqrt{\dfrac{1}{4}} \qquad \text{Square root property}$$

$$n - 2 = \pm\dfrac{1}{2} \qquad \text{Simplify.}$$

This means $n - 2 = \dfrac{1}{2}$ or $n - 2 = -\dfrac{1}{2}$. Solve both equations.

$$n - 2 = \dfrac{1}{2} \qquad\qquad n - 2 = -\dfrac{1}{2}$$

$$n = \dfrac{1}{2} + 2 \quad \text{or} \quad n = -\dfrac{1}{2} + 2 \qquad \text{Add 2 to each side.}$$

$$n = \dfrac{5}{2} \qquad\qquad n = \dfrac{3}{2}$$

The check is left to the student. The solution set is $\left\{\dfrac{3}{2}, \dfrac{5}{2}\right\}$.

b) $10y + 2y^2 = 3$

Step 1: Since the coefficient of y^2 is *not* 1, divide the whole equation by 2.

$$\dfrac{10y}{2} + \dfrac{2y^2}{2} = \dfrac{3}{2} \qquad \text{Divide by 2.}$$

$$5y + y^2 = \dfrac{3}{2} \qquad \text{Simplify.}$$

Step 2: The constant is on a side by itself. Rewrite the left side of the equation.

$$y^2 + 5y = \dfrac{3}{2}$$

Step 3: Complete the square: $\dfrac{1}{2}(5) = \dfrac{5}{2}$

$$\left(\dfrac{5}{2}\right)^2 = \dfrac{25}{4}$$

Add $\dfrac{25}{4}$ to both sides of the equation.

$$y^2 + 5y + \dfrac{25}{4} = \dfrac{3}{2} + \dfrac{25}{4}$$

$$y^2 + 5y + \dfrac{25}{4} = \dfrac{6}{4} + \dfrac{25}{4} \qquad \text{Get a common denominator.}$$

$$y^2 + 5y + \dfrac{25}{4} = \dfrac{31}{4} \qquad \text{Add.}$$

Step 4:   Factor: $\left(y + \dfrac{5}{2}\right)^2 = \dfrac{31}{4}$

$\uparrow$

$\dfrac{5}{2}$ is $\dfrac{1}{2}(5)$, the coefficient of y,
in the equation above.

Step 5: Solve using the square root property.

$$\left(y + \dfrac{5}{2}\right)^2 = \dfrac{31}{4}$$

$$y + \dfrac{5}{2} = \pm\sqrt{\dfrac{31}{4}} \qquad \text{Square root property}$$

$$y + \dfrac{5}{2} = \pm\dfrac{\sqrt{31}}{2} \qquad \text{Simplify.}$$

$$y = -\dfrac{5}{2} \pm \dfrac{\sqrt{31}}{2} \qquad \text{Add } -\dfrac{5}{2} \text{ to each side.}$$

Check the answers. The solution set is $\left\{ -\dfrac{5}{2} - \dfrac{\sqrt{31}}{2},\ -\dfrac{5}{2} + \dfrac{\sqrt{31}}{2} \right\}$.

[YOU TRY 3]   Solve by completing the square.

     a) $4c^2 + 8c - 45 = 0$ b) $2b^2 - 6b = 5$

ANSWERS TO [YOU TRY] EXERCISES

1) a) $w^2 + 2w + 1$; $(w + 1)^2$ b) $r^2 - 18r + 81$; $(r - 9)^2$

2) a) $\{-7, 3\}$ b) $\varnothing$ 3) a) $\left\{ -\dfrac{9}{2}, \dfrac{5}{2} \right\}$ b) $\left\{ \dfrac{3}{2} - \dfrac{\sqrt{19}}{2}, \dfrac{3}{2} + \dfrac{\sqrt{19}}{2} \right\}$

*Additional answers can be found in the Answers to Exercises appendix.

Objective 1: Complete the Square for an Expression of the Form $x^2 + bx$

1) What is a perfect square trinomial? Give an example.

2) In $x^2 - 9x + 14$, what is the

   a) quadratic term? x^2

   b) linear term? $-9x$

   c) constant? 14

Complete the square for each expression to obtain a perfect square trinomial. Then, factor.

Fill It In

Fill in the blanks with either the missing mathematical step or reason for the given step.

3) $y^2 + 18y$

   $\frac{1}{2}(18) = 9$ Find half of the coefficient of y.

   $9^2 = 81$ Square the result.

   $y^2 + 18y + 81$ Add the constant to the expression.

   The perfect square trinomial is $y^2 + 18y + 81$.

   The factored form of the trinomial is $(y + 9)^2$.

4) $c^2 - 5c$

   $\frac{1}{2}(-5) = -\frac{5}{2}$ Find half of the coefficient of c.

   $\left(-\frac{5}{2}\right)^2 = \frac{25}{4}$ Square the result.

   $c^2 - 5c + \frac{25}{4}$ Add the constant to the expression.

   The perfect square trinomial is $c^2 - 5c + \frac{25}{4}$.

   The factored form of the trinomial is $\left(c - \frac{5}{2}\right)^2$.

5) $a^2 + 12a$ 6) $b^2 + 8b$

7) $k^2 - 10k$ 8) $p^2 - 4p$

9) $g^2 - 24g$ 10) $z^2 + 26z$

11) $h^2 + 9h$ 12) $m^2 - 3m$

13) $x^2 - x$ 14) $y^2 + 7y$

Objective 2: Solve an Equation of the Form $ax^2 + bx + c = 0$ by Completing the Square

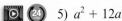

15) What are the steps used to solve a quadratic equation by completing the square? *Answers may vary.*

16) Can $x^3 + 12x + 20 = 0$ be solved by completing the square? Give a reason for your answer.
No, because the equation is not quadratic.

Solve by completing the square.

17) $x^2 + 6x + 8 = 0$ $\{-4, -2\}$ 18) $a^2 + 10a - 24 = 0$ $\{-12, 2\}$

19) $z^2 - 14z + 45 = 0$ $\{5, 9\}$ 20) $t^2 - 12t - 45 = 0$ $\{-3, 15\}$

21) $p^2 + 8p + 20 = 0$ $\varnothing$ 22) $c^2 - 2c + 37 = 0$ $\varnothing$

23) $y^2 - 11 = -4y$ $\{-2 - \sqrt{15}, -2 + \sqrt{15}\}$ 24) $k^2 - 1 = 8k$ $\{4 - \sqrt{17}, 4 + \sqrt{17}\}$

25) $x^2 - 10x = 3$ $\{5 - 2\sqrt{7}, 5 + 2\sqrt{7}\}$ 26) $-4b = b^2 - 14$ $\{-2 - 3\sqrt{2}, -2 + 3\sqrt{2}\}$

27) $2a = 22 + a^2$ $\varnothing$ 28) $w^2 + 39 = 6w$ $\varnothing$

29) $m^2 + 3m - 40 = 0$ $\{-8, 5\}$ 30) $h^2 - 7h + 6 = 0$ $\{1, 6\}$

31) $c^2 - 56 = c$ $\{-7, 8\}$ 32) $p^2 + 5p = -4$ $\{-4, -1\}$

33) $h^2 + 9h = -12$ 34) $q^2 - 3 = q$

35) $b^2 - 5b + 27 = 6$ $\varnothing$ 36) $g^2 + 3g + 11 = 4$ $\varnothing$

37) Can you complete the square on $2x^2 + 16x$ as it is given? Why or why not?
No, because the coefficient of x^2 is not 1.

38) What is the first thing you should do if you want to solve $3n^2 - 9n = 12$ by completing the square?
Divide both sides of equation by 3.

Solve by completing the square.

39) $4r^2 + 32r + 55 = 0$ 40) $4t^2 - 16t + 7 = 0$ $\left\{\frac{1}{2}, \frac{7}{2}\right\}$

41) $3x^2 + 39 = 30x$ $\{5 - 2\sqrt{3}, 5 + 2\sqrt{3}\}$ 42) $5p^2 + 30p = 10$ $\{-3 - \sqrt{11}, -3 + \sqrt{11}\}$

43) $7k^2 + 84 = 49k$ $\{3, 4\}$ 44) $10m = 2m^2 + 12$ $\{2, 3\}$

45) $54y - 6y^2 = 72$ 46) $8w^2 + 8w = 32$

47) $16z^2 + 3 = 16z$ $\left\{\frac{1}{4}, \frac{3}{4}\right\}$ 48) $5 - 16c = 16c^2$ $\left\{-\frac{5}{4}, \frac{1}{4}\right\}$

49) $3g^2 + 15g + 37 = 0$ $\varnothing$ 50) $7t^2 - 21t + 40 = 0$ $\varnothing$

51) $-v^2 - 2v + 35 = 0$ $\{-7, 5\}$

52) $-k^2 + 12k - 32 = 0$ $\{4, 8\}$

53) $(a - 4)(a + 10) = -17$ $\{-3 - 4\sqrt{2}, -3 + 4\sqrt{2}\}$

54) $(y + 5)(y - 3) = 5$ $\{-1 - \sqrt{21}, -1 + \sqrt{21}\}$

55) $n + 2 = 3n^2$ $\left\{-\frac{2}{3}, 1\right\}$ 56) $15 = m + 2m^2$ $\left\{-3, \frac{5}{2}\right\}$

57) $(5p + 2)(p + 4) = 1$ 58) $(3c + 4)(c + 2) = 3$

Solve each problem by writing an equation and completing the square.

59) The length of a rectangular portfolio is 7 in. more than its width. Find the dimensions of the portfolio if it has an area of 170 in^2.
width = 10 in., length = 17 in.

60) The area of a rectangular sign is 220 ft². Its width is 9 ft less than its length. What are the dimensions of the sign?   width = 11 ft, length = 20 ft

61) The area of a triangle is 60 cm². Its base is 1 cm less than twice its height. Find the lengths of the base and the height.   base = 15 cm, height = 8 cm

62) The length of the base of a triangle is 5 in. more than twice its height. Find the lengths of the base and the height if the area is 26 in².   base = 13 in., height = 4 in.

Find the lengths of the sides of each right triangle. Use the Pythagorean theorem, and solve by completing the square.

63)

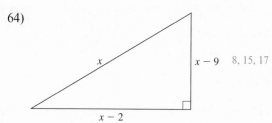

10, 24, 26

64)

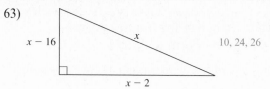

8, 15, 17

R Rethink

R1) Discuss any new patterns you noticed as you completed the exercises.

R2) How did they help you move through the exercises more quickly?

R3) Were there any problems you could not do? If so, write them down or circle them, then ask your instructor in class.

10.3 Solving Quadratic Equations Using the Quadratic Formula

P Prepare

O Organize

What are your objectives for Section 10.3?	How can you accomplish each objective?
1 Derive the Quadratic Formula	• Follow the steps taken to derive the **Quadratic Formula**. • Derive the quadratic formula in your notes by using the steps learned to complete the square. • Understand and *memorize* the **quadratic formula**.
2 Solve a Quadratic Equation Using the Quadratic Formula	• Complete the given examples on your own. • Complete You Trys 1–3.

W Work   Read the explanations, follow the examples, take notes, and complete the You Trys.

1 Derive the Quadratic Formula

In Section 10.2, we saw that any quadratic equation of the form $ax^2 + bx + c = 0$ ($a \neq 0$) can be solved by completing the square. We can develop another method, called the *quadratic formula,* for solving quadratic equations if we complete the square on $ax^2 + bx + c = 0$ ($a \neq 0$).

The steps we use to complete the square on $ax^2 + bx + c = 0$ are *exactly* the same steps we use to solve an equation like $2x^2 + 3x - 1 = 0$. We will do these steps side by side so that you can more easily understand the process.

Solve Each Equation for *x* by Completing the Square.

$$2x^2 + 3x - 1 = 0 \qquad | \qquad ax^2 + bx + c = 0$$

Step 1: The coefficient of the squared term must be 1.

$$2x^2 + 3x - 1 = 0 \qquad\qquad ax^2 + bx + c = 0$$

$$\frac{2x^2}{2} + \frac{3x}{2} - \frac{1}{2} = \frac{0}{2} \quad \text{Divide by 2.} \qquad \frac{ax^2}{a} + \frac{bx}{a} + \frac{c}{a} = \frac{0}{a} \quad \text{Divide by } a.$$

$$x^2 + \frac{3}{2}x - \frac{1}{2} = 0 \quad \text{Simplify.} \qquad x^2 + \frac{b}{a}x + \frac{c}{a} = 0 \quad \text{Simplify.}$$

Step 2: Get the constant on the other side of the equal sign.

$$x^2 + \frac{3}{2}x = \frac{1}{2} \quad \text{Add } \frac{1}{2}. \qquad\qquad x^2 + \frac{b}{a}x = -\frac{c}{a} \quad \text{Subtract } \frac{c}{a}.$$

Step 3: Complete the square.

$$\frac{1}{2}\left(\frac{3}{2}\right) = \frac{3}{4} \quad \tfrac{1}{2} \text{ of } x\text{-coefficient} \qquad \frac{1}{2}\left(\frac{b}{a}\right) = \frac{b}{2a} \quad \tfrac{1}{2} \text{ of } x\text{-coefficient}$$

$$\left(\frac{3}{4}\right)^2 = \frac{9}{16} \quad \text{Square the result.} \qquad \left(\frac{b}{2a}\right)^2 = \frac{b^2}{4a^2} \quad \text{Square the result.}$$

Add $\dfrac{9}{16}$ to both sides of the equation. | Add $\dfrac{b^2}{4a^2}$ to both sides of the equation.

$$x^2 + \frac{3}{2}x + \frac{9}{16} = \frac{1}{2} + \frac{9}{16} \qquad\qquad x^2 + \frac{b}{a}x + \frac{b^2}{4a^2} = -\frac{c}{a} + \frac{b^2}{4a^2}$$

$$x^2 + \frac{3}{2}x + \frac{9}{16} = \frac{8}{16} + \frac{9}{16} \quad \begin{matrix}\text{Get a common}\\\text{denominator.}\end{matrix} \qquad x^2 + \frac{b}{a}x + \frac{b^2}{4a^2} = -\frac{4ac}{4a^2} + \frac{b^2}{4a^2} \quad \begin{matrix}\text{Get a common}\\\text{denominator.}\end{matrix}$$

$$x^2 + \frac{3}{2}x + \frac{9}{16} = \frac{17}{16} \quad \text{Add.} \qquad x^2 + \frac{b}{a}x + \frac{b^2}{4a^2} = \frac{b^2 - 4ac}{4a^2} \quad \text{Add.}$$

Step 4: Factor.

$$\left(x + \frac{3}{4}\right)^2 = \frac{17}{16} \qquad\qquad\qquad \left(x + \frac{b}{2a}\right)^2 = \frac{b^2 - 4ac}{4a^2}$$

$$\uparrow \qquad\qquad\qquad\qquad\qquad\qquad \uparrow$$

$\frac{3}{4}$ is $\frac{1}{2}\left(\frac{3}{2}\right)$, the coefficient of *x*. | $\frac{b}{2a}$ is $\frac{1}{2}\left(\frac{b}{a}\right)$, the coefficient of *x*.

Step 5: Solve using the square root property.

$$\left(x + \frac{3}{4}\right)^2 = \frac{17}{16} \qquad\qquad \left(x + \frac{b}{2a}\right)^2 = \frac{b^2 - 4ac}{4a^2}$$

$$x + \frac{3}{4} = \pm\sqrt{\frac{17}{16}} \qquad\qquad x + \frac{b}{2a} = \pm\sqrt{\frac{b^2 - 4ac}{4a^2}}$$

$$x + \frac{3}{4} = \frac{\pm\sqrt{17}}{4} \quad \sqrt{16} = 4 \qquad\qquad x + \frac{b}{2a} = \frac{\pm\sqrt{b^2 - 4ac}}{2a} \quad \sqrt{4a^2} = 2a$$

$$x = -\frac{3}{4} \pm \frac{\sqrt{17}}{4} \quad \text{Subtract } \frac{3}{4}. \qquad\qquad x = -\frac{b}{2a} \pm \frac{\sqrt{b^2 - 4ac}}{2a} \quad \text{Subtract } \frac{b}{2a}.$$

$$x = \frac{-3 \pm \sqrt{17}}{4} \quad \begin{array}{l}\text{Same denominators,}\\\text{combine numerators.}\end{array} \qquad\qquad x = \frac{-b \pm \sqrt{b^2 - 4ac}}{2a} \quad \begin{array}{l}\text{Same denominators,}\\\text{combine numerators.}\end{array}$$

The result on the right is called the *quadratic formula*.

Definition The Quadratic Formula

The solutions of any quadratic equation of the form $ax^2 + bx + c = 0$ ($a \neq 0$) are

$$x = \frac{-b \pm \sqrt{b^2 - 4ac}}{2a}$$

This is called the **quadratic formula.**

W Hint

You can use the quadratic formula to solve *any* quadratic equation as long as it is in the form $ax^2 + bx + c = 0$.

Note

1) The equation to be solved *must* be written in the form $ax^2 + bx + c = 0$ so that a, b, and c can be identified correctly.

2) $x = \dfrac{-b \pm \sqrt{b^2 - 4ac}}{2a}$ represents the two solutions $x = \dfrac{-b + \sqrt{b^2 - 4ac}}{2a}$ and $x = \dfrac{-b - \sqrt{b^2 - 4ac}}{2a}$.

3) Notice that the fraction bar continues under $-b$ and does not end at the radical.

$$x = \frac{-b \pm \sqrt{b^2 - 4ac}}{2a} \qquad\qquad x = -b \pm \frac{\sqrt{b^2 - 4ac}}{2a}$$

   Correct   Incorrect

4) Using the $\pm$ when deriving the quadratic formula allows us to say that $\sqrt{4a^2} = 2a$.

5) The quadratic formula is a *very* important result and is one that is used often. *It should be memorized!*

2 Solve a Quadratic Equation Using the Quadratic Formula

W Hint

Are you writing out the steps as you read the example?

Solve using the quadratic formula.

a) $2x^2 + 3x - 1 = 0$ b) $3n^2 - 10n + 8 = 0$

Solution

a) Is $2x^2 + 3x - 1 = 0$ in the form $ax^2 + bx + c = 0$? *Yes.* Identify the values of a, b, and c, and substitute them into the quadratic formula.

$$a = 2 \qquad b = 3 \qquad c = -1$$

$$x = \frac{-b \pm \sqrt{b^2 - 4ac}}{2a} \qquad \text{Quadratic formula}$$

$$= \frac{-(3) \pm \sqrt{(3)^2 - 4(2)(-1)}}{2(2)} \qquad \text{Substitute } a = 2, b = 3, \text{ and } c = -1.$$

$$= \frac{-3 \pm \sqrt{9 - (-8)}}{4} \qquad \text{Perform the operations.}$$

$$= \frac{-3 \pm \sqrt{17}}{4} \qquad 9 - (-8) = 9 + 8 = 17$$

The solution set is $\left\{ \dfrac{-3 - \sqrt{17}}{4}, \dfrac{-3 + \sqrt{17}}{4} \right\}$. This is the same result we obtained when we solved this equation by completing the square at the beginning of the section.

b) Is $3n^2 - 10n + 8 = 0$ in the form $ax^2 + bx + c = 0$? *Yes.* Identify a, b, and c, and substitute them into the quadratic formula.

$$a = 3 \qquad b = -10 \qquad c = 8$$

$$n = \frac{-b \pm \sqrt{b^2 - 4ac}}{2a} \qquad \text{Quadratic formula}$$

$$n = \frac{-(-10) \pm \sqrt{(-10)^2 - 4(3)(8)}}{2(3)} \qquad \text{Substitute } a = 3, b = -10, \text{ and } c = 8.$$

$$n = \frac{10 \pm \sqrt{100 - 96}}{6} \qquad \text{Perform the operations.}$$

$$n = \frac{10 \pm \sqrt{4}}{6} \qquad \text{Simplify the radicand.}$$

$$n = \frac{10 \pm 2}{6} \qquad \text{Simplify } \sqrt{4}.$$

Find the two values of n, one using the plus sign and the other using the minus sign:

$$n = \frac{10 + 2}{6} = \frac{12}{6} = 2 \qquad \text{or} \qquad n = \frac{10 - 2}{6} = \frac{8}{6} = \frac{4}{3}$$

Check the values in the original equation. The solution set is $\left\{ \dfrac{4}{3}, 2 \right\}$.

YOU TRY 1

Solve using the quadratic formula.

a) $5p^2 - p - 3 = 0$ b) $3r^2 + r - 10 = 0$

If a quadratic equation is not in standard form, we must write it that way.

Solve using the quadratic formula.

a) $t^2 + 1 = 4t$ b) $2w(w + 3) = -5$

Solution

a) Is $t^2 + 1 = 4t$ in the form $ax^2 + bx + c = 0$? *No.* Before we can apply the quadratic formula, we must write it in that form.

$$t^2 - 4t + 1 = 0 \qquad \text{Subtract } 4t.$$

Identify a, b, and c: $a = 1$ $b = -4$ $c = 1$

$$t = \frac{-b \pm \sqrt{b^2 - 4ac}}{2a} \qquad \text{Quadratic formula}$$

$$t = \frac{-(-4) \pm \sqrt{(-4)^2 - 4(1)(1)}}{2(1)} \qquad \text{Substitute } a = 1, b = -4, \text{ and } c = 1.$$

$$t = \frac{4 \pm \sqrt{16 - 4}}{2} \qquad \text{Perform the operations.}$$

$$t = \frac{4 \pm \sqrt{12}}{2} \qquad \text{Simplify the radicand.}$$

$$t = \frac{4 \pm 2\sqrt{3}}{2} \qquad \text{Simplify the radical.}$$

$$t = \frac{2(2 \pm \sqrt{3})}{2} \qquad \text{Factor out 2.}$$

$$t = 2 \pm \sqrt{3} \qquad \text{Simplify.}$$

The solution set is $\{2 - \sqrt{3}, 2 + \sqrt{3}\}$.

b) Is $2w(w + 3) = -5$ in the form $ax^2 + bx + c = 0$? *No.* Distribute on the left side of the equation, then get all terms on the same side of the equal sign.

$$2w^2 + 6w = -5 \qquad \text{Distribute.}$$
$$2w^2 + 6w + 5 = 0 \qquad \text{Add 5.}$$

Identify a, b, and c: $a = 2$ $b = 6$ $c = 5$

$$w = \frac{-b \pm \sqrt{b^2 - 4ac}}{2a} \qquad \text{Quadratic formula}$$

$$w = \frac{-(6) \pm \sqrt{(6)^2 - 4(2)(5)}}{2(2)} \qquad \text{Substitute } a = 2, b = 6, \text{ and } c = 5.$$

$$w = \frac{-6 \pm \sqrt{36 - 40}}{4} \qquad \text{Perform the operations.}$$

$$w = \frac{-6 \pm \sqrt{-4}}{4} \qquad \text{Simplify the radicand.}$$

Since $\sqrt{-4}$ is not a real number, there is no real number solution to $2w(w + 3) = -5$. The solution set is $\varnothing$.

YOU TRY 2

Solve using the quadratic formula.

a) $y^2 + 4 = 6y$ b) $m(5m - 2) = -3$

W Hint

How could knowing a little bit about the discriminant help you check your answer?

The expression under the radical, $b^2 - 4ac$, in the quadratic formula is called the **discriminant.** Examples 1a) and 2a) show that if the discriminant is positive but not a perfect square, then the given equation has *two irrational solutions*. We see in Example 1b) that if the discriminant is positive and the square of an integer, then the equation has *two rational solutions* and can be solved by factoring. If the discriminant is negative, as in Example 2b), then the equation has *no real number solution*.

What if the discriminant equals 0? What does that tell us about the solution set?

EXAMPLE 3

In-Class Example 3

Solve $\dfrac{2}{15}w^2 - \dfrac{2}{3}w + \dfrac{5}{6} = 0$ using the quadratic formula.

Answer: $\left\{\dfrac{5}{2}\right\}$

Solve $\dfrac{2}{9}k^2 - \dfrac{2}{3}k + \dfrac{1}{2} = 0$ using the quadratic formula.

Solution

Is $\dfrac{2}{9}k^2 - \dfrac{2}{3}k + \dfrac{1}{2} = 0$ in the form $ax^2 + bx + c = 0$? *Yes.* However, working with fractions in the quadratic formula would be difficult. *Eliminate the fractions by multiplying the equation by 18, the least common denominator of the fractions.*

$$18\left(\frac{2}{9}k^2 - \frac{2}{3}k + \frac{1}{2}\right) = 18 \cdot 0 \qquad \text{Multiply by 18 to eliminate the fractions.}$$
$$4k^2 - 12k + 9 = 0$$

Identify a, b, and c: $\quad a = 4 \qquad b = -12 \qquad c = 9$

$$k = \frac{-b \pm \sqrt{b^2 - 4ac}}{2a} \qquad \text{Quadratic formula}$$

$$k = \frac{-(-12) \pm \sqrt{(-12)^2 - 4(4)(9)}}{2(4)} \qquad \text{Substitute } a = 4,\ b = -12,\ \text{and } c = 9.$$

$$k = \frac{12 \pm \sqrt{144 - 144}}{8} \qquad \text{Perform the operations.}$$

$$k = \frac{12 \pm \sqrt{0}}{8} \qquad \text{Simplify the radicand. The discriminant} = 0.$$

$$k = \frac{12 \pm 0}{8} = \frac{12}{8} = \frac{3}{2}$$

The solution set is $\left\{\dfrac{3}{2}\right\}$.

Example 3 illustrates that when the discriminant equals 0, the equation has *one rational solution*.

[YOU TRY 3]

Solve $\dfrac{3}{4}h^2 + \dfrac{1}{2}h + \dfrac{1}{12} = 0$ using the quadratic formula.

After multiplying the original equation in Example 3 by 18, we got the equation $4k^2 - 12k + 9 = 0$. Notice that we could have solved this equation by factoring:

$$4k^2 - 12k + 9 = 0$$
$$(2k - 3)^2 = 0 \qquad \text{Factor.}$$
$$2k - 3 = 0 \qquad \text{Use the square root property.}$$
$$2k = 3 \qquad \text{Add 3.}$$
$$k = \frac{3}{2} \qquad \text{Divide by 2.}$$

The solution set is $\left\{\dfrac{3}{2}\right\}$. This is the same as the result we obtained using the quadratic formula.

ANSWERS TO [YOU TRY] EXERCISES

1) a) $\left\{\dfrac{1 - \sqrt{61}}{10}, \dfrac{1 + \sqrt{61}}{10}\right\}$ b) $\left\{-2, \dfrac{5}{3}\right\}$

2) a) $\{3 - \sqrt{5}, 3 + \sqrt{5}\}$ b) $\varnothing$ 3) $\left\{-\dfrac{1}{3}\right\}$

E Evaluate **10.3** Exercises   Do the exercises, and check your work.

*Additional answers can be found in the Answers to Exercises appendix.

Objective 2: Solve a Quadratic Equation Using the Quadratic Formula

1) To solve a quadratic equation, $ax^2 + bx + c = 0$ ($a \neq 0$), for x, we can use the quadratic formula. Write the quadratic formula. $x = \dfrac{-b \pm \sqrt{b^2 - 4ac}}{2a}$

Find the error in each, and correct the mistake.

2) The solution to $ax^2 + bx + c = 0$ ($a \neq 0$) can be found using the quadratic formula

$$x = -b \pm \frac{\sqrt{b^2 - 4ac}}{2a}$$

The fraction bar should also be under $-b$:
$$x = \frac{-b \pm \sqrt{b^2 - 4ac}}{2a}$$

3) In order to solve $3x^2 - 5x = 4$ using the quadratic formula, a student substitutes a, b, and c into the formula in this way: $a = 3$, $b = -5$, $c = 4$.

$$x = \frac{-(-5) \pm \sqrt{(-5)^2 - 4(3)(4)}}{2(3)}$$

4) $\dfrac{-3 \pm 12\sqrt{5}}{3} = -1 \pm 12\sqrt{5}$

Solve using the quadratic formula.

5) $x^2 + 2x - 8 = 0$ $\{-4, 2\}$ 6) $p^2 + 8p + 12 = 0$ $\{-6, -2\}$

7) $6z^2 - 7z + 2 = 0$ $\left\{\dfrac{1}{2}, \dfrac{2}{3}\right\}$ 8) $2h^2 + h - 15 = 0$ $\left\{-3, \dfrac{5}{2}\right\}$

9) $k^2 + 2 = 5k$ 10) $d^2 = 5 - 3d$

11) $3w^2 = 2w + 4$ 12) $8r = 2 - 5r^2$

13) $y = 2y^2 + 6$ $\varnothing$ 14) $3v + 4v^2 = -3$ $\varnothing$

15) $m^2 + 11m = 0$ $\{-11, 0\}$ 16) $w^2 - 8w = 0$ $\{0, 8\}$

17) $2p(p - 3) = -3$ 18) $3q(q + 3) = 7q + 4$

19) $(2s + 3)(s - 1) = s^2 - s + 6$ $\{-1 - \sqrt{10}, -1 + \sqrt{10}\}$

20) $2n^2 + 2n - 4 = (3n + 4)(n - 2)$ $\{2 - 2\sqrt{2}, 2 + 2\sqrt{2}\}$

21) $k(k + 2) = -5$ $\varnothing$

22) $3r = 2(r^2 + 4)$ $\varnothing$

23) $(x - 8)(x - 3) = 3(3 - x)$ $\{3, 5\}$

24) $2(k + 10) = (k + 10)(k - 2)$ $\{-10, 4\}$

25) $8t = 1 + 16t^2$ $\left\{\dfrac{1}{4}\right\}$

26) $9u^2 + 12u + 4 = 0$ $\left\{-\dfrac{2}{3}\right\}$

27) $\dfrac{1}{8}z^2 + \dfrac{3}{4}z + \dfrac{1}{2} = 0$ $\{-3 - \sqrt{5}, -3 + \sqrt{5}\}$

28) $\dfrac{1}{6}p^2 + \dfrac{4}{3}p = \dfrac{5}{2}$ $\{-4 - \sqrt{31}, -4 + \sqrt{31}\}$

29) $\dfrac{1}{6}k + \dfrac{1}{2} = \dfrac{3}{4}k^2$ $\left\{\dfrac{1 - \sqrt{55}}{9}, \dfrac{1 + \sqrt{55}}{9}\right\}$

30) $\dfrac{1}{5} + \dfrac{1}{2}w = \dfrac{1}{5}w^2$ $\left\{\dfrac{5 - \sqrt{41}}{4}, \dfrac{5 + \sqrt{41}}{4}\right\}$

31) $0.8v^2 + 0.1 = 0.6v$ $\{0.25, 0.5\}$ 32) $0.2m^2 + 0.1m = 1.5$ $\{-3, 2.5\}$

33) $16g^2 - 3 = 0$ $\left\{-\dfrac{\sqrt{3}}{4}, \dfrac{\sqrt{3}}{4}\right\}$ 34) $49n^2 - 20 = 0$ $\left\{-\dfrac{2\sqrt{5}}{7}, \dfrac{2\sqrt{5}}{7}\right\}$

35) $9d^2 - 4 = 0$ 36) $4k^2 - 25 = 0$

37) $3(3 - 4r) = -4r^2$ $\left\{\dfrac{3}{2}\right\}$ 38) $5y(5y + 2) = -1$

39) $6 = 7h - 3h^2$ $\varnothing$ 40) $1 + t(2 + 5t) = 0$ $\varnothing$

41) $4p^2 + 6 = 20p$ $\left\{\dfrac{5 - \sqrt{19}}{2}, \dfrac{5 + \sqrt{19}}{2}\right\}$ 42) $4x^2 = 6x + 16$ $\left\{\dfrac{3 - \sqrt{73}}{4}, \dfrac{3 + \sqrt{73}}{4}\right\}$

Write an equation and solve. Give an exact answer, and use a calculator to round the answer to the nearest hundredth.

43) The hypotenuse of a right triangle is 1 inch less than twice the shorter leg. The longer leg is $\sqrt{23}$ inches long. Find the length of the shorter leg. $\dfrac{2 + \sqrt{70}}{3}$ in. or about 3.46 in.

44) The hypotenuse of the right triangle is 1 inch more than twice the longer leg. The length of the shorter leg is $\sqrt{14}$ in. Find the length of the longer leg. $\dfrac{-2 + \sqrt{43}}{3}$ in. or about 1.52 in.

Solve. Give an exact answer, and use a calculator to round the answer to the nearest hundredth.

45) An object is thrown upward from a height of 24 feet. The height h of the object (in feet) t seconds after the object is released is given by

$$h = -16t^2 + 24t + 24$$

     a) How long does it take the object to reach a height of 8 feet?   2 sec

     b) How long does it take the object to hit the ground? (Hint: When the object hits the ground, $h = 0$.) $\dfrac{3 + \sqrt{33}}{4}$ sec or about 2.19 sec

46) A ball is thrown upward from a height of 20 feet. The height h of the ball (in feet) t seconds after the ball is released is given by

$$h = -16t^2 + 16t + 20$$

     a) How long does it take the ball to reach a height of 8 feet?   1.5 sec

     b) How long does it take the ball to hit the ground? (Hint: When the ball hits the ground, $h = 0$.) $\dfrac{1 + \sqrt{6}}{2}$ sec or about 1.72 sec

Extension

47) What is the discriminant of a quadratic equation? It is the expression under the radical in the quadratic formula, $b^2 - 4ac$.

48) If the discriminant of a quadratic equation is positive but not a perfect square, what does that tell you about the solutions of the equation? The equation has two irrational solutions.

49) If the discriminant of a quadratic equation is positive and the square of an integer, what does that tell you about the solutions of the equation? The equation has two rational solutions.

50) If the discriminant of a quadratic equation is negative, what does that tell you about the solutions of the equation?   The equation has no real number solutions.

For each equation, find the value of the discriminant. Then, determine the number and type of solutions of each equation. *Do not solve the equation.*

51) $2n^2 - 7n + 3 = 0$ 25; two rational solutions

52) $w^2 - 8w + 5 = 0$ 44; two irrational solutions

53) $10p^2 - 9p + 3 = 0$ -39; no real number solutions

54) $4k^2 + 28k + 49 = 0$ 0; one rational solution

55) $-3x^2 = 4x - 1$ 28; two irrational solutions

56) $3t = 1 + 5t^2$ -11; no real number solutions

57) $-9 = 5h(5h + 6)$ 0; one rational solution

58) $6(y^2 - 2) = y$ 289; two rational solutions

R Rethink

R1) How could you use the quadratic formula to help you factor a more complicated quadratic expression?

R2) Do you think some of these exercises could have been solved using another method? Explain.

Putting It All Together

P Prepare

O Organize

What is your objective?	How can you accomplish the objective?
1 Decide Which Method to Use to Solve a Quadratic Equation	• While following Example 1 on your own, make a chart that will help you identify when to use each of the four methods. • Complete the given example on your own. • Complete You Try 1.

W Work

Read the explanations, follow the example, take notes, and complete the You Try.

We have learned four methods for solving quadratic equations.

Methods for Solving Quadratic Equations

1) Factoring

2) Square root property

3) Completing the square

4) Quadratic formula

While it is true that the quadratic formula can be used to solve *every* quadratic equation of the form $ax^2 + bx + c = 0$ ($a \neq 0$), it is not always the most *efficient* method. In this section, we will discuss how to decide which method to use to solve a quadratic equation.

1 Decide Which Method to Use to Solve a Quadratic Equation

EXAMPLE 1

In-Class Example 1

Solve.
a) $r^2 + 3r = 18$
b) $t^2 - 12t + 8 = 0$
c) $5p^2 + 2p + 4 = 0$
d) $(7y - 1)^2 - 3 = 0$

Answer: a) $\{-6, 3\}$
b) $\{6 - 2\sqrt{7}, 6 + 2\sqrt{7}\}$
c) $\varnothing$
d) $\left\{\dfrac{1 - \sqrt{3}}{7}, \dfrac{1 + \sqrt{3}}{7}\right\}$

Solve.

a) $c^2 - 3c = 28$

b) $x^2 - 10x + 18 = 0$

c) $2k^2 + 5k + 9 = 0$

d) $(5n + 3)^2 - 12 = 0$

Solution

a) Write $c^2 - 3c = 28$ in standard form: $c^2 - 3c - 28 = 0$

Does $c^2 - 3c - 28$ factor? *Yes. Solve by factoring.*

$$(c - 7)(c + 4) = 0$$

$c - 7 = 0$ or $c + 4 = 0$ — Set each factor equal to 0.

$c = 7$ or $c = -4$ — Solve.

The solution set is $\{-4, 7\}$.

b) To solve $x^2 - 10x + 18 = 0$, ask yourself, "Can I factor $x^2 - 10x + 18$?" *No.* We could solve this using the quadratic formula, but *completing the square* is also a good method for solving this equation. Why?

Completing the square is a good method for solving a quadratic equation when the coefficient of the squared term is 1 or −1 and when the coefficient of the linear term is even.

   We will solve $x^2 - 10x + 18 = 0$ by completing the square.

Step 1: The coefficient of x^2 is 1.

Step 2: Get the variables on one side of the equal sign and the constant on the other side: $x^2 - 10x = -18$

Step 3: Complete the square: $\dfrac{1}{2}(-10) = -5$

$$(-5)^2 = 25$$

Add 25 to *both* sides of the equation: $x^2 - 10x + 25 = -18 + 25$
$$x^2 - 10x + 25 = 7$$

Step 4: Factor: $(x - 5)^2 = 7$

Step 5: Solve using the square root property:

$$(x - 5)^2 = 7$$
$$x - 5 = \pm\sqrt{7}$$
$$x = 5 \pm \sqrt{7} \qquad \text{Add 5 to each side.}$$

   The solution set is $\{5 - \sqrt{7}, 5 + \sqrt{7}\}$.

Note

Completing the square works well when the coefficient of the squared term is 1 or −1 and when the coefficient of the linear term is *even* because when we complete the square in Step 3, we will not obtain a fraction. (Half of an even number is an integer.)

c) Ask yourself, "Can I solve $2k^2 + 5k + 9 = 0$ by factoring?" *No.* Completing the square would not be a very efficient way to solve the equation because the coefficient of k^2 is 2, and dividing the equation by 2 would give us $k^2 + \dfrac{5}{2}k + \dfrac{9}{2} = 0$.

   We will solve $2k^2 + 5k + 9 = 0$ using the quadratic formula.

Identify a, b, and c: $\quad a = 2 \quad\quad b = 5 \quad\quad c = 9$

$$k = \frac{-b \pm \sqrt{b^2 - 4ac}}{2a} \qquad \text{Quadratic formula}$$

$$k = \frac{-(5) \pm \sqrt{(5)^2 - 4(2)(9)}}{2(2)} \qquad \text{Substitute } a = 2, b = 5, \text{ and } c = 9.$$

$$k = \frac{-5 \pm \sqrt{25 - 72}}{4} = \frac{-5 \pm \sqrt{-47}}{4}$$

Since $\sqrt{-47}$ is not a real number, there is no real number solution to $2k^2 + 5k + 9 = 0$. The solution set is $\varnothing$.

d) Which method should we use to solve $(5n + 3)^2 - 12 = 0$?

We *could* square the binomial, combine like terms, then solve, possibly, by factoring or using the quadratic formula. However, this would be very inefficient. The equation contains a squared quantity and a constant.

We will solve $(5n + 3)^2 - 12 = 0$ using the square root property.

$$(5n + 3)^2 - 12 = 0$$
$$(5n + 3)^2 = 12 \qquad \text{Add 12 to each side.}$$
$$5n + 3 = \pm\sqrt{12} \qquad \text{Square root property}$$
$$5n + 3 = \pm 2\sqrt{3} \qquad \text{Simplify } \sqrt{12}.$$
$$5n = -3 \pm 2\sqrt{3} \qquad \text{Add } -3 \text{ to each side.}$$
$$n = \frac{-3 \pm 2\sqrt{3}}{5} \qquad \text{Divide by 5.}$$

The solution set is $\left\{ \dfrac{-3 - 2\sqrt{3}}{5}, \dfrac{-3 + 2\sqrt{3}}{5} \right\}$.

> **W Hint**
> Use your chart to help you decide which method to use to solve a quadratic equation.

[YOU TRY 1] Solve.

a) $3t^2 - 2 = -8t$ b) $m^2 + 36 = 13m$

c) $(4r - 1)^2 + 10 = 0$ d) $w^2 + 8w - 2 = 0$

ANSWERS TO [YOU TRY] EXERCISE

1) a) $\left\{ \dfrac{-4 - \sqrt{22}}{3}, \dfrac{-4 + \sqrt{22}}{3} \right\}$ b) $\{4, 9\}$ c) $\varnothing$ d) $\{-4 - 3\sqrt{2}, -4 + 3\sqrt{2}\}$

Putting It All Together Exercises

 Do the exercises, and check your work.

Additional answers can be found in the Answers to Exercises appendix.

Objective 1: Decide Which Method to Use to Solve a Quadratic Equation

Keep in mind the four methods we have learned for solving quadratic equations: *factoring, the square root property, completing the square,* and *the quadratic formula.* Solve the equations using one of these methods.

1) $f^2 - 75 = 0$ $\{-5\sqrt{3}, 5\sqrt{3}\}$

2) $t^2 - 8t = 4$ $\{4 - 2\sqrt{5}, 4 + 2\sqrt{5}\}$

 3) $a(a + 1) = 20$ $\{-5, 4\}$

4) $3m^2 + 7 = 4m$ $\varnothing$

5) $v^2 + 6v + 7 = 0$ $\{-3 - \sqrt{2}, -3 + \sqrt{2}\}$

6) $4u^2 - 5u - 6 = 0$ $\left\{ -\dfrac{3}{4}, 2 \right\}$

7) $3x(x + 3) = -5(x - 1)$ $\left\{ -5, \dfrac{1}{3} \right\}$

8) $5 = (p + 4)^2 + 7$ $\varnothing$

9) $n^2 + 12n + 42 = 0$ $\varnothing$

10) $\dfrac{1}{2}r^2 = \dfrac{3}{4} - \dfrac{3}{2}r$ $\left\{ \dfrac{-3 - \sqrt{15}}{2}, \dfrac{-3 + \sqrt{15}}{2} \right\}$

11) $12 + (2k - 1)^2 = 3$ $\varnothing$

12) $h^2 + 7h + 15 = 0$ $\varnothing$

13) $1 = \dfrac{x^2}{40} - \dfrac{3x}{20}$ $\{-4, 10\}$

14) $72 = 2p^2$ $\{-6, 6\}$

15) $b^2 - 6b = 5$ $\{3 - \sqrt{14}, 3 + \sqrt{14}\}$

16) $3m^3 + 42m = -27m^2$ $\{-7, -2, 0\}$

17) $q(q + 12) = 3(q^2 + 5) + q$ $\left\{ \dfrac{5}{2}, 3 \right\}$

18) $w^2 = 3w$ $\{0, 3\}$

19) $\dfrac{9}{c} = 1 + \dfrac{18}{c^2}$ $\{3, 6\}$

20) $2t(2t + 4) = 5t + 6$ $\left\{ \dfrac{-3 - \sqrt{105}}{8}, \dfrac{-3 + \sqrt{105}}{8} \right\}$

21) $(3v + 4)(v - 2) = -9$ $\varnothing$

22) $y^2 + 4y = 2$ $\{-2 - \sqrt{6}, -2 + \sqrt{6}\}$

23) $2r^2 + 3r - 2 = 0$ $\left\{ -2, \dfrac{1}{2} \right\}$

24) $(6r + 1)(r - 4) = -2(12r + 1)$ $\left\{ -\dfrac{2}{3}, \dfrac{1}{2} \right\}$

25) $5m = m^2$ $\{0, 5\}$

26) $6z^2 + 12z + 18 = 0$ $\varnothing$

27) $4m^3 = 9m$ $\left\{ -\dfrac{3}{2}, 0, \dfrac{3}{2} \right\}$

28) $\dfrac{8}{x^2} = \dfrac{1}{4} + \dfrac{1}{x}$ $\{-8, 4\}$

29) $2k^2 + 3 = 9k$ $\left\{ \dfrac{9 - \sqrt{57}}{4}, \dfrac{9 + \sqrt{57}}{4} \right\}$

30) $6v^2 + 3 = 15v$ $\left\{ \dfrac{5 - \sqrt{17}}{4}, \dfrac{5 + \sqrt{17}}{4} \right\}$

R1) Did you ever try a certain method and run into a problem? What did you learn from that?

R2) What is your favorite method when attempting to solve a quadratic equation? Why?

10.4 Graphs of Quadratic Equations

P Prepare

O Organize

What are your objectives for Section 10.4?	How can you accomplish each objective?
1 Graph Basic Parabolas	• Understand the definitions of *parabola* and *vertex*, and know the property **The Graph of a Quadratic Equation of the Form** $y = ax^2 + bx + c$. • Complete the given examples on your own. • Complete You Trys 1 and 2.
2 Graph a Quadratic Equation of the Form $y = ax^2 + bx + c$	• Review graphing parabolas by plotting points. • Learn the procedure for **Finding the Vertex of a Parabola of the Form** $y = ax^2 + bx + c$. • Learn the procedure for **Graphing a Parabola of the Form** $y = ax^2 + bx + c$. • Complete the given examples on your own. • Complete You Trys 3 and 4.

W Work

Read the explanations, follow the examples, take notes, and complete the You Trys.

1 Graph Basic Parabolas

In Chapter 3, we learned that the graph of a linear equation is a line. In this section, we will learn how to graph quadratic equations in two variables.

> **Property** The Graph of a Quadratic Equation of the Form
> $$y = ax^2 + bx + c$$
>
> The graph of a quadratic equation of the form $y = ax^2 + bx + c$ is called a **parabola.** The lowest point on a parabola that opens upward or the highest point on a parabola that opens downward is called the **vertex.**

The simplest form of a quadratic equation is $y = x^2$. To graph it, we will make a table of values.

EXAMPLE 1

Graph $y = x^2$.

Solution

Make a table of values. Plot the points, and connect them with a smooth curve.

W Hint

Use graph paper to graph the parabolas in this section.

$y = x^2$	
x	**y**
0	0
1	1
2	4
−1	1
−2	4

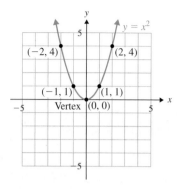

The graphs of quadratic equations where y is defined in terms of x open either upward or downward. Notice that this graph opens upward. The lowest point on a parabola that opens upward or the highest point on a parabola that opens downward is called the **vertex.** The vertex of the graph of $y = x^2$ is (0, 0). When graphing a quadratic equation by plotting points, it is important to locate the vertex.

[YOU TRY 1] Graph $y = x^2 - 3$.

EXAMPLE 2

Graph $y = -(x - 2)^2$.

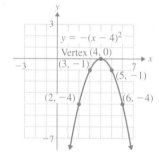

Solution

In Example 1, we said that it is important to locate the vertex of a parabola. Let's do that first. Begin by finding the x-coordinate of the vertex of this graph. *The x-coordinate of the vertex is the value of x that makes the expression being squared equal to zero.* The x-coordinate of the vertex of this parabola is 2.

Make a table of values, and use the x-coordinate of the vertex as the first x-value in the table. Then, we will choose a couple of values of x that are larger than 2 and a couple of values that are smaller than 2. Find the corresponding y-values, plot the points, and connect them with a smooth curve.

$y = -(x - 2)^2$	
x	**y**
2	0
3	−1
4	−4
1	−1
0	−4

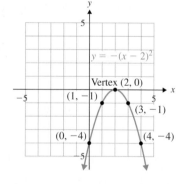

The vertex of this parabola is (2, 0). Notice that this time it is the *highest* point on the parabola because the graph opens downward.

[YOU TRY 2] Graph $y = -(x + 3)^2$.

Note

Every parabola has an **axis of symmetry.** In Example 1, the y-axis is the axis of symmetry, and in Example 2, the axis of symmetry is the vertical line $x = 2$. If you were to fold each graph along its axis of symmetry, one half of the graph would lie on top of the other. Each side is a mirror image of the other.

2 Graph a Quadratic Equation of the Form $y = ax^2 + bx + c$

How do we graph a more general parabola, one in the form $y = ax^2 + bx + c$? First, we must find the vertex. An important property of a parabola is that **if a parabola has two x-intercepts, then the x-coordinate of the vertex is always halfway between the x-coordinates of the x-intercepts of the parabola.** This will always be true and will help us derive a formula to find the vertex of a parabola with an equation like $y = -x^2 + 4x - 3$.

Recall from previous chapters that to find the x-intercepts of the graph of an equation, we let $y = 0$ and solve for x. Therefore, to find the x-intercepts of the graph of $y = ax^2 + bx + c$, let $y = 0$ and solve for x.

$$0 = ax^2 + bx + c$$

We can use the quadratic formula to solve for x, giving us the following x-coordinates of the x-intercepts:

$$x = \frac{-b - \sqrt{b^2 - 4ac}}{2a} \quad \text{and} \quad x = \frac{-b + \sqrt{b^2 - 4ac}}{2a}$$

Since the x-coordinate of the vertex of a parabola is halfway between the x-intercepts, we find the *average* of the x-coordinates of the x-intercepts to find the x-coordinate of the vertex.

$$\text{x-coordinate of the vertex} = \frac{1}{2}\left(\frac{-b - \sqrt{b^2 - 4ac}}{2a} + \frac{-b + \sqrt{b^2 - 4ac}}{2a}\right)$$

$$= \frac{1}{2}\left(\frac{-b - \sqrt{b^2 - 4ac} - b + \sqrt{b^2 - 4ac}}{2a}\right)$$

$$= \frac{1}{2}\left(\frac{-2b}{2a}\right)$$

$$= -\frac{b}{2a}$$

Procedure Finding the Vertex of a Parabola of the Form
$$y = ax^2 + bx + c \ (a \neq 0)$$

The **x-coordinate of the vertex of a parabola** written in the form $y = ax^2 + bx + c$ $(a \neq 0)$ is $x = -\dfrac{b}{2a}$. To find the y-coordinate of the vertex, substitute the x-value into the equation and solve for y.

When graphing a parabola, the first point we should locate is the vertex.

W Hint

Is there any more detail you would like to add to this procedure?

Procedure Graphing a Parabola of the Form $y = ax^2 + bx + c$

Step 1: Find the vertex. The x-coordinate of the vertex is $-\dfrac{b}{2a}$. Find the y-coordinate of the vertex by substituting the x-value into the equation.

Step 2: Find the y-intercept by substituting 0 for x and solving for y.

Step 3: Find the x-intercepts, if they exist, by substituting 0 for y and solving for x.

Step 4: Find additional points on the parabola using a table of values.

Step 5: Plot the points, and sketch the graph.

EXAMPLE 3

Graph $y = -x^2 + 4x - 3$.

In-Class Example 3

Graph $y = -x^2 - 2x + 3$.

Answer:

Vertex $(-1, 4)$ $y = -x^2 - 2x + 3$

Solution

Step 1: Find the vertex. First, find the x-coordinate of the vertex:

$$x = -\frac{b}{2a} = -\frac{4}{2(-1)} = \frac{4}{2} = 2 \qquad a = -1 \quad b = 4$$

Next, substitute $x = 2$ into the equation to find the y-coordinate of the vertex.

$$y = -x^2 + 4x - 3$$
$$y = -(2)^2 + 4(2) - 3 \qquad \text{Substitute 2 for } x.$$
$$y = -4 + 8 - 3 = 1$$

The vertex of the parabola is $(2, 1)$.

Step 2: Find the y-intercept by substituting 0 for x and solving for y:

$$y = -x^2 + 4x - 3$$
$$y = -(0)^2 + 4(0) - 3 = -3$$

The y-intercept is $(0, -3)$.

Step 3: Find the x-intercepts by substituting 0 for y and solving for x:

$$0 = -x^2 + 4x - 3 \qquad \text{Substitute 0 for } y.$$
$$0 = x^2 - 4x + 3 \qquad \text{Divide both sides by } -1.$$
$$0 = (x - 1)(x - 3) \qquad \text{Factor.}$$

$$x = 1 \qquad \text{or} \qquad x = 3 \qquad \text{Solve.}$$

The x-intercepts are $(1, 0)$ and $(3, 0)$.

Steps 4 and 5: Find additional points, plot the points, and sketch the graph.

	x	y
Vertex →	2	1
y-int. →	0	−3
x-int. →	1	0
x-int. →	3	0
	4	−3

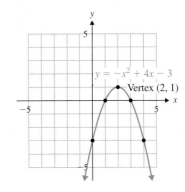

$\left[\text{ YOU TRY 3 }\right]$ Graph $y = -x^2 - 6x - 8$.

Sometimes we need to use the quadratic formula to find the x-intercepts of the graph.

EXAMPLE 4

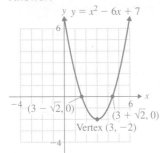

Graph $y = x^2 + 2x - 2$.

Solution

Step 1: Find the vertex. $x = -\dfrac{b}{2a} = -\dfrac{2}{2(1)} = -\dfrac{2}{2} = -1$ $a = 1 \quad b = 2$

Substitute $x = -1$ into the equation to find the y-coordinate of the vertex.

$$y = (-1)^2 + 2(-1) - 2 = 1 - 2 - 2 = -3$$

The vertex of the parabola is $(-1, -3)$.

Step 2: Find the y-intercept by substituting 0 for x and solving for y:

$$y = (0)^2 + 2(0) - 2 = -2$$

The y-intercept is $(0, -2)$.

Step 3: Find the x-intercepts by substituting 0 for y and solving for x:

$$0 = x^2 + 2x - 2$$

We will solve this equation using the quadratic formula.

$$x = \frac{-(2) \pm \sqrt{(2)^2 - 4(1)(-2)}}{2(1)} \qquad \text{Let } a = 1, b = 2, c = -2.$$

$$x = \frac{-2 \pm \sqrt{12}}{2} = \frac{-2 \pm 2\sqrt{3}}{2} = \frac{2(-1 \pm \sqrt{3})}{2} = -1 \pm \sqrt{3}$$

The x-intercepts are $(-1 - \sqrt{3}, 0)$ and $(-1 + \sqrt{3}, 0)$.

Steps 4 and 5: **Find additional points, plot the points, and sketch the graph.**

To plot the *x*-intercepts, we can either approximate them by hand or use a calculator.

	x	y
Vertex →	−1	−3
y-int. →	0	−2
x-int. →	$−1 − \sqrt{3} \approx −2.7$	0
x-int. →	$−1 + \sqrt{3} \approx 0.7$	0
	−2	−2

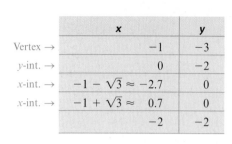

$$y = x^2 + 2x - 2 \quad \text{Vertex } (-1, -3)$$

[YOU TRY 4]

Graph $y = x^2 - 8x + 14$.

Notice in Example 3, $y = -x^2 + 4x - 3$, the value of *a* is negative ($a = -1$) and the parabola opens downward. In Example 4, the graph of $y = x^2 + 2x - 2$ opens upward and *a* is positive ($a = 1$). This is another characteristic of the graph of $y = ax^2 + bx + c$. If *a* is positive, the graph opens upward. If *a* is negative, the graph opens downward.

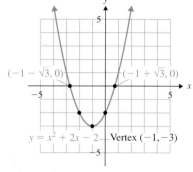

Using Technology

In Section 7.5, we said that the solutions of the equation $x^2 - x - 6 = 0$ are the *x*-intercepts of the graph of $y = x^2 - x - 6$. The *x*-intercepts are also called the zeros of the equation since they are the values of *x* that make $y = 0$. We can find the *x*-intercepts shown on the graphs by pressing 2nd TRACE and then selecting 2: zero. Move the cursor to the left of an *x*-intercept using the right arrow key and press ENTER. Move the cursor to the right of the *x*-intercept using the right arrow key and press ENTER. Move the cursor close to the *x*-intercept using the left arrow key and press ENTER. The result finds the *x*-intercepts $(-2, 0)$ and $(3, 0)$ as shown on the graphs below.

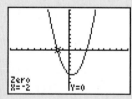 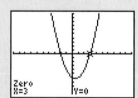

The *y*-intercept is found by graphing the equation and pressing TRACE 0 ENTER. As shown on the graph at the left below, the *y*-intercept for $y = x^2 - x - 6$ is $(0, -6)$.

The *x*-value of the vertex can be found using the vertex formula. In this case, $a = 1$ and $b = -1$, so $-\dfrac{b}{2a} = \dfrac{1}{2}$. To find the vertex on the graph, press TRACE, type 1/2, and press ENTER. The vertex is shown as $(0.5, -6.25)$ on the graph in the center below.

To convert the coordinates of the vertex to fractions, go to the home screen and select Frac from the Math menu. The vertex is then $\left(\dfrac{1}{2}, -\dfrac{25}{4}\right)$ as shown on the screen at the right below.

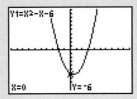

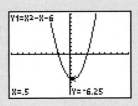

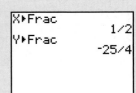

Find the *x*-intercepts, *y*-intercept, and vertex using a graphing calculator.

1) $y = x^2 + 2x - 3$ 2) $y = x^2 - 4x + 5$ 3) $y = (x - 1)^2 - 9$

4) $y = x^2 - 9$ 5) $y = x^2 - 4x + 4$ 6) $y = -x^2 + 6x - 10$

ANSWERS TO [YOU TRY] EXERCISES

1)

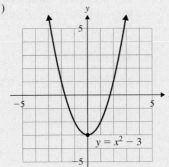

$y = x^2 - 3$

2)

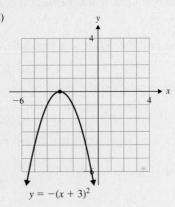

$y = -(x + 3)^2$

3)

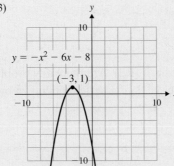

$y = -x^2 - 6x - 8$

$(-3, 1)$

4)

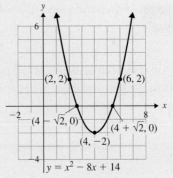

$(2, 2)$ $(6, 2)$

$(4 - \sqrt{2}, 0)$ $(4 + \sqrt{2}, 0)$

$(4, -2)$

$y = x^2 - 8x + 14$

ANSWERS TO TECHNOLOGY EXERCISES

1) *x*-intercepts $(1, 0)$, $(-3, 0)$; *y*-intercept $(0, -3)$; vertex $(-1, -4)$
2) no *x*-intercepts; *y*-intercept $(0, 5)$; vertex $(2, 1)$
3) *x*-intercepts $(-2, 0)$, $(4, 0)$; *y*-intercept $(0, -8)$; vertex $(1, -9)$
4) *x*-intercepts $(-3, 0)$, $(3, 0)$; *y*-intercept $(0, -9)$; vertex $(0, -9)$
5) *x*-intercept $(2, 0)$; *y*-intercept $(0, 4)$; vertex $(2, 0)$
6) no *x*-intercepts; *y*-intercept $(0, -10)$; vertex $(3, -1)$

*Additional answers can be found in the Answers to Exercises appendix.

Objective 1: Graph Basic Parabolas

1) What is the vertex of a parabola? It is the lowest point of a graph that opens upward or the highest point of a graph that opens downward.

2) What is the axis of symmetry of a parabola?
It is the line that cuts the parabola in half so that one half of the parabola is the mirror image of the other half.

Graph.

3) $y = x^2 + 1$

4) $y = x^2 + 2$

5) $y = -x^2$

6) $y = (x + 2)^2$

7) $y = -(x + 1)^2$

8) $y = -(x - 1)^2$

9) $y = x^2 - 5$

10) $y = -x^2 + 4$

11) $y = -(x - 3)^2$

12) $y = \frac{1}{2}x^2$

Objective 2: Graph a Quadratic Equation of the Form $y = ax^2 + bx + c$

13) Explain how to find the vertex of a parabola with equation $y = ax^2 + bx + c$.

14) How do you know whether the graph of $y = ax^2 + bx + c$ opens upward or downward?
If $a > 0$, the parabola opens upward. If $a < 0$, the parabola opens downward.

Sketch the graph of each equation. Identify the vertex.

15) $y = x^2 - 4$ $(0, -4)$

16) $y = -x^2 + 1$ $(0, 1)$

17) $y = -x^2 - 1$ $(0, -1)$

18) $y = x^2 - 5$ $(0, -5)$

19) $y = (x - 3)^2$ $(3, 0)$

20) $y = (x + 2)^2$ $(-2, 0)$

21) $y = x^2 - 4x + 3$ $(2, -1)$

22) $y = -x^2 + 6x - 5$ $(3, 4)$

23) $y = -x^2 + 2x + 3$ $(1, 4)$

24) $y = x^2 + 2x - 8$ $(-1, -9)$

25) $y = -x^2 - 2x + 4$ $(-1, 5)$

26) $y = x^2 + 4x + 1$ $(-2, -3)$

27) $y = x^2 + 6x + 11$ $(-3, 2)$

28) $y = -x^2 + 8x - 17$ $(4, -1)$

29) $y = -x^2 + 2x - 1$ $(1, 0)$

30) $y = x^2 + 6x + 9$ $(-3, 0)$

31) $y = 2x^2 - 4x$ $(1, -2)$

32) $y = -2x^2 - 8x - 6$ $(-2, 2)$

The solutions of $ax^2 + bx + c = 0$ are the x-intercepts of the corresponding quadratic equation $y = ax^2 + bx + c$. Use a graphing calculator to solve each equation in Exercises 33–38 by finding the x-intercepts of the graph of the corresponding equation, $y = ax^2 + bx + c$. Round your answer to the nearest hundredth where appropriate.

33) $5x^2 - 17x + 6 = 0$
$\{0.4, 3\}$

34) $4x^2 - 5x - 6 = 0$
$\{-0.75, 2\}$

35) $8x^2 + 38x + 9 = 0$
$\{-4.5, -0.25\}$

36) $20x^2 - 64x + 35 = 0$
$\{0.7, 2.5\}$

37) $x^2 + 2x - 6 = 0$
$\{-3.65, 1.65\}$

38) $x^2 + 10x + 22 = 0$
$\{-6.73, -3.27\}$

R Rethink

R1) What do you think the equation of a parabola that opens to the left would be? What about one that opens to the right?

R2) Are there any problems you could not do? If so, write them down or circle them and ask your instructor.

10.5 Introduction to Functions

P Prepare

What are your objectives for Section 10.5?

O Organize

How can you accomplish each objective?

What are your objectives for Section 10.5?	How can you accomplish each objective?
1 Define and Identify Relation, Function, Domain, and Range	• Learn the definitions of *relation, domain, range,* and *function,* and write them in your own words. • Complete the given example on your own. • Complete You Try 1.
2 Use the Vertical Line Test	• Understand the **Vertical Line Test,** and learn the procedure to use it effectively. • Complete the given example on your own.
3 Find the Domain and Range of a Relation from Its Graph	• Use the definitions from Objective 1 to help you understand Example 3. • Complete You Try 2.
4 Find the Domain of a Function Using Its Equation	• Follow the example to come up with your own procedure for **Finding the Domain of a Function,** and compare it to the procedure given. • Complete You Try 3.
5 Use Function Notation	• Write the definition of *function notation* in your own words. • Complete the given examples on your own. • Complete You Trys 4 and 5.
6 Solve Problems Using Functions	• Use meaningful letters to represent applied functions. • Complete the given example on your own.

 Work   **Read the explanations, follow the examples, take notes, and complete the You Trys.**

If you have a job and you earn $9.50 per hour, the amount of money you earn each week before deductions (gross earnings) depends on the number of hours you have worked.

Hours Worked	Gross Earnings
10 hours	$ 95.00
15 hours	$142.50
22 hours	$209.00
30 hours	$285.00

We can express these relationships with the ordered pairs

(10, 95.00)   (15, 142.50)   (22, 209.00)   (30, 285.00)

where the first coordinate represents the amount of time worked (in hours), and the second coordinate represents the gross earnings (in dollars).

We can also describe this relationship with the equation

$$y = 9.50x$$

where y is the gross earnings, in dollars, and x is the number of hours worked.

1 Define and Identify Relation, Function, Domain, and Range

If we form a set of ordered pairs from the ones listed on the previous page, we get a *relation:*

$$\{(10, 95.00), (15, 142.50), (22, 209.00), (30, 285.00)\}$$

Definition

A **relation** is any set of ordered pairs.

Definition

The **domain** of a relation is the set of all values of the first coordinates in the set of ordered pairs. The **range** of a relation is the set of all values of the second coordinates in the set of ordered pairs.

The domain of the given relation is $\{10, 15, 22, 30\}$. The range of the relation is $\{95.00, 142.50, 209.00, 285.00\}$.

The relation $\{(10, 95.00), (15, 142.50), (22, 209.00), (30, 285.00)\}$ is also a *function* because every first coordinate corresponds to *exactly one* second coordinate. A function is a very important concept in mathematics.

Definition

A **function** is a special type of relation. If each element of the domain corresponds to *exactly one* element of the range, then the relation is a function.

Relations and functions can be represented in another way—as a *correspondence* or a *mapping* from one set, the domain, to another, the range. In this representation, the domain is the set of all values in the first set, and the range is the set of all values in the second set.

EXAMPLE 1

In-Class Example 1

Identify the domain and range of each relation, and determine whether each relation is a function.
a) $\{(-4, -1), (-2, 1), (1, 5), (-2, 8), (6, 11)\}$
b) $\{(-9, -3), (-3, -1), (0, 0), (6, 2)\}$
c)

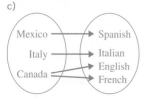

Identify the domain and range of each relation, and determine whether each relation is a function.

a) $\{(-3, -2), (4, 0), (5, 3), (5, -3)\}$

b) $\left\{(-4, -1), (-2, 0), (0, 1), \left(3, \dfrac{5}{2}\right), (4, 3)\right\}$

c)

Hint

Come up with your own
examples of a relation
and a function—they don't
have to have numbers!

Solution

a)   The *domain* is the set of first coordinates, {−3, 4, 5}. (We write the 5 in the set only once even though it appears in two ordered pairs.) The *range* is the set of second coordinates, {−2, 0, 3, −3}.

To determine whether or not this relation is a function, ask yourself, *"Does every first coordinate correspond to exactly one second coordinate?"* No. In the ordered pairs (5, 3) and (5, −3), the same first coordinate, 5, corresponds to two different second coordinates, 3 and −3. Therefore, this relation is *not* a function.

b)   The *domain* is {−4, −2, 0, 3, 4}. The *range* is $\left\{-1, 0, 1, \frac{5}{2}, 3\right\}$.

Ask yourself, "Does every first coordinate correspond to *exactly one* second coordinate?" *Yes.* This relation *is* a function.

c)   The *domain* is {Chicago, New York, San Diego}. The *range* is {Bears, Jets, Giants, Chargers}.

One of the elements in the domain, New York, corresponds to *two* elements in the range, Jets and Giants. Therefore, this relation is *not* a function.

[YOU TRY 1]   Identify the domain and range of each relation, and determine whether each relation is a function.

a)   {(−5, −3), (−4, 0), (2, 7), (8, 8), (10, 15)}

b)   {(−2, −13), (−2, 13), (0, −3), (1, −2)}

c)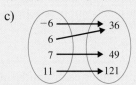

---

2   Use the Vertical Line Test

We know that a relation is a function if each element of the domain corresponds to *exactly one* element of the range.

If the ordered pairs of a relation are such that the first coordinates represent *x*-values and the second coordinates represent *y*-values (the ordered pairs are in the form (*x*, *y*)), then we can think of the definition of a function in this way:

Definition

A relation is a **function** if each *x*-value corresponds to exactly one *y*-value.

What does a function look like when it is graphed? Let's look at the graphs of the ordered pairs in the relations of Example 1a) and 1b).

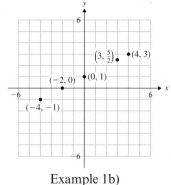

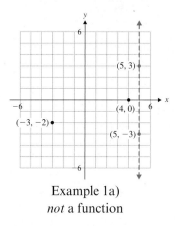

Example 1a)
not a function

Example 1b)
is a function

The relation in Example 1a) is *not* a function since the *x*-value of 5 corresponds to *two different y*-values, 3 and −3. Notice that we can draw a vertical line that intersects the graph in more than one point—the line through (5, 3) and (5, −3).

The relation in Example 1b), however, *is* a function—each *x*-value corresponds to only one *y*-value. Here we cannot draw a vertical line through more than one point on this graph.

This leads us to the **vertical line test** for a function.

Procedure The Vertical Line Test

If there is no vertical line that can be drawn through a graph so that it intersects the graph more than once, then the graph represents a function.

If a vertical line *can* be drawn through a graph so that it intersects the graph more than once, then the graph does *not* represent a function.

EXAMPLE 2

In-Class Example 2

Use the vertical line test to determine whether each graph represents a function.

a)

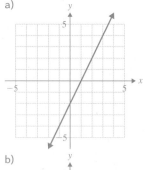

b)

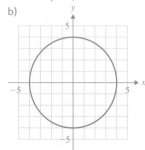

Answer: a) function
b) not a function

Use the vertical line test to determine whether each graph, in blue, represents a function.

a)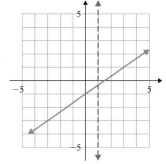

b)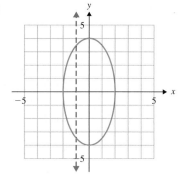

Solution

a) Anywhere a vertical line is drawn through the graph, the line will intersect the graph only once. *This graph represents a function.*

b) This graph fails the vertical line test because we can draw a vertical line through the graph that intersects it more than once. *This graph does* not *represent a function.*

We can identify the domain and range of a relation or function from its graph.

3 Find the Domain and Range of a Relation from Its Graph

In-Class Example 3

Identify the domain and range of each relation in In-Class Example 2.

Answer: a) domain: $(-\infty, \infty)$; range: $(-\infty, \infty)$
b) domain: $[-4, 4]$; range: $[-4, 4]$

Identify the domain and range of each relation in Example 2.

Solution

a)

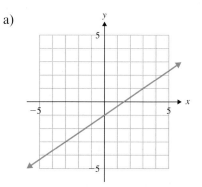

The arrows on the graph indicate that the graph continues without bound.

The domain of this function is the set of x-values on the graph. Since the graph continues indefinitely in the x-direction, the domain is the set of all real numbers. *The domain is* $(-\infty, \infty)$.

The range is the set of y-values on the graph. Since the arrows show that the graph continues indefinitely in the y-direction, *the range is* the set of all real numbers or $(-\infty, \infty)$.

b)

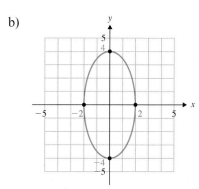

The set of x-values on the graph includes all real numbers from -2 to 2. *The domain is* $[-2, 2]$.

The set of y-values on the graph includes all real numbers from -4 to 4. *The range is* $[-4, 4]$.

Part (a) in Example 3 suggests the following point:

 Hint

Think about the graph of $y = b$, where b is a real number. Is $y = b$ a function?

Note

A linear equation of the form $y = mx + b$ is a function.

Equations of the form $x = a$ are vertical lines with undefined slope and are not functions.

[YOU TRY 2]

Use the vertical line test to determine whether each relation is also a function. Then, identify the domain and range.

a)

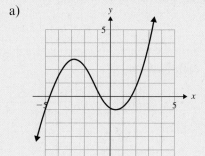

b)

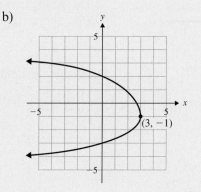

(3, −1)

4 Find the Domain of a Function Using Its Equation

We have seen how to determine the domain of a relation written as a set of ordered pairs, as a correspondence (or mapping), and as a graph. Next, we will discuss how to determine the domain of a relation written as an equation.

Sometimes, it is helpful to ask yourself, "Is there any number that *cannot* be substituted for x?"

EXAMPLE 4

In-Class Example 4

Determine the domain of each function.
a) $y = 6x - 9$ b) $y = x^3$
c) $y = \dfrac{1}{x}$

Answer: a) $(-\infty, \infty)$
b) $(-\infty, \infty)$
c) $(-\infty, 0) \cup (0, \infty)$

Determine the domain of each function.

a) $y = 3x - 1$ b) $y = x^2$ c) $y = \dfrac{1}{x}$

Solution

a) To determine the domain of $y = 3x - 1$, ask yourself, "Is there any number that *cannot* be substituted for x?" *No.* Any real number can be substituted for x, and $y = 3x - 1$ will be defined. The domain consists of all real numbers and can be written as $(-\infty, \infty)$.

b) Ask yourself, "Is there any number that *cannot* be substituted for x in $y = x^2$?" *No.* Any real number can be substituted for x. The domain is all real numbers, $(-\infty, \infty)$.

c) To determine the domain of $y = \dfrac{1}{x}$, ask yourself, "Is there any number that *cannot* be substituted for x?" *Yes. x cannot equal zero because a fraction is undefined if its denominator equals zero.* Any other number can be substituted for x, and $y = \dfrac{1}{x}$ will be defined.

The domain consists of all real numbers *except* 0. We can write the domain in interval notation as $(-\infty, 0) \cup (0, \infty)$.

Procedure Finding the Domain of a Function

To find the domain of a function written as an equation in terms of x:

1) Ask yourself, "Is there any number that *cannot* be substituted for x?"
2) If x is in the denominator of a fraction, determine what value of x will make the denominator equal 0 by setting the expression equal to zero. Solve for x. This x-value is *not* in the domain.

The domain consists of all real numbers that can be substituted for x.

[**YOU TRY 3**] Determine the domain of each function.

a) $y = x + 8$ b) $y = x^2 + 3$ c) $y = \dfrac{5}{x - 2}$

5 Use Function Notation

We can use *function notation* to name functions. If a relation is a function, then $f(x)$ can be used in place of y. $f(x)$ *is the same as* y.

For example, $y = x + 2$ is a function. We can also write $y = x + 2$ as $f(x) = x + 2$. *They mean the same thing.*

Definition

If y represents a function, then the **function notation** $f(x)$ is the same as y. We read $f(x)$ as "f of x," and we can say $y = f(x)$.

EXAMPLE 5

In-Class Example 5

a) Evaluate $y = -2x + 5$ for $x = 4$.   b) If $f(x) = -2x + 5$, find $f(4)$.

Answer: a) -3 b) -3

a) Evaluate $y = x + 2$ for $x = 5$.

b) If $f(x) = x + 2$, find $f(5)$.

Solution

a) To evaluate $y = x + 2$ for $x = 5$ means to substitute 5 for x and find the corresponding value of y.

$$y = x + 2$$
$$y = 5 + 2 \qquad \text{Substitute 5 for } x.$$
$$y = 7$$

When $x = 5$, $y = 7$. We can also say that the ordered pair $(5, 7)$ satisfies $y = x + 2$.

b) To find $f(5)$ (read as "f of 5") means to find the value of the function when $x = 5$.

$$f(x) = x + 2$$
$$f(5) = 5 + 2 \qquad \text{Substitute 5 for } x.$$
$$f(5) = 7$$

We can also say that the ordered pair $(5, 7)$ satisfies $f(x) = x + 2$ where the ordered pair represents $(x, f(x))$.

Note

Example 5 illustrates that evaluating $y = x + 2$ for $x = 5$ and finding $f(5)$ when $f(x) = x + 2$ are *exactly* the same thing. Remember, $f(x)$ is another name for y.

YOU TRY 4

a) Evaluate $y = 3x - 7$ for $x = -2$.   b) If $f(x) = 3x - 7$, find $f(-2)$.

Different letters can be used to name functions. $g(x)$ is read as "g of x," $h(x)$ is read as "h of x," and so on. Also, the function notation does *not* indicate multiplication; $f(x)$ does *not* mean f times x.

We can also think of a function as a machine—we put values into the machine, and the equation determines the values that come out. We can think of the function in Example 5b) like this:

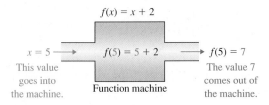

$f(x) = x + 2$

$x = 5$ ⟶
This value goes into the machine.

$f(5) = 5 + 2$
Function machine

$f(5) = 7$
The value 7 comes out of the machine.

Sometimes, we call evaluating a function for a certain value *finding a function value*.

EXAMPLE 6

In-Class Example 6

Let $f(x) = 4x - 9$ and $g(x) = x^2 - 5x - 4$. Find the following function values.
a) $f(8)$ b) $g(-3)$

Answer: a) 23 b) 20

Let $f(x) = 5x - 8$ and $g(x) = x^2 - 3x + 1$. Find the following function values.

a) $f(4)$ b) $g(-2)$

Solution

a) "Find $f(4)$" means to find the value of the function when $x = 4$. Substitute 4 for x.

$$f(x) = 5x - 8$$
$$f(4) = 5(4) - 8 = 20 - 8 = 12$$
$$f(4) = 12$$

We can also say that the ordered pair (4, 12) satisfies $f(x) = 5x - 8$.

b) To find $g(-2)$, substitute -2 for every x in the function $g(x)$.

$$g(x) = x^2 - 3x + 1$$
$$g(-2) = (-2)^2 - 3(-2) + 1 = 4 + 6 + 1 = 11$$
$$g(-2) = 11$$

The ordered pair $(-2, 11)$ satisfies $g(x) = x^2 - 3x + 1$.

W Hint

Does this process "feel" any different from what you have already been doing?

[YOU TRY 5] Let $f(x) = -6x + 5$ and $h(x) = x^2 + 9x - 7$. Find the following function values.

a) $f(1)$ b) $f(-3)$ c) $h(-2)$ d) $h(0)$

6 Solve Problems Using Functions

The domain of a function does not have to be represented by x. When using functions to solve problems, we often choose a more "meaningful" letter to represent a quantity. The same is true for naming the function.

A compact disk is read at 44.1 kHz (kilohertz). This means that a CD player scans 44,100 samples of sound per second on a CD to produce the sound that we hear. The function

$$S(t) = 44,100t$$

tells us how many samples of sound, $S(t)$, are read after t seconds. (www.mediatechnics.com)

a)   How many samples of sound are read in 20 sec?

b)   How many samples of sound are read in 1.5 min?

c)   How long would it take the CD player to scan 1,764,000 samples of sound?

d)   What is the smallest value t could equal?

Solution

a)   To determine how much sound is read in 20 sec, let $t = 20$ and find $S(20)$.

$$S(t) = 44,100t$$
$$S(20) = 44,100(20) \qquad \text{Substitute 20 for } t.$$
$$S(20) = 882,000 \qquad \text{Multiply.}$$

The CD player has read 882,000 samples of sound.

b)   To determine how much sound is read in 1.5 min, do we let $t = 1.5$ and find $S(1.5)$? *No.* Recall that t is in *seconds*. Change 1.5 min to seconds before substituting for t. We must use the correct units in the function.

$$1.5 \text{ min} = 90 \text{ sec}$$

Let $t = 90$ and find $S(90)$.

$$S(t) = 44,100t$$
$$S(90) = 44,100(90)$$
$$S(90) = 3,969,000$$

It has read 3,969,000 samples of sound.

c)   Since we are asked to determine *how long* it would take a CD player to scan 1,764,000 samples of sound, we will be solving for t. What do we substitute for $S(t)$? We substitute 1,764,000 for $S(t)$ and find t. That is, find t when $S(t) = 1,764,000$.

$$S(t) = 44,100t$$
$$1,764,000 = 44,100t \qquad \text{Substitute 1,764,000 for } S(t).$$
$$40 = t \qquad \text{Divide by 44,100.}$$

It will take 40 sec for the CD player to scan 1,764,000 samples of sound.

d)   Since t represents the number of seconds a CD has been playing, the smallest value that makes sense for t is 0.

A graphing calculator can be used to represent a function as a graph and also as a table of values. Consider the function $f(x) = 2x - 5$. To graph the function, press $\boxed{Y=}$, then type $2x - 5$ to the right of \Y1=. Press $\boxed{ZOOM}$ and select 6:ZStandard to graph the equation as shown on the left below. We can select a point on the graph. For example, press $\boxed{TRACE}$, type 4, and press $\boxed{ENTER}$. The point (4, 3) is displayed on the screen as shown below on the right.

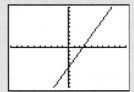 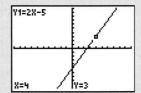

The function can also be represented as a table on a graphing calculator. To set up the table, press $\boxed{2nd}$ $\boxed{WINDOW}$, move the cursor after TblStart =, and enter a number such as 0 to set the starting x-value for the table. Enter 1 after ΔTbl = to set the increment between x-values as shown on the left below. Then press $\boxed{2nd}$ $\boxed{GRAPH}$ to display the table as shown on the right below.

 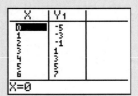

The point (4, 3) is represented in the table above as well as on the graph.

Given the function, find the function value on a graph and a table using a graphing calculator.

1) $f(x) = 3x - 4$; $f(2)$ 2) $f(x) = 4x - 1$; $f(1)$ 3) $f(x) = -3x + 7$; $f(1)$

4) $f(x) = 2x + 5$; $f(-1)$ 5) $f(x) = 2x - 7$; $f(-1)$ 6) $f(x) = -x + 5$; $f(4)$

ANSWERS TO $\boxed{\text{YOU TRY}}$ EXERCISES

1) a) domain: $\{-5, -4, 2, 8, 10\}$; range: $\{-3, 0, 7, 8, 15\}$; function   b) domain: $\{-2, 0, 1\}$; range: $\{-13, 13, -3, -2\}$; not a function   c) domain: $\{-6, 6, 7, 11\}$; range: $\{36, 49, 121\}$; function
2) a) function; domain: $(-\infty, \infty)$; range: $(-\infty, \infty)$ b) not a function; domain: $(-\infty, 3]$; range: $(-\infty, \infty)$ 3) a) $(-\infty, \infty)$ b) $(-\infty, \infty)$ c) $(-\infty, 2) \cup (2, \infty)$
4) a) $y = -13$ b) $f(-2) = -13$ 5) a) $f(1) = -1$ b) $f(-3) = 23$
c) $h(-2) = -21$ d) $h(0) = -7$

ANSWERS TO TECHNOLOGY EXERCISES

1) 2   2) 3   3) 4   4) 3   5) -9 6) 1

---

*Additional answers can be found in the Answers to Exercises appendix.

Mixed Exercises: Objectives 1–3

1) a) What is a relation?   any set of ordered pairs

   b) What is a function?   a relation in which each element of the domain corresponds to exactly one element of the range

   c) Give an example of a relation that is also a function.   Answers may vary.

2) Give an example of a relation that is *not* a function.
   Answers may vary.

Identify the domain and range of each relation, and determine whether each relation is a function.

 3) {(4, 10), (−4, 2), (1, −4), (8, −3)}
   domain: {−4, 1, 4, 8}; range: {−4, −3, 2, 10}; function

4) {(1, 3), (3, 4), (3, 2), (4, 5), (9, 1)}
   domain: {1, 3, 4, 9}; range: {1, 2, 3, 4, 5}; not a function

 5) {(9, −1), (25, −3), (1, 1), (9, 5), (25, 7)}
   domain: {1, 9, 25}; range: {−3, −1, 1, 5, 7}; not a function

6) {(−5, 1), (3, 3), (−2, 4), (6, −7), (5, −1)}
   domain: {−5, −2, 3, 5, 6}; range: {−7, −1, 1, 3, 4}; function

7) $\left\{(-4, -2), \left(-3, -\frac{1}{2}\right), \left(-1, -\frac{1}{2}\right), (0, -2)\right\}$
   domain: {−4, −3, −1, 0}; range: $\left\{-2, -\frac{1}{2}\right\}$; function

8) {(−4, 1), (4, 7), (2, 7), (−2, 8)}
   domain: {−4, −2, 2, 4}; range: {1, 7, 8}; function

9) {(−2.3, 6.2), (3.0, 7.8), (3.0, 3.1), (−4.1, −5.7)}
   domain: {−4.1, −2.3, 3.0}; range: {−5.7, 3.1, 6.2, 7.8}; not a function

10) {(5, 2), (6, 4), (1, 3), (0, 3)}
    domain: {0, 1, 5, 6}; range: {2, 3, 4}; function

11)

    domain: {Hawaii, New York, Miami};
    range: {State, City}; not a function

12)

    domain: {1, 2, 3, 4};
    range: {5, 10, 15, 20};
    function

 13)
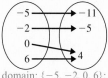
    domain: {−5, −2, 0, 6};
    range: {−11, −5, 4}; function

14)

    domain: {1, 16, 81};
    range: {1, 2, 3, −3}; not a function

15)
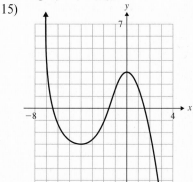
    domain: (−∞, ∞);
    range: (−∞, ∞);
    function

16)
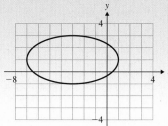
    domain: [−7, 1];
    range: [−1, 3];
    not a function

17)
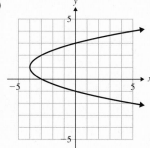
    domain: (−4, ∞);
    range: (−∞, ∞);
    not a function

18)
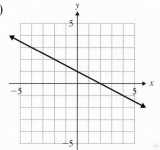
    domain: (−∞, ∞);
    range: (−∞, ∞);
    function

19)
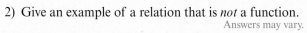
    domain: (−∞, ∞);
    range: (−∞, 6];
    function

20)
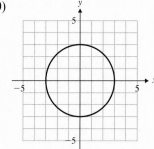
    domain: [−3, 3];
    range: [−3, 3];
    not a function

Objective 4: Find the Domain of a Function Using Its Equation

Determine the domain of each function.

21) $y = 3x + 1$ $(-\infty, \infty)$ 22) $y = -7x - 2$ $(-\infty, \infty)$

23) $y = -\dfrac{2}{5}x - 8$ $(-\infty, \infty)$ 24) $y = \dfrac{1}{4}x + 9$ $(-\infty, \infty)$

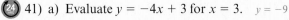

25) $y = x^2 - 4$ $(-\infty, \infty)$ 26) $y = -x^2 + 2$ $(-\infty, \infty)$

27) $y = x^3$ $(-\infty, \infty)$ 28) $y = 2x^4$ $(-\infty, \infty)$

29) $y = \dfrac{2}{x}$ $(-\infty, 0) \cup (0, \infty)$ 30) $y = \dfrac{9}{x}$ $(-\infty, 0) \cup (0, \infty)$

31) $y = \dfrac{12}{x - 5}$
$(-\infty, 5) \cup (5, \infty)$

32) $y = \dfrac{4}{x + 9}$
$(-\infty, -9) \cup (-9, \infty)$

33) $y = \dfrac{1}{x + 1}$
$(-\infty, -1) \cup (-1, \infty)$

34) $y = \dfrac{2}{x - 3}$
$(-\infty, 3) \cup (3, \infty)$

35) $y = \dfrac{6}{x - 20}$
$(-\infty, 20) \cup (20, \infty)$

36) $y = \dfrac{1}{x + 11}$
$(-\infty, -11) \cup (-11, \infty)$

Objective 5: Use Function Notation

37) What is the meaning of the notation $y = f(x)$?
y is a function, and y is a function of x.

38) Does $y = f(x)$ mean "$y = f$ times x"? Explain.
No; $f(x)$ is read as "f of x" and $y = f(x)$ means that y is a function of x.

39) a) Evaluate $y = 2x - 1$ for $x = 6$. $y = 11$

b) If $f(x) = 2x - 1$, find $f(6)$. $f(6) = 11$

40) a) Evaluate $y = -x - 5$ for $x = -3$. $y = -2$

b) If $f(x) = -x - 5$, find $f(-3)$. $f(-3) = -2$

41) a) Evaluate $y = -4x + 3$ for $x = 3$. $y = -9$

b) If $f(x) = -4x + 3$, find $f(3)$. $f(3) = -9$

42) a) Evaluate $y = -5x - 6$ for $x = -1$. $y = -1$

b) If $f(x) = -5x - 6$, find $f(-1)$. $f(-1) = -1$

For Exercises 43–52, let $f(x) = 2x - 11$ and $g(x) = x^2 - 4x + 2$, and find the following function values.

43) $f(4)$ -3 44) $f(1)$ -9 45) $f(0)$ -11

46) $f\left(\dfrac{11}{2}\right)$ 0 47) $g(2)$ -2 48) $g(-1)$ 7

49) $g(1)$ -1 50) $g(0)$ 2 51) $g\left(\dfrac{1}{2}\right)$ $\dfrac{1}{4}$

52) $g\left(-\dfrac{1}{3}\right)$ $\dfrac{31}{9}$

For each function f in Exercises 53–60, find $f(-2)$ and $f(5)$.

53) $f = \{(-1, 18), (-2, 12), (0, 4), (5, 4), (4, -3)\}$
$f(-2) = 12; f(5) = 4$

54) $f = \{(5, 10), (0, 7), (6, 4), (4, -5), (-2, -4)\}$
$f(-2) = -4; f(5) = 10$

55) $f = \left\{\left(-\dfrac{3}{4}, -1\right), (-2, 4), (5, 2), (8, 10)\right\}$
$f(-2) = 4; f(5) = 2$

56) $f = \left\{(-8, -1), \left(-2, \dfrac{3}{4}\right), (10, 8), (5, -1)\right\}$
$f(-2) = \dfrac{3}{4}; f(5) = -1$

57)
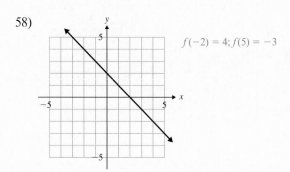
$f(-2) = -2; f(5) = 2$

58)
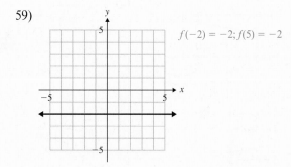
$f(-2) = 4; f(5) = -3$

59)
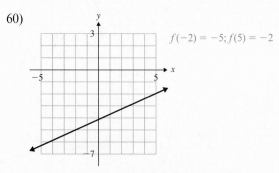
$f(-2) = -2; f(5) = -2$

60)

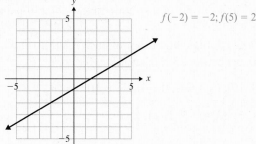
$f(-2) = -5; f(5) = -2$

61) $f(x) = -3x - 2$. Find x so that $f(x) = 10$. -4

62) $f(x) = -x + 7$. Find x so that $f(x) = -2$. 9

63) $h(x) = -\dfrac{3}{2}x - 5$. Find x so that $h(x) = 1$. -4

64) $g(x) = \dfrac{3}{4}x + 6$. Find x so that $g(x) = 12$. 8

Objective 6: Solve Problems Using Functions

65) The amount Fiona pays a babysitter, A (in dollars), for babysitting t hours can be described by the function $A(t) = 9t$.

   a) How much does she owe the babysitter for working 4 hr? $\$36.00$

   b) How much does she owe the babysitter for working 6.5 hr? $\$58.50$

   c) For how long could Fiona hire the sitter for $\$76.50$? 8.5 hr

   d) What is the slope of this linear function, and what does it mean in the context of this problem? $m = 9$; Fiona pays the babysitter $\$9$ per hour.

66) The number of miles, N, that Ahmed can drive in his hybrid car on g gallons of gas is given by the function $N(g) = 46g$.

   a) How far can Ahmed's car go on 5 gallons of gas? 230 miles

   b) How far can he drive on 11 gallons of gas? 506 miles

   c) How many gallons of gas would he need to drive 437 miles? 9.5 gallons

   d) What is the slope of this linear function, and what does it mean in the context of this problem? $m = 46$; Ahmed's car gets 46 miles per gallon.

67) A website sells concert tickets and charges a one-time $\$12.00$ service fee. The total cost, C (in dollars), of buying t tickets to a particular concert from this website is given by the function $C(t) = 59.50t + 12.00$.

   a) Find the cost of 2 tickets. $\$131.00$

   b) Find the cost of 4 tickets. $\$250.00$

   c) If Yoshiko paid $\$428.50$ for tickets from this website, how many did she buy? 7

68) When Ezra works an 8-hour day, he earns $\$90.40$ plus a 5% commission on everything he sells. Ezra's daily gross pay, P (in dollars), can be described by the function $P(s) = 0.05s + 90.40$, where s is the value of his sales, in dollars.

   a) How much would he earn if he sold $\$400.00$ worth of merchandise? $\$110.40$

   b) Find Ezra's gross pay if his sales totaled $\$290.00$. $\$104.90$

   c) If Ezra's gross pay on a Saturday was $\$143.40$, find the value of the merchandise he sold that day. $\$1060.00$

When skeletal remains are found at a crime scene, forensic scientists use formulas based on the person's sex, ethnicity, and bone length to estimate the person's height. Use the following formulas to answer the questions. (The *tibia* is a bone in the lower leg, and the *humerus* is a bone in the upper arm.) (http://forensics.rice.edu)

69) The height, h (in cm), of an Asian male in terms of the length of his tibia, t (in cm), is $h(t) = 2.39t + 81.45$. Find the height of an Asian man if his tibia is 41.6 cm long. Round to the nearest tenth of a cm. 180.9 cm (or about $5'11''$ tall)

70) The height, h (in cm), of a Caucasian male in terms of the length of his tibia, t (in cm), is $h(t) = 2.42t + 81.93$. Find the height of a Caucasian man if his tibia is 41.6 cm long. Round to the nearest tenth of a cm. 182.6 cm (or about $6'$ tall)

71) The height, h (in cm), of an African-American female in terms of the length of her humerus, L (in cm) is $h(L) = 3.08L + 64.67$. Find the height of an African-American woman if her humerus is 33.5 cm long. Round to the nearest tenth of a cm. 167.9 cm (or about $5'6''$ tall)

72) The height, h (in cm), of a Caucasian female in terms of the length of her humerus, L (in cm) is $h(L) = 3.36L + 57.97$. Find the height of a Caucasian woman if her humerus is 33.5 cm long. Round to the nearest tenth of a cm. 170.5 cm (or about $5'7''$ tall)

R1) How is evaluating an expression similar to evaluating a function for a given value?

R2) Is graphing a linear equation like $y = -2x + 5$ any different from graphing a linear function like $f(x) = -2x + 5$? Explain.

Group Activity — The Group Activity can be found online on Connect.

em  me What's Your Learning Style?

People learn in different ways. Depending on your personal learning style, visuals might be very useful to you as you study. On the other hand, you might take information in better if it is communicated to you by speech—say, in the context of a class lecture or study group. To discover how you learn best, read each of the following statements and rate them in terms of their usefulness to you as learning approaches. Base your ratings on your personal experiences and preferences, using the following scale:

1	=	Not at all useful
2	=	Not very useful
3	=	Neutral
4	=	Somewhat useful
5	=	Very useful

	1	2	3	4	5
1. Studying alone					
2. Studying pictures and diagrams to understand complex ideas					
3. Listening to class lectures					
4. Performing a process myself rather than reading or hearing about it					
5. Learning a complex procedure by reading written directions					
6. Watching and listening to film, computer, or video presentations					
7. Listening to a book or lecture on tape					
8. Doing lab work					
9. Studying teachers' handouts and lecture notes					
10. Studying in a quiet room					
11. Taking part in group discussions					
12. Taking part in hands-on classroom demonstrations					

	1	2	3	4	5
13. Taking notes and studying them later					
14. Creating flash cards and using them as a study and review tool					
15. Memorizing and recalling how words are spelled by spelling them "out loud" in my head					
16. Writing key facts and important points down as a tool for remembering them					
17. Recalling how to spell a word by seeing it in my head					
18. Underlining or highlighting important facts or passages in my reading					
19. Saying things out loud when I'm studying					
20. Recalling how to spell a word by "writing" it invisibly in the air or on a surface					
21. Learning new information by reading about it in a textbook					
22. Using a map to find an unknown place					
23. Working in a study group					
24. Finding a place I've been to once by just going there without directions					

To find your learning style, disregard your 1, 2, and 3 ratings. Add up your 4 and 5 ratings, using the following chart to link the statements to a learning style and to record your summed ratings:

Learning Style	Statements	Total (Sum) of Rating Points
Visual/verbal	1, 5, 9, 13, 17, and 21	
Visual/nonverbal	2, 6, 10, 14, 18, and 22	
Auditory/verbal	3, 7, 11, 15, 19, and 23	
Tactile/kinesthetic	4, 8, 12, 16, 20, and 24	

The total of your rating points for any given style will range from a low of 0 to a high of 30. The highest total indicates your main learning style. (Don't be surprised if you have a mixed style, in which two or more styles receive similar ratings.)

If you ranked highest in the **visual/verbal** learning style, you likely prefer information that is presented visually in a written format. You probably learn best when you have the opportunity to read about a concept.

If you ranked highest in the **visual/nonverbal** learning style, you probably learn most effectively when material is presented visually in a diagram or picture. Students with visual learning styles find it easier to see things in their "mind's eye"—to visualize a task or concept—than to be lectured about it.

If you ranked highest in the **auditory/verbal** learning style, you likely prefer listening to explanations rather than reading about them. Class lectures and discussions may be most beneficial for you, because you can easily take in the information that is being talked about.

If you ranked highest in the **tactile/kinesthetic** learning style, you probably like to learn by doing—touching, manipulating objects, and doing things. Students with this learning style often enjoy the act of writing because of the feel of a pencil or a computer keyboard—the tactile equivalent of "thinking out loud."

Chapter 10:   **Summary**

Definition/Procedure	Example

10.1 Solving Quadratic Equations Using the Square Root Property

The Square Root Property

Let k be a constant. If $x^2 = k$, then $x = \sqrt{k}$ or $x = -\sqrt{k}$.
(p. 606)

Solve $(t - 8)^2 = 5$.

$$t - 8 = \pm\sqrt{5} \qquad \text{Square root property}$$
$$t = 8 \pm \sqrt{5} \qquad \text{Add 8 to both sides.}$$

The solution set is $\{8 - \sqrt{5}, 8 + \sqrt{5}\}$.

10.2 Solving Quadratic Equations by Completing the Square

A **perfect square trinomial** is a trinomial whose factored form is the square of a binomial. **(p. 613)**

Perfect Square Trinomial	Factored Form
$g^2 + 6g + 9$	$(g + 3)^2$
$4t^2 - 20t + 25$	$(2t - 5)^2$

To find the constant needed to complete the square for $x^2 + bx$:

Step 1: Find half of the coefficient of x: $\quad\dfrac{1}{2}b$

Step 2: Square the result: $\quad\left(\dfrac{1}{2}b\right)^2$

Step 3: Add it to $x^2 + bx$: $\quad x^2 + bx + \left(\dfrac{1}{2}b\right)^2$ **(p. 613)**

Complete the square for $x^2 + 10x$ to obtain a perfect square trinomial. Then, factor.

Step 1: Find half of the coefficient of x: $\quad\dfrac{1}{2}(10) = 5$

Step 2: Square the result: $\quad 5^2 = 25$

Step 3: Add 25 to $x^2 + 10x$: $\quad x^2 + 10x + 25$

The perfect square trinomial is $x^2 + 10x + 25$.
The factored form is $(x + 5)^2$.

Solving a Quadratic Equation ($ax^2 + bx + c = 0$) by Completing the Square

Step 1: **The coefficient of the squared term must be 1.** If it is not 1, divide both sides of the equation by a to obtain a leading coefficient of 1.

Step 2: **Get the variables on one side of the equal sign and the constant on the other side.**

Step 3: **Complete the square.** Find half of the linear coefficient, then square the result. Add that quantity to *both* sides of the equation.

Step 4: **Factor.**

Step 5: **Solve using the square root property. (p. 615)**

Solve $x^2 + 8x + 9 = 0$ by completing the square.

$$x^2 + 8x + 9 = 0 \qquad \text{The coefficient of } x^2 \text{ is 1.}$$
$$x^2 + 8x = -9 \qquad \text{Get the constant on the other side}$$
$$\text{of the equal sign.}$$

Complete the square.

$$\frac{1}{2}(8) = 4$$
$$(4)^2 = 16$$

Add 16 to both sides of the equation.

$$x^2 + 8x + 16 = -9 + 16$$
$$(x + 4)^2 = 7 \qquad \text{Factor.}$$
$$x + 4 = \pm\sqrt{7} \qquad \text{Square root property}$$
$$x = -4 \pm \sqrt{7}$$

The solution set is $\{-4 - \sqrt{7}, -4 + \sqrt{7}\}$.

10.3 Solving Quadratic Equations Using the Quadratic Formula

The Quadratic Formula

The solutions of any quadratic equation of the form $ax^2 + bx + c = 0$ $(a \neq 0)$ are

$$x = \frac{-b \pm \sqrt{b^2 - 4ac}}{2a}$$

This is called the **quadratic formula. (p. 622)**

Solve $2x^2 - 9x + 8 = 0$ using the quadratic formula.
$$a = 2 \qquad b = -9 \qquad c = 8$$
Substitute the values into the quadratic formula.

$$x = \frac{-(-9) \pm \sqrt{(-9)^2 - 4(2)(8)}}{2(2)}$$

$$x = \frac{9 \pm \sqrt{81 - 64}}{4} = \frac{9 \pm \sqrt{17}}{4}$$

The solution set is $\left\{\dfrac{9 - \sqrt{17}}{4}, \dfrac{9 + \sqrt{17}}{4}\right\}$.

Definition/Procedure	Example

10.4 Graphs of Quadratic Equations

The graph of a quadratic equation of the form $y = ax^2 + bx + c$ is called a **parabola. (p. 631)**

The graph of $y = 2x^2 + x + 9$ is a parabola.

Graphing Basic Parabolas

The simplest form of a quadratic equation is $y = x^2$. **(p. 631)**

Graph $y = x^2$.

The lowest point on a parabola that opens upward or the highest point on a parabola that opens downward is called the **vertex. (p. 631)**

x	y
0	0
1	1
2	4
−1	1
−2	4

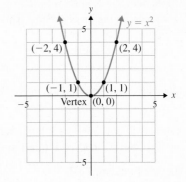

Graphing a Parabola of the Form $y = ax^2 + bx + c$ ($a \neq 0$)

Graph $y = x^2 + 4x + 3$.

Step 1: **Find the vertex.** The x-coordinate of the vertex is $x = -\dfrac{b}{2a}$. Find the y-coordinate of the vertex by substituting the x-value in the equation.

Step 1: **Find the vertex.**
$$x = -\frac{b}{2a} = -\frac{4}{2(1)} = -2$$
$$y = (-2)^2 + 4(-2) + 3 = 4 - 8 + 3 = -1$$

The vertex of the parabola is $(-2, -1)$.

Step 2: **Find the y-intercept** by substituting 0 for x and solving for y.

Step 2: **Find the y-intercept:** $y = (0)^2 + 4(0) + 3 = 3$
The y-intercept is $(0, 3)$.

Step 3: **Find the x-intercepts,** if they exist, by substituting 0 for y and solving for x.

Step 3: **Find the x-intercepts:**
$$0 = x^2 + 4x + 3$$
$$0 = (x + 3)(x + 1)$$
$$x = -3 \text{ or } x = -1$$
The x-intercepts are $(-3, 0)$ and $(-1, 0)$.

Step 4: **Find additional points** on the parabola using a table of values.

Step 4: **Find an additional point:** Let $x = -4$.
$$y = (-4)^2 + 4(-4) + 3 = 16 - 16 + 3 = 3$$

Step 5: **Plot the points, and sketch the graph. (p. 634)**

Another point on the graph is $(-4, 3)$.

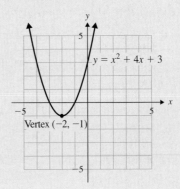

Definition/Procedure	Example

10.5 Introduction to Functions

A **relation** is any set of ordered pairs. A relation can also be represented as a correspondence or mapping from one set to another. **(p. 640)**

The **domain** of a relation is the set of values of the first coordinates in the set of ordered pairs.

The **range** of a relation is the set of all values of the second coordinates in the set of ordered pairs. **(p. 640)**

A **function** is a relation in which each element of the domain corresponds to exactly one element of the range. **(p. 640)**

Alternative definition: A relation is a **function** if each x-value corresponds to one y-value. **(p. 641)**

Relations:

a) $\{(-4, -12), (-1, -3), (3, 9), (5, 15)\}$

b)
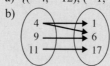

In a), the domain is $\{-4, -1, 3, 5\}$, and the range is $\{-12, -3, 9, 15\}$.

In b), the domain is $\{4, 9, 11\}$, and the range is $\{1, 6, 17\}$.

The relation in a) *is* a function.

The relation in b) *is not* a function.

The Vertical Line Test

If no vertical line can be drawn through a graph that intersects the graph more than once, then the graph represents a function.

If a vertical line *can* be drawn that intersects the graph more than once, then the graph does *not* represent a function. **(p. 642)**

This graph represents a function. Anywhere a vertical line is drawn, it will intersect the graph only once.

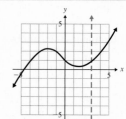

This graph is *not* the graph of a function. A vertical line can be drawn so that it intersects the graph more than once.

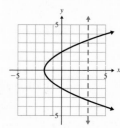

When determining the domain of a relation, it can be helpful to keep these tips in mind.

1) Ask yourself, "Is there any number that *cannot* be substituted for x?"

2) If x is in the denominator of a fraction, determine what value of x will make the denominator equal 0 by setting the expression equal to zero. Solve for x. This x-value is *not* in the domain. **(p. 644)**

Determine the domain of $f(x) = \dfrac{6}{x + 3}$.

$$x + 3 = 0 \qquad \text{Set the denominator} = 0.$$
$$x = -3 \qquad \text{Solve.}$$

When $x = -3$, the denominator of $f(x) = \dfrac{6}{x + 3}$ equals zero.

The domain contains all real numbers *except* -3. The domain of the function is $(-\infty, -3) \cup (-3, \infty)$.

$y = f(x)$ is called **function notation** and it is read as "y equals f of x."

Finding a function value means evaluating the function for the given value of the variable. **(p. 645)**

If $f(x) = 2x - 9$, find $f(4)$.

Substitute 4 for x and evaluate.

$f(4) = 2(4) - 9 = 8 - 9 = -1$

Therefore, $f(4) = -1$.

Chapter 10:   Review Exercises

*Additional answers can be found in the Answers to Exercises appendix.

(10.1) Solve using the square root property.

1) $w^2 = 81$ $\{-9, 9\}$

2) $p^2 - 50 = 0$ $\{-5\sqrt{2}, 5\sqrt{2}\}$

3) $3a^2 + 7 = 40$
$\{-\sqrt{11}, \sqrt{11}\}$

4) $(z + 6)^2 = 20$
$\{-6 - 2\sqrt{5}, -6 + 2\sqrt{5}\}$

5) $5 = (k - 4)^2 + 14$ $\varnothing$

6) $35 = (2y - 1)^2 + 7$ $\left\{\dfrac{1 - 2\sqrt{7}}{2}, \dfrac{1 + 2\sqrt{7}}{2}\right\}$

7) $\left(\dfrac{3}{4}c + 5\right)^2 - 2 = 14$ $\left\{-12, -\dfrac{4}{3}\right\}$

8) The power, P (in watts), in an electrical system is given by $P = \dfrac{V^2}{R}$, where V is the voltage, in volts, and R is the resistance, in ohms. Find the voltage if the power is 45 watts and the resistance is 5 ohms.   15 volts

(10.2) Complete the square for each expression to obtain a perfect square trinomial. Then, factor.

9) $x^2 + 18x$
$x^2 + 18x + 81; (x + 9)^2$

10) $n^2 - 20n$
$n^2 - 20n + 100; (n - 10)^2$

11) $z^2 - 3z$
$z^2 - 3z + \dfrac{9}{4}; \left(z - \dfrac{3}{2}\right)^2$

12) $y^2 + y$
$y^2 + y + \dfrac{1}{4}; \left(y + \dfrac{1}{2}\right)^2$

Solve by completing the square.

13) $d^2 + 10d + 9 = 0$
$\{-9, -1\}$

14) $v^2 - 8v + 11 = 0$
$\{4 - \sqrt{5}, 4 + \sqrt{5}\}$

15) $r^2 + 15 = 4r$ $\varnothing$

16) $x(x + 14) = -24$ $\{-12, -2\}$

17) $a^2 + 5a + 2 = 0$

18) $t^2 - 7 = 3t$
$\left\{\dfrac{3}{2} - \dfrac{\sqrt{37}}{2}, \dfrac{3}{2} + \dfrac{\sqrt{37}}{2}\right\}$

19) $4n^2 - 8n = 21$ $\left\{-\dfrac{3}{2}, \dfrac{7}{2}\right\}$

20) $2h - 7 = -2h^2$ $\left\{-\dfrac{1}{2} - \dfrac{\sqrt{15}}{2}, -\dfrac{1}{2} + \dfrac{\sqrt{15}}{2}\right\}$

(10.3)

21) We can use the quadratic formula to solve $ax^2 + bx + c = 0$ $(a \neq 0)$ for x.
Write the quadratic formula. $x = \dfrac{-b \pm \sqrt{b^2 - 4ac}}{2a}$

22) Correct the error in this simplification:
$$\dfrac{6 \pm 3\sqrt{10}}{3} = 6 \pm \sqrt{10}$$

Solve using the quadratic formula.

23) $x^2 + 7x + 12 = 0$
$\{-4, -3\}$

24) $3z^2 + z - 10 = 0$ $\left\{-2, \dfrac{5}{3}\right\}$

25) $2r^2 + 3r = -6$ $\varnothing$

26) $8n^2 + 3 = 14n$ $\left\{\dfrac{1}{4}, \dfrac{3}{2}\right\}$

27) $t(t - 10) = -7$ $\{5 - 3\sqrt{2}, 5 + 3\sqrt{2}\}$

28) $-1 = y(5 - 4y)$ $\left\{\dfrac{5 - \sqrt{41}}{8}, \dfrac{5 + \sqrt{41}}{8}\right\}$

29) $\dfrac{3}{8}p^2 + \dfrac{1}{2}p - \dfrac{1}{4} = 0$ $\left\{\dfrac{-2 - \sqrt{10}}{3}, \dfrac{-2 + \sqrt{10}}{3}\right\}$

30) $(2w + 3)(w + 4) = 6(w + 1)$ $\varnothing$

31) The hypotenuse of a right triangle is 1 inch less than three times the shorter leg. The longer leg is $\sqrt{55}$ inches long. Find the length of the shorter leg.   3 in.

Solve. Give an exact answer, and use a calculator to round the answer to the nearest hundredth.

32) An object is thrown upward from a height of 26 feet. The height h of the object (in feet) t seconds after the object is released is given by
$$h = -16t^2 + 22t + 26$$

a) How long does it take the object to reach a height of 6 feet?   2 sec

b) How long does it take the object to hit the ground? (Hint: When the object hits the ground, $h = 0$.)
$\dfrac{11 + \sqrt{537}}{16}$ sec or about 2.14 sec

Mixed Exercises

Keep in mind the four methods we have learned for solving quadratic equations: *factoring, the square root property, completing the square,* **and** *the quadratic formula.* **Solve the equations using one of these methods.**

33) $d^2 - 3d - 28 = 0$ $\{-4, 7\}$

34) $2y(y - 3) = -1$

35) $(3t - 1)(t + 1) = 3$

36) $\dfrac{1}{18}p^2 + \dfrac{7}{9}p + \dfrac{8}{3} = 0$ $\{-8, -6\}$

37) $(n - 9)^2 + 7 = 12$ $\{9 - \sqrt{5}, 9 + \sqrt{5}\}$

38) $(3a - 4)^2 + 15 = 11$ $\varnothing$

39) $6c^3 + 12c = 38c^2$ $\left\{0, \dfrac{1}{3}, 6\right\}$

40) $4m^2 - 25 = 0$ $\left\{-\dfrac{5}{2}, \dfrac{5}{2}\right\}$

41) $k - 2 = 5k^2 + 3k + 4$ $\varnothing$

42) $w^2 + 6w + 4 = 0$ $\{-3 - \sqrt{5}, -3 + \sqrt{5}\}$

43) $h^2 = 4h$ $\{0, 4\}$

44) $0.01x^2 + 0.09x + 0.12 = 0$ $\left\{\dfrac{-9 - \sqrt{33}}{2}, \dfrac{-9 + \sqrt{33}}{2}\right\}$

(10.4)

45) What do you call the graph of $y = ax^2 + bx + c$ $(a \neq 0)$?   parabola

46) Explain how to find the vertex of the graph of $y = ax^2 + bx + c$ $(a \neq 0)$.

47) What is the axis of symmetry of a parabola?

Sketch the graph of each equation. Identify the vertex.

48) $y = -x^2 + 4$ vertex: $(0, 4)$

49) $y = (x + 2)^2$ vertex: $(-2, 0)$

50) $y = x^2 + 4x + 3$ vertex: $(-2, -1)$

51) $y = -x^2 + 2x + 2$ vertex: $(1, 3)$

52) $y = x^2 - 6x + 11$ vertex: $(3, 2)$

(10.5) Identify the domain and range of each relation, and determine whether each relation is a function.

53) $\{(-6, 0), (5, 1), (8, 1), (10, 4)\}$ domain: $\{-6, 5, 8, 10\}$; range: $\{0, 1, 4\}$; function

54) $\{(-2, 4), (2, 4), (3, 9), (3, 16)\}$ domain: $\{-2, 2, 3\}$; range: $\{4, 9, 16\}$; not a function

55)

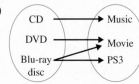

domain: $\{$CD, DVD, Blu-ray disc$\}$; range: $\{$Music, Movie, PS3$\}$; not a function

56)   domain: {−2, 1, 0}; range: {4, 3, 1}; function

57)

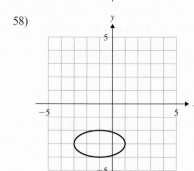

domain: $(-\infty, \infty)$; range: $(-\infty, \infty)$; function

58)

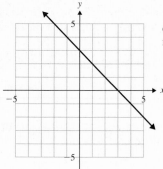

domain: $[-3, 1]$; range: $[-4, -2]$ not a function

Determine the domain of each function.

59) $y = 2x - 5$ $(-\infty, \infty)$

60) $y = \dfrac{1}{5}x + 6$ $(-\infty, \infty)$

61) $y = \dfrac{6}{x}$ $(-\infty, 0) \cup (0, \infty)$

62) $y = \dfrac{3}{x + 4}$

$(-\infty, -4) \cup (-4, \infty)$

63) $y = x^2 + 3$ $(-\infty, \infty)$

64) $y = x^3 - 9$ $(-\infty, \infty)$

For each function f, find f(2) and f(21).

65) $f = \{(0, 9), (-2, 5), (-1, -8), (2, 7)\}$ $f(2) = 7; f(-1) = -8$

66)

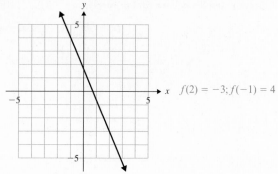

$f(2) = -3; f(-1) = 4$

67) Let $f(x) = -3x + 8$, $g(x) = x^2 - 9x + 4$. Find each of the following and simplify.

a) $f(3)$ −1

b) $f(-5)$ 23

c) $g(4)$ −16

d) $g(0)$ 4

68) $h(x) = 2x + 11$. Find x so that $h(x) = -7$. −9

69) In 2011, approximately 250 million photos were uploaded to Facebook each day. This can be modeled by the function $N(x) = 250x$, where N is the number of photos uploaded (in millions) to Facebook in x days. (www.facebook.com)

a) How many photos were uploaded to Facebook in 4 days? 1,000,000,000

b) How many photos were uploaded to Facebook in one week? 1,750,000,000

c) How long would it take for 3,000,000,000 photos to be uploaded? 12 days

Chapter 10: Test

1) Solve $3a^2 + 4 = 22$ using the square root property.
$\{-\sqrt{6}, \sqrt{6}\}$

2) The kinetic energy K, in joules, of an object is given by $K = \dfrac{1}{2}mv^2$, where m is the mass of the object, in kg, and v is the velocity of the object in m/sec. Find the velocity of a motorcycle with a mass of 180 kg and kinetic energy of 1440 joules. 4 m/sec

Solve by completing the square.

3) $m^2 - 12m + 26 = 0$
$\{6 - \sqrt{10}, 6 + \sqrt{10}\}$

4) $3k^2 + 15k + 9 = 0$

5) Elaine simplified the expression as follows:
$\dfrac{12 \pm 8\sqrt{5}}{4} = 3 \pm 8\sqrt{5}$. Explain her error, and simplify the expression correctly.

6) Solve $2p^2 + 9p + 5 = 0$ using the quadratic formula.
$\left\{\dfrac{-9 - \sqrt{41}}{4}, \dfrac{-9 + \sqrt{41}}{4}\right\}$

Solve using any method.

7) $(p - 7)^2 + 1 = 13$
$\{7 - 2\sqrt{3}, 7 + 2\sqrt{3}\}$

8) $\dfrac{1}{6}c^2 + \dfrac{2}{3}c - 2 = 0$
$\{-6, 2\}$

9) $(2k - 3)(2k + 5) = 6(k - 2)$
$\left\{\dfrac{1 - \sqrt{13}}{4}, \dfrac{1 + \sqrt{13}}{4}\right\}$

10) $n^2 = 3n - 10$ ∅

Solve. Give an exact answer, and use a calculator to round the answer to the nearest hundredth.

11) An object is thrown upward from a height of 32 feet. The height h of the object (in feet) t seconds after the object is released is given by $h = -16t^2 + 40t + 32$.

a) How long does it take the object to reach a height of 8 feet? 3 sec

b) How long does it take the object to hit the ground? (Hint: When the object hits the ground, $h = 0$.) $\dfrac{5 + \sqrt{57}}{4}$ sec

or about 3.14 sec

12) What do you call the graph of $y = ax^2 + bx + c$?
parabola

Sketch the graph of each equation. Identify the vertex.

13) $y = -(x - 1)^2$ vertex: (1, 0) 14) $y = x^2 + 2x - 3$
                                              vertex: (−1, −4)

15) $y = -x^2 + 4x - 1$ vertex: (2, 3)

16) What is a function? It is a relation in which each element of
the domain corresponds to exactly one element of the range.

**Identify the domain and range of each relation, and determine
whether each relation is a function.**

17) $\{(0, 0), (1, -1), (16, 2), (16, -2)\}$
domain: {0, 1, 16}; range: {0, −1, 2, −2}; not a function

18)

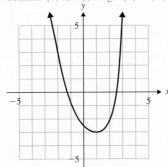

domain: $(-\infty, \infty)$;
range: $[-3, \infty)$; function

Determine the domain of each function.

19) $y = 9x + 2$ $(-\infty, \infty)$ 20) $y = \dfrac{5}{x + 8}$ $(-\infty, -8) \cup (-8, \infty)$

For each function f, find f(−3).

21) $f = \{(-5, 9), (-3, 4), (1, 0), (6, -3)\}$ 4

22)

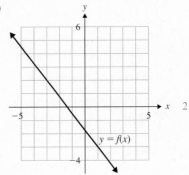

**Let $f(x) = -6x + 11$ and $g(x) = x^2 - 5x - 3$. Find each of the
following and simplify.**

23) $f(5)$ −19   24) $f\left(-\dfrac{3}{2}\right)$ 20

25) $g(-2)$ 11   26) $g(0)$ −3

27) A cell phone company has a plan that charges $9.99 per
month plus $0.06 per minute. The cost of this plan,
C (in dollars), can be described by the function
$C(x) = 0.06x + 9.99$, where x is the number of
minutes the phone is used.

a) Find the amount of a customer's bill if she used the
phone for 82 minutes during a month. $14.91

b) For how many minutes did she use the phone if her bill
was $16.83?   114 min

Chapter 10: Cumulative Review for Chapters 1–10

*Additional answers can be found in the Answers to Exercises appendix.

Perform the operations and simplify.

1) $\dfrac{5}{8} - \dfrac{2}{7}$ $\dfrac{19}{56}$

2) $\dfrac{12 - 56 \div 8}{(1 + 5)^2 - 2^4}$ $\dfrac{1}{4}$

3) Find the area and perimeter.

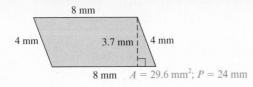

$A = 29.6$ mm²; $P = 24$ mm

4) Given this set of numbers,
$$\left\{ \sqrt{23}, -6, 14.38, \frac{3}{11}, 2, 5.\overline{7}, 0, 9.21743819 \ldots \right\}$$
list the

a) natural numbers   {2}   b) integers   {−6, 0, 2}

c) rational numbers   d) irrational numbers
                                   {$\sqrt{23}$, 9.21743819 …}

5) The lowest temperature on record in the state of Wyoming
is −66°F. Georgia's record low is 49° higher than
Wyoming's. What is the lowest temperature ever recorded
in Georgia? (www.weather.com)   −17°F

6) Use the rules of exponents to simplify.

a) $2^3 \cdot 2^2$ 32   b) $\left(\dfrac{1}{3}\right)^2 \cdot \left(\dfrac{1}{3}\right)$ $\dfrac{1}{27}$

c) $\dfrac{x^7 \cdot (x^2)^5}{(2y^3)^4}$ $\dfrac{x^{17}}{16y^{12}}$ d) $(6 - 8)^2$ 4

7) Evaluate.

a) $(-12)^0$ 1   b) $5^0 + 4^0$ 2

c) -6^{-2} $-\dfrac{1}{36}$ d) 2^{-4} $\dfrac{1}{16}$

8) Simplify by applying one or more of the rules of exponents.
The final answer should not contain negative exponents.

a) $(-3s^4t^5)^4$ $81s^{16}t^{20}$ b) $\left(\dfrac{d^{-4}}{d^{-9}}\right)^5$ d^{25}

c) $\left(\dfrac{3k^{-1}t}{5k^{-7}t^4}\right)^{-3}$ $\dfrac{125t^9}{27k^{18}}$ d) $\left(\dfrac{40}{21}x^{10}\right)(3x^{-12})\left(\dfrac{49}{20}x^2\right)$ 14

9) Write 0.0000575 in scientific notation. 5.75×10^{-5}

10) Simplify $\dfrac{8 \times 10^6}{2 \times 10^{13}}$ and write the final answer without an
exponent.   0.0000004

Solve each equation.

11) $8b - 7 = 57$ {8}

12) $6 - 5(4d - 3) = 7(3 - 4d) + 8d$ {all real numbers}

Solve using the Five Steps for Solving Applied Problems.

13) A number increased by nine is one less than twice the number. Find the number.   10

14) Jerome invested some money in an account earning 7% simple interest and $3000 more than that at 8% simple interest. After 1 year, he earned $915 in interest. How much money did he invest in each account? $4500 at 7% and $7500 at 8%

15) Solve $-15 < 4p - 7 \le 5$ and write the answer in interval notation. $(-2, 3]$

16) Complete the table of values for $4x - 6y = 8$.

x	y
2	0
0	$-\frac{4}{3}$
3	$\frac{2}{3}$
-4	-4

17) Graph $x - 2y = 6$ by finding the intercepts and at least one other point. $(6, 0), (0, -3); (2, -2)$ may vary

18) Write the *standard form* of the equation of the line given the following information. $5x + 3y = 6$

$$m = -\frac{5}{3} \text{ and } y\text{-intercept } (0, 2)$$

19) Solve the system of equations by graphing.

$$2x + 3y = 5$$

$$y = \frac{1}{2}x + 4 \quad (-2, 3)$$

Solve each system.

20) $\begin{aligned} 6x - y &= -3 \\ 15x + 2y &= 15 \end{aligned}$ $\left(\frac{1}{3}, 5\right)$

21) $\dfrac{3}{4}x - y = \dfrac{1}{2}$ $(2, 1)$

$-\dfrac{x}{3} + \dfrac{y}{2} = -\dfrac{1}{6}$

22) A car and a tour bus leave the same location and travel in opposite directions. The car's speed is 12 mph more than the speed of the bus. If they are 270 miles apart after $2\frac{1}{2}$ hours, how fast is each vehicle traveling?   car: 60 mph; bus: 48 mph

23) Graph the solution set of $x \le 4$ and $y \ge -\dfrac{3}{2}x + 3$.

24) Simplify $\left(\dfrac{3pq^{-10}}{2p^{-2}q^5}\right)^{-2} \cdot \dfrac{4q^{30}}{9p^6}$

Perform the indicated operation and simplify.

25) $\begin{array}{r} 5.8p^3 - 1.2p^2 + p - 7.5 \\ + 2.1p^3 + 6.3p^2 + 3.8p + 3.9 \\ \hline 7.9p^3 + 5.1p^2 + 4.8p - 3.6 \end{array}$

26) $\left(\dfrac{7}{4}k^2 + \dfrac{1}{6}k + 5\right) - \left(\dfrac{1}{2}k^2 + \dfrac{5}{6}k - 2\right)$ $\dfrac{5}{4}k^2 - \dfrac{2}{3}k + 7$

27) $-5(7w - 12)(w + 3)$ $-35w^2 - 45w + 180$

28) $(6q^2 + 2q - 35) \div (3q + 7)$ $2q - 4 - \dfrac{7}{3q + 7}$

Factor completely.

29) $t^2 - 2tu - 63u^2$ $(t + 7u)(t - 9u)$

30) $3g^2 + g - 44$ $(3g - 11)(g + 4)$

31) $a^2 + 16a + 64$ $(a + 8)^2$

32) Solve $(z + 2)^2 = -z(3z + 4) + 9$ $\left\{-\dfrac{5}{2}, \dfrac{1}{2}\right\}$

33) Find the length and width of the rectangle if its area is 28 cm^2.

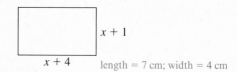

length = 7 cm; width = 4 cm

Perform the operation and simplify.

34) $\dfrac{3x^2 + 14x + 16}{15x + 40} \div \dfrac{11x + 22}{x - 5}$ $\dfrac{x - 5}{55}$

35) $\dfrac{10p + 3}{4p + 4} - \dfrac{8}{p^2 - 6p - 7}$ $\dfrac{10p^2 - 67p - 53}{4(p + 1)(p - 7)}$

36) $\dfrac{2}{5z^2} + \dfrac{9}{10z}$ $\dfrac{4 + 9z}{10z^2}$

37) Simplify $\dfrac{\dfrac{c}{c + 2} + \dfrac{1}{c^2 - 4}}{1 - \dfrac{3}{c + 2}}$ $\dfrac{c - 1}{c - 2}$

38) Solve $\dfrac{3k}{k + 9} = \dfrac{3}{k + 1}$. $\{-3, 3\}$

Simplify.

39) $\dfrac{\sqrt{200k^{21}}}{\sqrt{2k^5}}$ $10k^8$

40) $\sqrt{80} - \sqrt{48} + \sqrt{20}$ $6\sqrt{5} - 4\sqrt{3}$

41) $\dfrac{z - 4}{\sqrt{z} + 2}$ $\sqrt{z} - 2$

42) $\sqrt[3]{56p^{12}}$ $2p^4\sqrt[3]{7}$

Solve.

43) $\sqrt{6x + 9} - \sqrt{2x + 1} = 4$ {12}

44) $2c^2 - 11 = 25$ $\{-3\sqrt{2}, 3\sqrt{2}\}$

45) $n^2 + 8n - 3 = 0$ $\{-4 - \sqrt{19}, -4 + \sqrt{19}\}$

46) Graph $y = -x^2 + 2x + 3$. Identify the vertex. $(1, 4)$

Let $f(x) = 2x + 8$ and $g(x) = x^2 + 10x - 3$. Find each of the following.

47) $f(9)$ 26

48) $f(0)$ 8

49) $g(-5)$ -28

50) $g(3)$ 36

Appendix

A.1 Decimals

Objectives

1 Write a Decimal as a Fraction or Mixed Number
2 Add and Subtract Decimals
3 Multiply and Divide Decimals
4 Write a Fraction as a Decimal

Like a fraction, a *decimal* is a way to represent part of a whole.

Definition

A **decimal** is a number, containing a *decimal point,* that is another way to represent a fraction with a denominator that is a power of 10.

Examples: $0.1 = \dfrac{1}{10}$ Both of these are read as *one tenth.*

   $0.37 = \dfrac{37}{100}$ Both of these are read as *thirty-seven hundredths.*

Let's review place value in a decimal number. For example, here are the place-value names for the digits of 8,409,167.52387:

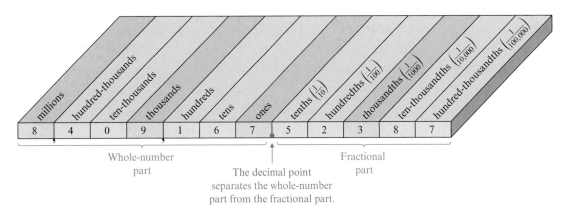

Whole-number part

The decimal point separates the whole-number part from the fractional part.

Fractional part

1 Write a Decimal as a Fraction or Mixed Number

Reading the decimal to ourselves will help us determine how to correctly write the decimal as a fraction. **The denominator of the fraction is the place value of the digit that is farthest to the right after the decimal point.**

EXAMPLE 1

Write each decimal as a fraction or mixed number in lowest terms.

a) 0.8 b) 0.0723 c) 2.164

Solution

a) We read 0.8 as *eight tenths*. So, $0.8 = \dfrac{8}{10}$.

8 is in the *tenths* place, so the denominator = 10.

Ask yourself, "Is $\dfrac{8}{10}$ in lowest terms?" No! Write it in lowest terms.

$$0.8 = \frac{8}{10} = \frac{8 \div 2}{10 \div 2} = \frac{4}{5}$$

b) The digit farthest to the right, 3, is in the *ten-thousandths* place. We read 0.0723 as *seven hundred twenty-three ten-thousandths.* Therefore,

$$0.0723 = \frac{723}{10,000}$$

3 is in the *ten-thousandths* place, so the denominator = 10,000.

Ask yourself, "Is $\dfrac{723}{10,000}$ in lowest terms?" Yes.

c) The digit farthest to the right, 4, is in the *thousandths* place. We read 2.164 as *two and one hundred sixty-four thousandths.* Therefore,

$$2.164 = 2\frac{164}{1000}$$

4 is in the *thousandths* place, so the denominator = 1000.

Is $\dfrac{164}{1000}$ in lowest terms? No. Write it in lowest terms.

$$2.164 = 2\frac{164}{1000} = 2\frac{164 \div 4}{1000 \div 4} = 2\frac{41}{250}$$

We could also write $2\dfrac{41}{250}$ as the improper fraction $\dfrac{541}{250}$.

[YOU TRY 1] Write each decimal as a fraction or mixed number in lowest terms.

a) 0.2 b) 0.0089 c) 4.375

2 Add and Subtract Decimals

To add or subtract *decimal* numbers, we **line up the decimal points**—that is, we line up the numbers in the tenths place, line up the numbers in the hundredths place, and so on. Then, add or subtract.

Procedure Adding or Subtracting Decimals

1) Write the numbers vertically so that the decimal points are lined up.

2) If any numbers are missing digits to the right of the decimal point, insert zeros. Then, add or subtract the same way we add or subtract whole numbers.

3) Place the decimal point in the answer *directly below* the decimal point in the problem.

Note
Using graph paper will help us line up the numbers correctly.

EXAMPLE 2

Add 13.27 + 6.198 + 5

Solution

Write the numbers vertically so that the decimal points are lined up. Remember that 5 can be written as 5.

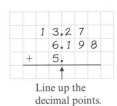

Line up the decimal points.

Insert zeros in the thousandths place of 13.27 and in three places to the right of the decimal point in 5. Then, add.

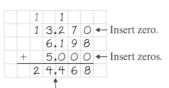

Line up the decimal point in the answer with the decimal points in the problem.

[YOU TRY 2]   Add 56.2 + 9 + 7.044

EXAMPLE 3

Subtract.

a) 59.7 − 23.661   b) Subtract 0.384 from 2

Solution

a) Line up the decimal points.

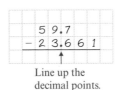

Line up the decimal points.

Insert zeros after the 7 in 59.7. Then, subtract.

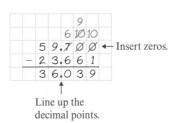

Line up the decimal points.

Remember, we can add zeros at the end of 59.7 because 59.7 is equivalent to 59.700.

---

b) Line up the decimal points.

Insert zeros and subtract.

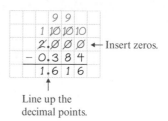← Insert zeros.

Line up the
decimal points.

Line up the
decimal points.

[**YOU TRY 3**] Subtract.

a) $30.4 - 17.82$ b) Subtract 6.093 from 14

---

3 Multiply and Divide Decimals

Procedure Multiplying Decimals

When **multiplying decimals,** we do *not* need to line up the decimal points. Line up the factors (the numbers being multiplied) on the right, and multiply them just as you would multiply whole numbers. The number of decimal places in the answer is the *total* number of decimal places in the factors.

 BE CAREFUL When we *multiply* decimals, we do **not** have to line up the decimal points. When we *add or subtract* decimals, we **must** line up the decimal points.

EXAMPLE 4

In-Class Example 4

Multiply.
a) 17.54×6.3 b) $(39.01)(25)$

Answer: a) 110.502
b) 975.25

Multiply.

a) 16.92×3.4 b) $(68.03)(45)$

Solution

a) Line up the numbers on the right. Do **not** line up the decimal points. **Multiply the numbers just like they were whole numbers.**

```
    1 6.9 2
  ×     3.4
    6 7 6 8
  5 0 7 6
  5 7 5 2 8
```

$16.92 \times 3.4 = 57.528$

Determine the total number of decimal places in the answer. This will be the total number of decimal places in the factors.

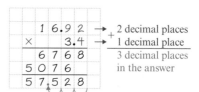

```
    1 6.9 2  →  2 decimal places
  ×     3.4  →  1 decimal place
    6 7 6 8     3 decimal places
  5 0 7 6       in the answer
  5 7.5 2 8
```

Start at the right side of the number and count 3 places to the left. Insert the decimal point.

---

b) Multiply the numbers just as if they were whole numbers. Determine the number of decimal places in the answer and insert the decimal point.

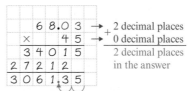

$(68.03)(45) = 3061.35$

YOU TRY 4

Multiply.

a) 61.85×4.7 b) $(24.89)(93)$

Sometimes, we have to insert zeros on the left side of the product to have the correct number of decimal places in the answer.

EXAMPLE 5

In-Class Example 5

Find 0.024×0.09

Answer: 0.00216

Find 0.038×0.07

Solution

Line up the numbers on the right. **Multiply the numbers just like they were whole numbers.**

Determine the total number of decimal places in the answer. This will be the total number of decimal places in the factors.

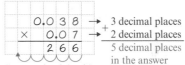

Start at the right side and count 5 places to the left. Insert zeros in the blank spaces.

There are 5 places to the right of the decimal point.

Write the final answer with a zero to the left of the decimal point, in the ones place.

$$0.038 \times 0.07 = 0.00266$$

Put a zero in the ones place.

YOU TRY 5

Find 0.092×0.04

Do you remember what we call the different parts of a division problem?

$$\begin{array}{r} 8 \leftarrow \text{Quotient} \\ \text{Divisor} \rightarrow 3\overline{)24} \leftarrow \text{Dividend} \end{array}$$

Now let's review how to divide decimals.

Procedure Dividing Decimals

1) When **dividing a decimal by a whole number**, like $0.548 \div 4$, write the decimal point in the quotient directly above the decimal point in the dividend. Divide as if the numbers were whole numbers.

2) When **dividing a number by a decimal**, like $1.22\overline{)37.149}$, move the decimal point to the **right** end of the divisor, 1.22, to make it a whole number. Move the decimal point in the dividend, 37.149, the same number of places to the right. Write the decimal point in the quotient directly above the decimal point in the dividend. Divide as if the numbers were whole numbers.

EXAMPLE 6

In-Class Example 6

Divide.
a) $0.738 \div 6$ b) $1.44\overline{)29.736}$

Answer: a) 0.123 b) 20.65

Divide.

a) $0.548 \div 4$ b) $1.22\overline{)37.149}$

Solution

a) We are **dividing a decimal by a whole number:** $0.548 \div 4$. Write the decimal point in the quotient directly above the decimal point in the dividend. Then, divide.

$$
\begin{array}{r}
0.1\ 3\ 7 \\
4\overline{)0.5\ 4\ 8} \\
-\ 4 \\
\hline
1\ 4 \\
-\ 1\ 2 \\
\hline
2\ 8 \\
-\ 2\ 8 \\
\hline
0
\end{array}
$$

b) In this problem, $1.22\overline{)37.149}$, we are **dividing a number by a decimal.**

Make the divisor a whole number.

$$\text{Divisor} \rightarrow 1.22\overline{)37.149} \qquad\qquad 122\overline{)3714.9}$$
$$\uparrow$$
$$\text{Dividend}$$

Move each decimal point 2 places to the right.

Begin the division process like we did before.

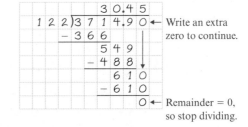

← No more digits to bring down

← Remainder $\neq 0$

We have reached the end of the dividend, the remainder is *not* zero, and there are no more digits to bring down.

Write an extra zero on the right end of the dividend and keep on dividing.

$$
\begin{array}{r}
3\ 0.4\ 5 \\
1\ 2\ 2\overline{)3\ 7\ 1\ 4.9\ 0} \\
-\ 3\ 6\ 6 \\
\hline
5\ 4\ 9 \\
-\ 4\ 8\ 8 \\
\hline
6\ 1\ 0 \\
-\ 6\ 1\ 0 \\
\hline
0
\end{array}
$$

← Write an extra zero to continue.

← Remainder = 0, so stop dividing.

The answer is 30.45.

[YOU TRY 6]

Divide.

a) $0.835 \div 5$ b) $1.74\overline{)35.931}$

If division does not work out evenly, we can approximate the answer to an indicated decimal place.

EXAMPLE 7

In-Class Example 7

Divide. Round the answer to the nearest hundredth.
$8 \div 1.3$

Answer: 6.15

Divide. Round the answer to the nearest hundredth. $9 \div 2.3$

Solution

Set up the division problem. $2.3\overline{)9}$

Because we must move the decimal point *one place* to the right, put a decimal point after the 9.

$$2.3\overline{)9.}$$ $$23\overline{)90.}$$

↑

Put a decimal point after the 9.

Move the decimal points 1 place to the right. Write a zero in the dividend, and put the decimal point in the quotient.

Divide.

```
        3.9 1 3  ← Because we are rounding to the
  2 3)9 0.0 0 0    hundredths place, carry out the
   - 6 9            division to the thousandths place.
     2 1 0
   - 2 0 7
         3 0
       - 2 3
           7 0
         - 6 9
             1    ← Remainder ≠ 0
```

Rounding 3.913 to the nearest hundredth, we get $9 \div 2.3 \approx 3.91$. The symbol $\approx$ means "is approximately equal to."

[YOU TRY 7]

Divide. Round the answer to the nearest hundredth. $8 \div 3.5$

We can multiply and divide decimals by powers of 10 in our heads if we remember these rules.

Procedure Multiplying and Dividing Numbers by a Power of 10

To **multiply** a number by a power of 10, **count** the number of zeros in the power of 10, **move the decimal point** in the number *to the right* the same number of spaces as the number of zeros in the power of 10. If necessary, add zeros as placeholders.

To **divide** a number by a power of 10, **count** the number of zeros in the power of 10, **move the decimal point** in the number *to the left* the same number of spaces as the number of zeros in the power of 10. If necessary, add zeros as placeholders.

EXAMPLE 8

In-Class Example 8

Perform the indicated operation.

a) 2.1745 · 100

b) 92.3 ÷ 1000

Answer: a) 217.45
b) 0.0923

Perform the indicated operation.

a) $7.9265 \cdot 100$

b) $61.8 \div 1000$

Solution

a) $7.9265 \cdot 100 = 792.65$ Move the decimal point *two* places to the *right*.

2 zeros

b) $61.8 \div 1000 = .0618 = 0.0618$ Move the decimal point *three* places to the *left*.

3 zeros

[YOU TRY 8]   Perform the indicated operation.

a) $4.9 \cdot 10,000$

b) $0.21 \div 100$

4 Write a Fraction as a Decimal

We will review two ways to write a fraction as a decimal. First, we will use long division. Recall that a fraction is one way to represent division. For example, $\frac{2}{3}$ means $2 \div 3$. **To write a fraction in decimal form, divide the numerator by the denominator.**

> **Procedure** Writing a Fraction as a Decimal Using Division
>
> To write a fraction as a decimal, divide the numerator by the denominator.

EXAMPLE 9

In-Class Example 9

Write each fraction as a decimal. a) $\frac{7}{4}$ b) $\frac{2}{3}$

Answer: a) 1.75
b) 0.6666… or $0.\overline{6}$

Write each fraction as a decimal.

a) $\frac{11}{4}$

b) $\frac{2}{3}$

Solution

a) $\frac{11}{4}$ means $11 \div 4$. We can change $\frac{11}{4}$ to a decimal by dividing.

$$
\begin{array}{r}
2.75 \\
4\overline{)11.00} \\
-8 \\
\hline
30 \\
-28 \\
\hline
20 \\
-20 \\
\hline
0
\end{array}
$$

← Write in zeros.

← Remainder = 0

$\frac{11}{4} = 2.75$

b) $\dfrac{2}{3}$ means $2 \div 3$. Divide.

```
      0.6 6 6 6
  3)2.0 0 0 0   ← Write in zeros, one by one.
   − 1 8
      2 0
    − 1 8
        2 0
      − 1 8
          2 0
        − 1 8
            2   ← The remainder will
                   always be 2.
```

Because the remainder will never equal zero and will *always* be 2, the exact decimal equivalent of $\dfrac{2}{3}$ is 0.6666... or the **repeating decimal** $0.\overline{6}$. (The bar over the 6 means that the 6 will repeat forever.) Therefore, we can write

$$\dfrac{2}{3} = 0.6666... \quad \text{or} \quad \dfrac{2}{3} = 0.\overline{6}$$

It is common to round repeating decimals. If we round the quotient to the nearest thousandth, we get $\dfrac{2}{3} \approx 0.667$.

[**YOU TRY 9**] Write each fraction as a decimal.

a) $\dfrac{7}{5}$ b) $\dfrac{4}{9}$

Recall that if a fraction already has a denominator that is a power of 10, we can immediately write it as a decimal. For example,

$$\dfrac{3}{10} = 0.3 \qquad \dfrac{7}{100} = 0.07 \qquad \dfrac{429}{1000} = 0.429$$

Therefore, if we can write a fraction as an *equivalent fraction* with a denominator of 10, 100, 1000, or another power of 10, then we can convert the fraction to a decimal without using long division.

EXAMPLE 10

Write each fraction as a decimal.

a) $\dfrac{3}{5}$ b) $\dfrac{11}{4}$

Solution

a) We can write $\dfrac{3}{5}$ with a denominator of 10: $\dfrac{3}{5} \cdot \dfrac{2}{2} = \dfrac{6}{10}$

Write $\dfrac{6}{10}$ as a decimal: 0.6   Therefore, $\dfrac{3}{5} = 0.6$.

b) Can we write $\dfrac{11}{4}$ as a fraction with a denominator of 10 or 100? *Yes*, we can write it with a denominator of 100.

$$\frac{11}{4} \cdot \frac{25}{25} = \frac{275}{100}$$

Write $\dfrac{275}{100}$ as a decimal: 2.75   So, $\dfrac{11}{4} = 2.75$.

Notice that this is the same as the result in Example 9a).

[YOU TRY 10] Write each fraction as a decimal.

a) $\dfrac{17}{20}$ b) $\dfrac{9}{5}$

Note

Not all fractions can be written as an equivalent fraction with a denominator that is a power of 10. In these cases, use division to write the fraction as a decimal.

ANSWERS TO [YOU TRY] **EXERCISES**

1) a) $\dfrac{1}{5}$ b) $\dfrac{89}{10,000}$ c) $4\dfrac{3}{8}$ 2) 72.244   3) a) 12.58 b) 7.907   4) a) 290.695 b) 2314.77

5) 0.00368   6) a) 0.167 b) 20.65   7) 2.29   8) a) 49,000 b) 0.0021   9) a) 1.4 b) $0.\overline{4}$
10) a) 0.85 b) 1.8

A.1 Exercises

*Additional answers can be found in the Answers to Exercises appendix.

Objective 1: Write a Decimal as a Fraction or Mixed Number

Identify the place value of each digit.

1) 413.7065   4—hundreds, 1—tens, 3—ones, 7—tenths,
0—hundredths, 6—thousandths, 5—ten thousandths

2) 1925.834   1—thousands, 9—hundreds, 2—tens, 5—ones,
8—tenths, 3—hundredths, 4—thousandths

Write each decimal in words.

3) 0.094   ninety-four thousandths

4) 38.76   thirty-eight and seventy-six hundredths

Write each word statement as a decimal.

5) one thousand seven and fifty-three hundredths   1007.53

6) twenty-six ten thousandths   0.0026

Round to the indicated place.

7) 56.4261; hundredths   56.43

8) 14.5382; tenths   14.5

9) 0.99118; tenths   1.0

10) 0.68854; thousandths   0.689

Write each decimal as a fraction or mixed number in lowest terms.

11) 0.7 $\dfrac{7}{10}$ 12) 0.3 $\dfrac{3}{10}$

13) 0.32 $\dfrac{8}{25}$ 14) 0.88 $\dfrac{22}{25}$

15) 0.075 $\dfrac{3}{40}$ 16) 0.006 $\dfrac{3}{500}$

17) 2.9554 $2\dfrac{4777}{5000}$ 18) 1.0425 $1\dfrac{17}{400}$

19) 16.008 $16\dfrac{1}{125}$ 20) 28.064 $28\dfrac{8}{125}$

Objective 2: Add and Subtract Decimals

21) Is it necessary to line up the decimal points when adding and subtracting decimals?   yes

22) In a whole number like 5, where is it understood that the decimal point goes?   after the 5 as in 5.

Perform the indicated operations.

23) $2.751 + 0.9 + 149.69$ 153.341

24) $56.07 + 7.7 + 8.4963$ 72.2663

25) $402.81 - 76.58$ 326.23

26) $310.52 - 63.27$ 247.25

27) Subtract 3.841 from 4.27 0.429

28) Subtract 5.976 from 6.12 0.144

29) $78.93 + 1519.0086 + 382 + 5.439$ 1985.3776

30) $3.25 + 8207 + 614.0909 + 49.26$ 8873.6009

31) $8 - 2.943$ 5.057

32) $5 - 3.817$ 1.183

33) Subtract 39.4 from 70.28 30.88

34) Subtract 24.6 from 50.19 25.59

Objective 3: Multiply and Divide Decimals

35) Explain, in your own words, how to multiply decimals. Answers may vary.

36) Explain, in your own words, how to divide a whole number by a number containing a decimal.
Answers may vary.

Multiply.

37) 43.7×2.59 113.183

38) 61.9×5.84 361.496

39) $(120.235)(64)$ 7695.04

40) $(140.485)(36)$ 5057.46

41) $0.317 \cdot 0.009$ 0.002853

42) $0.219 \cdot 0.008$ 0.001752

43) $(0.84)(5120.57)$ 4301.2788

44) $(0.76)(9038.13)$ 6868.9788

Divide.

45) $3\overline{)0.768}$ 0.256

46) $6\overline{)0.894}$ 0.149

47) $1.5\overline{)6.345}$ 4.23

48) $1.3\overline{)4.381}$ 3.37

49) $7.2 \div 0.04$ 180

50) $9.1 \div 0.07$ 130

51) $65.591 \div 30.65$ 2.14

52) $87.987 \div 20.85$ 4.22

53) $3 \div 2.4$ 1.25

54) $9 \div 7.2$ 1.25

55) $0.006\overline{)9}$ 1500

56) $0.005\overline{)8}$ 1600

Divide. Round the answer to the nearest hundredth.

57) $7\overline{)5.23}$ 0.75

58) $7\overline{)3.41}$ 0.49

59) $2.6\overline{)40.8}$ 15.69

60) $3.1\overline{)56.7}$ 18.29

61) $12 \div 4.5$ 2.67

62) $15 \div 2.7$ 5.56

63) Explain, in your own words, how to multiply a decimal number by 10,000.
Move the decimal point four places to the right.

64) Explain, in your own words, how to divide a decimal number by 1000.
Move the decimal point three places to the left.

Do the following operations in your head.

65) $0.71823 \cdot 10,000$ 7182.3

66) $0.53024 \cdot 1000$ 530.24

67) 96.4×100 9640

68) $81.9 \times 10,000$ 819,000

69) $1700 \times 100,000$ 170,000,000

70) $36,000 \times 10$ 360,000

71) $52.6 \div 100$ 0.526

72) $743.1 \div 1000$ 0.7431

73) $3.7 \div 10,000$ 0.00037

74) $8.2 \div 100$ 0.082

75) $95 \div 1000$ 0.095

76) $6 \div 10,000$ 0.0006

Objective 4: Write a Fraction as a Decimal

77) Describe two methods for writing $\dfrac{11}{20}$ as a decimal.
1) Divide the numerator by the denominator. 2) Write the fraction as an equivalent fraction with a denominator of 100, then write the fraction as a decimal.

78) Which method should you use to write $\dfrac{8}{9}$ as a decimal? Divide the numerator by the denominator. (This fraction cannot be written with a denominator that is a power of 10.)

Write each fraction as a decimal using long division. If the quotient is a repeating decimal, write it as such. If the answer is a nonrepeating, nonterminating decimal, round the answer to the nearest thousandth.

79) $\dfrac{2}{5}$ 0.4

80) $\dfrac{4}{5}$ 0.8

81) $\dfrac{31}{25}$ 1.24

82) $\dfrac{53}{20}$ 2.65

83) $\dfrac{27}{8}$ 3.375

84) $\dfrac{19}{8}$ 2.375

85) $\dfrac{8}{9}$ $0.\overline{8}$

86) $\dfrac{7}{9}$ $0.\overline{7}$

87) $\frac{15}{11}$ $1.\overline{36}$

88) $\frac{43}{11}$ $3.\overline{90}$

89) $\frac{9}{14}$ 0.643

90) $\frac{15}{17}$ 0.882

Write each fraction as a decimal by first writing the fraction as an equivalent fraction with a denominator that is a power of 10.

91) $\frac{1}{5}$ 0.2

92) $\frac{2}{5}$ 0.4

93) $\frac{9}{20}$ 0.45

94) $\frac{24}{25}$ 0.96

95) $\frac{337}{250}$ 1.348

96) $\frac{609}{200}$ 3.045

Mixed Exercises: Objectives 2 and 3

Perform the indicated operations. If a quotient is a repeating decimal, write it as such. If a quotient is a nonrepeating, nonterminating decimal, round the answer to the nearest thousandth.

97) $12.6 \div 0.03$ 420

98) $5.38 \times 10,000$ 53,800

99) $100.2 - 53.14$ 47.06

100) $7\overline{)0.29}$ 0.041

101) $(754.29)(18.6)$ 14,029.794

102) $23 - 19.851$ 3.149

103) $58.118 + 217.4 + 3.92 + 80$ 359.438

104) $507.4 - 286.53$ 220.87

105) $6\overline{)3.8}$ $0.6\overline{3}$

106) $641 \cdot 0.0058$ 3.7178

107) $97.6 \div 1000$ 0.0976

108) $1473.6 + 9 + 20.718 + 0.06$ 1503.378

109) $0.0007 \cdot 0.329$ 0.0002303

110) $27.9 \div 1.5$ 18.6

A.2 Percents

1 Convert Between Percents and Decimals

Objectives

1 Convert Between Percents and Decimals
2 Compute Percents Mentally
3 Solve Applied Problems Involving Percents

Percents are everywhere. A sign in a store might say, "*Save 30% off the original price.*" Or, we might read, "*The unemployment rate in the United States is 9%.*" But, what does percent mean?

Definition

Percent means *out of 100*.

If the unemployment rate in the United States is 9%, this means that 9 *out of* 100 people are unemployed. Therefore, we can write 9% as both a fraction and as a decimal.

$$9\% = \frac{9}{100} = 0.09$$

We can change percents directly to decimals by remembering that when we divide a number by 100, we move the decimal point two places to the left. For example,

$$9\% = \frac{9}{100} = 9 \div 100 = 0.09$$

Procedure How to Change a Percent to a Decimal

1) Remove the percent symbol.
2) Move the decimal point two places to the left.

Write each percent as a decimal.

a) 68%   b) 0.7%   c) 100%   d) 325%   e) $4\frac{1}{2}$%

Solution

a) $68\% = 68.\%$ Put the decimal point at the end of the number.

$= 0.68$ Remove the percent symbol, and move the decimal point two places to the left.

b) $0.7\% = 0.007$ Remove the percent symbol, and move the decimal point two places to the left.

c) $100\% = 100.\%$ Put the decimal point at the end of the number.

$= 1.00$ Remove the percent symbol, and move the decimal point two places to the left.

$= 1$

d) $325\% = 325.\%$ Put the decimal point at the end of the number.

$= 3.25$ Remove the percent symbol, and move the decimal point two places to the left.

e) $4\frac{1}{2}\% = 4.5\%$ Rewrite $4\frac{1}{2}$ as a decimal: $4\frac{1}{2} = 4.5$

$= 0.045$ Remove the percent symbol, and move the decimal point two places to the left.

Note

1) We see in Example 1c) that 100% = 1

2) Example 1d) shows that 325% = 3.25, a number that is greater than 1.

In general, we can say that a percent greater than 100 is equivalent to a decimal number greater than 1.

[YOU TRY 1]

Write each percent as a decimal.

a) 92%   b) 0.8%   c) 200%   d) 637%   e) $5\frac{3}{4}$%

If we remember the relationship between a decimal and a fraction, we can then write the decimal as a percent. For example, $0.41 = 41\%$ because

$$0.41 = \frac{41}{100} = 41\%$$

So, to change 0.41 to 41%, the decimal point has been moved *two places to the right* and we have put the percent symbol at the end of the number.

Procedure How to Change a Decimal to a Percent

1) Move the decimal point two places to the right.

2) Put the percent symbol at the end of the number.

EXAMPLE 2

Write each decimal as a percent.

a) 0.52   b) 0.034   c) 0.0089   d) 7.06   e) 1.5   f) 2

Solution

a) $0.52 = 52\%$ Move the decimal point two places to the right.
Put the % symbol at the end of the number.

b) $0.034 = 3.4\%$

c) $0.0089 = 0.89\%$

d) $7.06 = 706\%$

e) $1.5 = 150.$ Move the decimal point two places to the right.
$= 150\%$ Put the % symbol at the end of the number.

f) $2 = 2.$ Put the decimal point at the end of the number.
$= 200.$ Move the decimal point two places to the right.
$= 200\%$ Put the % symbol at the end of the number.

Note

1) Examples 2a)–2c) show that a number less than 1 is a percent less
than 100%.

2) Examples 2d)–2f) show that a number greater than 1 is a percent
greater than 100%.

3) We saw in Example 1c) that 1 = 100%.

[YOU TRY 2]

Write each decimal as a percent.

a) 0.62   b) 0.03   c) 0.0047   d) 8.95   e) 2.7   f) 4

2 Compute Percents Mentally

To find 10% of a number, we move the decimal point in that number one place to the *left*.
For example, 10% of $60 = 0.1 \times 60 = 6$.

To find percents that are multiples of ten, we can use the following procedure:

Procedure How to Find Percents That Are Multiples of Ten

If we know the value of 10% of a number,

1) find 20% of the number by multiplying the value by 2;

2) find 30% of the number by multiplying the value by 3;

3) find 40% of the number by multiplying the value by 4;

and so on.

EXAMPLE 3

In-Class Example 3

Perform each of the calculations mentally.
a) 10% of 40 b) 30% of 40
c) 5% of 40 d) 35% of 40

Answer: a) 4 b) 12
c) 2 d) 14

EXAMPLE 3

Perform each of the calculations mentally.

a) 10% of 80   b) 30% of 80   c) 5% of 80   d) 35% of 80

Solution

a) 10% of 80 = 8   Move the decimal point one place to the left.

b) In part a), we found that 10% of 80 = 8. To find 30% of 80, multiply that result by 3.

$$10\% \text{ of } 80 = 8$$
$$30\% \text{ of } 80 = 3 \cdot 8 = 24$$

30% of 80 = 24

c) Since 10% of 80 = 8, 5% of 80 is $\frac{1}{2}$ of that result.

5% is half of 10%
$$10\% \text{ of } 80 = 8$$
$$5\% \text{ of } 80 = \frac{1}{2} \cdot 8 = 4$$
Multiply 8 by $\frac{1}{2}$.

5% of 80 = 4

d) We want to think of 35% of 80 as 30% of 80 + 5% of 80.

$$35\% \text{ of } 80 = 30\% \text{ of } 80 + 5\% \text{ of } 80$$
$$= \quad 24 \quad + \quad 4$$
$$= 28$$

So, 35% of 80 = 28.

YOU TRY 3

Find

a) 10% of 120   b) 40% of 120   c) 5% of 120   d) 45% of 120

3 Solve Applied Problems Involving Percents

We can use these methods for computing percents mentally to solve everyday problems.

EXAMPLE 4

In-Class Example 4

Rashida's restaurant bill is $18.00. If she wants to leave a 15% tip, find the amount of the tip and the total amount of money Rashida will pay.

Answer: amount of tip: $2.70; total amount: $20.70

Yvonne's restaurant bill is $16.00. If she wants to leave a 15% tip, find the amount of the tip and the total amount of money Yvonne will pay.

Solution

Amount of the tip = 15% *of* $16.00

15% of $16.00 = 10% of $16.00 + 5% of $16.00
$$= \quad \$1.60 \quad + \quad \$0.80 \qquad 5\% \text{ of } \$16.00 = \$0.80 \text{ since } \frac{1}{2} \text{ of } \$1.60 = \$0.80.$$
$$= \$2.40$$

The amount of the tip is $2.40. The total amount Yvonne will pay is $16.00 + $2.40 = $18.40.

YOU TRY 4

Arturo wants to leave a 15% tip on the $28.00 restaurant bill. Find the amount of the tip and the final amount he pays.

Sometimes, rounding numbers we are working with will give us a ballpark figure of how much tip we should leave or how much money we are saving in a store. (After all, sometimes cash registers make mistakes!)

EXAMPLE 5

In-Class Example 5

A sign in a store says that all skateboards are 20% off. Shaun likes a skateboard with a regular price of $89.95. Approximately how much money will he save if he buys this item, and what is the sale price?

Answer: (Round $89.95 to $90.00.) Amount saved: $18.00; sale price: $72.00

A sign in a store says that all hoodies are 20% off. Vijay likes a hoodie with a regular price of $29.95. Approximately how much money will he save if he buys this item, and what is the sale price?

Solution

First, let's round $29.95 to $30.00 to make the number easier to work with.

Approximate amount of the discount = 20% of $30.00 = 2 · $3.00 = $6.00

Now, find the approximate sale price of the hoodie.

$$\text{Sale price} = \text{Original price} - \text{Amount of discount}$$
$$= \quad \$30.00 \quad - \quad \$6.00$$
$$= \$24.00$$

Vijay will save approximately $6.00, and the sale price will be about $24.00.

[YOU TRY 5]

Mary Beth wants to buy a new suitcase. The regular price is $79.89, and it is on sale for 40% off. Approximately how much money will she save, and what will be the approximate sale price?

Remember that in a statement like *find 10% of 70,* the *of* means multiply.

EXAMPLE 6

In-Class Example 6

Each month, 26% of Laura's income is used for rent. If she earns $2800 per month, how much does she pay in rent?

Answer: $728 per month

Each month, 28% of Hiroko's income is used for rent. If she earns $2400 per month, how much does she pay in rent?

Solution

To find the amount of Hiroko's rent, we must find 28% *of* $2400, so we must multiply:
28% *of* $2400 = 0.28 · $2400

Hiroko's rent is $672 per month.

[YOU TRY 6]

Mike earns $2350 per month, and his car payment is 8% of his income. Find his monthly car payment.

ANSWERS TO [YOU TRY] EXERCISES

1) a) 0.92   b) 0.008   c) 2   d) 6.37   e) 0.0575   2) a) 62%   b) 3%   c) 0.47%   d) 895%   e) 270%
f) 400%   3) a) 12   b) 48   c) 6   d) 54   4) $4.20; $32.20   5) $ 32.00; $48.00   6) $188

A.2 Exercises

*Additional answers can be found in the Answers to Exercises appendix.

Objective 1: Convert Between Percents and Decimals

1) What does *percent* mean? out of 100

Write each percent as a fraction and then as a decimal.

2) 37% $\frac{37}{100} = 0.37$

3) 91% $\frac{91}{100} = 0.91$

4) 74% $\frac{74}{100} = 0.74$

5) 18% $\frac{18}{100} = 0.18$

6) 2% $\frac{2}{100} = 0.02$

7) 6% $\frac{6}{100} = 0.06$

8) How do you change a percent *directly* to a decimal?
Remove the percent symbol, then move the decimal point two places to the left.

Write each percent as a decimal by moving the decimal point.

9) 53% 0.53

10) 12% 0.12

11) 40% 0.40 or 0.4

12) 80% 0.80 or 0.8

13) 7% 0.07

14) 5% 0.05

15) 0.3% 0.003

16) 0.6% 0.006

17) 8.1% 0.081

18) 9.4% 0.094

19) 100% 1

20) 400% 4

21) 220% 2.2

22) 350% 3.5

23) 0.04% 0.0004

24) 0.07% 0.0007

25) $1\frac{1}{2}\%$ 0.015

26) $5\frac{1}{2}\%$ 0.055

27) $18\frac{3}{4}\%$ 0.1875

28) $\frac{1}{4}\%$ 0.0025

29) Any percent greater than 100 is greater than which decimal number? 1

30) Any percent less than 100 is less than which decimal number? 1

Write each decimal as a percent.

31) 0.95 95%

32) 0.23 23%

33) 0.041 4.1%

34) 0.089 8.9%

35) 0.583 58.3%

36) 0.761 76.1%

37) 0.7 70%

38) 0.6 60%

39) 0.01 1%

40) 0.04 4%

41) 0.1 10%

42) 0.4 40%

43) 6.75 675%

44) 2.25 225%

45) 1.9 190%

46) 8.7 870%

47) 3 300%

48) 10 1000%

Objective 2: Compute Percents Mentally

Explain how to find the following percentages "in your head."

49) 10% of a number   Move the decimal point in the number one place to the left.

50) 60% of a number   Find 10% of the number, then multiply by 6.

51) 5% of a number   Find 10% of the number, then divide by 2.

52) 15% of a number   Find 10% of the number, and 5% of the number, then add those together.

Find each percentage "in your head."

53) a) 10% of 60 6   b) 20% of 60 12

   c) 70% of 60 42

54) a) 10% of 40 4   b) 30% of 40 12

   c) 80% of 40 32

55) a) 10% of 70 7   b) 40% of 70 28

   c) 110% of 70 77

56) a) 10% of 90 9   b) 60% of 90 54

   c) 120% of 90 108

Find each percentage "in your head."

57) 30% of 20 6

58) 40% of 50 20

59) 80% of 90 72

60) 70% of 80 56

61) 20% of 34 6.8

62) 20% of 42 8.4

63) 20% of $15.00 $3.00

64) 20% of $19.00 $3.80

65) 70% of $90.00 $63.00

66) 60% of $30.00 $18.00

Find each percentage "in your head."

67) a) 10% of 40 4   b) 5% of 40 2

   c) 15% of 40 6

68) a) 10% of 60 6   b) 5% of 60 3

   c) 15% of 60 9

69) a) 10% of 20 2   b) 5% of 20 1

   c) 35% of 20 7

70) a) 10% of 80 8   b) 5% of 80 4

   c) 45% of 80 36

Find each percentage "in your head."

71) 15% of 160   24

72) 15% of 180   27

73) 15% of 60   9

74) 15% of 40   6

75) 15% of $26.00 $3.90

76) 15% of 42 $6.30

77) 45% of 40   18

78) 35% of 60   21

79) 65% of 120   78

80) 85% of 80   68

Objective 3: Solve Applied Problems Involving Percents

Solve each problem.

81) Kanye buys concert tickets online for $220.00. The website charges an additional 10% handling fee. Find the amount of the handling fee and the total amount Kanye pays. $22.00; $242.00

82) Sidney buys tickets online for a hockey game for $240.00 plus a 10% handling fee. Find the amount of the handling fee and the total amount Sidney pays. $24.00; $264.00

83) Mark's restaurant bill is $14.00. If he wants to leave a 15% tip, find the amount of the tip and the total amount of money Mark will pay. $2.10; $16.10

84) A restaurant bill is $26.00. If Faviola wants to leave a 15% tip, find the amount of the tip and the total amount of money she will pay. $3.90; $29.90

85) Kim gets a haircut, style, and highlights for $120.00, and she wants to leave a 15% tip. Find the amount of the tip and the total amount she pays.
$18.00; $138.00

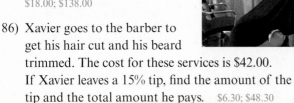

86) Xavier goes to the barber to get his hair cut and his beard trimmed. The cost for these services is $42.00. If Xavier leaves a 15% tip, find the amount of the tip and the total amount he pays. $6.30; $48.30

87) Mr. Ramsey rounds his family's restaurant bill to $70.00 and then leaves a 20% tip. Find the total cost of dinner. $84.00

88) Mrs. Child rounds her family's restaurant bill to $80.00 and then leaves a 20% tip. Find the total cost of dinner. $96.00

Solve each problem by first rounding the regular price to the nearest dollar.

89) Elzbieta finds a coat marked as 30% off its regular price of $89.99. Approximately how much money will she save if she buys the coat, and what is the sale price? $27.00; $63.00

90) Tadas sees a pair of sunglasses on sale for 40% off the regular price of $49.95. Find the approximate amount he will save as well as the sale price. $20.00; $30.00

91) Curtis finds a pair of jeans advertised as 25% off the regular price of $39.99. Find the approximate amount he will save as well as the sale price. $10.00; $30.00

92) Carrie wants to buy a tricycle that is 25% off the regular price of $59.99. Approximately how much will she save, and what is the sale price? $15.00; $45.00

93) A 32″ LCD HDTV is advertised for 20% off its regular price of $349.95. What is the approximate sale price? $280.00

94) A video game console is advertised for 30% off its regular price of $399.96. What is the approximate sale price? $280.00

Solve each problem. Find the *exact* amounts, not approximations.

95) A coffee shop served 400 customers one day, and 23% of them ordered decaffeinated coffee. How many customers ordered decaf coffee?   92

96) An apartment complex has 200 apartments, and 11% of them are empty. How many apartments are vacant?   22

97) Ting earns $3800 per month, and her student loan payment is 8% of this amount. Find her monthly student loan payment. $304.00

98) Kyle puts 4% of his monthly earnings into his savings account. If he makes $3600 per month, how much does he put into his savings account? $144.00

99) Before deductions, Manuel's paycheck for the last two weeks was $960.00. 4.5% of this amount was deducted for health insurance. How much was taken out for health insurance? $43.20

100) Before deductions, Susan's paycheck for the last two weeks was $1008.00. 13.5% of this was deducted for taxes. How much was taken out for taxes? $136.08

A.3 Factoring the Sum and Difference of Two Cubes

Objective

1 Factor the Sum and Difference of Two Cubes

In Section 7.4, we learned how to factor the difference of two squares like $x^2 - 4$. In this section, we will learn how to factor the sum and difference of two *cubes* like $k^3 + 27$ and $n^3 - 125$.

1 Factor the Sum and Difference of Two Cubes

Before we give the formulas for factoring the sum and difference of two cubes, let's look at two products.

$$(a + b)(a^2 - ab + b^2) = a(a^2 - ab + b^2) + b(a^2 - ab + b^2) \quad \text{Distributive property}$$
$$= a^3 - a^2b + ab^2 + a^2b - ab^2 + b^3 \quad \text{Distribute.}$$
$$= a^3 + b^3 \quad \text{Combine like terms.}$$

So, $(a + b)(a^2 - ab + b^2) = a^3 + b^3$, the sum of two cubes.

Let's look at another product:

$$(a - b)(a^2 + ab + b^2) = a(a^2 + ab + b^2) - b(a^2 + ab + b^2) \quad \text{Distributive property}$$
$$= a^3 + a^2b + ab^2 - a^2b - ab^2 - b^3 \quad \text{Distribute.}$$
$$= a^3 - b^3 \quad \text{Combine like terms.}$$

So, $(a - b)(a^2 + ab + b^2) = a^3 - b^3$, the difference of two cubes.

The formulas for factoring the sum and difference of two cubes, then, are as follows:

Formula Factoring the Sum and Difference of Two Cubes

$$a^3 + b^3 = (a + b)(a^2 - ab + b^2)$$
$$a^3 - b^3 = (a - b)(a^2 + ab + b^2)$$

Notice that each factorization is the product of a binomial and a trinomial.

Procedure To Factor the Sum and Difference of Two Cubes

Step 1: Identify a and b.

Step 2: Place them in the binomial factor and write the trinomial based on a and b.

Step 3: Simplify.

EXAMPLE 1

In-Class Example 1

Factor completely.
a) $p^3 + 8$ b) $t^3 - 27$

Answer:
a) $(p + 2)(p^2 - 2p + 4)$
b) $(t - 3)(t^2 + 3t + 9)$

Factor completely.

a) $k^3 + 27$ b) $n^3 - 125$

Solution

a) Use Steps 1–3 to factor.

Step 1: Identify a and b.

$$k^3 + 27$$

What do you cube to get k^3? k $(k)^3$ $(3)^3$ What do you cube to get 27? 3

So, $a = k$ and $b = 3$.

Step 2: Remember, $a^3 + b^3 = (a + b)(a^2 - ab + b^2)$.
Write the binomial factor, then write the trinomial.

Square a.   Product   Square b.
Same sign   of a and b

$$k^3 + 27 = (k + 3)[(k)^2 - (k)(3) + (3)^2]$$

Opposite sign

Step 3: Simplify: $k^3 + 27 = (k + 3)(k^2 - 3k + 9)$

b) **Step 1:** Identify a and b.

$$n^3 - 125$$

What do you cube to get n^3? n $(n)^3$ $(5)^3$ What do you cube to get 125? 5

So, $a = n$ and $b = 5$.

Step 2: Write the binomial factor, then write the trinomial. Remember,
$a^3 - b^3 = (a - b)(a^2 + ab + b^2)$.

Square a.   Product   Square b.
Same sign   of a and b

$$n^3 - 125 = (n - 5)[(n)^2 + (n)(5) + (5)^2]$$

Opposite sign

Step 3: Simplify: $n^3 - 125 = (n - 5)(n^2 + 5n + 25)$

[**YOU TRY 1**] Factor completely.

a) $m^3 + 1000$ b) $h^3 - 1$

EXAMPLE 2

Factor $64c^3 + 125d^3$ completely.

Solution

Step 1: Identify a and b.

$$64c^3 + 125d^3$$

What do you cube to get $64c^3$? $4c$ $(4c)^3$ $(5d)^3$ What do you cube to get $125d^3$? $5d$

So, $a = 4c$ and $b = 5d$.

Step 2: Write the binomial factor, then write the trinomial. Remember,
$a^3 + b^3 = (a + b)(a^2 - ab + b^2)$.

$$64c^3 + 125d^3 = (4c + 5d)[(4c)^2 - (4c)(5d) + (5d)^2]$$

Same sign · Square a. · Product of a and b · Square b. · Opposite sign

Step 3: Simplify: $64c^3 + 125d^3 = (4c + 5d)(16c^2 - 20cd + 25d^2)$.

YOU TRY 2 Factor $27p^3 - 64q^3$ completely.

Just as in the other factoring problems we've studied so far, the first step in factoring *any* polynomial should be to ask ourselves, *"Can I factor out a GCF?"*

EXAMPLE 3 Factor $5z^3 - 40$ completely.

In-Class Example 3

Factor $9c^3 - 9$ completely.

Answer: $9(c - 1)(c^2 + c + 1)$

Solution

"Can I factor out a GCF?" Yes. The GCF is 5.

$$5z^3 - 40 = 5(z^3 - 8)$$

Factor $z^3 - 8$. Use $a^3 - b^3 = (a - b)(a^2 + ab + b^2)$.

$$z^3 - 8 = (z - 2)[(z)^2 + (z)(2) + (2)^2]$$
$$(z)^3 - (2)^3 = (z - 2)(z^2 + 2z + 4)$$

Therefore, $5z^3 - 40 = 5(z^3 - 8)$
$$= 5(z - 2)(z^2 + 2z + 4)$$

YOU TRY 3 Factor completely.

a) $2t^3 - 54$ b) $2a^7 + 128ab^3$

As always, the first thing you should do when factoring is ask yourself, *"Can I factor out a GCF?"* and the last thing you should do is ask yourself, *"Can I factor again?"*

ANSWERS TO ⌈YOU TRY⌉ **EXERCISES**

1) a) $(m + 10)(m^2 - 10m + 100)$ b) $(h - 1)(h^2 + h + 1)$ 2) $(3p - 4q)(9p^2 + 12pq + 16q^2)$
3) a) $2(t - 3)(t^2 + 3t + 9)$ b) $2a(a^2 + 4b)(a^4 - 4a^2b + 16b^2)$

A.3 Exercises

Objective 1: Factor the Sum and Difference of Two Cubes

1) Find the following.

   a) 4^3 64

   b) 1^3 1

   c) 10^3 1000

   d) 3^3 27

   e) 5^3 125

   f) 2^3 8

2) If x^n is a perfect cube, then n is divisible by what number? 3

3) Fill in the blank.

   a) $(\_\_)^3 = m^3$ m

   b) $(\_\_)^3 = 27t^3$ $3t$

   c) $(\_\_)^3 = 8b^3$ $2b$

   d) $(\_\_)^3 = h^6$ h^2

4) If x^n is a perfect square *and* a perfect cube, then n is divisible by what number? 6

Complete the factorization.

5) $y^3 + 8 = (y + 2)($ $)$ $y^2 - 2y + 4$

6) $w^3 + 125 = (w + 5)($ $)$ $w^2 - 5w + 25$

7) $p^3 - 1000 = (p - 10)($ $)$ $p^2 + 10p + 100$

8) $n^3 - 64 = (n - 4)($ $)$ $n^2 + 4n + 16$

Factor completely.

9) $t^3 + 64$ $(t + 4)(t^2 - 4t + 16)$

10) $d^3 - 125$ $(d - 5)(d^2 + 5d + 25)$

11) $z^3 - 1$ $(z - 1)(z^2 + z + 1)$

12) $r^3 + 27$ $(r + 3)(r^2 - 3r + 9)$

13) $x^3 + 216$ $(x + 6)(x^2 - 6x + 36)$

14) $k^3 - 343$ $(k - 7)(k^3 + 7k + 49)$

15) $x^3 - \dfrac{1}{27}$ $\left(x - \dfrac{1}{3}\right)\left(x^2 + \dfrac{1}{3}x + \dfrac{1}{9}\right)$

16) $t^3 + \dfrac{1}{8}$ $\left(t + \dfrac{1}{2}\right)\left(t^2 - \dfrac{1}{2}t + \dfrac{1}{4}\right)$

17) $27m^3 - 125$ $(3m - 5)(9m^2 + 15m + 25)$

18) $64c^3 + 1$ $(4c + 1)(16c^2 - 4c + 1)$

19) $125y^3 - 8$ $(5y - 2)(25y^2 + 10y + 4)$

20) $27a^3 + 64$ $(3a + 4)(9a^2 - 12a + 16)$

21) $r^3 + t^3$ $(r + t)(r^2 - rt + t^2)$

22) $h^3 - k^3$ $(h - k)(h^2 + hk + k^2)$

23) $1000c^3 - d^3$ $(10c - d)(100c^2 + 10cd + d^2)$

24) $125v^3 + w^3$ $(5v + w)(25v^2 - 5vw + w^2)$

25) $8j^3 + 27k^3$ $(2j + 3k)(4j^2 - 6jk + 9k^2)$

26) $125m^3 - 27n^3$ $(5m - 3n)(25m^2 - 15mn + 9n^2)$

27) $64x^3 + 125y^3$ $(4x + 5y)(16x^2 - 20xy + 25y^2)$

28) $27a^3 - 1000b^3$ $(3a - 10b)(9a^2 + 30ab + 100b^2)$

29) $6c^3 + 48$ $6(c + 2)(c^2 - 2c + 4)$

30) $9k^3 - 9$ $9(k - 1)(k^2 + k + 1)$

31) $7v^3 - 7000w^3$ $7(v - 10w)(v^2 + 10vw + 100w^2)$

32) $216a^3 + 64b^3$ $8(3a + 2b)(9a^2 - 6ab + 4b^2)$

33) $3n^6 + 3$ $3(n^2 + 1)(n^4 - n^2 + 1)$

34) $4x^6 - 32$ $4(x^2 - 2)(x^4 + 2x^2 + 4)$

35) $h^6 - 64$ $(h + 2)(h - 2)(h^2 - 2h + 4)(h^2 + 2h + 4)$

36) $p^6 - 1$ $(p + 1)(p - 1)(p^2 - p + 1)(p^2 + p + 1)$

37) $(r - 2)^3 + 27$ $(r + 1)(r^2 - 7r + 19)$

38) $(x + 7)^3 + 8$ $(x + 9)(x^2 + 12x + 39)$

39) $(c + 4)^3 - 125$ $(c - 1)(c^2 + 13c + 61)$

40) $(p - 3)^3 - 1$ $(p - 4)(p^2 - 5p + 7)$

Instructor Answer Exercises

Chapter 1

Section 1.1

3) $\dfrac{1}{2}$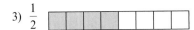

4) $\dfrac{1}{3}$

17) She multiplied the whole numbers and multiplied the fractions. She should have converted the mixed numbers to improper fractions before multiplying. Correct answer: $\dfrac{77}{6}$ or $12\dfrac{77}{6}$.

20) Convert the mixed numbers to improper fractions, then multiply the first fraction by the reciprocal of the second.

Section 1.2

4) 7 is a factor that appears five times in the product $7 \cdot 7 \cdot 7 \cdot 7 \cdot 7$.

Section 1.4

11)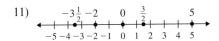

12)

13)

14)

Section 1.5

4)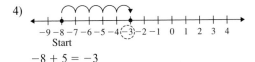

$-8 + 5 = -3$

5)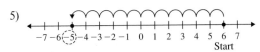

$6 - 11 = -5$

6)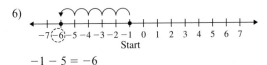

$-1 - 5 = -6$

7)

$-2 + (-7) = -9$

8)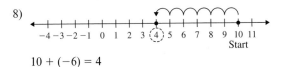

$10 + (-6) = 4$

Section 1.8

1)

Term	Coeff.
$7p^2$	7
$-6p$	-6
4	4

The constant is 4.

2)

Term	Coeff.
$-8z$	-8
$\dfrac{5}{6}$	$\dfrac{5}{6}$

The constant is $\dfrac{5}{6}$.

3)

Term	Coeff.
x^2y^2	1
$2xy$	2
$-y$	-1
11	11

The constant is 11.

4)

Term	Coeff.
w^3	1
$-w^2$	-1
$9w$	9
-5	-5

The constant is -5.

5)

Term	Coeff.
$-2g^5$	-2
$\dfrac{g^4}{5}$	$\dfrac{1}{5}$
$3.8g^2$	3.8
g	1
-1	-1

The constant is -1.

6)

Term	Coeff.
$121c^2$	121
$-d^2$	-1

There is no constant.

Chapter 1: Review Exercises

44)

90)

Term	Coeff.
$5z^4$	5
$-8z^3$	-8
$\frac{3}{5}z^2$	$\frac{3}{5}$
$-z$	-1
14	14

Chapter 1: Test

23)

25)

Term	Coeff.
$4p^3$	4
$-p^2$	-1
$\frac{1}{3}p$	$\frac{1}{3}$
-10	-10

Chapter 2

Section 2.2

1) **Step 1:** Clear parentheses and combine like terms on each side of the equation. **Step 2:** Isolate the variable. **Step 3:** Solve for the variable. **Step 4:** Check the solution.

Section 2.3

1) Eliminate the fractions by multiplying both sides of the equation by the LCD of all the fractions in the equation.

2) Eliminate the decimals by multiplying both sides of the equation by the appropriate power of 10.

63) **Step 1:** Read the problem until you understand it. **Step 2:** Choose a variable to represent an unknown quantity. **Step 3:** Translate the problem from English into an equation. **Step 4:** Solve the equation. **Step 5:** Check the answer in the original problem, and interpret the solution as it relates to the problem.

Section 2.6

3) Yes, a percent can be written as a fraction with a denominator of 100. For example, 25% can be written as $\frac{25}{100}$ or $\frac{1}{4}$.

Section 2.8

7)

8)

9)

10)

11)

12)

15)

16)

17)

18)

19)

20)

21)

22)

23)

24)

25)

26)

27)

28)

29)

30)

31)

32)

33)

34)

35)

36)

37) −3 −2 −1 0 1 2 3 38) −3 −2 −1 0 1 2 3

39) 8 9 10 11 12 13 14 40) −2 −1 0 1 2 3 4

41) −3 −2 −1 0 1 2 3 42) −3 −2 −1 0 1 2 3

43) −1 0 1 2 3 4 5 6

44) 0 1 2 3 4 5 6 7 8 9

45) −3 −2 −1 0 1 2 3

46) 10 11 12 13 14 15

47) −10 0 10 20 30 40

48) 0 2 4 6 8 10 12 14

53) −6 −5 −4 −3 −2 −1 0 1 2

54) −2 −1 0 1 2 3 4 5

55) −4 −3 −2 −1 0 1 2 3 4

56) −3 −2 −1 0 1 2 3

57) −4 −3 −2 −1 0 1 2 3 4

58) −4 −3 −2 −1 0 1 2 3 4

59) −4 −3 −2 −1 0 1 2 3 4

60) −6 −5 −4 −3 −2 −1 0 1 2

61) −6 −5 −4 −3 −2 −1 0 1 2 3 4 5

62) −5 −4 −3 −2 −1 0 1 2 3 4 5

63) 0 1 2 3 4 5 6 7

64) −3 −2 −1 0 1 2 3 4

65) −10 −8 −6 −4 −2 0 2 4 6 8

66) −5 −4 −3 −2 −1 0 1 2 3 4 5

67) 0 1 2 3 4 5 6 7

68) −5 −4 −3 −2 −1 0 1 2 3 4 5

69) 0 1 2 3 4 5 6 7 8

70) 0 1 2 3 4 5

71) 3 4 5 6 7 8 9 10 11

72) −4 −3 −2 −1 0 1 2 3 4

Chapter 2: Review Exercises

55) −4 −3 −2 −1 0 1 2 3 4

56) −4 −3 −2 −1 0 1 2 3 4

57) −3 −2 −1 0 1 2 3 4 5

58) −2 −1 0 1 2 3 4 5

59) −6 −5 −4 −3 −2 −1 0 1 2

60) 0 1 2 3 4 5 6 7 8 9

Chapter 2: Test

18) −4 −3 −2 −1 0 1 2 3 4

19) −1 0 1 2 3 4 5

Chapter 3

Section 3.1

39–42)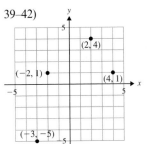

$(2, 4)$, $(-2, 1)$, $(4, 1)$, $(-3, -5)$

43–46)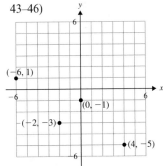

$(-6, 1)$, $(0, -1)$, $(-2, -3)$, $(4, -5)$

47–50)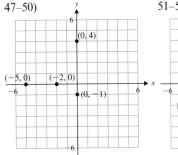

$(0, 4)$, $(-5, 0)$, $(-2, 0)$, $(0, -1)$

51–54)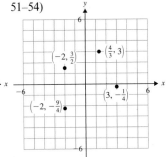

$\left(-2, \frac{3}{2}\right)$, $\left(\frac{4}{3}, 3\right)$, $\left(3, -\frac{1}{4}\right)$, $\left(-2, -\frac{9}{4}\right)$

55–56)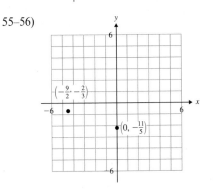

$\left(-\frac{9}{2}, -\frac{2}{3}\right)$, $\left(0, -\frac{11}{5}\right)$

61)

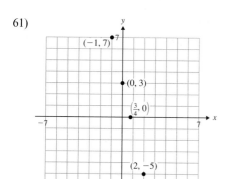

62)

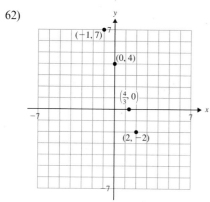

63)

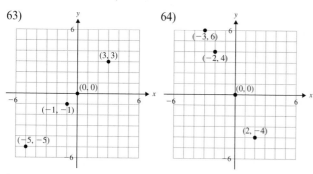

64)

65)

66)

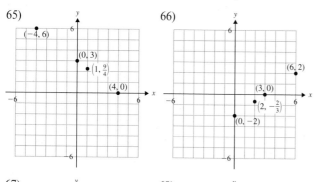

67)

68)

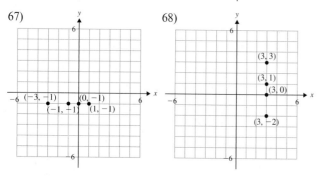

69)

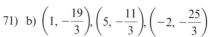

70)

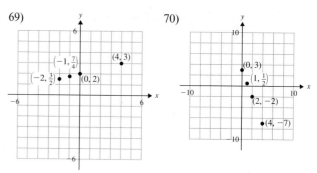

71) b) $\left(1, -\dfrac{19}{3}\right), \left(5, -\dfrac{11}{3}\right), \left(-2, -\dfrac{25}{3}\right)$

c) The x-values in part a) are multiples of the denominator of $\dfrac{2}{3}$. So, when you multiply $\dfrac{2}{3}$ by a multiple of 3 the fraction is eliminated.

82) b) In 2007, the average amount of time spent commuting to work was 28.5 minutes.

83) b)

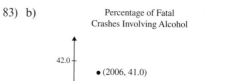

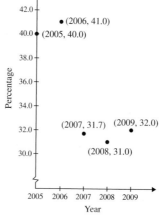

Percentage of Fatal
Crashes Involving Alcohol

84) b)

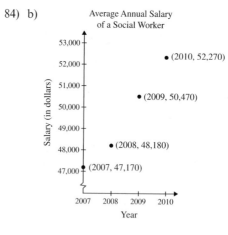

Average Annual Salary
of a Social Worker

85) b)

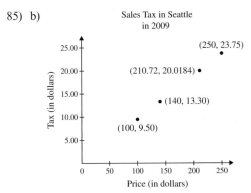

Sales Tax in Seattle in 2009

86) b)

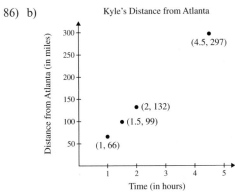

Kyle's Distance from Atlanta

Section 3.2

3)

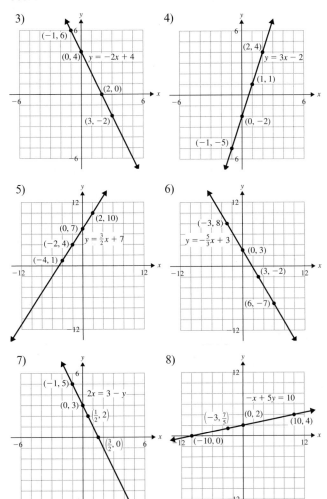

4)

5)

6)

7)

8)

9)

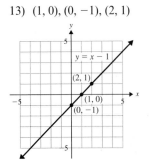

10)

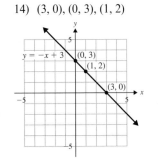

13) $(1, 0), (0, -1), (2, 1)$

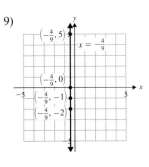

14) $(3, 0), (0, 3), (1, 2)$

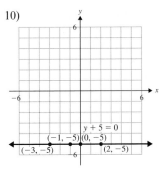

15) $(4, 0), (0, -3), (2, -\frac{3}{2})$

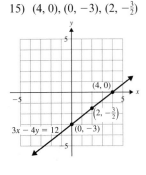

16) $(7, 0), (0, -2), (3, -\frac{8}{7})$

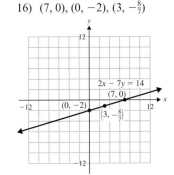

17) $(-2, 0), (0, -\frac{3}{2}), (2, -3)$

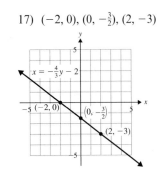

18) $(-5, 0), (0, 4), (-10, -4)$

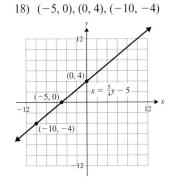

19) $(4, 0), (0, -8), (2, -4)$

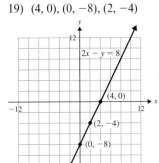

20) $(-2, 0), (0, -6), (-1, -3)$

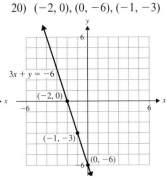

21) $(0, 0), (1, -1), (-1, 1)$

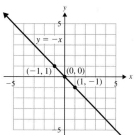

22) $(0, 0), (1, 3), (-1, -3)$

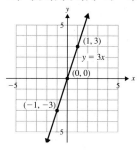

31) $\left(\frac{9}{4}, 0\right), (0, -9), (3, 3)$

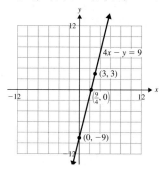

32) $\left(-5, 0\right), \left(0, -\frac{5}{3}\right), (4, -3)$

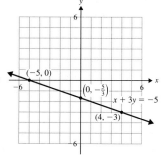

23) $(0, 0), (3, 4), (-3, -4)$

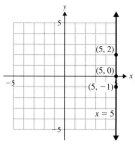

24) $(0, 0), (6, 5), (-6, -5)$

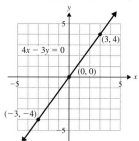

39) a)

x	y
0	0
4	5.16
7	9.03
12	15.48

$(0, 0), (4, 5.16),$
$(7, 9.03), (12, 15.48)$

b) (0, 0): If no songs are purchased, the cost is $0. (4, 5.16): The cost of downloading 4 songs is $5.16. (7, 9.03): The cost of downloading 7 songs is $9.03. (12, 15.48): The cost of downloading 12 songs is $15.48.

c)
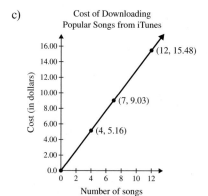

25) $(5, 0), (5, 2), (5, -1)$

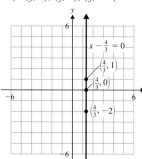

26) $(0, -4), (1, -4), (-3, -4)$

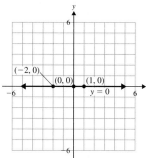

27) $(0, 0), (1, 0), (-2, 0)$

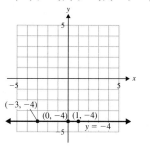

28) $(0, 0), (0, -1), (0, 2)$

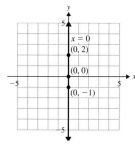

40) a)

x	y
0	0
0.5	50
1	100
1.5	150

$(0, 0), (0.5, 50),$
$(1, 100), (1.5, 150)$

b) (0, 0): The spring is not stretched if no force is applied. (0.5, 50): To stretch the spring 0.5 m, 50 N of force is needed. (1, 100): To stretch the spring 1 m, 100 N of force is needed. (1.5, 150): To stretch the spring 1.5 m, 150 N of force is needed.

c)
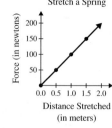

29) $\left(\frac{4}{3}, 0\right), \left(\frac{4}{3}, 1\right), \left(\frac{4}{3}, -2\right)$

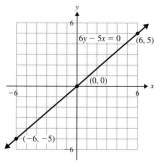

30) $(0, -1), (1, -1), (-3, -1)$

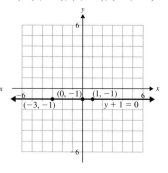

41) c)

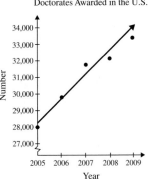

Number of Science and Engineering Doctorates Awarded in the U.S.

d) (0, 28,405); It looks like it is within 400 units of the number given by the equation.

42) c)

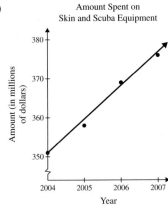

Amount Spent on Skin and Scuba Equipment

d) The y-intercept is 350.6. In 2004, approximately $350.6 mil was spent on skin and scuba diving gear. It looks like it is within about $1 mil of the plotted point.

Section 3.3

1) The slope of a line is the ratio of vertical change to horizontal change. It is $\dfrac{\text{Change in } y}{\text{Change in } x}$ or $\dfrac{\text{Rise}}{\text{Run}}$ or $\dfrac{y_2 - y_1}{x_2 - x_1}$, where (x_1, y_1) and (x_2, y_2) are points on the line.

15)

16)

39) d) $m = -0.1$; the number of injuries is decreasing by about 0.1 million, or 100,000 per year.

41)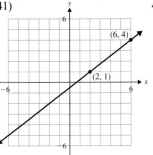

42)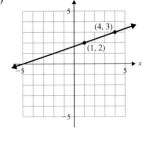

43)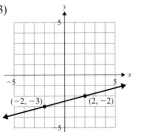

44)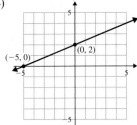

45)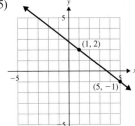

46)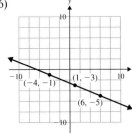

47)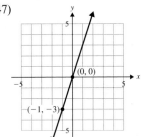

48)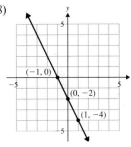

49)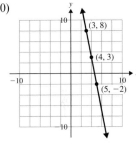

50)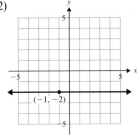

51)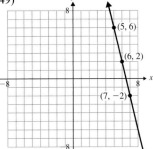

52)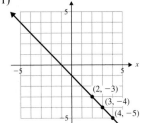

53)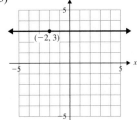

54)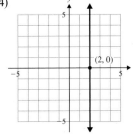

55)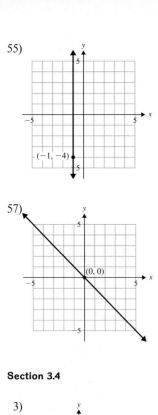

(−1, −4)

56)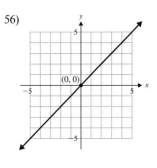

(0, 0)

9)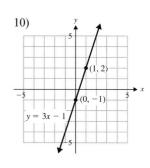

$y = -2x - 3$
(0, −3)
(1, −5)

10)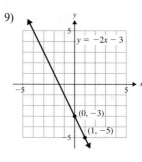

(1, 2)
(0, −1)
$y = 3x - 1$

57)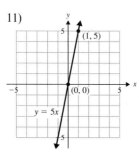

(0, 0)

11)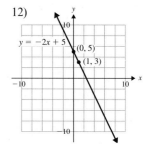

(1, 5)
(0, 0)
$y = 5x$

12)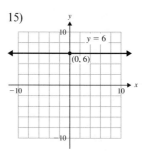

$y = -2x + 5$
(0, 5)
(1, 3)

Section 3.4

3)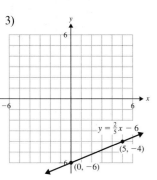

$y = \frac{2}{5}x - 6$
(5, −4)
(0, −6)

4)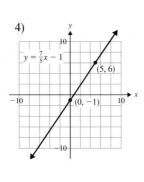

$y = \frac{7}{5}x - 1$
(5, 6)
(0, −1)

13)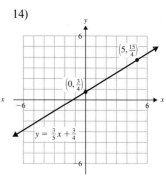

$y = -\frac{3}{2}x - \frac{7}{2}$
$\left(-2, -\frac{1}{2}\right)$
$\left(0, -\frac{7}{2}\right)$

14)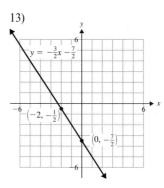

$\left(5, \frac{15}{4}\right)$
$\left(0, \frac{3}{4}\right)$
$y = \frac{3}{5}x + \frac{3}{4}$

5)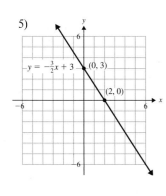

$y = -\frac{3}{2}x + 3$
(0, 3)
(2, 0)

6)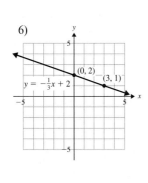

(0, 2)
$y = -\frac{1}{3}x + 2$
(3, 1)

15)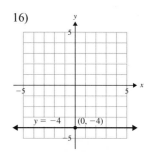

$y = 6$
(0, 6)

16)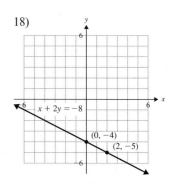

$y = -4$ (0, −4)

7)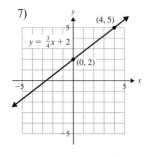

(4, 5)
$y = \frac{3}{4}x + 2$
(0, 2)

8)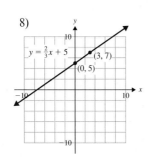

$y = \frac{2}{3}x + 5$
(3, 7)
(0, 5)

17)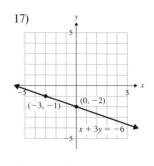

(−3, −1)   (0, −2)
$x + 3y = -6$

18)

$x + 2y = -8$
(0, −4)
(2, −5)

19)

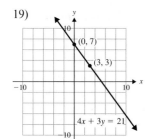

21) This cannot be written in slope-intercept form.

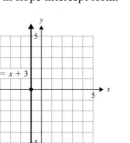

23) $y = -\frac{2}{3}x + 6$

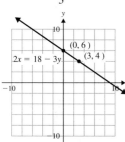

25) $y = -5$

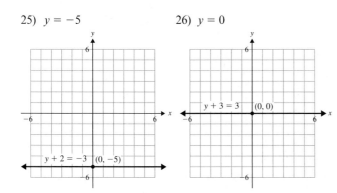

Section 3.5

102) b) The relationship between the Fahrenheit and Celsius scales is such that for every change of 9°F in the temperature, there is a change of 5°C.

20)

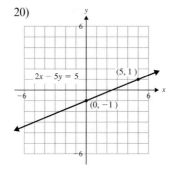

22) This cannot be written in slope-intercept form.

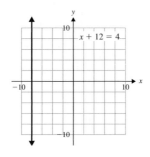

24) $y = \frac{4}{7}x + 2$

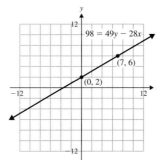

26) $y = 0$

Chapter 3: Review Exercises

11)

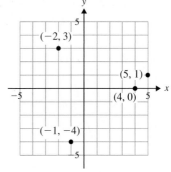

12)

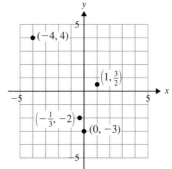

13) b)

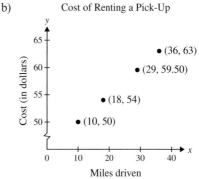

Cost of Renting a Pick-Up

15)

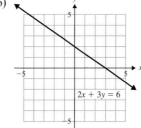

16)

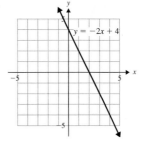

17) (2, 0), (0, −1);
    (4, 1) may vary.

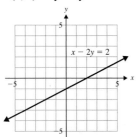

$x − 2y = 2$

18) (−1, 0), (0, 3);
    (−2, −3) may vary.

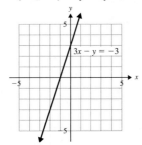

$3x − y = −3$

38)

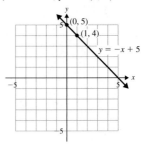

(−2, −3)

39) $m = −1$, y-int: (0, 5)

(0, 5)
(1, 4)
$y = −x + 5$

19) (2, 0), (0, 1);
    (−2, 2) may vary.

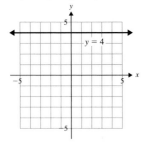

$y = −\frac{1}{2}x + 1$

20) (0, 0); (−1, 2), (1, −2)
    may vary.

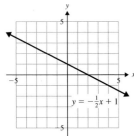

$2x + y = 0$

40) $m = 4$, y-int: (0, −2)

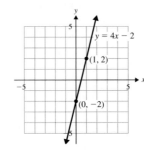

$y = 4x − 2$
(1, 2)
(0, −2)

41) $m = \frac{2}{5}$, y-int: (0, −6)

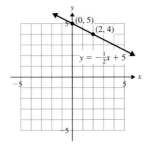

$y = \frac{2}{5}x − 6$
(5, −4)
(0, −6)

21) (0, 4); (2, 4),
    (−1, 4) may vary.

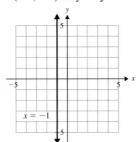

$y = 4$

22) (−1, 0); (−1, 2),
    (−1, −1) may vary.

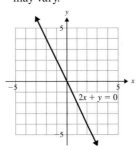

$x = −1$

42) $m = −\frac{1}{2}$, y-int: (0, 5)

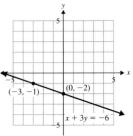

(0, 5)
(2, 4)
$y = −\frac{1}{2}x + 5$

43) $m = −\frac{1}{3}$, y-int: (0, −2)

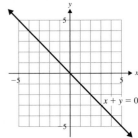

(−3, −1)
(0, −2)
$x + 3y = −6$

34)

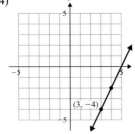

(3, −4)

35)

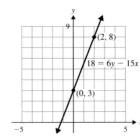

(−2, 2)

44) $m = \frac{5}{2}$, y-int: (0, 3)

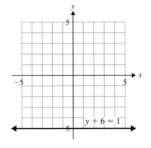

(2, 8)
$18 = 6y − 15x$
(0, 3)

45) $m = −1$, y-int: (0, 0)

$x + y = 0$

36)

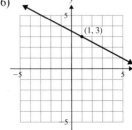

(1, 3)

37)

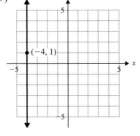

(−4, 1)

46) $m = 0$, y-int: (0, −5)

$y + 6 = 1$

Chapter 3: Test

2)

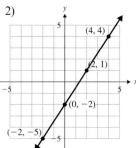

4) d)

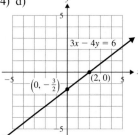

5)

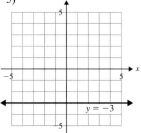

6)

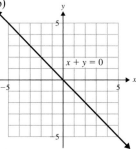

9)

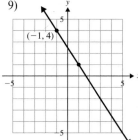

10)

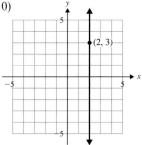

11) $y = \dfrac{3}{2}x - 5$

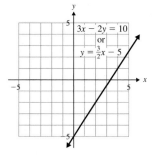

Chapter 3: Cumulative Review for Chapters 1–3

23)

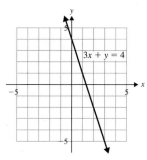

Chapter 4

Section 4.1

11) (3, 1)

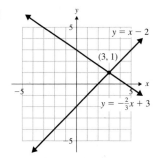

12) (2, 3)

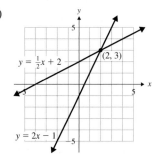

13) (2, 3)

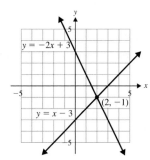

14) (2, −1)

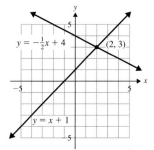

15) (4, −5)

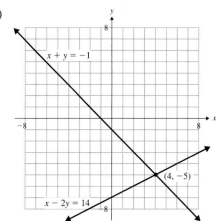

16) $(-3, -4)$

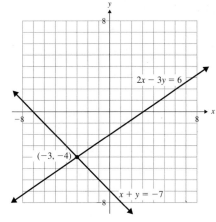

17) $(-1, -4)$

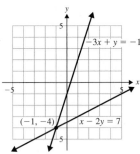

18) $(-4, 0)$

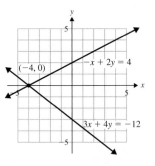

19) $\varnothing$; inconsistent system   20) $\varnothing$; inconsistent system

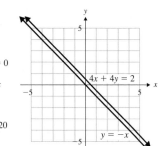

21) infinite number of
solutions of the form
$\left\{ (x, y) \left| y = \dfrac{1}{3}x - 2 \right. \right\};$
dependent equations

22) infinite number of solutions of the form
$\{(x, y) | x + y = 1\}$; dependent equations

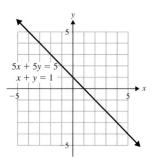

23) $(0, 2)$

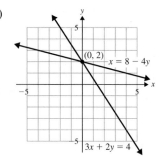

24) $(3, 3)$

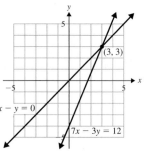

25) infinite number of
solutions of the form
$\{(x, y) | y = -3x + 1\};$
dependent equations

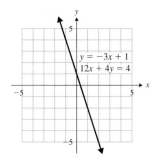

26) $\varnothing$; inconsistent system

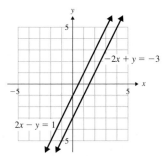

27) $(-2, 2)$

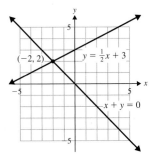

28) $(-2, 4)$

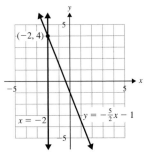

29) $(1, -1)$

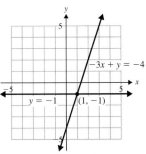

30) infinite number of solutions of the form $\{(x, y)|5x + 2y = 6\}$; dependent equations

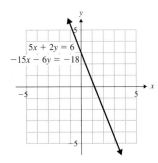

31) $\varnothing$; inconsistent system

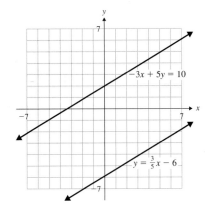

32) $(-1, -3)$

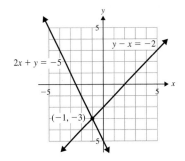

55) b) (2007, 1.4): In the year 2007, 1.4% of foreign students were from Saudi Arabia and 1.4% were from the United Kingdom.

Section 4.2

15) infinite number of solutions of the form $\{(x, y)|x - 2y = 10\}$

16) infinite number of solutions of the form $\{(x, y)|6x + y = -6\}$

21) infinite number of solutions of the form $\{(x, y)|-x + 2y = 2\}$

28) infinite number of solutions of the form $\{(x, y)|5x - 2y = 4\}$

37) infinite number of solutions of the form $\{(x, y)|y - \frac{5}{2}x = -2\}$

48) infinite number of solutions of the form $\{(x, y)|0.1x - 0.3y = -1.2\}$

55) c) (200, 120); If the cargo trailer is driven 200 miles, the cost would be the same from each company: $120.00.

d) If it is driven less than 200 miles, it is cheaper to rent from A+. If it is driven more than 200 miles, it is cheaper to rent from Rock Bottom Rental. If the trailer is driven exactly 200 miles, the cost is the same from each company.

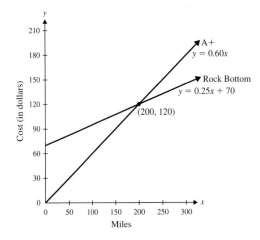

56) c) (6, 96); If the pressure washer is rented for 6 hours, the cost would be the same from each company: $96.00.

d) If the pressure washer is rented for less than 6 hours, it is cheaper to rent from Walsh Rentals. If it is rented for more than 6 hours, it is cheaper to rent from the Discount Company. If a pressure washer is rented for exactly 6 hours, the cost is the same from each company.

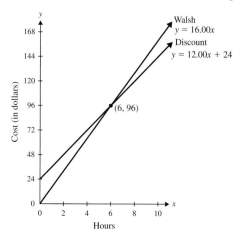

Section 4.3

19) infinite number of solutions of the form $\{(x, y)|9x - y = 2\}$

26) infinite number of solutions of the form $\{(x, y)|3x + 2y = 6\}$

30) infinite number of solutions of the form $\{(x, y)|4x - 9y = -3\}$

39) infinite number of solutions of the form $\left\{(x, y)\middle| y = \dfrac{2}{3}x - 7\right\}$

Chapter 4: Putting It All Together

1) Elimination method; none of the coefficients is 1 or -1; (5, 6).

2) Substitution; the first equation is solved for x and does not contain any fractions; (3, 5).

3) Since the coefficient of y in the second equation is 1, you can solve for y and use substitution. Or, multiply the second equation by 5 and use the elimination method. Either method will work well; $(\frac{1}{4}, -3)$.

4) Elimination method; none of the coefficients is 1 or -1; $(0, -\frac{2}{5})$.

5) Substitution; the second equation is solved for x and does not contain any fractions; (1, -7).

6) The second equation is solved for y, but it contains two fractions. Multiply this equation by 4 to eliminate the fractions, then write it in the form $Ax + By = C$. Use the elimination method to solve the system; (6, 4).

12) infinite number of solutions of the form $\{(x, y)|y = -6x + 5\}$

21) infinite number of solutions of the form $\{(x, y)|3x - y = 5\}$

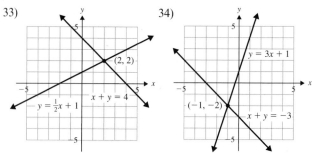

33)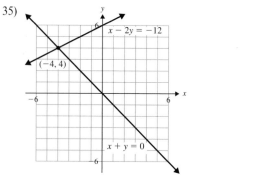

34)

35)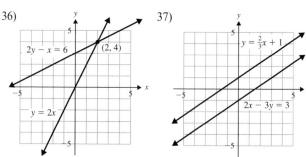

36)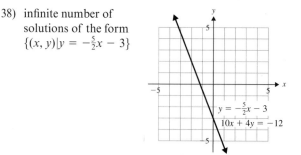

37)

38) infinite number of solutions of the form $\{(x, y)|y = -\frac{5}{2}x - 3\}$

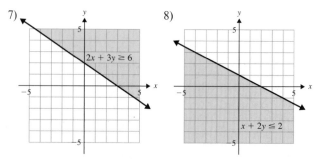

Section 4.5

7)

8)

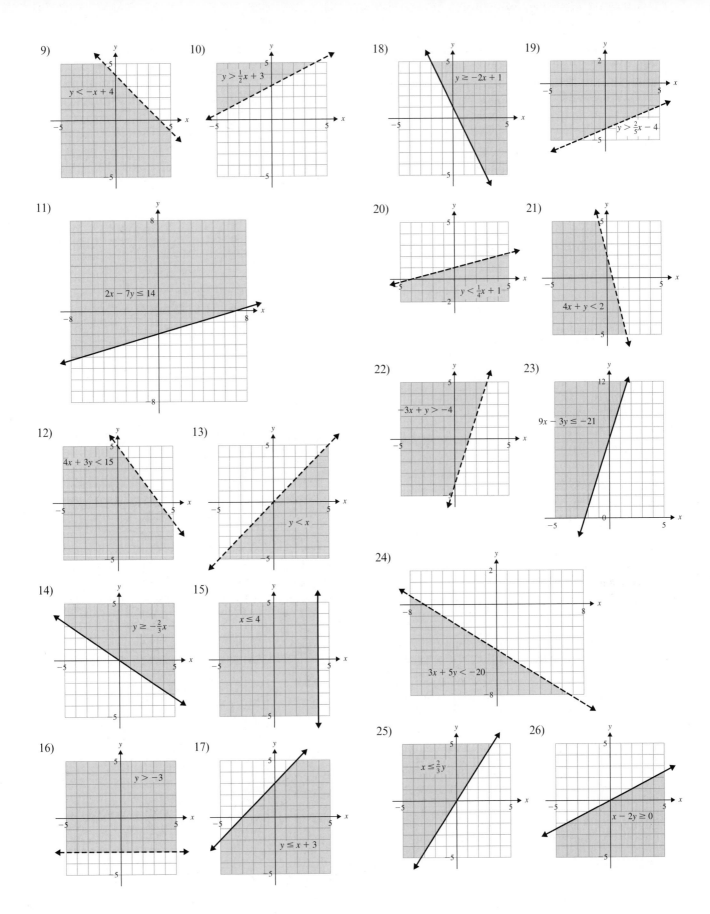

9) $y < -x + 4$

10) $y > \frac{1}{2}x + 3$

11) $2x - 7y \le 14$

12) $4x + 3y < 15$

13) $y < x$

14) $y \ge -\frac{2}{3}x -$

15) $x \le 4$

16) $y > -3$

17) $y \le x + 3$

18) $y \ge -2x + 1$

19) $y > \frac{2}{5}x - 4$

20) $y < \frac{1}{4}x + 1$

21) $4x + y < 2$

22) $-3x + y > -4$

23) $9x - 3y \le -21$

24) $-3x + 5y < -20$

25) $x \le \frac{2}{3}y$

26) $x - 2y \ge 0$

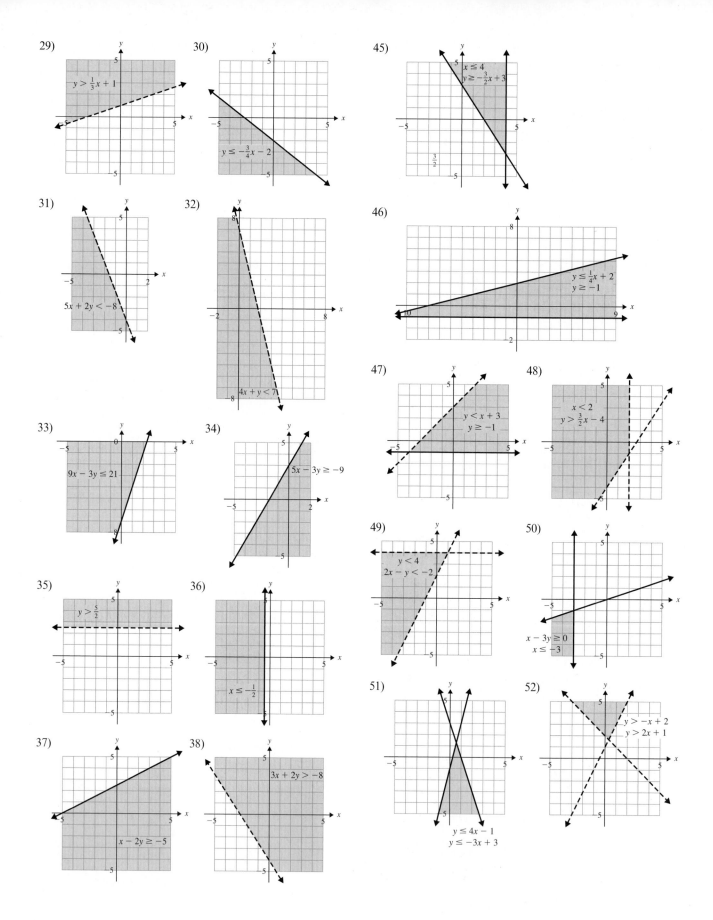

29) $y > \frac{1}{3}x + 1$

30) $y \leq -\frac{3}{4}x - 2$

31) $5x + 2y < -8$

32) $4x + y < 7$

33) $9x - 3y \leq 21$

34) $5x - 3y \geq -9$

35) $y > \frac{5}{2}$

36) $x \leq -\frac{1}{2}$

37) $x - 2y \geq -5$

38) $3x + 2y > -8$

45) $x \leq 4$
$y \geq -\frac{3}{2}x + 3$
$\frac{3}{2}$

46) $y \leq \frac{1}{4}x + 2$
$y \geq -1$

47) $y < x + 3$
$y \geq -1$

48) $x < 2$
$y > \frac{3}{2}x - 4$

49) $y < 4$
$2x - y < -2$

50) $x - 3y \geq 0$
$x \leq -3$

51) $y \leq 4x - 1$
$y \leq -3x + 3$

52) $y > -x + 2$
$y > 2x + 1$

53)

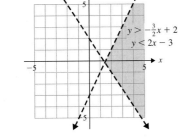

$y > -\frac{3}{2}x + 2$
$y < 2x - 3$

54)

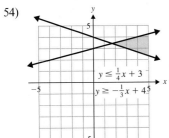

$y \le \frac{1}{4}x + 3$
$y \ge -\frac{1}{3}x + 4$

55)

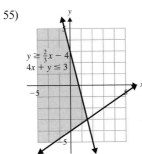

$y \ge \frac{2}{3}x - 4$
$4x + y \le 3$

56)

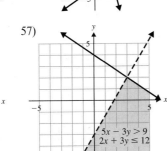

$y - 2x \le 1$
$y \ge -\frac{1}{5}x - 2$

57)

57)

$5x - 3y > 9$
$2x + 3y \le 12$

58)

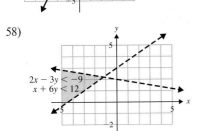

$2x - 3y < -9$
$x + 6y < 12$

59)

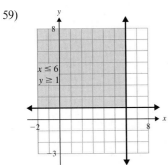

$x \le 6$
$y \ge 1$

60)

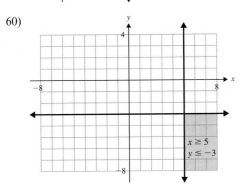

$x \ge 5$
$y \le -3$

61)

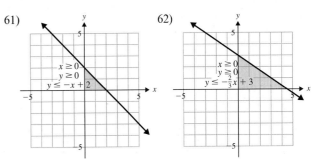

$x \ge 0$
$y \ge 0$
$y \le -x + 2$

62)

$x \ge 0$
$y \ge 0$
$y \le -\frac{2}{3}x + 3$

63)

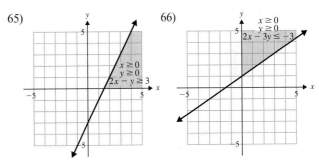

$x \ge 0$
$y \ge 0$
$y \le -3x + 4$

64)

$x \ge 0$
$y \ge 0$
$y \le -x + 1$

65)

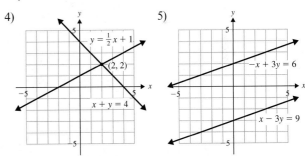

$x \ge 0$
$y \ge 0$
$2x - y \ge 3$

66)

$x \ge 0$
$y \ge 0$
$2x - 3y \le -3$

Chapter 4: Review Exercises

4)

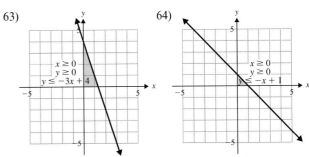

$y = \frac{1}{2}x + 1$
$(2, 2)$
$x + y = 4$

5)

$-x + 3y = 6$
$x - 3y = 9$

6) infinite number of solutions of the form
$\{(x, y) \mid -2x + y = -4\}$

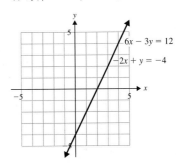

$6x - 3y = 12$
$-2x + y = -4$

7)

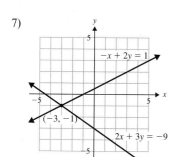
$-x + 2y = 1$
$(-3, -1)$
$2x + 3y = -9$

51)

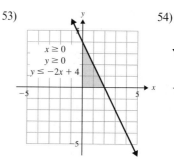
$-3x + y < 0$
$3x + 2y > 4$

52)

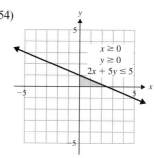
$x + 3y \le 3$
$x - y \ge 2$

10) b) (2008, 38); In 2008, Alabama and Ohio each had
38 million barrels of reserve crude oil.

c) 2005–2006; During this time, Alabama's reserve
decreased at a greater rate than at any other time on
the graph.

12) infinite number of solutions of the form
$\{(x, y)\,|\,y = -6x + 5\}$

53)

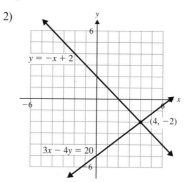
$x \ge 0$
$y \ge 0$
$y \le -2x + 4$

54)

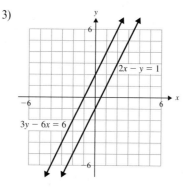
$x \ge 0$
$y \ge 0$
$2x + 5y \le 5$

42)

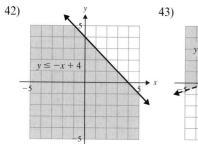
$y \le -x + 4$

43)

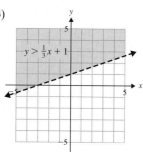
$y > \frac{1}{3}x + 1$

Chapter 4: Test

2)

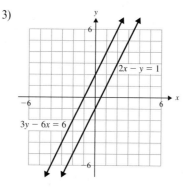
$y = -x + 2$
$3x - 4y = 20$
$(4, -2)$

44)

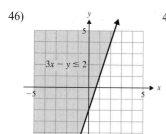
$x > 2$

45)

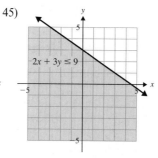
$2x + 3y \le 9$

46)

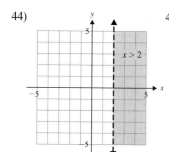
$3x - y \le 2$

47)

$y > 0$

3)

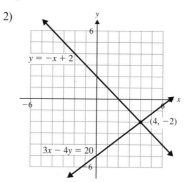
$2x - y = 1$
$3y - 6x = 6$

4) c) Virginia from Sept. 2008 to Sept. 2009; During this time,
Virginia had the largest increase in its unemployment
rate of all of the times represented on the graph.

49)

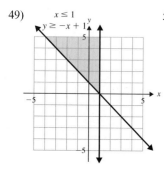
$x \le 1$
$y \ge -x + 1$

50)

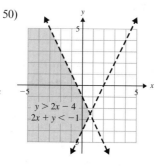
$y > 2x - 4$
$2x + y < -1$

19)

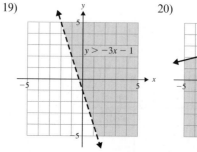
$y > -3x - 1$

20)

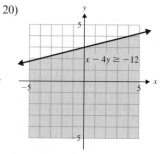
$x - 4y \ge -12$

21)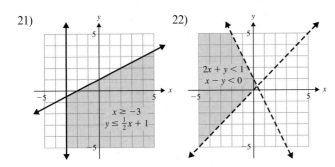

22)

$2x + y < 1$
$x - y < 0$

$x \geq -3$
$y \leq \frac{1}{2}x + 1$

Chapter 4: Cumulative Review for Chapters 1–4

15)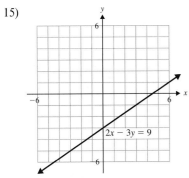

$2x - 3y = 9$

25)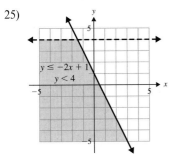

$y \leq -2x + 1$
$y < 4$

Chapter 5

Section 5.1A

23) $(3 + 4)^2 = 49$, $3^2 + 4^2 = 25$. They are not equivalent because when evaluating $(3 + 4)^2$, first add $3 + 4$ to get 7, then square the 7.

24) $(7 - 3)^2 = 16$, $7^2 - 3^2 = 40$. They are not equivalent because when evaluating $(7 - 3)^2$, first subtract $7 - 3$ to get 4, then square the 4.

26) No. $(-2)^4 = (-2) \cdot (-2) \cdot (-2) \cdot (-2) = 16$;
$-2^4 = -1 \cdot 2^4 = -16$

Chapter 6

Section 6.1

1) Yes; the coefficients are real numbers and the exponents are whole numbers.

2) Yes; the coefficients are real numbers and the exponents are whole numbers.

3) No; one of the exponents is a negative number.

4) Yes; the coefficient is a real number and the exponent is a whole number.

5) No; two of the exponents are fractions.

6) No; there is a variable in the denominator.

19)

Term	Coeff.	Degree
$3y^4$	3	4
$7y^3$	7	3
$-2y$	-2	1
8	8	0

Degree of polynomial is 4.

20)

Term	Coeff.	Degree
$6a^2$	6	2
$2a$	2	1
-11	-11	0

Degree of polynomial is 2.

21)

Term	Coeff.	Degree
$-4x^2y^3$	-4	5
$-x^2y^2$	-1	4
$\frac{2}{3}xy$	$\frac{2}{3}$	2
$5y$	5	1

Degree of polynomial is 5.

22)

Term	Coeff.	Degree
$3c^2d^2$	3	4
$0.7c^2d$	0.7	3
cd	1	2
-1	-1	0

Degree of polynomial is 4.

77) No. If the coefficients of the like terms are opposite in sign, their sum will be zero.
Example: $(3x^2 + 4x + 5) + (2x^2 - 4x + 1) = 5x^2 + 6$

Section 6.2

103) No. The order of operations tells us to perform exponents, $(r + 2)^2$, before multiplying by 3.

Section 6.4

39) No. For example, $\dfrac{12x + 8}{3x} = 4 + \dfrac{4}{3x}$. The quotient is not a polynomial because one term has a variable in the denominator.

Chapter 6: Review Exercises

1)

Term	Coeff.	Degree
$7s^3$	7	3
$-9s^2$	-9	2
s	1	1
6	6	0

Degree of polynomial is 3.

2)

Term	Coeff.	Degree
a^2b^3	1	5
$7ab^2$	7	3
$-2ab$	-2	2
$9b$	9	1

Degree of polynomial is 5.

79) $-\dfrac{5}{3}x^3y^2 - 4xy^2 + 1 - \dfrac{5}{4y^2}$

Chapter 6: Cumulative Review for Chapters 1–6

9) x-int: $(8, 0)$; y-int: $(0, -3)$ 10)

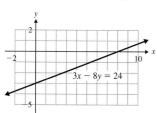

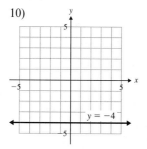

Chapter 7

Section 7.4

7) The middle term does not equal $2(2a)(-3)$. It would have to equal $-12a$ to be a perfect square trinomial.

8) Only one term, x^2, is a perfect square, so it can't be a perfect square trinomial.

63) $\left(\dfrac{1}{3}t + \dfrac{5}{2}\right)\left(\dfrac{1}{3}t - \dfrac{5}{2}\right)$ 64) $\left(\dfrac{4}{3}x + \dfrac{1}{7}\right)\left(\dfrac{4}{3}x - \dfrac{1}{7}\right)$

Section 7.5

5) If the product of two quantities equals 0, then one or both of the quantities must be zero.

6) No. The zero product rule says that we must also let the other factor, m, equal zero so that $m = 0$ is another solution. The solution set is $\{0, 8\}$.

9) $\left\{\dfrac{3}{2}, 10\right\}$ 10) $\left\{-9, \dfrac{4}{5}\right\}$ 27) $\left\{\dfrac{4}{3}, 2\right\}$

28) $\left\{-\dfrac{5}{2}, -1\right\}$ 29) $\left\{-\dfrac{3}{4}, \dfrac{2}{3}\right\}$ 30) $\left\{-\dfrac{1}{4}, \dfrac{5}{2}\right\}$

37) $\left\{-\dfrac{7}{10}, \dfrac{7}{10}\right\}$ 38) $\left\{-\dfrac{9}{2}, \dfrac{9}{2}\right\}$ 39) $\left\{-\dfrac{6}{5}, -1\right\}$

40) $\left\{-\dfrac{3}{2}, \dfrac{4}{3}\right\}$

Section 7.6

44) g) The 10-in. shell would need to be 410 ft farther horizontally from the point of explosion than the 3-in. shell.

6)

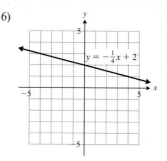

Chapter 8

Section 8.1

9) Set the denominator equal to zero and solve for the variable. That value cannot be substituted into the expression because it will make the denominator equal to zero.

10) No. A rational expression equals zero when its numerator equals zero. If $x = 0$, then the numerator equals 9. If x is a positive or a negative number, then x^2 is positive and $x^2 + 9$ is positive, too. Therefore, $x^2 + 9$ cannot equal zero; it is always positive.

24) a) $\dfrac{11}{6}$

b) never undefined—any real number may be substituted for m

71) possible answers:

$\dfrac{-u-7}{u-2}, \dfrac{-(u+7)}{u-2}, \dfrac{u+7}{2-u}, \dfrac{u+7}{-(u-2)}, \dfrac{u+7}{-u+2}$

72) possible answers:

$\dfrac{-8y+1}{2y+5}, \dfrac{1-8y}{2y+5}, \dfrac{-(8y-1)}{2y+5}, \dfrac{8y-1}{-2y-5}, \dfrac{8y-1}{-(2y+5)}$

73) possible answers:

$\dfrac{-9+5t}{2t-3}, \dfrac{5t-9}{2t-3}, \dfrac{-(9-5t)}{2t-3}, \dfrac{9-5t}{-2t+3}, \dfrac{9-5t}{3-2t}, \dfrac{9-5t}{-(2t-3)}$

74) possible answers:

$-\dfrac{w-6}{4w-7}, \dfrac{6-w}{4w-7}, \dfrac{w-6}{7-4w}, -\dfrac{6-w}{-4w+7}$

75) possible answers:

$-\dfrac{12m}{m^2-3}, \dfrac{12m}{-(m^2-3)}, \dfrac{12m}{-m^2+3}, \dfrac{12m}{3-m^2}$

76) possible answers:

$-\dfrac{9x+11}{18-x}, \dfrac{9x+11}{x-18}, \dfrac{-(9x+11)}{18-x}, \dfrac{9x+11}{-(18-x)}$

Section 8.2

14) $\dfrac{6k^2}{(5k+6)^3}$ 32) $\dfrac{1}{2(x-7)(x+4)^2}$ 34) $\dfrac{2a-5}{4a^2}$

63) $\dfrac{9}{8n^6} = \dfrac{27}{24n^6}$; $\dfrac{2}{3n^2} = \dfrac{16n^4}{24n^6}$

64) $\dfrac{5}{8a} = \dfrac{25a^4}{40a^5}$; $\dfrac{7}{10a^5} = \dfrac{28}{40a^5}$

65) $\dfrac{6}{4a^3b^5} = \dfrac{6a}{4a^4b^5}$; $\dfrac{6}{a^4b} = \dfrac{24b^4}{4a^4b^5}$

66) $\dfrac{3}{x^3y} = \dfrac{15y^4}{5x^3y^5}$; $\dfrac{6}{5xy^5} = \dfrac{6x^2}{5x^3y^5}$

67) $\dfrac{r}{5} = \dfrac{r^2 - 4r}{5(r - 4)}$; $\dfrac{2}{r - 4} = \dfrac{10}{5(r - 4)}$

68) $\dfrac{t}{5t - 1} = \dfrac{7t}{7(5t - 1)}$; $\dfrac{8}{7} = \dfrac{40t - 8}{7(5t - 1)}$

69) $\dfrac{3}{d} = \dfrac{3d - 27}{d(d - 9)}$; $\dfrac{7}{d - 9} = \dfrac{7d}{d(d - 9)}$

70) $\dfrac{5}{c} = \dfrac{5c + 10}{c(c + 2)}$; $\dfrac{4}{c + 2} = \dfrac{4c}{c(c + 2)}$

71) $\dfrac{m}{m + 7} = \dfrac{m^2}{m(m + 7)}$; $\dfrac{3}{m} = \dfrac{3m + 21}{m(m + 7)}$

72) $\dfrac{z}{z - 4} = \dfrac{z^2}{z(z - 4)}$; $\dfrac{5}{z} = \dfrac{5z - 20}{z(z - 4)}$

73) $\dfrac{a}{30a - 15} = \dfrac{2a}{30(2a - 1)}$; $\dfrac{1}{12a - 6} = \dfrac{5}{30(2a - 1)}$

74) $\dfrac{7}{24x - 16} = \dfrac{21}{24(3x - 2)}$; $\dfrac{x}{18x - 12} = \dfrac{4x}{24(3x - 2)}$

75) $\dfrac{9}{k - 9} = \dfrac{9k + 27}{(k - 9)(k + 3)}$; $\dfrac{5k}{k + 3} = \dfrac{5k^2 - 45k}{(k - 9)(k + 3)}$

76) $\dfrac{6}{h + 1} = \dfrac{6h + 42}{(h + 1)(h + 7)}$; $\dfrac{11h}{h + 7} = \dfrac{11h^2 + 11h}{(h + 1)(h + 7)}$

77) $\dfrac{3}{a + 2} = \dfrac{9a + 12}{(a + 2)(3a + 4)}$; $\dfrac{2a}{3a + 4} = \dfrac{2a^2 + 4a}{(a + 2)(3a + 4)}$

78) $\dfrac{b}{6b - 5} = \dfrac{b^2 - 9b}{(6b - 5)(b - 9)}$; $\dfrac{8}{b - 9} = \dfrac{48b - 40}{(6b - 5)(b - 9)}$

79) $\dfrac{9y}{y^2 - y - 42} = \dfrac{18y^2}{2y(y + 6)(y - 7)}$; $\dfrac{3}{2y^2 + 12y} = \dfrac{3y - 21}{2y(y + 6)(y - 7)}$

80) $\dfrac{12q}{q^2 - 6q - 16} = \dfrac{24q^2}{2q(q + 2)(q - 8)}$; $\dfrac{4}{2q^2 - 16q} = \dfrac{4q + 8}{2q(q + 2)(q - 8)}$

81) $\dfrac{c}{c^2 + 9c + 18} = \dfrac{c^2 + 6c}{(c + 6)^2(c + 3)}$; $\dfrac{11}{c^2 + 12c + 36} = \dfrac{11c + 33}{(c + 6)^2(c + 3)}$

82) $\dfrac{z}{z^2 - 8z + 16} = \dfrac{z^2 + 8z}{(z - 4)^2(z + 8)}$; $\dfrac{9z}{z^2 + 4z - 32} = \dfrac{9z^2 - 36z}{(z - 4)^2(z + 8)}$

83) $\dfrac{11}{g - 3}$ already has the LCD. $\dfrac{4}{3 - g} = -\dfrac{4}{g - 3}$

84) $\dfrac{6}{n - 9}$ already has the LCD. $\dfrac{1}{9 - n} = -\dfrac{1}{n - 9}$

85) $\dfrac{4}{3x - 4} = \dfrac{12x + 16}{(3x + 4)(3x - 4)}$; $\dfrac{7x}{16 - 9x^2} = -\dfrac{7x}{(3x + 4)(3x - 4)}$

86) $\dfrac{12}{5k - 2} = \dfrac{60k + 24}{(5k + 2)(5k - 2)}$; $\dfrac{4k}{4 - 25k^2} = -\dfrac{4k}{(5k + 2)(5k - 2)}$

87) $\dfrac{2}{z^2 + 3z} = \dfrac{6z + 18}{3z(z + 3)^2}$; $\dfrac{6}{3z^2 + 9z} = \dfrac{6z + 18}{3z(z + 3)^2}$; $\dfrac{8}{z^2 + 6z + 9} = \dfrac{24z}{3z(z + 3)^2}$

88) $\dfrac{4}{w^2 - 4w} = \dfrac{28w - 112}{7w(w - 4)^2}$; $\dfrac{6}{7w^2 - 28w} = \dfrac{6w - 24}{7w(w - 4)^2}$; $\dfrac{11}{w^2 - 8w + 16} = \dfrac{77w}{7w(w - 4)^2}$

89) $\dfrac{t}{t^2 - 13t + 30} = \dfrac{t^2 + 3t}{(t + 3)(t - 3)(t - 10)}$; $\dfrac{6}{t - 10} = \dfrac{6t^2 - 54}{(t + 3)(t - 3)(t - 10)}$; $\dfrac{7}{t^2 - 9} = \dfrac{7t - 70}{(t + 3)(t - 3)(t - 10)}$

90) $-\dfrac{2}{a + 2} = -\dfrac{2a^2 - 6a + 4}{(a + 2)(a - 2)(a - 1)}$; $\dfrac{a}{a^2 - 4} = \dfrac{a^2 - a}{(a + 2)(a - 2)(a - 1)}$; $\dfrac{15}{a^2 - 3a + 2} = \dfrac{15a + 30}{(a + 2)(a - 2)(a - 1)}$

Section 8.4

17) b) Multiply the numerator and denominator of $\dfrac{4}{9b^2}$ by $2b^2$, and multiply the numerator and denominator of $\dfrac{5}{6b^4}$ by 3.

18) b) Multiply the numerator and denominator of $\dfrac{8}{x - 3}$ by x, and multiply the numerator and denominator of $\dfrac{2}{x}$ by $x - 3$.

45) No. If the sum is rewritten as $\dfrac{9}{x-6} - \dfrac{4}{x-6}$, then the LCD $= x - 6$. If the sum is rewritten as $\dfrac{-9}{6-x} + \dfrac{4}{6-x}$, then the LCD is $6 - x$.

47) $\dfrac{6}{q-4}$ or $-\dfrac{6}{4-q}$ 　　48) $\dfrac{4}{z-9}$ or $-\dfrac{4}{9-z}$

49) $\dfrac{26}{f-7}$ or $-\dfrac{26}{7-f}$ 　　50) $\dfrac{9}{a-b}$ or $-\dfrac{9}{b-a}$

51) $\dfrac{8-x}{x-4}$ or $\dfrac{x-8}{4-x}$ 　52) $\dfrac{-m-11}{m-5}$ or $\dfrac{m+11}{5-m}$

55) $\dfrac{3(1+2u)}{2u-3v}$ or $-\dfrac{3(1+2u)}{3v-2u}$

56) $\dfrac{3(c+3)}{11b-5c}$ or $-\dfrac{3(c+3)}{5c-11b}$

57) $-\dfrac{2(x-1)}{(x+3)(x-3)}$ 　58) $-\dfrac{4(y+5)}{(y+8)(y-8)}$

59) $\dfrac{3(3a+4)}{(2a+3)(2a-3)}$ 　60) $\dfrac{3(4b+5)}{(3b+5)(3b-5)}$

Section 8.5

1) Method 1: Rewrite it as a division problem, then simplify.

$$\dfrac{2}{9} \div \dfrac{5}{18} = \dfrac{2}{\cancel{9}_{1}} \cdot \dfrac{\cancel{18}^{2}}{5} = \dfrac{4}{5}$$

Method 2: Multiply the numerator and denominator by 18, the LCD of $\dfrac{2}{9}$ and $\dfrac{5}{18}$. Then simplify.

$$\dfrac{\cancel{18}^{2}\left(\dfrac{2}{9}\right)}{\cancel{18}_{1}\left(\dfrac{5}{18}\right)} = \dfrac{4}{5}$$

2) Method 1: Subtract the fractions in the numerator and add the fractions in the denominator. Then rewrite the complex fraction as a division problem and simplify.

$$\dfrac{\dfrac{3}{2} - \dfrac{1}{5}}{\dfrac{1}{10} + \dfrac{3}{5}} = \dfrac{\dfrac{15}{10} - \dfrac{2}{10}}{\dfrac{1}{10} + \dfrac{6}{10}}$$

$$= \dfrac{\dfrac{13}{10}}{\dfrac{7}{10}}$$

$$= \dfrac{13}{10} \div \dfrac{7}{10}$$

$$= \dfrac{13}{\cancel{10}} \cdot \dfrac{\cancel{10}}{7}$$

$$= \dfrac{13}{7}$$

Method 2: Multiply the numerator and denominator by 10, the LCD of all of the fractions. Simplify.

$$\dfrac{10\left(\dfrac{3}{2} - \dfrac{1}{5}\right)}{10\left(\dfrac{1}{10} + \dfrac{3}{5}\right)} = \dfrac{15 - 2}{1 + 6} = \dfrac{13}{7}$$

Chapter 8: Review Exercises

7) a) $-\dfrac{5}{3}, 2$

b) never undefined—any real number may be substituted for m

8) a) 3

b) never undefined—any real number may be substituted for d

17) possible answers:

$$\dfrac{-4n-1}{5-3n}, \dfrac{-(4n+1)}{5-3n}, \dfrac{4n+1}{3n-5}, \dfrac{4n+1}{-5+3n}, \dfrac{4n+1}{-(5-3n)}$$

18) possible answers:

$$\dfrac{-u+8}{u+2}, \dfrac{8-u}{u+2}, \dfrac{u-8}{-u-2}, \dfrac{-(u-8)}{u+2}, \dfrac{u-8}{-(u+2)}$$

46) $\dfrac{4k^2 - 24k}{(k-6)(k-9)}$ 　47) $\dfrac{6z}{z(2z+5)}$

51) $\dfrac{8c}{c^2 + 5c - 24} = \dfrac{8c^2 - 24c}{(c-3)^2(c+8)}$;

$\dfrac{5}{c^2 - 6c + 9} = \dfrac{5c+40}{(c-3)^2(c+8)}$

53) $\dfrac{7}{2q^2 - 12q} = \dfrac{7q+42}{2q(q+6)(q-6)}$;

$\dfrac{3q}{36 - q^2} = -\dfrac{6q^2}{2q(q+6)(q-6)}$;

$\dfrac{q-5}{2q^2 + 12q} = \dfrac{q^2 - 11q + 30}{2q(q+6)(q-6)}$

Chapter 8: Cumulative Review for Chapters 1–8

6) x-int: $\left(\dfrac{3}{2}, 0\right)$; y-int: $(0, -2)$

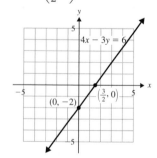

Chapter 9

Section 9.1

1) False; the $\sqrt{}$ symbol means to find only the positive square root of 100. $\sqrt{100} = 10$

33) 3.3

$\sqrt{11}$

0 1 2 3 4 5 6 7 8 9

34) 1.7

$\sqrt{3}$

0 1 2 3 4 5

35) 1.4

$\sqrt{2}$

0 1 2 3 4 5 6 7 8 9

36) 2.2

$\sqrt{5}$

0 1 2 3 4 5 6 7 8 9

37) 5.7

$\sqrt{33}$

2 3 4 5 6 7

38) 6.2

$\sqrt{39}$

2 3 4 5 6 7

39) 7.4

$\sqrt{55}$

3 4 5 6 7 8 9

40) 8.5

$\sqrt{72}$

5 6 7 8 9 10

Section 9.2

97) (i) Its radicand will not contain any factors that are perfect cubes.
   (ii) The radicand will not contain fractions.
   (iii) There will be no radical in the denominator of a fraction.

98) (i) Its radicand will not contain any factors that are perfect fourth powers.
   (ii) The radicand will not contain fractions.
   (iii) There will be no radical in the denominator of a fraction.

Section 9.5

61) He multiplied incorrectly: $\dfrac{9}{\sqrt[3]{2}} \cdot \dfrac{\sqrt[3]{2}}{\sqrt[3]{2}} = \dfrac{9\sqrt[3]{2}}{\sqrt[3]{4}}$.

The correct way is $\dfrac{9}{\sqrt[3]{2}} \cdot \dfrac{\sqrt[3]{4}}{\sqrt[3]{4}} = \dfrac{9\sqrt[3]{4}}{\sqrt[3]{8}} = \dfrac{9\sqrt[3]{4}}{2}$.

110) $\dfrac{3\sqrt{42} + 3\sqrt{30} - \sqrt{14} - \sqrt{10}}{2}$

Chapter 9: Review Exercises

8) 5.8

$\sqrt{34}$

0 1 2 3 4 5 6 7 8 9

Chapter 9: Cumulative Review for Chapters 1–9

6)

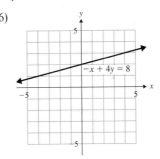

$-x + 4y = 8$

Chapter 10

Section 10.1

35) $\left\{ -\dfrac{\sqrt{10}}{3}, \dfrac{\sqrt{10}}{3} \right\}$

36) $\left\{ -\dfrac{\sqrt{11}}{4}, \dfrac{\sqrt{11}}{4} \right\}$

37) $\left\{ -\dfrac{\sqrt{55}}{5}, \dfrac{\sqrt{55}}{5} \right\}$

38) $\left\{ -\dfrac{\sqrt{21}}{3}, \dfrac{\sqrt{21}}{3} \right\}$

57) $\left\{ \dfrac{10 - 2\sqrt{6}}{3}, \dfrac{10 + 2\sqrt{6}}{3} \right\}$

58) $\left\{ \dfrac{7 - 3\sqrt{7}}{2}, \dfrac{7 + 3\sqrt{7}}{2} \right\}$

59) $\left\{ \dfrac{-8 - 5\sqrt{5}}{5}, \dfrac{-8 + 5\sqrt{5}}{5} \right\}$

60) $\left\{ \dfrac{-5 - 2\sqrt{11}}{4}, \dfrac{-5 + 2\sqrt{11}}{4} \right\}$

61) $\left\{ \dfrac{-3 - 3\sqrt{6}}{2}, \dfrac{-3 + 3\sqrt{6}}{2} \right\}$

62) $\left\{ \dfrac{1 - 3\sqrt{10}}{6}, \dfrac{1 + 3\sqrt{10}}{6} \right\}$

65) $\left\{ -\dfrac{2}{5}, \dfrac{6}{5} \right\}$

66) $\left\{ -2, \dfrac{4}{3} \right\}$

67) $\left\{ -\dfrac{5}{3}, -\dfrac{2}{3} \right\}$

68) $\left\{ -\dfrac{1}{4}, \dfrac{7}{4} \right\}$

69) $\left\{ \dfrac{11 - \sqrt{14}}{2}, \dfrac{11 + \sqrt{14}}{2} \right\}$

70) $\left\{ \dfrac{-8 - 2\sqrt{2}}{5}, \dfrac{-8 + 2\sqrt{2}}{5} \right\}$

Section 10.2

1) a trinomial whose factored form is the square of a binomial; examples may vary.

5) $a^2 + 12a + 36; (a + 6)^2$

6) $b^2 + 8b + 16; (b + 4)^2$

7) $k^2 - 10k + 25; (k - 5)^2$

8) $p^2 - 4p + 4; (p - 2)^2$

9) $g^2 - 24g + 144; (g - 12)^2$

10) $z^2 + 26z + 169; (z + 13)^2$

11) $h^2 + 9h + \dfrac{81}{4}; \left(h + \dfrac{9}{2} \right)^2$

12) $m^2 - 3m + \dfrac{9}{4}; \left(m - \dfrac{3}{2} \right)^2$

13) $x^2 - x + \dfrac{1}{4}; \left(x - \dfrac{1}{2} \right)^2$

14) $y^2 + 7y + \dfrac{49}{4}; \left(y + \dfrac{7}{2} \right)^2$

33) $\left\{-\dfrac{9}{2} - \dfrac{\sqrt{33}}{2}, -\dfrac{9}{2} + \dfrac{\sqrt{33}}{2}\right\}$

34) $\left\{\dfrac{1}{2} - \dfrac{\sqrt{13}}{2}, \dfrac{1}{2} + \dfrac{\sqrt{13}}{2}\right\}$ 39) $\left\{-\dfrac{11}{2}, -\dfrac{5}{2}\right\}$

45) $\left\{\dfrac{9}{2} - \dfrac{\sqrt{33}}{2}, \dfrac{9}{2} + \dfrac{\sqrt{33}}{2}\right\}$

46) $\left\{-\dfrac{1}{2} - \dfrac{\sqrt{17}}{2}, -\dfrac{1}{2} + \dfrac{\sqrt{17}}{2}\right\}$

57) $\left\{-\dfrac{11}{5} - \dfrac{\sqrt{86}}{5}, -\dfrac{11}{5} + \dfrac{\sqrt{86}}{5}\right\}$

58) $\left\{-\dfrac{5}{3} - \dfrac{\sqrt{10}}{3}, -\dfrac{5}{3} + \dfrac{\sqrt{10}}{3}\right\}$

Section 10.3

3) The equation must be written as $3x^2 - 5x - 4 = 0$ before identifying the values of a, b, and c. $a = 3$, $b = -5$, $c = -4$;

$$x = \frac{-(-5) \pm \sqrt{(-5)^2 - 4(3)(-4)}}{2(3)}$$

4) You cannot divide only the -3 by 3.

$$\frac{-3 \pm 12\sqrt{5}}{3} = \frac{3(-1 \pm 4\sqrt{5})}{3} = -1 \pm 4\sqrt{5}$$

9) $\left\{\dfrac{5 - \sqrt{17}}{2}, \dfrac{5 + \sqrt{17}}{2}\right\}$

10) $\left\{\dfrac{-3 - \sqrt{29}}{2}, \dfrac{-3 + \sqrt{29}}{2}\right\}$

11) $\left\{\dfrac{1 - \sqrt{13}}{3}, \dfrac{1 + \sqrt{13}}{3}\right\}$

12) $\left\{\dfrac{-4 - \sqrt{26}}{5}, \dfrac{-4 + \sqrt{26}}{5}\right\}$

17) $\left\{\dfrac{3 - \sqrt{3}}{2}, \dfrac{3 + \sqrt{3}}{2}\right\}$

18) $\left\{\dfrac{-1 - \sqrt{13}}{3}, \dfrac{-1 + \sqrt{13}}{3}\right\}$

35) $\left\{-\dfrac{2}{3}, \dfrac{2}{3}\right\}$ 36) $\left\{-\dfrac{5}{2}, \dfrac{5}{2}\right\}$ 38) $\left\{-\dfrac{1}{5}\right\}$

Section 10.4

3)
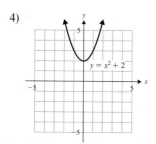

4)
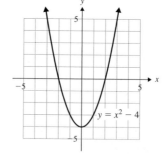

5)
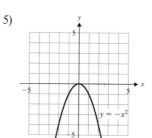

6)
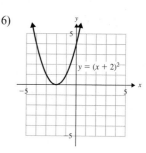

7)
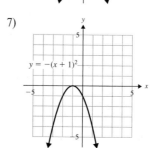

8)
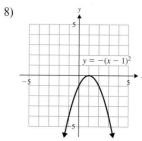

9)
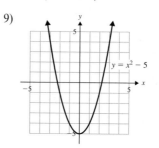

10)
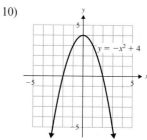

11)
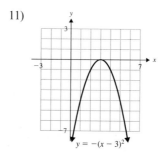

12)
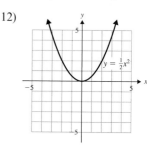

13) The x-coordinate of the vertex is $-\dfrac{b}{2a}$. Substitute that value into the equation to find the y-coordinate of the vertex.

15) Vertex: $(0, -4)$

16) Vertex: $(0, 1)$

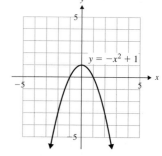

17) Vertex: (0, −1)

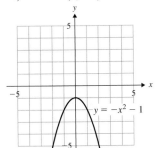

$y = -x^2 - 1$

18) Vertex: (0, −5)

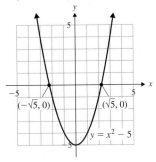

$(-\sqrt{5}, 0)$ $(\sqrt{5}, 0)$ $y = x^2 - 5$

25) Vertex: (−1, 5)

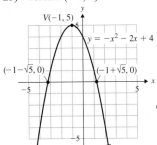

$V(-1, 5)$ $y = -x^2 - 2x + 4$ $(-1-\sqrt{5}, 0)$ $(-1+\sqrt{5}, 0)$

26) Vertex: (−2, −3)

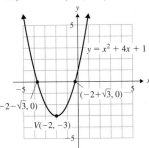

$y = x^2 + 4x + 1$ $(-2+\sqrt{3}, 0)$ $(-2-\sqrt{3}, 0)$ $V(-2, -3)$

19) Vertex: (3, 0)

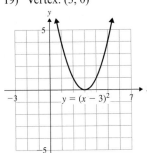

$y = (x - 3)^2$

20) Vertex: (−2, 0)

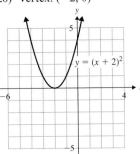

$y = (x + 2)^2$

27) Vertex: (−3, 2)

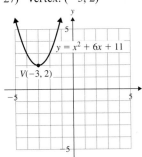

$y = x^2 + 6x + 11$ $V(-3, 2)$

28) Vertex: (4, −1)

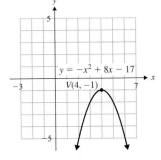

$y = -x^2 + 8x - 17$ $V(4, -1)$

21) Vertex: (2, −1)

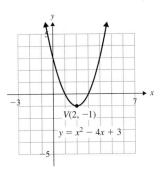

$V(2, -1)$ $y = x^2 - 4x + 3$

29) Vertex: (1, 0)

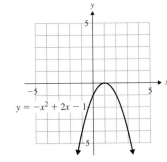

$y = -x^2 + 2x - 1$

30) Vertex: (−3, 0)

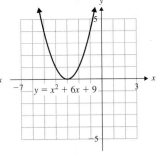

$y = x^2 + 6x + 9$

22) Vertex: (3, 4)

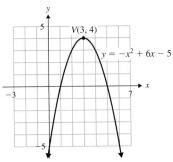

$V(3, 4)$ $y = -x^2 + 6x - 5$

31) Vertex: (1, −2)

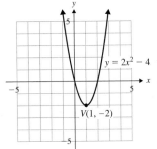

$y = 2x^2 - 4$ $V(1, -2)$

32) Vertex: (−2, 2)

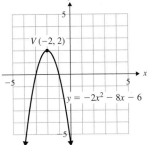

$V(-2, 2)$ $y = -2x^2 - 8x - 6$

23) Vertex: (1, 4)

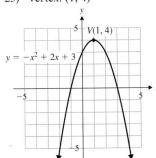

$V(1, 4)$ $y = -x^2 + 2x + 3$

24) Vertex: (−1, −9)

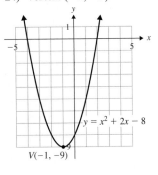

$y = x^2 + 2x - 8$ $V(-1, -9)$

Chapter 10: Review Exercises

17) $\left\{ -\dfrac{5}{2} - \dfrac{\sqrt{17}}{2}, -\dfrac{5}{2} + \dfrac{\sqrt{17}}{2} \right\}$

22) $\dfrac{6 \pm 3\sqrt{10}}{3} = \dfrac{3(2 \pm \sqrt{10})}{3} = 2 \pm \sqrt{10}$

34) $\left\{ \dfrac{3 - \sqrt{7}}{2}, \dfrac{3 + \sqrt{7}}{2} \right\}$

35) $\left\{\dfrac{-1-\sqrt{13}}{3}, \dfrac{-1+\sqrt{13}}{3}\right\}$ 36) $\{28, 26\}$

46) The x-coordinate of the vertex is $-\dfrac{b}{2a}$. Substitute that value into the equation to find the y-coordinate of the vertex.

47) It is the line that cuts the parabola in half so that one half of the parabola is the mirror image of the other half.

48) Vertex: $(0, 4)$

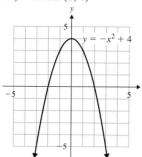

49) Vertex: $(-2, 0)$

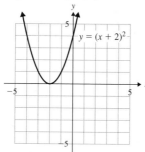

50) Vertex: $(-2, -1)$

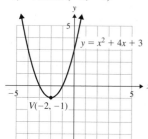

51) Vertex: $(1, 3)$

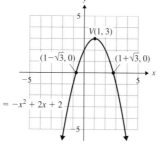

52) Vertex: $(3, 2)$

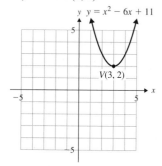

Chapter 10: Test

4) $\left\{-\dfrac{5}{2}-\dfrac{\sqrt{13}}{2}, -\dfrac{5}{2}+\dfrac{\sqrt{13}}{2}\right\}$

5) She divided only the 12 in the numerator by the 4 in the denominator. The correct way to simplify is

$$\dfrac{12 \pm 8\sqrt{5}}{4} = \dfrac{4(3 \pm 2\sqrt{5})}{4} = 3 \pm 2\sqrt{5}.$$

13) Vertex: $(1, 0)$

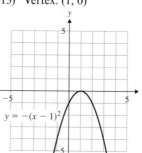

14) Vertex: $(-1, -4)$

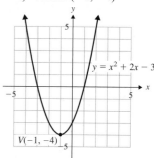

15) Vertex: $(2, 3)$

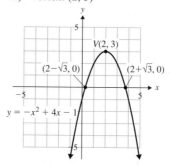

Chapter 10: Cumulative Review for Chapters 1–10

4) c) $\left\{-6, 14.38, \dfrac{3}{11}, 2, 5.\overline{7}, 0\right\}$

17) $(6, 0), (0, -3); (2, -2)$ may vary

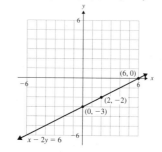

19) $(-2, 3)$

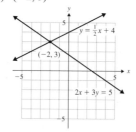

23)

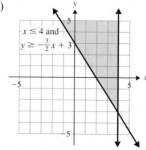

46) Vertex: $(1, 4)$

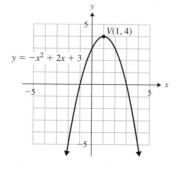

Photo Credits

Page 1: © Jill Braaten; **p. 2:** © Andrea Laurita/Getty RF; **p. 16:** © Vol. 61 PhotoDisc/Getty RF; **p. 17:** © BananaStock/Punchstock RF; **p. 85:** © EP100/PhotoDisc RF/Getty RF; **p. 86:** © Rubberball/Getty RF; **p. 118:** © Photo by Keith Weller/USDA; **p. 119(top):** © Big Cheese Photo/Punchstock RF; **p. 119(bottom):** © Ingram Publishing/ Superstock RF; **p. 131:** © Lawrence M. Sawyer/Getty RF; **p. 138:** © The McGraw-Hill Companies Inc./Ken Cavanagh Photographer; **p. 141:** © The McGraw-Hill Companies, Inc./Jill Braaten, photographer; **p. 148:** © Blend Images/Jupiterimages RF; **p. 149(top):** © Jeremy Woodhouse/Blend Images RF; **p. 149(bottom left):** © John Lund/Drew Kelly/Blend Images RF; **p. 149(bottom right):** © BananaStock/PunchStock RF; **p. 150:** © LuciannePashley/age fotostock RF; **p. 158:** © Getty RF; **p. 160(left):** © S. Meltzer/PhotoLink/Getty RF; **p. 160(right):** © DynamicGraphics/ Jupiter Images RF; **p. 167:** © Corbis RF; **p. 168:** © Sherri Messersmith; **p. 169:** © Royalty-Free/Corbis; **p. 170:** © Design Pics/Don Hammond; **p. 171:** © Jill Braaten; **p. 172:** © Ingram Publishing RF; **p. 186:** © S. Meltzer/PhotoLink/ Getty RF; **p. 194:** © Digital Vision RF; **p. 198:** © Royalty-Free/Corbis; **p. 217:** © Brooklyn Productions/Corbis RF; **p. 226:** © Adam Gault/Getty RF; **p. 230:** © SeidePreis/Getty RF; **p. 241:** © Jupiterimages/Comstock/Getty RF; **p. 242:** © Rubberball/Getty RF; **p. 282(left):** © The McGraw-Hill Companies, Inc.; **p. 283:** © Burke/Triolo/ BrandX RF; **p. 284(left):** © BrandX/ Corbis RF; **p. 284(right):** © Digital Vision/Getty RF; **p. 304:** © The McGraw-Hill Companies, Inc./Ken Cavanagh, photographer; **p. 305:** imac/ Alamy RF; **p. 306(left):** © Thinkstock/ Jupiterimages RF; **p. 306(right):** © Sandra Ivany/BrandX/Getty RF; **p. 307:** © C. Thatcher/Getty RF; **p. 308:** © JGI/ Tom Grill/Blend Images/Getty RF; **p. 338:** © Vol. 16 PhotoDisc RF/Getty; **p. 339:** © Ingram Publishing/Alamy RF; **p. 343:** © Digital Vision RF; **p. 345:** © Jupiterimages/Comstock/Getty RF; **p. 346:** © Sri Maiava Rusden/Getty RF; **p. 385:** © Royalty-Free/Corbis; **p. 386:** © Image Source/Getty RF; **p. 439:** © The McGraw-Hill Companies, Inc./John Flournoy, photographer; **p. 451:** © Ian Hooton/SPL/Getty RF; **p. 452:** © AID/a.collectionRF/Getty RF; **p. 512:** © Daisuke Morita/Getty RF; **p. 519:** © Dynamic Graphics/JupiterImages RF; **p. 522:** © Corbis RF; **p. 524:** © Stockbyte/ Getty RF; p. 525: © PhotoAlto RF; **p. 535:** © C Squared Studios/Getty RF; **p. 537:** Getty/Stockbyte RF; **p. 538:** © Andersen Ross/Getty RF; **p. 594:** © Royalty-Free/Corbis; **p. 602:** © Comstock/ Getty RF; **p. 603:** © ImageBazaar/Getty RF; **p. 611:** © Philip Coblentz/Brand X Pictures/PictureQuest RF; **p. 612:** © Purestock/Getty RF; **p. 651:** © Ingram Publishing/Alamy RF; **p. A18(top right):** © Randy Faris/Corbis RF; **p. A18(left):** © Jack Hollingsworth/ BrandX/Getty RF; **p. A18(bottom right):** © Glowimages/Getty RF

Index

solving
    by elimination, 262–268, 271–272, 300
    by graphing, 246–249, 299
    by substitution, 255–257, 271–272,
      299–300
Systems of linear inequalities
    solution set of, 291
    solving by graphing, 291–293, 302

T

Terms
    in algebraic expressions, identifying,
      69–70, 80
    like. *See* Like terms
    in polynomial expressions, degree of,
      349, 379
    radical expressions with two, rationalizing
      denominators of, 580–581
Test point, for linear inequalities in two
    variables, 288–290, 301–302
Textbooks, reading, 86
Time log, 172
Time management, 172
Time style, 232
Timetable, weekly, 172
To-do list, 172
Trapezoid
    area of, 26–28
    perimeter of, 26–28
Trial and error, factoring trinomials by,
    407–410, 446
Triangles
    acute, 25–26
    area of, 26–28
    equilateral, 25–26, 78
    identifying, 25–26
    isosceles, 25–26, 78
    measure of angles of, 25, 78
      applied problems with, 124–125
    obtuse, 25–26
    perimeter of, 26–28
    right, 25–26
    scalene, 25–26, 78
    similar, 139–140

Trinomials
    definition of, 351
    difference of two squares
      definition of, 415
      factoring, 415–418, 446
    factoring
      difference of two squares,
        415–418, 446
      in form $ax^2 + bx + c(a \neq 1)$, 406–410,
        445–446
      in form $x^2 + bx + c$, 400–403, 445
      by grouping, 406–410
      perfect square trinomials, 413–415, 446
      skills for, 398–399
      by trial and error, 407–410, 446
      in two or more variables, 403
    perfect square
      definition of, 413
      factoring, 413–415, 446

U

Unit price, 136

V

Variable
    definition of, 60, 79
    in exponent bases, with zero exponent,
      322–323, 341
    finding unknown, 121–122
    isolating, 90
    solving formula for, 127–130, 164
    solving rational equations for,
      505–506, 531
Variation
    constant of, 517
    direct
      applied problems with, 517–519
      definition of, 517
Vertex, of parabola, 631, 633, 655
Vertical angles
    applied problems with, 125–126
    definition of, 24

Vertical line
    equation of, 192–193
    graphing, 192–193, 234
    slope of, 204–205, 234
    writing equation of, 223–224, 236
Vertical line test, 641–642, 656
Visuals, creating, 603
Volume
    definition of, 29
    formulas for, 29–30

W

Weekly timetable, 172
Whole numbers
    definition of, 5
    in set of real numbers, 37
Work problems, 515–517, 531–532

X

x-axis, in Cartesian coordinate system, 178
x-intercept
    definition of, 190, 233
    of linear equation in two variables,
      graphing from, 190, 233

Y

y-axis, in Cartesian coordinate system, 178
y-intercept
    definition of, 190, 233
    of linear equation in two variables
      graphing from, 190, 233
      writing equation given slope and, 221

Z

Zero
    as denominator, in rational expression,
      456–458, 487
    as exponent, 320, 341
      with variable bases, 322–323, 341
    as product, 424, 446
Zero product rule, 424, 446